Large Meteorite Impacts and Planetary Evolution II

Edited by

Burkhard O. Dressler
Lunar and Planetary Institute
3600 Bay Area Boulevard
Houston, Texas 77058
USA

and

Virgil L. Sharpton
Lunar and Planetary Institute
3600 Bay Area Boulevard
Houston, Texas 77058
USA

SPECIAL PAPER

339

1999

Copyright © 1999, The Geological Society of America, Inc. (GSA). All rights reserved. GSA grants permission to individual scientists to make unlimited photocopies of one or more items from this volume for noncommercial purposes advancing science or education, including classroom use. Permission is granted to individuals to make photocopies of any item in this volume for other noncommercial, nonprofit purposes provided that the appropriate fee ($0.25 per page) is paid directly to the Copyright Clearance Center, 222 Rosewood Drive, Danvers, MA 01923, USA, phone (978) 750-8400, http://www.copyright.com (include title and ISBN when paying). Written permission is required from GSA for all other forms of capture or reproduction of any item in the volume including, but not limited to, all types of electronic or digital scanning or other digital or manual transformation of articles or any portion thereof, such as abstracts, into computer-readable and/or transmittable form for personal or corporate use, either noncommercial or commercial, for-profit or otherwise. Send permission requests to GSA Copyrights.

Copyright is not claimed on any material prepared wholly by government employees within the scope of their employment.

Published by The Geological Society of America, Inc.
3300 Penrose Place, P.O. Box 9140, Boulder, Colorado 80301

Printed in U.S.A.

GSA Books Science Editor Abhijit Basu

Library of Congress Cataloging-in-Publication Data

Large meteorite impacts and planetary evolution II / edited by B.O.
   Dressler and V. L. Sharpton.
      p. cm. — (Special paper ; 339)
   Papers from the International Conference on Large Meteorite
Impacts and Planetary Evolution, September 1 through September 3,
1997, Sudbury, Ontario.
   Includes bibliographical references and index.
   ISBN 0-8137-2339-6
   1. Meteorite craters—Congresses. 2. Cratering—Congresses.
3. Planets—Origin—Congresses. 4. Impact—Congresses.
I. Dressler, Burkhard O. II. Sharpton, Virgil L.
III. International Conference on Large Meteorite Impacts
and Planetary Evolution (1997 : Sudbury, Ont.) V. Series: Special
papers (Geological Society of America) ; 339.
QB754.8.L37   1999
551.3'97—dc20   94-34436
                                                             CIP

Cover: The New Quebec Crater is located in New Quebec, Canada, at lat 61°17′N and long 73°40′W. The crater is 3.44 km in diameter and has an age of 1.4 ± 0.1 Ma. Photograph taken July 1998. Courtesy of George Burnside, Manotick, Ontario, Canada.

10 9 8 7 6 5 4 3 2 1

Dedication

The geosciences community engaged in large-body impact research can count among itself an appreciable number of researchers who dedicate their scientific careers to the fascinating effects that impacts have had on the evolution of the planets of our solar system and on how they have influenced the paths life has taken on Earth. R. S. Dietz and E. M. Shoemaker, two scientists who, over several decades, were leading advocates of the significance of meteorite impacts in the history of Earth and the other planets of our solar system, and participated in the 1992 Sudbury Conference on Large Meteorite Impacts and Planetary Evolution, are no longer among us. It is to these two leaders of impact geology, and to Colonel Paul Barringer, generous supporter of impact research, that the authors of this volume dedicate this publication.

Contents

Preface viii

INVESTIGATIONS OF TERRESTRIAL IMPACT STRUCTURES

1. Anatomy of the Popigai impact crater, Russia 1
   V. L. Masaitis, M. V. Naumov, and M. S. Mashchak

*2. Popigai impact structure (Arctic Siberia, Russia): Geology, petrology, geochemistry,
   and geochronology of glass-bearing impactites* 19
   Sergei Vishnevsky and Alessandro Montanari

*3. Morokweng impact structure, South Africa: Geologic, petrographic, and isotopic results,
   and implications for the size of the structure* 61
   Wolf Uwe Reimold, Christian Koeberl, Franz Brandstätter, F. Johan Kruger, Richard A. Armstrong,
   and Cornelis Bootsman

*4. Ni- and PGE-enriched quartz norite impact melt complex in the Late Jurassic
   Morokweng impact structure, South Africa* 91
   M. A. G. Andreoli, L. D. Ashwal, R. J. Hart, and J. M. Huizenga

5. Slate Islands, Lake Superior, Canada: A mid-size, complex impact structure 109
   B. O. Dressler, V. L. Sharpton, and P. Copeland

6. Lycksele structure in northern Sweden: Result of an impact? 125
   H. Thunehed, S.-Å. Elming, and L. J. Pesonen

*7. Lake Karikkoselkä impact structure, central Finland: New geophysical
   and petrographic results* 131
   Lauri J. Pesonen, Seppo Elo, Martti Lehtinen, Tarmo Jokinen, Risto Puranen, and Liisa Kivekäs

*8. Seismic expression of the Chesapeake Bay impact crater: Structural and
   morphologic refinements based on new seismic data* 149
   C. Wylie Poag, Deborah R. Hutchinson, Steven M. Colman, and Myung W. Lee

9. Gravity signature of the Teague Ring impact structure, Western Australia 165
   Jeffrey B. Plescia

10. BP and Oasis impact structures, Libya, and their relation to Libyan Desert Glass 177
    Begosew Abate, Christian Koeberl, F. Johan Kruger, and James R. Underwood, Jr.

11. Mjølnir Structure, Barents Sea: A marine impact crater laboratory 193
    Filippos Tsikalas, Steinar Thor Gudlaugsson, Jan Inge Faleide, and Olav Eldholm

EFFECTS AND PRODUCTS OF IMPACT CRATERING

12. Shock-induced effects in natural calcite-rich targets as revealed by X-ray powder diffraction 205
Roman Skála and Petr Jakeš

13. Carbon isotope study of impact diamonds in Chicxulub ejecta at Cretaceous-Tertiary boundary sites in Mexico and the Western Interior of the United States 215
R. M. Hough, I. Gilmour, and C. T. Pillinger

14. Formation of a flattened subsurface fracture zone around meteorite craters 223
Yevgeny Zenchenko and Vsevolod Tsvetkov

15. Effect of erosion on gravity and magnetic signatures of complex impact structures: Geophysical modeling and applications 229
J. Plado, L. J. Pesonen, and V. Puura

16. Impact crises, mass extinctions, and galactic dynamics: The case for a unified theory 241
Michael R. Rampino

17. Late Archean impact spherule layer in South Africa that may correlate with a Western Australian layer 249
Bruce M. Simonson, Scott W. Hassler, and Nicolas J. Beukes

THE CHICXULUB STRUCTURE—A THREE-DIMENSIONAL VIEW

18. Deep seismic reflection profiles across the Chicxulub crater 263
D. B. Snyder and R. W. Hobbs

19. Near-surface seismic expression of the Chicxulub impact crater 269
John Brittan, Joanna Morgan, Mike Warner, and Luis Marin

20. Morphology of the Chicxulub impact: Peak-ring crater or multi-ring basin? 281
Joanna Morgan and Mike Warner

21. Upper crustal structure of the Chicxulub impact crater from wide-angle ocean bottom seismograph data 291
G. L. Christeson, R. T. Buffler, and Y. Nakamura

NEW DEVELOPMENTS IN SUDBURY GEOLOGY

22. Sudbury Structure 1997: A persistent enigma 299
B. O. Dressler and V. L. Sharpton

23. Sudbury Breccia distribution and orientation in an embayment environment 305
John S. Fedorowich, Don H. Rousell, and Walter V. Peredery

24. Impact diamonds in the Suevitic Breccias of the Black Member of the Onaping Formation, Sudbury Structure, Ontario, Canada 317
V. L. Masaitis, G. I. Shafranovsky, R. A. F. Grieve, F. Langenhorst, W. V. Peredery, A. M. Therriault, E. L. Balmasov, and I. G. Fedorova

25. The Green Member of the Onaping Formation, the collapsed fireball layer of the Sudbury impact structure, Ontario, Canada 323
M. E. Avermann

Contents

26. Carbonaceous matter in the rocks of the Sudbury Basin, Ontario, Canada 331
Ted E. Bunch, Luann Becker, David Des Marais, Anne Tharpe, Peter H. Schultz,
Wendy Wolbach, Daniel P. Glavin, Karen L. Brinton, and Jeffrey L. Bada

**27. Origin of carbonaceous matter, fullerenes, and elemental sulfur in rocks
of the Whitewater Group, Sudbury impact structure, Ontario, Canada** 345
D. Heymann, B. O. Dressler, J. Knell, M. H. Thiemens, P. R. Buseck, R. B. Dunbar, and
D. Mucciarone

28. Isotopic evidence for a single impact melting origin of the Sudbury Igneous Complex 361
A. P. Dickin, T. Nguyen, and J. H. Crocket

**29. Sudbury Igneous Complex: Simulating phase equilibria and in situ differentiation
for two proposed parental magmas** 373
Alexey A. Ariskin, Alexander Deutsch, and Markus Ostermann

30. Sudbury impact event: Cratering mechanics and thermal history 389
B. A. Ivanov and A. Deutsch

**31. Emplacement geometry of the Sudbury Igneous Complex:
Structural examination of a proposed impact melt-sheet** 399
E. J. Cowan, U. Riller, and W. M. Schwerdtner

32. Structural evolution of the Sudbury impact structure in the light of seismic reflection data 419
D. E. Boerner and B. Milkereit

33. Summary: Development of ideas on Sudbury geology, 1992–1998 431
A. J. Naldrett

Index 443

Preface

Since the Sudbury 1992 Conference on Large Meteorite Impacts and Planetary Evolution, research into impact geology has proceeded at an accelerated pace. In 1994, worldwide attention was focused on the spectacular impact of the Shoemaker-Levy comet fragments on Jupiter, convincing remaining skeptics that meteorite impacts are a reality and may pose a potential threat to Earth and human civilization. The number of known terrestrial impact structures has risen to >150, and recently the Morokweng Structure, a fourth very large structure, possibly 300 km in diameter, has been discovered in northern South Africa. The other large structures are the Chicxulub (Mexico), Sudbury (Canada), and Vredefort (South Africa) structures. Investigations on various aspects of large-body impacts continue to be carried out. Among them are those that relate to cratering mechanics, killing mechanisms related to impact, geochemistry of impact melts, carbon including fullerenes in impact breccia deposits, the effects of impact on geologic materials, and impact modeling.

The Ontario Geological Survey, Sudbury, and the Lunar and Planetary Institute, Houston, convened a second conference on Large Meteorite Impacts and Planetary Evolution, September 1–3, 1997, once again in Sudbury, Ontario, Canada, and referred to as Sudbury 1997. Sponsors of the conference were the Barringer Crater Company, Falconbridge Ltd., the Geological Survey of Canada, Inco Ltd., the Lunar and Planetary Institute, the Ontario Ministry of Northern Development and Mines through the Ontario Geological Survey, and the Quebec Ministère des Ressources Naturelles. Approximately 140 scientists from 18 countries attended Sudbury 1997 to present and discuss results of recent field and laboratory investigations. About 80 talks and 30 posters were presented. The sessions were grouped around the following themes: the Sudbury Impact Structure, Diamonds and Carbonaceous Matter: Sudbury and Other Impact Structures, Terrestrial Impact Structures, the Chicxulub Structure, Effects of Planetary Impact, and Geophysical Constraints on the Character of Chicxulub and Other Large Impact Structures. In addition, a well-attended, excellent public lecture entitled "Strange World: Radar Encounters with Earth-Approaching Asteroids" was presented at the Science North Museum by Steven J. Ostro of the Jet Propulsion Laboratory, Pasadena, California. The scientific talks and posters presented in Sudbury provided important new insights into various aspects of impact cratering. It is for this reason that the organizers decided to communicate the conference results to a wider audience in this Special Paper.

Without the dedication of the members of both the scientific and local organizing committees, the conference would not have been the success it was. Members of the Scientific Organizing Committee were A. Deutsch, University of Münster; B. O. Dressler, Chair, Lunar and Planetary Institute, Houston, Texas; B. M. French, Smithsonian Institution, Washington, D.C.; R. A. F. Grieve, Geological Survey of Canada, Ottawa, Ontario; G. W. Johns, Ontario Geological Survey, Sudbury, Ontario; and V. L. Sharpton, Lunar and Planetary Institute, Houston, Texas.

B. O. Dressler
V. L. Sharpton
Volume Editors

Anatomy of the Popigai impact crater, Russia

V. L. Masaitis, M. V. Naumov, and M. S. Mashchak
Karpinsky Geologic Research Institute (VSEGEI), Sredny pr., 74, 199106 St.-Petersburg, Russia; e-mail: vsegei@wplus.mail.net

ABSTRACT

Based on results of geologic mapping, geophysical observations, and drilling, three-dimensional geologic models of the multi-ring Popigai impact structure (D = 100 km), as well as of its morphostructural elements and large bodies of impact rocks, are constructed.

The true bottom of the crater is uneven and composed of disturbed target rocks. Its blocky relief was formed before deposition of impactites and allogenic breccia that form "pseudo-layered" sequences (1 km thick and more) consisting of relatively thin, sub-horizontal lenticular and sheet-shaped bodies. The distribution of molten and fragmented masses in the crater interior is characterized by asymmetry and probably was caused by an oblique impact.

The extent and shape of allogenic monomict and polymict lithic breccia, suevite, and tagamite bodies were traced in detail in three sectors of the crater (30–250 km^2) and to a depth of about 0.5–1.5 km. These areas represent well-studied parts of main concentric morphostructural elements of the crater: the annular uplift, the annular trough, and the outer zone of deformed target rocks. The morphology and inner structure of certain bodies are characterized by a combination of radial and concentric patterns.

Analysis of various cross sections, of the distribution of masses of certain complex impact rock bodies, their inner structures, and interactions shows that they were formed by the ejection of separate jets and currents that differed in composition, impulses, and trajectories. Although the formation of the main mass of impact rocks was simultaneous, the mixing of melted and crushed material occurred only at the initial stages of mass movement and was insignificant during the main phases of transportation and deposition.

INTRODUCTION

Numerous models of terrestrial impact cratering and the origin of impact melt bodies have been proposed (for example, Basilevsky et al., 1983; Melosh, 1989). However, these models usually are not based on adequate geologic data sets. Such a data set exists for the well-preserved Popigai impact crater, Eastern Siberia. It represents one of the best examples of complex terrestrial impact structures. Its principal geologic features were described by Masaitis et al. (1975, 1980, 1998) and Masaitis (1984, 1994). Some constraints on impact processes can be inferred from observations made at the Popigai Crater.

Popigai Crater originated 35.7 ± 0.2 Ma (Bottomley et al., 1997) at the northeast edge of the Anabar shield. The crater diameter is about 100 km, but weakly deformed and fractured crystalline basement and sedimentary cover rocks are traceable to a distance of about 75 km from the center of the ring structure. The crater is filled by various kinds of allogenic lithic breccias and melted rocks—tagamites and suevites (Figure 1). Shocked target rocks and their minerals may occur as inclusions in these breccias and impactites.

The main geologic units found within the crater are (1) target rocks—sedimentary and crystalline, partly parautochthonous; (2) allogenic lithic breccia, subdivided into coarse (mega-

---

Masaitis, V. L., Naumov, M. V., and Mashchak, M. S., 1999, Anatomy of the Popigai impact crater, Russia, *in* Dressler, B. O., and Sharpton, V. L., eds., Large Meteorite Impacts and Planetary Evolution II: Boulder, Colorado, Geological Society of America Special Paper 339.

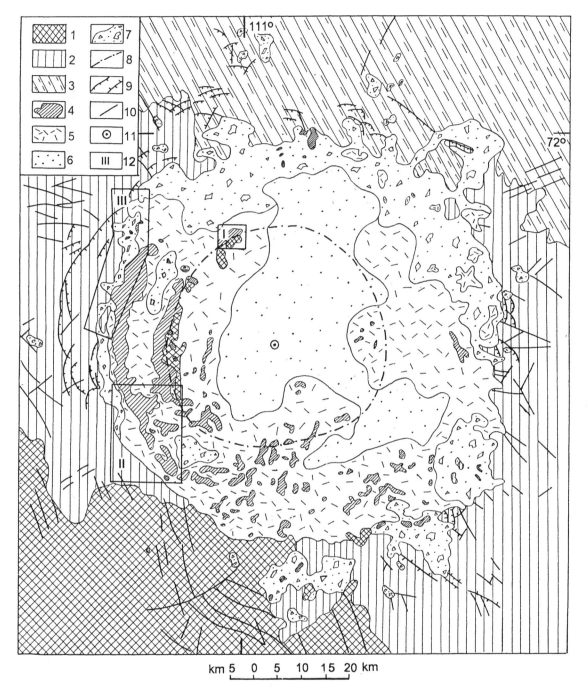

Figure 1. Sketch geologic map of Popigai Crater (Quaternary deposits not shown). 1, Archean crystalline rocks; 2, Upper Proterozoic and lower Paleozoic sedimentary rocks; 3, upper Paleozoic and Mesozoic sedimentary and igneous rocks; 4, tagamites; 5, suevites; 6–7, allogenic lithic breccia; 6, fine-grained (coptoclastites), 7, megabreccia and blocky breccia; 8, axis of the annular uplift; 9, thrusts and upthrusts; 10, faults; 11, geometric center of the crater; 12, areas, described in detail (I, Mayachika Upland; II, Balagan River basin; III, Sogdoku Upland).

breccia, usually crystalline, blocky polymict breccia, and blocky sedimentary breccia) and fine grained (coptoclastite). The megabreccia in places is cemented by impact melt rock—tagamite; (3) impactites—suevites and tagamites. Subdivision and nomenclature of certain rock species are made according to a proposal worked out previously (Masaitis et al., 1975; Raikhlin et al., 1980).

Crystalline rocks are represented mainly by Archean biotite-, garnet-, pyroxene-, and in places cordierite- and sillimanite-bearing gneisses; schist, calciphyre, granite may also present.

Minor Upper Proterozoic and Triassic dolerites also occur. Terrigenous and carbonaceous rocks of Upper Proterozoic, lower and upper Paleozoic, and Cretaceous age represent the uppermost target. Allogenic lithic breccia consists of blocks and fragments of all types of target rocks set in a fine-grained matrix, in places with an admixture of impact glass particles and bombs (<10%). Fine-grained breccia originated at the moment of cratering due to comminution and ejection of loose Cretaceous and upper Paleozoic sands and sandstones and are distinguished as a separate kind of allogenic breccia.

The rocks, which contain more than 10% of chilled or crystallized impact melt, are considered as impactites. Tagamites consist of glassy or crystalline matrix with numerous inclusions. Two varieties of tagamites may be distinguished: high temperature (HT) and low temperature (LT). They are distinct by some specific petrographic features (Masaitis, 1994; Masaitis et al., 1998), the main ones including crystallinity of matrix, composition of main liquidus phases, Fe^{2+}/Fe^{3+} ratio, and degree of thermal transformations of clasts. Suevites consist of bombs, lapilli- and ash-sized particles of impact glass, and fragments of crystalline and sedimentary target rocks in a clastic matrix of the same components. Suevites are subdivided into two varieties: one is composed mainly of glass fragments or vitroclasts (V-suevites). The other is composed mainly of fragments of crystalline or sedimentary rocks and their minerals; these rocks are called C-suevites and S-suevites, respectively. Based on common classifications of pyroclastic rocks, these suevites may be also regarded as agglomerate-, lapilli- or ash-sized varieties.

The geologic bodies differentiated in this study are composed mostly of certain kinds of breccia and impactites. Tagamites are in general homogeneous, but in places contain a large number of blocks of shocked gneisses. Allogenic lithic breccia and suevites may include irregularly distributed, small bodies of tagamites, not exceeding 10–20% by volume. Fine-grained, melt-free breccia (coptoclastite) may alternate with small suevite lenses on a scale of 1–20 m. The contacts between bodies of differing composition and structure may be sharp, but commonly they are transitional. Nevertheless, the individual bodies made of certain rock types are clearly distinguishable on outcrop and in drill core.

The modeling of the impact structure including the three-dimensional shape of specific geologic bodies within the structure is based mostly on geologic mapping and detailed drilling in local areas (the depths of boreholes are from some tens of meters to as large as 1.5 km). More than 500 drill holes were studied in detail. Some geophysical observations on gravity and magnetic fields and on electric resistivity were taken into account in the modeling process.

MAIN STRUCTURAL ELEMENTS

The main elements of the inner structure of Popigai Crater are a central depression filled in by impactites and allogenic impact breccia, an annular uplift of disturbed crystalline target rocks (authigenic breccia, parautochthon), an annular trough filled in by coarse allogenic breccia and impactites, and an outer flat slope of crater and zone of deformed target rocks—crystalline and sedimentary. Shallow, radial troughs are excavated on this slope and filled in by breccia and impactites as well (Fig. 1). The diameters of these four structural elements are D_1 = 35–37 km (outer edge), D_2 = 45 km (axis), D_3 = 56–58 km (axis), and D_4 = 100 km (outer limit corresponding to the rim diameter).

These elements are well expressed in the three-dimensional relief model of the true crater bottom made of crystalline and, in part, sedimentary rocks underlying the allogenic rocks (Figure 2). It is based on geologic and topographic maps, drilling data, and gravity observations (Figure 3). The latter show an annular gravity high surrounded by an annular gravity low, which correspond to main structural elements, as listed above. Thus, the crater has a multi-ring inner structure.

A weak gravity high exists in the crater center and a flat central uplift is suspected to exist in the central depression. The depth of central depression may be ~2 km, which corresponds to the thickness of allogenic breccia and impactites. The annular trough has a similar depth in its southeastern sector but is more shallow in the northwestern part (about 1,000 m). The true bottom (including the annular uplift) is uneven, having a relief of some hundreds of meters.

Impactites and allogenic breccia occupy an area of about 5,000 km^2 and their total volume is about 4,800 km^3. In general, the center of gravity of entire mass of melted rocks is slightly displaced, relatively to geometric center of the crater, to the southwest. This displacement may be the result of an oblique impact.

Three segments of the crater, ranging in size from 30 to 250 km^2 and to depths of 0.5–1.5 km were studied in considerable detail. They represent parts of the main concentrical morphostructural elements of the crater: annular uplift, annular trough, and outer zone of deformed target rocks (Figure 1), as described below. They represent a continuous radial structural section from the inner slope of the annular uplift up to the outer edge of the impact crater.

There are only limited data on the distribution and shape of allogenic breccia and impactites bodies in the central depression. This is the result of limited outcrop in this central zone of the impact structure, and only the upper part of the sequence was penetrated here by rare and shallow drill holes. Generally, fine-grained clastic breccias (coptoclastites) occur in this area; they contain thin lenses of suevites (mostly S-suevites) and small irregular bodies of tagamites. One of drill holes locating at 15 km westward from the crater center penetrated 800-m sequence composed of tagamites and suevites, but it did not reach the crater floor.

ANNULAR UPLIFT: MAYACHIKA UPLAND

The shocked crystalline basement is well exposed in the northwestern sector of the impact crater (Fig. 1), especially in the Mayachika Upland, making up an isolated group of small hills,

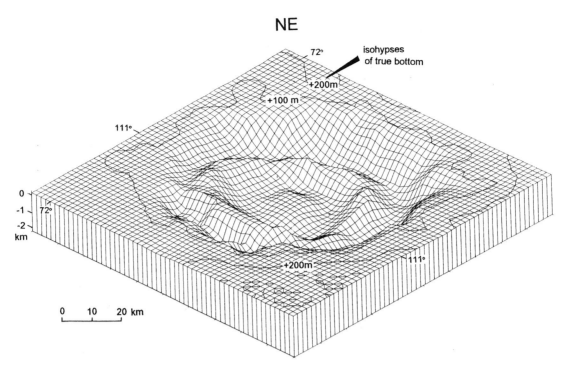

Figure 2. Sketch relief map of the true bottom of Popigai Crater. Principal features of the relief are shown: central depression with suspected central uplift, annular uplift, and annular trough transient to the outer zone of deformed crystalline and sedimentary rocks.

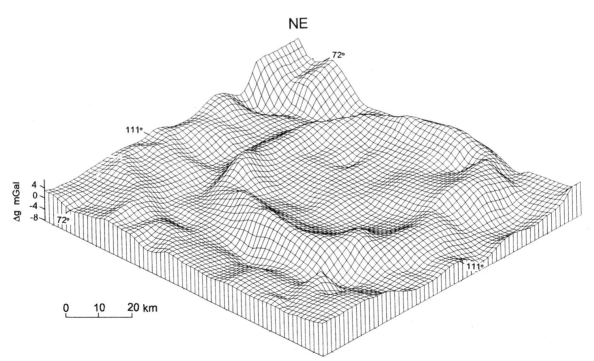

Figure 3. Residual gravity Bouguer anomalies (after removal of regional field) of central and northwest parts of Popigai Crater. Annular gravity high corresponds to the annular uplift of crystalline rocks.

surrounded by a swampy, glacial plain. The uplift (its axis has a northeastern strike) is composed of parautochthonous shocked gneisses and overlain in part by coarse allogenic polymict breccia, which are partially cemented by tagamite at the outer northwestern slope, and by suevites at the inner southeastern slope of the annular uplift. The shock features of rock-forming minerals of these crystalline rocks are indicative of shock pressure of about 8–10 GPa. Lens-like and sheet-like bodies of tagamites, suevites, and fine-grained, glass-free allogenic breccia overlie the annular crystalline uplift, but their sequences on its inner and outer slopes are different (Figures 4 and 5).

The relief of parautochthonous crystalline rocks of the outer slope is steeper than on the inner slope, but on both sides it is irregular (Figure 6). Elongated blocky uplifts and depressions, parallel and perpendicular to the axis of the annular uplift, are characteristic of the outer slope. The amplitude of these displacements may be estimated at 100–200 m. It is interesting to note that the relief of Mayachika Upland inherits the original shape of this large part of the annular uplift, because they are coated by allogenic breccia and impactites on all sides. This uplifted block was only slightly modified by erosion.

The thickness of crystalline megabreccia overlapping parautochthonous rocks is insignificant in the upper part of the steep outer slope (50 m), but it increases abruptly to as much as 340 m at the base of the annular uplift and remains probably the same in the annular trough. In the middle and lower parts of this slope, megabreccia is cemented by tagamites that grade into an extended tagamite (LT-tagamite) body overlapping this breccia. The volumetric ratio megablocks/tagamite matrix in the megabreccia is about 4:1. The base of this lower tagamite sheet gently dips northwestward, although its thickness varies abruptly in the northeast-southwest direction. Both megabreccia and lower tagamite sheet are split near the foot of the slope, where they surround a thick lens of polymict allogenic breccia that is composed mostly of blocks of sedimentary rocks.

The main mass of impact melt-rocks of the outer slope of the annular uplift consists of a lenticular body of S-suevites overlain by an upper tagamite sheet. These units lie in a semicircular depression between the outcropping parautochthonous basement of the annular ring and the crystalline megabreccia whose roof forms a steep morphologic rise of 750 m/km.

It is important to note that C-suevite deposits are thickest

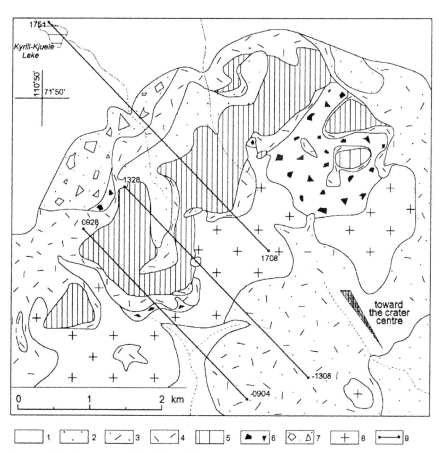

Figure 4. Geologic map of the Mayachika Upland. 1, Quaternary deposits; 2, coptoclastites; 3, S-suevites; 4, C-suevites and V-suevites; 5, tagamites; 6–7, allogenic lithic breccia; 6, crystalline megabreccia; 7, sedimentary blocky breccia; 8, gneisses (parautochthon); 9, profiles of drill holes.

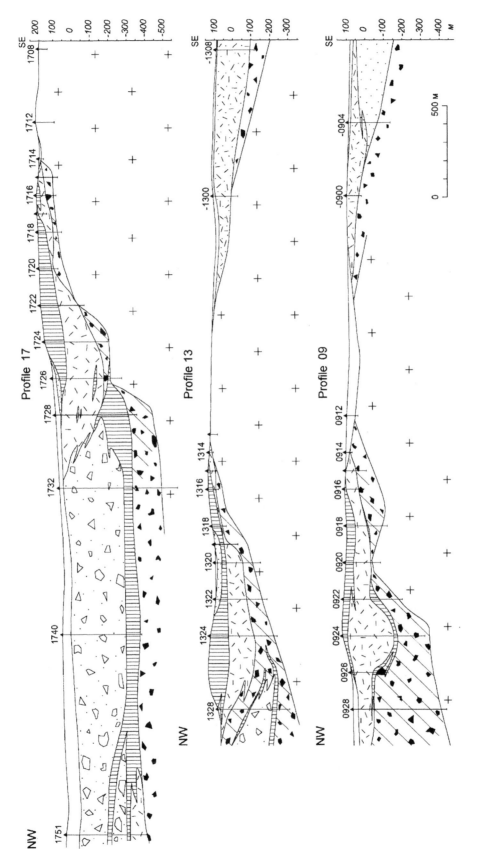

Figure 5. Geologic sections along profiles of drill holes shown in Figure 4. Striated areas indicate the presence of impactites in the matrix of the crystalline megabreccia.

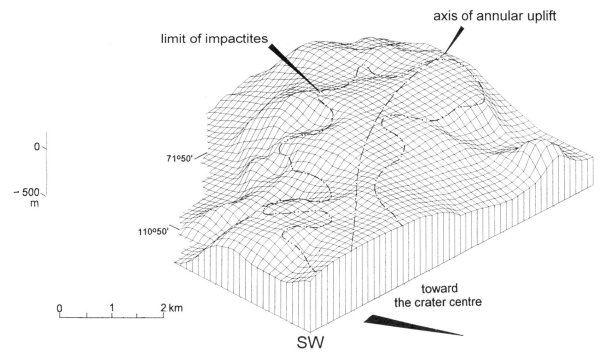

Figure 6. Sketch relief map of the surface of crystalline rocks (parautochthon) on the Mayachika Upland. The surface at elevations lower than –500 m is not shown.

where they are overlain by thickest LT-tagamites, while there are no such correlations with the lower tagamite sheet. The zones of maximum thickness of the C-suevite body have radial trends, while maximum thicknesses of the upper tagamite body have meridional trends (Figure 7). High variability in thickness of suevites in a semicircular depression is caused by the uneven base, reflecting an appropriate relief of the uplifted parautochthon basement. In turn, the variability of the thickness of the upper tagamite body is a result of its uneven original roof modified by recent erosion.

Comparing the hypsometric levels of the upper and lower surfaces of lens-like and sheet-like bodies located at the base of the sequence with those at the top shows that from below the dispersion of hypsometry of these surfaces decreases by a factor of four times, meaning that the upper bodies are more flat than the lower ones. In addition, the axes of gravity mass for the upper units are displaced toward the crater center relative to the lower ones. Thus the centripetal tendency of this distribution exists.

On the inner slope of the annular uplift, suevites occur directly on shocked gneisses and on the allogenic crystalline megabreccia. They are distinct from suevites described above and mainly made of the S-suevite variety interlayered by lenses of coptoclastites. These coptoclastites are considered to represent the uppermost member of the sequence of ejected and deposited material in this area because they occur as sags in the roof of the upper tagamite body. On the inner slope of the annular uplift, the relief of the parautochthonous basement and the distribution of impact lithologies are less well confined due to the lack detailed data.

ANNULAR TROUGH: BALAGAN RIVER BASIN

The structure of the annular trough was studied in detail in the southwestern sector of Popigai Crater (Balagan River basin) where the landscape is a hilly plain with a relief of ~100–200 m and where tagamite, suevite, and minor allogenic breccia outcrop in the valleys of Balagan River and some other rivers. Only in the northeast and southwest parts of this area do exposures of parautochthonous target rocks occur.

A cross section of the annular trough in the Balagan River basin is shown in Figure 8. This area embraces the outer slope of the annular uplift, the annular trough of 13–16 km in width, and its outer slope. Here, a thick complex impact melt sheet was discovered and mapped by deep drilling (Figure 9). Maximum total thickness of impactites and allogenic breccia in the central part of the trough can reach 1.4 km. These rocks overlie parautochthonous (partly disturbed and displaced) crystalline rocks, which in some places alternate with large blocks of Upper Proterozoic quartzite. The shock level of crystalline rocks attenuates outward from 8–10 GPa at the annular uplift up to <5 GPa at the outer slope of the trough where no microscopic shock effects were observed. The outer slope of the annular uplift (its width here is about 3 km) is twice as steep as the inner one. Farther outward from the uplift, the crater bottom gently subsides, reaching its minimum hypsometric level (lower than –1,000 m sea level) about 5–6 km from the outer edge of the trough. Thus the maximum depth of the trough is displaced toward its outer slope. The detailed relief of the crater bottom is presumed to be more irregular, and may be considered as blocky.

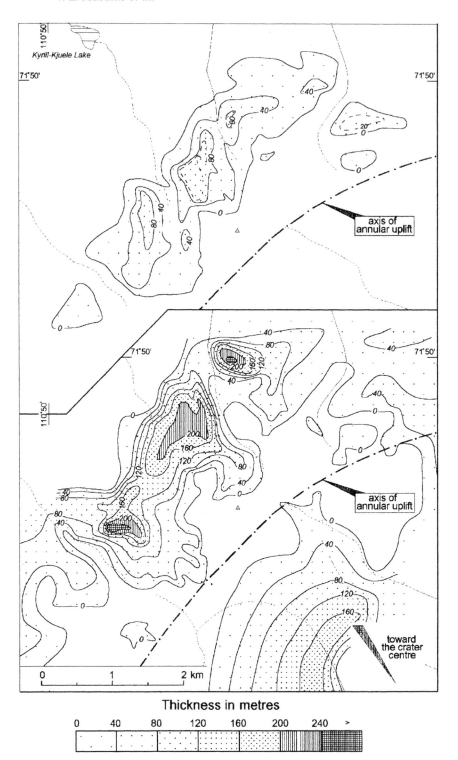

Figure 7. Isopach maps of the upper tagamite body (top map), and of the suevites (bottom map), Mayachika Upland. Main contour interval, 40 m.

The total ratio of the estimated volumes of allogenic lithic breccia, tagamite to suevite in the annular trough, is 12:4:1. Near the annular uplift and in the central part of the trough the sequence is composed mainly of impactites. Close to the outer edge, suevites and tagamites wedged out gradually, and allogenic lithic breccias are the predominant type of rock here.

Allogenic lithic polymict breccia is the lowermost unit of the trough sequence. It is confined to outer part of the annular trough and probably reaches about 1,000 m in thickness. In its lower part, this breccia is composed of blocks of sedimentary, rarely crystalline, rocks cemented together by fine-grained, polymict, clastic material (coptoclastite). Upward in the section, the admix-

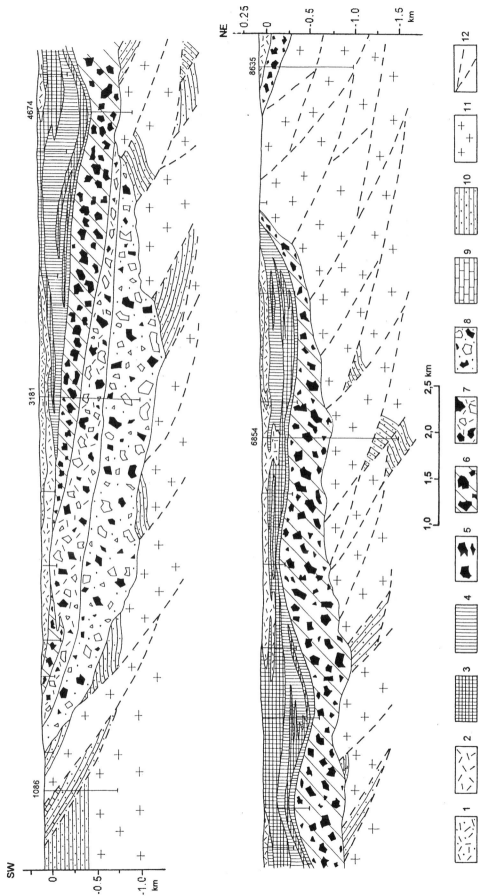

Figure 8. Geologic profile through the annular trough, Balagan River basin. 1, 2, suevites: 1, S-suevites; 2, C-suevites and V-suevites; 3–4, tagamites: high-temperature (3) and low-temperature (4) varieties; 5–8, allogenic lithic breccia; 5, monomict crystalline megabreccia; 6, the same, cemented by impactites; 7, polymict blocky breccia, cemented by impactites; 8, polymict blocky breccia, cemented by coptoclastites; 9, Lower Cambrian carbonate rocks; 10, Upper Proterozoic quartzites; 11, Archean gneisses (parautochthon); 12, thrusts, upthrows.

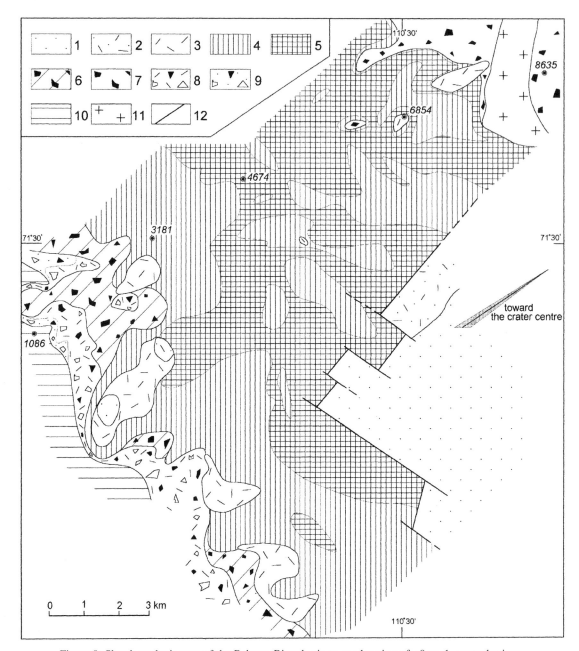

Figure 9. Sketch geologic map of the Balagan River basin at an elevation of +0 m. 1, coptoclastites; 2, S-suevites; 3, C-suevites and V-suevites; 4–5, tagamites: low-temperature (4) and high-temperature (5) varieties; 6–9, allogenic lithic breccia 6, crystalline megabreccia, cemented by impactites; 7, crystalline megabreccia, 8, polymict blocky breccia, cemented by impactites; 9, polymict blocky breccia, cemented by coptoclastites; 10, deformed sedimentary and partially crystalline rocks of the outer slope of the annular trough; 11, Archean gneisses (parautochthon); 12, faults.

ture of glass particles increases as does the content of crystalline blocks, and the cement matter of the breccia may be considered as suevite matrix.

Close to the annular uplift, parautochthonous rocks of true crater floor are overlain by a megabreccia of shocked (up to more than 20 GPa) crystalline rocks, cemented by crushed material of the same, or by tagamites and suevites. The megabreccia has variable thickness (up to 400–500 m in the central part of the trough), due not only to its uneven base but also to the irregular upper surface of this breccia. In places the roof of megabreccia rises above altitude +0 m, whereas in the adjoining area it may occur at –500 to –600 m sea level (Figure 9). These irregularities may be the result of an accumulation of large blocks; on the other hand, they may be due to erosion caused by radial flows of impact melt. The latter process occurred mainly near the outer edge where the borders of certain rock units are lobate in plan. Both the base (upper surface of polymict breccia) and the roof of crystalline megabreccia rise steeply in an outward direction (updip) here. These surfaces are characterized by several radially oriented grooves and ridges extending from the margin of the trough, at a distance of about 4 km (Figure 9) and partly reflecting the appropriate relief of the parautochthonous crater floor. Tagamite cement contributes about 13% of the total volume of crystalline megabreccia on average, although locally the amount may reach up to 50%. Upward in the section, the megabreccia grades to a continuous tagamite sheet comprising four-fifths of the upper part of the rock sequence.

The southeastern part of the Balagan River basin differs distinctly in its geologic structure from the region just described. A continuous field of coptoclastites grading to S-suevites northward and containing lenses of small bodies of suevites and tagamites was mapped here. Beneath a 228-m-thick coptoclastite layer, tagamites were encountered in a drill hole at an elevation of –133 m. The occurrence of tagamite at this depth is probably the result of post impact downfaulting, with the amplitude above 100 m.

The thickness of the tagamite sheet exceeds 600 m. This sheet spreads from the annular uplift up to the outer edge of the annular trough, locally (Figure 10). In our reconstructions we consider this as the central part of the area where the base of the sheet is reached by drill holes. Outside this segment northwestward and southeastward, the thickness of tagamites diminishes but may still be about 150–200 m.

The volume of tagamite body in this area is more than 10

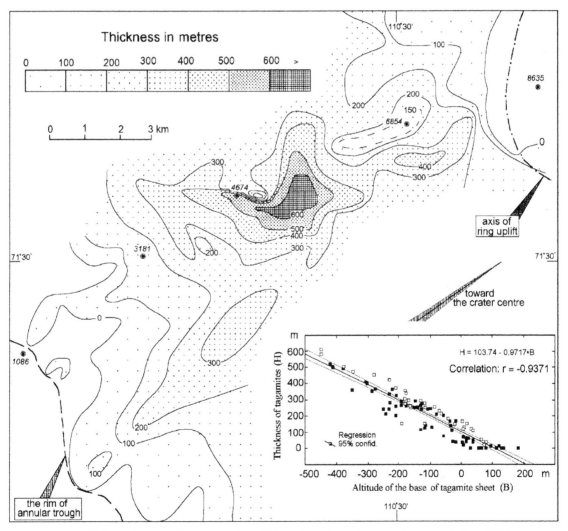

Figure 10. Isopach map of the tagamite sheet, Balagan River basin. Diagram shows correlation of thickness of tagamites (black squares, true thickness; open squares; apparent thickness) and elevation of the base of the tagamite sheet. The diversion of open squares from the regression line to higher value of thickness shows that the initial thickness of tagamites in these places was greater before erosion. Contour interval, 100 m.

times greater than in the Mayachika Upland. Besides, here the relief of the base and the roof of the tagamite sheet are significantly different. The base arises in a generally outward direction, but it has a more complicated relief in detail (Figure 11). The main features are several radial and near-radial trenches 1–3 km wide, separated by accumulations of large blocks of crystalline megabreccia that are uplifted locally to 450 m above the base of the tagamite sheet. These trenches could be traced to the outer margin of the annular trough and contain the bulk of the volume of tagamites. Although farther toward the outer edge of the annular trough, the tagamite sheet quickly thins; in some trenches it is still more than 100 m thick (for example, at a distance only 0.5 km from the margin).

Thus the thickness of tagamite sheet is correlated entirely with the relief of its base (Figure 10). However thickening of the tagamite sheet occasionally results in a relative rising of its upper surface; moreover, this elevation was more significant before erosion, this fact being a consequence of the correlation diagram (Figure 10). The roof of the tagamite sheet (Figure 12) is characterized by a general subsidence in centrifugal directions as well as by a combination of radial and concentrical elements of its structure. The radial elements appear as chains of narrow (0.5–1.5 km) deep (up to 150 m and more) depressions that do not coincide with trenches excavated in the base of tagamite sheet near the uplift, but coincide with them close to the outer slope.

A more precise examination of the tagamite sheet revealed some interesting internal structural patterns. Some subhorizontal bodies composed of LT- or HT-tagamites could be distinguished in its interior, in places with thick transition zones (up to 100 m in vertical section) between them. These simple bodies are not continuous and are variable in thickness (Figure 8); nevertheless, such a subdivision is inferred from a statistically recurring sequence. The LT variety contributes 58% to the total volume of the tagamite sheet and has a more uniform space distribution. The HT-tagamites commonly cause an abrupt increase in sheet thickness and local elevation of the roof. In radial directions, HT-tagamite wedge-out much closer to the annular uplift than LT-tagamites. Although the LT/HT volume ratio, in general, increases in centrifugal direction (appropriate correlation coefficient, r = 0.63), its correlation with the main appearance features (e.g., thickness, reliefs of the base and roof) of the tagamite sheet is insignificant. Consequently, the variability of tagamites originated before deposition of impact melt in the annular trough and is not related to postimpact processes.

About 1–2% of the tagamite sheet volume is made up by blocks of shocked crystalline rocks. Apart from areas near the annular uplift, the blocks are most common in narrow bands parallel but displaced southward relative to the position of maximum-thickness tagamites; in general, crystalline blocks are more common in the frontal zones of above-mentioned trenches excavated in the crystalline breccia base of the sheet.

Minor occurrences of tagamites are present in overlying suevites. They form lenses up to 45 m thick and more, and contribute less than 1% of the total volume of tagamites in the annular trough. Yet, in the upper part of the tagamite sheet (up to 100 m in vertical section), suevite lenses up to 40 m thick occur rarely. Tagamite lenses in suevites are found in depressions of the roof of the sheet and decrease regularly outward from the annular uplift. In contrast, suevite bodies in the tagamite sheet appear in zones where its roof is uplifted; they occur mainly in the outer part of the annular trough. The majority of tagamite bodies occurring in the suevite sequence undoubtedly belong to the uppermost

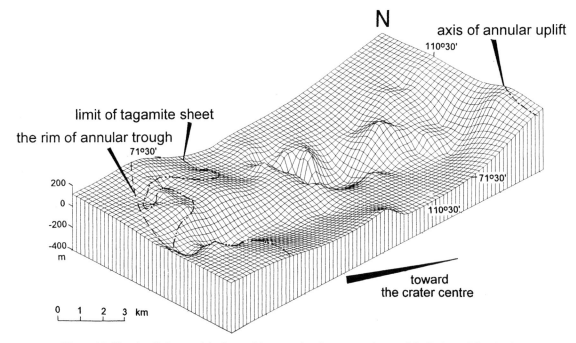

Figure 11. Sketch relief map of the base of the tagamite sheet, central part of the Balagan River basin.

varieties of the tagamite sheet, a fact confirmed by the similar correspondence of tagamite varieties in both settings.

Suevites overlie tagamites everywhere in the area with the exception the uppermost tagamite sheet (elevation above +150 m); they are inferred to have formed a continuous cover before erosion. The average thickness of preserved suevite cover is 42 m, with a maximum of 196 m. Despite erosion, the basic regularities of the inner structure of suevite cover are preserved. This suevite sheet is unhomogeneous due to local variations in the composition of clastic matter, so that all the three above-mentioned suevite types, namely V-, C-, and S-suevites, could be distinguished. The significant negative correlation coefficients between thicknesses of various suevite types show the sharp lateral distinction in vertical sections of entire suevite sequence from place to place.

As a whole, both the base and the erosional top of this suevite sheet dip southwestward away from the annular uplift to the edge of the annular trough; thus its thickness is mainly determined by the relief of the base, i.e., the roof of the tagamite sheet (correlation coefficient r = –0.82). Thus, the spatial distribution of suevites is characterized in general by a combination of radial and concentrical elements. However, the distribution of certain

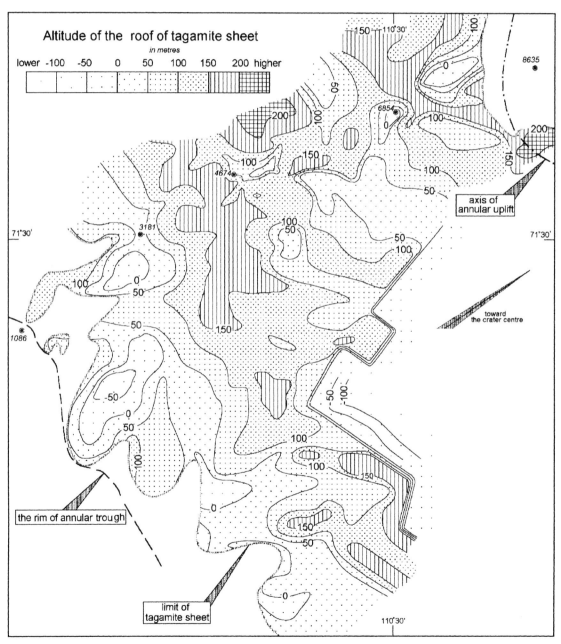

Figure 12. Hypsometric map of the roof of the tagamite sheet, Balagan River basin. The eastern part of the sheet has subsided along the system of faults. Contour interval 50 m.

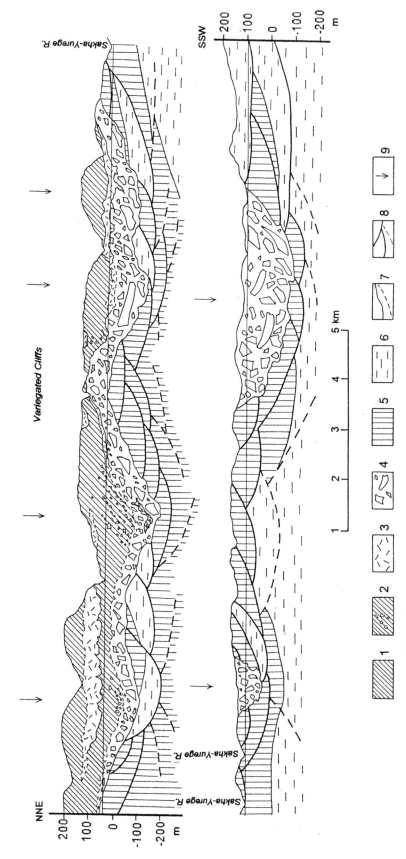

Figure 13. Geologic profile along the outer northwest edge of Popigai Crater, Sogdoku Upland. 1, tagamites; 2, tagamites; enriched by blocks of shocked gneisses; 3, suevites; 4, polymict megabreccia and blocky breccia; 5, Lower Cambrian carbonate rocks; 6, Upper Proterozoic quartzites; 7, geologic borders, traced and supposed; 8, thrusts, upthrows; 9, positions of axes of radial troughs. Straight line at an elevation above sea level of +50–100 m shows local base level of erosion.

suevite types is different. V-suevites and C-suevites are relatively evenly distributed. The V-suevites occur mainly in the areas where the thickest tagamite sheet occurs, in the central and near-uplift parts of the annular trough. C-suevites are the predominant variety recorded near the outer rim. On the contrary, discontinuous, narrow near-radial depressions in the roof of the tagamite sheet are filled mainly by S-suevites usually directly overlying tagamite; V-suevites (in inner parts of the annular trough) and C-suevites (in its outer parts) underlie S-suevites only in areas where the base of suevite cover has subsided. Finally, small (less than 0.5 km^2 in area) bodies of coptoclastites up to 132 m thick occur locally in these trenches.

In the southeastern faulted block (Figure 9), S-suevites alternating with predominant coptoclastites form a continuous cover expanding southeastward parallel to the principal structural elements of the crater. This block probably represents the uppermost part of the impact rock sequence that was removed by erosion elsewhere in the Balagan area.

OUTER EDGE OF THE TROUGH: SOGDOKU UPLAND

The Sogdoku Upland is located on the northwest flank of Popigai Crater between the continuous field of impactite and allogenic breccia and the deformed Upper Proterozoic and Lower Cambrian sequences. All these rocks are exposed at a distance about 35 km along river valleys parallel to western slopes of the Sogdoku Upland and its continuation to the south—the Suon-Tumul Upland. On the panorama sketch of Figure 13, the character of disturbed basement and its morphology are shown. The geologic bodies and their interrelations below the straight line (which is the local base level of erosion) are inferred according to data of geologic mapping and rare drill holes in the region. Sedimentary rocks of the crater edge (units 5 and 6 of Figure 13) generally gently dip northward and are disturbed by multiple thrusts and upthrows, arcuate in plan. Their planes dip eastward and southeastward; displacement amplitudes are evaluated at the least as in kilometers (Figure 14). Minor folds and faults occur here. In general, the base of allogenic blocky breccia and overlapping impactites has a lower hypsometric position in the north than in the south.

Radial troughs (they are peculiar morphostructural elements of the second order), which are excavated in deformed sedimentary rocks, are filled by blocky allogenic breccia, and partly by suevites and tagamites. There are six such troughs in the section, incised into sedimentary rocks down to a depth of about 100–300 m below the average level of the crater floor. One of the radial troughs located in the circular disturbed zone (Figure 15) was mapped in detail (Masaitis et al., 1975, Danilin, 1982). The section displays the inner structure of this trough filled in by blocky allogenic lithic breccia and C- and V-suevites with minor tagamites. The roof of the allogenic breccia is characterized by ridges and grooves parallel to the radial trough axis. In the front part of the trough the polymict allogenic breccia thins out and agglomeratic C- and V-suevites rest directly on upthrown sedimentary rocks. Thus, outward displacement of the upper rock units in the vertical section of the radial trough manifests a centrifugal tendency.

The Sogdoku Upland may be considered as a narrow plateau made up of tagamites and suevites, with a total maximal thickness that may reach 150–200 m. The tagamite sheet thickens to the north, split into two by a suevite lens. Similar lenses occur in this sheet in the southern part of the area (Figure 14). In a southward direction this sheet connects with the northwestern part of the thick tagamite sheet of the Balagan River basin described above. The base of the complex impactite body is irregular; this body thickens along the axes of radial troughs. In places, underlying allogenic lithic breccia "protrudes" into the tagamite sheet and is traced on the flat surface of the Sogdoku Upland. In the lower part of the tagamite sheet, numerous inclusions of shocked crystalline rock may occur; in these cases tagamites grade into megabreccia cemented by tagamites. The remnants of the suevite cover of the tagamite sheet are traceable in places; they may be regarded as sags (Figure 13), most of which are composed of V-suevites as well as lenses in the northern part of the section. In general, with transition from the lower part of the sequence to the upper one, there occurs a flattening of the base of lenticular and sheeted breccia bodies and impactite bodies, which are subhorizontal in the upper part.

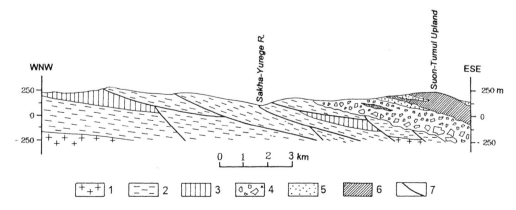

Figure 14. Geologic profile through the western edge of Popigai Crater, southern part of the Sogdoku Upland. 1, Archean gneisses; 2, Upper Proterozoic quartzites; 3, Lower Cambrian carbonate rocks; 4, allogenic polymict megabreccia and blocky breccia; 5, suevites; 6, tagamites; 7, thrusts, upthrows.

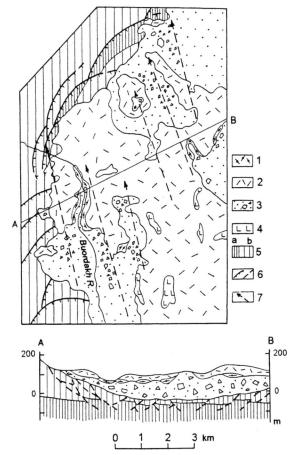

Figure 15. Sketch geologic map of the Buordakh radial trough. 1, Agglomeratic C-suevites; 2, lapilli V-suevites; 3, allogenic polymict blocky breccia; 4, tagamites; 5, sedimentary rocks a, Upper Proterozoic quartzites; b, Lower Cambrian carbonate rocks; 6, thrusts, upthrows in the map and on the section (suspected); 7, axes of the roof ridges of allogenic breccia.

CONCLUDING REMARKS

Popigai Crater may be regarded as a multi-ring impact structure, with the main concentric elements being expressed in the relief of true bottom, namely; central depression possibly with a flat central uplift, annular uplift, annular trough, and an outer crater zone. With more detailed consideration, radial and concentrical structural patterns of second and higher orders were recognized both in the relief of the crater bottom (in particular, radial troughs on the outer edge of the crater) and in the internal structure of certain impact rocks sequences.

The true bottom (crater floor) is composed of crystalline and sedimentary parautochthonous rocks and displays irregular relief caused by displacements of large blocks. These displacements were formed during the stage of early modification of the impact process before the deposition of ejected crushed and melted fallback material. The complicated structure of the parautochthon crater floor is partly reflected in the morphology of certain bodies of impact rocks including those occurring in the upper parts of sequences. The distribution of molten and fragmented masses in the crater interior is characterized by asymmetry probably caused by an oblique impact. Both parautochthon crater floor and impactites and allogenic lithic breccia are partly disturbed by local postimpact faulting.

Impactites (tagamites and suevites) and coarse- and fine-grained allogenic breccia form thick (up to 1 km and more) sequences in the crater, which may be considered as "pseudolayered" because they consist of subhorizontal lenticular and sheet-shaped geologic bodies. In particular, it should be noted, that in general sequence, several tagamite lenses or sheets divided by suevite layers occur. In some cases distinct types of impactites and allogenic lithic breccia have gradual transitions between each other, but transition zones are thin in comparison with the thickness of specific geologic bodies. They are mostly homogeneous, but not infrequently show complicated combinations of different lithologies. The largest tagamite sheets (up to 600 m thick) are located on the outer slope of the annular uplift and in the surrounding annular trough. Commonly, they fill concentrically elongated depressions in the relief of underlying allogenic breccia. In the deposited sequence the uppermost tagamite bodies are more flat in appearance compared to the lower member. Comparison of the tagamite sheets of the Mayachika Upland, Balagan River basin, and Sogdoku Upland reveal distinct features, including the character of the roof, interrelations between the relief of the roof, the relief of the base with the thickness of the tagamite sheets, and variation of thickness. These features are believed to be the result of differences in the amount of melt, and, consequently, of its thermal capacity, viscosity, and the like.

The morphology and the inner structure of certain rock bodies are characterized by a combination of radial and concentric patterns. The radial patterns have their arm-shaped forms in ridge-like relief of their bases, and show specific distributions of such features as clasts and other components. The significance of these patterns becomes less distinct, from the bottom to the top of thick sequences, as well as laterally in a centrifugal direction. In the same directions the variability of hypsometry of boundaries between certain sheet-like bodies and the thickness of transition zones become less obvious. The concentric patterns are reflected in such ways as the distribution of thicknesses of certain bodies, in the positions of their outer borders as well as of their gravity centers of mass (in general, higher in the section they are displaced to the center of the crater), and in morphology of boundaries of upper units.

The spatial differentiation of rock varieties was caused by distinct composition of ejected separate flows or currents. We believe that mixing of melted and brecciated material, which was ejected along different trajectories, occurred only during the initial stages of material movement and was insignificant during the main phases of transportation and deposition. At the same time, the existence of transition zones indicates simultaneous deposition of the main mass of melted and brecciated rocks. Only the fine-grained breccia (coptoclastite) and minor suevites were deposited late, mostly from clouds of dust and melt particles. It is possible to distinguish two main tendencies in the spatial interrelations of the frontal parts of different geologic units higher in the section of thick sequences. Centripetal and centrifugal depositional environments can be distinguished. The first environment is confined to the annular trough, the second to radial troughs near the rim of the crater. The depositional environments were caused by differences in impulses of ejected currents and their trajectories.

Although the models and structural patterns described above are based on a study of relatively small areas of the entire impact structure, the results can be extrapolated to most of Popigai Crater. The similarity of constitution of these areas to the more extended territories is based on the similarity of their geologic settings.

The interpretation of our data resulted in three-dimensional models of impact rock bodies within the Popigai impact crater. The data provide a better knowledge of various mechanisms of breccia and melt-rock deposition. Combined with drilling data and geophysical studies of other terrestrial impact structures, this work will eventually lead to a better understanding of impact cratering processes.

ACKNOWLEDGMENTS

The drilling program was carried out under the supervision of L. M. Zaretsky, N. A. Donov, and V. T. Kirichenko; we thank them all. We are also grateful to A. I. Raikhlin, V. A. Maslov, and A. N. Danilin for some drill-core data. Assistance with the preparation of the manuscript by S. M. Gadasina, A. T. Maslov, and T. I. Vasiljeva is also acknowledged. M. Dence, B. Dressler, and S. Kieffer provided thorough and constructive reviews of the manuscript, which led to its considerable improvement.

REFERENCES CITED

Basilevsky, A. T., Ivanov, B. A., Florensky, K. P., Yakovlev, O. I., Feldman, V. I., and Granovsky, L. B., 1983, Impact craters on the Moon and planets: Moscow, Nauka Press, 200 p. (in Russian).

Bottomley, R., Grieve, R., York, D., and Masaitis, V., 1997, The age of the Popigai impact event and its relation to events in the stratigraphic column: Nature, v. 388, p. 365–368.

Danilin, A. N., 1982, On characteristics of inner structure of thick sequences of allogenic breccia and suevites in large astroblemes: Meteoritika, v. 40, p. 102–106 (in Russian).

Masaitis, V. L., 1984, Giant meteorite impact: some models and their consequences *In* Modern ideas of theoretical geology: Leningrad, Nedra Press, p.151–179 (in Russian).

Masaitis, V. L., 1994, Impactites of the Popigai impact structure, *in* Dressler, B. O., Grieve, R. A. F., and Sharpton, V. L., eds., Large meteorite impacts and planetary evolution: Geologic Society of America Special Paper 293, p. 153–162.

Masaitis, V. L., Mikhailov, M. S., and Selivanovskaya, T. V., 1975, The Popigai meteorite crater: Moscow, Nauka Press, 124 p. (in Russian).

Masaitis, V. L., Danilin, A. N., Mashchak, M. S., Raikhlin, A. I., Selivanovskaya, T. V., and Shadenkov, Ye. M., 1980, The geology of astroblemes: Leningrad, Nedra Press, 231 p. (in Russian).

Masaitis, V. L., Mashchak, M. S., Raikhlin, A. I., Selivanovskaya, T. V., and Shafranovsky, G. I., 1998, Diamond-bearing impactites of the Popigai astrobleme: St. Petersburg, VSEGEI Press, 180 p. (in Russian).

Melosh, H. J., 1989, Impact cratering: geologic process: New York, Oxford University Press, 245 p.

Raikhlin, A. I., Selivanovskaya, T. V., and Masaitis, V. L., 1980, Rocks of terrestrial impact craters: problems of classification, in Proceedings, 11th Lunar and Planetary Science Conference, Abstracts: Houston, Texas, Lunar and Planetary Institute, p. 911–913.

MANUSCRIPTED ACCEPTED BY THE SOCIETY DECEMBER 16, 1998

Popigai impact structure (Arctic Siberia, Russia): Geology, petrology, geochemistry, and geochronology of glass-bearing impactites

Sergei Vishnevsky
Institute of Mineralogy and Petrology, 3 University Prospect, Novosibirsk-90, 630090, Russia

Alessandro Montanari
Osservatorio Geologico di Coldigioco, 62020, Frontale di Apiro, Italy

ABSTRACT

The Popigai basin in central Arctic Siberia has a diameter of about 100 km, and is the largest known Cenozoic impact structure on the Earth. Its size, undeformed outline, well-preserved state, good exposure, and wide variety of target lithologies and impactites make it a prime object for the study of large impact structures.

We describe Popigai impactites following an original classification and genetic interpretation, focusing on differences and contrasts with other published works. Popigai impactites exhibit a reversed stratigraphy in the crater margin, as is commonly found in other impact craters, but the upper impactoclastic fill within the crater contains abundant fragments of low-shocked, soft Mesozoic rocks derived from the uppermost layers of the target. This suggests a complex origin of the structure that does not satisfy a simple cratering model. This and other unusual aspects of the Popigai impact formations are here explained with the hypothesis of a "dynamic barrier" in the expanding explosion cloud. This barrier is represented by a torus-shaped screen of vertically ejected megablocks of Mesozoic rocks above marginal and middle zones of the growing crater, which would have controlled the process of mixing, transport, and deposition of ejected material.

Two petrographic groups of impact melt rocks and suevite glasses are recognized in the Popigai: cryptocrystalline tagamites and pristine suevite glasses and microcrystalline tagamites and recrystallized suevite glasses. Petrographic and geochemical analyses indicate that water inherited from the target rocks had an important role in the evolution of these different impact melts. Most of impact melt rocks and suevite glasses are here interpreted as products of low-temperature, quickly cooled impact melt saturated with fragments of crystalline rocks, and derived from marginal parts of the melting zone. We propose that high-temperature impact melt derived from the inner part of the melting zone was excavated, disintegrated, and disseminated outside the crater, and sporadic occurrences of homogeneous clast-free glasses in some suevites within the crater are the sole preserved evidence of this high-temperature melt.

Globular glasses in some microcrystalline tagamites were produced by a selective mobilization of Si, K, and volatiles, accompanied by loss of Fe, Ca, and Mg, and are here interpreted as products of early "impact anatexis" of crystalline target rock inclusions that occurred in the final stage of the impact shock-loaded state, or immediately after it.

Results from iridium instrumental nuclear activation analyses of Popigai impactites indicate that some impact melt rocks are contaminated by the impactor.

Radioisotopic dates from Popigai impact glasses range from 81 to 8 Ma, suggesting the unpredictable influence of target rock inclusions, and/or diagenetic alteration effects on age determinations. Nevertheless, the bulk of radioisotopic dates and several fission-track age determinations indicate a mid-Cenozoic age for the Popigai impact. Modern radioisotopic laser-spot geochronology on the right samples is needed to test the hypothesis that the Popigai impact event is the source for an impactoclastic air-fall layer globally found in upper Eocene marine sediments.

INTRODUCTION

The Popigai structure, centered at 71°38′ N, 111°12′ E in central Arctic Siberia (Fig. 1), has a diameter of about 100 km, and is the largest known Cenozoic impact structure on Earth. The impact origin of this structure was first recognized by Masaitis (1970), and documented with detailed petrographic, structural, and geochemical investigations by a number of Russian researchers (e.g., Masaitis, 1983, 1984, 1994; Masaitis and Selivanovskaya, 1972; Masaitis et al., 1971, 1972, 1975, 1980; Vishnevsky, 1978, 1981, 1994a; Vishnevsky and Palchik, 1975; Vishnevsky et al., 1975a,b).

An extensive drilling project of more than 450 drill holes, some of them to a depth of 1.5 km, was carried out by the Regional Krasnoyarsk Geological Survey between 1973 and 1978 for the prospecting of impact diamonds. About 1,000 tons of drill cores were obtained and stored in the Siberian taïga near the drilling sites (Vishnevsky, 1994b,c).

This impact crater is unusual because of its large size, undeformed outline, relatively young age, well-preserved state, good exposure, and various types of target rocks that are easily recognizable in various impactites (Vishnevsky, 1994b). Moreover, Popigai exhibits a wide variety of impact rocks and formations, some of them with no known equivalent in other terrestrial impact craters. Popigai is included in the UNESCO (United Nations Environmental, Scientific and Cultural Organization) provisional global list of geologic heritage as a first-class site of scientific and general cognitive interest (Cowie, 1991).

Some apparently contradictory geologic and petrologic features of Popigai impactites have been described by Vishnevsky (1994a). One is a lack of reversed stratigraphy in the suevite sequence and the presence of soft Mesozoic target rock blocks (up to 80 m in size) in the uppermost part of the suevite column within the inner part of the crater; these rocks were derived from upper target stratigraphy, and, according to simple cratering models (e.g.,

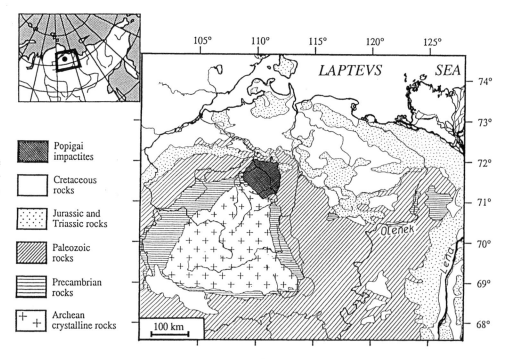

Figure 1. Geographic location of the Popigai impact structure and simplified geology of surrounding areas.

Gault et al., 1968; Maxwell, 1977; Ivanov, 1979) these blocks should have been removed, transported, and deposited outside the crater and covered by suevites. Another feature is the two spatially different sources of material in suevites that correspond to a sedimentary source characterized by unshocked or mildly shocked material from the crater's near-margin inner zone, and a basement source characterized by highly shocked gneisses, derived from the central zone of the crater. A third feature is evidence of intense mixing between different types of suevites that took place when the material was in a plastic or even lithified state; as a result, very complex relations between different suevite deposits are observed, including "breccia-in-breccia" relations, in which fragments of one suevite are cemented by another type of suevite.

These features are not easily explained by simple models of ballistic and fall-out deposition of suevites. Breccia-in-breccia relations and the lack of unequivocable evidence of the projectile in impactites were, among others, the reasons that led several Russian investigators to propose a volcanic origin for the Popigai impact structure (e.g., Polyakov and Trukalev, 1974; Trukhalev and Solov'ev, 1981; Mezhvilk, 1981; Vaganov et al., 1985; Marakushev et al., 1993).

Popigai has no obvious analogs among other known large impact structures on Earth (e.g., Manicouagan, Sudbury, Vredefort, Kara, Chicxulub, Morokweng) that are either deeply eroded, partially exposed, or completely buried and therefore known only from geophysical investigations and limited drill core data. On the other hand, well-preserved and thoroughly studied younger craters such as the Ries in southern Germany (24 km in diameter), may not be comparable with Popigai because of their smaller size. As a result, it is difficult to describe some important features of the Popigai impact formations strictly using classifications and definitions commonly used in the international literature (e.g., Stöffler and Grieve, 1994). We present here our original classification of Popigai impactites mainly based on the work of Vishnevsky (1978, 1981, 1994a, in Russian), recently summerized in English by Vishnevsky and Montanari (1994).

In this chapter, we summarize the results of 25 yr of studies conducted at the Institute of Mineralogy and Petrology in Novosibirsk on the geology of the Popigai impact structure that have never been published in the international literature. In particular, we describe the petrology, geochemistry, and geochronology of its glass-bearing impactites. A geologic map of the structure is shown in Figure 2A, while the locations of all the outcrops and samples mentioned herein are shown in Figure 2B.

TARGET AND CRATER GEOLOGY

Target

The target of the Popigai impact structure are Archean crystalline rocks of the Anabar Shield, and overlying Proterozoic to Mesozoic sedimentary sequences. Archean basement rocks are represented by the Verkhne-Anabar, and the Khapchan series (Rabkin, 1959; Lutz, 1964, 1985), which are exposed to the south of the crater. These rocks belong to the amphibolite and granulite facies of regional metamorphism, and are strongly deformed and folded about a west-northwest strike, with steep or nearly vertical dips of gneiss sheets. The folding is associated with faulting and linear mylonite zones.

The Verkhne-Anabar Series is up to 10 km thick and represents the lower, terrigenous metamorphosed member of the Archean Geosyncline Zone (Lutz, 1985). It is mainly composed of leuco- and mesocratic hypersthene, and two-pyroxene gneisses and plagiogneisses, as well as amphibole- and biotite-gneisses. Minor units of quartzite, biotite-garnet-sillimanite-, and cordierite-gneisses are also present.

The Khapchan Series is up to 5 km thick and is a metamorphosed, carbonate flysch sequence of the upper part of the Archean Geosyncline Zone. It consists of biotite-garnet-hypersthene gneisses (commonly graphite-bearing), two-pyroxene gneisses, and plagiogneisses. Biotite-amphibole gneisses, marbles, calcite-pyroxene-scapolite-quartz rocks, and scapolite-diopside schists are also present in this unit.

The average composition of Anabar Shield gneisses (Rabkin, 1959), compared with the average compositions of other Archean shields (Poldervaart, 1955), and the Popigai impact melt rocks, or tagamites (Masaitis et al., 1975), are shown in Table 1.

Basic and ultrabasic rocks, and various granites of Archean age, are present as igneous bodies in the crystalline basement. Metabasalts are common in the Verkhne-Anabar Series, comprising up to 10–20% of its volume, whereas they are rare in the Khapchan Series. These basic bodies are mainly represented by amphibole schists, two-pyroxene schists, and other types of schist, and occur as sheets, lenses, vein-like bodies, and isolated blocks. These rocks are of special interest because they represent a possible source for Ni, Co, and Cr anomalies detected in Popigai impactites (see below). The bulk geochemical compositions of some of these basic rocks are shown in Table 2, while the contents of Ni, Co and Cr are shown in Table 3.

Ultrabasic rocks, being a minor constituent of the crystalline basement, occur in gneisses as sheets 10–100 m thick, and up to 1–2 km long. They consist of amphibolized and serpentinized peridotites and pyroxenites.

Granitoids of various compositions (granodiorites, granosyenites, granites, and alaskites), morphology (irregular or sheet-, lens-, and vein-like bodies), age, and size comprise up to 10% of the gneiss volume.

The sedimentary cover has a total thickness of 1.5–1.7 km, and begins with a 500-m-thick, upper Proterozoic sequence of gray and pinkish-gray to brick-red quartz arenites, dolomites, and sandstones. These rocks are exposed west and southeast of the crater. Both the Archean units and Proterozoic strata are intersected by gabbro-diabase and diabase dikes of Proterozoic age. Gray and yellowish-gray Cambrian dolomites, marls, and limestones, with a total maximum thickness of 850 m, are exposed along the western and eastern margins of the crater. Along the northern side of the structure, dark Permian sandstones are interbedded with coal-bearing siltstones and clays

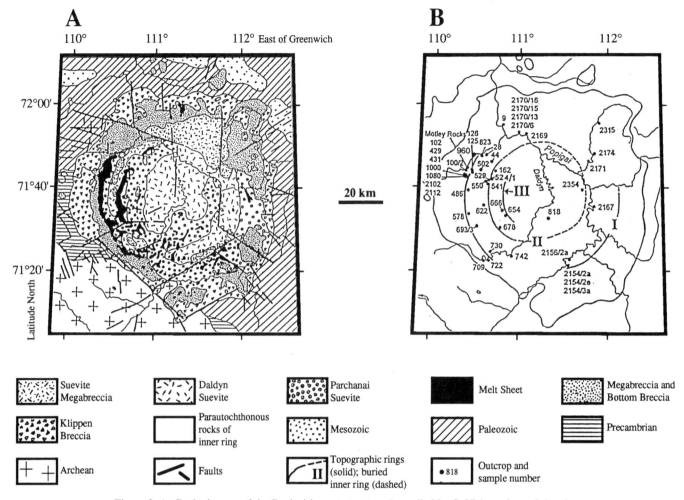

Figure 2. A, Geologic map of the Popigai impact structure (compiled by S. Vishnevsky and data by M. T., Kirjushina, E. I., Podkopaev, L. P., Smirnov, A. I., Trukhalev, K. S., Zaburdin, and V. L., Masaitis, among others. B, Sample and outcrop locations (indicated by numbers) described in the text.

with a total thickness of 150 m. Numerous dolerite dikes of Permian-Triassic age intrude all the above-mentioned target rocks. Rare sill-like bodies of these rocks are also present. A Mesozoic unit, sporadically exposed northeast in the nearest vicinity of the crater, includes some Triassic volcano-sedimentary formations, none of which are found in Popigai impactites, some Jurassic marine and continental formations, fragments of which were observed in Popigai suevites, and Cretaceous siltstones, sandstones, coal-bearing sands, and clays. The preimpact thickness of the Cretaceous member in the area now occupied by the northeastern part of the crater was at least 150 m (Vishnevsky et al., 1995a). These rocks are the most important components of the Popigai suevite formations.

Crater morphology and general geology

The Popigai structure is a circular basin characterized by a complex topography. The central and eastern sectors are occupied by flat lake and swamp plains with elevations of 20–60 m a.s.l. The western and southern sectors are topographically more complicated, with elevations up to 230 m a.s.l. A system of hills is arranged in three concentric chains, forming incomplete topographic rings, which are separated from each other by river valleys (Figs. 2B, 3, and 4). Rings I and III are of secondary origin: erosional for ring I, and complex for ring III, which resulted from a combination of slumping of melt masses, isostatic relaxation, and erosion (Vishnevsky, 1980).

The basin is surrounded by a smooth plateau, with relics of a Paleogene peneplane surface. This plateau reaches elevations of 300–400 m a.s.l. and is cut by deep and narrow river valleys. The fluvial network in the basin and its vicinity has, in general, a radial-concentric pattern.

Both surface and buried geology of the Popigai impact structure are characterized by a concentric arrangement (Figs. 2A and 4). This structure, which lacks a topographic outer rim, is separated from slightly deformed surrounding areas by a complex system of dislocations such as uplifted blocks, centrifugal thrusts, klippens, and overturned thrusts. As with the Ries (Pohl et al.,

TABLE 1. CHEMICAL COMPOSITIONS OF SOME REPRESENTATIVE ARCHEAN TARGET ROCKS OF THE ANABAR SHIELD COMPARED WITH AVERAGE COMPOSITIONS OF ARCHEAN SHIELD ROCKS AND POPIGAI IMPACTITES

	Biotite-garnet Gneiss*	Biotite-hypersthene Garnet Gneiss*	Garnet Granulite*	Biotite-hypersthene Plagiogneiss[†]	Hypersthene Gneiss[†]	Hypersthene Plagiogneiss[†]	Anabar Shield Average[§]	Archean Shield Average**	Popigal Tagamites[‡]
(wt. %)									
SiO_2	70.54	59.08	72.29	69.70	63.24	63.91	64.4	64.4	63.13
TiO_2	1.0	0.79	0.14	0.42	0.52	0.46	0.6	0.6	0.76
Al_2O_3	12.09	19.20	14.22	15.07	17.64	17.16	16.6	15.5	14.68
Fe_2O_3	4.11	0.63	2.00	2.03	1.46	1.87	2.4	1.8	1.99
FeO	2.46	7.65	2.89	2.29	3.33	3.71	3.1	2.8	4.97
MnO	0.08	0.05	0.25	0.04	0.06	0.15	0.1	0.1	0.08
MgO	2.96	2.97	1.27	1.87	2.34	2.58	2.3	2.0	3.82
CaO	2.75	5.15	1.96	3.50	5.77	5.54	4.2	3.8	3.43
Na_2O	0.67	2.79	3.40	3.02	3.39	3.45	3.1	3.5	1.96
K_2O	1.93	1.68	1.20	1.17	1.84	0.98	3.0	3.3	2.72
P_2O_3	....	0.37	0.03	0.14	0.13	0.10	....	....	....
H_2O[§§]	....	0.08	0.19	0.07	0.08	0.06	....	....	....
L.O.I.	0.11	....	0.61	0.36	0.49	0.53	....	....	2.31
Total	98.70	100.44	100.45	99.68	100.29	100.50	97.1	97.8	99.85

*Khapchan Series after Rabkin, 1959.
[†]Verkhne-Anabar Series after Rabkin, 1959.
[§]After Rabkin, 1959, without water and L.O.I.
**After Poldervaart, 1955, without water and L.O.I.
[‡]After Masaitis et al., 1975.
[§§]Hygroscopic water extracted at 105°C.

TABLE 2. CHEMICAL COMPOSITIONS OF SOME METAVOLCANIC MAFIC ARCHEAN TARGET ROCKS
(After Lutz, 1985)

	Khapchan Series		Granulites***	Verkhne-Anabar Series			Anabar Shield Average[§]	Popigal Tagamites Average[§§]
	Hypersthene Gneiss*	Hypersthene Gneiss**		Hypersthene Gneiss[†]	Hypersthene Gneiss[††]	Hypersthene Gneiss[†††]		
(wt. %)								
SiO_2	56.50	65.33	68.13	54.73	60.16	65.46	64.4	63.13
TiO_2	1.10	0.91	0.47	0.80	0.73	0.68	0.6	0.76
Al_2O_3	17.88	14.80	15.49	15.71	16.95	15.54	16.6	14.68
Fe_2O_3	3.80	1.88	1.77	2.31	2.19	1.59	2.4	1.99
FeO	5.24	4.51	2.65	6.60	5.20	4.18	3.1	4.97
MnO	0.15	0.11	0.08	0.15	0.13	0.08	0.1	0.08
MgO	2.48	2.81	1.37	5.53	3.49	1.19	2.3	3.82
CaO	6.28	4.73	3.18	8.58	5.18	3.54	4.2	3.43
Na_2O	3.79	2.07	3.09	3.44	3.42	3.23	3.1	1.96
K_2O	1.60	1.71	2.80	0.94	1.35	2.22	3.0	2.72
P_2O_5	0.19	0.15	0.09	0.17	0.24	0.18	....	....
H_2O+L.O.I.	....	....	....	....	....	....	....	2.31
Total	99.01	99.01	99.12	98.96	99.04	97.89	97.1	99.85

Note: Ni, Co, and Cr compositions in ppm for analyses for Khapchan and Verkhne-Anabar series rocks are reported in Table 7.
*, **, ***, and [†], [††], [†††] correspond to samples indicated with the same symbols in Table 3.
[§]After Rabkin, 1959, without water and L.O.I.
[§§]After Masaitis et al., 1975.

TABLE 3. Ni, Co, AND Cr CONTENTS IN
VARIOUS POPIGAI TARGET ROCKS

Rock Type	Ni	Co	Cr
(ppm)			
Peridotite	1000	600	1500
Pyroxenite	1000	600	1000
Amphibole-two-pyroxene-garnet schist[§]	46	47	100
Amphibole-two-pyroxene-garnet schist[§]	70	35	200
Hypersthene gneiss[†]	80	30	165
Hypersthene gneiss[††]	32	21	115
Hypersthene gneiss[†††]	16	16	50
Hypersthene gneiss	15	19	50
Hypersthene gneiss	10	15	18
Granulite	15	11	17
Granulite	6	6	20
Amphibole-pyroxene schist[§§]	90	50	320
Hypersthene gneiss[*]	16	20	210
Hypersthene gneiss[**]	15	22	150
Granulite[***]	10	7	100
Granulite	10	4	50
Average Anabar Shield gneisses[¥]	27	13	80
Average Permian-Triassic dolerites[¥¥]	220	71	390
Average terrestrial mafic rocks[-]	160	45	200
Average Popigal tagamites[--]	85	13	110

[*], [**], [***], and [†], [††], [†††] correspond to samples indicated with the same symbols in Table 2.
[§]From the Verkhne-Anabar Series.
[§§]From the Khapchan Series.
[¥]After Masaitis and Raikhlin, 1989.
[¥¥]In the immediate vicinity of the Anabar Shield, after Kopylova and Oleinikov, 1973.
[-]After Vinogradov, 1962.
[--]After Masaitis and Raikhlin, 1989.

Figure 3. Panoramic view of topographic ring I in the western part of the Popigai Crater basin looking north from the top of Motley Rock. The western cliffs of the Khara-Khaia Mountain (151 m above sea level.; on the right), which is made up of the Melt Sheet Formation, dominate the Rassokhe River valley, about 120 m below (on the left). This valley separates ring I from the border of the crater basin and surrounding plateau (up to 300 m above sea level) seen at the horizon of the picture.

1977), the boundary between slightly deformed surroundings and the crater itself is represented by this tectonic rim.

Structurally, the Popigai is characterized by the following five main features (Figs. 2A and 4): (1) a deeply buried central uplift recognizable only in gravity and magnetic data (Masaitis, et al., 1980); (2) an inner ring of Archean basement rocks, 47–50 km in diameter, exposed in the northwest and west sectors, and buried in other places where it was recognized through geophysical investigations; (3) a central or inner basin 2.5 km deep, bounded by the inner ring; (4) an annular trough 1.3–2 km deep that surrounds the inner ring; and (5) a tectonic rim zone. A zone of slightly deformed rocks surrounding the crater is recognizable to a maximum distance of 70 km from the crater's center following topographic analyses and remote sensing (Masaitis, 1994). A concentric arrangement of the crater periphery is complicated by a system of shallow radial troughs, up to 15 km in length, which were formed by centrifugal flows of displaced material from the inner part of the crater locally arranged like festoons.

Target rocks surrounding the crater exhibit attenuating faulting (Masaitis, 1994). Impact diatremes and horsts are locally found in sedimentary terrains in the west, northwest, and southeast surroundings of the crater, 3–5 km from its tectonic rim. In west and northwest outskirts, these dislocations are localized in subhorizontal Cambrian rocks. These are represented by Pastakh (Fig. 5) and Ed'en-Yurege diatremes with elliptical elongated shapes, up to 3.5 km in length, which are expressed by topographic depressions, and are filled with a chaotic mixture of fragments of various sizes, from blocks up to 700 m in size down to pebble size. This clastic material is derived from all country rock types, from Archean gneisses to Triassic dolerites. Horsts from 300 m to 2 km in size are here represented by Proterozoic rocks. The elongation of these structures follows the regional west-northwest faulting trend. Except for "gries" breccia, which is a good field evidence of weak shock metamorphism, and which was described as such by Preuss (1969) in the Ries Crater, no other evidence of shock metamorphism is recognizable here (Vishnevsky, 1986). Sometimes, impact horsts are found in klippen breccia close to the tectonic rim of the crater. Diatremes and

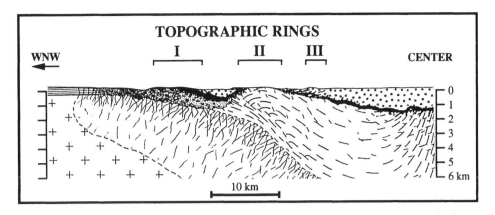

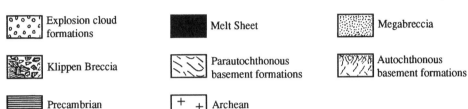

Figure 4. Radial cross section of the Popigai impact structure interpreted and synthetized from original field data and other published information (e.g., Masaitis et al., 1975, 1980).

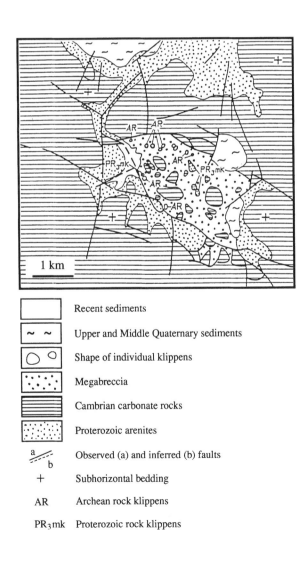

horsts provide evidence for subvertical shock impulses transmitted from the deep interior of the growing crater.

Proximal ejecta deposits outside the crater are not very widespread and have a distinct asymmetry in their distribution. Only small, isolated patches of ejecta are known in the western, southeastern, and eastern outskirts of the crater, whereas in the northern and southern surroundings there are several large patches of ejecta deposits. These patches are found in valleys as well as in elevated areas and hills.

The scarcity of proximal ejecta deposits is a particular feature of Popigai that is discussed below. Distal ejecta deposits are even more elusive than proximal deposits but apperently more widespread, as indicated by a strewn field of impact diamonds that can be traced at distances up to 500 km from the crater center (Fig. 6A) (Vishnevsky et al., 1995b). Impact diamonds are also widespread in impact melt rocks within the crater (Fig. 6B).

IMPACT FORMATIONS

The classification of Popigai impact formations (Fig. 7) is based on their geologic setting, stage of shock metamorphism, stratigraphic provenance, petrographic composition, and macroscopic evidence of transport mode (i.e., ballistic vs. turbulent). Following the nomenclature commonly used in the international literature (e.g., Stöffler and Grieve, 1994), we apply the term "impact" for all the geologic formations produced by the impact. As a first rank of classification, the Popigai impact formations are

Figure 5. Geologic map of the Pastakh impact diatreme located on sedimentary terrains about 3 km west of the Popigai structure (after Vishnevsky, 1986).

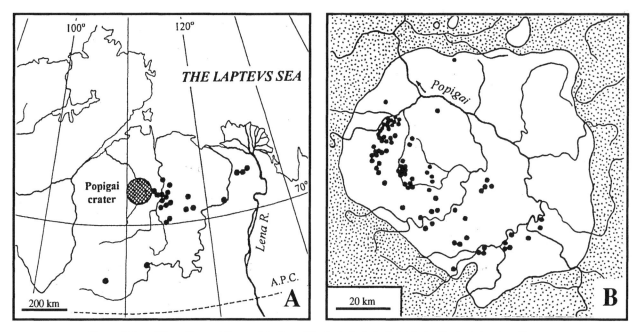

Figure 6. Localities of impact diamond findings (black circles). A, Outside the Popigai structure (the so-called Yukatite strewn field after Vishnevsky et al., 1995b); B, within the impact structure (after Vishnevsky et al., 1997).

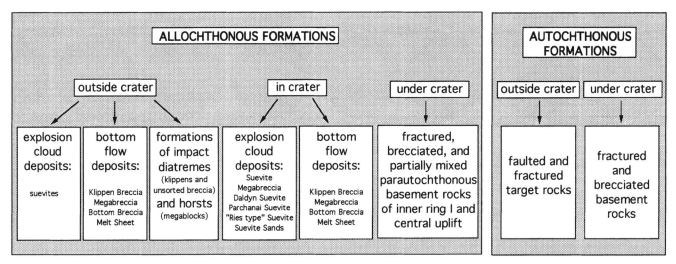

Figure 7. Classification of Popigai impact formations according to Vishnevsky (1978, 1981, 1994a) and Vishnevsky and Montanari (1994).

divided into "allogenic" or "allochthonous" (i.e., impactites were displaced from their original position during cratering), and "authochthonous" (i.e., impactites remained more or less *in situ* during cratering).

A second rank of classification takes into account the spatial position of the rocks with respect to the tectonic border of the crater and its basement. Accordingly, impact formations are defined as proximal ejecta deposits outside the crater, crater fill inside the crater, and basement formations (Fig. 7).

A third rank of subdivision is based on the position of a formation within the crater fill and its contact relations with other formations. This distinction is crucial for interpreting the geology and cratering mechanics of the Popigai. Accordingly, two formation groups can be recognized. A first group includes formations composed of material we interpret as transported, mixed, and deposited in a nonballistic mode (bottom flow deposits); a second group includes impact formations whose material is here interpreted as ballistically transported (explosion cloud deposits) (Fig. 7; see also discussion below).

The first group comprises klippen breccia, megabreccia, tuff-like breccias, and impact melt rocks (tagamites). These impactites may be referred to as "base surge formations", inas-

TABLE 4. MAIN IMPACT FORMATIONS OF THE POPIGAI CRATER

Impact Formations	Particle/Body Size	Stratigraphic Provenance	Shock Metamorphism*	Geological Setting	Texture
Faulted; fractured, and brecciated autochthonous rocks	....	All stratigraphic units	Stage 0 <5 GPa	Undisplaced crater basement; adjacent area outside the crater.	Monomict
Faulted, fractured, and brecciated parautochthonous	....	Crystalline rocks	Stages 0–II <35 GPa	Displaced rocks of crater basement (Inner ring and central uplift)	Monomict
Impact diatremes[†] and horsts	As much as 2 km	Crystalline to Triassic rocks	Stages 0–1 <10 GPa	Outside the crater; in margin part of klippen zone	Polymict for diatremes, monomict for horsts
Klippen Breccia	>100 m to several km	All stratigraphic units	Stages 0–I <10 GPa	Klippen zone; crater basement in annular trough; outside the crater; among other bottom formations	Polymict for klippen zone
Megabreccia	>1 to 100 m	All stratigraphic units	Stages 0–II <35 GPa	Megabreccia zone and filling of annular trough; irregular bodies outside the crater, in klippen zone and among other bottom formations.	Polymict
Bottom Breccia	<0.001 to 1 m	All stratigraphic units	Stages 0–IV <100 GPa	Irregular bodies outside the crater, in klippen and megabreccia zones, and in the Melt Sheet layer.	Polymict
Melt Sheet	As much as hundreds $km^{3†}$	Crystalline rocks	Stage IV 55–100 GPa	Continuous layer and irregular bodies of tagamite among other formations inside and outside the crater.	Polymict (impact glass + rock and mineral fragments)
Explosion cloud formations	<0.001 to 80 m	All stratigraphic units	Stages 0–IV <100 GPa	Uppermost filling of central basin and annular trough; irregular bodies in the upper Melt Sheet and other bottom formations in and rarely outside the crater.	Polymict
Dikes[†]	0.01 m to several m	All stratigraphic units	Stages 0–IV <100 GPa	Crater basement; klippen and megabreccia blocks; suevites.	Polymict

*Stages of shock metamorphism after Stöffler, 1971.
[†]Body size or volume.

much they are made of material transported in the form of dense, near-surface, turbulent, centrifugal megaflows. However, because impact melt rocks are particularly abundant in this group, the term "bottom flow formations" (Vishnevsky, 1978, 1981, 1994a) is, in our opinion, more applicable here than "base surge formations," which would exclude the presence of large volumes of transported melt. Tuff-like, coarse- to fine-grained breccias belonging to this group of impact formations are here referred to as "bottom suevite" and "bottom fragmental breccia," depending on whether impact glass is present.

A second group of impact formations is here referred to as "explosion cloud deposits," and includes all impact products we interpret as ballistically transported (i.e., suevites and suevite megabreccia; see discussion below).

A special group of allogenic formations outside the crater is represented by impact diatremes (i.e., chaotic mixtures of klippens and unsorted breccias) (see Fig. 5), and horsts (Vishnevsky, 1986). Trukhalev and Solov'ev (1981) described some of these diatremes as cryptovolcanic objects. Another special group of allogenic formations is represented by parautochthonous basement rocks that constitute the inner ring and the central uplift. Finally, dikes of various origin, not shown in Figure 7 complete the rank of Popigai allogenic formations. The geologic setting, texture, stratigraphic provenance, shock metamorphism, and particle or body size characteristics of the main Popigai impact formations are shown in Table 4.

Dikes in the Popigai impact structure are of various genetic types. Some of them repersent injections of impact melt into shocked and brecciated gneisses of the inner ring, and megabreccia fragments (Masaitis, 1994). Similar dikes are expected to be common in the buried crater basement within the central uplift and the inner basin. Pseudotachylite dikes are also known in some klippens and megabreccia fragments of crystalline rocks. Other dikes were intruded during transport of bottom centrifugal flow material. For example, dikes of injected impact melt are known among the klippen fragments of Proterozoic rocks in the western sector of the klippen zone (Vishnevsky, 1978). Another group of dikes was formed during deposition of suevite megabreccia: subvertical dikes of Daldyn suevite are present in large fragments of soft Cretaceous rocks, and were formed probably by upward ejection of mobile material in subsiding masses (Vishnevsky, 1994a). Finally, postcratering neptunian dikes are known in some suevites and tagamites (Mashchak and Fedorova, 1987).

Bottom formations

Klippen Breccia Formation. This formation constitutes a ~10-km-wide belt around the crater, and is made of displaced klippens of all the target rocks once present in a given sector of the cratering area, from 100 m up to several kilometers in size. Except for the larger size of the blocks (klippens), this formation may be compared to the "megablock zone" of the Ries Crater. However, close to the tectonic rim of the crater, the variety of klippen lithologies is limited to the target rocks exposed in the immediate surroundings. The direction of displacement is clearly centrifugal, and locally klippens are arranged in sickle-like patterns with their concave sides toward the center of the crater. In some other places, only simple uplift, folding, and rotation of klippens are evident. The klippen rocks are fractured or brecciated, but their original stratigraphy is preserved. The intensity of block fragmentation and mixing of klippens increases toward the center of the crater. The transition from Klippen Breccia to Megabreccia occurs either as a gradual decrease of the size of the fragments, or as an interfingering zone between the two formations.

Megabreccia Formation. This formation (Fig. 8A) is a polymict breccia containing clasts of all target rocks that were present in a given sector of the cratering area; it is made of fragments 1–100 m in size. This formation constitutes another concentric belt that overlays the inner part of the Klippen Breccia belt. Megabreccia fragments exhibit various grades of shock metamorphism with maxima of stage II (Stöffler, 1971) in crystalline rocks. Stishovite was first found in a block of Archean crystalline rock in this formation by Vishnevsky et al. (1975a). Hard Paleozoic and Proterozoic rock fragments commonly exhibit "gries" texture and contain rare shatter cones, whereas soft Mesozoic fragments show no visible traces of shock metamorphism. In most cases, fragments in the Megabreccia Formation are cemented by bottom suevites and fragmental breccias

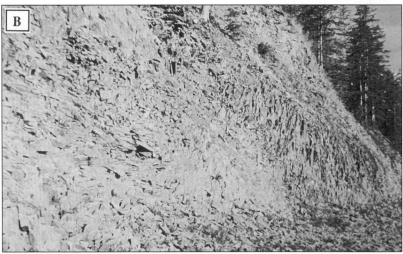

Figure 8. A, Tagamite cap and underlying megabreccia, representing bottom flow formations (Motley Rocks outcrop in the northwest sector of the crater margin; see Fig. 2B). The cap was emplaced by gravity flow of impact melt. B, Outcrop of a large body of tagamite representing part of the Melt Sheet, exposed in the southwest sector of the crater periphery along the Daldyn River bank (outcrop no. 722).

(see below); as a geologic body, this formation is similar to Ries Bunte Breccia. However, the Ries Bunte Breccia lacks impact glass that is common in the matrix of the Popigai Megabreccia; and it is mostly found outside the crater.

Bottom Breccia Formation. This formation (Fig. 7) is a tuff-like polymict breccia made up of two main components: fragments (<1 mm to 1 m in size) of comminuted target rocks, and particles of impact glass. Sand is the dominating grain size of target rock clasts contained in the Bottom Breccia. A general feature of impact glass particles in this formation is their low porosity. These breccias are found as discrete bodies, which justify their distinction as a separate impact formation. However, the same kind of polymict tuff-like breccia is an important matrix component in the Megabreccia and sometimes in the Klippen Breccia Formations. Bottom Breccia Formation rocks are texturally and compositionally variable from place to place; variations from one kind to another may be either gradational or sharp. As in the Megabreccia, shock metamorphism is more intense in crystalline clasts, and it is low grade or absent in sedimentary clasts. However, crystalline clasts in the Bottom Breccia Formation exhibit higher grades of shock metamorphism than in the Megabreccia, up to stages III–IV of Stöffler (1971), as is demonstrated by the presence of coesite (Vishnevsky et al., 1974; 1975b).

Two types of rocks in the Bottom Breccia Formation are distinguished on the basis of their stratigraphic provenance and range of shock metamorphism.

Type I is a loose or weakly to moderately lithified impactite where sedimentary fragments are significantly more abundant than crystalline ones. Impact glass is not a discriminatory characteristic for this type of impactite. In most cases it is lacking, while in some other cases rounded fragments of impact glass comprise up to 30% of the rock volume. In most cases, this type of impactite constitutes the matrix in megabreccia, and it also forms discrete bodies within or in the uppermost part of the Megabreccia Formation.

Type II, the second type of impactite in the Bottom Breccia Formation, is a moderately to strongly lithified rock, dominantly made of finely comminuted or partially molten crystalline rocks with a minor component of sedimentary clasts. Coarser crystalline rock clasts are less common. Sizable fragments of other target lithologies are rare or even absent in some places. Glass is present mainly in the form of schlierens impregnating the clastic matrix. Clastic glass particles are common in some cases. The abundance of glass varies broadly, from 20–30% to 80–90% of the rock volume. This type of rock sometimes constitutes the matrix in the Megabreccia Formation, but more commonly it forms irregular bodies distributed in the uppermost part of the formation. Type II Bottom Breccia is in contact with the overlying Melt Sheet Formation (see below), and exhibits complex relationship with it (interfingering and isolated inclusions in the lower part of the Melt Sheet). This Bottom Breccia is similar in some respects to the Ries crystalline breccia described by Pohl et al. (1977), but differs from it in its presence of impact glass (Fig. 9).

Following the classification of Stöffler and Grieve (1994), Bottom Breccia rocks free of impact glass would be classified as fragmental breccia, whereas glass-bearing rocks would be considered as "suevite." However the features and spatial location of the rocks described above require distinction between "standard" suevite and the glass-bearing rocks of type II Bottom Breccia; for this reason we describe them as "bottom suevites" in this chapter; similarly, type I of Bottom Breccia Formation is here described as "bottom fragmental breccia."

Melt Sheet Formation. This formation is essentially made of impact melt rock, concentrated in the uppermost part of bottom flow deposits (Figs. 2A, 4, and 8A, B), and represented by a stratum up to 600 m thick (Masaitis, 1994) that gently dip toward the center of the crater. In some exposures within the inner ring, the Melt Sheet Formation lies directly on the crater basement. In other parts of the inner crater, numerous intrusions of tagamite into the suevite cover indicate that most or all the bottom of the inner crater is covered by this formation and related impact melt rocks. Vertical and lateral relations with surrounding formations are very complex, and indicate turbulent mixing during transport and emplacement.

Figure 9. Bottom Breccia Formation type II. The rock is strongly lithified, contains Archean gneiss fragments (light patches), and it is cross-cut by two subhorizontal fracture systems. Small tagamite bodies (dark elongated areas marked with "t"), with swirled and complex morphology, indicate turbulent mixing of the material in plastic or molten state (outcrop no. 960).

Explosion cloud formations

In the annular trough and in the inner crater, tagamites of the Melt Sheet and other bottom formations are covered by various explosion cloud formations made of different types of suevites (Figs. 2A and 4). In general, all of them may be described in accordance with Stöffler and Grieve (1994) as a polymict clastic breccia composed of lithic and mineral clasts of variously shocked target rocks, and cogenetic particles of impact glass. However, at least five explosion cloud formations can be recognized in the Popigai impact structure (Vishnevsky, 1978, 1992, 1994a): Parchanai Formation, Daldyn Formation; Suevite Sand Formation; Suevite Megabreccia Formation; and "Ries-Type" Suevite Formation. Except for the last formation (see below), all other suevite formations are composed of non- or low-shocked Mesozoic (mostly Cretaceous) clastic material and various amounts of impact glass derived from crystalline rocks. Unshocked or moderately shocked clasts of sedimentary Proterozoic and Paleozoic target rocks are less common than Mesozoic clasts, whereas strongly shocked crystalline rock fragments are rare.

With the exception for the Suevite Megabreccia Formation, the grain size of target rock clasts in Popigai suevites is generally less than 2 mm. Coarser fragments of target rocks are less common, whereas single fragments up to several meters across are rare. The size of the glass fragments varies in the different types of suevites (see below). Suevite glasses from explosion cloud deposits, and especially those of the Parchanai Formation, are generally pumice and constitute a macroscopic criterion for recognizing these suevites from those deposited from the bottom flow.

Parchanai Formation. This formation, named after the Parchanai River where it is extensively exposed (see Vishnevsky, 1978, 1994a), is a well-lithified rock composed of 50–70 vol% lapilli-size glass fragments. Coarser fragments and bomb-size glass bodies are rare. The rest of this suevite formation is a matrix made up of comminuted (mainly Mesozoic) target rocks. Lapilli and coarse glass particles were deposited in a solid state, whereas bombs up to several decimeters in size show clear signs of deformation, indicating that they were in a plastic state when deposited. Intense alteration of the glass is a common feature in these suevites.

Large bodies of suevite occupy the lower part of the suevite cover in the annular trough and in the inner basin of the crater (Fig. 2A). The contact between Parchanai suevites and the underlying tagamites is characterized in some places of the inner ring by a transitional zone (Vishnevsky and Pospelova, 1988), which is represented by a mixture of irregular and swirled tagamite and suevite bodies up to several tens of meters in size. Rare tagamite bodies 2 m or less is size are also disseminated in the Parchanai Formation in the central basin of the crater. Within the first topographic ring, small irregular bodies of Parchanai suevites are present in the uppermost part of the Melt Sheet, and in the upper part of the Megabreccia (see, for example, the case of Motley Rock in Fig. 8A).

Daldyn Formation. This unit named after the Daldyn River where this formation is exposed (Vishnevsky, 1994a), is a poorly to moderately lithified rock made of suevites compositionally similar to those of the Parchanai Formation. However, in contrast to Parchanai suevites, the impact glasses here are almost always fresh. Another distinction is that glass fragments in Daldyn suevites are smaller, mainly in the ash-size range. Rare glass bombs are also fresh, and sometimes exhibit evidence of plastic deformation after deposition. Voluminous masses of Daldyn suevites occupy the upper part of the suevite cover in the inner basin and annular trough (Fig. 2A). This type of suevite often constitutes the matrix of suevite megabreccia.

Suevite Sand Formation. This formation is a loose or poorly lithified polymict impact rock mostly made of sand-size clasts of Cretaceous target rocks with minor amount of clasts of other target lithologies derived from marginal areas of the crater. Ash-size particles of quenched, pure impact glass constitute 3–5% to 15–20% of the volume of the rock. Low-stage shock metamorphism is seldomly seen in sedimentary rock fragments, whereas strong shock metamorphism is evident in rare gneiss fragments. Suevite sands were deposited in a cold state, and are associated with other explosion cloud deposits at the top of the suevitic column as bodies in Daldyn suevites, as well as matrix in the suevite megabreccia.

Suevite Megabreccia Formation. This formation is a chaotic mixture of mainly Cretaceous target rock fragments, 1–80 m in size, cemented by suevite sand and Daldyn-type suevite. More rarely, lumps of other sedimentary target rocks are included in the breccia. Fragments of megabreccia display rare "gries" breccia and shatter cones indicating weak shock metamorphism. This suggests that lumps of the megabreccia were excavated from middle and marginal areas of the growing crater. The overall texture of the megabreccia matrix is very complex, with evidence of turbulent mixing in a dense lithoid state between various kinds of the suevites (Fig. 10A–D). Other traces of strong mechanical activity indicate forced movement of soft target rock fragments in the matrix (Fig. 11A-B).

Ries type Suevite Formation. This formation is a strongly lithified polymict breccia made up of clastic matrix (up to 70 vol%), and ash- and lapilli-size fragments of quenched fresh glass. The clastic matrix is made up of finely comminuted, target rock fragments generally less than 1 mm in size, that are mixed with various amounts of glass microparticles and fragments of Mesozoic rocks, including coal. An important component in the matrix is represented by granoclasts of crystalline rocks. This formation is characterized by a dominating proportion of material derived from shocked crystalline rocks, as in the Ries fall-back suevite; it differs from all other suevites described above. It has a limited occurrence, and is found stratigraphically above the Melt Sheet Formation within the inner ring zone (see Fig. 2B, point no. 678), but its contact relations with tagamites and other suevite formations in this part of the crater are not exposed.

Proximal impact formations

In western and northwestern areas outside the crater, there are several small isolated bodies of impactites. The largest of these measures 700 × 1,100 m, and it is found on subhorizontal Cambrian rocks in the immediate vicinity of the crater's tectonic rim. It is made up of a megabreccia (all country rocks including blocks of Archean gneisses) with a coarse, glass-bearing matrix. Impact glass is present in the form of irregular schlieren-like particles of several millimeters to several centimeters in size. In other localities at distances of 5–10 km from the tectonic rim of the crater, small groups of Archean gneiss blocks were deposited on Cambrian rocks in valleys and on hillside slopes.

In a locality 11 km from the southwestern tectonic rim, a small (~0.5 km across) isolated patch of tagamite associated with irregular suevite bodies is present on a watershed, resting on Archean gneisses. Large patches of ejecta deposits are localized on Cambrian and Permian terrains in valleys, on hillside slopes, and on watersheds up to 20 km from the northern rim of the crater. They are made up of megabreccia and single klippens of various country rocks with a matrix of bottom fragmental breccia and suevite. Single bodies of bottom glass-bearing breccia and tagamite are also present. Proximal ejecta deposits in south and southeast areas outside the crater were briefly described by Masaitis (1994).

Autochthonous and parautochthonous formations

The parautochthonous basement, which constitutes the inner ring of the Popigai impact structure, is exposed in the western sector of the inner ring (Figs. 2A and 4). The basement is composed of intensely dislocated crystalline rocks. They exhibit evidence for shock metamorphism up to stage II of Stöffler (1971), including development of planar deformation features (PDFs) in quartz. From place to place, the basement of this part of the inner

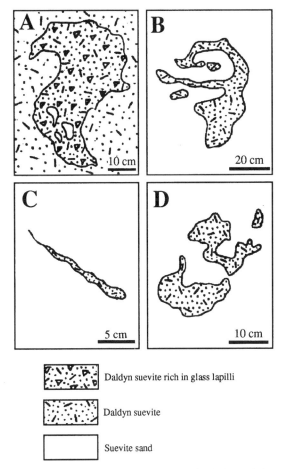

Figure 10. Suevite megabreccia: traces of turbulent mixing between different suevite bodies in lithoid state. A, Sketch from outcrop no. 2170; B and C, from outcrop no. 2315; D, from outcrop no. 2374. Glass-rich Daldyn suevite body in Figure 10A contains fragments <1 cm of Proterozoic arenites; three small inclusions of suevite sand are present in the lower part of this body.

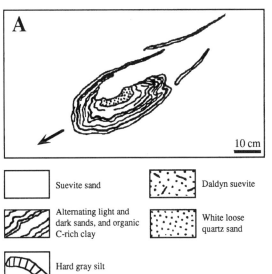
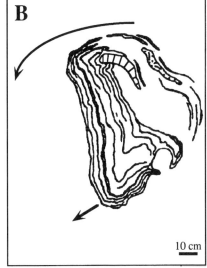

Figure 11. Suevite Megabreccia Formation: two examples (A) and (B) of forced movement of Cretaceous rock fragments in lithoid matrix made of suevite sand (outcrop no. 2315).

ring is covered either by various rocks of the Melt Sheet and Bottom Breccia or by Parchanai suevite. Moreover, in some cases, Parchanai suevites were deposited in the spaces between dislocated blocks of the crystalline basement. Slumping of the tagamite melt and molten masses of bottom suevite is observed on the slopes of the ring. These features indicate a fast uplifting of more than 1.5 km of the inner ring during cratering.

PETROGRAPHY OF TAGAMITES AND SUEVITE GLASSES

Two types of glass-bearing impactites occur at the Popigai structure: tagamites and suevites. In this section, we describe the petrography of different types of Popigai tagamites and of Daldyn suevite. Tagamites and suevites alike are essentially impact breccias, for they both contain various amounts of impactoclasts (mineral or lithic, with or without evidence of shock metamorphism) derived from the comminution of target rocks. The main difference is that, in tagamites, the impact glass constitutes the matrix, whereas in suevites the matrix is essentially clastic, and impact glass is present as particles.

Among all the suevite formations described above, we focus on Daldyn because it best exemplifies suevite as an impact breccia, and it contains unaltered glass. Except for the "Ries-type" suevite, which is briefly touched in this section, suevites from other impact formations are not dealt with here because they either contain scarce glass or are completely altered. Further petrographic information on these Popigai suevites can be found in Masaitis (1983, 1994), Masaitis et al. (1975, 1980), and Vishnevsky (1978, 1981, 1992, 1994a).

Tagamites

Tagamite (from the Tagamy hills of the Popigai basin) is the name proposed by Masaitis et al. (1971) for rocks formed after impact melting of Archean crystalline basement in the central part of the Popigai Crater. Thereafter, this term has been broadly used in Russian literature also for other craters. Thus, tagamite may be regarded as a synonym for impact melt rock (or impact melt breccia) commonly used in the international literature (e.g., Stöffler and Grieve, 1994). This impact melt rock contains 5–30 vol% fragments of parental crystalline rocks dispersed in a glass matrix.

Irregular bodies of tagamite are sometimes disseminated in various bottom formations, but the bulk of this rock is concentrated in the Melt Sheet Formation (Figs. 2A, 4, and 8A,B).

Two types of tagamites can be recognized in the Popigai. The first type is a dark gray (sometimes with a bluish tinge), very hard cryptocrystalline rock with conchoidal fractures. The second type is a dark gray, fragile, microcrystalline rock with irregular fracturing. Microcrystalline tagamite is subdivided into single-melt rocks, whose matrix is made up of one kind of glass only, and globular rocks that display glass globules dispersed in a glassy matrix of different composition (Fig. 12A–C).

The spatial relations between cryptocrystalline and microcrystalline tagamites within the Melt Sheet are very complex and often characterized by close interfingering between bodies of various sizes. Globular tagamites are exposed on topographic rings I and II, in the marginal and middle parts of the crater. They comprise up to 10–15% of the Melt Sheet Formation in the western sector of the crater (Vishnevsy and Pospelova, 1984).

Most of the clasts in tagamites range in size between 0.02 and 3 mm. Coarser fragments, in the range of centimeters to decimeters, are much less abundant, whereas large fragments one to several meters across are very rare. The <2-mm fraction is made up mainly of clasts of quartz and feldspar, with subordinate amounts of biotite, pyroxene, zircon, and a few other accessory minerals. Coarser fragments are mostly represented by crystalline rocks, whereas clasts of other target lithologies are very rare in these impactites.

The bulk geochemistry of Popigai tagamites (see below), and the paucity of other than crystalline rock inclusions, indicates that mineral clasts in these rocks were derived from the crystalline basement. Therefore, all the fragments included in Popigai tagamites are here referred to as Archean rock fragments.

Relatively large Archean rock fragments, up to several decimeters in size, often exhibit traces of intensive plastic deformation such as folding and bending. These Archean rock fragments display various stages of shock metamorphism (Fig. 13A,B). However, these features are often poorly preserved due to postimpact annealing and recrystallization within the cooling melt, especially in cryptocrystalline tagamites. Nevertheless, three groups of large Archean rock fragments are recognized by Vishnevsky and Pospelova (1986) in unaltered microcrystalline tagamite on the basis of their shock metamorphism (Table 5). Archean rock fragments of group I, which show low-stage shock metamorphism, were clearly incorporated in the tagamite melt from outside the melting zone.

Small Archean rock fragments, ranging from 0.02 to 1–2 mm in size, are represented by mineral clasts, such as quartz and feldspar, and constitute the dominating part of the fragments in both crypto- and microcrystalline tagamites. Other mineral fragments of crystalline rocks, such as pyroxene, biotite, zircon, and others, are rare. Mineral clasts in tagamites also exhibit various stages of shock metamorphism (Figs. 13,C–F).

High stages of shock metamorphism in quartz are represented by lechatelierite schlierens (fused silica glass), and by angular grains of diaplectic quartz glass. Fresh forms of these products, like those shown in Figure 13A, are rare in tagamites and are usually replaced by new generations of quartz (Fig. 13C) or by semi-transparent fine quartz-cristobalite aggregates with ballen structure similar to those described by French et al. (1970), Vishnevsky et al. (1974), and Carstens (1975) (Fig. 13B). Grains of fresh shocked quartz, partially transformed into diaplectic glass (such as those shown in Fig. 14A,B) are also rare in tagamites. Lower stages of shock metamorphism represented by quartz grains with PFs (planar fractures) and PDFs (planar deformation features) are present in some cases, whereas fragments of

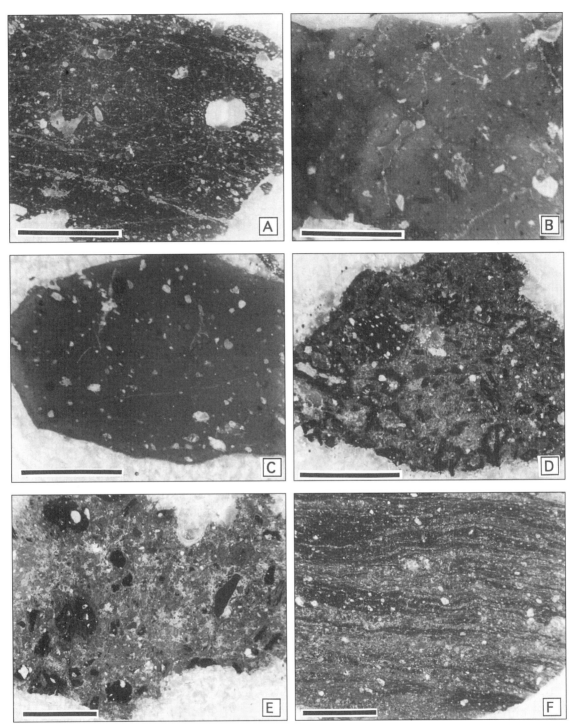

Figure 12. Popigai impact rocks described in the text as seen in polished slabs: A, Altered cryptocrystalline tagamite: sample no. 730; B, slightly-altered microcrystalline tagamite: sample no. 429; C, globular tagamite, very weakly altered: sample no. 102; D, Ries-type suevite containing slightly altered pristine glass fragments, sample no. 678; E, Daldyn suevite with strongly altered glass fragments (clasts and rims around large fragments), sample no. 709. F, Slightly altered fragment of banded glass from a suevite bomb; dark bands are made of recrystallized glass, whereas light bands are pristine glass; sample no. 2154/2a. Scale bars, 1 cm.

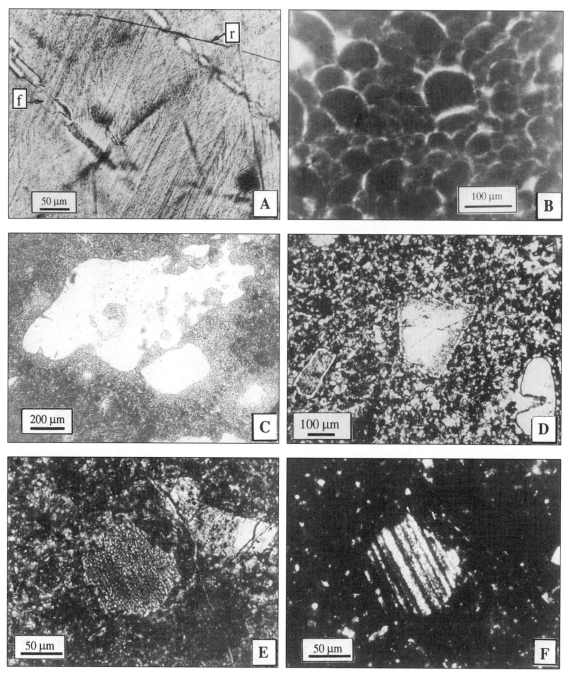

Figure 13. Shock metamorphism in rock and mineral fragments of Archean gneisses contained in Popigai impact melt rocks. A, Diaplectic quartz collected from a 25-cm Archean gneiss fragment in a cryptocrystalline tagamite, showing a birefringence as low as 0.001 (iron-gray colors under crossed polars), and displaying two sets of PDFs; needle-like inclusions of crystalline rutile (r) and chains of fused feldspar (f) are also present. After Stöffler and Langenhorst (1994), this sample documents shock pressures ranging between 33 and 40 GPa (sample no. 578, parallel polars). B, Diaplectic quartz glass annealed to fine-grained quartz-cristobalite aggregates with "ballen" structure, contained in a 20-cm fragment of shocked graphite-bearing quartz from cryptocrystalline tagamite. This fragment contains also aggregates of impact diamond paramorphs along parental graphite (cubic and hexagonal lonsdaleite phases), with impurity of chaoite (Vishnevsky and Pal'chik, 1975) (sample no. 486, parallel polars). C, Schlieren blob of lechatelierite replaced by quartz, in microcrystalline tagamite (sample no. 28, parallel polars). D, Fragments of shocked feldspar in microcrystalline tagamite, partially (in center), or completely (on the right) replaced by fine-grained new generations of feldspar with simultaneous extinction of crystallites, and surrounded by rim overgrowths of pure feldspar (sample no. 1000, parallel polars). E and F, Fine-grained, new generation feldspar replacing an Archean feldspar fragment in a microcrystalline tagamite, and inheriting the polysynthetic twinning of the precursor mineral (sample no. 1000, in parallel (E) and crossed (F) polars). As in Figure 8D, these aggregates with preferred orientation of crystallites were probably developed from diaplectic feldspar glass.

TABLE 5. SHOCK METAMORPHISM OF CRYSTALLINE INCLUSIONS IN POPIGAI TYPE IIB TAGAMITE
(After Vishnevsky and Pospelova, 1986)

Shock metamorphism (after Stöffler, 1971)	Type of parental rock Fragment size Sample Number	Rock forming minerals: shock metamorphism, annealing, and recrystallization				Other features
		Quartz	Feldspar	Garnet	Biotite	
Group 1 Stage Ib (low shock)	Biotite-garnet gneiss 5 m No. 2102	Several sets of PFs and PDFs	PDFs and partial transition to diaplectic feldspar glass; the glass is recrystallized into fine aggregates with parallel extinction and with non-vitrified parts of the mineral.	Cryptocrystalline aggregates of olivine and cordierite* which replace garnet of pyrope (36.8%)-almandine (63.2%) composition. The garnet morphology is preserved.	Replacement by fine intergrowths of opaques, sanidine (?), and cordierite (?). Original biotite morphology is preserved.	Interstitial films of impact anatectic glass (see text) which comprises as much as 2.3% of the rock volume. See composition in Table 9, analyses 16 and 18. Glass exhibits coronas of non-Ca ferrihypersthene (53% ferrosilite) at the contact with garnet ghost.
Group 2 Stage II-III; (moderate shock)	Biotite garnet gneiss 30 cm No. 2112	Diaplectic glass recrystallized into cryptograins of quartz-cristobalite aggregates* with "ballen" structure	Diaplectic plagioclase(?) glass recrystallized into fine aggregates of new feldspar generation with simultaneous extinction. Fused (K?) feldspar recrystallized into fine aggregates with random extinction.	Replacement by fine intergrowths of olivine (74% fayalite) and cordierite (43% Fe species) composition. The garnet morphology is preserved.	Replacement by fine intergrowths of opaques, sanidine (?), and cordierite (?). Original biotite morphology is preserved.	Broad patches of impact anatectic glass, which comprise as much as 15–20% of the rock volume. See composition in Table 9, analyses 19 and 20. Coronas of ferrohypersthene (62% ferrosilite) around ghosts of quartz crystals.
Group 3 Stage III: (strong shock)	Ghost gneiss inclusion 20 cm No. 431	Diaplectic quartz glass replaced by fine aggregates of quartz, cristobalite, and tridymite*	Partial or complete homogenization into mixed glass; schlieren-like ghosts of percursor mineral in the form of aggregates of feldspar microliths with random extinction.	Ghosts of precursor mineral in aggregates of fine hercynite crystals (Mg species).	Homogenization into mixed glass.	Mixed glasses in spaces between percursor quartz grains, and ghosts of other minerals. Coronas of coarse ferrihypersthene around ghosts of quartz crystals, pyroxene trichites, and euhedral ilmenite flakes. Impact anatectic glass only as a border around gneiss inclusions; see composition in Table 9, analyses 10, 11, 14, and 15.

* = determination by x-ray diffractometry.

the mineral with no visible traces of shock metamorphism are common.

As for quartz, shocked feldspar fragments may range from recrystallized aggregates of fused monomineralic glass (schlieren-like, fine aggregates of crystallites with random extinction), to more or less recrystallized diaplectic feldspar glass with parallel extinction of crystallites, often inheriting the twin structures of the precursor mineral (Fig. 13D–F). Low-shocked feldspar with planar microdeformation structures is rarely found in tagamites, whereas feldspar lacking traces of shock metamorphism is common.

Other Archean rock minerals, such as pyroxene and biotite, are less common in tagamites and are usually opacitized, with relics of the precursor mineral in the cores. Some completely opacitized grains may represent shocked garnet, recrystallized to cryptocrystalline aggregates of olivine and cordierite similar to those described in Table 5 for shocked garnet in gneiss inclusions. Single grains of zircon are rare, and show no evidence of shock metamorphism by optical means.

Tagamite types described above exhibit clear petrographic differences at a microscopic scale, and are characterized by various degrees of glass recrystallization and alteration (Vishnevsky, 1978, 1981; Vishnevsky and Montanari, 1994). Cryptocrystalline rocks (Fig. 15A) exhibit a glassy matrix with cryptocrystals or very fine crystallites, <10 μm in size. Single-melt microcrystalline tagamite is characterized by a glass matrix which is made

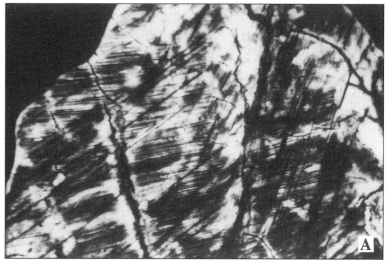

Figure 14. Intergrowth of several shocked quartz crystals in a 2-mm gneiss fragment contained in a pristine glass fragment in Daldyn suevite. This shocked quartz is partially transformed into low birefringent silica (dark areas), with several sets of planar microstructures produced by shock pressures of 25–35 GPa (according to Grieve et al., 1996). Grain 1 on the left contains two sets of PFs, parallel to ω {10$\bar{1}$3} orientation (PF1 - c-axis∧⊥ = 24°, and PF2 - c-axis∧⊥ = 22° (tabular ω{103} c-axis∧⊥ = 22°56′), and one set of PDFs, parallel to π {10$\bar{1}$2} orientation (measured c-axis∧⊥ = 34° vs. tabular value π {10$\bar{1}$2} c-axis∧⊥ = 32°25′); (crossed polars; sample no. 678).

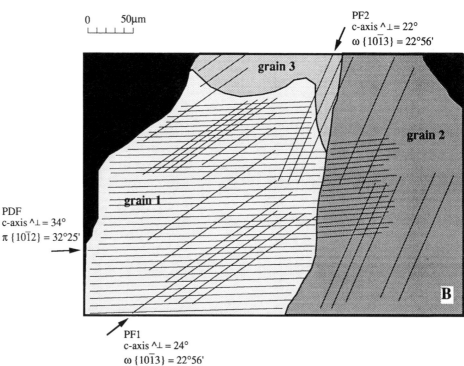

up of one kind of glass only, partially recrystallized into well-developed crystal aggregates (crystallites 10–200 μm in size) (Fig. 15B) coarser than those in cryptocrystalline tagamites. Globular tagamite is similar to the previous one, but is characterized by two coexisting types of glass, one of which is in the form of globules dispersed in the glass matrix (Fig. 15C).

Quartz fragments in microcrystalline tagamites are often surrounded by coronas of relatively coarse hypersthene crystals, and an outer border of Fe-, Ca-, and Mg-depleted residual glass (Fig. 15G). Grains of recrystallized diaplectic feldspar glass are also bordered by a new generation of a clear feldspar (Figs. 13D and 15D). Together with coarser crystallites, these features are the evidence of a well-developed diffusion of components during crystallization in microcrystalline tagamite.

The glass matrix in all tagamites often exhibits evident flow structures. The mass of crystallites is represented by feldspar and pyroxene (in cryptocrystalline tagamites, both minerals are detectable by x-ray diffractometry only). Accessory minerals are represented by opaques and fine troilite-pyrrhotite globules, translucent brown-reddish euhedral flakes of ilmenite, and in some cases by cordierite and hercynite. The residual glass of the matrix in both types of microcrystalline tagamites is fresh and has a refractive index <1.537; when visible, in cryptocrystalline tagamite, this residual glass appears altered. The glass globules in microcrystalline tagamites are fresh, with a refractive index <1.537, and contain less then 10% crystallites of ilmenite, feldspar, and pyroxene. These globules are sharply outlined and have no reaction or diffusion borders with the surrounding matrix, and their elongated shapes indicated that they were in a liquid state during deformation. Thus, these globules were formed before the crystallization of the matrix melt, and remained in a liquid state throughout the crystallization of the enclosing impact melt.

A common feature of Popigai tagamites is a close co-existence, even at a microscopic scale, of Archean rock mineral fragments exibiting a broad range of shock metamorphism, from lechatelierite, to diaplectic quartz and feldspar glasses, to fresh grains of quartz and feldspar. Fresh quartz fragments, decorated by coronas of pyroxene crystallites (Fig. 15G), are also present.

Cryptocrystalline tagamite is usually altered to various clays and other secondary minerals, whereas microcrystalline tagamites are more resistant to alteration. In some cases, close interfingering between crypto- and microcrystalline tagamites results in banded fabrics, similar to the interfingering observed at macro- and megascale within the Melt Sheet Formation. The varieties of tagamites investigated in this study are summarized in Table 6.

Suevite glasses

Suevite glass from the Daldyn and "Ries-Type" Formations described in this chapter is found in various size ranges, from microscopic particles to bombs several decimeters across, whereas the bulk of the glass material is of ash-lapilli size, i.e., up to 10 mm (Fig. 12D,E). In impact cloud formations, small particles of glass were quenched during flight and deposited in a solid state. On the other hand, glass bombs sometimes display traces of plastic deformation after deposition. The glass in these suevites is usually fresh, although in some cases it appears altered as shown in Figure 12E.

As with tagamites, suevite glasses contain various amounts (from none or a single fragment per glass particle, up to 25 vol% of the whole glass) of Archean rock fragments, 0.02–5 mm in size, most of which are single-mineral clasts, mainly quartz and feldspar. Fragments of other minerals such as biotite and pyroxene are rare. These fragments exhibit various stages of shock metamorphism. For instance, in the same suevite glass, quartz grains may be found free of visible traces of shock metamorphism, and/or as diaplectic quartz with PFs and PDFs, and/or as diaplectic quartz glass. Schlieren blobs of lechatelierite are also common here. An example of diaplectic quartz partially transformed into isotropic diaplectic quartz glass is shown in Figure 14A–B. The full range of shock metamorphism, from apparently unshocked grains to diaplectic and fused glass, is also common in feldspar. Flakes of biotite, if present, are opacitized; rare fragments of fractured pyroxene are also found in these suevite glasses.

In contrast to tagamites, diaplectic and fused glasses of quartz and feldspar are often fresh or slightly recrystallized. However, in some cases, schlierens of lechatelierite are completely replaced by quartz-cristobalite aggregates. Coesite was found in one recrystallized lechatelierite blob by Vishnevsky et al. (1975b).

As with Popigai tagamites, the glasses from suevites can be divided into two groups: pristine (nonrecrystallized) glasses (Fig. 15D), and recrystallized glasses (Fig. 15E). Pristine glass is more common than recrystallized glass in suevites. Strong crystallization in Daldyn suevites sometimes leads to the formation of a dark, translucent rock. Crystallites are very fine and often indistinguishable by optical means. When resolved under the microscope, they are represented by 10–20-μm laths of feldspar, and possibly pyroxene. Both types of glasses coexist in the same suevite body, even at the scale of a hand sample. The geochemical affinity of the suevite glasses and tagamites in terms of bulk geochemistry is described in the next section.

Both suevite glasses may show various degrees of vesicularity, from low porosity to pumice. Flow structures and heterogenity are common features in both glass types. However, some suevites from Daldyn Formation are characterized by optically homogeneous glass fragments with rare Archean rock fragments, if present at all.

In some cases, pristine and recrystallized glasses are finely interbanded (see Figs. 12F and 15F). This unusual glass exhibits a fabric similar to banded glasses found in Zhamanshin and Elgygytgyn Craters, as well as in Muong Nong–type tektites (indochinites) (Vishnevsky and Pospelova, 1986). Fast quenching of the glasses led to preservation of flow structures that indicate unstable microflow along the boundary between melts similar to those found in quenched glasses from the Bottom Breccia (Fig. 15H).

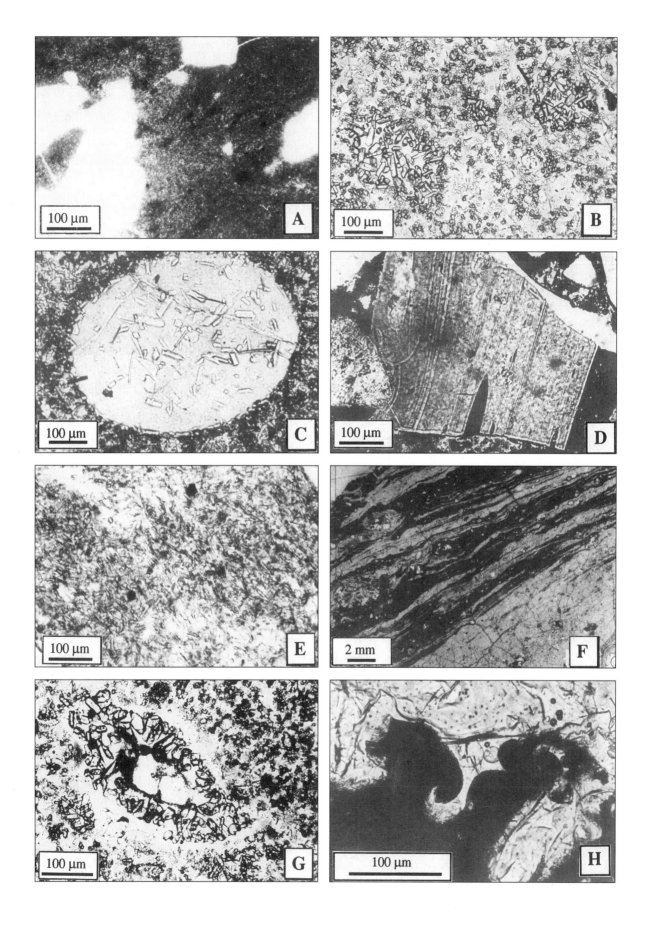

Mushroom-like structures such as those shown in Figure 15H were also produced in shock experiments with different kinds of metals (Vishnevsky and Staver, 1985). Banded glasses are rare in Popigai suevites and were found only in the Daldyn Formation (outcrop nos. 2154, 2156, 2167, and 2170).

Microglobules of troilite-pyrrhotite are common in all types of the suevite glasses as accessories. Additionally, banded suevitic glass contains minor amounts of minute ilmenite and rutile crystals, and globules of pentlandite and native iron (Vishnevsky and Pospelova, 1988).

A brief description of suevite glasses investigated in this study is given below.

Pristine suevite glass. Several fragments of this glass from sample no. 678 ("Ries-Type" Suevite Formation), and sample no. 709 (Daldyn Formation) include mineral clasts of Archean rocks up to 25 vol%, and vesicles up to 5 vol% of the whole glass. The glass is pale-yellowish, has a refractive index >1.537, and contains 15–120 µm accessory globules of troilite-pyrrhotite. Archean rock fragments, ranging in size from 0.02 to 2 mm, are represented by quartz and feldspar without visible traces of shock metamorphism, and by schlierens of lechatelierite, partly or completely replaced by quartz-crystobalite aggregates. Alteration in sample no. 709 is well developed, and represented by clay minerals replacing the glass along microfractures, and in rims around glass fragments up to 200 µm wide; small particles of glass in this sample are completely altered. Vesicles in the glass are filled with chalcedony. Alteration is also evident in sample no. 678 and represented by up to 20-µm rims of clay replacement around glass fragments.

Recrystallized suevite glass. Several small fragments of this glass from sample no. 678 ("Ries-Type" Suevite Formation) and a glass from sample no. 2156/2a (a bomb from the Daldyn Formation) contain up to 20 vol% (no. 678), and up to 10–15 vol% (no. 2156/2a) clasts of Archean rocks; vesicles comprise <5% and 7–10% of the glass volume, correspondingly.

Glass in sample no. 678 has a relatively low (3–15 vol%) content of crystallites, except for a single fragment, which shows incipient to well-developed recrystallization into semi-transparent aggregates. Glass in sample no. 2156/2a (index of refraction <1.537) exhibits heterogeneous flow structures, and has up to 70–75 vol% of the crystallites, represented by needle-like feldspar laths (up to 10–15 µm m in length), and 3–7-µm xenomorphic inclusions of (probable) pyroxene. Accessory troilite-pyrrhotite globules, up to 15 mm in diameter, and other unidentified opaque minerals are always present in all these glasses.

Mineral fragments of Archean rocks, 0.02–2 mm in size, are mainly represented by quartz and feldspar grains, which exhibit evidence of various stages of shock metamorphism comparable to the other glasses described above, with common schlieren blobs of lechatelierite, some of which show partial replacement by fine-grained quartz-cristobalite aggregates. Accessory opacitized biotite flakes and diaplectic pyroxene are also present.

Glass alteration is not evident in sample no. 2156/2a, whereas in sample 678, it shows slight alteration represented by 10–20-µm rims of cryptocrystalline aggregates of clay minerals around the fragments. The matrix in sample no. 678 also shows local patches of Fe-hydroxides. Chloritization and other kinds of alterations are probably present in this sample.

Banded glass. Several fragments of this glass, represented by sample nos. 2154/2a, 2154/3a, and 2170/16 (all of them fragments of large glass bombs), and one small fragment from sample no. 2167, contain varying amounts of mineral clasts of Archean rocks, ranging from 5–7% up to 20–30% of the rock volume. In some cases, the bands of recrystallized glass contain more fragments than the pristine glass bands by a factor of two or three. The vesicles vary from rare occurrences up to 30% of the glass volume, as in the case of frothy glass sample no. 2170/16.

The overall characteristics of both glasses in this banded facies, and of the Archean mineral grains, are similar to those described above for individual samples of the pristine and recrystallized glass. Alteration varies from sample to sample: in no. 2154/3a it is practically absent, whereas in nos. 2154/2a and 2170/16, alteration is represented by 10–70-µm rims of cryptocrystalline aggregates of clay minerals in vesicles and around the glass fragments. Strong alteration in sample no. 2167 is rep-

Figure 15. Main petrographic features of impact glasses from Popigai tagamites and suevites (parallel polars). A, cryptocrystalline tagamite (sample no. 162) characterized by a dark semi-transparent matrix of recrystallized glass made of cryptocrystalline aggregates of pyroxene and feldspar; light area on the left is a schlieren of lechatelierite. B, Microcrystalline tagamite (sample no. 126) characterized by partial crystallization of the melt into coarser aggregates of euhedral hypersthene and plagioclase, with fresh residual glass in the interstices. Coronas of coarser pyroxene crystals (clusters in the lower left and upper right corners), similar to those shown in Figure 7G, are partially cut by the thin section plane. C, Globular type of microcrystalline tagamite (sample no. 1080) characterized by globules of impact anatectic glass (center of the image) disseminated in a matrix of pyroxene, feldspar, and other mineral crystallites, with residual interstitial glass. D, Fragment of quenched, pristine, heterogeneous glass with a fluidal linear texture, incorporated in a fine-grained clastic matrix (sample no. 742 from a Daldyn suevite). E, Quenched, unaltered, recrystallized glass with crystallites of pyroxene and felpdspar (sample no. 2154/2e). F, Fragment of a quenched banded glass bomb from Daldyn suevite showing fine alternation of clear pristine and dark recrystallized glasses (sample no. 2154/2a). G, "Coarse" hypersthene overgrowth with a rim of Fe-, Mg-, and Ca-depleted glass around quartz fragment (center of the image) in a microcrystalline tagamite (sample no. 125). H, Mushroom-like microstructures indicating unstable boundary microflow between the original melts now represented by pristine (clear) and recrystallized (dark) glasses in a schlieren glass body from Bottom Breccia (sample no. 9).

TABLE 6. VARIETIES OF POPIGAI TAGAMITES

Tagamite Type	Sample Number	Vesicles Vol.% of Whole Rock	Alteration of Whole Rock	Archean Rock Fragments	Main Crystallites	Rock Matrix Accessories	Residual Glass
Cryptocrystalline	730	<5	Evident	Size range: 0.02–4 mm 10–15 vol.% of whole rock	~60 vol.% of matrix size range: 2–7 μm	Troilite-pyrrhotite size range: 5–10 μm	Altered, ~40 vol.% of matrix
Single-melt microcrystalline	44, 125, 429, 541	~5	Evident along microfractures	Size range: 0.02–10 mm 7–15 vol.% of whole rock	40–80 vol.% of matrix 1 to 3 types of pyroxene ranging in size to as much as 100 μm plagioclase: 10–30 μm laths; K-feldspar: 40–60 μm subequant crystals	Troilite-pyrrhotite, cordierite, and ilmenite size range: 5-30 μm wider size range for opaques	Fresh, pale-brownish, 20–60 vol.% of matrix
Globular microcrystalline	102, 126	Rare	Evident along microfractures	Size range: 0.02–1 mm ~10 vol.% of whole rock	40–50 vol.% of matrix 2 types of pyroxene in the size range of 15–70 μm plagioclase: 15–40 μm laths K-feldspar: 40–60 μm subequant crystals	Troilite-pyrrhotite, cordierite, and ilmenite size range: 7–20 μm wider size range for opaques	Fresh, pale-yellowish, 50–60 vol.% of matrix
Banded tagamite	502, 818	Rare	Evident	Size range: 0.02–3 mm cryptocrystalline tagamite: 25–30 vol.% of whole rock; microcrystalline tagamite: 10–20 vol.% of whole rock	Cryptocrystalline tagamite: as much as 80 vol.% of matrix size range: 2–7 μm microcrystalline tagamite: 50–70 vol.% of matrix feldspar 15–40 μm laths pyroxene 20–60 μm crystals	Troilite-pyrrhotite size range: 7–20 μm, wider size range for opaques	Cryptocrystalline tagamite: altered, <20 vol.% of matrix microcrystalline tagamite: altered, 30–50 vol.% of matrix

Note: All Archean fragments exhibit common unshocked quartz and feldspar, recrystallized diaplectic quartz and feldspar glass, and recrystallized lechatelierite, and rare non-recrystallized diaplectic quartz and feldspar with PFs and PDFs recrystallized fused feldspar glass, and opacitized biotite and pyrox .

resented by complete chloritization of the glass, and by 0.1–0.2-μm cryptograin rims around the fragment probably made of chlorite-montmorillonite.

GEOCHEMISTRY AND MINERALOGY OF TAGAMITES AND SOME SUEVITE GLASSES

Analytical methods and techniques

Analytical techniques and methods used in this study comprise wet chemistry, instrumental neutron activation analyses (INAA), and electron microprobe, gas chromatography, and carbon-hydrogen-nitrogen (CHN) studies. Bulk major-element compositions of the rocks were determined at the United Institute of Geology, Geophysics, and Mineralogy (UIGGM) in Novosibirsk by wet chemistry techniques, using the methods of A. K. Boiko, A. G. Pilipenko, E. M. Gel'man and others (Hel'man and Starobina, 1972; Gusev et al., 1981). Analyses yielded total sums of 99.9 ± 0.8 wt%.

Electron microprobe analyses were performed at UIGGM in Novosibirsk with a Camebax electron microprobe (some analyses were also made with a JXA-5A microprobe). Silicate minerals and glasses were analyzed at 20-kV accelerating voltage; beam currents of 20 nA were used throughout. The beam diameter was 4–5 μm. For matrix analyses, the beam was defocussed up to a diameter of 15–20 μm. Standard ZAF correction procedures were used. Relative errors of the determinations were estimated to be about 1 rel% for Si and Al, and about 2–3 rel% for Fe, Mg, Ca, Na, and K. The limits of detection were 0.02–0.03 wt% for Ti, Cr, Mn, and P. Details of the method are described in Lavrent'ev et al. (1974).

Gas chromatographic analyses were performed at UIGGM in Novosibirsk, with a device constructed by N. Yu. Osorgin (Osorgin, 1990a,b). This device can analyze the following thermally extracted gases: CO_2, H_2O, CO, H_2, O_2 + Ar, N_2, H_2S, and

TABLE 7. CHEMICAL COMPOSITIONS OF POPIGAI IMPACT MELT ROCKS COMPARED TO AVERAGE COMPOSITIONS OF ANABAR SHIELD ROCKS

Sample	SiO_2	TiO_2	Al_2O_3	FeO	Fe_2O_3	MnO	CaO	MgO	Na_2O	K_2O	P_2O_5	H_2O*	L.O.I	Total
(wt. %)														
Cryptocrystalline tagamite														
730	63.3	0.75	14.60	5.20	1.00	0.08	3.60	3.20	2.23	2.25	0.09	0.40	2.56	99.26
Banded cryptocrystalline and single glass microcrystalline tagamites														
818	65.5	0.71	14.30	4.10	2.20	0.08	3.40	3.00	1.96	2.60	0.09	0.50	1.88	100.32
502	61.4	0.78	15.40	3.30	3.30	0.07	3.20	3.20	2.06	2.73	0.06	0.68	3.06	99.24
Single glass microcrystalline tagamite														
44	62.4	0.81	15.00	3.60	2.90	0.07	3.40	3.20	2.04	2.55	0.09	0.64	2.62	99.32
125	61.7	0.73	14.90	5.32	0.99	0.06	3.08	3.34	2.16	2.71	0.06	0.81	3.77	99.63
429	62.9	0.75	15.10	5.64	0.63	0.06	3.19	3.26	2.21	2.68	0.09	0.62	2.80	99.93
541	63.4	0.75	15.40	5.57	0.81	0.07	2.96	3.18	2.23	2.62	0.11	0.30	2.52	99.92
Globular microcrystalline tagamite														
102	61.8	0.72	15.00	5.39	1.01	0.06	3.19	3.18	2.34	2.81	0.07	0.55	3.85	99.77
126	62.9	0.73	15.00	5.50	0.99	0.07	2.96	3.34	2.05	2.70	0.06	0.56	2.64	99.50
Banded suevite glass														
2154/2a	63.4	0.75	14.90	4.80	2.80	0.07	4.10	3.40	1.94	2.63	0.06	0.16	1.66	100.67
2154/3a	64.4	0.68	15.00	4.89	0.87	0.07	4.21	3.58	1.86	2.64	0.05	0.10	1.36	99.71
Pristine suevite glass														
2170/13	64.1	0.69	14.70	5.21	1.26	0.07	3.96	3.90	1.97	2.40	0.05	0.32	1.28	99.91
2170/15	63.5	0.67	14.50	5.10	1.53	0.07	3.83	3.66	1.86	2.28	0.06	0.53	2.19	99.78
Recrystallized suevite glass														
2156/2a	64.0	0.72	14.90	5.46	0.63	0.06	3.96	3.50	2.03	2.68	0.07	0.12	1.86	99.99
2154/2e	62.3	0.68	14.70	5.21	1.21	0.07	3.96	3.58	1.94	2.45	0.08	0.28	3.06	99.52
2170/6	62.9	0.71	13.50	3.88	3.39	0.06	2.06	2.74	1.44	1.08	0.05	2.99	4.48	99.28
†	63.13	0.76	14.68	4.97	1.99	0.08	3.43	3.82	1.96	2.72	....	2.31	....	99.85
§	64.4	0.6	16.6	3.1	2.4	0.1	4.2	2.30	3.10	3.00	....	....		97.10
**	64.4	0.6	15.2	2.8	1.8	0.1	3.8	2.00	3.50	3.30	3.5	....		97.80

Note: All analyses were performed by "wet chemistry."
*Hygroscopic water extracted at 105°C, except for analysis indicated by † where H_2O and L.O.I. were combined.
†Popigai tagamites average after Masaitis et al., 1975.
§Anabar shield average after Rabkin, 1959, without water and L.O.I.
**Archean shields as a whole, average after Poldervaart, 1955, without water and L.O.I.

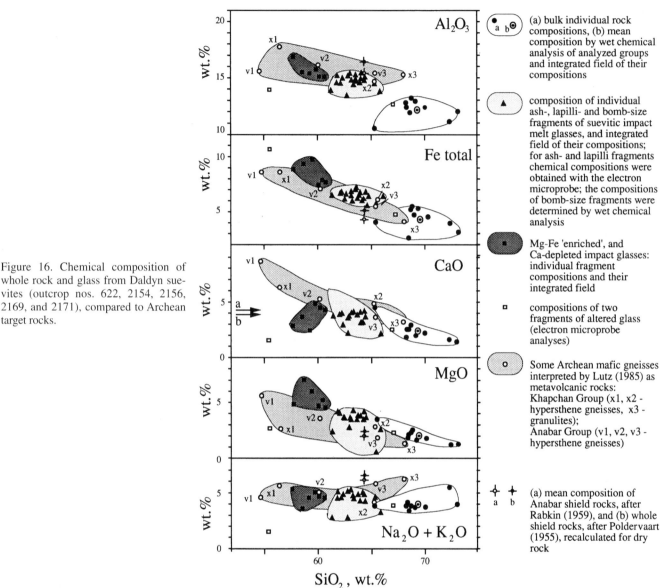

Figure 16. Chemical composition of whole rock and glass from Daldyn suevites (outcrop nos. 622, 2154, 2156, 2169, and 2171), compared to Archean target rocks.

$CH_4C_4H_{10}$ group of hydrocarbons. This device was calibrated for water using standard mineral samples, such as cordierite (1.34 ± 0.1 wt% H_2O) and quartz (0.0624 wt% H_2O). The quality of determinations is routinely controlled before and after the analyses by background determination of water in the device. The relative precision of the analyses for each extracted gas are the following: H_2O, 15–16%; CO_2, 9–10%; H_2, 22-25%; N_4, 9–10%; and CH_4, 7–12% CHN analyses (Hel'man et al., 1987) were performed on sample aliquots of 0.6–0.7 ± 0.0001 mg each at the Institute of Organic Chemistry (Novosibirsk) with a Hewlett-Packard CHN-analyzer (Model 185). The device is calibrated daily by burning of a standard material. The relative precision of the determinations is ±0.3%.

Instrumental neutron activation analyses were made by Frank Asaro at the Lawrence Berkeley National Laboratory (LBNL) on 50-mg sample aliquots, using standards of DIN0-one (Alvarez et al., 1982a), and Standard Pottery (Perlman and Asaro, 1969, 1971). Irradiation was performed in the nuclear reactor at the University of Missouri, Columbia; calibrations and measurements of unknowns were made with the Luis W. Alvarez iridium coincidence spectrometer (LWAICS), designed specifically to measure Ir abundances (Asaro et al., 1987, 1988; Michel et al., 1990).

Micro- and cryptocrystalline aggregates in impactites were analyzed by standard powder x-ray diffractometry.

Results

The bulk geochemical compositions (by wet chemistry) of tagamites and suevite glasses reported in this chapter are in good

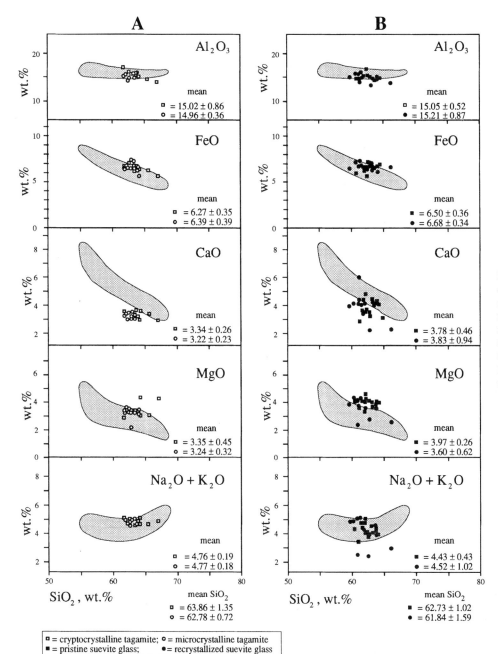

Figure 17. Bulk geochemistry of Popigai crypto- and microcrystalline tagamites (11 and 15 wet chemical analyses, respectively, in column A) and pristine and recrystallized glasses (21 and 14 electron microprobe analyses, respectively, in column B), compared to the composition of basic members of Archean target gneisses (after Lutz, 1985; outlined fields).

agreement with the data of Masaitis et al. (1975), as shown in Table 7. The same data for individual ash-, lapilli-, and bomb-size glass fragments, and bulk composition of Daldyn suevite are shown in Figure 16. The principal compositional feature of tagamite and suevite glasses of all size ranges is their close similarity to the average composition of Archean target gneisses, as well as to Archean rock complexes in general. Such similarity clearly indicates the Archean source of the Popigai impact melt, as it was originally shown by Masaitis et al. (1975, 1980). However, some suevitic glass fragments are enriched in Fe and Mg compared to the average of Anabar Shield rocks or their basic members, probably reflecting either some local heterogeneity in the gneiss complex, or a possible source from ultrabasic members of the target.

Bulk geochemical compositions of the Popigai tagamites, pristine and recrystallyzed glasses, and banded glasses, compared with the basic members of the Archean target gneisses, are shown in Table 7 and in Figures 16–18. A common feature for these glasses is their close compositional similarity. Slight compositional variations in suevitic glasses may be due to incomplete homogenization of the melt. However, despite their geochemical similarity, cryptocrystalline tagamite and pristine suevite glasses

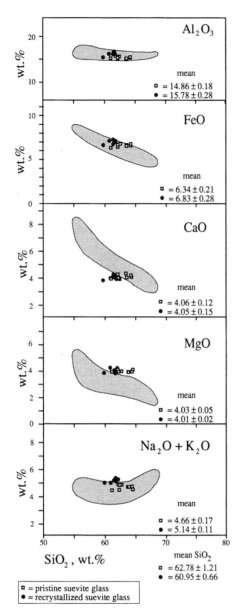

Figure 18. Bulk geochemistry of Popigai banded glasses, compared with the composition of basic members of Archean target gneisses (after Lutz, 1985; outlined fields); nine microprobe analyses for pristine glass, and eight microprobe analyses for recrystallized glass.

TABLE 8. THERMAL STEP EXTRACTION OF WATER
(GAS CHROMATOGRAPHY)
FROM POPIGAI IMPACT MELT ROCKS

Temperature °C	H_2O, rel.% total extraction	
	Cryptocrystalline tagamite and pristine suevite glass	Microcrystalline tagamite and recrystallized suevite glass
100–500	41.6	79.60
500–700	41	17.25
700–800	17.4	3.15

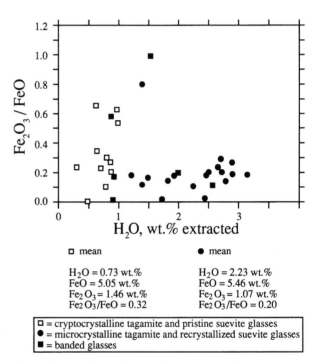

Figure 19. Fe_2O_3/FeO - H_2O plot for cryptocrystalline tagamite and pristine suevite glasses on one hand, and microcrystalline tagamites and recrystallized suevite glasses on the other (11 and 18 analyses, respectively), and for banded tagamites and suevite glasses (6 analyses), analyzed by wet chemistry (for iron oxidation state), and by step-heating gas chromatography (for H_2O content).

on the one hand, and microcrystalline tagamite and recrystallized suevite glasses on the other, clearly differ in three main characteristics (Vishnevsky, 1996a,b) (see Table 8; Fig. 19): (1) loss on ignition (LOI; on the average, 1.97 wt% for cryptocrystalline tagamite and pristine glasses vs. 2.71 wt% for microcrystalline tagamite and recrystallized suevite glasses); (2) direct determination of H_2O content (0.73 wt% vs. 2.23 wt%, respectively); and (3) ferric to ferrous ratio (0.29 vs. 0.19, respectively). These glasses also differ in the way they release water during step heating (Table 8). Apparently, cryptocrystalline tagamite and pristine suevite glasses are more retentive of water than microcrystalline tagamite and recrystallized suevite glasses.

Products of glass recrystallization (pyroxene and feldspar) in cryptocrystalline tagamites are detected by x-ray diffraction only. In microcrystalline tagamites, hypersthene (Fs_{46-50}), andesine plagioclase, and K-feldspar are found as major mineral components, whereas ilmenite, cordierite, hercynite, globules of Ni-enriched troilite-pyrrhotite, pentlandite, metal iron, globular or euhedral crystals of rutile, and other minerals are found as accessories. Traces of a well-developed diffusion are observed in glasses of microcrystalline tagamites (Fig. 15G), and are represented by relatively coarse hypersthene (Fs_{62}) overgrowths (up to 200 μm), and rims of residual glass depleted in Fe, Ca, and Mg.

In comparison to the bulk rock composition, felsic interstitial glass between crystallites in the matrix of microcrystalline tagamites (refractive index, <1.537) is strongly depleted in Fe (total FeO as low as 1.5–1.6 wt%), CaO (as low as 0.5–0.6 wt%), and MgO (as low as 0.05–0.14 wt%), but enriched in SiO_2 (up to 69–73 wt%), K_2O (up to 2.6–3.9 wt%), and possibly fluids, as is suggested by low microprobe totals around 90–91.5 wt% (Table 9).

Compared to the bulk rock composition, globules of immiscible glass in the matrix of globular tagamite are enriched in SiO_2 (up to 77 wt%), K_2O (up to 6 wt%), and H_2O (up to 5–7 wt%); and depleted in Na_2O (down to 0.2–2 wt%), FeO (1.1–2.2 wt%), CaO (0.2–0.5 wt%), and especially in MgO (down to 0.01–0.2 wt%). Concentrations of TiO_2 (0.5–1 wt%), and Al_2O_3 (10.9–12.3 wt%) are similar to those known for tagamites (Table 9) (see also Vishnevsky and Pospelova, 1986). Interstitial residual glass in microcrystalline tagamite matrix and impact anatectic glass from immiscible globules in globular tagamites are compositionally very similar. Recrystallization in suevite glasses is mainly represented by feldspar and possibly pyroxene, neither of which was investigated by microprobe.

TABLE 9. CHEMICAL COMPOSITIONS OF IMPACT ANATECTIC GLASS IN BI-MELT TAGAMITES (INCLUDING GLOBULES IN MATRIX, RIMS, AND INTERSTITIAL SEGREGATIONS IN GNEISS FRAGMENTS), AND OF RESIDUAL MATRIX GLASS

Analysis	Sample	SiO_2	TiO_2	Al_2O_3	FeO*	MgO	CaO	Na_2O	K_2O	Total	H_2O+CO_2†	Total§
(wt. %)												
1	430	74.84	0.52	11.79	1.52	0.17	0.34	1.14	4.71	94.83	....	....
2	431	74.88	0.49	11.48	1.36	0.15	0.31	1.36	4.40	94.43	6.71**	101.14
3	1000	75.19	0.49	12.11	1.83	0.22	0.46	2.02	5.65	97.97	....	....
4	1080	74.36	0.63	11.24	1.18	0.14	0.25	0.97	4.34	93.11	7.66**	100.77
5	2102/1	73.22	0.20	11.61	1.08	0.00	0.53	0.46	3.49	90.59	9.40**	99.99
6	2124/1	77.09	1.04	12.09	1.13	0.02	0.23	0.24	5.15	96.99		
7	2124/1	74.83	0.79	10.87	1.06	0.01	0.21	0.22	4.04	92.03	7.42‡	....
8	2128	72.93	0.62	11.18	1.44	0.04	0.37	0.37	3.50	90.45		
9	2128	74.00	0.75	11.27	1.40	0.03	0.36	0.45	3.65	91.91	8.00‡	....
Average		74.57	0.61	11.52	1.33	0.09	0.34	0.80	4.32	93.58		
10	431/1	70.88	0.47	12.31	2.19	0.02	0.31	0.14	2.01	88.33	....	....
11	431/1	70.72	0.52	12.07	2.16	0.01	0.31	0.14	2.33	88.31	....	....
12	2112/1	74.73	0.51	11.40	0.98	0.00	0.27	0.30	3.26	91.45	9.50**	100.95
13	2112/1	72.17	0.47	11.81	1.39	0.01	0.28	0.63	3.89	90.65		100.15
14	431/1	72.89	0.12	11.72	1.05	0.03	0.26	0.23	2.52	88.82	....	....
15	431/1	71.92	0.15	12.03	1.55	0.02	0.31	0.59	4.30	90.87	....	....
Average		72.23	0.37	11.89	1.56	0.02	0.29	0.34	3.05	89.75		
16	2102/1	75.49	0.77	11.63	1.78	0.19	0.42	1.49	4.79	96.56	....	....
17	2102/1	75.71	0.62	11.85	1.52	0.10	0.45	1.48	4.70	96.43	....	....
18	2102/1	73.63	0.44	11.24	1.34	0.03	0.49	0.40	3.70	91.27	....	....
19	2112/1	73.80	0.20	11.32	1.60	0.02	0.27	0.24	2.58	90.03	....	....
20	2112/1	72.25	0.42	11.52	1.28	0.00	0.29	0.31	3.60	89.67	....	....
Average		74.18	0.49	11.51	1.50	0.07	0.39	0.78	3.87	92.79		
21	100/2	72.91	0.91	12.09	1.49	0.05	0.62	0.50	2.61	91.18	....	....
22	100/2	89.40	0.84	12.02	1.57	0.14	0.56	1.36	3.89	89.78	....	....
23	100/2	72.06	0.85	11.92	1.59	0.08	0.56	0.62	3.32	91.00	....	....
24	100/2	70.77	0.92	12.30	1.59	0.09	0.54	0.67	3.50	90.38	....	....
25	100/2	72.03	0.95	11.99	1.83	0.08	0.54	0.68	3.39	91.29	....	....

Note: Data by Vishnevsky and Pospelova, 1986, microprobe determinations; analyses 1 through 9 are from globules in matrix; analyses 10 through 15 are from borders around gneiss fragments; analyses 16 through 18 are from interstitial glass in gneiss fragments; analyses 19 through 23 are from felsic residual glass in the matrix of tagamite.
*As total iron.
†Separate determination by CHN analyses.
§As total from microprobe and CHN analyses.
**CHN analysis on the same globule.
‡CHN analysis on other glob

TABLE 10. WHOLE ROCK INAA ANALYSES OF Fe AND SOME TRACE ELEMENTS IN POPIGAI IMPACTITES

Sample (%) Lab ID	Tagamite				Recrystallized Suevite		Error in Standard Abundances
	Cryptocrystalline 730	Cryptocrystalline 2354	Microcrystalline 125	Globular 102	2167	74034	
	S15-227	S15-226	S15-228	S15-229	S15-231	S15-230	
(ppm)							
Ir(ppb)	0.2 ± 0.03	0.08 ± 0.04	1.55 ± 0.05	1.49 ± 0.06	0.58 ± 0.07	0.09 ± 0.03	2.0
Ca	98	102	120	140	108	79	4.9
Co	18.2	18.8	20.8	21.3	20.7	13.9	1.1
Cr	114	117	148	142	141	89	3.9
Cs	0.97	0.52 ± 0.01	0.54 ± 0.01	0.56 ± 0.01	0.47 ± 0.02	0.69	6.6
Eu	1.53	1.58	1.62	1.58	1.65	1.24	2.6
Fe (wt.%)	4.98	4.94	4.89	5.07	5.60	3.88	1.2
Hf	7.9	7.3 ± 0.1	7.8 ± 0.2	7.7	7.7	5.7	7.1
Ni	68 ± 5	72 ± 5	131 ± 5	136 ± 6	90 ± 9	51 ± 4	2.5
Rb	88 ± 3	85 ± 4	88 ± 3	98 ± 5	88 ± 8	67 ± 3	1.9
Sb	<0.11	<0.09	<0.08	<0.11	<0.14	0.06 ± 0.04	2.9
Sc	17.6	18.2	17.4	18.0	19.9	13.3	1.6*
Se	0.22 ± 0.02	0.18 ± 0.02	0.24 ± 0.02	0.22 ± 0.03	0.14 ± 0.03	0.13 ± 0.02	25
Ta	0.61 ± 0.02	0.61 ± 0.03	0.53 ± 0.02	0.59 ± 0.02	0.62 ± 0.02	0.50 ± 0.01	2.8
Th	11.5	11.8	17.3 ± 0.2	21.5	12.3	8.9	2.8*
Zn	133 ± 2	68 ± 3	73 ± 3	76 ± 3	83 ± 2	51 ± 3	15

Note: Abundances are in parts per million (ppm) except for those of Fe, which are in weight percent (wt.%), and Ir, which are in parts per billion (ppb, i.e., 10^{-9} grams of Ir per gram of rock). Errors are the best estimates of 1σ in the precision of the measurement. Where no error is given, the precision is estimated at 1 percent. Accuracies must include the errors in the abundances in the standard used, and these are given in the last column. The Ir standard DINO-1 (Alvarez et al., 1982a) was used to calibrate Zn and Se, as well as Ir. The abundances of Ir, Zn, and Se were taken as 315 ppb (Alvarez et al., 1982a), 457 ppm, and 27.6 (F. Asaro, personal communication, April 1997) ppm respectively. Standard Pottery (Periman and Asaro, 1971) was used to calibrate all other elements in this table. The abundances of Cr, Ni, and Rb in Standard Pottery were taken as 102 (Alvarez, et al., 1981), 279 (Alvarez et al., 1981) and 65.5 (F. Asaro, personal communication, April 1997) ppm, respectively. All other abundances in Standard Pottery were taken from Perlman and Assaro, 1971.
*Although the accuracies of the measurement of Sc and Th abundances in Standard Pottery were given as 1.6 and 2.8 percent, respectively, in Perlman and Asaro, 1971, Frank Asaro stated (personal communication, April 1997) that comparison with the measurements of two other INAA groups suggest that Sc and Th values are 6–10 percent too high.

The concentrations of some rare earth and trace elements in tagamite and suevite glasses are shown in Table 10. All the samples are enriched in Ir (0.08–1.56 ppb) in respect to the average concentration of 0.007 ppb detected in two samples of ordinary Archean gneisses by Masaitis and Raikhlin (1986, 1989). Similar element enrichments compared to Archean gneisses are observed for Ni, Co, and Cr (52–136, 14–22 and 89–142 ppm in impactites vs. 27, 13 and 80 ppm in gneisses, respectively) (see Tables 3 and 10). However, basic and ultrabasic units of the Archean basement, as well as the compositions of basic rocks in the Permian-Triassic dolerite dike complexes (Table 3), show Ni, Co, and Cr compositions similar to those of the analyzed Popigai impactites (see discussion below).

In contrast with the data by Masaitis and Raikhlin (1989), our data show that Ir and Ni contents in the analyzed rocks correlate well (Fig. 20), and that, although weaker, a correlation exists also between Ir-Co and Ir-Cr. Our data for Ni and Cr contents (Table 10) agree with results by Masaitis and Raikhlin (1986, 1989), but our Ir values, ranging from 0.08 to 1.56 ppb, are two or three orders of magnitude lower than the maximum value of 570 ppb reported by Masaitis (1994) for a Popigai tagamite. However, because the extraordinarily high concentration of Ir found by Masaitis (1994) in a tagamite sample is comparable to that of pure meteoritic matter (e.g., Crocket, 1981), we are compelled to consider this result as doubtful, and recommend that it needs to be confirmed by replicate analysis before accepting it as evidence for meteoritic contamination.

The abundances of Ir, Ni, and Cr in our samples do not correlate with those of chalcophiles elements (Ag, Se, Zn). Poor or no correlation is found between Ir and some lithophiles (Rb and Ce), and there is no correlation between Ir-Ta and Ir-Cs.

GEOCHRONOLOGY

The rich and well-diversified fossil fauna (including bivalves, cephalopods, and other marine organisms) and flora (pollen) recognized in numerous clasts contained in Popigai impact rocks (e.g., Kirjushina, 1959; Smirnov, 1962; Vishnevsky et al., 1995a) indicate that the Popigai impact is younger than Cenomanian (i.e., <94–92 Ma). About 70 samples of presumably Cenozoic sands and clays from Popigai suevites collected for micropaleontologic study resulted in no detection of any recog-

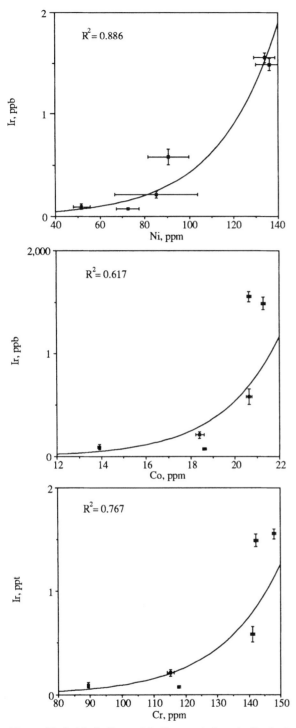

Figure 20. Ir-Ni, Ir-Co, and Ir-Cr correlations in Popigai impactites (cf. Table 10).

TABLE 11. RADIOISOTOPIC AND FISSION TRACK DATES FROM POPIGAI IMPACTITES

Rock Type	Sample	Technique	Age (ma)	Reference*
?	?	K/Ar	28.8 ± 9	1
Tagamite	3752	K/Ar	~40–50	2
Tagamite	3753	K/Ar	~45	2
Tagamite	3654	K/Ar	~40	2
Tagamite	100/2	K/Ar	54 ± 1	3
Tagamite	524/1	K/Ar	38 ± 5	3
Tagamite	529	K/Ar	71 ± 5	3
Tagamite	550	K/Ar	81.5 ± 3	3
Tagamite	639/3	K/Ar	42 ± 6	3
Tagamite	654	K/Ar	57 ± 2	3
Tagamite	666	K/Ar	64.5 ± 6	3
Tagamite	818	K/Ar	35 ± 5	3
Tagamite	823	K/Ar	47 ± 3	3
Tagamite	828m	K/Ar	~43	4
Tagamite	840a	K/Ar	~41	4
Tagamite	1619v	K/Ar	~29	4
Tagamite	1713a	K/Ar	~46	4
Tagamite	1798a	K/Ar	~31	4
Suevite glass	Vv	K/Ar	~42	4
?	?	F.T.	38.7†	4
?	?	F.T.	30.5 ± 1.2	5
Suevite glass	74034	$^{40}Ar/^{39}Ar$	65.4 ± 0.3	6
Tagamites	§	$^{40}Ar/^{39}Ar$	35.7 ± 0.8§	7
Tagamite	102 matrix	$^{40}Ar/^{39}Ar$	26.0 ± 1.6	8
Tagamite	102 glass blob	$^{40}Ar/^{39}Ar$	8.1 ± 1.2	8
Tagamite	102 glass blob	$^{40}Ar/^{39}Ar$	25.3 ± 0.8	8
Banded glass	2170/16	$^{40}Ar/^{39}Ar$	37.09 ± 0.19	8
Banded glass	2170/16	$^{40}Ar/^{39}Ar$	40.6 ± 0.5	8

*1 = Vasil'eva, 1968, in Masaitis et al., 1968; 2 = Firsov, 1970; 3 = Firsov and Vishnevsky, 1972, Institute of Geology and Geophysics, Novosibirsk, unpublished data; 4 = Komarov and Raikhlin, 1976, Institute of Geology and Geophysics, Novosibirsk, unpublished data; 5 = Storzer and Wagner, 1979; 6 = Deino et al., 1991; 7 = Bottomley et al., 1997; 8 = Deino et al., 1995.
†Average from numerous measurements.
§Average age from 5 runs.

nizable fossil. Therefore, although clasts of sand and clay contained in Popigai suevites are almost certainly derived from the preimpact Cenozoic sedimentary cover of the Anabar Shield, no paleontologic age younger than Late Cretaceous was ever determined in Popigai impactites.

Early K/Ar dates of the Popigai melt rocks (Vasil'eva, 1968, in Masaitis et al., 1975; Firsov, 1970; L. V. Firsov and S. A. Vishnevsky, 1972, unpublished, in Vishnevsky and Montanari, 1994; Komarov and Raikhlin, 1976) yielded ages ranging from 81.5 to about 29.0 Ma (Table 11). The probable reasons for such a wide scatter of radioisotopic dates are that they were obtained from rock samples containing various amounts of precursor Archean target rock fragments, and/or that the analyzed rocks have lost some radiogenic argon as a result of postimpact alteration. In the first case, the dates would appear older than the real, unknown age of the impact melt, although Bogard et al. (1988) maintained that contamination from impact-shocked inclusions would be in any case small, whereas in the second case the dates would appear younger than the age of impact melt formation.

A set of fission-track dates by Komarov and Raikhlin (1976) ranging between 32.4 and 44.0 Ma, (average, 38.7 Ma) strongly supports a Cenozoic age for this impact event. Additional fission-track dating of Popigai impact melt rocks by Storzer and Wagner (1979) yielded an age of 30.5 ± 1.2 Ma, further supporting a Cenozoic age for this impact (Table 11).

More recent results obtained with the $^{40}Ar/^{39}Ar$ laser fusion and step-heating techniques (Deino et al., 1991; Bottomley et al., 1997) (see Table 11) show similar age inconsistencies as in previous Russian K/Ar datings. $^{40}Ar/^{39}Ar$ analyses by Deino et al. (1995) on hand-picked, apparently pure and fresh glass fragments from a globular tagamite surprisingly produced more inconsistent ages: six runs from the matrix and a blob of immiscible glass yielded poorly defined plateaus with ages ranging between 8 and 25 Ma—the youngest ages ever obtained from Popigai melt rocks. On the other hand, banded glass from suevite sample no. 2170/16 yielded an age of 37.7 ± 0.3 Ma (Deino et al., 1995). Of all the radioisotopic dates shown in Table 11, those by Bottomley et al. (1997) appear as the most consistent and reproducible, and indicate a late Eocene age for the Popigai impact event (35.7 ± 0.8 Ma).

Proximal impactoclastic air-fall beds in marine sequences attributable to the Popigai impact have yet to be found. On the other hand, late Eocene marine sediments throughout the world contain two closely spaced layers with clear evidence for large, distal impacts (see Montanari et al., 1993, and references therein, for a detailed review). The youngest level, which seems to be restricted to the Gulf of Mexico, the Caribbean, and the East Coast of North America, contains microtektites and shocked quartz, but no Ir anomaly. Microtektites contained in this layer yielded an $^{40}Ar/^{39}Ar$ age of 35.5 ± 0.3 Ma (Obradovich et al., 1989), whereas compositionally identical tektite fragments from the late Eocene Bath Cliff section in Barbados yielded an $^{40}Ar/^{39}Ar$ age of 35.4 ± 0.6 Ma (Glass et al., 1986). This impactoclastic air-fall layer has been correlated with the recently discovered 85–90-km diameter Chesapeake Bay Crater (e.g., Poag et al., 1994; Koeberl et al., 1996).

The second layer, which in some Caribbean deep-sea cores is found 25 cm below the microtektite-bearing level (Glass and Burns, 1987), displays a worldwide distribution and is characterized by clinopyroxene microkrystites, Ni-rich spinel (Pierrard et al., 1996), and shocked quartz (Clymer et al., 1996; Langenhorst, 1996), associated with an Ir anomaly in the order of a few tens of parts per billion (Alvarez et al., 1982b; Keller et al., 1987; Montanari et al., 1993). In the Global Stratotype Section and Point for the Eocene/Oligocene Boundary (GSSP) at Massignano (Italy), this layer is represented by a 5-cm-thick marly horizon containing altered microkrystites and Ni-spinel (Pierrard et al., 1998), shocked quartz (Clymer et al. 1996; Langenhorst, 1996), and Ir (Montanari et al., 1993). At Massignano, this horizon is dated at 35.5 ± 0.3 Ma from tightly interpolated $^{40}Ar/^{39}Ar$, K/Ar, Rb/Sr, and U/Th dates obtained from numerous biotite-rich volcaniclastic layers (e.g., Montanari et al., 1993). Is this Ir-rich impactoclastic air-fall layer the distal, worldwide expression of the Popigai impact? At present, radioisotopic dates from Popigai impact melt rocks cannot give a credible answer to this question because of their evident inconsistency and wide scatter. However, this question (Montanari et al., 1993; Clymer et al., 1996, Langenhorst, 1996) could be taken as a strong working hypothesis for further interdisciplinary research on both the Popigai impact structure, and the late Eocene, Ir-enriched impactoclastic layer. The detailed information on the petrography and geochemistry of Popigai glass-bearing impactites documented in this work may be useful for the correct choice of material for further radioisotopic dating of the Popigai event.

DISCUSSION

General interpretation of impact formations and cratering

The classification of impactites, and the general interpretation of Popigai geology presented here, differ from previous models proposed by Masaitis et al. (1975, 1980) and Masaitis (1994). Some remarks on these contrasting models were discussed by Vishnevsky (1978, 1981, 1994a). There are at least three main contrasting points in these models that are critical for the interpretation of the general geology of this impact crater, as well as the dynamics of the impact process.

A first controversial point is about the nature of the Archean rocks in the inner ring. We interpret them as parautochthonous crystalline rocks that were dislocated by block fragmentation, fracturing, and autoclastic brecciation. These rocks underwent plastic megaflow centrifugal displacement for short distances from their original location, and experienced a rapid uplift during cratering up to 1.5 km. In contrast, Masaitis et al. (1975, 1980) interpreted them as autochthonous formations formed in situ during the cratering process.

The second critical point regards the definition of the Popigai crater-fill formations. We divide them into two groups (Vishnevsky, 1978, 1981, 1994a; Vishnevsky and Montanari, 1994): those transported, mixed, and deposited in a nonballistic manner (deposits of the bottom centrifugal flow), and those deposited after ballistic transport (the explosion cloud deposits). Masaitis (1994) and Masaitis et al. (1975, 1980) divided them into three categories: "allogenic breccia" (all coarse clastic formations, including part of our suevite megabreccia), tagamites, and suevites.

The third critical point is about the interpretation of suevites in terms of their provenance and emplacement. All Popigai suevites are considered by Masaitis et al. (e.g., 1975, 1980) exclusively as explosion cloud deposits. However, the glass-bearing rocks of our Bottom Breccia Formation, which are genuine suevites, according to Stöffler and Grieve (1994), are not derived from the explosion cloud, but rather from the near-bottom, turbulent mixing, transport, and deposition of the impact base surge. Masaitis and co-workers (e.g., 1975, 1980) ignored the spatial distribution of these rocks, which are covered by the tagamites of the Melt Sheet Formation; this is why they reached a different interpretation of the geology and cratering mechanism for the Popigai.

A similar problem exists for the Bunte Breccia in the Ries Crater, which stimulated a controversy between Pohl et al. (1977)

and Hörz et al. (1977), who proposed a ballistic transport for megablocks, and Chao (1977a,b), who proposed a roll-and-glide mode of transport for the same material.

These different classifications and genetic interpretations of impactites lead to substantially different interpretations of some other important features of Popigai geology and cratering. For instance, Vishnevsky (1978, 1981, 1994a) interpreted exposures of the impact melt rocks within topographic ring I in the western quadrant of the crater (Figs. 2A and 4) as the marginal part of a massive "melt sheet" deposited on top of other bottom formations. A contrasting interpretation viewed these impactites as either a cover of tagamite extruded from a deeply buried impact melt sheet over a suevite layer (Masaitis et al., 1975), or as a huge splash-like mass, which traveled through the atmosphere in ballistic trajectories, together with suevitic material directly derived from the explosion cloud (Masaitis et al., 1980). Similarly, tagamite and breccia matrix between klippens and fragments of the Megabreccia formation were interpreted by Masaitis et al. (1975, 1980) as ballistically deposited products, whereas Vishnevsky (1978, 1981, 1994a) explained these relations as the result of dense bottom flow mixing.

Another problem concerns the terminology and definitions of the Popigai suevites. In fact, the term suevite was originally applied to Ries polymict impact breccias containing aerodynamically shaped glass masses by a number of authors (e.g., Preuss, 1969; Horz, 1965, 1982; Stahle, 1972). Later, the term suevite was extended to various tuff-like impact breccias, including those that were deposited during cratering from dense bottom turbulent flows in different craters. As Dence (1971, p. 5554) pointed out, "In these breccias, the fresh or recrystallized glasses show intricate intrusive relationship with the enclosing fragmental rocks, indicating that the masses of melt were incorporated into breccia while still hot and mobile. In some cases, breccia fragments are entrained between layers of glass to give a banded fabric. Such breccias have also been called 'suevites' but the term is more properly applied only for the rocks like the Ries material, in which the glassy masses are of the Fladen type."

Following these considerations by Dence (1971), Vishnevsky (1978, 1981, 1992, 1994a) never applied the term suevite to the rocks of the Bottom Breccia Formation because of their clear petrographic and genetic distinction from those rocks named as suevites in Ries and some other impact craters: they were referred to as bottom impact breccias, not suevites. In short, it was confusing and useless in the study of Popigai rocks, and in the interpretation of cratering, to categorize all glass-bearing polymict impact breccias under the umbrella label of suevite. Vishnevsky's classification considers, along with petrographic criteria, the source materials and the stratigraphic position of these breccias within the Popigai Crater.

A similar approach was recently applied in the description of impact breccias of the Manson structure by Witzke and Anderson (1996), who stated that the nomenclature of Stöffler and Grieve (1994) does not adequately characterize impact breccias in this structure. We can follow the conclusion by Witzke and Anderson (1996, p.142), that "further studies, both theoretical and empirical, are recommended before a comprehensive genetic classification scheme for impactites is advanced."

Theoretical modeling of impact cratering allows estimation of some energetic, physical, and dimensional parameters of the Popigai impact event. Using the modeling by Dence et al. (1977), a total energy of about 6.2×10^{22} J for the Popigai impact event was estimated by Vishnevsky (1994a), which led to the result of a diameter of about 5 km for a stony projectile, in the case of a vertical impact at about 25 km/s (Shoemaker, 1977). Estimates by Masaitis (1984) yielded a total impact energy of about 1.7×10^{23} J, and a 7.8 km diameter for a stony projectile. Using the formula of Ivanov et al. (1982), the resulting shock-wave attenuation along the vertical axis of the impact is here estimated at pressures (GPa)/depths (km) of 100/11.5, 60/13.9, 50/14.9, 35/17, 20/20.9, and 10/27. Using the modeling of Grieve (1987), an axial distribution of residual postshock temperatures of the gneisses is here estimated at temperatures (°C)/depths (km): 2530/11.5; 1390/13.9; 1000/14.9; 450/17; 150/20.9 and 100/27. Using the modeling of Melosh (1989), the diameter of the Popigai transient cavity would be about 50–60 km, and about 15–20 km deep, with a maximum depth of excavation of about 5–6 km. The uplift during gravity collapse of the transient cavity would be up to 9 km. In less than 10 s, shock-wave phenomena were extinct, while the excavation process took about 70 s (estimate by Vishnevsky, 1994a, using the formula by Florensky et al., 1983).

Maximum velocities for the bulk of vaporized and molten ejecta were proposed in the range from 4 km/s (Ivanov, 1976) up to 7.3 km/s (Kieffer and Simonds, 1980). A voluminous number of impact melts were generated in the central part of the crater within a ~14-km radius hemispherical zone defined by an attenuating shock wave isobar of 55–60 GPa. The total volume of melt produced in the Popigai was about 1,750 km^3 (Masaitis, 1984).

Omitting initial jetting at the contact of the projectile with the target, and phenomena of cumulative centripetal movement of vaporized material at the initial stage of the excavation, two modes of material transport that took place at the origin of Popigai impact structure may be envisioned on the basis of a simple cratering model: (1) low-angle, or even subhorizontal, centrifugal transport of material in the form of bottom (or base surge) flow, which led to the origin of the bottom formations described above; and (2) high-angle, cone-shaped centrifugal ejection, which led to the origin of explosion cloud formations in the upper part of the crater fill. In the first case, the complex relationships among different bottom formations indicate that transport and deposition was turbulent and nonballistic. In the second case, the monotonous composition and gradual transitions between suevite formations indicate that they were transported and deposited in a ballistic mode. Some features of suevite megabreccia are regarded separately, and are discussed below.

During the Popigai impact event, the front of the bottom flow would have prograded outward for 10–15 s before starting to interact with the hot, strongly shocked, and fast-moving material derived from inner parts of the crater (Vishnevsky, 1992). Moving klippens and megabreccia were mixed with, injected, and overlapped by comminuted and moderately shocked target material (bottom fragmental breccia and suevite). Soon after, these moving mixtures were reached and overlapped by high-speed flows of comminuted and strongly shocked, partially and/or totally molten material (bottom suevite and tagamites, respectively). Part of this fast-moving turbulent material was injected into, and mixed with, the underflowing impactoclastic mass, but the bulk of it was emplaced as a cover on other bottom formations, and led to the origin of the Melt Sheet Formation.

In some places, tongues of bottom flow impactites were emplaced outside the tectonic rim of the crater. In some other places inside the crater, coarse bottom formations were not covered by molten masses (e.g., the Motley Rock in Fig. 8A), and contain no or rare irregular tagamite inclusions.

At the final stages of cratering, still-moving bottom flow masses were covered by suevites and suevite megabreccia. The bulk of the explosion cloud material was deposited inside the crater to form a thick, layered suevitic cover, in the annular trough and central basin. The suevite megabreccia is found in the uppermost part of the cover in the annular trough only. In some places, irregular suevite bodies were admixed with the upper part of the molten masses that will form the Melt Sheet, with the bottom megabreccia in the crater margin, and other impact deposits outside the crater. Being localized mainly in the lower part of the suevite cover (Fig. 2A), Daldyn suevites came from the lower part of the explosion cloud, and were the first to be deposited in a relatively hot state within the crater. Parchanai and Suevite Sands Formations, which occupy the upper part of the cover, came from the upper part of the explosion cloud, and were deposited in a relatively cold state.

A rapid uplift of the inner ring during cratering complicated the emplacement and deposition of bottom and explosion cloud formations, and led to redistribution of the impactites. In some places, Daldyn suevites were emplaced directly on top of the parautochthonous gneiss breccia of the inner ring.

Complex relations between tagamites and suevites at the contact between the Melt Sheet and the suevite cover on the slope of the inner ring, indicate that there was a gradual transition between bottom centrifugal flow and explosion cloud.

The sequences of events during Popigai cratering is similar in some respects to the Ries, as indicated by a similar reverse stratigraphy of impactites in marginal parts of the crater and outside it. However, stratigraphic provenance of Popigai suevites shows lack of reverse stratigraphy inside the crater. Dominating Mesozoic lithologies and suevite megabreccia in the central part of the crater are the evidence of very steep trajectories of the ejecta derived from loose sedimentary rocks making up the top of the target rock sequence. Only rare occurrences of Ries-type suevites, similar to the Ries fall-back suevites, are present in the Popigai impact structure.

Popigai suevites and the hypothesis of a dynamic barrier

The occurrence of weakly or non shocked clasts of soft Mesozoic rocks in impactites inside the crater contrasts with simple cratering models. In fact, following these models (e.g., Gault et al., 1968; Maxwell, 1977; Ivanov, 1979), unshocked or slightly shocked material derived from the upper layers of the target is expected to have been ejected outside the crater. Unshocked or weakly shocked fragments of Cretaceous rocks (<5 GPa) in the suevite megabreccia indicate that they originated not less than 20 km from the center of the crater. Therefore, a high-angle to subvertical ejection is required to explain the present location of suevite megabreccia within the crater (e.g., outcrop no. 2171, 22 km from the center of the crater) (Fig. 2B).

The hypothesis of a dynamic barrier in the explosion cloud was proposed by Vishnevsky (1992, 1994a) in the attempt to explain this and other features of Popigai suevites. It takes into account the low density of target Mesozoic rocks (~1.8–2 g/cm^3), and their thickness (up to 150 m). All other sedimentary target rocks are characterized by higher densities (up to 2.6 g/cm^3), similar to those of most Archean basement rocks (~2.8 g/cm^3).

The dynamic barrier hypothesis is based on the supposition that a density contrast in target rocks and the surface attenuation of the shock wave led to a subsurface lag of the shock wave front passing through soft, surficial Mesozoic rocks at a certain distance from the center of the impact. This hypothesis improves the well-known simple cratering model of Gault et al. (1968), and explains the high-angle to vertical ejection of loose Mesozoic rocks (Fig. 21). The ejection itself, or its combination with near-surface spallation of target-rock fragments described by Melosh (1989), originated a torus-shaped cloud of low-shocked Mesozoic rock fragments vertically rising above the middle and marginal zones of the growing crater.

This cloud would function as a dynamic barrier for high-speed ejecta derived from other target rocks. The conical plume of these high-velocity ejecta, which is predicted by simple cratering models, was screened in its middle part by the rising torus-shaped cloud of Mesozoic rock lumps. This hypothesis also explains why proximal suevites are scarce outside the Popigai Crater.

The colliding of high-velocity ejecta of strongly shocked crystalline rocks with the torus-shaped cloud of Mesozoic fragments caused their intensive mixing and a secondary shock of the whole ejecta mass. Such mixing can explain the coexistence, in Popigai suevites, of material derived from spatially different zones of shock metamorphism and excavation: weakly or nonshocked Mesozoic clasts from marginal zones of the crater, and strongly shocked gneiss clasts and impact glass particles derived from its core. During this aerial collision, part of the suevite material was condensed and compacted to a dense lithoid or even lithified state, and mixed with lumps of unshocked Mesozoic target rocks. Seventy seconds of flight (Vishnevsky, 1994a) was sufficient for chilling the melt fraction of the cloud, and for lithification of some condensed masses. Turbulent interaction between dense plastic suevitic material and fragmentation of

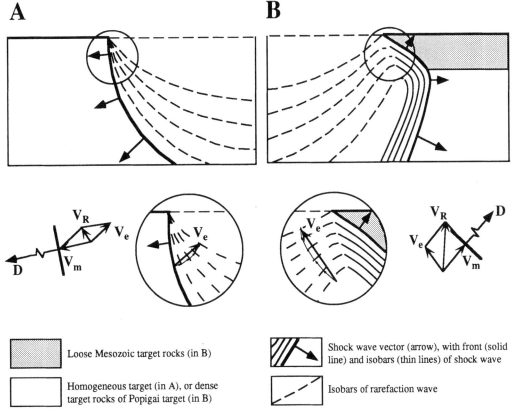

Figure 21. Simple model of crater excavation after Gault et al. (1968) in homogeneous target (A), and its application to the supposed subsurface lag of shock wave front in soft Mesozoic rocks of Popigai target (B). D = shock wave velocity; V_m = particle velocity behind the shock wave front; V_e = expansion velocity of compressed rock in rarefaction waves, and its elementary components ΔV_e orthogonal to isobars of rarefaction wave at every progradation increment; V_R = resultant particle velocity after shock.

lithified masses also took place during the collision at the dynamic barrier, and during the collapse of the explosion cloud. As a result, suevite mixtures (Fig. 10), and forced injections of Mesozoic rock lumps in dense lithoid material, like those shown in Figure 11, as well as breccias-within-breccias relations were formed during deposition. At the collapse of the explosion cloud, the finer fraction of this material was deposited as Suevite Sands and Daldyn Formations, while mixtures with large blocks of Mesozoic rocks were emplaced as Suevite Megabreccia.

During and after the late stages of cratering, the redistribution of impactites took place by gravity flows of molten masses on slopes of the inner ring, and tagamite melt was deposited locally on various suevite formations in the annular trough. Gravity flows occurred also in marginal areas of the crater, where the tagamite cover was emplaced directly on megabreccia (Fig. 8A). In many places, especially in the south sector of the crater, buried impact melt generated numerous intrusions into the overlying suevites. Block faulting tectonism, which is still active today, played an important role in the postimpact modification of the Popigai impact structure. This tectonic activity is presently expressed by a mosaic of structural blocks with different elevations.

Evolution of impact melts

Tagamites exhibit different degrees of melt crystallization. However, these impactites share a variety of common features, such as source, bulk geochemistry, same target-rock inclusions with a broad spectrum of shock metamorphism, and same geologic setting.

Tagamite differentiation was attributed to low (for cryptocrystalline rocks) and high (for microcrystalline rocks) initial temperatures of the impact melt (Masaitis, 1994). However, this thermal interpretation may be applicable for large, separate bodies of melt with different temperatures. In this case, crystallization rates would be different in the two types of melt before heat exchange leads to thermal homogenization of the entire rock mass. However, in the case of small bodies of one kind of melt incorporated in a large body of the other (which is a common occurrence in the Melt Sheet at Popigai), or of dense interfingering and even microscale banding of two different melts, thermal homogenization will occur quickly, and crystallization will develop homogeneously throughout the molten body. Thus, the thermal model of Masaitis (1994) cannot explain these cases.

A similar problem arises in explaining the petrologic differences between pristine and recrystallized suevite glasses. They also share a common history, same source and bulk geochemistry, equal conditions of melting, transport, quenching, and geologic setting, but they are clearly different from each other on the basis of their crystallinity.

We propose (Vishnevsky and Montanari, 1994; Vishnevsky, 1996a,b) that the water originally contained in melts controlled the crystallization rate of impact glasses: low water content inhibits crystallization in cryptocrystalline tagamites, whereas a relatively high water content catalyzes crystallization in microcrystalline tagamites. This hypothesis, which we call "hygrogenic" (from old Greek *hygro* = humid) is supported by the different LOI in impact glasses (in average, 1.97 wt% for cryptocrystalline tagamites and pristine glasses, vs. 2.71 wt% for microcrystalline tagamites and recrystallized suevite glasses), as well as by direct water determinations in two types of impactites (0.73 wt% for cryptocrystalline tagamites and pristine suevite glasses vs. 2.23 wt% for microcrystalline tagamites and recrystallized suevite glass; see Vishnevsky, 1996a,b) (Fig. 18). Thus, crystal-free, dry, pristine suevite glass is the quenched equivalent of cryptocrystalline tagamite, whereas recrystallized wet suevite glass corresponds to microcrystalline tagamite.

In agreement with our hygrogenic model of impact melt crystallization, the banded microfabric of suevite glass, as well as banded fabrics in some other impactites, indicate that water-rich and water-poor portions of the original melt were not homogenized, in some places, during the early stages of impact melt formation and turbulent mixing.

Residual glass in microcrystalline tagamite (but not in cryptocrystalline rocks), is strongly depleted in Ca, Fe, and Mg, and enriched in Si, K, and volatiles, possibly $H_2O + CO_2$. As a result, cryptocrystalline tagamite appears more intensely altered than microcrystalline tagamite because the "acid"-depleted glass is here more resistant to hydrothermal processes compared with the more basic glass in cryptocrystalline tagamite. Alteration clays provide higher Fe_2O_3/FeO ratios in cryptocrystalline than in microcrystalline tagamite (0.29 vs. 0.19, respectively) (see Vishnevsky, 1996a,b) (Fig. 19). Step-heating water extraction in gas chromatographic analyses in the two types of impact glasses from tagamites and suevites shows essentially different dynamics (Table 8): water extraction is more pronounced in cryptocrystalline tagamite and pristine suevite glass than in microcrystalline tagamite and recrystallized suevite glass. We presume that a portion of this water is derived from montmorillonite (100°–500°C), chlorite (500°–800°C), and from other hydromicas. This is probably due to stronger alteration in cryptocrystalline tagamite and pristine suevite glass compared to microcrystalline tagamite and recrystallized suevite glass, and would indicate that the former were initially "dryer" than what they appear to be in our analyses: hydromicas provide additional water of hydrothermal origin.

Different water contents in impactites were probably inherited from Archean target rocks that comprise both granulite and amphibolite facies. These rocks differ from each other in water content, inasmuch as water-bearing minerals like cordierite, biotite, and amphibole are uncommon or totally absent in granulites, whereas they are abundant in amphibolite facies. This is one primary source of water in target Archean rocks. Furthermore, fluid inclusions in rock-forming minerals in granulites contain less water than in rocks of amphibolite facies. In addition to this constitutional water, the shallower parts of the target (i.e., 3–3.5 km depth) are more porous than deeper parts, and consequently richer in water. Thus, Archean target gneisses are significantly different in their water content, being a complex assemblage of water-rich and water-poor rocks, blocks, layers, fragments, and dislocation zones.

Fragments of Archean rocks in Popigai impact glasses are supposedly derived from the same zone where the melt was originated, and then they were disseminated in the impact melt by turbulent mixing during cratering (Vishnevsky and Pospelova, 1986). Thus, these clasts are here considered as residual fragments of partially molten target rocks, and they are not exotic clasts derived from outside the melting zone, nor minerals formed later in the crystallizing melt. In our view, there is no other realistic mechanism that could produce such a pervasive saturation of unmelted, and variously shocked, clasts in large melt volumes. The same mechanism can also explain Archean rock fragments in suevitic glass.

The constant presence of Archean clasts in the Popigai impact melt rocks, including those that are weakly shocked, indicates that the melt was derived from outer, colder parts of the melting zone. In inner parts of the melting zone close to vaporization, the impact melts were hotter and more homogeneous than in marginal zones, and contained no or rare target-rock clasts.

We have no objective means to precisely estimate the temperature boundary between these relatively cool and hot melts, but we can infer that the integrated temperature of "cold" melts after homogenization of hot spots by thermodiffusion, was lower than the melting points of quartz (~1,700°C), or feldspar (~1,300°–1,400°C). As for the hot melt, it is probably represented by homogeneous clast-poor or clast-free fragments of impact glass that are sometimes found in suevites. The rare occurrence of these glasses in Popigai suevites may indicate that most of the hot melt from the inner part of the melting zone was dispersed in the explosion cloud, and ejected together with vaporized material from the crater to form distal impactoclastic deposits.

Immiscible impact glass and "impact anatexis"

The microcrystalline globular tagamite is especially interesting because it contains sparse 0.2–3-mm globules of fresh glass (Fig. 15C) dispersed in the matrix glass. The globules constitute 1.2–1.6% of the rock, and exhibit a geochemistry different from that of the enclosing tagamite matrix (see Table 9). Vishnevsky and Pospelova (1984) proposed two alternate sources for this material. The first is that this immiscible glass is a product of melting of target alaskite, whereas the second possibility envisions this material as the result of shock separation of components during the impact melting stage.

Studies focused on microcrystalline tagamites (Vishnevsky and Pospelova, 1984, 1986) show that the globular glass was derived from shocked Archean gneiss fragments. The most abundant generation of this parental melt is found in the fragments affected by shock metamorphism of stages II and III of Grieve et al. (1996) (Table 5). At stage III, when almost complete mixing of all mineral phases other than quartz takes place in Archean fragments, not much parental globule glass is produced. At higher stages of shock metamorphism, Archean fragments reach total fusion, and globular glass is not produced at all.

Glass segregations in Archean gneiss fragments include interstitial films between mineral grains, borders around fragments, and injections of the glass into the matrix of the tagamite, and are compositionally very similar to the glass composing the globules (Table 9). These forms of segregations indicate that the interstitial glass is the result of anatectic melting of the fragments. The close chemical similarity between this interstitial glass and the globules, along with the textural characteristics mentioned above, indicates that the globules represent isolated blobs of interstitial anatectic glass intruded into the surrounding tagamite melt.

This specific melting has been referred to as "impact anatexis" by Vishnevsky and Pospelova (1986) and Vishnevsky et al. (1987), and seems to be produced by a combination of high deformation rate, high excitation of atoms in the crystal lattice, and other extreme conditions common in the impact shock process. A result of this process is a selective mobilization of Si, K, and volatiles, and $H_2O + CO_2$, accompanied by a loss of Na, Ca, Fe, and Mg. The "impact anatexis" is consistent with a model of highly excited solids that increases the rate of component diffusion up to 15 orders of magnitude (Panin, 1986). In this unusual, short-lived but extremely excited state, the components of solid matter become as mobile as they would be in a gas state.

Intense turbulent flow at a microscopic scale in the tagamite melt is required for such a fine disintegration and widespread dispersion of the anatectic glass. This condition is reached either in impact melts during the final stages of shock pressure release, or immediately after the unloading of the melt (Vishnevsky and Staver, 1985). Soon after the pressure release, intense turbulent flow is quickly eliminated by internal viscosity friction.

Masaitis and Raikhlin (1985) contended that the glass globules in microcrystalline tagamites are the result of postimpact eutectic melting of gneiss inclusions incorporated into the hot tagamite melt. Accordingly, films, borders, and schlierens of eutectic glass can be considered as products of "a later impact anatexis" (Vishnevsky and Pospelova, 1986) resulting from postshock thermal metamorphism. However, this explanation is hard to reconcile with the immiscible glass globules pervasively dispersed in the rock matrix. Such a thorough dispersion can be attributed only to an extremely fluid, turbulent melt. Thus, we propose that formation and dispersal of the glass globules occurred before crystallization of the molten matrix that, initially, had a different chemical composition, similar to that of the bulk (crystallized) tagamite. This precrystallization compositional difference is the premise for immiscibility. If dispersal of globules in the matrix has occurred later (by some unkown mechanisms), i.e., after partial crystallization of the matrix melt, compositional similarity between globules and residual melt would lead to miscibility: the sharp boundaries of the globules indicate that no interaction (homogenization, contamination) ever occurred between them and the surrounding matrix melt.

The duration of shock compression is an important factor in early impact anatexis, which may be a feature common to large impacts only. That is why no analogues of globular tagamite, such as the one described here, are known in impact structures other than the Popigai. Nevertheless, possible traces of early impact anatexis in other impact structures may be represented by potassium-enriched glass borders around lechatelierite schlierens, such as those described by Stahle (1972) in Ries suevites, and by Dressler et al. (1997) in Wanapitei impact glasses.

Examples of immiscible glass spheroids in impactites from West Clearwater Lake Crater (Dence et al., 1974) are the result of a process different from early impact anatexis. These spheroids were formed by mixing between liquid silica glasses with contrasting compositions, and other Fe-, Mg-, and water-rich melts. Immiscible glasses recently described in the Zhamanshin (Fel'dman and Sazonova, 1993) and Logoi (Glazovskaya et al., 1993) Craters are the products of true liquid equilibria in impact melts of various compositions, and have no analogy with the impact anatectic glass herein described for Popigai microcrystalline tagamite.

The portions of "acid" (70–77 wt% SiO_2), potash-rich (up to 6 wt% K_2O), and water-saturated (up to 6–7 wt% H_2O) "granite" melt derived from Popigai target gneisses as a result of the early impact anatexis (see Table 9 and description in the petrography section above) may be of particular petrologic concern. The composition of this melt is a confirmation of the geochemical affinity of Si, K, and $H_2O + CO_2$ fluid in impactites that had been already known in static geology and experiments and comparable to the anatectic high-potassium acid rocks of the earliest Earth, as proposed by Vishnevsky and Popov (1987) and Vishnevsky (1993).

Trace element geochemistry and the nature of the impactor

First attempts to find the signatures of the extraterrestrial projectile in Popigai impactites were focused on broadly widespread, minute (3–5 to 200 µm) globules of Ni-enriched (0.5–12 wt% Ni) troilite-pyrrhotite (Masaitis and Sysoev, 1975a,b; Vishnevsky, 1975), as well as rare, fine globules and segregations of native iron (Vishnevsky, 1975), and native nickel (Masaitis and Sysoev, 1975b). Further investigations by Vishnevsky and Pospelova (1988) led to the discovery of globules of pentlandite (37 wt% Fe; 31 wt% S; 22 wt% Ni; 2 wt% Cu; and traces of such elements as Ti, Cr, and Co) in banded glasses from suevites. Similar Ni-bearing troilite-pyrrhotite globules (0.9–2 wt% Ni) were also found in Ries glasses (El Goresy et al., 1968), and were interpreted as possible remnants of the Ries projectile.

The Ni-bearing minerals in Popigai impact glasses were

interpreted either as direct evidence of the projectile (Masaitis and Sysoev, 1975a), or as meteoritic matter, whose cosmic source would be confirmed by a comparative study of opaque minerals in target source rocks (Vishnevsky, 1975). Soon after, it was shown by Vishnevsky (1976) that similar Ni-enriched accessory troilite-pyrrhotite globules containing up to 6 wt% Ni are common and widespread in Archean target gneisses. This piece of evidence leads us to the conclusion that the meteoritic source of the Ni-enriched troilite-pyrrhotite globules in Popigai impactites is doubtful.

Other attempts in identifying traces of the projectile in Popigai impactites consisted of the quest for anomalous concentrations of the so-called "meteoritic elements" such as Ni, Co, Cr, and Ir (Masaitis and Raikhlin, 1986, 1989; Masaitis, 1994). The apparent enrichments in Ni, Co, and Cr in impactites with respect to ordinary target gneisses led Masaitis and Raikhlin (1986 and 1989) to the conclusion that these elements represent a contamination from the extraterrestrial projectile. On the basis of these data, the same authors estimated the amount of meteoritic material in tagamites and suevites ranging from 0.01–0.5 to 1–4 wt% of the impactite volume, and suggested that the impactor was an ordinary chondrite. Nevertheless, the possible derivation of Ni, Co, and Cr from basic target rocks (Table 3) was not taken into consideration by Masaitis and Raikhlin (1986, 1989) in their interpretation. However, the background contents for comparison were obtained only from a limited collection of ordinary gneisses (41 samples for Ni, Co, and Cr; Masaitis and Raikhlin, 1986, 1989), inadequately representing the whole suite of target rocks. In reality, the target crystalline basement from which the large volume of impact melt ($\sim$1,750 km^3) was generated is of complex composition; it contains, in addition to ordinary gneisses, metabasalts comprising up to 10–20% of Verkhne-Anabar Series, dolerite dikes, and rare ultrabasic rocks.

The average Ir abundance in Popigai impactites is 0.064 ppb (Masaitis and Raikhlin, 1986, 1989). Further analyses of Popigai impactites yielded Ir concentrations of up to 5.71 ppb. After comparing these high concentrations with background Ir contents in ordinary gneisses, (0.007 ppb on average; data by Masaitis and Raikhlin, 1986,1989), these authors proposed (Masaitis and Raikhlin, 1986, 1989) a contamination of Popigai impactites by meteoritic material. They also pointed out that Ni/Ir ratios in chondrites and Popigai impactites are different from each other. This discrepancy was explained by a different volatility of Ni and Ir during the impact process (Masaitis and Raikhlin, 1989).

In general, the Ir content in common rocks of crystalline shields is very low ($\sim$0.018 ppb on average) (Nazarov, 1985), and it is similar to the average value of 0.007 ppb found in two gneiss samples by Masaitis and Raikhlin (1986, 1989). However, Nazarov (1995) showed that the range of Ir concentrations in shield rocks varies broadly, from 0.001 to 1 ppb. The same range of Ir abundances, from 0.001 to 1 ppb, is common for basic terrestrial rocks (Crocket, 1979). If compared with Ir concentrations in Popigai impactites (data from Masaitis and Raikhlin, 1986, 1989; see also our determinations in Table 6), target basic rocks of the Anabar Shield may provide at least some of the Ir found in Popigai impactites.

In ultrabasic rocks, Ir abundances usually range from 1 to 5 ppb, with maxima up to 20 ppb (Crocket, 1979, 1981), and rarely up to 20–80 ppb, as was found in some Newfoundland ultrabasites by Page and Talkington (1984). Therefore, the ultrabasic members of Archean target complex, despite their limited extent, may provide most of the Ir known in Popigai impactites, including the high concentration of 5.7 ppb found in glass veins from shocked gneisses by Masaitis and Raikhlin (1989). For comparison, the Ir content in C1–C2 chondrites is 518 ppb on the average (Crocket, 1981).

Ir is known as a good indicator for meteoritic contamination in impact melt rocks, and several terrestrial impact structures show clear enrichments (e.g., Palme, 1982). Recently, very uneven Ir enrichments were also reported in impactites from the giant Chicxulub Crater in Yucatán (Schuraytz et al., 1994, 1996). These values have so far not been confirmed by other laboratories (Rocchia et al., 1994; Claeys et al., 1995), although the presence of an extraterrestrial component in Chicxulub melt rocks is supported by Re-Os isotope systematics (Koeberl et al., 1994). Re-Os systematics was also applied in detection of meteoritic components in Ivory Coast tektites (Koeberl and Shirey, 1993).

All the samples we analyzed contain anomalous concentrations of Ir (0.08–1.56 ppb), which are one to three orders of magnitude higher than background abundances measured in ordinary gneisses by Masaitis and Raikhlin (1989). The analyzed rocks show no evidence of ultrabasic target components, and their bulk geochemistry indicates a gneiss source for the melt. Thus, the anomalous concentrations of Ir have to be attributed to an meteoritic source. However, we may conclude that meteoritic contamination in Popigai impactites varies widely from place to place, ranging up to two orders of magnitude.

Another side of the problem of projectile contamination in impactites in large impact structures is often underestimated, having to do with the kinematics of disruption and dispersion of the impacting object. W. von Englehardt (personal communication in Bunch and Cassidy, 1972) supposed that the lack of projectile traces in the Ries impactites may be due to the deposition of contaminated material in the top of the suevite column, which was completely eroded in postimpact times. A similar model of deposition of projectile matter in the top horizons of allogenic crater fill was also proposed for the Popigai by Vishnevsky and Pospelova (1984).

A recent numerical simulation of the Popigai impact shows that dispersal of the projectile was essentially controlled by a cumulative explosion effect, producing vertical jets of vaporized matter (vapor density ranging from 0.0009 to 0.4 g/cm^3) at velocities as high as 14.6 km/s (Vishnevsky et al., 1996). A similar vertical vapor jet model was proposed by Melosh (1989, see Fig. 5.6 therein). As much as 3×10^9 t of vaporized projectile matter ($\sim$1.5% of the impactor mass) was ejected at velocities of >11.2 km/s, enough to escape the Earth, and as much as 1.7×10^{10} t of the same matter ($\sim$8.5% of meteorite mass) was ejected at velocities between 2 and 11.2 km/s, high enough to leave the crater along ballistic trajectories (Vishnevsky et al., 1996). According to this model and in agreement with the model proposed by Alvarez (1995) on the ballistics of the Chicxulub impact ejecta, distal dispersion of meteoritic matter and strongly shocked and vaporized target rocks would have formed an impactoclastic air-fall bed probably of global distribution, and possibly containing a detectable extraterrestrial chemical signature (i.e., an Ir anomaly). The rest of the projectile matter (~90% of meteorite mass) was ejected at velocities between 0.1 and 2 km/s; most of it was deposited in a proximal ejecta blanket and on top of crater-fill formations. A large portion of these allogenic impact products enriched with extraterrestrial material were quickly eroded after the impact.

According to these considerations, a wide range of factors control the migration processes for disrupted projectile matter, and not all of them may result in a broad dissemination of projectile matter in the crater deposits. We propose that bolide-contaminated impactites were deposited in the uppermost breccia deposits and were likely subject to postimpact erosion.

CONCLUSIONS

The particular structure of the target, with soft Mesozoic rocks overlying harder lithologies, and the large scale of the event are important factors controlling the Popigai cratering process. Allogenic crater-fill impact formations are divided into turbulent centrifugal bottom flow deposits (impact base surge), and ballistically transported deposits of the explosion cloud. The first group of formations include Klippen Breccia, Megabreccia, Bottom Breccia, and Melt Sheet; the second group is represented by various suevite formations. Bottom flow deposits exhibit a reversal stratigraphy in peripheral zones of the crater and in areas outside the crater. Some characteristics of ballistically deposited impactites cannot easily be explained by a simple cratering model, and indicate a high-angle ejection of Mesozoic target rocks. The hypothesis of a dynamic barrier in the explosion cloud proposed in this chapter can explain stratigraphic provenance, rate of shock metamorphism, and some other unusual features of Popigai suevites and suevite megabreccia inside the crater, and their paucity outside the crater. This hypothesis envisions a complex structure of the explosion cloud, with a torus-shaped screen of vertically ejected megablocks of Mesozoic rocks above marginal and middle zones of the growing crater.

Popigai tagamite matrices are of two types: cryptocrystalline (crystallites <10 µm in size) and microcrystalline (crystallites 10–200 µm in size). Impact glass fragments from suevites are either pristine or recrystallized. Rock and mineral fragments of crystalline target rocks with various degrees of shock metamorphism are abundant in both tagamites and suevite glasses. Spatial relations between tagamite bodies, as well as microbanding in tagamites and suevite glasses, are very complex and indicate turbulent mixing.

All tagamites and suevite glasses investigated here, like those studied earlier by others, are similar to each other in terms of bulk geochemistry, which reflects the average composition of crystalline target rocks. However, cryptocrystalline tagamites and their suevitic equivalents (i.e., pristine glasses) contain less water (0.73 wt% on average) than microcrystalline tagamites and recrystallized suevite glasses (2.23 wt% on average). We propose that differences in water content in the melt had an important role in the crystallization of these different impact melts. The water in different impact melts was inherited from more or less wet Archean target rocks, which can explain megascale interfingering of bodies of different tagamites. Finer interfingering and banding of tagamites and other impact glasses may reflect heterogeneity of target rocks at a smaller scale.

Clasts of crystalline rocks, which are always present in Popigai impact glasses, are remnants of partially molten target rocks derived from the same melting zone in which the melt originated. We suggest that the melt for both types of tagamite and suevite glasses was also derived from marginal parts of the melting zone. It was a low-temperature melt, quickly cooled by thermodiffusion to a final integrated temperature, which was probably much lower then the melting points of quartz and feldspar (i.e., ~1,300°–1,700°C). Only homogeneous clast-free glass fragments from some suevites may be regarded as formed from high-temperature melts with integrated temperature >1,700°C. This hot melt was derived from inner parts of the melting zone.

Globules of impact anatectic glass dispersed in some microcrystalline tagamites, and derived from shocked gneisses, may be regarded as a result of component separation during the final stage of shock loading or immediately after it. This process provided selective mobilization of Si, K, and $H_2O + CO_2$ accompanied by loss of Fe, Ca, and Mg. A possible cause of this early impact anatexis may have been a short-lived, strongly excited state of compressed gneiss with a super-high rate of component diffusion. This process has nothing to do with the origin of various globular textures in impactites derived from mixtures of monomineral melt or liquation processes. Globular microcrystalline tagamites appear to be an exclusive feature of the Popigai impact structure, although traces of early impact anatexis may also be present in some other craters (e.g., Ries, Wanapitei).

The identification of the projectile in Popigai impactites is still an open question. There is no conclusive evidence that enrichments in Ni, Co, and Cr were derived from either the impactor itself or from basic and ultrabasic target rocks. However, our Ir INAA of selected Popigai impactites indicates that, in fact, the impact melt was contaminated to a certain degree by the projectile.

Numerous attempts to date the Popigai impact event led to widely scattered and inconsistent results, which we suspect were due to unsuitable impactite samples. We predict that fresh and thoroughly degassed glasses of microcrystalline tagamites are most likely to yield reliable radioisotopic dates. These rocks, as well as fresh suevite glasses, are also expected to be suitable for reliable fission-track age determinations. Products of high-temperature and partially vaporized impact melts, which are probably present in some Daldyn suevites, are perhaps the best material for radioisotopic and fission-track dating.

ACKNOWLEDGMENTS

This study was undertaken at the Osservatorio Geologico di Coldigioco and was supported by Coldigioco Research Fund; by a scientific exchange grant from the Eurpean Science Foundation (to S.V.); by the Network on Impact Cratering and Evolution of Planet Earth (D. Stöffler, chairman); and by the Association pour la Recherche et le Developpement des Methodes et Processus Industriels (ARMINES), Ecole des Mines des Paris.

We thank Bernard Beaudoin (Paris) and Philippe Claeys (Berlin) for critical reviews of an early draft of the manuscript, and Frank Asaro (Lawrence Berkeley National Laboratory) for having performed the instrumental neutron activation analyses on the Popigai impactites presented in this work. We also express our sincere gratitude to Burkhard Dressler, Bevan French, Christian Koeberl, and Uwe Reimold for their careful and patient reviews of the original manuscript. Their comments, suggestions, and criticism allowed us to improve considerably not only the form but also the substance of this work, overcoming language and conceptual/cultural barriers that, prior to this effort, prevented the senior author from properly communicating his knowledge of the Popigai to the international scientific community.

REFERENCES CITED

Alvarez, L. W., Asaro, F., Goulding, F. S., Landis, D. A., Madden, N. W., and Malone, D. F., 1988, Instrumental measurement of iridium abundances in the part-per-trillion range following neutron activation: American Chemical Society, l96th National Meeting, Los Angeles, CA. September 25–30, 1988: Washington D.C., Abstracts, NUCL-30.

Alvarez, W., Claeys, P., and Kieffer, S. W., 1995, Emplacement of Cretaceous-Tertiary boundary shocked quartz from Chicxulub crater: Science, v. 269, p. 930–935.

Alvarez, W., Alvarez, L. W., Asaro, F., and Michel, H. V., 1982a, Current status of the impact theory for the terminal Cretaceous extinction, in Silver, L. T., and Schultz, P. H., eds., Geological implications of impacts of large asteroids and comets on the Earth: Geological Society of America Special Paper 190, p. 305–315.

Alvarez, W., Asaro, F., Michel, H. V., and Alvarez, L. W., 1982b, Iridium anomaly approximately synchronous with terminal Eocene extinctions: Science, v. 216, p. 886–888.

Asaro, F., Alvarez, L. W., Alvarez, W., and Michel, H. V., 1987, Operation of the iridium coincidence spectrometer: studies in the middle Miocene and near the Cenomanian/Turonian boundary: Beijing, China, International Geological Correlation Program, Abstracts, p. 65.

Bogard, D., Horz, F., and Stöffler, D.,1988, Loss of radiogenic argon from shocked granite clasts in suevite deposits from the Ries crater: Geochimica et Cosmochimica Acta, v. 52, p. 2639–2649.

Bottomley, R. J., Grieve, R. A. F., York, D., and Masaitis, V., 1997, The age of the Popigai impact event and its relations to events at the Eocene/Oligocene boundary: Nature, v. 388, p. 365–368.

Bunch, T. E., and Cassidy, W. A., 1972, Petrographic and electron microscope study of the Monturaqui impactites: Contribution to Mineralogy and Petrology, v. 36, p. 95–112.

Carstens, H., 1975, Thermal history of impact melt rocks in the Fennoscandian Shield: Contribution to Mineralogy and Petrology, v. 59, p. 145–155.

Chao, E. C. T., 1977a, Preliminary interpretation of the 1973 Ries drill core: Geologica Bavarica, v. 75, 421–441.

Chao, E. C. T., 1977b, Impact cratering phenomena for the Ries multiring structure based on constraints of geological, geophysical and petrological studies and the nature of impacting body, in Roddy, D. J., Pepin, R. O., and Merrill, R. B., eds., Impact and explosion cratering: New York, Pergamon Press, p. 405–424.

Claeys, P., Asaro, F., Rocchia, R., Hildebrand, A. R., Grajales, J. M., and Cedillo, E., 1995, Ir abundances in the Chicxulub melt-rock: European Science Foundation, 4th International Workshop on The effects of impacts on the evolution of the atmosphere and biosphere with regard to short- and long-term changes, Ancona, Italy, May 12–18, 1995, Abstracts, p. 55–57.

Clymer, A. Bice, D. M., and Montanari, A., 1996, Shocked quartz from the late Eocene: impact evidence from Massignano, Italy: Geology, v. 24, p. 483–486.

Cowie, J. W., 1991, UNESCO World Heritage Convention. Report on Task Force Meeting, Paris, February 1991: New York, UNESCO, 37 p.

Crocket, J. H., 1979, Platinum-group elements in basic and ultrabasic rocks: a survey: Canadian Mineralogist, v. 17, p. 391–402.

Crocket, J. H., 1981, Geochemistry of the platinum-group elements: in Cabri, L. S., ed., Platinum-group elements: mineralogy, geology, recovery: Montreal, Canadian Institute of Mining and Metallurgy Special Volume 23, p. 47–65.

Deino, A., Garvin, J. B., and Montanari, A., 1991, K/T Ar-Ar age for the Popigai impact?: Lunar and Planetary Science Conference, Abstracts: Houston, Texas, Lunar and Planetary Institute, p. 297-298.

Deino, A. L., Montanari, A., and Vishnevsky, S., 1995, $^{40}Ar/^{39}Ar$ dating of glass from the Popigai impact crater, Russia: European Science Foundation, 4th International Workshop on The effects of impacts on the evolution of the atmosphere and biosphere with regard to short- and long-term changes, Ancona, Italy, May 12–18, 1995, Abstracts, p. 66–67.

Dence, M. R., 1971, Impact melts: Journal of Geophysical Research, v. 76, p. 5552–5565.

Dence, M. R., Engelhardt, W. von, Plant, A. G., and Walter, L. S., 1974, Indications of fluid immiscibility in glass from West Clearwater Lake impact crater, Quebec, Canada: Contribution to Mineralogy and Petrology, v. 46, p. 81–97.

Dence, M. R., Grieve, R. A. F., and Robertson, P. B., 1977, Terrestrial impact structures: principal characteristics and some energy considerations, in Roddy, D. J., Pepin, R. O., and Merrill, R. B., eds., Impact and explosion cratering: New York, Pergamon Press, p. 247–275.

Dressler, B. O., Crabtree, D., and Shuraytz, B.C., 1997, Incipient melt formation and devitrification at the Wanapitei impact structure, Ontario, Canada: Meteoritics and Planetary Science, v. 32, p. 249–258.

El-Goresy, A., Fechtig, H., and Ottemann, J., 1968, Opaque minerals in impactite glasses, in French, B. M., and Short, N. M., eds., Shock metamorphism of natural materials: Baltimore, Maryland, Mono Book Corp., p. 531–554.

Fel'dman, V. I., and Sazonova, L. V., 1993, Formation and cooling conditions for impact melts in Zhamanshin astroblema: Petrology, v. 1, p. 596–614 (in Russian).

Firsov, L. V., 1970, The Paleogenic basaltic rocks in the Popigai graben (the Anabar Schield): Doklady Akademii Nauk SSSR, v. 194, p. 664–666 (in Russian).

Florensky, K. P., Basilevsky, A. T., Ivanov, B. A., and others, 1983, Shock craters on the Moon and other planets: Moscow, Nauka Press, 200 p. (in Russian).

French, B. V., Hartung, J. B., Short, N. M., and Dietz, R. S., 1970, Tenoumer crater, Mauritania: age and petrologic evidence for origin by meteorite impact: Journal of Geophysical Research, v. 75, p. 4396–4406.

Gault, D. E., Quaide, W. L., and Oberbeck, V. R., 1968, Impact cratering mechanics and structures; in French, B. M., and Short, N. M., eds., Shock metamorphism of natural materials: Baltimore, Maryland, Mono Book Corp., p. 87–99.

Glass, B. P., and Burns, C. A., 1987, Late Eocene glass-bearing spherules: two layers or one?: Meteoritics, v. 22, p. 265–279.

Glass, B. P., Hall, C. M., and York, D., 1986, $^{40}Ar/^{39}Ar$ laser-probe dating of

North American tektite fragments from Barbados and the age of the Eocene/Oligocene boundary: Chemical Geology, v. 59, p. 181–189.

Glazovskaya, L. I., Parfenova, O. V., and Il'inkevitch, G. I., 1993, Impactites of Logoi astroblema: Petrology, v. 1, p. 634–644 (in Russian).

Grieve, R. A. F., 1987, Terrestrial impact structures: Annual Review of Earth and Planetary Science, v. 15, p. 245–270.

Grieve, R. A. F., Langenhorst, F., and Stöffler, D., 1996, Shock metamorphism of quartz in nature and experiment: significance in geoscience: Meteoritics and Planetary Science, v. 31, p. 6–35.

Gusev, G. M., Guletskaya, E. S., and Kozyrev, N. L., 1981, Calculative methods of elements' content determinations at photometric analyses of minerals and materials: Physical and chemical methods of investigations in geology, in Proceedings, Novosibirsk Institute of Geology and Geophysics, no. 450: Novosibirsk; Nauka Press, p. 189–192 (in Russian).

Hel'man, E. M., and Starobina, I. Z., 1972, Determination of rock-forming elements (silicates and carbonates) by photometric method: Analytical methods at geochemical investigations, in Proceedings, 4th Geochemical Conference, Leningrad, May 18–20, 1970: Leningrad, Mining Institute Press, p. 50–52 (in Russian).

Hel'man, E. M., Terent'eva, E. A., Shanina, T. M., and Kiparenko, L. M., 1987, Methods of numerical organic elemental analyses: Moscow, Khimiya Press, 296 p. (in Russian).

Horz, F. P., 1965, Unterschungen am Riesglasern: Beitrage zur Mineralogie und Petrographie, v. 11, p. 621–661.

Horz, F., 1982, Ejecta of the Ries crater, Germany, in Silver, L. T., and Schultz, P. H., eds., Geological implications of impacts of large asteroids and comets on the Earth: Geological Society of America Special Paper 190, p. 39–55.

Horz, F., Hall, H., Huttner, R., and Oberbeck, V. R., 1977, Shallow drilling in the "Bunte Breccia" impact deposits, Ries crater, Germany: in Roddy, D. J., Pepin, R. O., and Merrill, R. B., eds., Impact and explosion cratering: New York, Pergamon Press, p. 425–448.

Ivanov, B. A., 1976, The effect of gravity on cratering formation: thickness of ejecta and concentric basins, in Proceedings, 7th, Lunar and Planetary Science Conference, v. 3: New York, Pergamon Press, p. 2947–2965.

Ivanov, B. A., 1979, A simple model of cratering: Meteoritika, v. 38, p. 68–85 (in Russian).

Ivanov, B. A., Basilevsky, A. T., and Sazonova, L. T., 1982, On the origin of central uplifts in meteoritic craters: Meteoritika, v. 40, p. 67–81 (in Russian).

Keller, G., D'Hondt, S. L., Orth, C. J., Gilmore, J. S., Oliver, P. Q., Shoemaker, E. M., and Molina, E., 1987, Late Eocene impact microspherules: stratigraphy, age, and geochemistry: Meteoritics, v. 22, p. 25–60.

Kieffer, S. W., and Simonds, C. H., 1980, The role of volatiles and lithology in the impact cratering processes: Reviews on Geophysics and Space Physics, v. 18, p. 143–181.

Kirjushina, M. T., 1959, Meso-Cenozoic volcanites at the northern margin of the Siberian platform: Izvestija Akademii Nauk SSSR, Geology, n. 1, p. 50–55 (in Russian).

Koeberl, C., and Shirey S. B., 1993, Detection of a meteoritic component in Ivory Coast tektites with rhenium-osmium isotopes: Science, v. 261, p. 595–598.

Koeberl, C., Sharpton, V. L., Schuarytz, B. C., Shirey, S. B., Blum, J. D., and Marin, L. E., 1994, Evidence for a meteoritic component in impact melt rock from the Chicxulub structure: Geochimica et Cosmochimica Acta, v. 58, p. 1679–1684.

Koeberl, C., Poag, C. W., Reimold, W. U., and Brandt, D., 1996, Impact origin of the Chesapeake Bay structure and source of the North American tektites: Science, v. 271, p. 1263–1266.

Komarov, A. N., and Raikhlin, A. I., 1976, Comparative investigation of the age of impact rocks by fission track, and K/Ar methods: Doklady Akademii Nauk SSSR, v. 228, p. 673–676 (in Russian).

Kopylova, A. G., and Oleinikov, B. V., 1973, The distribution of Ni, Co, Cr, V and Cu in rocks and minerals of differentiated trapp intrusions, in Geology and geochemistry of the eastern part of the Siberian Platform: Moscow, Nauka Press, p. 164–191 (in Russian).

Langenhorst, F., 1996, Characteristics of shocked quartz in late Eocene impact ejecta from Massignano (Ancona, Italy): clues to shock conditions and source crater: Geology, v. 24, p. 487–490.

Lavrent'ev, Yu. G., Pospelova, L. N., Sobolev, N. V., and Malikov, Yu. I., 1974, The determination of rock-forming minerals by X-ray spectra microanalyses with electron probe: Zavodskaya Laboratoria, v. 40, p. 657–661 (in Russian).

Lutz, B. G., 1964, The petrology of granulite facies of the Anabar Shield: Moscow, Nauka Press, 124 p. (in Russian)

Lutz, B. G., 1985, The metamorphism in the mobile belts of the Early Earth: Moscow, Nauka Press, 216 p. (in Russian).

Marakushev, A. A., Bogatyrev, O. S., Fenogenov, A. D., Panyakh, N. A., and Fedosova, S. P., 1993, Impactogenesis and volcanism: Petrologia, v. 1, p. 571–595 (in Russian).

Masaitis, V. L., 1970, A short review on the igneous rocks of the Siberian Platform, in Minerageniya Sibirskoi Platformy: Moscow, Nedra Press, p. 42–62 (in Russian).

Masaitis, V. L., ed., 1983, Structures and textures of explosive breccias and impactites, in Proceedings, Karpinsky Geological Institute, new seria, n. 316: Leningrad, Nedra Press, 159 p. (in Russian).

Masaitis, V. L., 1984, Giant meteorite impacts: Some models and their consequences: in Rundkvist, D. V., ed., Modern ideas of theoretical geology: Moscow, Nedra Press, p. 151–179 (in Russian).

Masaitis, V. L., 1994, Impactites from Popigai crater, in Dressler, B. O., Griver, R. A. F., and Sharpton, V. L., eds., Large meteorite impacts and planetary evolution: Geological Society of America Special Paper 293, p. 153–162.

Masaitis, V. L., and Raikhlin, A. I., 1985, Immiscibility of impact and pyrometamorphic melt: Meteoritika, v. 44, p. 159–163 (in Russian).

Masaitis, V. L., and Raikhlin, A. I., 1986, The Popigai crater was formed by an ordinary chondrite impact: Doklady Akademii Nauk SSSR, v. 286, p. 1476–1478 (in Russian).

Masaitis, V. L., and Raikhlin, A. I., 1989, The traces of projectile in various rock types of impact craters: Meteoritika, v. 48, p. 161–170 (in Russian).

Masaitis, V. L., and Selivanovskaya, T. V., 1972, The shock-metamorphosed rocks and impactites of the Popigai crater: Zapiski Vsesoyuznogo Mineralogicheskogo Obstchestva, pt. 101, p. 383–393 (in Russian).

Masaitis, V. L., and Sysoev, A. G., 1975a, The meteoritic matter in Popigai impactites: Pis'ma v Astronomichesky Zhurnal, v. 1, p. 43–47 (in Russian).

Masaitis, V. L., and Sysoev A. G., 1975b, Nickel-bearing iron sulphides and native nickel in suevites of Popigai crater: Zapiski Vsesoyuznogo Mineralogicheskogo Obstchestva, pt. 104, p. 204–208 (in Russian).

Masaitis, V. L., Mikhailov, M. V., and Selivanovskaya, T. V., 1971, The Popigai meteoritic crater: Sovetskaya Geologia, no. 6, p. 143–147 (in Russian).

Masaitis, V. L., Futergendler, S. I., and Gnevushev, M. A., 1972, Diamonds in impactites of Popigai crater: Zapiski Vsesoyuznogo Mineralogicheskogo Obstchestva, pt. 101, p. 108–113 (in Russian).

Masaitis, V. L., Mikhailov M. V., and Selivanovskaya T. V., 1975, The Popigai impact crater: Moscow, Nauka Press, 124 p (in Russian).

Masaitis, V. L., Mastchak, M. S., Selivanovskaya, T. V., Raikhlin, A. I., and Danilin, A. N., 1980, The Popigai astroblema, in Pogrebitsky, Yu. E., ed., Geology of astroblemes: Leningrad, Nedra Press, p. 114–130 (in Russian).

Mashchak, M. S., and Fedorova, I. G., 1987, Composition and origin conditions for clastic dikes in tagamites of Popigai astrobleme: Meteoritika, v. 46, p. 124–127 (in Russian).

Maxwell, D. E., 1977, Simple Z-model of cratering, ejection and overturned flap, in Roddy, D. J., Pepin, R. O., and Merrill, R. B., eds., Impact and explosion cratering: New York, Pergamon Press, p. 1003–1008.

Melosh, H. J., 1989, Impact cratering, a geologic process: New York, Oxford University Press, Oxford Monographs on Geology and Geophysics, no. 11, 245 p.

Mezhvilk, A. A., 1981, The Popigai volcano-tectonic structure: Izvestia Akademii Nauk SSSR, geology, no. 6, p. 19–24 (in Russian).

Michel, H. V., Asaro, F., Alvarez, W., and Alvarez, L. W., 1990, Geochemical studies or the Cretaceous-Tertiary boundary in ODP holes 689B and 690C: ODP Proceedings, Scientific Results, v. 113, p. 159–168.

Montanari, A., Asaro, F., Michel, H. V., and Kennett, J. B., 1993, Iridium anomalies of late Eocene age at Massignano (Italy), and ODP Site 689B (Maud Rise, Antarctic): Palaios, v. 8, p. 420–437.

Nazarov, M. A., 1985, The types of Moon rocks and minerals carried by "Moon" seria of Automatic Lunar Landing Stations (Some consequences on the origin of the Moon's and Earth's crust), [Cand. of Sci. thesis]: Moscow, Vernadsky Institute of Geochemistry and Analytical Chemistry, 27 p. (in Russian).

Nazarov, M. A., 1995, The geochemical evidence of large impact events in geological record of the Earth [abs. Ph.D. thesis]: Moscow, Vernadsky Institute of Geochemistry and Analytical Chemistry, 48 p. (in Russian).

Obradovich, J. N., Snee, L. W., and Izett, G. A., 1989, Is there more than one glassy impact layer in the late Eocene?: Geological Society of America Abstracts with Programs, v. 2, p. A134.

Osorgin, N. Yu., 1990a, Chromatographic analyses of gases from minerals (methodology, technique and metrology): Novosibirsk, Institute of Geology and Geophysics Press, no. 11, 32 p. (in Russian).

Osorgin, N. Yu., 1990b, Chromatographic device for simulating determinations of organic and inorganic gasses, in Simonov, V. A., ed., Thermobarogeochemistry of mineral formation processes: Novosibirsk, Institute of Geology and Geophysics Press, no. 1, p. 129–140 (in Russian).

Page, N. J., and Talkington, R. W., 1984, Palladium, platinum, rhodium, ruthenium and iridium in peridotites and chromitites from ophiolite complexes of Newfoundland: Canadian Mineralogist, v. 22, p. 137–149.

Palme, H., 1982, Identification of projectiles of large terrestrial impact craters and some implications for the interpretation of Ir-rich Cretaceous/Tertiary boundary layers, in Silver, L. T., and Schultz, P. H., eds., Geological implications of impacts of large asteroids and comets on Earth: Geological Society of America Special Paper 190, p. 223–233.

Panin, V. E., 1986, A new field of solid state physics: Izvestia Vuzov SSSR, Physics, v. 30, p. 3–8 (in Russian).

Perlman, I., and Asaro, F., 1969, Pottery analysis by neutron activation: Archaeometry, v. 11, p. 21–52.

Perlman, I., and Asaro, F., 1971, Pottery analysis by neutron activation, in Brill, R. H., ed., Science and archeology: Cambridge, Massachusetts, MIT Press, p. 182–195.

Pierrard, O., Robin, E., and Rocchia, R., 1996, The stratigraphic distribution of Ni-rich spinel crystals near the Eocene-Oligocene boundary: 17th Regional African European Meeting of Sedimentology: Stax, Tunisia, International City of Sedimentologists, p. 209–210.

Pierrard, O., Robin, E., Rocchia, R., and Montanari, A., 1998, Extraterrestrial Ni-rich spinel in upper Eocene sediments from Massignano, Italy: Geology, v. 26, p307–310.

Poag, C. W., Powars, D. S., Poppe, L. J., and Mixon, R. B., 1994, Meteoroid mayhem in Ole Virginy: source for the North American tektite strewn field: Geology, v. 22, p. 691–694.

Pohl, J., Stöffler, D., Gall H., and Ernstson, K., 1977, The Ries impact crater, in Roddy, D. J., Pepin, R. O., and Merrill, R. B., edts., Impact and explosion cratering: New York, Pergamon Press, p. 343–404.

Poldervaart, A., 1955, Chemistry of the Earths' crust, in Poldervaart, A., ed., Crust of the Earth: Geological Society of America Special Paper 62, p. 119–144.

Polyakov, M. M., and Trukhalev, A. I., 1974, The Popigai volcano-tectonic ring structure: Izvestia Akademii Nauk SSSR, Geology, no. 4, p. 85–94 (in Russian).

Preuss, E., 1969, Einführung in die Ries–Forschung: Geologica Bavarica, v. 61, p. 22–25.

Rabkin, M. I., 1959, The geology and petrology of the Anabar Shield: Proceedings, Arctic Geology Research Institute (NIIGA), v. 87, 164 p. (in Russian).

Rocchia, R., Claeys, P., and Robin, E., 1994, Lack of Ir enrichment in the Chicxulub crater impact sheet: Eos transactions, American Geophysical Union, Fall Meeting Supplement, v. 75, p. 416.

Schuraytz, B. C., Sharpton, V. L., and Marin, L. E., 1994, Petrology of impact-melt rocks at the Chicxulub multiring basin, Yucatan, Mexico: Geology, v. 22, p. 868–872.

Schuraytz, B. C., Lindstrom, D. J., Marin, L. E., Martinez, R. R., Mittlefehldt, D. W., Sharpton, V. L., and Wentworth, S. J., 1996, Iridium metal in Chicxulub impact melt: forensic chemistry on the K/T smoking gun: Science, v. 271, p. 1573–1576.

Shoemaker, E. M., 1977, Astronomically-observed crater-forming projectiles, in Roddy, D. J., Pepin, R. O., and Merrill, R. B., eds., Impact and explosion cratering: New York, Pergamon Press, p. 617–628.

Smirnov, L. P., 1962, Stratigraphy of Cretaceous continental deposits of Popigai basin, in Collection of articles on geology and petroleum and gas potential of Arctica: Proceedings, Arctic Geology Research Institute (NIIGA), v. 121, p. 29–43 (in Russian).

Stahle, V., 1972, Impact glasses from suevites of the Nordlingen-Ries: Earth and Planetary Science Letters, v. 17, p. 275–293.

Stöffler, D., 1971, Progressive metamorphism and classification of shocked and brecciated crystalline rocks at impact craters: Journal of Geophysical Research, v. 76, p. 5541–5551.

Stöffler, D., and Grieve, R. A. F., 1994, Classification and nomenclature of impact metamorphic rocks, a proposal to the IUGS Subcommission on the Systematics of metamorphic rocks: European Science Foundation, Network on Impact Cratering and the Evolution of Planet Earth, in Montanari, A., and Smit, J., eds., Post-Östersund Newsletter, Osservatorio Geologico di Coldigioco, p. 9–15.

Stöffler, D., and Langenhorst, F., 1994, Shock metamorphism of quartz in nature and experiment: I. Basic observations and theory: Meteoritics, v. 29, p. 155–181.

Storzer, D., and Wagner, G., 1979, Fission-track dating of El'gygytgyn, Popigai and Zhamanshin craters: no source for Australasian or North American tektites: Meteoritics, v. 14, p. 541–542.

Trukhalev, A. I., and Solov'ev, I. A., 1981, Manifestation of Mesozoic-Cenozoic volcanism on the North-Eastern slope of the Anabar Shield: Geologia i Geofizika, v. 22, no. 5, p. 54–60 (in Russian).

Vaganov, V. I., Ivankin, P. F., Kropotkin, P. N., Trukhalev, A. I., Semeneko, N. P., Tsymbal, S. N., Tatarintsev, V. I., Glukhovsky, M. Z., and Bulgakov, E. A., 1985, Explosive ring structures on the shields and platforms: Moscow, Nedra Press, 200 p. (in Russian),

Vinogradov, A. P., 1962, Average elemental abundancies in main types of igneous rocks of the Earth's crust: Geokhimia, no. 7, p. 552–571 (in Russian).

Vishnevsky, A. S., Balagansky I. A., and Vishnevsky, S. A., 1996, Computer simulation of the Popigai impact event (compression and initial excavation stages), and some consequences on global dispersion of projectile and tektite glasses: in Drobne, K., Gorican, S., Kotnik, B., eds., International Workshop on "The Role of Impact Processes in the Geological and Biological Evolution of Planet Earth," Postojna, Slovenia, September 27–October 2, 1996: Geology of West Slovenia, Field Guide Ljubljana, Slovenia, p. 95–96, Abstracts.

Vishnevsky, S. A., 1975, Some opaque minerals in impact glasses from the Popigai structure: Methodology and methodics of geological and geophysical investigations in Siberia: Novosibirsk, Institute of Geology and Geophysics Press, p. 105–115 (in Russian).

Vishnevsky, S. A., 1976, On the origin of Ni-enriched troilite-pyrrhotite in Popigai impactites: Geologia i Geofizika, no. 7, p. 110–112 (in Russian).

Vishnevsky, S. A., 1978, The features of the rocks and the origin of the Popigai structure [Ph.D. thesis]: Novosibirsk, Institute of Geology and Geophysics, 273 p. (in Russian).

Vishnevsky, S. A., 1980, On the nature of the complex craters' morphostructure, in Dolgov, Yu. A., ed., The interaction of meteoritic matter with the Earth: Novosibirsk, Nauka Press, p. 54–66 (in Russian).

Vishnevsky, S. A., 1981, Impactites of giant complex meteoritic craters, in Marakushev, A. A., ed., Impactites: Moscow, Moscow University Press, p. 171–184 (in Russian).

Vishnevsky, S. A., 1986, A margin part of Popigai astroblema: impact diatremes and horsts, and a new interpretation of deep interior of the structure, in Dolgov, Yu. A., ed, The cosmic matter and the Earth: Novosibirsk, Nauka Press, p. 131–159 (in Russian).

Vishnevsky, S. A., 1992, Suevites of the Popigai astroblem: some paradoxes and false secondary relations, (preprint): Novosibirsk, United Institute of Geology, Geophysics and Mineralogy Press, 53 p. (in Russian).

Vishnevsky, S. A., 1993, Terrestrial planets interior differentiation: the earliest Si-K-fluid ($H_2O + CO_2$) separation by impact? European Geophysical Society, General Assembly, 18th, Wiesbaden, May 3–7, 1993. Pt. III, Space and Planetary Sciences: Annales Geophysicae, v. 11, suppl. III, p. C449.

Vishnevsky, S. A., 1994a, The suevite megabreccia: a new type of explosion cloud deposits at the Popigai astroblema. I. General characteristics (preprint): Novosibirsk, United Institute of Geology, Geophysics and Mineralogy Press, 66 p. (in Russian).

Vishnevsky, S. A., 1994b, The Popigai astroblema, a possible site of our geological heritage: characteristics and aspect, problems and ideas of conservation, in Proceedings, 1st, International Symposium on the Conservation of Our Geological Heritage, Digne-les-Bains, France, June 11–16, 1991: Memories de la Societé geologique de France, nouvelle serie, no. 165, p. 71–74.

Vishnevsky, S. A., 1994c, The Popigai boring cores project: a history, present state, proposals and perspectives: in European Science Foundation, 3rd International Workshop on "Shock wave behavior of solids in nature and experiments," Limoges, France, September 18–21, 1994, Feildtrip guide, p. 70.

Vishnevsky, S. A., 1996a, Two groups of Popigai impact glasses: a result of initial water content in target rocks: in International Conference on Natural Glasses, 3rd, Jena, Germany, March 21–23, 1996: Jena, Germany, Friedrich-Schiller-Universitat Press, Abstracts, p. 14.

Vishnevsky, S. A., 1996b, Two groups of Popigai impact glasses: a result of initial water content in target rocks: Chemie der Erde, v. 56, p. 493–497.

Vishnevsky, S., and Montanari, A., 1994, Petrography and geochemistry of Popigai impact melt rocks: a selection of material for potentially reliable radioisotopic dating: European Science Foundation, Network on Impact Cratering and Evolution of Planet Earth, in Montanari, A., and Smit, J., eds., Post-Östersund Newsletter, Osservatorio Geologico di Coldigioco, p. 16–27.

Vishnevsky, S. A., and Pal'chik, N. A., 1975, Graphite in the rocks of Popigai structure: destruction and transformation to other phases of carbon system: Geologia i Geofizika, no. 1, p. 67–75 (in Russian).

Vishnevsky, S. A., and Popov, N. V., 1987, The possible model of origin of the earliest high-potassium acid rocks of the Earth, in Popov, N. V., ed., Processes of metamorphism on the shields and folding belts: models of evolution. Novosibirsk: Institute of Geology and Geophysics Press, p. 58–67 (in Russian).

Vishnevsky, S. A., and Pospelova, L. N., 1984, Some petrologic and geochemical features of the impact processes, in Dolgov, Yu. A., ed., The meteoritic research in Siberia (75th anniversary of the Tunguska event): Novosibirsk, Nauka Press, p. 156–191 (in Russian).

Vishnevsky, S. A., and Pospelova L. N., 1986, The impact anatexis of the shocked gneisses at the Popigai astroblema, [in Russian]: in Dolgov, Yu. A., ed., Meteoritic matter and Earth: Novosibirsk, Nauka Press, p. 117–131 (in Russian).

Vishnevsky, S. A., and Pospelova, L. N., 1988, The fluid regime of the impactites: dense fluid inclusions in the high-silica glasses and their petrological significance (preprint): Novosibirsk, Institute of Geology and Geophysics Press, no. 16, 53 p. (in Russian).

Vishnevsky, S. A., and Staver, A. M., 1985, Some distinctive features of deformation and melting in impact metamorphism: Geologia i Geofizika, v. 26, no. 2, p. 22–30 (in Russian).

Vishnevsky, S. A., Kovaleva, L. T., and Pal'chik, N. A., 1974, Coesite in the rocks of Popigai structure: Geologia i Geofizika, v. 26, no. 6, p. 140–145 (in Russian).

Vishnevsky, S. A., Beisel, A. L., and Zakharov, V. A., 1995a, The Popigai crater: a guide to regional paleogeography: Meteoritics, v. 30, p. 591–592.

Vishnevsky, S., Afanas'ev, V., Koptil', V., and Montanari, A., 1995b, Popigai distal ejecta deposits: evidence by the strewn field of impact diamonds, European Science Foundation, 4th International Workshop on the effects of impacts on the evolution of the atmosphere and biosphere with regard to short- and long-term changes: Ancona, Italy, May 12–18, 1995, Abstracts, p. 160–162.

Vishnevsky, S. A., Dolgov, Yu. A., Kovaleva, L. T., and Pal'chik, N. A., 1975a, Stishovite in the rocks of the Popigai structure: Doklady Academii Nauk SSSR, v. 221, p. 1167–1169 (in Russian).

Vishnevsky, S. A., Dolgov, Yu. A., Kovaleva, L.T., and Pal'chik, N. A., 1975b, Stishovite in the rocks of the Popigai structure: Geologia i Geofizika, no. 0, p. 149–156 (in Russian).

Vishnevsky, S. A., Popov, N. V., and Pospelova, L. N., 1987, Bi-melt structures in impactites; origin and role, in 2nd International Conference on Natural Glasses, Prague, September 21–23, 1987: Prague, Czechoslovakia, Karlov University Press, p. 68.

Vishnevsky, S., Afanas'ev, V., Argunov, K. P., and Pal'chik, N. A., 1997, Impact diamonds: their features, origin, and significance: Novosibirsk, United Institute of Geology, Geophysics, and Mineralogy Press, no. 835, 110 p.

Witzke, B. J., and Anderson, R. R., 1996, Sedimentary-clast breccias of the Manson impact structure, in Koeberl, C., and Anderson, R. R., eds., The Manson impact structure, Iowa: anatomy of an impact crater: Geological Society of America Special Paper 302, p. 115–144.

Manuscript Received by the Society December 16, 1998

Morokweng impact structure, South Africa: Geologic, petrographic, and isotopic results, and implications for the size of the structure

Wolf Uwe Reimold
Impact Cratering Research Group, Department of Geology, University of the Witwatersrand, P.O. Wits, Johannesburg 2050, South Africa

Christian Koeberl
Institute of Geochemistry, University of Vienna, Althanstrasse 14, A-1090 Vienna, Austria; e-mail: christian.koeberl@univie.ac.at

Franz Brandstätter
Mineralogisch-Petrographische Abteilung, Naturhistorisches Museum, P.O. Box 417, A-1014 Vienna, Austria

F. Johan Kruger
Hugh Allsopp Laboratory, Bernard Price Institute, University of the Witwatersrand, Johannesburg 2050, South Africa

Richard A. Armstrong
Research School of Earth Sciences, Australian National University, Canberra, ACT 0200, Australia

Cornelis Bootsman
Department of Geography and Environmental Studies, University of the Northwest, P.O. Box X2046, Mmabatho 2735, South Africa

ABSTRACT

A large impact structure has been identified around Morokweng in the Northwest Province of South Africa. The central aeromagnetic anomaly of this structure is caused by a ~30-km-wide melt body. Textural, mineralogical, and chemical analyses of many drill-core samples of this melt rock are reported. Shock metamorphism in granitoids drilled below the melt body confirms the origin of this melt rock by impact melting. The age of the Morokweng impact melt rock is well-constrained at 145 Ma from single zircon U-Pb isotopic analysis. This age is indistinguishable from the biostratigraphic age of the Jurassic-Cretaceous boundary. Rb-Sr isotopic data of whole-rock and mineral specimens of impact melt rock show that the melt-clast system was not completely isotopically equilibrated. Secondary alteration caused, at least partial, open system behavior. Petrographic studies of the granitoids below the melt body suggest that this ~50-m-wide zone represents a megabreccia and not, as previously suggested, Archean basement. Analysis of a suite of dike breccia samples from this zone revealed the presence of injected impact melt rock, cataclastic breccia, and altered melt breccia, which could have resulted either from impact melting of highly shock-metamorphosed material or from frictional melting. The occurence of several distinct breccia types does not allow us to classify these dike breccias summarily as pseudotachylite, as earlier workers were prone to do. New surface and subsurface information, especially with regard to the stratigraphic thicknesses of the preimpact Dwyka Group and postimpact Kalahari Group in the region, suggest that the Morokweng impact structure could originally have been at least 200 km wide.

Reimold, W. U., Koeberl, C., Brandstätter, F., Kruger, F. J., Armstrong, R. A., and Bootsman, C., 1999, Morokweng impact structure, South Africa: Geologic, petrographic, and isotopic results, and implications for the size of the structure, *in* Dressler, B. O., and Sharpton, V. L., eds., Large Meteorite Impacts and Planetary Evolution II: Boulder, Colorado, Geological Socicty of America Special Paper 339.

INTRODUCTION

A large, near-circular aeromagnetic anomaly in the region around the town of Morokweng, approximately centered at 23°32′E and 26°20′S, close to the border with Botswana in the Northwest Province of South Africa (Fig. 1), had long been interpreted as the expression of an intrusive body (e.g., Stettler, 1987; discussion in Corner, 1994). However, reinterpretation of regional gravity and aeromagnetic images by Corner (1994) and Corner et al. (1996, 1997; also Reimold et al., 1996), as well as first reports of impact characteristic shock metamorphic effects in rocks from the Morokweng area (Andreoli et al., 1995, 1996a,b; Corner et al., 1996, 1997; Koeberl et al., 1997a,b), demonstrated the presence of a large meteorite impact structure. The size of the Morokweng impact structure is still the subject of debate, but previous workers agree that it must have been larger than 70 km, which is the diameter of the present composite magnetic anomaly. Some estimates (e.g., Andreoli et al., 1995; Corner et al., 1997) proposed original diameters in excess of 300 km.

Koeberl et al. (1997a) and Hart et al. (1997) presented petrographic, chemical, and isotopic data for an impact melt body (which is also the cause of the magnetic anomaly) in the Morokweng structure. Preliminary chemical work by these authors revealed significant levels of siderophile elements. Koeberl et al. (1997a) demonstrated the presence of approximately 5% by weight of a meteoritic component in this impact melt rock. Both Koeberl et al. (1997a) and Hart et al. (1997) obtained near-identical U-Pb single zircon ages for the Morokweng impact melt rock, of 145 ± 0.8, 144.7 ± 0.7, and 146.2 ± 1.5 Ma, respectively. This age is indistinguishable from the currently accepted age for the Jurassic-Cretaceous boundary, which is placed at 145 Ma at the base of the Berriasian (Harland et al., 1990; Gradstein et al., 1994). Hart et al. (1997) and Koeberl et al. (1997a–c) discussed the implications of the discovery of a large impact structure at or near the J–K boundary.

After the initial recognition of the Morokweng impact structure and its siderophile-element–enriched melt rocks, limited exploration activity resulted, as many other large impact structures known on earth are associated with economic mineral and hydrocarbon deposits. For example, the Vredefort Structure in South Africa is located in the center of the gold- and uranium-rich Witwatersrand basin. The Chicxulub Structure was discovered in the course of hydrocarbon exploration in Mexico, and the enormous nickel and platinum group element (PGE) deposits of the Sudbury Structure in Canada are legendary. However, the levels of siderophile element enrichment in the Morokweng melt body and low abundance of sulfide minerals soon discouraged further exploration. Nevertheless, the record shows that large impact structures may be of outstanding economic value (e.g., Masaitis and Grieve, 1994; Reimold, 1996).

The discovery of the large Morokweng impact structure is significant, as it underlines the necessity to continue the search for still-unknown impact structures on the African continent. The current impact crater record for Africa stands at 18 (Koeberl, 1994; Vincent and Beauvilain, 1996; Master et al., 1996). However, the majority of these impact structures measure only a few kilometers in diameter. The only large impact structures known in Africa are the Vredefort Structure, of $2,023 \pm 4$-Ma age and probably some 250 km original diameter (Reimold and Gibson, 1996), the Morokweng Structure, and the ca. 24 km-wide Highbury Structure (cf. Koeberl, 1994). The current number of documented impact craters in Africa is approximately a factor of two lower than that for other cratons. Theoretically, several other, and presumably rather large, impact structures remain to be discovered in Africa.

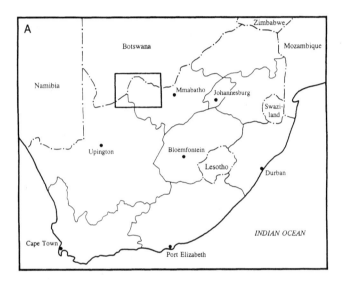

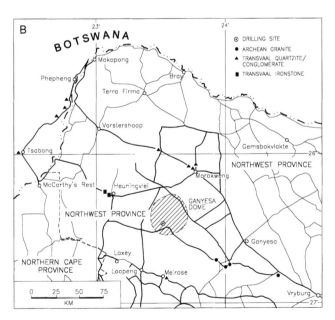

Figure 1. A, Locality of the Morokweng Structure in the Northwest Province of South Africa. B, Location of the central aeromagnetic anomaly (cf. Corner et al., 1997) on the so-called Ganyesa Dome southwest of the town of Morokweng.

Our group has carried out extensive petrographic, chemical, and chronologic work on the Morokweng impact melt body and underlying basement lithologies. Here we report the petrograpic and chronologic findings, whereas micropetrographic and geochemical results will be presented by Koeberl and Reimold (in prep.).

REGIONAL GEOLOGY

The area over the Morokweng aeromagnetic anomaly and its environs is extensively covered by Tertiary and Quaternary sands and calcrete of the Kalahari Group (Fig. 2) (SACS, 1980). Outcrops and subcrops of major bedrock lithologies are scarce (cf. Corner et al., 1997). Limited subsurface geologic information is available from shallow hydrogeologic drilling (Geological Survey of South Africa, 1974). In Figure 2 the regional surface and subsurface geologic knowledge, as well as interpretation of geophysical data after Corner et al. (1997), have been compiled. Some outcrops are located to the north-northwest of the town of Morokweng, where subvertically oriented Black Reef Quartzite from near the base of the ca. 2.25–2.5-Ga Transvaal Supergroup (SACS, 1980) is exposed in outcrops generally only a few tens of meters wide. Black Reef conglomerate is prominently exposed as a ridge just east of the township of Melrose (cf. Fig. 4 of Corner et al., 1997). Locally, Transvaal dolomite, sometimes intercalated with chert bands up to 60 cm wide, is present, especially in the vicinity of Morokweng town (e.g., along the road to Vorstershoop). Metasedimentary rocks, occurring to the west of the Ganyesa Dome, are correlated with the Griqualand West and Olifants-

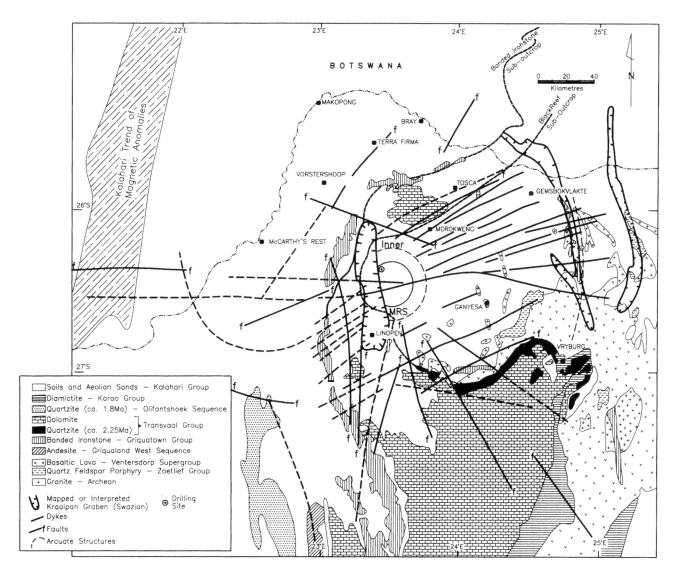

Figure 2. Regional geology of the study region based on a combination of surface geological information, geophysical analysis, and hydrogeological drilling (modified after Corner et al., 1997).

hoek Supergroups (SACS, 1980), which are thought to be post-Bushveld (<2.05 Ga) in age (Cheney and Winter, 1995).

A few patchy, but sometimes extensive (hundreds of meters), exposures of Archean basement granitoids occur some 50–70 km southeast of the Ganyesa Dome, especially in the Moshaweng stream bed along the road from Loopeng (also known as Linopen) to Ditshipeng. These rocks are massive and show no signs of major deformation. Some veinlets of cataclasite that are up to several meters long, but never more than a few centimeters wide, are present in places (Fig. 3A,B). It appears from the limited observations permitted on these two-dimensional exposures that the breccia zones generally have subvertical attitude. Thin-section studies of such cataclasite did not reveal any evidence of shock deformation. As discussed by Corner et al. (1997), it is unclear whether the limited deformation should be associated with the Morokweng event, or is the result of regional tectonics.

Several major aeromagnetic anomalies in the region around the Ganyesa Dome have been associated with greenstone, banded ironstone, metavolcanics, and ultramafic rocks of the Archean Kraaipan Group (Corner et al., 1997, p. 358; Zimmermann, 1994; Cornell et al., 1996; Anhaeusser and Walraven, 1997; Walraven and Martini, 1995). Patchy subcrop of such strata is indicated on the regional geologic map (Geological Survey of South Africa, 1974), but outcrops were not detected during our surface exploration.

About 35 km west of the Ganyesa Dome, the undulating topography of terrane covered by the Kalahari Group changes dramatically. Thick successions of Griqualand West Sequence rocks form a prominent ridge-and-valley morphology, dominated by massive banded ironstone formations, in association with cherts and quartzites. Where it was possible to obtain geometric data, it appears that these strata are consistently oriented subhorizontally. In one of these ironstone formations, near the village of Heuningvlei (Fig. 1B), Corner et al. (1997) detected a parautochthonous quartzite breccia containing quartz grains with up to several sets of planar deformation features (PDFs).

The drilling information compiled by the Geological Survey of South Africa (1974) indicates diamictite occurrences of the Dwyka Group of the Permian-Jurassic Karoo Supergroup (see, e.g., Reimold et al., 1997) in an arcuate region around the Ganyesa Dome. Geophysical studies (Corner et al., 1997; Hart et al., 1997) emphasize the presence of mafic intrusives, mainly

Figure 3. A, A ca. 1.3-m-wide exposure of Archean basement in the Moshaweng streambed southeast of the Ganyesa Dome. A fault-controlled breccia veinlet occurs here, which also shows a number of offshoots from the main generation vein. Although this occurrence of breccia macroscopically resembles pseudotachylite, microscopic analysis unambiguously proved that this material is cataclasite. For scale, knife is 8.5 cm long. B, Another occurrence of cataclasite from the same locality. Here the fault-controlled breccia occurs in a zone of criss-crossing veinlets up to 30 cm wide, which could be referred to as a small network breccia.

in the form of dikes, in the region around Morokweng. These intrusives are believed to represent the ca. 2.7-Ga Ventersdorp Supergroup, the Ongeluk Lava (an equivalent of the about 2.24-Ga Hekpoort Andesite of the eastern parts of the Transvaal basin; Walraven and Martini, 1995) of the Transvaal Supergroup, and 180-Ma dolerite of the Karoo Supergroup.

Several years ago, three shallow boreholes were drilled into the south-central area of the aeromagnetic anomaly (Fig. 4). Borehole MWF03 was drilled to a depth of 130.3 m, MWF04 to 189.3 m, and MWF05 to 271.3 m. These boreholes represent a

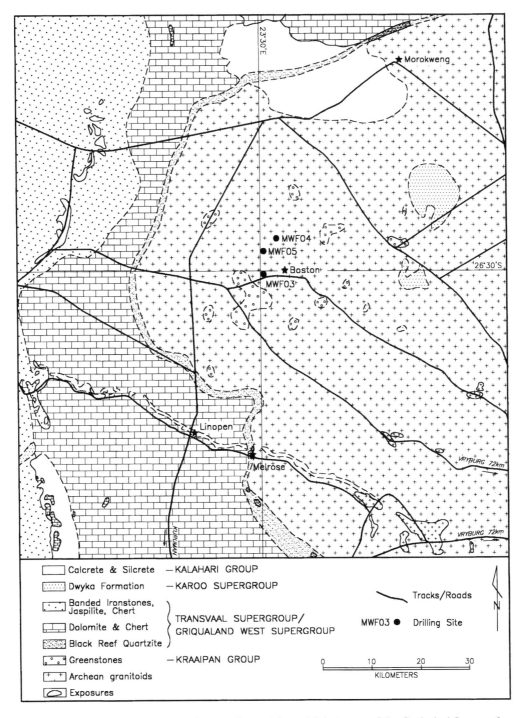

Figure 4. Geology of part of the Ganyesa Dome (after published maps of the Geological Survey of South Africa, 1974; based on hydrogeologic percussion drilling) with locations of the three drill cores into the Morokweng melt body (see text for detail).

circa 6.5-km-long profile from the southern edge of the anomaly (MWF03), via MWF05, to its central part (MWF04). Koeberl et al. (1997) and Hart et al. (1997) have provided preliminary descriptions of these boreholes. A more extensive characterization is presented in this chapter.

DRILL CORE STRATIGRAPHY AND SAMPLING

All three boreholes first penetrated up to 100 m of Kalahari Group sediments, most prominently calcrete (Fig. 5; Table 1). Only in the case of drill core MWF05 was it possible to study the whole calcrete succession, but it can be categorically stated that the lower zone (at least several meters) of this sedimentary package carries clasts of the underlying dark brown melt rock. Hart et al. (1997) described a ca. 50-cm-wide zone at the base of the Kalahari strata consisting of conglomerate with pebbles of what they termed "quartz norite" (equivalent to the melt rock) and granite from the underlying basement in boreholes MWF05 and MWF03. Hart et al. (1997) further reported on the presence of shock-metamorphosed quartz, with planar deformation features (PDFs) of crystallographic orientations characteristic of shock, in granite pebbles from this layer. All melt rock clasts within Kalahari sediments are strongly altered and contain much secondary carbonate.

Kalahari sediments are underlain by a brownish rock type of igneous appearance, which in all drill cores is capped by a substantial zone of strong alteration (Fig. 5). Only in borehole MWF05 was the lower contact of this melt rock intersected, at a depth of 225 m, where granitic basement was reached. The other two holes were terminated in melt rock. Consequently, an approximate thickness of 130 m can be estimated for this body. The altered upper parts of the melt rock are characterized by strong secondary hydrothermal alteration in the form of pervasive formation of secondary minerals, but also by the presence of veinlets of secondary mineral assemblages, especially carbonate, which may form zones up to 40 cm wide. Secondary biotite is present, and much of the feldspar (predominantly plagioclase) has undergone saussuritization or contains secondary carbonates.

About 15 m into the melt rock, alteration has decreased significantly. The fresh melt rock appears homogeneous, with exception of sporadic changes in grain size (from coarse grained to fine or medium grained), relative abundance of clasts, and local (i.e., on a decimeter to meter scale) changes in mineralogic composition (typified by varied ratios of felsic and mafic mineral abundances). Though rather abundant, clasts are often difficult to discern, due to their apparent marginal assimilation by the melt rock. Rock types represented by clasts are equally difficult to distinguish macroscopically, but appear to be dominated by mafic lithologies. Obvious granitoid clasts are rare. With exception of the lowest part of the melt rock, the maximum clast size is 15 cm, but the vast majority of clasts is smaller than a few centimeters. Some mafic clasts have reaction rims of up to several millimeters wide.

TABLE 1. MOROKWENG DRILL CORE DESCRIPTIONS

MWF04 to

Depth	Description
34.5 m	Well-laminated calcrete.
36.5	Top to melt rock, strongly altered; many joints with secondary carbonate.
50.6	Altered melt rock.
189.3	Melt rock: minor subhorizontal veining, spotted with inclusions (including relatively rare granitoid and many mafic inclusions – as much as 10 cm in size); slight color changes in certain areas (brown to lighter brown), occasional silica-filled amygdales; generally massive and dense, i.e., no brecciation or jointing.

MWF05 to

Depth	Description
~100 m	Crumbly (some core loss) calcrete containing some granitoid clasts and, toward the base, some clasts of altered melt rock.
~115	Strongly weathered melt rock with several, as much as 40 cm wide, carbonate layers.
213.82	Clast- and vesicle-poor melt rock; a few carbonate-filled veins.
224.8	Granitoid clast-rich melt rock (clast intersections as much as 45 cm long).
271.29	Granitoids of different types, with intrusive relationships; a pegmatoidal component, some aplitic veins; intercalated with <2 cm wide breccia veinlets around 240.98 and 263.8 m depths, as well as ca. 30 cm-wide-breccia zones at 265.3 and 270 m depths

MWF03 to

Depth	Description
7 m	Crumbly calcrete.
34.31	Calcrete with large, extensively weathered melt rock clasts.
43	Weathered melt rock, with some subhorizontal carbonate veining.
59	Weathered melt rock, but slightly darker than above.
130.25	Melt rock, relatively clast-poor, but with a number of mafic clasts, as well as relatively fewer granitoid-derived ones; some subhorizontal and low-angle veining (mainly carbonate fill).

In particular, in drill cores MWF04 and MWF03, the melt rock is traversed by a number of veinlets of secondary materials, in which carbonate and epidote minerals are prominent. In addition, several centimeter-wide "pockets" or "pods" of relatively coarse-grained material occur in certain zones (see Fig. 5). Andreoli et al. (1995) mentioned the existence of several "flows" within this melt body, without providing supporting evidence. We assume that their observation may be related to the distinctive changes in grain size that are noted in certain zones.

From 213.82 to 224.8-m depth in drill core MWF05, the melt rock contains numerous intersections of granitoid clasts up to 45 cm long. Granitoid rock types represented are the same ones that occur in the more massive granitoids intersected below the melt body. Grain sizes are varied in these generally granitic rocks, ranging from pegmatoidal to very fine grained aplitic. Composition also varies, but granitic compositions prevail and

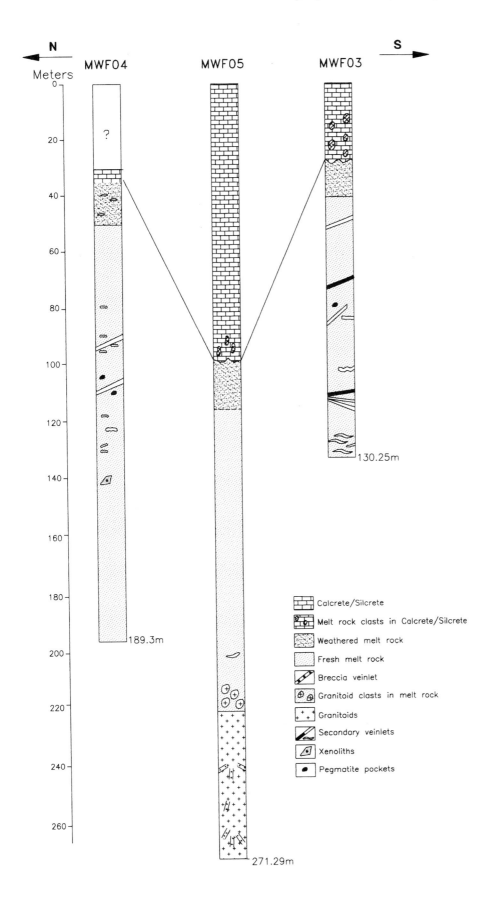

Figure 5. Stratigraphy of the three Morokweng drill cores (see text for details).

granodiorite is scarce. In some places these lithologies apparently form intrusive relationships. In contrast to Andreoli et al. (1995), who referred to the lower margin of the melt body as the contact with the Archean granitic basement, and to Hart et al. (1997, p. 29), who reported a "clear, chilled margin against the underlying basement granite," we prefer to interpret this zone as representing the upper part of a continuously granitoid-enriched breccia zone (compare with the Discussion below). In this zone the melt rock–granitoid ratio seems to decrease toward lower depths.

The section from 224.8 m depth to the termination of this borehole at 271.29 m includes granitoids that appear massive, but are transected by several types of breccia veinlets (here termed "dike breccias"). These breccias were also mentioned by Andreoli et al. (1995) and Hart et al. (1997) who compared them to "pseudotachylite." With regard to the definition of "pseudotachylite" as *friction melt rock*, the drill core does not permit us to observe a relationship between faulting and breccia formation (e.g., no displacements are obvious); therefore the superficial, macroscopic appearance may be rather misleading (compare, for example, Reimold, 1995, 1998) and needs to be supplemented by micropetrographic analysis (compare below).

Throughout the lower granitoid zone, the rock is obviously strongly thermally overprinted. Macroscopic study alone indicates that grain shapes and interrelationships have become diffuse, suggestive of thermal alteration. It is unfortunate that drilling was suspended at a depth of 271.29 m. However, the evidence at hand can be interpreted as indication that, at this depth, no coherent basement has been encountered, but rather a breccia zone with downward increasing clast sizes.

Samples for detailed petrographic and chemical analysis were taken from all three drill cores. The samples represent all parts, as well as all lithologies, intersected in these boreholes. Altogether 77 samples were analyzed, with additional subsamples (for example, selected clasts or duplicate specimens) studied as well. A drill core stratigraphic log with sampling depth indications is provided herein dealing with the detailed geochemical data (Koeberl and Reimold, in prep.).

ANALYTICAL PROCEDURES

We studied normal as well as polished thin sections from the drill core samples with the petrographic microscope. Initial petrographic observations were mentioned by Reimold et al. (1996) and Koeberl et al. (1997a,b). Details on major and trace element chemical analyses and age dating are given in Koeberl et al. (1997a) and Koeberl and Reimold (1998). Selected samples were subjected to detailed scanning electron microscope and energy dispersive x-ray analysis at the Naturhistorisches Museum in Vienna. The instrument used was a JEOL JSM-6400 scanning electron microscope (SEM) with an energy dispersive spectrometer. In addition, quantitative wavelength dispersive electron microprobe analysis of mineral constituents of melt rock samples was conducted in Vienna with an ARL-SEMQ instrument, using standard procedures.

Several melt rock specimens were selected for Rb-Sr isotopic analysis in an attempt to utilize this technique for the dating of melt rock formation and evaluation of the isotopic systematics of this intriguing lithology. Pyroxene and plagioclase separates were obtained from two melt rock samples with standard mineral separation techniques (including heavy liquid and magnetic separation), followed by handpicking.

The Rb-Sr analytical methods utilized are essentially the same as those described by Smith et al. (1985) and Brown et al. (1989). A single-collector Micromass 30 mass spectrometer and a multiple collector VG354 mass spectrometer at the Hugh Allsopp Laboratory, University of the Witwatersrand, were employed to analyze Rb and Sr isotopic concentrations, respectively. Uncertainty estimates are 0.6, 1.5, 1.6, and 0.005 rel% on Rb and Sr concentrations, and $^{87}Rb/^{86}Sr$ and $^{87}Sr/^{86}Sr$ ratios, respectively. Processing and regression of the isotopic data was carried out using the GEODATE version 2.2 package developed by Egglington and Harmer (1991), applying the decay constant of $1.42 \times 10^{-11} y^{-1}$. Error augmentation followed Method I of Egglington and Harmer (1991). Age calculations were performed with 1.5% errors on the $^{87}Rb/^{86}Sr$ ratios, and with 0.0001 blanket errors on the $^{87}Sr/^{86}Sr$ ratios.

Total method blanks did not exceed 0.5 ng for Rb and 8.5 ng for Sr, values that are insignificant with respect to the concentrations observed in the samples. Thus, blank corrections were obsolete. The average $^{87}Sr/^{86}Sr$ ratio determined for the E&A standard was 0.70800 ± 4 (1 σ standard deviation), and for the SRM987 standard 0.71022 ± 3. Accuracy of Rb concentration measurements was monitored through periodic measurement of the SRM607 standard, and tracer solution calibration is regularly carried out against in-house gravimetric standards.

PETROGRAPHIC STUDIES

Petrographic information for the samples is given in Table 2, and photomicrographs representing all rock types are shown in Figures 6 through 14.

Morokweng melt rock

Andreoli et al. (1995) referred to the Morokweng melt rock as a charnockite. Hart et al. (1997), in contrast, discussed a quartz norite. Reimold and Koeberl (1997) recognized distinct similarities between this rock type and the Vredefort Granophyre (e.g., Reimold et al., 1990; Therriault et al., 1996, 1997) and, consequently, termed it "Morokweng Granophyre" (see Fig. 6A–E).

The dominant mineral constituents of this lithology are plagioclase, ortho- and clinopyroxene, quartz, alkali and K-feldspar, and opaque minerals. Hart et al. (1997) provided modal percentages for these minerals (in the same order from plagioclase to K-feldspar: 60, 20, 10, and 5%). Whereas the dominance of the first two minerals is certainly correct, we find that mineralogically this melt rock body is highly variable. In

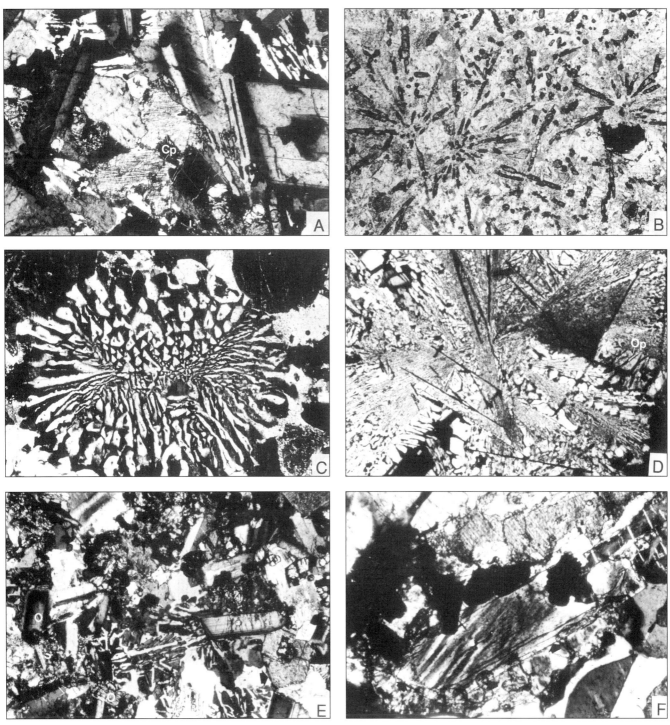

Figure 6. A, Typical matrix texture of medium-grained Morokweng melt rock. Note tabular, twinned clinopyroxene (Cp). Plagioclase blades display marginal absorption and melting in central parts. Sample MO16; width of field of view; 3.4 mm, crossed polarizers. B, Spherulitic growth of hypersthene, radiating outward from completely annealed microclasts; interstices between pyroxene crystals filled with micropegmatitic growths. Sample MO59C, 3.4 mm wide, parallel polarizers. C, Patch of micropegmatite in partially melted, 1-cm large granitic clast. Sample MO57-1, 3.4 mm wide, crossed polarities. D, Patch of granophyric growth with needle-shaped orthopyroxene (Op) and ilmenite between garben-textured micropegmatite. Granitic clast MO32C, 3.4 mm wide, crossed polarities. E, Another part of recrystallized melt in clast MO32C, with large, twinned plagioclase blades, small prismatic orthopyroxene (O) crystals, and numerous oxidic relics after disassociated mafic minerals—probably biotite. The arrow marks a clinopyroxene clast. 3.4 mm wide, crossed polarities. F, Kinkbanded biotite clast (center) with overgrowths of biotite (arrows, lower and left margin of this grain), as well as twinned orthopyroxene (upper part) in largely recrystallized clast in melt rock sample MO30B; ca. 1.1 mm wide, crossed polarities.

TABLE 2. PETROGRAPHIC DETAILS OF MOROKWENG DRILL CORE SAMPLES

Sample	Depth (m)	Characterization
MWF04		
24	39.70–39.90	Strongly altered (secondary carbonate) melt rock; heterogeneous distribution of plagioclase and pyroxene.
25	41.35–41.50	As 24.
27	48.25–48.45	Coarse-grained granular to intersertal melt rock.
28	48.95–49.07	One cm mafic clast in coarse-grained intersertal melt rock, partially surrounded by a garben-textured felsic, granophyric melt rock zone.
29	49.22–49.32	Melt rock with granophyric patches alternating with more coarse-grained subophitically textured areas.
30	50.56–50.71	Typical coarse-grained melt rock, with mafic clasts 30A and 30B (30B contains a biotite clast with secondary biotite overgrowth).
31	55.65–55.80	Locally altered melt rock (veins and patches of secondary carbonate).
32C	57.18–57.23	Granitic clast in melt rock: completely recrystallized, only recognizable as clast by lack of pyroxene crystals and presence of abundant ilmenite blades.
33	58.27–59.44	Typical coarse-bladed melt rock.
34	64.30–64.55	Similar to melt rock 33, but somewhat smaller plagioclase crystals.
35	78.80–79.00	Coarse-grained melt rock (contains large 'precursor' biotite and gabbro clast).
36	83.87–84.15	Coarse-grained melt rock, partially with large feldspar blades, locally with more tabular feldspar crystals; relic plagioclase with overgrowths.
37	90.35–90.65	As 36.
38	107.00–107.08	As 36 and 37.
39	115.80–116.00	As 36, but locally altered (chlorite, smectite), relic biotite, primary K-feldspar with overgrowths.
40	117.70–118.00	Melt rock as 36, plus a porphyritic mafic clast.
43	182.20–182.40	Typical coarse-grained melt rock.
44	193.90–194.25	Typical coarse-grained melt rock with a sub-mm mafic clast.
MWF05		
45	119.00–119.40	Typical coarse-grained melt rock with lots of secondary biotite.
46	131.85–132.00	Coarse-grained melt rock with many remnants of precursor feldspar; somewhat altered, much secondary biotite in patches.
47	143.87–144.00	Similar to melt rock 46, but slightly more altered.
48A+B	167.25–167.38	Medium-grained melt rock with several mafic and felsic clasts.
49A	178.40–178.49	Felsic (granitoid-derived) clast in typical medium-grained melt rock; some remnants of PDFs in quartz of clast.
49B	175.00–175.20	Typical melt rock with completely annealed mafic clasts.
51	183.40–183.60	Medium-grained melt rock, cut by several mm-wide quartz veinlets; lots of oxides from decomposition of primary pyroxene and biotite; large gabbro clast with tabular pyroxene.
52	187.50–187.63	Relatively finer-grained, more granular melt rock; a 1 cm wide unannealed granitic clast; a large poikilitic orthopyroxene clast (Karoo-derived?).
MWF05 (continued)		
53	190.45–190.65	Medium-grained, strongly altered melt rock; rather heterogeneous with alternating quartz-rich patches and others characterized by typical feldspar blades and short-prismatic pyroxene.
53B	190.45	Cm-sized felsic clast, internally melted and annealed to microgranophyre.
54	197.30–197.41	Variably textured (coarse-grained, bladed to fine- or medium-grained) melt rock; in patches finer-grained (probably after felsic clasts).
55	200.85–201.00	Fine- to medium-grained, in patches coarse-grained melt rock.
57	209.90–210.05	Typical fine- to medium-grained melt rock, with a 1 cm wide patch of very fine-grained granophyre; a second thin section contains a quartzitic, fully annealed clast with biotite-rich reaction rim.
58	214.32–214.44	Part of a large granitoid clast, which was partially melted.
59A	216.60–216.85	Large granitic clast with dioritic inclusion; more mafic parts are melted and annealed; patches of relic feldspar recognizable; partially oxidized biotite; relic K-feldspar with plagioclase overgrowths.
59C	218.40–218.98	Melt rock around clast 59A; in places spherulitic texture; contains some fully annealed pseudo-quartzitic clasts.
60	222.90–223.05	Contact of melt rock and large (dm) granite clast; typical melt rock, but contains numerous fully recrystallized clasts (as much as several mm wide); quite strongly altered; large clast is nearly fully annealed and recrystallized; a few quartz grains with PDFs.
62A,B	225.60–226.00	Medium- to coarse-grained granite, at contact with pegmatitic granitoid; completely recrystallized; a few relic PDFs in quartz and K-feldspar; ballenquartz (probably after diaplectic quartz glass); a second section consists of largely annealed pegmatoid, with relics of shocked quartz, K-feldspar, and alkali feldspar with PDFs.
63A	231.00–231.28	Strongly annealed, in places, melted granite; some diaplectic quartz glass.
63	232.35–232.60	Strongly annealed and locally melted granite; remnants of shocked quartz and plagioclase with PDFs; a second thin section shows the contact between an aplitic breccia vein and granite; many feldspar grains in the aplite have single or multiple sets of PDFs.
64	234.65–235.00	Fine- to medium-grained granitoid, strongly annealed and even recrystallized, but in patches still original material with relic microcline and (oxidized) biotite.
65	238.65–238.83	Partially annealed granite; less shock deformation than in 63; some PDFs in K-feldspar, local melting and recrystallization.
66	249.80–249.95	As 65, locally aplitic, fine-grained and K-feldspar rich; less annealing than in previous samples; biotite partially oxidized, ca. 20 vol% of melted material; no relic grains with PDFs found.
67	252.40–252.60	A: shocked, rather fine-grained, and strongly annealed biotite-granite. B: contact between fine-grained biotite-granite and pegmatoidal phase (remnants of primary plagioclase and of some quartz with PDFs). C: contact between annealed granite and 2 cm wide breccia vein. D: annealed fine-grained granite.

TABLE 2. PETROGRAPHIC DETAILS OF MOROKWENG DRILL CORE SAMPLES (continued)

Sample	Depth (m)	Characterization	Sample	Depth (m)	Characterization
MWF05 (continued)			**MWF03**		
68	260.35–260.58	Coarse-grained granite; remnants of quartz, K-feldspar, and alkali feldspar with PDFs.	2	ca. 12.35	Clast of altered melt rock in calcrete.
69	265.30–265.65	A series of thin sections across host rock and 10-cm-wide breccia vein: 69: upper contact between typical coarse-grained melt rock and dike breccia, with much clastic material, some of which contains quartz with PDFs and shows local melting. A: Dike breccia fully altered, type unknown; some quartz clasts with PDFs. B: breccia of unknown type, could be melt rock; evidence of melting and devitrification of glass in center. C: similar to B, but definitely melt breccia with fine-grained intersertal texture. D: lower contact to host rock (largely annealed granite).	4	ca. 17.90	Slightly less altered melt rock clast in calcrete.
			8	34.20–34.31	Altered melt rock (strongly replaced by carbonate).
			10	49.70–49.79	Somewhat fresher, but still altered melt rock (similar to granular Vredefort Granophyre).
			12	63.00–63.12	Typical coarse-grained melt rock.
			13	69.03–69.20	As melt rock 12.
			14	73.20–73.35	Granophyric melt rock; remnants of felsic precursor minerals; macroscopically clast-bearing; PDFs in microscopic mineral clast relics.
			16	78.73–78.91	Similar to melt rock 14, but more tabular pyroxene clasts.
70	267.75–257.80	Dike breccia-granite contact. A: fine-grained, completely recrystallized breccia, similar to that of 69; some shocked clasts, e.g., a partially melted quartz clast with PDFs (melting after shock deformation!!!). B: contact granite-breccia; granite largely annealed, some diaplectic quartz glass. C: largely annealed, partially shock melted granite; relics of shocked feldspar and quartz with PDFs.	17	88.18–88.40	Similar to melt rock 16, but more newly grown biotite laths.
			19	99.90–101.10	Typical coarse-grained melt rock with large felsic clast (pseudo-quartzitic); near the contact between the clast and melt rock the melt rock becomes much finer-grained; also several cm-sized annealed mafic clasts.
71	270.00–279.35	Completely annealed granitoid. A: slightly darker phase: relatively many remnants of dissociated mafic minerals; PDFs in a relic plagioclase grain. B: relatively lighter phase: less opaque minerals (or remnants thereof) than in A and less annealing in general; among the 15 vol% relic grains, many show PDFs, especially K-feldspar relics.	21	119.00–119.08	As melt rock 17.
			22	127.23–127.33	Coarse-grained (feldspar blades) melt rock.

many of our samples, biotite is another important mineral (up to 15 vol%). Some individual samples are dominated by more than 50 vol% of pyroxene, whereas others are predominantly felsic.

Texturally, the melt rock samples are very variable as well (Table 2). They may closely resemble ophitic to subophitic igneous rocks, or consist entirely of micropegmatite (Koeberl et al., 1997a). This granophyric intergrowth consists mainly of quartz and various feldspar minerals, but may also contain needle to lath or more stubby shaped, prismatic orthopyroxene (Fig. 6C,D). Coarser grained melt rock samples characteristically display large blades or more equant (tabular) plagioclase crystals and prismatic pyroxene. Tabular crystals of pyroxene are often twinned. Relatively more fine-grained specimens exhibit either the same crystal shapes at reduced grain sizes, or may indicate relatively rapid cooling resulting in fine-grained intersertal or spherulitic textures.

Several specimens with spherulitic growths of needle-shaped orthopyroxene (Fig. 6B) closely resemble the spherulitic type of impact melt rock (Vredefort Granophyre) from the Vredefort Structure. Sample MO28 from drillcore MWF04 contains melt rock with a unique textural appearance: very fine-grained, needle-like orthopyroxene and ilmenite form a microspinifex texture, which could also be described as garben-texture.

Thus, the textural types identified include all the textures described from Vredefort Granophyre, which represents confirmed impact melt rock (Koeberl et al., 1996a). The composi-

tions of matrix plagioclase are not very variable and range from nearly pure albite to oligoclase compositions (Fig. 7A). Matrix alkali feldspar compositions are more variable and range from Ab_{20} to Ab_{77}. Pyroxene compositions measured on matrix crystals are uniform at $Wo_{40}En_{43}Fs_{17}$ and $Wo_4En_{58}Fs_{38}$, respectively. All analyzed pyroxene crystals are internally homogeneous and unzoned. Clinopyroxene in the matrix often displays corroded shapes and narrow overgrowths of a Ca-bearing (1–2.5 wt%) amphibole, which also contains significant Ni concentrations. All pyroxene crystals analyzed contain appreciable amounts of Ni. However, the small matrix orthopyroxenes contain up to 0.20 wt% Ni and, thus, are relatively enriched in comparison with the more platy clinopyroxene crystals which, at maximum, contain 0.12 wt%. These lines of evidence suggest that the clinopyroxene grains are relics of mafic precursor rocks to the Morokweng melt rock.

Biotite occurs in two forms: first, as medium- to coarse-grained tabular crystals, often with idiomorphic or subidiomorphic shapes; and second, as narrow overgrowths on pyroxene crystals, mafic lithic clasts, and primary biotite. In addition, biotite is a component in many mafic inclusions, where it often appears distinctly bleached and/or partially oxidized, and may also exhibit kinkbands. Bleached biotite crystals also can be observed occasionally as mineral clasts in the matrix of this melt rock.

Accessory minerals identified are magnetite and titanomagnetite, some ilmenite, chromium- and nickel-rich spinel, rutile

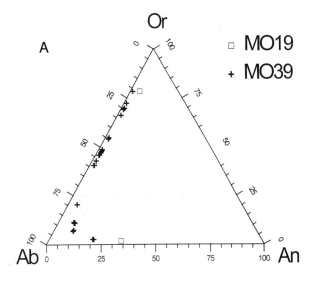

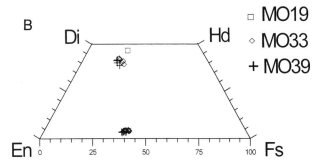

Figure 7. A, Ternary feldspar diagram with results of electron microprobe analyses of matrix plagioclase and alkali feldspar. B, Electron microprobe analytical results for ortho- and clinopyroxene in Morokweng melt rock. The orthopyroxene crystals analyzed are matrix pyroxene, whereas the clinopyroxene component represents clasts derived from a mafic precursor lithology.

(which is often intergrown with ilmenite), and monazite, zircon, baddeleyite, chalcocite, and trevorite (with about 25 wt% NiO). The presence of baddeleyite is clear evidence of the high-temperature origin of this melt rock. Hart et al. (1997) reported additional Ni minerals, such as millerite (NiS), bunsenite (NiO), Ni-rich ilmenite, liebenbergite (a Ni-rich olivine phase), and willemseite (a Ni-talc mineral). They stated that these minerals may occur in nodules or veinlets, and that one such veinlet also carried a 3-μm-sized Pt nugget.

Zircon crystals observed by us are all euhedral and transparent. They do not contain cores of an older generation of zircon, are undeformed, and not metamict. No shocked zircon crystals have been identified to date. Because of the igneous nature of the zircon in the Morokweng melt body, the U-Pb single zircon dating results by Koeberl et al. (1997b) and Hart et al. (1997) are interpreted as marking the age of the melt body.

We commonly found skeletal magnetite within irregularly shaped pyroxene, which is intimately intergrown with gra-

nophyric matrix. Large spinel crystals show a lamellar internal structure consisting of Ni-enriched zones (25 wt% NiO), Ni-poor zones, and rutile exsolution lamellae.

Besides the quite strongly altered zones along the upper contacts to Kalahari Group sediments discussed above, in which sericitization of plagioclase, carbonate deposition, and oxidation of pyroxene is pervasive, the bulk of the melt rock is generally fresh. However, plagioclase may be locally sericitized or saussuritized, and pyroxene can be uralitized, saussuritized, or chloritized, or show overgrowths of amphibole and/or secondary biotite. Where K-feldspar is prominent, it may also be altered and larger crystals appear pitted. Locally, complete replacement of matrix by carbonate has occurred. In very narrow zones along cross-cutting secondary veinlets, filled with quartz or carbonate, enhanced alteration of matrix minerals is present. This, however, rarely exceeds distances of a few millimeters from the veinlets. Several samples displayed a more patchy alteration style, in which areas up to several millimeters wide in the matrix were affected by the formation of secondary clay minerals, such as illite, smectite, or chlorite. Especially fine-grained samples displayed such alteration, or zones around large, originally granitic clasts. It appears likely that volatiles derived from the thermally altered clasts contributed to the formation of these secondary minerals. Some clasts are completely altered to secondary, phyllosilicate-, and/or carbonate-rich assemblages.

The population of lithic clasts (cf. Fig. 8A–F) is dominated by gabbro- and granitoid-derived inclusions, whereby the latter are mainly represented by granitic and dioritic clasts. Macroscopically mainly mafic inclusions can be discerned, but thin-section studies revealed that granitoid-derived clasts are a major, if not the major, component of the clast population. In addition to microscopically easily recognizable granitic and dioritic clasts, we recognized numerous partially assimilated smaller clasts from such precursor rocks.

In contrast to lithic clasts, it is much more difficult to recognize mineral clasts. That they exist in the micropegmatitic parts of the matrix is obvious from occasional observation of shock-metamorphosed quartz clasts. The following observations can also be utilized to identify remnants of clasts in the matrix: corroded margins of feldspar, quartz, or mafic mineral grains, due to partial absorption by the melt matrix; the presence of deformation effects, such as fracturing or remnants of planar deformation features; the presence of thermal alteration effects, such as partial annealing of quartz or feldspar, or decomposition of mafic minerals to oxide-bearing secondary mineral assemblages; the presence of apparently exotic mineral grains (e.g., microcline derived from a granitic precursor rock); distinct zonation of large feldspar crystals, which is unlikely in this apparently overall rather quickly cooled material; and often subtle changes in grain size across a clast/matrix boundary. Overall, the problems of recognition of clastic material in the melt matrix of Morokweng melt rock are the same as for Vredefort Granophyre. Both melt rocks

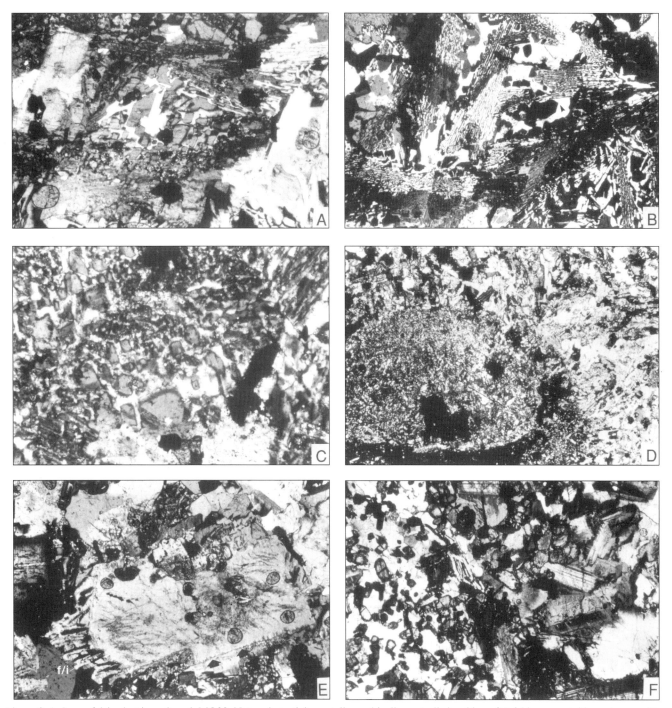

Figure 8. A, Large felsic clast in melt rock MO28: Note substantial, crystallographically controlled melting of K-feldspar crystal in center, resulting in so-called checkerboard texture (e.g., Reimold, 1982). Where melting has progressed even further, micropegmatitic growths have developed (especially in lower, central part of image). Several other K-feldspar (lower right) and perthitic alkali feldspar (lower right) crystals have not been melted and appear undeformed, but have overgrowths of matrix feldspar. Width, 3.1 mm, crossed nicols. B, Another area of the clast in MO28: Garben-textured micropegmatite as the result of large-scale melting of this area. Quartz crystals between garben areas have been completely recrystallized with subgrains now forming 120° triplepoints (arrow). Width, 3.4 mm, crossed nicols. C, High magnification photograph of a partially melted feldspar crystal in the clast of MO28. Dark areas are oxide remnants after a mafic mineral. Width, 1.1 mm, crossed nicols. D, Completely annealed mafic clast in melt rock sample MO19. The melt rock is of the fine-grained subophitic variety, with relatively small granophyric proportions. The clast consists of microcrystals of plagioclase and pyroxene, and is surrounded by a reaction corona that is enriched in biotite crystals and oxides, and is partially invaded by matrix crystals. Width, 3.4 mm, crossed nicols. E, A large K-feldspar clast with marginal resorption texture (arrow, lower rim), as well as overgrowth of matrix-derived feldspar and ilmenite (f/i, left margin). In the upper part of the photo, a pocket of micropegmatite and partially melted feldspar crystals is visible. Round features are polishing-derived epoxy blobs. Width of image, 3.4 mm, crossed nicols. F, Ophitic-textured melt rock (right side) MO4, with completely recrystallized mafic clast composed of plagioclase, orthopyroxene, and magnetite. Width, 3.4 mm, crossed nicols.

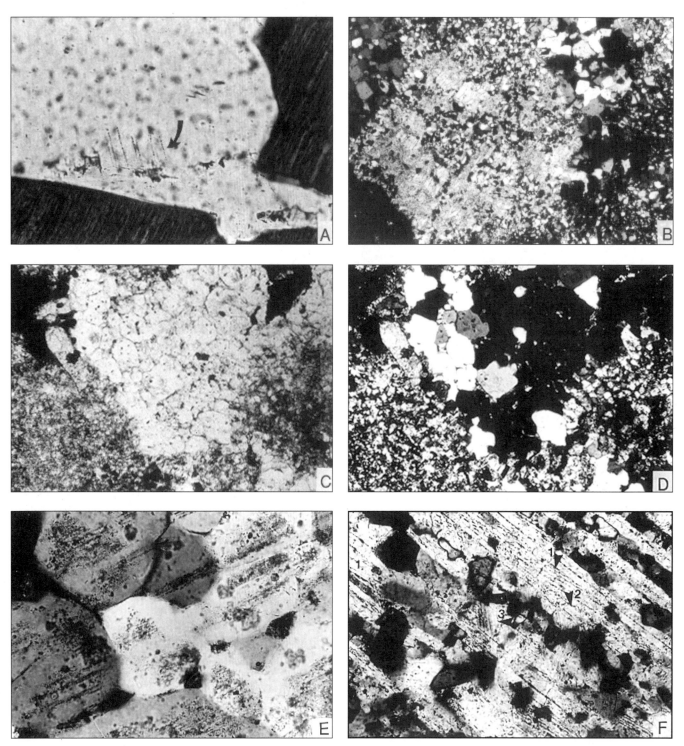

Figure 9. A, Granitic clast in melt rock sample MO32, containing a quartz grain with remnants of a basal set of PDFs (arrow) in the form of a planar fluid inclusion trail (grains in extinction are perthitic alkali feldspar). Width: 355 µm, crossed nicols. B, Partially annealed plagioclase crystal (containing remnants of two sets of PDFs that are not visible at this magnification) in otherwise largely recrystallized granite MO67B (a pegmatoidal granitoid type) from the zone below the melt body. Width, 3.4 mm, crossed nicols. C, Ballenquartz (probably after diaplectic quartz glass) in partially annealed granite MO70B, close to the contact with a breccia vein. Note: Some subgrains not in extinction could have formed from a part of this crystal that had actually been melted. Width: 2.2 mm, parallel nicols. D, As in (C), but crossed nicols; note isotropic parts of this partially recrystallized ballenquartz aggregate. E, Recrystallized quartz in pegmatoidal granite MO63. Note remnants of originally continuous planar deformation features, now as segmented sets of planar fluid inclusion trails of, for PDFs, characteristic spacings of only a few micrometers; 355 µm wide, crossed nicols. F, PDFs of three different crystallographic orientations (arrows) in plagioclase of granitoid MO71A. Plagioclase was partially melted. Width, 565 µm, crossed nicols.

conspicuously resemble each other, not only with regard to melt textures, but also with regard to thermal alteration and assimilation of clasts (cf. also Figure 9A).

Matrix texture varies on a decimeter- or centimeter-scale. We believe that the highly variable textures observed within these drill cores are, to a large extent, the result of differential melting of precursor material as well as of differential cooling times in clast-poor and clast-rich parts of the melt body. Some early reports on the Morokweng melt rock (Andreoli et al., 1995; Andreoli and Ashwal, 1996) described the melt body as composed of several flows or displaying numerous examples of cross-cutting relations between at least four successive pulses of hypersthene-bearing intrusives. We observed no evidence in support of these suggestions. Some samples of distinctly varied grain size, on a scale of several hundreds of micrometers to several millimeter, were eventually recognized as large granitic clasts up to several centimeters wide, that had been partially annealed.

We distinguished several types of mafic, generally gabbroic, clasts. The main type is a medium- to coarse-grained gabbro, which is dominated by plagioclase and pyroxene, and texturally closely resembles Karoo dolerite or the Anna's Rust Sheet type of 1,100-Ma-old intrusive from the Vredefort Structure (e.g., Pybus, 1995). However, originally finer grained intrusive rocks do occur in the region around Morokweng, as discovered in the course of our surface studies (Corner et al., 1997; Koeberl and Reimold, in prep.). As mentioned earlier, magnetic dikes are a prominent feature of the subsurface geology in the Morokweng region (Corner et al., 1997). Overall, it is likely that Archean Kraaipan Group, e.g., Anheusser and Walraven, 1997, Ventersdorp Supergroup (2.7 Ga), Transvaal Supergroup (ca. 2.25 Ga; e.g., Ongeluk Lava; Cornell et al., 1996), and Karoo Supergroup (180 Ma) intrusives and extrusives were part of the preimpact crust at Morokweng.

Both granitoid- and mafic rock-derived clasts are mostly annealed (with regard to both the ratio between annealed and thermally unaltered clasts and to the recrystallization degree within individual clasts). Textures resulting from this pervasive thermal alteration (Fig. 9) are often very similar to those known from clasts in the Vredefort Granophyre and in other impact melt rocks (e.g., French et al., 1989; Reimold et al., 1990). Often, granitic clasts are completely annealed to irregularly shaped or roundish patches of quartzitic material set into fine-grained melt matrix, that is very similar to general melt matrix. However, it may contain areas with many small aggregates of oxidic material, which is the obvious relic of thermally decomposed mafic precursor minerals (mostly biotite, with some amphibole). Gabbroic clasts may also be completely annealed to aggregates of finest grained plagioclase and pyroxene. In some such cases, the original (precursor rock) texture is still recognizable.

Such completely recrystallized lithic clasts may show a slight decrease in grain size (of the newly formed microcrystals) toward the contact with surrounding melt matrix. Where the sizes and shapes of the newly formed crystals are similar to those of matrix minerals, the outline of the clast may be gleaned only from slight textural changes, for example, of crystal orientation, crystal shape, or distribution of oxidic remnants of mafic precursor minerals within such a patch. This phenomenon has been widely described from impact melt rocks (e.g., Reimold, 1982); such clasts are known as "ghost clasts."

The Morokweng melt rock also contains a large number of clasts that are rimmed by distinct reaction rims (e.g., Fig. 8D). Such clasts may be of (the annealed type) quartzitic, granitic, or gabbroic composition. Reaction rims are mostly composed of an enrichment of small orthopyroxene crystals that occur in narrow zones up to several hundred micrometers wide, together with quartz, with albite or K-feldspar, typically some apatite and/or zircon, and—significantly—small lath-shaped biotite. Such "coronae" are well-known from clasts in impact melt breccias from many confirmed impact structures (e.g., Koeberl et al., 1996b, and references therein).

Numerous felsic, lithic, and mineral clasts were scanned at high microscope magnification for the possible presence of shock metamorphic effects. Only in a limited number of primary mineral grains in felsic clasts (e.g., sample MO49A) did we find shock metamorphic deformation in the form of remnants of planar deformation features (mostly only one set, but several times two or three sets per grain; e.g., Fig. 9A). That up to three sets of PDFs were preserved is surprising, in the light of the discussion of preferential annealing of PDFs of high-order crystallographic orientations (e.g., Grieve et al., 1990). Perhaps the thermal overprint on these clasts was too short to cause all high-order PDFs to become annealed.

In addition, a few very small (<65 μm) quartz crystals in granophyric matrix patches, which by the nature of their irregular and corroded looking shapes were thought to represent clasts, exhibited short remnants of (generally short) single or double sets of planar deformation features. Other quartz or quartz-bearing clasts contain densely spaced *planar* trails of microinclusions, which we interpret as relics of PDFs.

SEM studies of melt matrix

We studied in detail four areas of typical matrix in melt rock sample MO39 (also utilized for Rb-Sr isotopic analysis) with the scanning electron microscope and electron microprobe (Fig. 10A–F). Figure 10A shows an area of typical coarse-grained melt matrix. Main mineral constituents are large tabular crystals of alkali feldspar and albite/oligoclase, orthopyroxene (some with overgrowths of chlorite), and aggregates of ilmenite and rutile, sometimes overgrown by thin magnetite seams. Accessory minerals identified are apatite and zircon. Figure 10B shows matrix dominated by plagioclase, with dark interstices of quartz, and intergrown with short-prismatic matrix orthopyroxene. Also visible are several clinopyroxene crystals, which are, by the nature of their irregular shapes and apparent marginal resorption, interpreted as clasts. Other phases identified in this

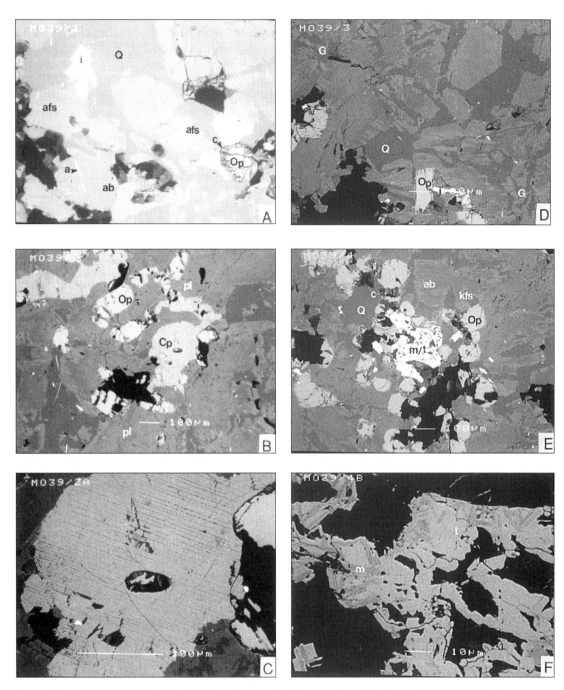

Figure 10. A–F, Backscattered electron images of matrix areas in melt rock sample MO39, illustrating the important matrix constituents and their textural relationships. See text for detailed description. Q = quartz; afs = alkali feldspar; a = apatite; ab = albite; c = chlorite; Op = orthopyroxene; Cp = clinopyroxene; pl = plagioclase; G = granophyric groundmass; i = ilmenite; mt = magnetite; t = trevorite; m/t = magnetite/trevorite intergrowths; kfs = K-feldspar. Scale bars are printed onto individual images.

area are rutile and hematite, apatite needles (in matrix quartz), and crystallites of chrome-spinel.

One of the tabular clinopyroxene crystals is shown in Figure 10C. The saw-tooth outline, presumably the result of marginal melting, is obvious. The crystal displays perfect cleavage parallel to (001) and mechanical twinning (possibly shock-induced), along (010). In matrix area 3 (Fig. 10D), we analyzed a series of alkali feldspars: they are either (K,Na)-feldspar of variable K/Na ratios, or alkali feldspar with up to 1.9 wt% CaO. This photograph also shows short prismatic orthopyroxene, matrix quartz, some quartz-alkali feldspar granophyric intergrowths (lower right), and some (light) laths of ilmenite or magnetite. In area 4 (Fig. 10E), a large intergrowth of magnetite and trevorite was encountered, in addition to the normal matrix silicate minerals. A part of this magnetite-trevorite assemblage is magnified in Figure 10F (trevorite appearing slightly lighter than magnetite).

Granitoids below the melt rock

Granitoids drilled below the melt rock in borehole MWF05 (Fig. 9B–F, and 11A,B) are petrographically similar to the granitoid clasts contained in the lowest part of the melt body (Table 1). Clearly, these granitoids represent a suite of mutually intrusive rocks. Granite containing small (several centimeters in size) dioritic xenoliths is present. In addition, several granitic to granodioritic lithologies occur, which are cross-cut by fine-grained aplitic and quartz-feldspar pegmatitic veins.

All these rocks, over the whole extent from a depth of about 225 to 271.29 m, are strongly thermally altered. Quartz has been nearly completely converted to fine- to medium-grained aggregates of quartz. Often the individual crystals of the medium-grained aggregates have euhedral shapes and grain boundaries that intersect at 120° angles. Many such grains have up to three sets of planar fluid inclusion trails in their interior parts, and frequently such planar trails in adjacent grains are perfectly aligned. Obviously, these euhedral crystals have grown at the expense of a (perhaps brecciated) precursor crystal that contained relatively continuous and long planar deformation features, of which the fluid inclusion trails are the product of segmental annealing. Very similar annealing textures and annealed planar fluid inclusion trails occur in the granitoids of the central part of the Vredefort Dome (compare, for example, Leroux et al., 1994, and references therein).

A number of our granitoid specimens also contain relics of PDFs in plagioclase, K-feldspar, and alkali feldspar crystals. Diaplectic quartz glass and quartz with ballen texture occur as well. Microcline may be completely recrystallized, but some

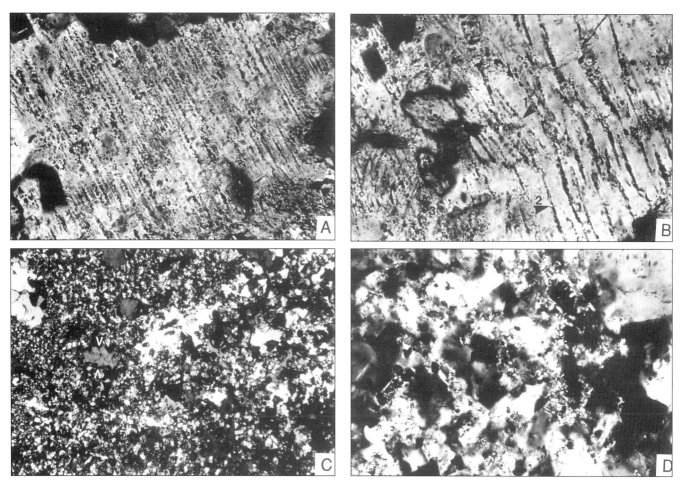

Figure 11. A, Coarse-grained quartz-feldspar granitoid MO68: planar fluid inclusion trails in quartz. Width, 355 μm, crossed nicols. B, Multiply oriented sets (arrows) of planar fluid inclusion trails, after PDFs, in alkali feldspar of sample MO68. Width, 355 μm, crossed nicols. C, Completely annealed matrix of breccia MO69A. Note that a few clasts are only partially annealed, which might indicate marginal melting (arrows). 3.4 mm wide, crossed nicols. D, Higher magnification view of the breccia shown in (C): it is not possible to resolve whether the matrix to this breccia consists of, at least partially, melted or clastic material, or both. Clearly, recrystallization was efficient, and a lot of tiny particles of secondary minerals are present, as well. Width, 355 μm, crossed nicols.

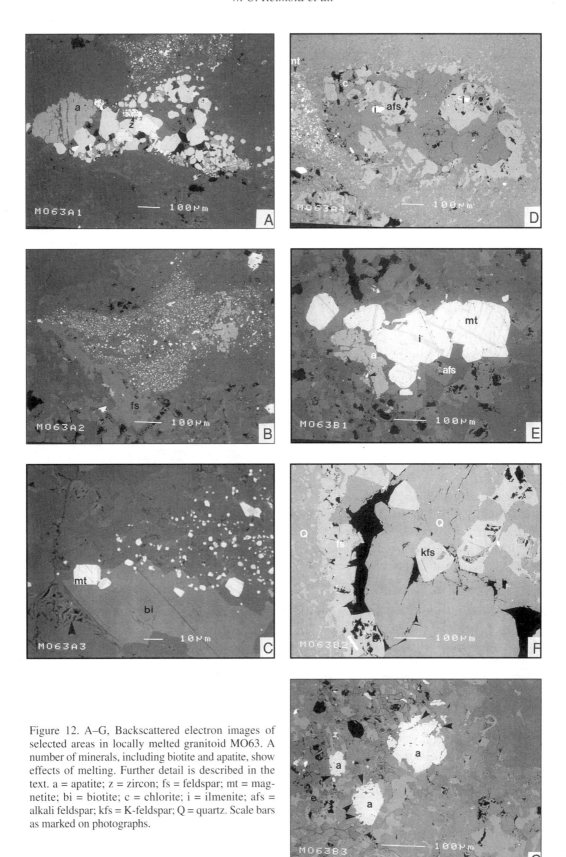

Figure 12. A–G, Backscattered electron images of selected areas in locally melted granitoid MO63. A number of minerals, including biotite and apatite, show effects of melting. Further detail is described in the text. a = apatite; z = zircon; fs = feldspar; mt = magnetite; bi = biotite; c = chlorite; i = ilmenite; afs = alkali feldspar; kfs = K-feldspar; Q = quartz. Scale bars as marked on photographs.

unannealed patches are still recognizable. Biotite in this zone below the melt body has been largely thermally decomposed and reduced to patches of oxide relics. Very rarely, relic grains with kinkbanding were seen. In addition, a few patches with brownish glass, probably after biotite and/or amphibole, occur. Locally, we obtained the impression that material was melted (compare the chapter on SEM studies). However, most of these patches are also fine-grained recrystallized, probably as the result of contact metamorphic overprint in the vicinity of the melt body.

The degree of annealing and partial melting of granitoid material does not seem to decrease from the top to the bottom of this zone. In fact, the sample collected at 270.00 m depth is still thoroughly annealed. It consists of two different "lithologies": a relatively dark one, with a high proportion of brownish, oxidic remnants of thermally dissociated mafic minerals, and a lighter one, with a comparatively higher proportion of K-feldspar and quartz. About 15 vol% of this material is composed of primary material, mainly K-feldspar, many grains of which display PDFs. Equally, it is not possible to identify a change in the degree of shock deformation from the top to the bottom of this ca. 50-m-wide zone. It is difficult to study this aspect because of the high proportion of annealed and even melted material. Study of the remaining relic grains indicates no decrease in the number of quartz grains with multiple sets of PDFs. Similarly, at least some diaplectic quartz glass is present throughout this zone.

SEM studies of granitoids from below the melt body

The area shown in Figure 12A shows how strongly individual minerals have been deformed (e.g., the zircon grain in the center or the apatite grain at left—which is also transected by alkali feldspar). The zircon grain appears as if it is partially melted. This whole "pocket" seems to be the result of local melting of precursor minerals (e.g., ilmenite/magnetite from biotite or amphibole). The mottled area at center top is magnified in Figure 12B. A schlierig area with numerous scattered magnetite particles is located between partially melted quartz and feldspar minerals (note lower left and compare with Figure 10A). Figure 12C shows a large biotite crystal, which is marginally melted (lower left, arrow) to form new, but much smaller and elongated, biotite crystallites. The light phase is magnetite with ilmenite exsolution lamellae, which, together with the vesicular quartz observed in the area just above the relic biotite crystal, are the result of thermal decomposition of the biotite.

Figure 12D records a second area with melt, which is rich in magnetite particles (middle left). Medium gray alkali feldspar is partially melted and transected by "channels" of slightly darker (SiO_2 enriched) material. Vesicular areas, most likely also the result of melting, are partially filled with secondary chlorite spherulites. Figure 11E shows a large magnetite aggregate with ilmenite lamellae, next to a brecciated apatite crystal with a small monazite inclusion (arrow). Main dark phases are quartz and alkali feldspar, the latter locally appearing corroded (melted?). In Figure 12F, several K-feldspar crystals show evidence of local transformation to vesicular, and partially altered, melt. At the left margin, a larger area of partially melted quartz and feldspar is visible. Marginally and more pervasively melted apatite is shown in Figure 12G.

Four areas were studied in detail in granitoid sample MO69. Area 1 shows completely annealed quartz (Fig. 13A, right side). Interestingly, the newly formed equigranular and euhedral subgrains are surrounded by a melt film, which, in backscattered electron images, is distinctly lighter than quartz. In fact, its brightness level is similar to that of the K-feldspar that forms a seam along the contact of the quartz aggregate and the melted/recrystallized alkali feldspar at lower left. This originally large alkali feldspar grain has been dissociated to a mixture of albite, K-feldspar, and quartz, an assemblage that is similar to the average micropegmatite of the Morokweng melt rock.

At the upper left, a rounded (marginally melted?) relic quartz grain is visible. Several idiomorphic ilmenite and Fe-oxide crystals are also shown (very bright) and could well have crystallized from the K-feldspar–rich melt. Figure 13B shows another recrystallized and locally melted quartz grain (upper left), which again contains melt seams of obvious affinity to K-feldspar. In the adjacent mixture of quartz and K- and alkali feldspar, some muscovite was analyzed as well. In the lower right portion of the image, several melt pockets are seen, which contain many small ilmenite and Fe-oxide particles. It is thought that these areas represent the melting products of mafic precursor minerals (biotite?).

Figure 13C shows several thermally altered quartz grains (quartz aggregates of the type discussed in Fig. 13A, B), as well as a large K-feldspar crystal, the central part of which has been melted to heterogeneous melt of, locally, either K- or Na-enrichment. Some lateral exchange of melted materials must have taken place to result in these different melt compositions. Figure 13D shows myrmekitic growths (after obviously melted precursor minerals), which consist of ilmenite (light) containing a few (even lighter) monazite inclusions, K-feldspar, and plagioclase of albite or oligoclase composition. Vugs (possibly the result of volatile production during the breakdown of Ti-bearing mafic minerals) are partially filled with secondary chlorite needles.

Three more areas were studied in granitoid sample MO70A. Area 1 (Fig. 14A) is a melt pocket containing a dark matrix composed of small areas of either relatively Ca-rich plagioclase (marked 1) or K-feldspar–like (marked 2; 2 wt% Na_2O and 0.8 wt% CaO) melt, forming a myrmekitic intergrowth. In separate parts of this melt pocket, one can observe either magnetite with ilmenite exsolution lamellae (upper right), apatite needles, or a series of complex oxides (Si, Ca, Ce, Nd, Fe, (Ti)-oxide), monazite, and phosphorus-bearing phases (P, Th, La, Ce, O), all of which are REE (rare earth element)-rich, plus apatite and rutile needles.

Area 2 (Fig. 14B) shows an intergrowth of dark melt with plagioclase relic grains (on the left side) and a series of larger

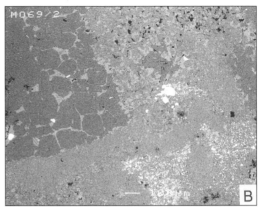

Figure 13. A–D, Backscattered electron images obtained in breccia MO69, showing various melting phenomena. See text for details. Q = quartz; a = alkali feldspar; k = K-feldspar; m = melt; i = ilmenite; c = chlorite.

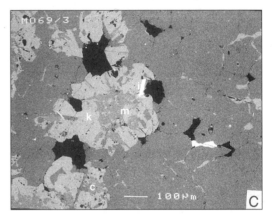
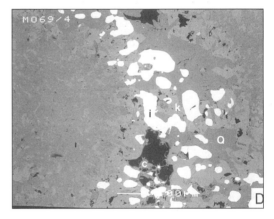

quartz and alkali feldspar crystals, which appear to be the result of annealing (equigranular and euhedral shapes, with 120° triplepoints). The large alkali feldspar grain near the center of the image encloses a euhedral biotite crystal, which appears to be melted in its central part. It is interesting that the alkali feldspar crystal is fractured in such a way that it could be interpreted as a result of decomposition of biotite, leading to the production of overpressure from released volatiles, which then exceeded the strength of the surrounding alkali feldspar crystal.

Figure 14C shows a large primary biotite crystal between locally melted quartz crystals. The biotite is marginally melted (arrow, upper left margin, interstice next to euhedral magnetite crystal on left side, and upper right margin). Ilmenite and magnetite intergrowths, in this case, are likely primary components. Also identified were small ZnS and chalcopyrite particles.

Breccia veins

Andreoli et al. (1995) mentioned that breccias and recrystallized pseudotachylite veins whose widths range between few millimeters and 10 cm occur in the granitoids (cf. also Andreoli et al., 1996a). Hart et al. (1997) stated that the granitoids below the melt body are brecciated and cut by veins of fine-grained recrystallized breccia that resemble pseudotachylite.

In this study of the MWF05 drill core a number of <2-cm-wide breccia veinlets were identified between depths of 240.97 m and 263.8 m, as well as approximately 30-cm-wide breccia zones at depths of 265.30 and 270 m (cf. Fig. 11C,D). An aplite breccia veinlet from the 232.35-m depth seems to be the result of shock brecciation of this particular granitoid intrusion. Perhaps the contact between the two adjacent granitoid types was exploited as a fault during shock compression, leading to the formation of a cataclastic breccia.

Sample 67C (252.4 m depth) represents a 2-cm-wide breccia vein. The contact to the adjacent, completely annealed granitoid is highly irregular and, at thin section scale, could be interpreted as an intrusive rather than a frictional contact. Clasts in the breccia are relics of the minerals found in the host granitoid (outside the annealed zone), but contain a rather high proportion of mafic relic grains. The matrix to the breccia is completely crystalline, consisting mainly of finest grained quartz, K- or alkali feldspar, an unidentified mafic mineral (biotite?), and specks of opaques. Whether the matrix originally was clastic and has been completely recrystallized or whether this breccia represents a melt breccia (either impact melt or friction melt = pseudotachylite) cannot be determined.

We studied a series of samples from the wide breccia zone

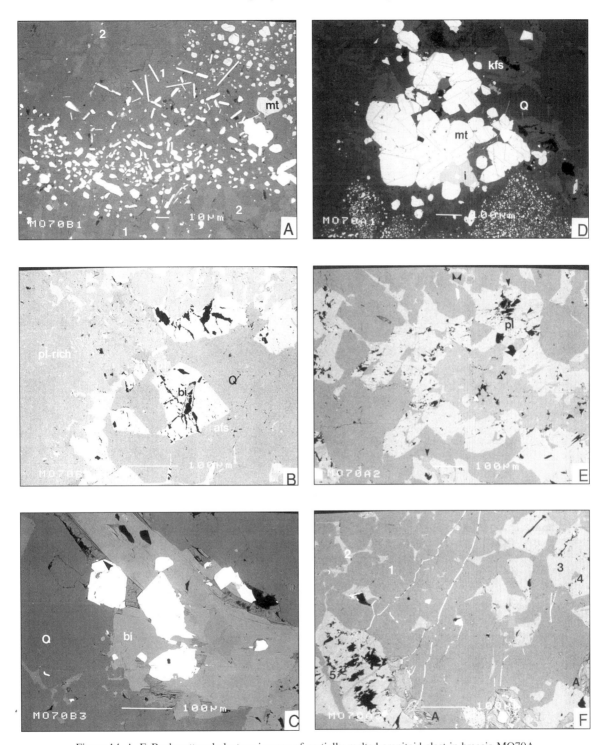

Figure 14. A–F, Backscattered electron images of partially melted granitoid clast in breccia MO70A. See text for detail. Q = quartz; kfs = K-feldspar; i = ilmenite; mt = magnetite; pl = plagioclase; bi = biotite; A = amphibole; afs = alkali feldspar. Numbers as explained in text.

at 265.30-m depth (samples 69A–D). Two breccia varieties occur within this zone (Table 2): at the upper contact to granitoid host rock, typical melt rock was encountered. This breccia is in contact with a type of dike breccia, the true nature of the matrix of which unfortunately could not be assessed. Secondary phyllosilicates are ubiquitous, and the vein material also contains oxidic relic material from the breakdown of mafic precursor minerals. Furthermore, the breccia is well-annealed. Most vein margins studied are schlierig and may represent clastic material. However, high-resolution work on the center of this vein revealed crystallization textures, for example, in the form of spherulitic or garben growth of a mafic mineral (pyroxene or amphibole?). This seems to be an indication that at least some melting occurred. If this is the correct interpretation, it still remains to be clarified whether this case is the product of frictional or impact melting (cf. Reimold, 1995, 1998). However, undoubtedly typical melt rock is part of this zone, which is an important observation with regard to the nature of the whole granitoid zone intersected below the melt body.

A third dike breccia (samples 70A–C) was sampled at depths of 267.75–268.00 m. As in the case of sample 69, breccia 70 consists of a very fine grained crystalline (or recrystallized) matrix, containing granitoid-derived (i.e., locally derived) clasts, some of which clearly display shock deformation indicative of a range of moderate (single to multiple sets of PDFs) to high (diaplectic quartz glass, mineral melting) shock pressures.

It was hoped that a chemical study of host rock-breccia pairs would assist in identifying the mode of origin of these breccia pairs (Koeberl and Reimold, in prep.). However, the results are ambiguous. Whereas some significant chemical differences between breccias and respective host rocks were determined, they do not define uniform trends that would indicate specific modes of origin.

SEM investigation of a breccia vein

A granitoid clast in the breccia vein at 267.75-m depth (samples MO70) was studied with the SEM (Fig. 14D–F). Figure 14D shows an aggregate of euhedral magnetite crystals with ilmenite exsolution lamellae, which is surrounded by locally melted quartz and K-feldspar crystals. It also contains a larger ilmenite crystal. The melt pockets in the bottom part of the image contain numerous small Fe-oxide particles, which contain up to 1 wt% SiO_2 and up to 0.8 wt% Al_2O_3. Again, it appears likely that these melt pockets are derived from melting of mafic precursor minerals.

Another area in this sample (Fig. 14E) consists of large, generally angular relic grains of quartz, K-feldspar, and plagioclase, as well as irregularly shaped, sometimes amoebal, areas that are interpreted as locally produced melt (arrows). Some of these apparent melt pockets contain idiomorphic crystals of ilmenite and pyrite (light grain in center). In area 3 (Fig. 14F), a very large, completely recrystallized quartz grain (marked 1), with K-feldspar-like melt interstitial to subgrains (2), is flanked by quartz-melt (3) invaded relic K-feldspar (4) crystals (upper right), as well as a marginally melted (5, saw-tooth margins) K-feldspar grain and deformed (fractures) amphibole (lower left and bottom right). Both amphibole crystals are partially converted to calcite-silica aggregates. Late-stage fractures cutting across this sample are filled with pyrite, and pyrite also occurs in the form of euhedral crystals in amphibole and K-feldspar–like melt. However, as mentioned before, sulfide minerals are extremely rare in the melt body.

GEOCHEMICAL DATA

Since 1995, our group has analyzed 24 melt-rock samples, as well as 17 samples of granitoids and breccia dikes from below the melt-rock body and 7 surface samples (for a discussion of surface sampling, see Corner et al., 1997), for major and trace element abundances. A detailed discussion of these results will be presented in a separate publication (C. Koeberl and W. U. Reimold, unpublished data, 1998). However, several overall valid conclusions need to be presented here also, as they—combined with the petrographic findings—are pertinent to the discussion of the origin of the Morokweng melt body.

Disregarding a small number of obviously altered melt-rock samples, the Morokweng melt body is extremely homogeneous. Variations for major elements do not exceed 2–5 rel%. With respect to many elements analyzed, the chemical composition, is also remarkably similar to that of the Vredefort Granophyre (Reimold and Koeberl, 1997). As this impact melt rock, the Morokweng melt rock contains relatively high proportions of CaO (on average, 3.41 wt%), MgO (3.70 wt%), and Fe_2O_3 (5.87 wt%), for an average SiO_2 content of 65.75 wt%. Siderophile elements are consistently enriched (variation between samples is less than a factor of 2) in the melt rock samples in comparison with rocks of such major element composition (granodioritic to dioritic), with average values for Cr of 440 ppm, 50 ppm for Co, 780 ppm for Ni, and 32 ppb for Ir. No variations with depth and no differences between drill cores were noted. As discussed by Koeberl et al. (1997a,b), it can be excluded that mafic to ultramafic country rocks could have contributed more than 2 ppb of Ir. In light of isotopic evidence for the melt rock cited by these authors, it also must be excluded that mantle-derived sources could be responsible for these high siderophile and Ir concentrations. Abundances of the other platinum group elements (determined by neutron activation analysis after NiS fire assay) (Robert et al., 1971) confirmed a near-chondritic composition of this component. After correction for indigenous components, Koeberl et al. (1997a,b) concluded that a meteoritic (chondritic) component of 2–5% was present in the Morokweng melt rock.

ISOTOPE DATA

Results of first Rb-Sr and Sm-Nd isotopic measurements on Morokweng melt rock were reported in abstract form by Andreoli et al. (1995). These authors reported whole-rock Rb-Sr isotopic data for five melt rock samples, which allegedly yielded an errorchron of 1.4 ± 0.2 Ga, for an initial ratio of 0.716. Their Sm-Nd isotopic data did not allow calculation of age information due to insufficient data spread, but T_{CHUR} model ages were of the order of 3 Ga, and T_{DM} model ages of 3.2 Ga. Thus, these preliminary data obviously do not represent the age of the impact.

More recently, full publications by Koeberl et al. (1997a) and Hart et al. (1997) presented single zircon U-Pb data for zircons isolated from Morokweng impact melt rock. Koeberl et al. (1997a) reported U-Th-Pb isotopic compositions on the zircons that were obtained by ion microprobe (SHRIMP) analysis, following techniques described by Williams and Claesson (1987) and Claoué-Long et al. (1995), among others. Common Pb corrections, which were made using the measured ^{207}Pb/^{206}Pb and ^{206}Pb/^{238}U values (see Claoué-Long et al., 1995, and references therein), were almost negligible at 0.1% of the total ^{206}Pb in the analysis (which would result in an average correction of the calculated ^{206*}Pb/^{238}U ages of <0.2 Ma). The measurements allow calculation of a ^{206}Pb/^{238}U age of 146.2 ± 1.5 Ma and a ^{208}Pb/^{232}Th age of 144.7 ± 1.9 Ma. This age represents the crystallization age of the zircons in the melt. The quality of these data is obvious from the histograms of Fig. 15. Both numbers exclude the analyses of two zircons on statistical grounds (excess scatter; cf. Koeberl et al., 1997a). However, inclusion of these numbers would not change the final age significantly (resulting in a statistically less valid overall age of 147.5 ± 1.4 Ma). As discussed before (Koeberl et al., 1997a,b; Koeberl and Reimold, 1997), this age is indistinguishable from the age of the Jurassic-Cretaceous (J-K) boundary (see below).

In addition, we carried out preliminary Rb-Sr isotopic analysis on a number of melt-rock samples (Table 3; Fig. 16). If all data (except the MO43 results, which were obtained later) are regressed, an age of 267.5 ± 59.3 Ma is obtained (I_{Sr} = 0.725958 ± 3, at 95% confidence level), but the regression is an errorchron (MSWD = 21.15). If the MO39 plagioclase data are removed, the age becomes 306.5 ± 32.5 Ma for a similar initial ratio of 0.725720 ± 3; however in this case, the MSWD value is only 5.62. From the petrographic study it is known that sample MO39 is not quite fresh. The relatively high Rb content of this mineral separate suggests that probably some K-feldspar was contained in this sample, a phase particularly susceptible to alteration.

The five whole-rock samples display significant scatter in their isotopic compositions. They have very similar Sr concentrations, but vary with regard to their Rb contents. Four whole-rock data are very similar, but the Rb concentration in sample

TABLE 3. MOROKWENG MELT ROCK Rb-Sr ISOTOPE DATA

Sample*	Rb (ppm)	Sr (ppm)	^{87}Rb/^{86}Sr	^{87}Sr/^{86}Sr†
MO16wr	80.47	365.9	0.6376	0.728361 ± 12
MO19wr	110.6	363.0	0.8834	0.729776 ± 10
MO22wr	71.61	378.4	0.5486	0.728251 ± 6
MO33wr	80.43	351.4	0.6636	0.728456 ± 8
MO39wr	82.10	373.5	0.6373	0.728382 ± 12
MO39px	1.948	36.45	0.2300	0.726801 ± 53
MO39pl	91.21	472.4	0.8311	0.728541 ± 16
MO43wr	69.5	346.5	0.5815	0.728418 ± 10
MO43px	27.35	280.8	0.2824	0.729265 ± 7
MO43Fsp	89.08	433.6	0.5956	0.728739 ± 17
MO43Fspl.	88.15	374.5	0.6824	0.729000 ± 18

*Wr = whole rock; px = pyroxene; pl = plagioclase; Fsp = feldspar; Fspl. = leached feldspar.
†Uncertainties in the ^{87}Sr/^{86}Sr isotope ratio are 1σ and refer to the last digits.

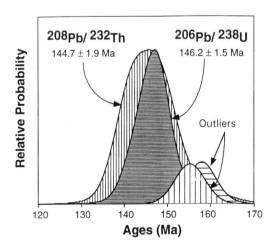

Figure 15. Relative probability plot of the ^{208}Pb/^{232}Th and ^{206}Pb/^{238}U age data calculated from the single zircon SHRIMP analysis dataset presented by Koeberl et al. (1997a). Agreement between the two independent isotopic systems is evident. The small curves represent the outliers, which were excluded from the calculation of the averages on statistical grounds on the basis of excess scatter of the data. Inclusion of these data would not change the resulting age beyond the presently calculated standard deviation.

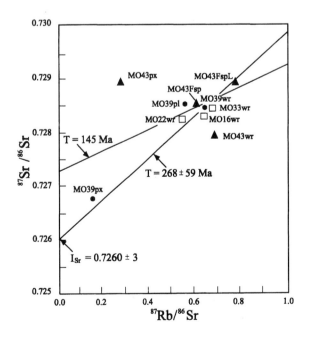

Figure 16. Rb-Sr isochron diagram for selected whole-rock and two mineral separate specimens (refer to text for discussion). Black circles indicate sample MO39; black triangles, sample MO43 (L = leached feldspar sample); open squares, other whole rock samples.

MO19 is about 30 ppm higher than in the other samples. It appears likely that this discrepancy is the result of the inclusion of clastic precursor rock material.

The overall Rb-Sr isotopic results indicate an age for the Morokweng melt rock, which is significantly younger than the 1.4-Ga age suggested by the results of Andreoli et al. (1995), who also have encountered significant scatter in their data. This kind of behavior of Rb-Sr isotopic systematics has been shown to be quite characteristic for impact melt-rock samples (e.g., Reimold, 1982; see also Deutsch and Schärer, 1994). Petrographic analysis indicates that the melt-rock matrix consists of an intimate mixture of melt and relic precursor rock material. Compared to the well constrained zircon U-Pb age of the Morokweng melt body, the Rb-Sr age information is clearly wrong and the Rb-Sr age of 267 Ma too high. Our preferred interpretation for this result is that isotopic homogenization between melt phase and clast component at 145 Ma ago was incomplete. Unfortunately the choice of sample MO39 was not an optimum one, as this sample is now known for substantial secondary alteration. For this reason, a much less altered melt-rock sample, MO43, was selected for mineral separation and additional Rb-Sr isotopic analysis. Besides a whole-rock sample, we analyzed pyroxene and feldspar (mostly plagioclase) fractions. In addition, an aliquot of the feldspar separate was leached in dilute HCl to investigate the effect of alteration phases (Table 3).

The results are straightforward: the considerable scatter of the four new data from sample MO43 is further evidence of open system behavior with respect to the Rb-Sr isotope systematics of the Morokweng melt rock. GEODATE regression of the whole suite of 11 data results in an errorchron corresponding to an age of 146 Ma, but this age carries a 1σ error of 75 Ma, and is thus of little significance. However, the initial $^{87}Sr/^{86}Sr$ ratio of 0.72730 ± 70 is similar to the earlier regression result and confirms the crustal nature of this melt rock.

DISCUSSION

Petrographic constraints on the origin of the Morokweng Structure

Much evidence has been presented that leaves no doubt about the origin of the Morokweng melt rock by impact melting. The fact that nearly every investigated sample contains either relics of granitoid clasts (in melt rock) with impact-characteristic shock metamorphic effects, such as single or, especially, multiple, sets of planar deformation features (PDFs), diaplectic quartz glass and ballenquartz textures, local and pervasive melting, or crystals with such shock deformation (in the case of granitoids from below the melt body), provides convincing evidence in favor of impact. This result further supports the earlier conclusion by Andreoli et al. (1995), Corner et al. (1997), Hart et al. (1997), and Koeberl et al. (1997a) that an impact structure occurs in the region of the Ganyesa Dome in Northwest Province of South Africa.

Many of the petrographic observations made on matrix and clasts of Morokweng impact melt rock are identical to observations made on Vredefort impact melt rock, the Vredefort Granophyre. The distinct thermal alteration and melting-related textures are consistent with the high-temperature (superheating) origin of impact melt. The transitional stages from local melting in clasts to completely melted (and recrystallized) matrix confirms that these large bodies of impact melt rock form from intensely homogenized clastic material, which, either in small parts or comprehensively, has been subjected to the high shock pressures and temperatures produced in the impact melting zone relatively close to the point of impact. It follows that the chemical homogeneity observed typically in impact melt bodies is a phenomenon which characterizes such melt rocks on a decimeter scale. Sampling of smaller specimens would result in considerable scatter of chemical data, as is obvious from the more felsic to more mafic centimeter-sized areas described above. The limited variation in chemical compositions of melt rock discussed by Koeberl et al. (1997a) are caused by this effect. These small, heterogeneous areas consist of either chemically heterogeneous melt or represent relic clasts or more or less melted clastic material.

In the case of Morokweng, the observed clast population in the melt rock indicates that the precursor materials for the melt rock must have comprised one or more granitoid components, as well as a gabbroic component characterized by large tabular clinopyroxene crystals (a suitable component could be 180-Ma-old Karoo dolerite) and other more fine-grained mafic components. Koeberl and Reimold (in prep.) performed chemical analyses for some of the regionally occurring granitic and mafic intrusive rocks, and mixing calculations aimed at reproducing the composition of the Morokweng impact melt rock, and thus better constraining the indigenous component to the siderophile element content of the impact melt.

We have noted that the chemical composition of Morokweng impact melt rock is very similar to that of the Vredefort Granophyre (Reimold and Koeberl, 1997). This is viewed as a coincidence, due to a similar composition of the target region (in which granitoids formed the main component), which was overlain by both quartzitic and carbonate rocks (in the case of Morokweng belonging to the Transvaal Supergroup, in the case of Vredefort to the Transvaal and Witwatersrand Supergroups), and intruded or partially covered by mafic intrusives or extrusives. Much of the carbonate target component would have been vaporized/dissociated and not mixed into the impact melt phase. As outlined below, it is likely that the Morokweng Structure represents one of the largest impact structures known on Earth. If this was not the case, and a much smaller area had been melted to form the Morokweng impact melt body, the probability of obtaining the same chemical composition for this mixture and the Vredefort Granophyre would have been much lower.

Breccia Formation in granitoids below the melt body

The petrographic data reported here, as well as chemical data (Koeberl and Reimold, in prep.), for a series of breccia veinlets in the granitoids drilled below the Morokweng impact melt body demonstrated that the granitoids at this level are not only transected by impact melt, they also contain other breccia types. Whereas it is difficult to definitely identify the nature of any of these veinlets due to extensive thermal alteration of the available material, we can positively state that evidence for the existence of both melt breccia (in the form of crystallization products within vein matrix as well as extensive indication of melting on clasts within such veins) and cataclasite exists. However, it is not possible to pinpoint whether the melt breccia was formed as (1) impact melt, which was intruded into the granitoids; (2) locally formed impact melt; or (3) friction-produced pseudotachylite. It is equally impossible to decide whether the cataclasites represent lithic impact breccia or fault-generated clastic breccia.

Consequently, the earlier reports of pseudotachylite veinlets in Morokweng drill cores (Andreoli et al., 1995, 1996a; Hart et al., 1997) are based on a superficial morphologic resemblance to tectonically (fault-related) produced friction melts (Reimold, 1995, 1998) and are not correct. These same authors also believed that the granitoids intersected in drill core MWF05 represented basement, i.e., presumably coherent crater floor. In contrast, our findings of at least some impact melt breccia in veins cutting these granitoids suggest that the solid crater floor was not reached in this borehole. Rather, a brecciated, granitoid-clast-rich zone was encountered, which could be referred to as a "megabreccia zone," as some of the granitoid intersections are a few to tens of meters wide. This reinterpretation of the MWF05 granitoid zone is important to the following discussion of scaling of the Morokweng impact structure, particularly with regard to the regional geologic setting. One of the inherent problems is the unknown depth of erosion of the Morokweng impact structure.

Constraints on the size of the Morokweng impact structure

Since Corner (1994) first proposed that geophysical anomalies centered on the Ganyesa Dome could represent the central part of a large impact structure, the size of the structure has been debated. Early size estimates ranged from 70 km (Andreoli et al., 1995) to >300 km (Andreoli et al., 1996a; Corner et al., 1997; Hart et al., 1997), based on different interpretations of the aeromagnetic and gravity data. Especially because the age of the Morokweng Structure coincides with that of the Jurassic-Cretaceous boundary, it is important to determine the size of this structure for the purpose of better assessing the potential global effects of this impact event.

First-order size estimation is based on the aeromagnetic expression of the Ganyesa Dome. A 30-km-wide inner, positive magnetic anomaly is surrounded by a magnetically "quiet" zone of 20-km width. Corner et al. (1997) proposed that the regional Bouguer gravity and aeromagnetic signatures indicated ring anomalies of about 300–340 km diameter. Geologic studies on the surface, have been severely hindered by lack of satisfactory exposures. However, the following facts have been established: Black Reef quartzite, sampled about 10 km outside of the Ganyesa Dome aeromagnetic expression, is unshocked. In contrast, a parautochthonous breccia sample from Heuningvlei, ca. 45 km west of the center of the structure, revealed quartz grains with multiple sets of PDFs. A large number of Archean granite specimens obtained in situ ca. 70 km from the center of the Morokweng Structure, are completely unshocked, but limited structural deformation (cataclasite veining) might or might not be related to the formation of this impact structure. The Black Reef quartzite north of Morokweng is oriented subvertical, whereas the Heuningvlei exposures of Transvaal banded ironstone have a subhorizontal attitude.

In addition, limited subsurface geologic information is available from hydrogeologic drilling results, compiled by one of us (C.B.) in the course of a study of the development of the paleodrainage pattern of the Molopo River. Estimated thicknesses of Kalahari Group sediments in the region around the Morokweng Structure indicate that two semi-annular troughs trend around the aeromagnetic anomaly (Fig. 17). The inner trough follows the outline of the anomaly, whereas the outer one occurs at a radial distance of about 100 km. A similar compilation of occurrences and thicknesses of Dwyka sediments (presumably diamictite; compare, for example Reimold et al., 1997) indicates significant occurrences at ca. 35 km east of the center, and again at 100 km to the west (Fig. 18). Whether these strata indeed represent Dwyka diamictite or are Morokweng impact breccia filling a crater moat needs to be further investigated.

A schematic geologic cross section for the region of the Morokweng Structure, based on surface geology and the limited drilling information discussed above, suggests an approximately 200-km-wide impact structure (Fig. 19). In this interpretation, the central geophysical anomalies (70 km diameter) represent a central uplift feature. With the exception of the regionally well-constrained thicknesses of Dwyka and Kalahari Group rocks, and of the Morokweng melt body, lithologic thicknesses (e.g., of Transvaal Supergroup and Griqualand West/Olifantshoek Group strata) are not known. The unshocked granitic basement exposed outside these anomalies would correspond to a deep-level exposure of crater floor, which, in turn, necessitates rather deep erosion of the whole impact structure. This does not seem to be easily reconcilable with the existence of a still at least 130-m-thick impact melt body in the center of the structure, which presumably extends over the entire width of the central, positive magnetic anomaly. It is possible that the central part of this structure is considerably more complex than suggested in this reconstruction. Without further geophysical and drilling constraints, we cannot determine how thick the megabreccia zone is, whether the basement in the central area represents crater

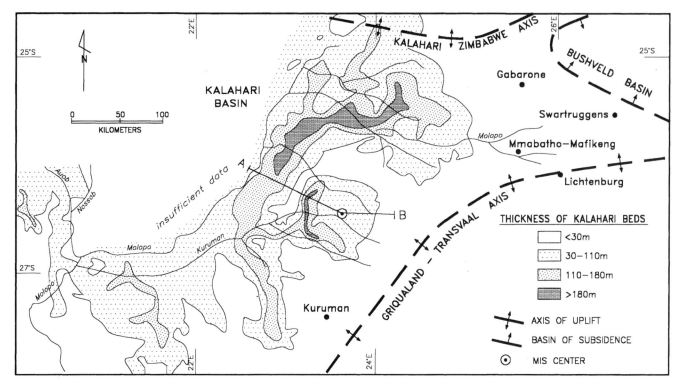

Figure 17. Regional thickness of Kalahari Group sediments in the region around the Morokweng Structure, compiled from hydrogeologic drilling results summarized on open-file geologic maps. MIS = Morokweng Impact Structure. The Griqualand-Transvaal Axis is a feature that marks enhanced uplift of the southern part of this region. As discussed in the text, it is not unlikely that the whole MIS could have been tilted toward the northwest, to an uncertain extent. Profile A-B corresponds to schematic cross section shown in Fig. 19.

floor, or whether the central structure corresponds to a collapsed central uplift or peak ring feature.

At the western edge of this profile, extensive Matsap Quartzite (Griqualand West Group) occurs. The cross section shown in Figure 19 is asymmetrical, and it appears that a relatively deeper level of erosion may be exposed on the southeastern side of the structure. In the case of the Vredefort impact structure, a similar asymmetry has been explained by tilting of the whole structure toward the northwest (e.g., Henkel and Reimold, 1998; Reimold and Gibson, 1996).

Implications for mass extinctions

The radiometric age of the Morokweng impact structure is indistinguishable from recently interpolated stratigraphic ages of the Jurassic-Cretaceous (J-K) boundary. This apparent coincidence suggests a causal relationship. However, the Morokweng situation is somewhat different from that of the Cretaceous-Tertiary (K-T) boundary, where a well-defined and firmly dated impact-derived layer (with shocked minerals and siderophile element anomalies) marks the boundary (compare papers in Silver and Schultz, 1982; Sharpton and Ward, 1990; Ryder et al., 1996) and has led to the discovery of a large impact structure of equivalent age. There is no equally well-defined stratotype for the J-K boundary, and there are only a few well-documented boundary locations in the world. Whereas there is a report of platinum group element enrichment at a J-K boundary site in Siberia, with associated microspherules (Zhakarov et al., 1993), these data are not yet confirmed, and even the identity of this layer in Siberia with the J-K boundary is uncertain (Rampino and Haggerty, 1996). Thus, at the present time no well-defined distal impact ejecta have been identified at the J-K boundary (Koeberl and Reimold, 1997).

Until recently, even the placement of the J-K boundary was not agreed on, as the cross-calibration of the biostratigraphic time scale with the paleomagnetic time scale was hampered by uncertainties in the numerical calibration of each time scale (Remane, 1991). However, paleomagnetic analyses within the last 15 yr (e.g., Lowrie and Channell, 1983; Ogg and Lowrie, 1986) have provided a more solid basis for global correlation of the J-K boundary event. Several authors have recently placed the boundary between the Tithonian (Uppermost Jurassic) and the Berriasian (Lowermost Cretaceous) stages at approximately 145 Ma (Harland et al., 1990: 145.5 ± 2.5 Ma; Gradstein et al., 1994: 144.2 ± 2.9 Ma).

Whereas earlier workers did not assign much importance to the J-K boundary (cf. Remane, 1991), Raup and Sepkoski (1986), Sepkoski (1992), and Sepkoski (cited in Rampino and Haggerty, 1996) now cite the J-K extinction as one of the large extinction events. Rampino and Haggerty (1996) interpreted

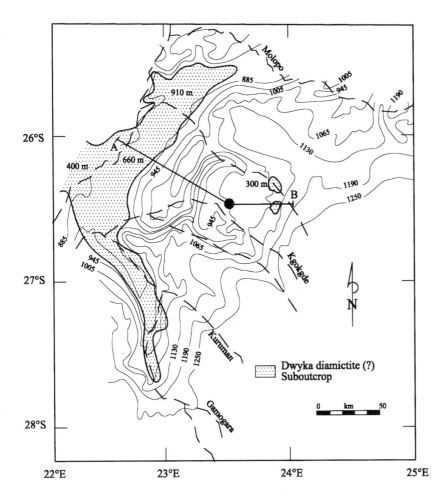

Figure 18. Occurrences and thicknesses (subcrops) of Dwyka Group strata (stippled areas) of the Karoo Supergroup in the region of the Morokweng impact structure. The figure also contains contour lines representing pre-Kalahari Group (i.e., at ca. 70 Ma) topography (meters above mean sea level). Confirmation of whether the mapped Dwyka strata are diamictite of the Karoo Supergroup or instead represent impact breccia is vital for further assessment of the erosion level in this region and the geologic situation underneath and around the melt body. Black dot indicates approximate center of aeromagnetic anomaly. Profile A-B corresponds to schematic cross section shown in Figure 19.

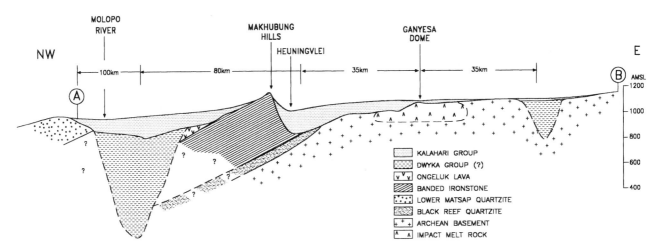

Figure 19. Schematic geologic cross section through the Morokweng impact structure between the Molopo River bed in the northwest and an allegedly Dwyka-filled trough to the east of the Ganyesa Dome (cf. with Fig. 17 and 18). This interpretation is largely based on geophysical, limited drilling, and very poor surface geologic findings. Lower contacts are not constrained, as only shallow drilling information is available. Just west of Heuningvlei is the banded ironstone outcrop where Corner et al. (1997) detected a breccia with shocked quartz.

data from these papers to indicate that the J-K extinction is the fifth-largest extinction event in the last 300 m.y., as recorded by marine genera. Furthermore, reptiles (including dinosaurs) show a distinct peak in extinction at the J-K boundary (e.g., Bakker, 1978; Benton, 1988; Rampino and Haggerty, 1996). Some evidence has been cited to suggest that the extinction events occurred at slightly different times in the Boreal and Tethyan provinces (cf. Ogg and Lowrie, 1986; M. Rampino, personal communication, 1997). For example, Gradstein and Ogg (1996) placed the boreal J-K boundary (offshore Norway) between the Volgian and the Ryazanian stages at an interpolated age of 142 ± 2.6 Ma. It is interesting to note that the top of the Volgian offshore Norway is marked by impact ejecta from the nearby 40-km-diameter Mjølnir impact structure (Dypvik et al., 1996). However, it is difficult to use biostratigraphic correlations to determine if the J-K boundary represents a global event, because of limited faunal exchange between different paleobiogeographic provinces.

At the end-Jurassic, the central Atlantic Ocean was in its initial rifting stages (Fig. 20). Janssen et al. (1995) noted that when final breakup between southern Africa and South America began in the latest Jurassic, there was a change in the direction of the tensional stress in southeastern Africa. It remains to be established if the Morokweng impact event had any relevance for this process. The formation of a large impact crater of about 200 km diameter, which is one of the largest such impact events presently documented on Earth, must have had severe environmental and biologic consequences (e.g., Toon et al., 1997). It may be speculated if the lesser extinction rate at the J-K boundary compared to the K-T boundary could be the result of a difference in target rocks. While the bolide that formed the Chicxulub Crater in Mexico hit a sequence of crustal rocks overlain by carbonate and evaporite rocks, leading to the release of enormous quantities of CO_2, SO_2, and SO_3 (with subsequent formation of acid rain and long-term climate changes; compare papers in Ryder et al., 1996), the Morokweng impact—of comparable size—occurred mainly in granitoids.

SUMMARY AND CONCLUSIONS

The Morokweng melt rock formed by bulk melting of granitoid and gabbroic precursor materials, remnants of which are still ubiquitous throughout the melt body. Alteration textures of clasts and the intimate relationship between remnants of clastic material and crystals grown from the melt phase are consistent with observations made at numerous impact melt-rock occurrences, including the very similar range of textures of the Vredefort Granophyre (cf. also Reimold and Koeberl, 1997). The following evidence demonstrates that the Morokweng melt-rock represents bona fide impact melt rock: (1) the occurrence of characteristic shock metamorphic effects (such as single and multiple sets of planar deformation features) in many lithic and mineral clasts from throughout the melt body; (2) the obvious shock deformation of granitoid from below the melt body, which includes shock effects covering the whole range of shock pressures between ca. 10 and 45 GPa (PDFs, diaplectic glasses, mineral melting); and (3) the confirmed presence of a small meteoritic component, as determined by siderophile element analysis (Koeberl et al., 1997a).

Petrographic and chemical analysis of several breccia veinlets in granitoid from the lower parts of drill core MWF05 (Koeberl and Reimold, in prep.) revealed that at least one of the studied breccias represents injected impact melt, and another cataclasite (sample 63). In the other cases it is ambiguous whether the vein fillings represent recrystallized clastic material, of either local (cataclasite) or foreign (injected clastic impact breccia) origin, or melt breccia of either local friction-related "pseudotachylite" or either local or foreign (injected impact melt breccia) origin. To describe such breccias indiscriminately as "pseudotachylite" is unjustified and should be avoided until an origin by friction along faults has been properly established.

The age of the Morokweng impact melt rock is well-constrained by the results of single zircon U-Pb isotopic analyses. The age of 145 Ma coincides with several recent age estimates of the Jurassic-Cretaceous boundary and implies the necessity to further study the possible relationship between this impact event, the boundary itself, and the biotic changes associated with the boundary. In addition to the situation at the K-T boundary, this is the second coincidence between a large impact event and a major stratigraphic boundary. The J-K biostratigraphic record is not as well studied as the K-T record,

Figure 20. Estimated position (star) of the Morokweng impact structure in a schematic paleogeographic map at about 145 Ma (after TME program, Sageware Corp., 1990).

however, although there is evidence of some marine extinctions (as well as among reptiles) at or near the boundary.

Current constraints on the size of the Morokweng impact structure are restricted to the limited surface geologic information available, drill core data from the area of the central magnetic anomaly, and stratigraphic information regarding the distribution of pre- and post-Morokweng strata of the Dwyka and Kalahari Groups. Whereas earlier geophysical work suggested that the size of the Morokweng impact structure could have exceeded 300 km, the data presented here favor a diameter of approximately 200 km. At this size, Morokweng is still a member of the Big Four, together with Chicxulub, Vredefort, and Sudbury, and consequently deserves special attention. Further drilling and geophysical surveying are required to better document the subsurface structure (for example, extent of central uplift or peak ring) of this large impact structure and to determine the level of erosional exposure.

ACKNOWLEDGMENTS

The director of the Council of Geoscience, Pretoria, permitted access to the Morokweng drillcores, and D. Grobler of the Council's office in Mmabatho facilitated sampling of the cores. Dona Jalufka, Lyn Whitfield, and Di Du Toit assisted with drafting, and Henja Czekanowska provided valuable photographic assistance. Mike Dence and Wylie Poag are acknowledged for critical and constructive reviews of the manuscript. This work was supported by a grant from the Foundation for Research Development of South Africa (to W. U. R.), and by the Austrian Fonds zur Förderung der wissenschaftlichen Forschung Project Y58-GEO (to C. K). This is the University of the Witwatersrand Impact Cratering Research Group Publication No. 6.

REFERENCES CITED

Andreoli, M. A. G., Ashwal, L. D., 1996, Petrology, geochemistry, and structures of melt rocks in the Morokweng impact, southern Kalahari, South Africa; in Anniversary Conference, 12th: Geological Society of South Africa, Tectonics Division, Abstract, 1 p. (unnumbered).

Andreoli, M. A. G., Ashwal, L. D., Hart, R. J., Smith, C. B., Webb, S. J., Tredoux, M., Gabrielli, F., Cox, R. M., and Hambleton-Jones, B. B., 1995, The impact origin of the Morokweng ring structure, southern Kalahari, South Africa, in Centennial Geocongress: Johannesburg, Geological Society of South Africa, Abstract, p. 541–544.

Andreoli, M. A. G., Huizenga, J. M., and Mokgatlha, K., 1996a, Shock metamorphism in the Morokweng impact structure, southern Kalahari, South Africa: implications for planar features in the Vredefort structure in Anniversary Conference 12th: Pretoria, Geological Society of South Africa, Tectonic Division, Abstract, 1 p. (unnumbered).

Andreoli, M. A. G., Ashwal, L. D., Hart, R. J., and Tredoux, M., 1996b, The charnockitic rocks of the sand-covered Morokweng impact structure, southern Kalahari, South Africa: evidence for a possible impact melt origin [abs.]: Lunar and Planetary Science, v. 27, p. 29–30.

Anhaeusser, C. R., and Walraven, F., 1997, Polyphase crystal evolution of the Archaen Kraaipan granite-greenstone terrane, Kaapvaal Craton, South Africa: Johannesburg, Univerisity of the Witwatersrand Economic Geology Research Unit, Information Circular 313, p. 1–27.

Bakker, R. T., 1978, Dinosaur feeding behaviour and the origin of flowering plants: Nature, v. 274, p. 661–663.

Benton, M. J., 1988, Mass extinctions in the fossil record of reptiles: Paraphyly, patchiness, and periodicity (?), in Larwood, G. P., ed., Extinction and survival in the fossil record: Oxford, United Kingdom, Oxford University Press, p. 269–294.

Brown, R. W., Allsopp, H. L., Bristow, J. W., and Smith, C. B., 1989, Improved precision of Rb-Sr dating of kimberlitic micas: an assessment of a leaching technique: Chemical Geology, v. 79, p. 125–136.

Carstens, H., 1975, Thermal history of impact melt rocks in the Fennoscandian Shield: Contributions to Mineralogy and Petrology, v. 50, p. 145–155.

Cheney, E. S., and Winter, H. de la R., 1995, The late Archean to Mesoproterozoic major unconformity-bounded units of the Kaapvaal Province of southern Africa: Precambrian Research, v. 74, p. 203–223.

Claoué-Long, J. C., Compston, W., Roberts, J., and Fanning, C. M., 1995, Two Carboniferous ages: a comparison of SHRIMP zircon dating with conventional zircon ages and $^{40}Ar/^{39}Ar$ analysis, in Berggren, W. A., Kent, D. V., Aubry, M.-P., and Hardenbol, J., eds., Geochronology, time scales and global stratigraphic correlation, SEPM Special Publication 54, p. 3–21.

Cornell, D. H., Schütte, S. S., and Eglington, B. L., 1996, The Ongeluk basaltic andesite formation in Griqualand West, South Africa: submarine alteration in a 2222 Ma Proterozoic sea: Precambrian Research, v. 79, p. 101–123.

Corner, B., 1994, Crustal framework of the Kaapvaal Province from geophysical data, in Conference on Proterozoic Crustal and Metallogenic Evolution: Windhoek, Geological Survey of Namibia, Abstract, p. 9.

Corner, B., Reimold, W. U., Brandt, D., and Koeberl, C., 1996, Evidence for a major impact structure in the Northwest province of South Africa—The Morokweng impact structure. (abs.): Lunar and Planetary Science, v. 27, p. 257–258.

Corner, B., Reimold, W. U., Brandt, D., and Koeberl, C., 1997, Morokweng impact structure, Northwest Province, South Africa: Geophysical imaging and some preliminary shock petrographic studies: Earth and Planetary Science Letters, v. 146, p. 351–364.

Deutsch, A., and Schärer, U., 1994, Dating terrestrial impact events: Meteoritics, v. 29, p. 301–322.

Dypvik, H., Gudlaugsson, S. T., Tsikalas, F., Attrep, M., Ferrell, R. E., Krinsley, D. H., Mørk, A., Faleide, J. I., and Nagy, J., 1996, Mjølnir structure: an impact crater in the Barents Sea: Geology, v. 24, p. 779–782.

Egglington, B. M., and Harmer, R. E., 1991, GEODATE: a program for the processing and regression of isotope data using IBM compatible microcomputers: CSIR Manual EMA-H 9101: Pretoria, South Africa, C S I R, 57 p.

French, B. M., Orth, C. J., and Quintana, C. R., 1989, Iridium in the Vredefort bronzite granophyre: impact melting and limits on a possible extraterrestrial component; in Proceedings 19th Lunar and Planetary Science Conference: Houston, Texas, Lunar and Planetary Institute, p. 733–744.

Geological Survey of South Africa, 1974, Geological map, sheet 2622 Morokweng: Pretoria, South Africa, Geological Survey of South Africa, scale 1:250,000.

Gradstein, F. M., and Ogg, J., 1996, A Phanerozoic time scale: Episodes, v. 19, no. 1-2, p. 3–5.

Gradstein, F. M., Agterberg, F. P., Ogg, J. G., Hardenbol, J., van Veen, P., Thierry, J., and Huang, Z., 1994, A Mesozoic time scale: Journal of Geophysical Research, v. 99, p. 24051–24074.

Grieve, R. A. F., Coderre, J. M., Robertson, P. B., and Alexopoulos, J., 1990, Microscopic planar deformation features in quartz of the Vredefort structure: anomalous but still suggestive of an impact origin: Tectonophysics, v. 171, p. 185–200.

Harland, W. B., Armstrong, R. L., Cox, A. V., Craig, L. E., Smith, A. G., and Smith, D. G., 1990, A geologic time scale 1989, Oxford, United Kingdom, Cambridge University Press, 263 p.

Hart, R. J., Andreoli, M. A. G., Tredoux, M., Moser, D., Ashwal, L. D., Eide, E. A., Webb, S. J., and Brandt, D., 1997, Late Jurassic age for the Morokweng impact structure, southern Africa: Earth and Planetary Science Letters, v. 147, p. 25–35.

Henkel, H., and Reimold, W. U., 1998, Integrated geophysical modeling of a giant, complex impact structure: anatomy of the Vredefort structure, South Africa: Tectonophysics, v. 287, p. 1–20.

Janssen, M. E., Stephenson, R. A., and Cloetingh, S., 1995, Temporal and spatial correlations between changes in plate motions and the evolution of rifted basins in Africa: Geological Society of America Bulletin, v. 107, p. 1317–1332.

Koeberl, C., 1994, African meteorite impact craters: Characteristics and geological importance: Journal of African Earth Sciences, v. 18, p. 263–295.

Koeberl, C., and Reimold, W. U., 1997, The large Jurassic-Cretaceous boundary age Morokweng (South Africa) impact structure—the search for ejecta [abs.]: Geological Society of America Abstracts with Programs, v. 29, no. 6, p. A-80.

Koeberl, C., Reimold, W. U., and Shirey, S. B., 1996a, Re-Os isotope and geochemical study of the Vredefort Granophyre: clues to the origin of the Vredefort structure, South Africa: Geology, v. 24, p. 913–916.

Koeberl, C., Reimold, W. U., Kracher, A., Träxler, B., Vormaier, A., and Körner, W., 1996b, Mineralogical, petrographical, and geochemical studies of drill core samples from the Manson impact structure, Iowa, in Koeberl, C., and Anderson, R. R., eds., The Manson impact structure, Iowa: anatomy of an impact crater: Geological Society of America Special Paper 302, p. 145–219.

Koeberl, C., Armstrong, R. A., and Reimold, W. U., 1997a, Morokweng, South Africa: a large impact structure of Jurassic-Cretaceous boundary age: Geology, v. 25, p. 731–734.

Koeberl, C., Reimold, W. U., and Armstrong, R. A., 1997b, The Jurassic-Cretaceous boundary impact event: the Morokweng impact structure, South Africa [abs.]; in Conference on Large Meteorite Impacts and Planetary Evolution (Sudbury 97): Houston, Texas, Lunar and Planetary Institute, Contribution 922, p. 28–29.

Leroux, H., Reimold, W. U., and Doukhan, J. C., 1994, A T.E.M. investigation of shock metamorphism in quartz from the Vredefort dome, South Africa: Tectonophysics, v. 230, p. 223–239.

Lowrie, W., and Channell, J. E. T., 1983, Magnetostratigraphy of the Jurassic-Cretaceous boundary in the Maiolica limestone (Umbria, Italy): Geology, v. 12, p. 44–47.

Masaitis, V. L, and Grieve, R. A. F., 1994, The economic potential of terrestrial impact craters: International Geology Reviews, v. 36, p. 195–151.

Master S., Reimold W. U., and Brandt D., 1996, Evidence for shock metamorphic origin of multiply-striated joint surfaces (MSJS) in sandstones of the Sinamwenda meteorite impact structure, Zimbabwe [abs.]: Lunar and Planetary Science, v. 27, p. 827–828.

Ogg, J. G., and Lowrie, W., 1986, Magnetostratigraphy of the Jurassic/Cretaceous boundary: Geology, v. 14, p. 547–550.

Pybus, G. Q. J., 1995, Geological and mineralogical analysis of some mafic intrusions in the Vredefort Dome, Central Witwatersrand Basin [M.S. thesis]: Johannesburg, South Africa, University of the Witwatersrand, 282 p.

Rampino, M. R., and Haggerty, B. R., 1996, Impact crises and mass extinctions: a working hypothesis, in Ryder, G., Fastovsky, D., Gartner, S., eds., The Cretaceous-Tertiary event and other catastrophes in Earth history, Geological Society of America Special Paper 307, p. 11–30.

Raup, D. M., and Sepkoski, J. J., Jr., 1986, Periodic extinctions of families and genera: Science, v. 231, p. 833–836.

Reimold, W. U., 1982, The impact melt rocks of the Lappäjärvi meteorite crater, Finland: petrography, Rb-Sr, major and trace element geochemistry: Geochimica et Cosmochimica Acta, v. 46, p. 1203–1225.

Reimold, W. U., 1995, Pseudotachylite—generation by friction melting and shock brecciation?—a review and discussion: Earth-Science Reviews, v. 39, p. 247–264.

Reimold, W. U., 1996, Impact cratering—a review, with special reference to the economic importance of impact structures and the Southern African impact crater record: Earth, Moon and Planets, v. 70, p. 21–45.

Reimold, W. U., 1998, Exogenic and endogenic breccias: a discussion of major problematics: Earth-Science Reviews, v. 43, p. 25–47.

Reimold, W. U., and Gibson, R. L., 1996, Geology and evolution of the Vredefort impact structure, South Africa: Journal of African Earth Sciences, v. 23, p. 125–162.

Reimold, W. U., and Koeberl, C., 1997, Mineralogical-geochemical comparison of the Vredefort and Morokweng Granophyre [ext. abs.]; in MINSA Symposium: Pretoria, Mineralogical Society of South Africa, p. 26–28.

Reimold, W. U., Horsch, H., and Durrheim, R. J., 1990, The "Bronzite"Granophyre from the Vredefort structure—a detailed analytical study and reflections on the origin of one of Vredefort's enigmas, in Proceedings, 20th Lunar and Planetary Science Conference: Houston, Texas, Lunar and Planetary Institute, p. 433–450.

Reimold, W. U., Corner, B., Koeberl, C., and Brandt, D., 1996, The Morokweng structure in the Northwest Province of South Africa: remnant of a large impact structure? [abs.]: Meteoritics, v. 31, p. A114.

Reimold, W. U., von Brunn, V., and Koeberl, C., 1997, Are diamictites impact ejecta?—no supporting evidence from South African Dwyka Group diamictite: Journal of Geology, v. 105, p. 517–530.

Remane, J., 1991, The Jurassic-Cretaceous boundary: problems of definition and procedure: Cretaceous Research, v. 12, p. 447–453.

Robert, R. V. D., van Wyk, K. E., and Palmer, R., 1971, Concentration of the noble metals by a fire-assay technique using nickel sulphide as the collector: Johannesburg, South Africa, National Institute for Metallurgy Report 1371, 14 p.

Ryder, G., Fastovsky, D., and Gartner, S., eds., 1996, The Cretaceous-Tertiary event and other catastrophes in Earth history: Geological Society of America Special Paper 307, 576 p.

SACS (South African Committee for Stratigraphy), 1980, Stratigraphy of South Africa, Pt. 1; in Kent, L. E., compiler, Lithostratigraphy of the Republic of South Africa, South West Africa/Namibia and the Republics of Bophutatswana, Transkei and Venda: Pretoria, South Africa. Department of Mining and Energy Affairs, Government Printer, 690 p.

Sepkoski, J. J., Jr., 1992, A compendium of fossil marine animal families (second edition): Milwaukee Public Museum Contributions to Biology and Geology 83, 156 p.

Sharpton, V. L., and Ward, P. D., eds., 1990, Global catastrophes in Earth history: Geological Society of America Special Paper 247, 631 p.

Silver, L. T., and Schultz, P. H., eds., 1982, Geological implications of impacts of large asteroids and comets on the Earth: Geological Society of America Special Paper 190, 528 p.

Smith, C. B., Allsopp, H. L., Kramers, J. D., Hutchinson, G., and Roddick, J. C., 1985, Emplacement ages of Jurassic-Cretaceous South African kimberlites by the Rb-Sr method on phlogopite and whole-rock samples: Transactions of the Geological Society of South Africa, v. 88, p. 249–266.

Stettler, E. H., 1987, Preliminary interpretation of aeromagnetic and gravity data covering a section of the Thlaping-Thlaro-Ganyesa districts in Bophutatswana; in 14th Colloquium on African Geology: Berlin, Abstracts, p. 339–342.

Therriault, A. M., Reimold, W. U., and Reid, A. M., 1996, Field relations and petrography of the Vredefort Granophyre: South African Journal of Geology, v. 99, p. 1–22.

Therriault, A. M., Reimold, W. U., and Reid, A. M., 1997, Geochemistry and impact origin of the Vredefort Granophyre: South African Journal of Geology, v. 100, p. 115–122.

Toon, O. B., Zahnle, K., Morrison, D., Turco, R. P., and Covey, C., 1997, Environmental perturbations caused by the impacts of asteroids and comets: Reviews of Geophysics, v. 35, p. 41–78.

Vincent P., and Beauvilain A., 1996, Découverte d'un nouveau cratère d'impact météoritique en Afrique: l'astrobleme de Gweni-Fada (Ennedi, Sahara du Tchad): Comptes Rendus de l'Academie des Sciences, v. 323 (II), p. 987–997.

Walraven, F., and Martini, J., 1995, Zircon Pb-evaporation age determinations of the Oak Tree Formation, Chuniespoort Group, Transvaal Sequence: implications for Transvaal-Griqualand West basin correlations: South African Journal of Geology, v. 98, p. 58–67.

Williams, I. S., and Claesson, S., 1987, Isotopic evidence for the Precambrian provenance and Caledonian metamorphism of high-grade paragneisses from the Seve Napes, Scandinavian Caledonides. II. Ion microprobe zircon U-Th-Pb: Contributions to Mineralogy and Petrology, v. 97, p. 205–217.

Zhakarov, V. A., Lapukhov, A. S., and Shenfil, O. V., 1993, Iridium anomaly at the Jurassic-Cretaceous boundary in northern Siberia: Russian Journal of Geology and Geophysics, v. 34, p. 83–90.

Zimmermann, O. T., 1994, Aspects of the geology of the Kraaipan Group in the Northern Cape Province and the Republic of Bophutatswana [M.S. thesis]: Johannesburg, South Africa, University of the Witwatersrand, 145 p.

MANUSCRIPT ACCEPTED BY THE SOCIETY DECEMBER 16, 1998

A Ni- and PGE-enriched quartz norite impact melt complex in the Late Jurassic Morokweng impact structure, South Africa

M. A. G. Andreoli
Earth and Environmental Technology Division, Atomic Energy Corporation of South Africa, P.O. Box 582, Pretoria, 0001, South Africa; Schonland Research Center, University of the Witwatersrand, P.O. Box 3, Wits, 2050, South Africa

L. D. Ashwal
Department of Geology, Rand Afrikaans University, P.O. Box 524, Auckland Park, 2006, South Africa

R. J. Hart
Schonland Research Center, University of the Witwatersrand, P.O. Box 3, Wits, 2050, South Africa; on secondment from the Council for Geoscience, P. Bag X112, Pretoria, 0001, South Africa

J. M. Huizenga
Department of Geology, Rand Afrikaans University, P.O. Box 524, Auckland Park, 2006, South Africa

ABSTRACT

The 145 ± 0.8-Ma Morokweng impact structure in north-central South Africa (diameter, ≥140 km) is almost entirely covered by post-Cretaceous sand and continental sediments. Boreholes from the center of the structure intersect texturally complex, ≥170-m-thick pyroxene-bearing rocks, the most typical being a medium-grained quartz norite. These rocks, interpreted as an impact melt sheet, present a chilled contact with basement granitoids that are affected by very intense shock and thermal metamorphism. In places, veins and dikes of heterogeneous and/or fine-grained quartz norite, pyroxene-bearing granitoids or pegmatoids, and granophyre cross-cut the impact melt sheet. Constituent minerals in the quartz norite include plagioclase (An_{17-48}), orthopyroxene (En_{55-60}), minor subophitic augite (En_{38-42}, Wo_{40-45}), granophyric quartz–K-feldspar intergrowths, and numerous opaque accessories. Rare nodules and veinlets consist of Ni- and Cu-sulfides (millerite, bornite, chalcopyrite, chalcocite), Ni-rich oxides (trevorite, Ni-rich ilmenite, bunsenite), Ni-rich silicates (willemseite, liebenbergite), minor selenides, and traces of native platinum. The quartz norite is characterized by an intermediate to acid bulk chemistry (SiO_2 = 59–66 wt%, MgO = 2.4–4.9 wt%), poorly fractionated rare earth elements, and a striking enrichment in siderophile elements (avg. Ir = ~20 ppb; Ni = 480 ppm; Cr = 360 ppm) when compared to mantle-derived rocks. Morokweng differs from the majority of other impact structures as its impact melt rocks shows some degree of differentiation, and strong siderophile element enrichment including high Ni contents in silicates, oxides and sulfides. In particular, the sulfide-oxide–rich assemblages have nonchondritic platinum group element (PGE) patterns and abundances ranging from 2 to 20 times the chondrite values. The complex crystallization of the Morokweng quartz norite is consistent with the proposed large size of the Morokweng impact structure.

Andreoli, M. A. G., Ashwal, L. D., Hart, R. J., and Huizenga, J. M., 1999, A Ni- and PGE-enriched quartz norite impact melt complex in the Late Jurassic Morokweng impact structure, South Africa, *in* Dressler, B. O., and Sharpton, V. L., eds., Large Meteorite Impacts and Planetary Evolution II: Boulder, Colorado, Geological Society of America Special Paper 339.

INTRODUCTION

The Morokweng structure, located in north-central South Africa, was first recognized as a distinctive cluster of magnetic anomalies on the regional aeromagnetic maps of the southern Kalahari (inset, Fig. 1) (Corner et al., 1986; Stettler, 1987). These authors interpreted such features as an intrusive, low-density plug-like body ($D = 30$ km) of magnetic syenite or granite. More recent works (Heard et al., 1992; Andreoli et al., 1995; Hart et al., 1997; Corner et al., 1997; Koeberl et al., 1997) have reinterpreted the Morokweng magnetic anomaly as the core of a meteorite impact structure on the basis of several lines of evidence. First, partly recrystallized planar deformation features in quartz and other shock features indicative of meteorite impact (Alexopoulos et al., 1988; Huffman et al., 1991; Stöffler and Langenhorst, 1994) are present in basement granite and pebbles from the Morokweng area (Andreoli et al., 1995; Hart et al., 1997; Corner et al., 1997). Second, the Morokweng magnetic anomaly (Hart et al., 1997; Corner et al., 1997) shows marked similarities with the signature of other known impact sites such as Ries (Pilkington and Grieve, 1992), Vredefort (Hart et al., 1995) and other large multi-ring structures. Third, boreholes drilled in the center of the anomaly intersected intermediate to acid igneous-textured rocks interpreted as impact melt rocks in view of the anomalously high concentrations of Ni and Ir (Andreoli et al., 1995; Hart et al., 1997). The U-Pb ages for zircons and Ar-Ar ages for primary biotite recovered from these rocks are 145 ± 0.8 and 144 ± 4 Ma, respectively (Hart et al., 1997). The age of ~145 Ma, later confirmed by Koeberl et al. (1997), is interpreted as the time of crystallization of the impact melt rocks and, thus, the time of impact.

In this chapter, we present more comprehensive petrographic, mineralogic and geochemical data on samples from a number of borehole cores drilled close to the center of the Morokweng structure. These data, consistent with an impact origin of the Morokweng structure, show that the associated impact melt rocks exhibit an unusually complex crystallization history, leading to incipient fractionation, reminiscent of the in situ differentiation that characterizes the Sudbury Igneous Complex (Ostermann, 1996). In addition, we report on the first occurrence, in impact melt rocks, of a suite of Ni-silicates, Ni-oxides, Ni-sulfides and selenides.

GEOLOGIC SETTING

The Morokweng impact structure is almost entirely covered by the Kalahari Group, a broad inland-dipping basin of Late Cretaceous to Cenozoic continental sediments (inset, Fig. 1). This sequence consists largely of wind-blown sand, calcrete, silcrete, sandstone, and up to 180 m of conglomerate (Dingle et al., 1983). Until recently, little was known of the pre-Kalahari geology of the Morokweng area, and the published geologic maps have been compiled from rare exposures and percussion hole chips (Smit, 1977; Levin et al., 1985; Saggerson and Turner, 1992). Apart from the impact-related rocks (Hart et al., 1997), the basement in the Morokweng area is represented by Archean granitic rocks, with scattered banded iron formations, mafic and acid volcanics, and rare ultramafic rocks of the Archean Kraaipan Group (Stettler, 1987; Smit, 1977; Zimmerman, 1994). The domal structure of the basement complex and the overlying Late Archean-Paleoproterozoic strata is represented in Figure 1. These supracrustal rocks comprise quartzite, dolomite, slate, volcanics, and iron ± manganese beds of the 2,300–2,070-Ma Griqualand West Supergroup (thickness = 3.9 ± 1.6 km) (SACS, 1980). These, in turn, are overlain by the younger, largely quartzitic Volop Group and other members that constitute the >1780-Ma Olifantshoek Sequence (thickness = >4.9 km) (SACS, 1980).

Ground water and prospecting boreholes have also revealed two pre-Kalahari (Late Cretaceous) basins, one floored by diamictite in broad arcs approximatively concentric to the Morokweng structure (Fig. 1) (Smit, 1977). This diamictite, reputed to be Permian and of glacial origin (Visser, 1983), can reach a thickness of 600 m (R. Heard, personal communication). In addition, the Morokweng area was intruded by a swarm of prominent preimpact mafic dikes with an east-northeast trend, of which little is known in terms of age, chemistry, and origin (Stettler, 1987; Corner et al., 1997). The geologic map of the southern Kalahari clearly portrays the multi-ring nature of the Morokweng structure, characterized by a prominent Archean basement uplift with a diameter of ~140 km, previously known as the Ganyesa dome (GD in Fig. 1) (Hunter and Hamilton, 1978). This subcircular structure is recognizable as an incomplete ring on Landsat TM and National Oceanic and Atmospheric Administration satellite images, as well as a prominent Bouguer gravity anomaly low ($D = ~170$ km) (S. Webb, personal communication). The same circular outline is further accentuated on the satellite images by the arcuate orientation of the Late Mesozoic to Cenozoic drainage (Fig. 1). The outer limit of the Morokweng structure is not clearly defined: an image of geoid anomalies in southern Africa shows that the Morokweng structure matches a circular ($D = ~180$–200 km), ~3-m-deep depression of the geoid (S. Webb, personal communication). The arched outcrops of Archean cover rocks and the surface drainage trends north of Vryburg (Fig. 1) suggest a diameter of ~260 km, whereas Corner et al. (1997) and Hart et al. (1997) inferred a diameter of ~340 km from the arcuate shape of certain magnetic anomalies further afield.

The Morokweng impact structure is further characterized by the presence of several large sub-concentric and radial faults. The latter include the notheast-trending ~100-km Morokweng fault (MF, in Fig. 1) and the southeast-trending ~200 km (~90 m throw) Bothitong-Reivilo fault (BRF, in Fig. 1). In a number of cases, dikes have been emplaced along these faults, the most prominent of which are the east-trending ~600-km-long Machavie and related dikes (MD in Fig. 1) (Corner et al., 1986). Newly acquired airborne magnetic data clearly demonstrate the postimpact Cretaceous age of these dikes (M. Andreoli, unpublished data).

A Ni- and PGE-enriched quartz norite impact melt complex, Morokweng impact structure, South Africa

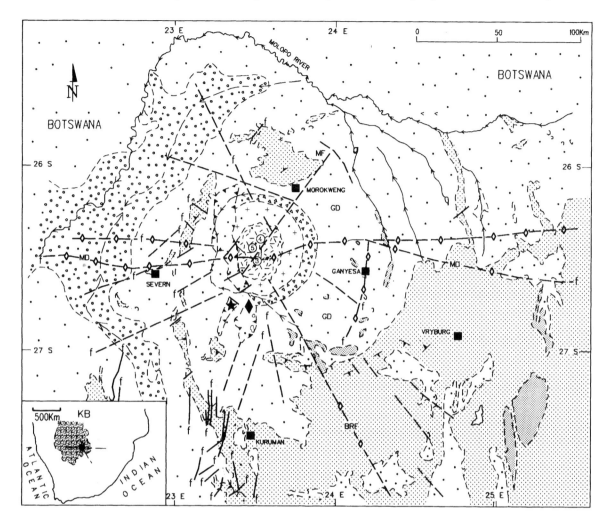

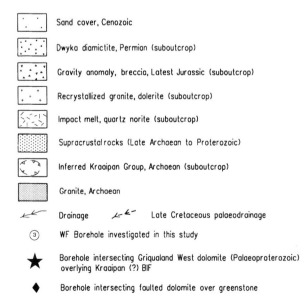

LEGEND

- Sand cover, Cenozoic
- Dwyka diamictite, Permian (suboutcrop)
- Gravity anomaly, breccia, Latest Jurassic (suboutcrop)
- Recrystallized granite, dolerite (suboutcrop)
- Impact melt, quartz norite (suboutcrop)
- Supracrustal rocks (Late Archaean to Proterozoic)
- Inferred Kraaipan Group, Archaean (suboutcrop)
- Granite, Archaean
- Drainage   Late Cretaceous palaeodrainage
- ③ WF Borehole investigated in this study
- ★ Borehole intersecting Griqualand West dolomite (Palaeoproterozoic) overlying Kraaipan (?) BIF
- ◆ Borehole intersecting faulted dolomite over greenstone
- Dyke, post-Jurassic   Fault
- Strike and dip of beds (dip <20°)

Figure 1. Generalized geology of the Morokweng impact structure showing localities and features, including borehole cores, as described in the text. GD = Ganyesa dome; MF = Morokweng fault; MD = Machavie dike; BRF = Bothitong-Reivilo fault. Inset: location of the Morokweng structure in relation to the southern part of the Kalahari basin (KB). Geology adapted from Saggerson and Turner (1992); Levin et al. (1985); Hart et al. (1997); Andreoli et al. (1995); Smit (1977); Corner et al. (1997); Stettler (1987); Dingle et al. (1983) and A. J. Siebrits (personal communication).

LITHOSTRATIGRAPHY OF THE BOREHOLE CORES

The only direct information on the Morokweng impact comes from three boreholes drilled into the central magnetic anomaly (Figs. 1 and 2) (Brinn, 1991; Hart et al., 1997). The basement beneath the Late Cretaceous–Cenozoic Kalahari cover consists of homogeneous, medium-grained crystalline rocks of superficial igneous appearance (Fig. 3A), previously referred to as diorite by Brinn (1991), as quartz norite by Hart et al. (1997), and as impact melt by Koeberl et al. (1997). In this chapter we prefer to retain the term quartz norite in accordance to Streckeisen's (1976) criteria, without reference to any specific genetic model. The Morokweng quartz norite hosts occasional, widely

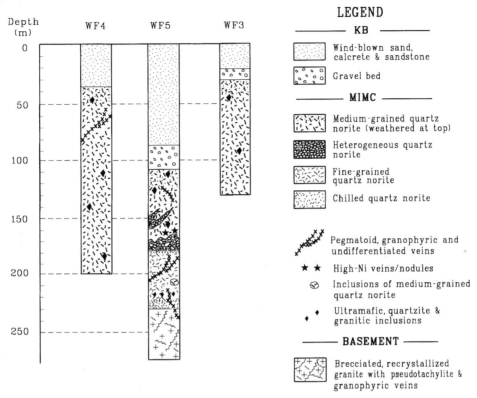

Figure 2. Generalized stratigraphy of boreholes WF 03, -04, and -05 from localities shown in Figure 1, KB = Kalahari beds; MIMC = Morokweng impact melt complex (modified from Hart et al., 1997).

spaced inclusions (up to 5 cm across) of ultramafic to mafic rocks and, less commonly, quartzite. In borehole WF 05, the medium-grained quartz norite is intruded along sharp, irregular contacts by a large composite igneous body (Figs. 2 and 3B–3D). The top 30 m of this intrusion largely consists of a heterogeneous quartz norite comprising fine-grained melanocratic norite inclusions of irregular shape (up to 5 cm across) that are tightly packed within a fine-grained leucogranite matrix (Figs. 3C, 3D). This heterogeneous igneous rock grades, over an interval of several meters, into a fine-grained, more homogeneous quartz norite (Figs. 2 and 3E). Inclusions in this fine-grained quartz norite are common, although less abundant than in the heterogeneous type above. The clasts comprise both basement rocks (quartzite, dolerite, pyroxenite, serpentinite, and finely banded pyroxene-magnetite hornfels) and medium-grained quartz norite (cf. Fig. 2). Closer to the underlying basement granite, the fine-grained quartz norite develops a chilled, microcrystalline texture.

The entire sequence of noritic rocks is cut by secondary, narrow (1–5 cm wide), steeply dipping veins (Fig. 2) comprising leucocratic quartz norite (Fig. 3E); pyroxene-bearing granite; miarolitic pyroxene-bearing pegmatoids (Fig. 3C); miarolitic, fine-grained granophyre; and rare sulfide-, and oxide-rich assemblages (Fig. 2).

The basement granitoids in WF 05 (224–271.3 m) consist of massive to weakly foliated, medium- to coarse-grained leucogranites with scattered pegmatitic lenses, and contain ubiquitous small (~1–2 mm³) open vesicles that are more abundant (up to

TABLE 1. ESTIMATED MODES OF QUARTZ NORITES (BOREHOLE WF 05)

| Column* | 1 | 2 | 3 | 4 |
Type†	M.g.	Het.	F.g.	Chil.
Plg§	60	56	58	60
Qtz	10	12	9	10
Kfp	5	6	4	5
Opx	12	12	12	10
Cpx	4	2	7	5
Bt	5	7	5	5
Op	4	5	5	5
M.C.I.	25	26	29	25
Opx/Bt	2.4	1.7	2.4	2
Opx/Cpx	3	6	1.7	2
M.C.I./Op	6.2	5.2	5.8	5

*Col. 1, average of samples: N5 (134.5 m), LA-137 (136.5 m), N4 (140 m), LA-141 (140.5 m), LA-161 (160.9 m), LA-172 (172 m). Col. 2, average of samples LA-174 (173.7 m), N2 (181 m), LA-186 (185.7 m). Col. 3, average of samples N-1 (195 m), LA-197 (196.7 m), LA-213 (213 m), LA-216 (216.3 m). Col. 4, sample LA-224 (224 m).
†M.g. medium-grained; Het., heterogeneous; F.g., fine-grained; Chil., chilled.
§Plg = plagioclase; Qtz = quartz; Kfp = K-feldspar; Opx = orthopyroxene; Cpx = clinopyroxene; Bt = biotite; Op = opaque; M.C.I. = modal color index.

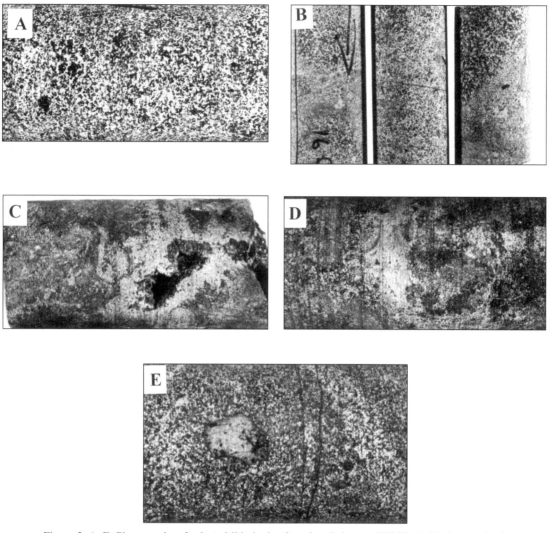

Figure 3. A–E, Photographs of selected lithologies from borehole core WF 05. A, Medium-grained quartz norite (134 m). B, Vein of fine-grained quartz norite crosscutting medium-grained quartz norite (160.9 m). C, Pegmatoid vein with druse in heterogeneous quartz norite (170 m). D, Heterogeneous quartz norite (185 m). E, Fine-grained, inclusion-bearing quartz norite cross-cut by a more leucocratic veinlet (197 m). Core width, 4.5 cm.

10 vol%) where the rock is more melanocratic. The granitoids are cut by networks of annealed fractures, microshears (width = 1–5 mm), breccias, gray recrystallized pseudotachylite, and posttectonic granophyre veins (width = <1 mm–15 cm). These veins are of unusually low density (~2.3 g/cm³) due to a porous texture of interconnecting, small miarolitic cavities lined by euhedral crystals (~1–2 mm in length) of quartz and feldspar.

MINERALOGY

Quartz norite

The mafic mineral content of the quartz norite is typical of an intermediate rock with an average modal color index of ~20 (range = 13–30) (Table 1), whereas true norite (or gabbro) must have 35–65% mafic minerals. It might be appropriate to classify the igneous rocks of the Morokweng impact structure with the charnockitic suite (Streckeisen, 1976), and specifically as charnoenderbites, because of the ubiquitous presence of plagioclase, quartz, K-feldspar, and orthopyroxene (Table 1). However, we consider the charnockitic terminology to be misleading, as it is generally applied to granitoids found in metamorphic terrains of high metamorphic grade.

The textures of the Morokweng quartz norites are varied, but the most abundant rocks in boreholes WF 03, -04, and -05 consist of lath-shaped, randomly oriented plagioclase and smaller, generally euhedral orthopyroxene crystals (Fig. 4A). The same rocks also comprise ~20–25 vol% of interstitial micropegmatitic intergrowths of quartz and K-feldspar, the latter commonly with a micrographic texture. Subordinate clinopyroxene is always inter-

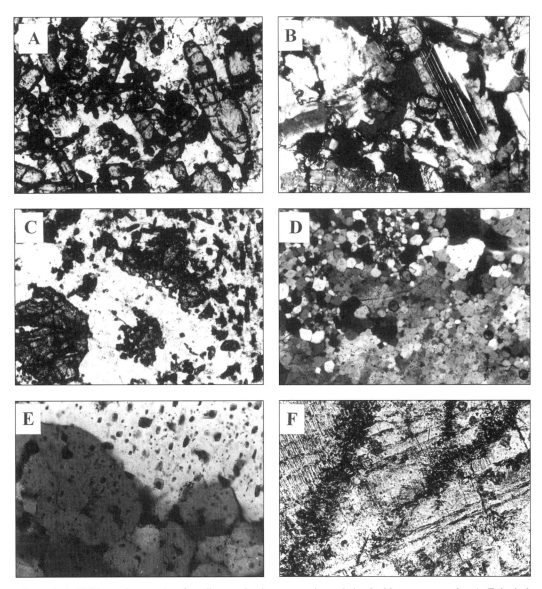

Figure 4. A–H, Photomicrographs of medium-grained quartz norite and shocked basement granite. A, Euhedral orthopyroxene in a quartz-feldspar groundmass. B, Zoned, subhedral plagioclase (cross-polarized light). C, Contact between a medium-grained (left) and an intrusive fine-grained quartz norite (right); note the grain-size difference between the older, euhedral phenocrysts and the more rapidly cooled, flow-aligned pyroxenes on the right. D, Mosaic-like fabric of square-section β-quartz in a recrystallized quartz grain from the basement granite. E, detail of (D), showing devitrified glass inclusions in recrystallized quartz grains. F, Plagioclase grain with dark, broad deformation bands, traces of albite-twin planes (orientation, northeast), and two sets of planar fractures (orientations: north-northwest and east). Fields of view; A–E, = 1 mm; F = 0.3 mm.

stitial, with subophitic texture and numerous partings. Although there is considerable variability in grain size within and between the rocks examined, the bulk of the melt sheet contains plagioclase crystals that average between 0.6 and 1.0 mm in length (Fig. 4B). Minor and accessory constituents typically comprise opaques and less common biotite and hornblende.

Electron microprobe analyses indicate that the plagioclase is ubiquitously zoned from a calcic andesine core to an oligoclase rim (An_{17-48}) (Table 2). Some plagioclase crystals consist of a euhedral core, surrounded by a highly zoned rim (Fig. 4B) that in some case grades into interstitial micropegmatite. The data presented in Table 2 indicate that the difference in feldspar composition between quartz norite types is relatively limited, although fine-grained quartz norites tend to have the plagioclase with the highest An content. Occasional plagioclase grains characterized by fractures and anomalous extinction patterns possibly represent clastic fragments from the target rocks. The K_2O content of the plagioclase is uniformly low (1–5 mol% Or), whereas the Fe con-

TABLE 2. ELECTRON MICROPROBE ANALYSES OF PLAGIOCLASE FROM QUARTZ NORITES (BOREHOLE WF05)*

Type†	M.g.	M.g.	M.g. (c)	M.g. (r)	Het. (r)	Het. (c)	F.g.	Chil.
Depth (m)	160.9	172	172	172	176.6	176.6	213–216§	224
SiO_2	56.30	54.91	58.22	62.89	63.45	55.59	55.22	58.15
Al_2O_3	27.86	28.69	25.86	23.11	21.15	28.48	27.64	26.96
FeO	0.28	0.27	0.16	0.11	0.56	0.26	0.29	0.23
CaO	9.10	10.24	7.5	4.27	3.5	9.94	9.71	7.81
Na_2O	6.73	6.0	7.25	9.65	8.88	5.89	5.84	7.33
K_2O	0.37	0.34	0.52	0.52	0.93	0.25	0.35	0.53
Total**	100.64	100.45	99.51	100.55	98.47	98.41	99.05	100.30
Na	0.58	0.52	0.63	0.83	0.78	0.52	0.51	0.64
K	0.02	0.02	0.03	0.03	0.05	0.01	0.02	0.03
Ca	0.44	0.49	0.36	0.02	0.17	0.49	0.47	0.37
Fe	0.01	0.01	0.01	0.00	0.02	0.01	0.01	0.01
Al	1.47	1.52	1.37	1.20	1.12	1.43	1.48	1.39
Si	2.52	2.47	2.62	2.78	2.86	2.55	2.51	2.60
Total‡	5.05	5.04	5.02	5.05	5.00	5.01	5.0	5.04
Ab	0.56	0.50	0.62	0.78	0.78	0.51	0.52	0.61
An	0.42	0.48	0.35	0.19	0.17	0.48	0.46	0.36
Or	0.02	0.02	0.03	0.03	0.05	0.01	0.02	0.03

*Microprobe analyses were carried out on a Cameca Camebax instrument, with operating conditions of 15 kV and 15 na. Natural standards and conventional ZAF procedures were used.
†Type: M.g. = medium-grained; Het. = heterogeneous; F.g. = fine-grained; Chil. = chilled; c = core; r = rim.
§Average of samples LA-213 and LA-216.
**MgO not detected; MnO <0.09.
‡Number of cations on the basis of 8 oxygen atoms.

tent is within the range of most analyses of terrestrial plagioclase (Smith and Brown, 1989).

Orthopyroxene ($Wo_{3.3}$ En_{55-60}) and the subordinate Ca-rich pyroxene (Wo_{40-46} En_{38-12}) (Table 3) are characteristic minerals of the quartz norite, yet are also present in the pegmatoidal veins that cut the quartz norite. The pyroxenes in the veins are set in a matrix of graphic alkali feldspar, quartz and large (up to ~2 cm wide) miarolitic cavities (Fig. 3C). Like the feldspars, the pyroxenes from the different quartz norites tend to be chemically similar, apart from minor variations in the Mg* (=100 Mg/(Mg + Mn + Fe)) values of the Ca-pyroxenes that progressively decrease with depth. Distinctive features of all the pyroxenes are their appreciable content of Ni (max. NiO in opx, 0.28 wt%; in cpx, 0.14 wt%; Table 3) (Koeberl et al., 1997) and Cr (max. Cr_2O_3 in opx, 0.29 wt%; in cpx, 0.5 wt%; Table 3). The NiO values generally exceed those in pyroxenes from most of the ultramafic and mafic rocks quoted by Deer et al. (1978, Tables 3–6, p. 34–50). Moreover, pyroxenes with similar Cr_2O_3 values are generally restricted to Mg-rich ultramafic rocks (Deer et al., 1978). In addition to these groundmass pyroxenes, there are also others, less common, that are larger, cross-cut by fractures (Fig. 4C) and may represent xenocrysts from the basement.

In boreholes WF 03, -04, and -05, hydrous constituents such as red-brown biotite, hornblende, and serpentine are minor, late-stage replacement products of mafic silicates and Fe-Ti-oxides. However, some biotite was also found in equilibrium with the orthopyroxene (Hart et al., 1997). Accessory minerals include elongate crystals of apatite (<1 vol%), 4–5 vol% of evenly disseminated (width = 1–2 mm) ilmenite-magnetite grains (the latter often strongly altered to hematite and maghemite), and scattered specks of sulfides. Grains of (Cr-, Ni-rich) spinel, rutile, monazite, chalcocite, trevorite, zircon, baddeleyite, and skeletal Cr- and Ni-magnetite (as possible exsolutions in pyroxenes) were also mentioned in the quartz norite by Koeberl et al. (1997).

Within the quartz norite there are also rare lumps and veins (width: 0.5–3 cm; max. length ≥6 cm) consisting of unusual Ni-sulfides, Ni-oxides, Ni-rich silicates (Tables 4 and 5), and minor amounts of Cu-selenides detected—on a qualitative basis—by ion (proton) probe. The vein investigated was subvertical, with a millerite-bornite-chalcopyrite-rich core, a trevorite ± magnetite-rich mantle, and a hematite-stained ~1-mm-wide rim composed of symplectitic lamellae of quartz + trevorite/Ni-magnetite ± millerite. The latter assemblage appears in textural equilibrium with the pyroxenes of the surrounding quartz norite. Finally, close to the contact with the granite basement, there are finer-grained zones (e.g., at depths of 186 and 224 m) that in hand specimen present a typical chilled appearance. In this rock, plagioclase averages

TABLE 3. ELECTRON MICROPROBE ANALYSES OF PYROXENES FROM QUARTZ NORITES
(BOREHOLE WF 05)*

Type†	M.g.	Het.	F.g.	Chil.	M.g.	Het.	F.g.	F.g.	Chil.
Depth (m)	160.9	176.6	213	224	160.9	176.6	213	216	224
Mineral	Opx	Opx	Opx	Opx	Cpx	Cpx	Cpx	Cpx	Cpx
SiO_2	53.26	51.7	52.8	52.7	52.69	52.83	52.93	52.4	52.41
TiO_2	0.16	0.26	0.21	0.33	0.12	0.40	0.23	0.22	0.55
Al_2O_3	0.35	0.51	0.27	n.d.§	0.47	1.19	0.26	0.11	0.89
Cr_2O_3	0.20	0.28	0.29	0.08	0.38	0.5	0.17	0.2	0.34
FeO	24.14	25.83	23.49	25.68	9.26	11.47	11.41	10.94	11.97
MnO	0.48	0.54	0.42	0.52	0.15	0.29	0.22	0.29	0.31
NiO	n.d.	0.28	n.d.	n.d.	n.d.	0.14	n.d.	n.d.	n.d.
MgO	21.15	19.71	21.27	19.33	14.57	14.58	13.98	13.58	13.36
CaO	1.66	1.51	1.57	1.75	22.08	19.31	21.67	21.28	20.3
Na_2O	0.14	0.04	0.2	0	0.28	0.44	0.37	0.25	0.25
Total	101.54	100.66	100.52	100.39	100.00	101.15	101.24	99.25	100.38
Si	1.976	1.959	1.976	1.994	1.971	1.960	1.972	1.987	1.968
Ti	0.004	0.007	0.006	0.009	0.003	0.011	0.006	0.006	0.016
Al	0.015	0.023	0.012	n.d.	0.021	0.052	0.011	0.005	0.039
Cr	0.006	0.008	0.009	0.002	0.011	0.015	0.005	0.006	0.01
Fe	0.749	0.818	0.735	0.812	0.289	0.358	0.356	0.347	0.376
Mn	0.015	0.017	0.013	0.017	0.005	0.009	0.007	0.009	0.010
Ni	n.d.	0.009	n.d.	n.d.	n.d.	0.004	n.d.	n.d.	n.d.
Mg	1.170	1.113	1.187	1.090	0.812	0.806	0.777	0.767	0.748
Ca	0.066	0.061	0.063	0.071	0.885	0.767	0.865	0.865	0.817
Na	0.014	0.003	0.015	0.00	0.020	0.032	0.027	0.018	0.018
Total**	4.015	4.018	4.016	3.995	4.017	4.012	4.026	4.010	4.002
Mg*‡	60.50	57.1	61.3	57.0	73.0	69.0	68.1	68.3	66.0
Wo	3.3	3.1	3.2	3.6	44.6	39.8	43.3	`43.7	42.1
En	59	55.9	59.8	55.2	40.8	41.8	38.9	38.8	38.5
Fs	37.7	41.1	37.0	41.2	14.6	18.4	17.8	17.5	19.4

*Instruments and experimental conditions as in Table 2.
†Type: M.g. = medium-grained; Het. = heterogeneous; F.g. = fine-grained; Chil. = chilled.
§n.d. = not determined.
**Number of cations on the basis of 6 atoms.
‡Mg* = 100 Mg/(Mg + Fe + Mn).

0.1–0.4 mm in length, and orthopyroxene needles have grown radially in rosettes ~1 mm across.

Basement granite

The Archean granitoid rocks at the base of borehole WF 05 are generally highly recrystallized. Quartz grains visible in hand specimen have been invariably replaced by mosaic-like aggregates of smaller grains (Fig. 4D) with the bipyramidal shape of β-quartz. The cores of these secondary grains are often darkened by dense clouds and/or intersecting trails of transparent to brownish, devitrified glass inclusions with rounded to negative crystal shapes (diameter, 30–50 μm) (Fig. 4E). Commonly included in these same inclusions are one or more smaller inclusions of anisotropic crystals and what appears to be gas bubbles. The glassy nature of the inclusions that form the trails (Fig. 4E) was determined under an optical microscope fitted with a U.S. Geological Survey heating-freezing stage. Observed between -200 and +200°C, the bubbles within the inclusions did not change their size (10–30 vol% of the inclusion), nonspherical shape, transparency, and position. Such properties indicate that the inclusions in the grains of quartz originally consisted of silicate melt/glass, probably lechatelierite (Carstens, 1975; Clocchiatti, 1975; Touret and Frezzotti, 1993), and that the bubbles contain no water. Laser-Raman microspectrometry with a Microdil-Dilor 28 instrument (cf. Burke and Lustenhouwer, 1987) further indicated that the bubbles within the glass inclusions contain no CO_2, CH_4, N_2, nor H_2S. These observations, and the results of fluid density calculations for fluid entrapment in the silicate melt after the impact (T = ~1000°–2000°C; P = ~2 kb) (cf. Saxena and Fei, 1987), indicate that the bubbles are now probably filled by air.

Most plagioclase grains in the basement granite have been extensively replaced by secondary aggregates of fine-grained, granoblastic grains, some of which display a crude spherulitic

TABLE 4. ELECTRON MICROPROBE ANALYSES OF Ni-OXIDES AND Ni-SILICATES (BOREHOLE WF 05)*

Mineral† Depth (m)	Bunsenite 170.6	Liebenbergite 170.6	Trevorite 170.6	Ilmenite(-Ni) 170.6	Willemseite 170.6
SiO_2	0.70	26.5	n.d.§	n.d.	54.12
TiO_2	n.d.	n.d.	n.d.	49.81	n.d.
FeO	1.17	5.70	72.21	38.21	7.39
NiO	99.40	62.20	25.60	11.08	21.85
MgO	n.d.	3.52	n.d.	n.d.	13.18
Total	101.27	97.92	97.81	99.10	96.54
Si	0.008	0.94	n.d.	n.d.	3.93
Ti	n.d.	n.d.	n.d.	1.94	n.d.
Fe^{3+}	n.d.	n.d.	16.150	n.d.	n.d.
Fe^{2+}	0.012	0.17	1.688	1.65	0.45
Ni	0.965	1.76	6.083	0.48	1.27
Mg	n.d.	0.19	n.d.	n.d.	1.44
Total**	0.985	3.06	23.92	4.05	7.09
O	1.000	4.00	32.00	6.00	12.00
Ni/(Ni+Fe+Mg)	0.988	0.830	0.783	0.218	0.402

*Microprobe operating conditions as in Table 2.
†Standard mineral formulas are as follows: bunsenite, NiO; liebenbergite, Ni_2SiO_4; trevorite, $Ni_{0.8}Fe_{2.2}O_4$; ilmenite (-Ni), $Fe_{0.8}Ni_{0.2}TiO_3$; willemseite $(Ni,Mg)_3Si_4O_{10}(OH)_2$ (Tredoux et al., 1989).
§n.d. = not determined.
**Cation totals based on number of oxygen atoms as indicated below.

TABLE 5. ELECTRON MICROPROBE ANALYSES OF SULFIDES AND NATIVE METAL (BOREHOLE WF 05)*

Mineral† Depth (m)	Millerite 170.3 and 170.6	Bornite 170.6	Chalcopyrite 170.6	Chalcocite 170.6	Platinum 170.6
Fe	0.43	11.3	30.35	0.48	0.00
Ni	63.77	0.13	1.39	0.05	0.00
Co	0.17	n.d.§	n.d.	n.d.	0.00
Cu	n.d.	62.1	33.8	75.3	0.00
S	34.66	23.2	31.1	19.7	0.00
Pt	n.d.	n.d.	n.d.	n.d.	100.0
Total	99.03	96.73	96.64	96.72**	100.0‡
Fe	0.021	0.912	1.120	0.013	0
Ni	1.004	0.009	0.049	n.d.	0
Co	0.003	n.d.	n.d.	n.d.	0
Cu	n.d.	4.415	1.097	1.928	0
Pt	n.d.	n.d.	n.d.	n.d.	1.000
S	1.000	4.000	2.000	1.000	0

*Microprobe operating conditions as in Table 2.
†Standard mineral formulas are as follows: millerite, NiS; bornite, Cu_5FeS_4; chalcocite, Cu_2S; chalcopyrite, $CuFeS_2$.
§ n.d. = not determined.
**Total includes 1.19% MgO.
‡Total includes traces of Pd.

texture. The occasional relics of the original plagioclase display anomalous extinction patterns akin to mosaicism (Officer and Carter, 1991; Dressler, 1984), and are cross-cut by cracks, anastomosing fracture bands, and multiple sets of intersecting planar features (Fig. 4F). The original crystals of alkali (K, Na-) feldspar are largely replaced by fine-grained mosaics of smaller, granoblastic alkali feldspar grains. In rare cases, however, these aggregates present traces of intersecting planar fractures. Primary biotite crystals are invariably replaced by pseudomorphs of spinels, chlorite, secondary biotite and minute crystals of (serpentinized) orthopyroxene. Apatite and zircon are intensely deformed, fractured, and in places recrystallized into fine-grained, mosaic-like aggregates. Small, euhedral prisms of apatite and zircon were also observed.

Granophyre veins

The veins of miarolitic granophyre that cross-cut quartz norite and basement granite vary in width between less than 1 mm and ~30 cm. The same granophyre constitutes interstitial pockets or rims at the contact between recrystallized quartz and feldspar grains, as well as the matrix of brecciated metagranite. The main constituents of the granophyre veins are fine- to medium-grained quartz and alkali feldspar crystals that occur as micrographic lamellar intergrowths that commonly display a crude radial texture. Small (2–3 mm wide) blades of hematite are a minor constituent in the veins that cross-cut the quartz norite.

Biotite, carbonate, serpentinized orthopyroxene ± colloform chlorite, opaques, and rare zircons distinguish the granophyric veins in the basement granite.

GEOCHEMISTRY

Samples of 15 quartz norites and 6 granitoids were taken from borehole WF 05; clasts, inclusions, and veins were avoided. Whole-rock analyses of these samples were obtained by standard x-ray fluorescence (XRF) and instrumental neutron activation analysis (INAA) techniques for 10 major/minor elements and 23 trace elements. Selenium was determined separately by XRF on two samples, and confirmed qualitatively by INAA (Table 6).

Detailed consideration of the data indicates small but consistent differences between the medium-grained quartz norites (Column M.g. in Table 7), the matrix of the heterogeneous quartz norite (Column Het. In Table 7), and the underlying fine-grained quartz norites (Column F.g. in Table 7). The compositional differences between the main quartz norite types are clearly demonstrated when the rocks are classified according to the relative proportions of normative quartz, K-feldspar, and plagioclase (Fig. 5). In this diagram, the medium-grained, chilled, and heterogeneous quartz norites plot in the granodiorite (charnoenderbite) field, whereas all but one of the fine-grained rocks are more feldspathic, less silicic, and plot in the quartz-monzodiorite (quartz jotunite) field. In accordance with this observation, the medium-grained quartz norites display a normative color index (N.C.I. =

TABLE 6. WHOLE ROCK ANALYSES OF QUARTZ NORITES (BOREHOLE WF 05)									
Sample	N-5	LA-137	N-4	LA-141	LA-161	N-3	V-170.3	LA-172	LA-174
Type*	M.g.	M.g.	M.g.	M.g.	M.g.	M.g.	Vein	M.g.	Het.
Depth (m)	134.5	136.5	140	140.5	160.9	166.5	170.3	172	173.7
(wt. %)									
SiO_2	64.65	64.27	65.42	64.14	65.78	64.38	12.0†	64.63	67.33
TiO_2	0.43	0.38	0.43	0.37	0.47	0.43	n.d.§	0.41	0.52
Al_2O_3	13.24	13.18	13.38	13.17	12.77	13.03	1.0	13.03	13.37
Fe_2O_3**	1.23	3.01	1.17	2.61	3.16	1.17	18.4	3.15	3.12
FeO	4.42	2.78	4.22	3.07	2.36	4.21	n.d.	2.41	1.51
MnO	0.09	0.08	0.09	0.1	0.07	0.09	n.d.	0.08	0.03
MgO	4.05	3.71	3.93	3.92	3.15	4.04	n.d.	3.58	2.36
CaO	3.49	3.38	3.38	3.31	3.25	3.36	n.d.	3.16	2.9
Na_2O	3.88	4.39	3.86	4.65	3.97	3.58	n.d.	4.04	4.28
K_2O	2.05	2.04	2.14	2.2	2.28	2.21	n.d.	2.31	2.42
P_2O_5	0.12	0.09	0.1	0.05	0.08	0.12	n.d.	0.11	0.09
H_2O	n.d.	1.53	n.d.	1.27	1.35	n.d.	n.d.	1.46	1.31
L.O.I.	1.67	0.59	1.75	0.51	0.66	2.09	11.8	0.66	0.69
S	n.d.	n.d.	n.d.	n.d.	n.d.	n.d.	30.0	n.d.	n.d.
Total	99.32	99.48	99.87	99.41	99.37	99.71	101.9‡	99.05	99.95
(ppm)									
V	72	72	71	72	69	75	n.d.	72	70
Cr	427	415	415	415	333	414	n.d.	398	208
Co	36	41	35	41	39	42	45.0E2	51	27
Ni	535	900	519	577	500	634	273.0E3	513	205
Cu	22	46	20	38	33	30	99.0E2	30	39
Zn	58	63	60	64	60	59	n.d.	55	60
Rb	73	64	79	70	97	75	n.d.	83	99
Sr	404	382	392	387	330	374	n.d.	356	287
Ba	424	471	444	471	463	439	n.d.	490	407
Y	17	17	18	19	17	18	n.d.	19	22
Zr	112	110	120	112	118	123	n.d.	114	128
Nb	6.5	7	6	8	8	6.3	n.d.	8	9
U	1.7	n.d.	1.8	n.d.	n.d.	1.9	n.d.	n.d.	n.d.
Th	4.5	n.d.	4.5	n.d.	n.d.	4.7	n.d.	n.d.	n.d.
La	19.01	n.d.	18.96	n.d.	n.d.	18.57	n.d.	n.d.	n.d.
Ce	34.92	n.d.	34.11	n.d.	n.d.	35.65	n.d.	n.d.	n.d.
Sm	3.71	n.d.	3.63	n.d.	n.d.	3.89	n.d.	n.d.	n.d.
Eu	0.63	n.d.	0.71	n.d.	n.d.	0.68	n.d.	n.d.	n.d.
Tb	0.57	n.d.	0.55	n.d.	n.d.	0.61	n.d.	n.d.	n.d.
Sc	13.8	n.d.	13.4	n.d.	n.d.	13.5	n.d.	n.d.	n.d.
Ta	0.64	n.d.	0.65	n.d.	n.d.	0.68	n.d.	n.d.	n.d.
W	0.54	n.d.	0.35	n.d.	n.d.	0.49	n.d.	n.d.	n.d.
Cs	3.3	n.d.	3.4	n.d.	n.d.	4.1	n.d.	n.d.	n.d.
(ppb)									
Ir	23	n.d.	n.d.	n.d.	n.d.	22	75.1E2§§	n.d.	n.d.
Au	2.8	n.d.	2.3	n.d.	n.d.	1.8	30.8E2	n.d.	n.d.
Cr/Sc	31	n.d.	31	n.d.	n.d.	30.7	n.d.	n.d.	n.d.
Ir/Au	8.2	n.d.	n.d.	n.d.	n.d.	12.2	2.4	10.1	7.6
Ni/Co	14.9	22	14.8	14.1	12.8	15.1	60.7	10.05	7.59

20.2) that is appreciably lower than that of the fine-grained rocks (N.C.I. = 27.8; Table 7).

When the trace elements are considered, the Morokweng quartz norites are in general significantly enriched in Cr (mean = 358 ppm) and siderophile elements such as Ni (mean = 485 ppm), Co (mean = 41 ppm), Ir (mean = 20 ppb) when compared to typical igneous rocks of comparable bulk composition, such as granodiorite and andesite (ranges for the latter are: Cr = 10–16 ppm; Ni = 8.8–75 ppm; Co = 6.6–47 ppm; Ir = 0.2–0.3 ppb) (Govindaraju, 1989). The siderophile element content of the Morokweng rocks is generally comparable to that of mafic to ultramafic lithologies (de Wit and Tredoux, 1988), whereas Ir exceeds typical concentrations in mantle-derived ultramafic rocks such as dunite, peridotite and komatiite (Ir = 0.69–6.9 ppb) (Tredoux et al., 1989;

TABLE 6. WHOLE ROCK ANALYSES OF QUARTZ NORITES (BOREHOLE WF 05) (continued)

Sample Type* Depth (m)	N-2 Het. 181	LA-186 Het. 185.7	N-1 F.g. 195	LA-197 F.g. 196.7	LA-213 F.g. 213	LA-216 F.g. 216.3	LA-224 Chil. 224	Mean	S.D.
(wt. %)									
SiO_2	62.84	64.24	59.32	59.93	60.68	64.27	61.99	63.59	2.23
TiO_2	0.54	0.50	0.79	0.77	0.61	0.42	0.64	0.51	0.13
Al_2O_3	14.1	13.74	13.43	13.18	13.28	12.86	13.3	13.27	0.33
Fe_2O_3**	1.32	3.21	1.80	4.55	4.12	3.58	4.36	2.77	1.18
FeO	4.75	2.55	6.46	3.59	3.25	2.17	2.46	3.35	1.28
MnO	0.09	0.07	0.14	0.11	0.1	0.07	0.08	0.09	0.02
MgO	3.81	3.15	4.88	4.7	4.5	3.86	3.68	3.82	0.63
CaO	4.23	3.95	5.26	5.12	4.24	3.16	3.87	3.74	0.71
Na_2O	4.02	4.55	3.5	3.91	4.4	4.44	3.91	4.09	0.34
K_2O	1.52	1.85	1.67	1.95	1.97	2.54	1.71	2.05	0.29
P_2O_5	0.11	0.09	0.16	0.14	0.09	0.1	0.12	0.10	0.03
H_2O	n.d.	0.87	n.d.	0.87	1.15	1.3	1.58	1.27	0.25
L.O.I.	1.48	0.72	1.53	0.72	0.67	0.92	1.34	1.07	0.52
S	n.d.	n.d.	n.d.	n.d.	n.d.	n.d.	n.d.	n.d.	n.d.
Total	98.81	99.52	98.94	99.59	99.12	99.7	99.09	99.73	n.d.
(ppm)									
V	92	76	146	133	108	72	101	87	24.7
Cr	292	234	306	315	384	430	385	358	73
Co	32	28	39	48	61	52	51	41	10
Ni	363	312	361	364	479	541	480	485	161
Cu	24	40	47	69	50	41	43	38	13
Zn	73	71	81	78	71	64	76	66	8
Rb	49	60	71	71	73	83	89	76	13
Sr	393	359	340	331	349	335	353	358	31
Ba	368	300	338	374	435	445	470	423	55
Y	18	21	24	27	21	20	23	20	3
Zr	112	100	119	116	108	121	129	116	8
Nb	4.9	7	5.3	7	7	9	8	7	1
U	1.5	n.d.	1.3	n.d.	n.d.	n.d.	n.d.	1.6	0.2
Th	3.4	n.d.	3.2	n.d.	n.d.	n.d.	n.d.	4.1	0.7
La	16.24	n.d.	16.53	n.d.	n.d.	n.d.	n.d.	17.86	1.36
Ce	31.42	n.d.	33.28	n.d.	n.d.	n.d.	n.d.	33.88	1.63
Sm	3.65	n.d.	4.06	n.d.	n.d.	n.d.	n.d.	3.79	0.18
Eu	0.73	n.d.	0.84	n.d.	n.d.	n.d.	n.d.	0.72	0.08
Tb	0.61	n.d.	0.72	n.d.	n.d.	n.d.	n.d.	0.55	0.15
Sc	18.2	n.d.	26.7	n.d.	n.d.	n.d.	n.d.	17.12	5.72
Ta	0.59	n.d.	0.63	n.d.	n.d.	n.d.	n.d.	0.64	0.03
W	n.d.	n.d.	2.74	n.d.	n.d.	n.d.	n.d.	1.03	1.1
Cs	3.6	n.d.	4.5	n.d.	n.d.	n.d.	n.d.	3.78	0.51
(ppb)									
Ir	16	n.d.	18	n.d.	n.d.	n.d.	n.d.	19.7	n.d.
Au	1.4	n.d.	3.2	n.d.	n.d.	n.d.	n.d.	2.3	n.d.
Cr/Sc	16	n.d.	11.5	n.d.	n.d.	n.d.	n.d.	20.9	n.d.
Ir/Au	11.3	n.d.	5.6	n.d.	n.d.	n.d.	n.d.	8.5	n.d.
Ni/Co	11.34	11.1	9.3	7.6	7.6	10.4	9.4	11.8	n.d.

Note: Major and trace elements from V to Nb have been determined by X-ray fluorescence spectroscopy of fused glass disks, with Na analyzed on pressed pellets. Analytical uncertainty for most trace elements by XRF is ±5% or more. Elements U to Au have been determined by INAA (Instrumental Neutron Activation Analysis), following the method of Erasmus et al., 1977.
*Type: M.g. = medium-grained; Het. = heterogeneous; F.g. = fine-grained; Chil. = chilled.
†Semi-quantitative determination for SiO_2, Al_2O_3, Fe_2O_3, and L.O.I.
§n.d. = not determined.
**Samples N-1 to N-5: $FeO/(FeO+Fe_2O_3) = 0.8$ (assumed).
‡Total includes values for Co, Ni, and Cu as indicated below.
§§Other PGE (all data in ppm): Os = 9.5; Ru = 9.9; Pt = 2.7; Pd = 23 (Andreoli et al., 1997); sample V-170.3 includes Ag = 36 ppm; Se = 75 ppm (determined by X-ray fluorescence).

TABLE 7. NORMATIVE MINERAL AND SELECTED TRACE ELEMENT COMPOSITION OF QUARTZ NORITES (BOREHOLE WF 05)

Type*	M.g.	S.D.†	Het.§	F.g.
Qz**	20.1	2.24	24.94	13.36
Or	13.24	0.67	14.64	11.31
Ab	35.22	3.06	36.98	34.15
An	11.5	1.73	10.36	13.46
Di	4.1	1.31	3.08	8.55
Hy	11.65	3.59	4.58	12.37
Mt	3.28	1.43	3.50	5.18
Il	0.81	0.07	0.05	1.41
Ap	0.23	0.06	0.22	0.32
Total	100.13		99.31	100.11
(ppm)				
Ni	597	141	205	401
Cr	402	32	208	335
An %‡	24.70	4.27	21.9	28.3
N.C.I.§§	20.18	1.67	12.5	27.83

Note: Norms calculated from mean of analyses listed in Table 6.
*Rock types as in Table 1.
†S.D. = standard deviation (7 analyses).
§Average of samples N1, La-197, and LA-213.
**Qz = quartz; Or = orthoclase; Ab = albite; An = anorthite; Di = diopside; Hy = hypersthene; Mt = magnetite; Il = ilmenite; Ap = apatite.
‡Percentage of anorthite in plagioclase.
§§N.C.I. = normative color index = Di + Hy + Mt + Il + Ap.

enriched in Cr, Ni, and the PGE when compared to the fine-grained types. This observation is intriguing because the content of Cr, Ni, and the PGE tends to be higher in mafic and ultramafic rocks with the lowest silica content (Govindaraju, 1989). In addition, the medium-grained quartz norites have a rather uniform composition, whereas the fine-grained rocks show progressive changes with depth: Ba and K decrease upward whereas V, TiO_2, and N.C.I. increase (Fig. 6). These trends indicate that the fine-grained quartz norite is more feldspathic near the base and grades upward toward less feldspathic, pyroxene-richer types. There are also small but consistent downward decreases in Cr/Sc (from 31 to 11.5) and less regular changes in Ir/Au (from 12.2 to 2.4) and Ni/Co (from 7.6 to 60.7; see Table 6).

Compared to the overlying quartz norites, the basement granitoids are richer in alumina, silica, alkalis and other LILEs (large ion lithophile elements) such as Rb, Ba, Th, Zr, and the light REE (rare earth elements) (Fig. 7; Tables 6 and 8). In addition, the granitic rocks display more variable and fractionated REE patterns (avg. La/Tb = 7.3) compared to the quartz norites (avg. La/Tb = 4.2) (Fig. 7). The quartz norites are also enriched in U, Cs, and W (avg. U = 1.6 ppm; Cs = 3.8 ppm; W = 1.0 ppm) (Table 6) relative to the granitic rocks (avg. U = 0.7 ppm; Cs = 2.2 ppm; W = 0 ppm) (Table 8), despite being significantly more mafic in composition.

DISCUSSION

Shock metamorphism

The petrographic study of the basement granitoids provides evidence that shock metamorphism affected all constituent minerals. The cores of many quartz grains are decorated by intersecting planar trails of glass inclusions, probably lechatelierite (Fig. 4E) (Carstens, 1975). The presence of numerous air-filled bubbles in most of the glass inclusions is taken to indicate heterogeneous trapping, i.e., that a gas phase (subsequently leaked) was present together with the melt. These glass inclusions probably

Tredoux and McDonald, 1996). Siderophile elements (PGE) in samples of quartz norite exhibit a near-chondritic pattern (R. J. Hart, unpublished data), with a slight enrichment in Re, Au, Pt, and Pd (elements with the lowest melting point) relative to the more refractory Os, Ir, and Ru (cf. Koeberl et al., 1997).

Small, yet appreciable, differences in trace element distribution can also be identified between the various types of quartz norites. The medium-grained rocks are both more silicic and

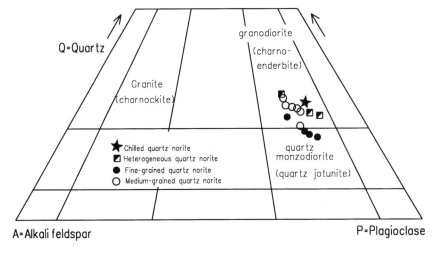

Figure 5. Quartz–K-feldspar–plagioclase ternary diagram showing normative compositions of quartz norite from WF 05, in relation to Streckeisen's (1976) classification for plutonic rocks with >5 vol% hypersthene.

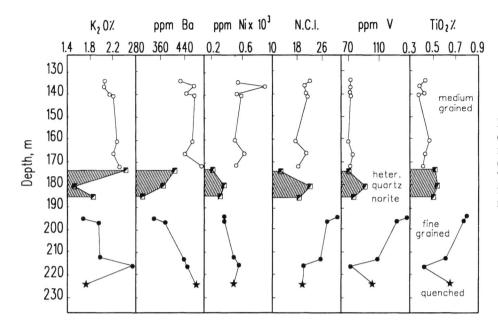

Figure 6. Plot showing changes with depth of selected major and trace elements, and of the normative color index (N.C.I.) in borehole WF 05 (data after Table 6; symbols after Fig. 5). Shaded field outlines heterogeneous quartz norites; Ni-rich nodules/veins, not plotted.

represent relics of previously continuous, shock-generated intracrystalline glass lamellae (Stöffler and Langenhorst, 1994). Plagioclase (Fig. 4F) offers additional evidence for shock metamorphism in the form of multiple (up to three), intersecting sets of planar features or anomalous extinction patterns resembling mosaicism (cf. French, 1968; Dressler, 1984).

K-feldspar has commonly been converted to a mosaic of smaller crystals, whereas all primary biotite crystals have been replaced by symplectitic aggregates of silicates (including very small, secondary biotite) and opaques, probably spinels. Textures similar to these, e.g., breakdown of hydrous minerals, and undeformed micrographic/microgranophyric veinlets ± orthopyroxene pseudomorphs, have been reported in the granitic rocks from the center of Vredefort (Schreyer, 1983; Hart et al., 1991), and in shocked breccia clasts from the Sudbury complex (Dressler, 1984). At Vredefort, such textures have been interpreted as evidence of high-temperature conditions outlasting shock metamorphism (Schreyer, 1983; Hart et al., 1991). The origin of the microcavities in the metagranitic rocks remains uncertain; however, vesiculated and highly shocked country rocks were found as inclusions in impact melt bombs from Ries, Germany (Newsom et al., 1990), and in parautochthonous and allochthonous rocks from the Manicouagan impact, Canada (Floran et al., 1978).

Experimental studies demonstrate that diaplectic silica glass is formed along planar features in quartz grains subjected to the shock wave of hypervelocity meteorite impacts at P = 30–38 GPa (Sharpton and Grieve, 1990; Da Silva et al., 1990; Huffman et al., 1991; Stöffler and Langenhorst, 1994, and references therein). Crystallization of euhedral/subhedral β-quartz from a diaplectic quartz glass (Fig. 4D), extensive resorption of the glass inclusions in secondary quartz grains (Fig. 4E), and the ubiquitous presence of undeformed, orthopyroxene-bearing granophyric textures/veins

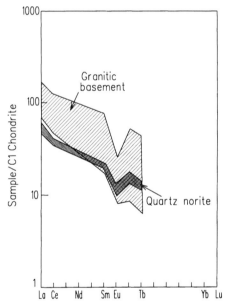

Figure 7. Chondrite-normalized REE abundances of quartz norites (n = 5) in relation to field (light diagonal rule) of basement granitoids (n = 6).

point to a high-temperature static event in the basement rocks (T = >750°C; cf. Schreyer, 1983). We interpret this event to reflect the thermal effect of the overlying impact melt sheet.

Petrogenesis of the quartz norite

Compositions of impact melt rocks from large craters, including Manicouagan, Ries, West Clearwater, Mistastin, Popigai, and Vredefort show a remarkable homogeneity (Grieve et al.,

TABLE 8. WHOLE ROCK ANALYSES OF BASEMENT GRANITOIDS

Sample	G-4	G-3	G-6	G-2	G-1	G-5	Mean	S.D.*
Depth (m)	227	231.8	244.5	246.4	258.2	270.5		
(wt. %)								
SiO_2	72.49	70.75	71.91	71.4	71.92	67.00	70.91	2.00
TiO_2	0.18	0.19	0.22	0.25	0.21	0.55	0.27	0.14
Al_2O_3	14.22	15.62	14.16	14.32	14.01	14.89	14.54	0.61
FeO	1.39	1.5	1.56	1.85	1.79	3.59	1.95	0.82
MnO	0.04	0.03	0.03	0.06	0.05	0.08	0.05	0.02
MgO	0.41	0.4	0.38	0.55	0.47	1.02	0.54	0.24
CaO	1.67	1.89	1.68	1.9	1.35	2.44	1.82	0.36
Na_2O	4.52	5.16	4.53	4.62	3.72	4.73	4.55	0.47
K_2O	3.46	3.49	3.54	3.13	4.85	3.27	3.62	0.62
P_2O_5	0.07	0.03	0.1	0.1	0.04	0.28	0.10	0.09
H_2O	0.52	0.49	0.46	0.57	0.44	1.03	0.58	0.22
Total	98.97	99.55	98.57	98.75	98.85	98.88	98.93	n.d.†
(ppm)								
V	22.9	25.6	22.4	33.9	33.5	44.5	30.47	8.5
Cr	8	6	9	13	9	14	9.8	3
Co	5	5	6	6	5	9	6	1.5
Ni	3.7	2.4	3.0	4.7	3.4	8.4	4.3	2.16
Cu	0	0	0	0	0	8.2	1.4	n.d.
Zn	35	39	43	48	64	128	59	35
Rb	161	162	154	144	163	152	156	7.4
Sr	411	360	491	520	448	543	462	69
Ba	590	732	783	670	796	702	712	76
Y	12	11	11	12	18	47	18.5	4.2
Zr	128	144	150	165	168	238	165	38
Nb	4	3.8	5	6.4	6	18.3	7.2	5.5
U	0.7	0.5	0.8	0.5	0.5	1.3	0.7	0.3
Th	5	4.5	6	5.1	10.5	13.2	7.4	3.6
La	28.2	20.8	23.3	32.4	46.3	56.9	34.6	14.1
Ce	46.34	37.51	41.56	58.38	85.98	126.72	66.1	34.5
Sm	3.77	2.71	3.05	3.66	6.1	13.92	5.5	4.3
Eu	0.71	0.8	0.55	0.78	0.91	1.77	0.92	0.43
Tb	0.42	0.3	0.31	0.42	0.66	1.97	0.68	0.64
Sc	2.9	2.9	2.6	4.0	4.3	7.9	4.1	2
Ta	0.38	0.31	0.49	0.63	0.71	3.0	0.92	1.03
W	0	0	0	0	0	0	0	0
Cs	2.7	2.7	1.8	1.6	2.2	2.4	2.2	0.5
(ppb)								
Ir	b.d.§	b.d.	b.d.	b.d.	<2	b.d.	b.d.	n.d.
Au	0.5	0.3	0.2	0.4	<5	<2	n.d.	n.d.
Cr/Sc	2.76	2.07	3.46	3.25	2.09	1.83	2.39	n.d.

Note: Analytical techniques and procedures as described in Table 6.
*S.D. = standard deviation.
†n.d. = not determined.
§b.d. = below detection limit.

1977; Floran et al., 1978; Reimold et al., 1990). The Morokweng quartz norites are also reasonably homogeneous, as indicated by the low variability of silica (avg. $SiO_2 = 63.59 \pm 2.16$ wt% (1σ)) and by the limited spread in the chondrite-normalized REE data (Fig. 7). Many of the quartz norites from the Morokweng impact also contain clearly recognizable clastic material ranging in size from about 3 to 4 cm, down to the limit of microscopic resolution (~10 µm). In particular, the texture of the heterogeneous quartz norite (Fig. 3D) strongly suggests that this rock represents a breccia in which many of the clasts were molten during emplacement. Typically, however, the medium- and fine-grained quartz norites of Morokweng show only a few scattered inclusions (Fig. 3A). In this they differ from well-documented impact melt rocks from most terrestrial and lunar craters that typically contain lithic and/or mineral clasts and may be considered breccias with igneous matrices (e.g., Simonds et al., 1978a; Warner et al., 1973). The quartz norites from Morokweng perhaps compare best to impact melt rocks from craters at Manicouagan (thickness [h] ≥230 m) (Floran et al., 1978), Mistastin Lake (h, ≥80 m) (Grieve, 1975), and Brent (h, 34 m) (Grieve, 1978). At these impact sites the proportion of the lithic fragments decreases to zero in the upper or central parts of the impact melt sheets. Yet, even these coarser grained rocks include recognizable mineral fragments, e.g., refractory calcic plagioclase cores (of xenocrystic origin), overgrown by more sodic feldspar (Floran et al., 1978; Grieve, 1978). In contrast, the finer grained clastic material of the Morokweng quartz norite appears to have been disaggregated, and original boundaries can be difficult to identify due to reaction with the host melt. We propose that the ubiquitous and dramatic continuous zoning of plagioclase (Fig. 4B) has taken place around cores of possible xenocrystic origin.

Perhaps the strongest geochemical argument in support of an impact melt origin for the quartz norites is represented by the exceptionally high siderophile element (especially Os, Ir, and Ru) content in these rocks (Fig. 8). One explanation for such anomalous siderophile element enrichment would invoke contamination of the impact melt (quartz norite) by the impacting bolide (Masaitis, 1994; Palme, 1982; Schuraytz et al., 1996; Hart et al., 1997). Based on the assumption that Cr, like Ir, is largely of meteoritic origin, Koeberl et al. (1997) proposed that the Morokweng quartz norite comprises 2–5 wt% of a chondritic impactor. However, these values may be perhaps revised upward, e.g., to 10 wt%, if Ni-PGE-rich assemblages (such as those listed in Tables 4 and 5) were a minor, although persistent, component of the quartz norite of Morokweng. The alternate hypothesis that the impact melt rocks received their siderophile element endowment principally from mantle-derived melts, perhaps as suggested for the Sudbury Igneous Complex (cf. Dressler et al., 1996; Pye et al., 1984, and references therein), is discounted on the basis of the PGE data. Even in the extreme, theoretical case that the impact melt rocks were ultramafic in character, with a large proportion of mantle-derived (e.g., komatiitic) component, a refractory siderophile element such as Ir would probably occur in concentrations between 3 and 20 times lower than the values measured in the quartz norite (cf. Table 6; Fig. 8). It is then clear that if the impact melt rocks comprise a mantle contribution, its signature is at present masked by the swamping effect of the meteoritic component. In view of these considerations, and of the absence of oucrops in the target area (Fig. 1), we consider as premature any attempt to derive the composition of the quartz norite by mixing calculations that use only selected members of the stratigraphic column (cf. Koeberl et al., 1997).

Anomalies in the impact melt model

The data presented in this chapter reveal a number of differences between the quartz norites of Morokweng and the impact melt rocks from other large impact structures such as Chicxulub

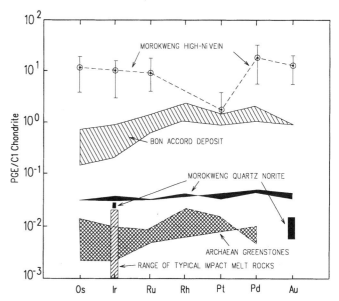

Figure 8. Chondrite-normalized PGE abundances for Morokweng high-Ni vein (averages with ranges; cf. Table 6) (Andreoli et al., 1997; R. J. Hart, unpublishlished data), and quartz norite (black bars indicate data from Table 6; black field represents data from Koeberl et al., 1997). Range of typical impact melt rocks comprises 90% of published (n = 170) Ir determinations in terrestrial impact melt rocks (Bazilevskiy et al., 1984, and references therein; Schmidt et al., 1997; Evans et al., 1993; Palme et al., 1978; Pernicka et al., 1996; French et al., 1988; Masaitis, 1994; Williams, 1994; Schuraytz et al., 1994). Fields of Ni-rich ore from the Bon Accord deposit and of typical greenstones and komatiites are also indicated (Tredoux et al., 1989; Tredoux and McDonald, 1996).

(Schuraytz et al., 1994, 1996), Manicouagan (Grieve and Floran, 1978), and Mistastin Lake Craters (Grieve, 1975). The multiple intrusions by a number of distinct igneous types, namely fine-grained, leucocratic, and/or heterogeneous quartz norite, pegmatoids, granophyre, sulfide-oxide rich veins, and pyroxene granite (Figs. 2 and 3B–E) are not commonly documented at other impact structures. Complex boundaries between melt zones of different crystallinity and fragment saturation are known, however, at large (D = >30 km) impact craters, such as Manicouagan (Floran et al., 1978; Currie, 1972; Grieve and Floran, 1978), West Clearwater (Simonds et al., 1978b), Popigai (Masaitis, 1994), and Sudbury (Grieve et al., 1991).

The high Ni content of the silicates (Tables 3 and 4) and the progressive increase in the normative color index and in certain interelement ratios (e.g., Cr/Sc, Ni/Co, and Ir/Au) (Table 6), from the bottom to the top of the fine-grained quartz norite (Fig. 6), are also atypical. With few exceptions (e.g., the Mistastin Lake impact melt rocks [Grieve, 1975], the Manicouagan impact melt rocks [Grieve and Floran, 1978], and the Sudbury igneous complex [Pye et al., 1984; Dressler et al., 1996; Grieve et al., 1991; Corfu and Lightfoot, 1996; Ostermann, 1996]), even the slowest cooled, clast-poor upper portions of the largest terrestrial impact melt sheets show no evidence for crystal accumulation or progressive compositional changes with depth (Ashwal et al., 1982).

The medium-grained quartz norite of Morokweng hosts the

first reported occurrence of nodules and veins with a complex, high-Ni mineral assemblage consisting chiefly of millerite (NiS), liebenbergite (Ni_2SiO_4), willemseite (($Ni,Mg)_3Si_4O_{10}(OH)_2$), bunsenite (NiO), trevorite ($Ni_{0.8}Fe_{2.2}O_4$), Ni-ilmenite ($Ni_{0.2}Fe_{0.8}TiO_3$), minor Cu-rich selenides, and native platinum. (Tables 4 and 5). The petrogenesis of this and related assemblages is not discussed here, but will be the topic of a subsequent paper. Preliminary data for the same millerite-trevorite vein indicate that the PGE are significantly fractionated and not chondritic because Pt is depleted and Pd enriched relative to Os, Ir, and Ru (Fig. 8; Table 6) (Andreoli et al., 1997). If the PGE are of chondritic derivation (Koeberl et al., 1997), then significant postimpact fractionation, as well as concentration, is required to produce the pattern observed in the vein. The possible role of subsolidus differentiation is further emphasized by the presence of millerite and selenides, minerals generally absent in high-temperature sulfide deposits (e.g., Sudbury), but characteristic of skarns and of a variety of lower temperature environments (e.g., hydrothermal deposits, epithermal veins, coal) (Leutwein, 1974; Simon and Essene, 1996; Anthony et al., 1990).

Finally, the enrichment of siderophile element (Ni, Co, PGE) in the quartz norites (Table 6; Fig. 8) may imply special conditions during the impact because Ir concentrations >1 ppb are in general restricted to impact melt rocks from craters with D <25 km (Palme, 1982). Published Ir data show that more than 90% of impact melt rock from structures >1.5 km (e.g., excluding the Wabar Crater, Saudi Arabia, and Meteor Crater, U.S.) have Ir ≤15 ppb (Fig. 8). Among the large impact structures, very high Ir contents (Ir, 550–570 ppb) are known only from the base of the Popigai melt sheet (Masaitis, 1994). With the possible exception of the Sudbury impact-related rocks, the Morokweng quartz norites and associated veins therefore represent the most PGE-enriched impact melt rocks known (Fig. 8; Table 6).

CONCLUSIONS

The present study is consistent with the impact melt origin of the quartz norites at the core of the Morokweng structure, as proposed in previous works (Andreoli et al., 1995; Corner et al., 1997; Hart et al., 1997; Koeberl et al., 1997). The available field, remote sensing, and geophysical data (S. Webb, personal communication) indicate that the Morokweng impact melt rocks form a shallow lens near the center of a much broader, possibly multiring, structure with an apparent diameter of ≥140 km (Fig. 1) (Andreoli et al., 1995; Corner et al., 1997; Hart et al., 1997). Elsewhere in the world, the identification of multi-ring basins (D = >100 km) remains controversial, as these features may be limited to the Sudbury (Grieve et al., 1991; Spry and Thompson, 1995), Chicxulub (Sharpton et al., 1993), Popigai (Masaitis, 1994), and Vredefort (Hart et al., 1995; Moser, 1997) impact structures. Morokweng is also characterized by the presence of several major radial and concentric dikes and/or faults that probably coincide with, or at least postdate, the impact event (Fig. 1). The petrographic, mineralogic and geochemical complexity of the Morokweng quartz norites stands in contrast with the majority of other terrestrial impact melt rocks, which are generally homogeneous and cryptocrystalline (e.g., partly glassy to basaltic) (Ashwal et al., 1982; Masaitis, 1994; Floran et al., 1978; Schuraytz et al., 1994).

A number of unresolved questions remain, such as what caused the unusual enrichment of the quartz norites in siderophile elements, the unique Ni mineralogy (Tables 4 and 5), the highly anomalous nonchondritic PGE pattern of the millerite-trevorite–rich vein (Table 6; Fig. 8), and what controlled the incipient differentiation of the quartz norites (Fig. 6). In particular, we have the paradox that the high-Ni assemblages (Tables 4 and 5) comprise selenides and millerite, more typical of low-temperature mineralizations (Leutwein, 1974; Simon and Essene, 1996; Anthony et al., 1990), in close association to bunsenite (Ni-oxide), trevorite (Ni-Fe-oxide), and liebenbergite (Ni-olivine) that may record the ultra-high temperatures (~2000°C) (cf. Tredoux et al., 1989) prevailing early in the evolution of the impact melt. As a consequence, Morokweng probably represents the impact structure with the highest degree of extraterrestrial contamination known to date (Fig. 8). Although the body of igneous-textured rocks at Morokweng is obviously much smaller and more deeply eroded than that of Sudbury, there may be some similarities worthy of further consideration.

ACKNOWLEDGMENTS

We thank N. Beukes, R. Hargraves, J. Touret, D. Broad, R. Heard, E. Stettler, and S. Webb for discussions. Comments by R. A. F. Grieve and M. Ostermann, our referees, were particularly valuable. We also thank I. McDonald, R. Cox, B. Corner, E. A. J. Burke, K. Mogkhatlha, J. Bester, D. H. Corbett, J. Malherbe, M. Le Grangie, N. Day, and the staff of Pelindaba's SAFARI-1 reactor for technical and logistic support. We are grateful to S. Ellis and A. J. Siebrits for providing borehole and geophysical data, and to the Faculty of Earth Sciences, Vrije Universiteit, Amsterdam, and WACOM for access to their Raman laser spectrometry facilities. We also thank the Management of the Atomic Energy Corporation of South Africa for financial support and permission to publish this contribution. The Foundation for Research Development partly supported Ashwal's work in this study.

REFERENCES CITED

Alexopoulos, J. S., Grieve, R. A. F., and Robertson, P. B., 1988, Microscopic lamellar deformation features in quartz: discriminative characteristics of shock generated varieties: Geology, v. 16, p. 796–799.

Andreoli, M. A. G., Ashwall, L. D., Smith, C. B., Webb, S. J., Tredoux, M., Gabrielli, F., Cox, R. M., and Hambleton-Jones, B. B., 1995, The impact origin of the Morokweng ring structure, southern Kalahari, South Africa: Centennial Congress, Geological Society of South Africa, Johannesburg, Abstracts, v. 1, p. 541–544.

Andreoli, M. A. G., Hart, R. J., Ashwal, L. D., and Tredoux, M., 1997, Nickel and platinum-group-element-enriched quartz norite in the Latest Jurassic Morokweng impact structure, South Africa, in Conference on Large Meteorite Impacts and Planetary Evolution (Sudbury 1997): Houston,

Texas, Lunar and Planetary Institute, Contribution No. 922, Abstracts, p. 3.

Anthony, J. W., Bideaux, R. A., Bladh, K. W., and Nichols, M. C., 1990, Handbook of mineralogy. Vol. I, Elements, sulfides, sulfosalts: Tucson, Arizona, Mineral Data Publishing, 588 p.

Ashwal, L. D., Warner, J. L., and Wood, C A., 1982, SNC meteorites: evidence against an asteroidal origin, in Proceedings, 13th, Lunar and Planetary Science Conference: Journal of Geophysical Research, Pt. 1, v. 87 suppl., p. A393–A400.

Bazilevskiy, A. T., Feldman, V. I., Kapustkina, S. G., and Kolesov, G. M., 1984, The iridium distribution in rocks from terrestrial impact craters: Geokhimiya, v. 6, p. 781–790 (in Russian).

Brinn, C., 1991, Preliminary report on the investigation of a magnetic anomaly at Water Fouché, Kudumane/Ganyesa District, Bophuthatswana: Pretoria, South African Council for Geoscience Report (unpublished), 40 p.

Burke, E. A. J., and Lustenhouwer, W. J., 1987, The application of a multi-channel laser Raman Microscope (Microdil 28) to the analyses of fluid inclusions: Chemical Geology, v. 61, p. 11–17.

Carstens, H., 1975, Thermal history of impact melt rocks in the Fennoscandian shield: Contributions to Mineralogy and Petrology, v. 50, p. 145–155.

Clocchiatti, R., 1975, Les inclusions vitreuses des cristaux de quartz. Étude optique, thermo-optique et chimique. Applications géologiques: Bulletin Societé Géologique Française, v. 164, p. 229–242.

Corfu, F., and Lightfoot, P. C., 1996, U-Pb geochronology of the sublayer environment, Sudbury igneous complex, Ontario: Economic Geology, v. 91, p. 1263–1269.

Corner, B., Durrheim, R. J., and Nicolaysen, L. O., 1986, The structural framework of the Witwatersrand basin as revealed by gravity and magnetic data: Congress Geological Society of South Africa, 21st, Johannesburg, Abstracts, p. 27–30.

Corner, B., Reimold, W. U., Brandt, D., and Koeberl, C., 1997, Morokweng impact structure, Northwest Province, South Africa: geophysical imaging and shock petrographic studies: Earth and Planetary Science Letters, v. 146, p. 351–364.

Currie, K. L., 1972, Geology and petrology of the Manicouagan resurgent caldera, Quebec: Geological Survey of Canada Bulletin 198, 153 p.

Da Silva, S. L., Wolff, J. A., and Sharpton, V. L., 1990, Explosive volcanism and associated pressures; implications for models of endogenically shocked quartz, in Sharpton, V. L., and Ward, P. D., eds., Global catastrophes in Earth history; a multidisciplinary conference on impacts, volcanism, and mass mortality: Geological Society of America Special Paper 247, p. 139–145.

Deer, W. A., Howie, R. A., and Zussman, J., 1978, Rock-forming minerals. Vol. 2A, Single-chain silicates (2nd ed.): Harlow, United Kingdom, Longman Group, 668 p.

de Wit, M. J., and Tredoux, M., 1988, PGE in the 3.5 Ga Jamestown Ophiolite Complex, Barberton Greenstone belt, with implication for PGE distribution in simatic lithosphere, in Prichard, H. M., Potts, P. J., Bowles, J. F. W., and Cribb, S., eds., Geoplatinum 87: New York, Elsevier, p. 319–341.

Dingle, R. V., Siesser, W. G., and Newton, A. R., 1983, Mesozoic and Tertiary geology of southern Africa: Rotterdam, Netherlands, A. A. Balkema, 375 p.

Dressler, B. O., 1984, The effects of the Sudbury event and the intrusion of the Sudbury igneous complex on the footwall rocks of the Sudbury Structure, in Pye, E. G., Naldrett, A. J., and Giblin, P. E., eds., The geology and ore deposits of the Sudbury structure: Ontario Geological Survey Special Volume 1, p. 97–136.

Dressler, B. O., Weiser, T., and Brockmeyer, P., 1996, Recrystallized impact glasses of the Onaping Formation and the Sudbury igneous complex, Sudbury structure, Ontario, Canada: Geochimica et Cosmochimica Acta, v. 60, p. 2019–2036.

Erasmus, C. S., Fesq, H. W., Kable, E. J. D., Rasmussen, S. E., and Sellschop, J. P. F., 1977, The NIMROC samples as reference materials for neutron activation analysis: Journal of Radioanalytical Nuclear Chemistry, v. 39, p. 323–334.

Evans, N. J., Gregoire, D.C., Grieve, R. A. F., Goodfellow, W. D., and Veizer, J., 1993, Use of platinum-group elements for impactor identification: terrestrial impact craters and Cretaceous-Tertiary boundary: Geochimica et Cosmochimica Acta, v. 57, p. 3093–3104.

Floran, R. J., Grieve, R. A. F., Phinney, W. C., Warner, J. L., Simonds, C. H., Blanchard, D. P., and Dence, M. R., 1978, Manicouagan impact melt, Quebec 1. Stratigraphy, petrology and chemistry: Journal of Geophysical Research, v. 83, p. 2737–2759.

French, B. M., 1968. Sudbury structure, Ontario: some petrographic evidence for an origin by meteorite impact, in French, B. M., and Short, N. M., eds., Shock metamorphism of natural materials: Baltimore, Maryland., Mono Book Corp., p. 383–412.

French, B. M., Orth, C. J., and Quintana, L. R., 1988, Iridium in the Vredefort bronzite granophyre: impact melting and limits on a possible extraterrestrial component, in Proceedings, 19th, Lunar and Planetary Science Conference, Abstracts: Houston, Texas, Lunar and Planetary Institute, p. 356–357.

Govindaraju, K., 1989, 1989 compilation of working values and sample description for 272 geostandards: Geostandards Newsletter, v. 13, p. 1–113.

Grieve, R. A. F., 1975, Petrology and chemistry of the impact melt sheet at Mistastin Lake crater, Labrador: Geological Society of America Bulletin 86, p. 1617–1629.

Grieve, R. A. F., 1978, The melt rocks at Brent crater, Ontario, Canada, in Proceedings, 9th, Lunar and Planetary Science Conference: Geochimica et Cosmochimica Acta, suppl. 6, p. 2579–2608.

Grieve, R. A. F., and Floran, R. J., 1978, Manicouagan impact melt, Quebec—2, Chemical interrelations with basement and formational processes: Journal of Geophysical Research, v. 83, B6, p. 2761–2771.

Grieve, R. A. F., Dence, M. R., and Robertson, P. B., 1977, Cratering process: as interpreted from the occurrence of impact melts, in Roddy, D. J., Pepin, R. O., and Merrill, R. B., eds., Impact and explosion cratering: New York, Pergamon Press, p. 791–814.

Grieve, R. A. F., Stöffler, D., and Deutsch, A., 1991, The Sudbury structure: controversial or misunderstood?: Journal of Geophysical Research, v. 96, E5, p. 22753–22764.

Hart, R. J., Andreoli, M. A. G., Reimold, W. U., and Tredoux, M., 1991, Aspects of the dynamic and thermal metamorphic history of the Vredefort cryptoexplosion structure: implications for its origin: Tectonophysics, v. 192, p. 313–331.

Hart, R. J., Hargraves, R. B., Andreoli, M. A. G., Tredoux, M., and Doucouré, C. M., 1995, Magnetic anomaly near the center of the Vredefort structure: implications for impact-related magnetic signatures: Geology, v. 23, p. 277–280.

Hart, R. J., Andreoli, M. A. G., Tredoux, M., Moser, D., Ashwal, L. D., Eide, E. A., Webb, S. J., and Brandt, D., 1997, Late Jurassic meteorite impact in southern Africa: Earth and Planetary Science Letters, v. 147, p. 25–35.

Heard, R., Andersen, N. J. B., Andreoli, M. A. G., Hennop, F., and Schoeman, R. P., 1992, The Kalahari project. An investigation of the geology of an area bounded by 22º East, the Botswana border, and the edge of the Kalahari beds: Pretoria, Atomic Energy Corporation of South Africa, Report GEA-996, 18 p.

Huffman, A. R., Brown, J. M., Carter, N. L., and Reimold, W. U., 1991, The microstructural response of quartz and feldspar under shock loading at variable temperatures: Journal of Geophysical Research, v. 98, B12, p. 22171–22197.

Hunter, D. R., and Hamilton, P. J., 1978, The Bushveld complex, in Tarling, D. H., ed., Evolution of the Earth's crust: London, Academic Press, p. 107–173.

Koeberl, C., Armstrong, R. A., and Reimold, W. U., 1997, Morokweng, South Africa: A large impact structure of Jurassic-Cretaceous boundary age: Geology, v. 25, p. 731–734.

Leutwein, F., 1974, Selenium, in Wedephol, K. H., ed., Handbook of Geochemistry: Berlin, Springer-Verlag, v. II-3, 34 F, p. 1–5.

Levin, M., Hambleton-Jones, B. B., and Smit, M. C. B., 1985, Uranium in the groundwater of the Kalahari region south of the Molopo River, in Hutchins, D. G., and Lynam, A. P., eds., Proceedings, Seminar on the

Mineral Exploration of the Kalahari: Geological Survey of Botswana Bulletin 29, p. 234–250.

Masaitis, V. L., 1994, Impactites from Popigai crater, in Dressler, B. O., Grieve, R. A. F., and Sharpton, V. L., eds., Large meteorite impacts and planetary evolution: Geological Society of America Special Paper 293, p. 153–162.

Moser, D. E., 1997, Dating the shock wave and thermal imprint of the giant Vredefort impact, South Africa: Geology, v. 25, p. 7–10.

Newsom, H. E., Graup, G., Iseri, D. A., Geissman, J. W., and Keil, K., 1990, The formation of the Ries Crater, West Germany; evidence of atmospheric interactions during a large cratering event, in Sharpton, V.L., and Ward, P. D., eds., Global catastrophes in Earth history; an interdisciplinary conference on impacts, volcanism, and mass mortality: Geological Society of America Special Paper 247, p. 195–206.

Officer, C. B., and Carter, N. L., 1991, A review of the structure, petrology, and dynamic deformation characteristics of some enigmatic terrestrial structures: Earth Science Reviews, v. 30, p. 1–49.

Ostermann, M, 1996, Die Geochemie der Impaktschmelzdecke (Sudbury Igneous Complex) im Multiring-Becken Sudbury [Ph.D. thesis]: Muenster, Westfaelische Wilhelms-Universität, 168 p.

Palme, H., 1982, Identification of projectiles of large terrestrial impact craters and some implications for the interpretation of Ir-rich Cretaceous/Tertiary boundary layers: Geological Society of America Special Paper 190, p. 223–233.

Palme, H., Janssens, M.-J., Takahashi, H., Anders, E., and Hertogen, J., 1978, Meteoritic material at five large impact craters: Geochimica et Cosmochimica Acta, v. 42, p. 313–323.

Pernicka, E., Kaether, D., and Koeberl, C., 1996, Siderophile element concentrations in drill core samples from the Manson crater, in Koeberl, C., and Anderson, R. R., eds., The Manson impact structure, Iowa: anatomy of an impact crater: Geological Society of America Special Paper 302, p. 325–330.

Pilkington, M., and Grieve, R. A. F., 1992, The geophysical signature of terrestrial impact craters: Reviews of Geophysics, v. 30, p. 161–181.

Pye, E.G., Naldrett, A.J., and Giblin, P.E., eds., 1984, The geology and ore deposits of the Sudbury Structure: Ontario Geological Survey Special Volume 1, 603 p.

Reimold, W. U., Horsch, H., and Durrheim, R. J., 1990, The "bronzite"-granophyre from the Vredefort structure —a detailed analytical study and reflections on the genesis of one of Vredefort's enigmas, in Proceedings, 20th, Lunar and Planetary Science Conference: Houston, Texas, Lunar and Planetary Institute, p. 433–450.

Saggerson, E. P., and Turner, L. M., compilers, 1992, Metamorphic map of the Republics of South Africa, Bophuthatswana, Venda and Ciskei and the Kingdoms of Lesotho and Swaziland: Pretoria, Council for Geoscience, scale 1:1,000,000, 4 sheets.

Saxena, S. K., and Fei, Y., 1987, Fluids at crustal pressures and temperatures: Contributions to Mineralogy and Petrology, v. 95, p. 370–375.

Schmidt, G., Palme, H., and Kratz, K.-L., 1997, Highly siderophile elements (Re, Os, Ir, Ru, Rh, Pd, Au) in impact melts from three European impact craters (Saaksjarvi, Mien, and Dallen): clues to the nature of the impacting bodies: Geochimica et Cosmochimica Acta, v. 61, p. 2977–2987.

Schreyer, W., 1983, Metamorphism and fluid inclusions in the basement of the Vredefort dome, South Africa: guidelines to the origin of the structure: Journal of Petrology, v. 24, p. 26–47.

Schuraytz, B. C., Sharpton, V. L., and Marín, L. E., 1994, Petrology of impact-melt rocks at the Chicxulub multi-ring basin, Yucatán, Mexico: Geology, v. 22, p. 868–872.

Schuraytz, B. C., Lindstrom, D. J., Marín, L. E., Martinez, R. R., Mittlefehldt, D. W., Sharpton, V. L., and Wentworth, S. J., 1996, Iridium metal in Chicxulub melt: forensic chemistry on the K-T smoking gun: Science, v. 271, p. 1573–1576.

Sharpton, V. L., and Grieve, R. A. F., 1990, Meteorite impacts, cryptoexplosions, and shock metamorphism; a perspective on the evidence at the K/T boundary, in Sharpton, V. L., and Ward, P. D., eds., Global catastrophes in Earth history; a multidisciplinary conference on impacts, volcanism, and mass mortality: Geological Society of America Special Paper 247, p. 301–318.

Sharpton, V.L., Burke, K., Camargo-Zanoguera, A., Hall, S. A., Lee, S. D., Marin, L. E., Suárez-Reynoso, G., Quezada-Muñeton, J. M., Spudis, P. D., and Urrutia-Fucugauchi, J., 1993, Chicxulub multi-ring impact basin: size and other characteristics derived from gravity analysis: Science, v. 261, p. 1564–1567.

Simon, G., and Essene, E. J., 1996, Phase relations among selenides, tellurides, and oxides: I. Thermodynamic properties and calculated equilibria: Economic Geology, v. 91, p. 1183–1208.

Simonds, C. H., Floran, R. J., McGee, P. E., Phinney, W. C., and Warner, J. L., 1978a, Petrogenesis of melt rocks, Manicouagan impact structure, Quebec: Journal of Geophysical Research, v. 83, p. 2773–2788.

Simonds, C.H., Phinney, W.C., McGee, P.E., and Cochran, A., 1978b, West Clearwater, Quebec impact structure, Part 1: Field geology, structure and bulk chemistry, in Proceedings, 9th, Lunar and Planetary Science Conference: Houston, Texas, Lunar and Planetary Institute, v. 2, p. 2633–2658.

Smit, P. J., 1977, Die geohidrologie in die opvanggebied van die Moloporivier in die noordelike Kalahari [Ph.D. thesis]: Bloemfontein, South Africa, University of the Orange Free State, 354 p.

Smith, J. V., and Brown, W. I., 1989, Feldspar minerals. Vol. I; Crystal structures, physical, chemical, and microtextural properties (2nd ed.): Heidelberg, Germany, Springer-Verlag, 828 p.

South African Committee for Stratigraphy (SACS), Kent, L. E., compiler, 1980, Stratigraphy of South Africa. Part 1, Pretoria. Geological Survey of South Africa Handbook 8, 690 p.

Spry, J. G., and Thompson, L. M., 1995, Friction melt distribution in a multiring impact basin: Nature, v. 373, p. 130–132.

Stettler, E. H., 1987, Preliminary interpretation of aeromagnetic and gravity data covering a section of the Thlaping-Thlaro-Ganyesa districts in Bophuthatswana, in Matheis, G., and Schandelmeier, H., eds., Current research in African earth sciences, 14th Colloquium on African Geology: Rotterdam, Netherlands, A. A. Balkema, p. 339–342.

Stöffler, D., and Langehorst, E., 1994, Schock metamorphism of quartz in nature and experiment: 1. Basic observation and theory: Meteoritics, v. 29, p. 155–181.

Streckeisen, A., 1976, To each rock its proper name: Earth-Sciences Review, v. 12, p. 1–35.

Touret, J. L. R., and Frezzotti, M. L., 1993, Magmatic remanents in plutonic rocks: Bulletin Societé Géologique Française, v. 164, p. 229–242.

Tredoux, M., de Wit, M. J., Hart, R. J., Armstrong, R. A., Lindsay, N. M., and Sellschop, J. P. F., 1989, Platinum group elements in a 3.5 Ga nickel-iron occurrence: possible evidence of deep mantle origin: Journal of Geophysical Research, v. 94, p. 795–813.

Tredoux, M., and McDonald, I., 1996, Komatiite WITS-1, low concentration non metal standard for the analysis of non-mineralized samples: Geostandards Newsletter, v. 20, p. 267–276.

Visser, J. N. J., 1983, An analysis of the Permo-Carboniferous glaciation in the marine Kalahari basin, southern Africa: Palaeogeography, Palaeoclimatology, Palaeoecology, v. 44, p. 295–315.

Warner, J. L., Simonds, C. H., and Phinney, W. C., 1973, Apollo 16 rocks: classification and petrogenetic model, in Proceedings, 4th, Lunar and Planetary Science Conference: Geochimica et Cosmochimica Acta, suppl.4, vol. 1, p. 481–504.

Williams, G. E., 1994, Acraman, a major impact structure from the Neoproterozoic of Australia, in Dressler, B. O., Grieve, R. A. F., and Sharpton, V. L., eds., Large meteorite impacts and planetary evolution: Geological Society of America Special Paper 293, p. 209–224.

Zimmermann, O. T., 1994, Aspects of the geology of the Kraaipan Group in the Northern Cape Province and the Republic of Bophuthatswana [M.S. thesis]: Johannesburg, South Africa, University of the Witwatersrand, 148 p.

MANUSCRIPT ACCEPTED BY THE SOCIETY DECEMBER 16, 1998

Slate Islands, Lake Superior, Canada: A mid-size, complex impact structure

B. O. Dressler and V. L. Sharpton
Lunar and Planetary Institute, 3600 Bay Area Boulevard, Houston, Texas 77058, USA
P. Copeland
Department of Geosciences, University of Houston, 4800 Calhoun Road, Houston, Texas 77204, USA

ABSTRACT

The target rocks of the 30–32-km diameter Slate Islands impact structure in northern Lake Superior, Canada, are Archean supracrustal and igneous rocks and supracrustal Proterozoic rocks. Shatter cones, pseudotachylites, impact glasses, and microscopic shock metamorphic features were formed during the contact and compression phase of the impact process, followed, during excavation and central uplift, by polymict, clastic matrix breccias in the uplifted target, and by allogenic fall-back breccias (suevite and bunte breccia). Monomict, autoclastic breccias were mainly observed on Mortimer Island and the other outlying islands of the archipelago and were probably generated relatively late in the impact process (central uplift and/or crater modification). The frequency of low index planar shock metamorphic features in quartz was correlated with results from shock experiments to estimate shock pressures experienced by the target rocks. The resulting shock attenuation plan across the archipelago is irregular, probably because the shock wave did not expand from a point or spherical source, and because of the destruction of an originally more regular shock attenuation plan during the central uplift and crater modification stages of the impact process. No impact melt rock bodies have been positively identified on the islands. An impact melt may be present in the annular trough around the islands, though and—based on a weighted mixture of target rocks—may have an intermediate-mafic composition. No such impact melt was found on the archipelago. An ^{40}Ar-^{39}Ar release spectrum of a pseudotachylite provides an age of about 436 Ma for the impact structure, substantiating age constraints based on various stratigraphic considerations.

INTRODUCTION

The Slate Islands archipelago is located in northern Lake Superior, ~10 km south of Terrace Bay and ~150 km east of Thunder Bay, Ontario (Fig. 1). It is 7–8 km in diameter and represents the partially eroded central uplift of a complex impact structure 30–32 km in diameter (Halls and Grieve, 1976; Dressler et al., 1995). The size estimate is based on bathymetric information from around the island group outlining an annular trough and a crater rim at a radial distance of about 15–16 km from the center of the central uplift (Halls and Grieve, 1976). This is substantiated by our interpretation of the reprocessed northern part of the Great Lakes International Multidisciplinary Program on Crustal Evolution (GLIMPCE) (Mariano and Hinze, 1994) seismic Line A in Figure 2 (courtesy of B. Milkereit, Geological Survey of Canada, 1994). Evidence for an origin of the structure by asteroid or comet impact is provided by shatter cones, a variety of microscopic shock metamorphic features, impact glasses, and strong

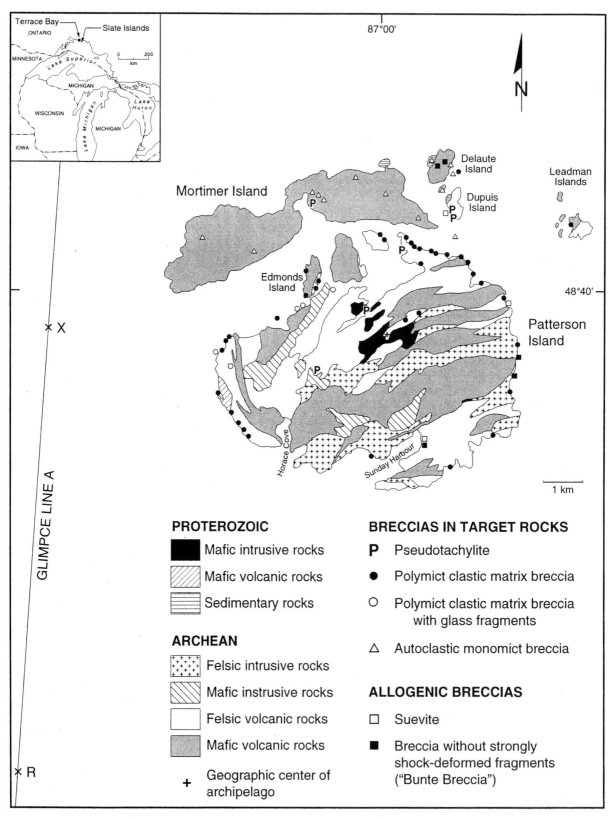

Figure 1. Central uplift of Slate Islands Impact Structure. General geology (Sage, 1991), location of various impact breccias (Dressler and Sharpton, 1997), and location of the northern part of seismic Line A of GLIMPCE (Mariano and Hinze, 1994). Location of geographic center of archipelago is approximate only and different from that of the area of highest shock pressure (Fig. 5). For location of points R and X on the seismic Line A of GLIMPCE, compare with Figure 2 and its caption.

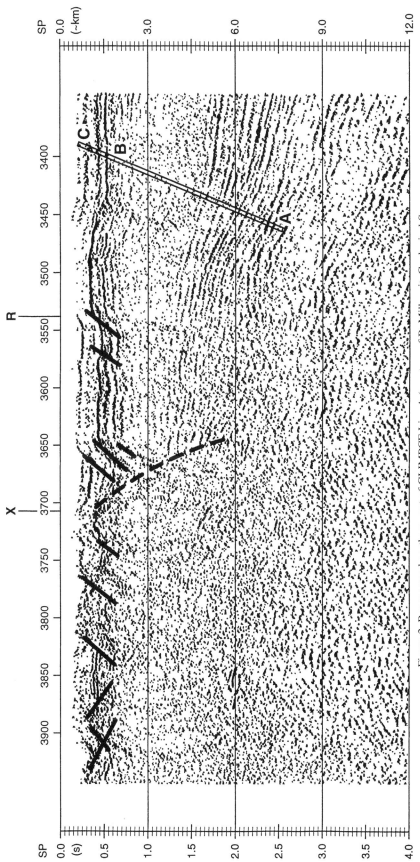

Figure 2. Reprocessed northern part of GLIMPCE Line A (courtesy of B. Milkereit, Geological Survey of Canada, Ottawa, 1994). Left vertical axis is seconds of two-way-time; right vertical axis is approximate depth in kilometers; horizontal axis shows shot points. X is the westward projection of the approximate center of the central uplift. The distance from the center of the central uplift (approximate geographic center of archipelago) to R is 15–16 km. R lies approximately where the rim of the structure is placed based on bathymetry. Compare with Figure 1. The strong reflections at 0.5 s probably represent arenites of the Jacobsville Formation and not multiple reflections of the lake bottom, which is at relatively shallow depth in the area investigated here. AB: Keweenawan basalt; BC: Jacobsville Formation.

target rock brecciation (Halls and Grieve, 1976; Grieve and Robertson, 1976; Dressler and Sharpton, 1997). Sage (1991) mapped the regional geology in considerable detail and recognized a wide spectrum of Archean and Proterozoic target rock units (Fig. 1). Among them are ~2.7-Ga felsic to mafic metavolcanics, metasediments, and felsic to mafic intrusive rocks. Mudstones, cherts, and carbonate-hematite ironstones of the ~1.8-Ga Animikie Group and basalts and minor siltstones and sandstones of the 1.1–1.2-Ga Keweenawan Supergroup represented the upper target stratigraphy. The Archean rocks were subjected to greenschist facies, the Proterozoic ones to subgreenschist to greenschist facies regional metamorphism (Sage, 1991). Fragments of a distinctive, buff to pinkish-brown quartz arenite occur in impact breccias and are believed to be derived from the Jacobsville Formation which, according to Card et al. (1994), is ~800 m.y. old.

The Slate Islands impact structure has been selected for detailed investigations of rock units and macroscopic and microscopic deformations related to the impact process (the locations of the various breccias are shown in Fig. 1). Excellent, large and continuous exposures along the shores of the islands of the archipelago, which are considerably better than in most other terrestrial impact structures, allow detailed investigation of rock units related to impact. We believe this resulted in well-founded interpretations of our field and laboratory observations. Here we attempt to relate brecciation (Dressler and Sharpton, 1997) and shock metamorphism (Dressler et al., 1998) to the various phases of the impact process and to construct an impact model integrating all observations made in the field and laboratory. Details on the petrography, sample locations, and investigative methods applied are presented in Dressler and Sharpton (1997) and in Dressler et al. (1998). Age constraints based on Slate Islands stratigraphy and preliminary radiometric age determinations are also presented.

SLATE ISLANDS IMPACT STRUCTURE AND ITS EFFECTS

Approximately 150 impact structures are presently known on Earth (Grieve et al., 1995), and each year a few more are added to this number. Comet and asteroid impact has become an accepted planetologic process over the last 30 yr following recognition of a vast number of circular structures produced by impact on the solid planets of the solar system. There are simple craters, complex craters, and multi-ring basins. Simple craters on Earth are as large as 2–4 km in diameter and are bowl-shaped. The larger complex craters, such as the Slate Islands structure, have a central uplifted peak or peak ring in their centers, surrounded by an annular depression. The largest impact structures, the so-called multi-ring basin, are characterized by structural rings surrounding the crater. They are known from the Earth's moon and several planets of the solar system. On Earth four large basins exist: the Chicxulub (180–300 km diameter in Mexico; Sharpton et al., 1993; Hildebrand et al., 1995), the Morokweng (possibly 200–>300 km diameter in South Africa; Andreoli et al., this volume; Reimold et al., this volume), the Sudbury (180–250 km diameter in Canada; Deutsch et al., 1994, and references therein), and the Vredefort (~190–335 km diameter in South Africa; Therriault et al., 1993, 1995, and references therein) structures.

Over the last 30 yr, many field and laboratory studies have advanced our understanding of terrestrial impact processes. Here we wish to build on and enhance this understanding by documenting effects of the Slate Islands impact structure on the target units, and assign them to the various stages of the impact process, namely, contact and compression, excavation and central uplift, and crater modification (Melosh, 1989; 1997). Excavation and central uplift represent the most destructive, in part overlapping, phases following contact and compression. Central uplift collapse and isostatic readjustment are believed to be less destructive.

Preimpact target

The Slate Islands archipelago lies close to the northern rim of the North American midcontinental rift system of Proterozoic (~1.1 Ga) age (Sims et al., 1980), which is characterized by mafic rocks of the Keweenawan Supergroup. They overlie 1.8–1.9-Ga Animikie Group ironstones and siltstones, which in turn horizontally overlie deformed and metamorphosed Archean supracrustal and igneous rocks (Sage, 1991) that are ~2.7 Ga (Ontario Geological Survey, 1992). No Proterozoic supracrustal rocks are known to occur on the mainland north of the archipelago; however, a >5-km-thick, more or less flat-lying sequence of Keweenawan and possibly Animikie rocks is present south of Slate Islands beneath the waters of Lake Superior. This is based on GLIMPCE seismic profiles (Mariano and Hinze, 1994) and the reprocessed northern part of GLIMPCE Line A in Figure 2 (courtesy of B. Milkereit, Geological Survey of Canada, 1994). We do not know the preimpact thickness of the Proterozoic rocks in the target area, i.e., we do not know how rapidly the Proterozoic sequences thinned toward the northern rim of the midcontinental rift along GLIMPCE Line A. Therefore, the thickness of the Proterozoic sequences in Figure 3A is only an approximation.

Contact and compression

The contact and compression stage of the impact process (Fig. 3B) is the briefest of the three phases, lasting only few times longer than it takes for the asteroid or comet, arriving at a cosmic velocity of ~15–70 km/s, to penetrate its own diameter into the target rock. At this time the kinetic energy and momentum of the projectile is transferred to the target rock. The initial, compressional shock wave travels with supersonic speed through the target and forces it into downward and outward motion, resulting in the formation of a "transient crater" (Melosh, 1989). It travels faster through the target than the excavation flow, imprinting its effects onto the target before the target rocks become involved in the excavation. The effects due to the immediate passage of the shock wave can be recognized in the parautochthonous target

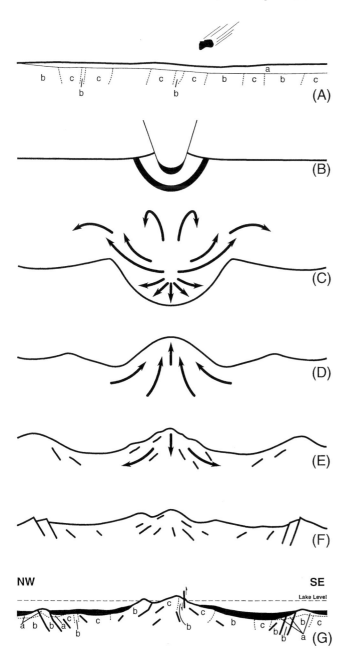

Figure 3. Formation of the Slate Islands impact structure: A, preimpact target; B, contact and compression; C, excavation; D, central uplift; E, central uplift collapse and modification; F, final structure; G, present structure; black areas indicate impact melt overlain by allogenic breccias (assumed, not shown in D–F). a, Proterozoic and younger supracrustal rocks. Deformed Archean greenstone assemblage (assumed in annular trough); b, mafic metavolcanics, minor metasediments and intrusive rocks; c, Intermediate and felsic metavolcanics, minor metasediments, and intrusive rocks.

rocks and in rock fragments that become incorporated into various breccias during subsequent excavation, central uplift, and crater modification stages of the impact event. In general, the supersonic shock wave causes evaporation and vitrification of target material close to ground zero (Pierazzo et al., 1997) and the

Figure 4. Planar deformation features in vein quartz, southern Patterson Island. Crossed polarizers. Scale, 0.2 mm.

formation of microscopic shock metamorphic features and shatter cones around the zone of melting (French and Short, 1968; Stöffler and Langenhorst, 1994, and references therein). Melt-matrix pseudotachylite veins[1] are also formed during the compressional stage of the impact process; we have observed pseudotachylite clasts and rock fragments cut by pseudotachylite veinlets within breccias formed during the subsequent excavation and central uplift stage of the impact process.

A detailed study of about 270 rock thin sections from all across the Slate Islands archipelago allowed us to document various microscopic shock metamorphic features in parautochthonous target rocks and breccia components (Dressler et al., 1998). We recognized scarce maskelynite in a diabase target rock of central Patterson Island (Fig. 1), shock-induced planar fractures and planar deformation features (PDFs) in quartz (Fig. 4), PDFs in feldspar and kink bands in micaceous minerals, the latter not necessarily diagnostic of the passage of a shock wave (Dressler et al., 1998). These shock features were observed mainly in the target rocks of Patterson Island (Dressler et al., 1995; Dressler et al., in 1998). They formed very early in the impact process as they occur in mineral and rock fragments in pseudotachylite veins, also believed to form very early, and in clasts of breccias formed during the subsequent excavation/central uplift stage (Dressler and Sharpton, 1997). The large number of specimens collected on the various islands of the archipelago allowed a detailed study of the distribution of shock metamorphic features across the central uplift. We compared the crystallographic orientation of PDFs in quartz with results from calibrated shock recovery experiments (Hörz, 1968; Müller and Défourneaux, 1968; Reimold and Hörz, 1986; Langenhorst, 1994; Langenhorst and Deutsch, 1994), applying

---

[1] "Pseudotachylite" is here used as a strictly descriptive term for tachylite-like, dark-colored, fine-grained to aphanitic, inclusion-bearing rock regardless of the specific mode of formation. It forms dikes or irregularly shaped bodies and has a glassy, igneous or clastic matrix.

the universal stage method of Engelhardt and Bertsch (1969) on 20 quartz grains with PDFs per sample. We assigned shock pressures to >100 specimens, resulting in the shock attenuation plan of Figure 5. To arrive at this figure we computed average shock pressure values for all data points across the archipelago by assigning values of 7.5, 15.0, and 22.5 GPa to individual planar microstructures parallel to the c (0001), $\varpi\{10\bar{1}3\}$ and $\pi\{10\bar{1}2\}$ crystallographic orientations, respectively. These values represent averages based on various shock experiments (Hörz, 1968; Müller and Défourneaux, 1968; Langenhorst, 1994; Stöffler and Langenhorst, 1994, and references therein). The resulting shock attenuation plan is remarkably irregular and nonconcentric. However, from the areas of highest shock values in central Patterson Island to the western shore of this islands there is a somewhat more regular shock pressure decrease with a gradient of 4.5 GPa/km (Dressler et al., 1998). The results are also plotted in Figure 6, showing a little pronounced decrease in shock pressures from the central area of highest shock (Fig. 5). A shock attenuation plan based on onset pressures, i.e., the first appearance of PDFs parallel to c, ϖ, and π, is also irregular, non-concentric around a shock center assumed to lie in central Patterson Island (Dressler et al., 1998). The irregularity of the shock attenuation plan is regarded as mainly the results of two effects (Sharpton and Dressler, 1997a; Dressler et al., 1998). First, the shock wave did not expand from a point or spherical source. The projectile responsible for the formation of a 30–32 km impact structure (see above) had an estimated diameter of 1–1.5 km and probably was not spherical; this estimate is based on the assumption that the projectile was an asteroid traveling at an average asteroid velocity of ~20–25 km/s. Second, destruction of a shock attenuation plan, that originally was probably more less concentric, occurred during subsequent stages of the impact process (see below). A low-angle, oblique impact would also result in a noncircular shock attenuation plan.

Shock waves of >4 ± 2 GPa result in the formation of shatter cones in crystalline rocks (Roddy and Davis, 1977), also during the compression phase. Shatter cones (Fig. 7A,B) occur on all islands of the archipelago (Stesky and Halls, 1983), indicating that a shock pressure of about 3 GPa specifies the minimum shock pressure that all Slate Islands target rocks were subjected to. We have observed shatter-coned fragments in polymict impact breccias (see below). This is evidence that these conical features, like PDFs, form early during the impact process. Some very spectacular, >10-m-long, shatter cones are exposed on the islands (Fig. 7A) (Sharpton et al., 1996). To our knowledge, they are the largest known anywhere on Earth. Shatter cones with apices pointing in opposite or diverging directions occur, but are not common (Fig. 7B).

Close to ground zero, projectile and target rock experience the highest shock pressures leading to the vaporization and melting of projectile and target material, also very early in the impact process. By analogy with other impact structures such as the Popigai Crater (Masaitis, 1994), melts may form a sheet around the archipelago beneath the waters of Lake Superior, possibly overlain by suevitic breccias (Fig. 3G). Vitrified rock fragments become involved in the brecciation and excavation processes and are found in various polymict breccias on the islands (see below). Likewise, we have observed pseudotachylite-bearing clasts in polymict breccias, suggesting that pseudotachylites also form during the earliest stages of the impact process (Dressler and Sharpton, 1997). These pseudotachylites, however, commonly contain quartz fragments with PDFs, suggesting that melt-matrix pseudotachylite formation is somewhat slower than PDF formation.

Excavation and central uplift

According to the usual impact model, the compressional shock wave created during contact of the projectile with the target expands and eventually weakens into an elastic wave no longer capable of inducing permanent phase changes to rocks and minerals. While the shock and elastic waves expand, the impact crater opens by a much slower excavation flow (Fig. 3C) (Melosh, 1989). A gravity-dominated crater on Earth of the size of the Slate Islands Crater is excavated in a time span of about 1 min. to a depth of ~1.5 km (Melosh, 1989; 1997). Immediately after the formation of the transient crater, and possibly somewhat overlapping with it, elastic rebound and gravity force the central portion of the crater into uplift, resulting in a "central uplift" or central "peak ring" (Fig. 3D). At the Slate Islands impact structure, the uplift possibly reached a height of ~3 km (Grieve et al., 1981; Grieve, 1987). Terraces are formed at the rims of the transient crater, enlarging the crater.

Excavation and ejection result in the formation of allogenic breccias within the crater itself and in the surrounding area. At Slate Islands, due to erosion, no ejected breccias are preserved outside the impact structure north of the archipelago on the mainland. However, relatively small deposits have been noted on the flanks of the central uplift: at Sunday Harbour and the eastern shore of Patterson Island, on Dupuis and on Delaute Islands. They are probably also present in the annular trough around the central uplift. Allogenic, suevitic breccias (e.g., at Sunday Harbour) contain shock metamorphosed Archean and Proterozoic rock and mineral clasts and 3–16 vol% of altered, shard-like glass fragments (Dressler and Sharpton, 1997). The absence of aerodynamically shaped glass fragments from the suevite,[2] coupled with the position of the suevite deposits within the impact structure, characterizes the glass-bearing breccias as "crater suevite" or "fall-back suevite." At the suevite type location, the Ries Crater in Germany, a complex crater some 7 km smaller in diameter than the Slate Islands structure, ~270 m of fall-back suevite have been encountered in a drill hole near the center of the crater (Engelhardt, 1997, and references therein).

At the Slate Islands structure we do not know the original

---

[2] In this chapter we wish to restrict the terms "suevite and "suevitic breccia" to mean allogenic fall-back and ejecta deposits.

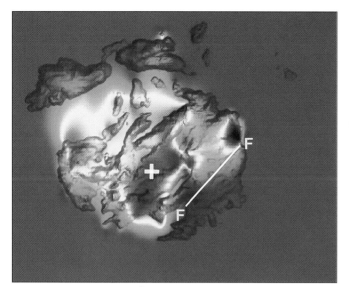

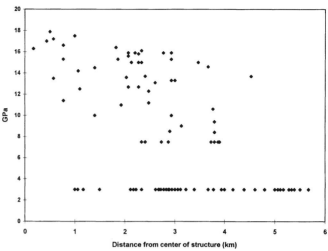

Figure 5. Shock attenuation across the central uplift of the Slate Islands impact structure. Note the irregular, nonconcentric plan and the low pressure area of southeastern Patterson Island. The boundary between this area and a high pressure area just to the west is a relatively straight line extending from Sunday Harbour to the east-central shore of Patterson Island. We interpret this line as a fault along which weakly shocked rocks were uplifted from greater depth during the central uplift stage of the impact process. Highest mean shock pressures of up to 18 GPa are depicted in red. Lower pressures are shown in green and blue. F = Assumed fault; plus sign indicates center of area of highest shock pressure. (After Dressler et al., 1998).

Figure 6. Shock attenuation across the central uplift of the Slate Islands impact structure. The sharp gaps between 7.5 and 3 GPa and less than 3 GPa are the result of the shock pressure calibration method applied. No planar microstructures diagnostic for pressures <7.5 GPa exist. We assigned a shock pressure of 3 GPa to quartz-bearing samples that are shatter-coned but do not exhibit planar microstructures.

Figure 7. Slate Islands shatter cones. Most apices point upward and more or less toward the center of the impact structure (Stesky and Halls, 1983). In a few places they point in two directions, opposite or not. A, shatter cones >10 m long in intermediate metavolcanic rocks; northwestern Patterson Island. B, shatter cones with apices pointing in two, not opposite, directions in Keweenawan metabasalt; eastern Patterson Island. Coin has a diameter of 1.9 cm.

thickness of allogenic breccias nor the exact amount of central uplift above the floor of the annular trough, and therefore cannot easily compare our observations with those at Ries Crater and estimate the amount of erosion that affected the Slate Islands fallback breccia deposits. However, based on the preservation of some fall-back breccias on the archipelago and a comparison of the thickness of breccia deposits at the ~24-km-diameter Ries Crater, we tentatively estimate that erosion at the Slate Islands structure probably did not amount to more than 100–350 m since the formation of the impact crater. Allogenic breccias that do not contain glass fragments or strongly shock metamorphosed mineral and rock fragments at Sunday Harbour and on Delaute Island have been compared with "bunte breccia" deposits at Ries Crater (Dressler and Sharpton, 1997). These Slate Islands breccias consist of clasts derived mainly from Proterozoic, i.e., stratigraphically upper, target rocks. No contacts with suevite have been observed. Within the Ries crater, bunte breccia, an allogenic, clastic matrix breccia consisting of clasts derived mainly from the upper target stratigraphy occurs in the megablock zone (e.g., drill hole Wörnitzostheim) beneath the suevite, but appears not to be very important volumetrically there. It contains virtually no impact glass and strongly shocked target rock fragments. Large and thick deposits of it, however, occur outside the Ries Crater, also beneath suevite.

Bunte breccia and suevite are not the only breccias formed during the excavation/central uplift stage of the impact process. Decompression, especially during the central uplift when target rocks are in a relatively cohesionless state of dilation, allows for brecciation of the parautochthonous target rock of the central uplift and surrounding crater floor. Large, irregularly shaped breccia bodies and dikes between and within target rock blocks are formed, making up an estimated ~15–25% of the rocks that underlie the archipelago (Dressler and Sharpton, 1997). These breccias are polymict, clastic matrix breccias (Fig. 8A–E), have sharp contacts with their host rocks, and contain clasts exhibiting shock metamorphic features, such as PDFs in quartz and feldspar. Some contain scarce, altered glass shards of shock metamorphic origin. All known target rocks have been noted as breccia components, and up to seven clasts of different composition may occur in one breccia body. Rock fragments with PDF-bearing minerals and glass shards may occur in breccia dikes where host rocks apparently did not experience shock pressures high enough for vitrification or the formation of PDFs. These observations all attest to vigorous mixing of clasts and substantial downward and/or outward clast transport mainly away from the center of the impact up to and possibly >2 km. This is the distance between breccia dikes on the Leadman Islands that contain clasts with PDFs and the eastern shore of Patterson Island where the parautochthonous target rocks are strongly shock-metamorphosed. They contain quartz with PDFs. The host rocks of the polymict, clastic matrix breccias on the Leadman Islands and on several islands closer to Patterson Island, however, do not contain these shock features.

The polymict, clastic breccias are assigned by us to the excavation/central uplift stage of the impact process when, as stated above, the target rocks were in a relatively cohesionless state, allowing mixing and turbulent transport of breccia components over considerable distances. The breccias contain fragments with shock features, fragments of pseudotachylite or of rocks cut by pseudotachylite veinlets (Fig. 8C), glass shards, and shatter-coned fragments, which were all formed during the early compressional stage. This stage takes considerably less time (<<1 s) than formation of the transient crater and crater excavation (Melosh, 1989). Central uplift collapse may contribute to further polymict, clastic matrix brecciation, but we believe to a considerable lesser degree.

Crater modification

Following excavation and central uplift, the central peak in large, complex impact structures may collapse, leading to the formation of an internal peak ring or rings (Grieve, 1987; Melosh, 1989) (Fig. 3E,F). At the Slate Islands we observe a single peak structure only. We do not know if the present height of the peak is the result of stratigraphic uplift only or of uplift followed by partial collapse of the central peak and erosion. The uplift is accompanied by inward sliding of terraces at the transient crater rim, which is probably not instantaneous but continues for several minutes or tens of minutes followed by longtime isostatic readjustment across the structure. These late movements and reactions to seismic energy imparted by the impact lead to faulting and minor brecciation. No detailed information on the annular trough and rim of the structure is available. What is known from around the archipelago is solely based on bathymetry and on seismic Line A of GLIMPCE (Fig. 2).

On the outlying islands of the archipelago, on Mortimer and Delaute Islands, and on a small island east of the latter island are monomict, autoclastic breccias that have formed relatively late, during the crater modification and possibly also the central uplift stage of the impact process. In these breccias we have noted pseudotachylites and polymict, clastic matrix breccia dikes that were formed earlier during the central uplift and compression stages of the impact process, respectively. They have been affected by this relatively late, autoclastic brecciation and faulting. These autoclastic breccias are clast-supported and have transitional borders with their host rocks (Dressler and Sharpton, 1997). Fragments are angular, somewhat rotated, and set in a fine, autogene matrix. We did not recognize any diagnostic features, such as polymict, clastic matrix breccia dikes or pseudotachylite dikes cutting across monomict, autoclastic breccia bodies, that would allow us to assign these breccias to another than a late stage of the impact process. Explosive injection of polymict breccia dikes into target rocks removed from, but originating in, the central crater may occur prior to cessation of strong seismic vibration in the surrounding area that is thought to be responsible for the autoclastic brecciation. These autoclas-

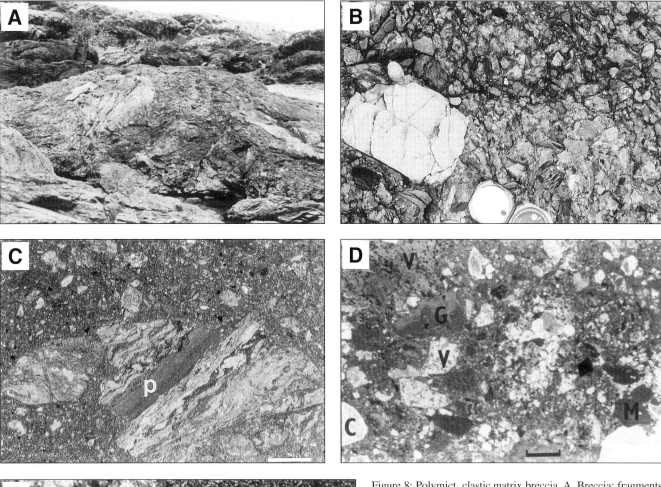

Figure 8: Polymict, clastic matrix breccia. A, Breccia: fragments are Archean supracrustal rocks and Proterozoic metavolcanics. West shore of Patterson Island. Exposure in foreground is ~5 m long. B, Breccia: fragments are angular to subrounded, consisting of Archean mafic and felsic metavolcanic and igneous rocks in a greenish-gray, clastic matrix. From east shore of Patterson Island. C, Photomicrograph of polymict, clastic matrix breccia. Various rock and mineral clasts are set in a fine clastic matrix. One is clast cut by a pseudotachylite veinlet (p). Plane polarized light. Scale, 0.5 mm. D, Polymict, clastic matrix breccia. Rock and mineral clast in a fine-grained, clastic matrix. From the east shore of Patterson Island. Electronically scanned thin section. Scale, 3.5 mm. C = chert; G = metagabbro, M = metabasalt; V = intermediate metavolcanic rock. E, Polymict, clastic matrix breccia dikelet in fine-grained diabase. Clasts of intermediate metavolcanics (most common), quartz, plagioclase, host diabase (arrows), and pseudotachylite in fine clastic matrix. Plane polarized light. Dikelet is about 3 mm wide.

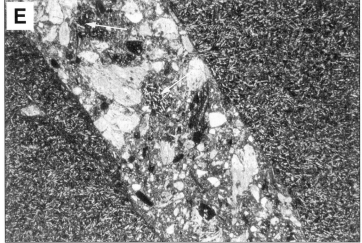

tic fragmental rocks appear to be very scarce on Patterson Island. We believe that late stage movement and seismic vibration on this island during the waning phases of the impact process were possibly accommodated by earlier formed, polymict, clastic matrix breccia bodies, leading in places to the formation of breccia-in-breccia clasts within these early breccias (Dressler and Sharpton, 1997).

Melt Rocks and Age Constraints

Impact melts occur in three forms in terrestrial impact structures (Dence, 1971; Engelhardt, 1972; Grieve et al., 1977; Phinney and Simonds, 1977; Floran et al., 1978; Grieve and Ber, 1994; Dressler et al., 1997): (1) as glassy or recrystallized bombs or fragments in allogenic breccia deposits inside and outside the structure, (2) as sheets and lenses of igneous rocks within the structure, and (3) as dikes intruded into the basement target rocks of the structure. At the Slate Islands structure we have observed altered glass fragments in allogenic breccias and, rarely, in polymict, clastic matrix breccia dikes. Large melt bodies are not exposed on the archipelago, but we speculate that they may be present around the island group beneath the waters of Lake Superior (Fig. 3G). The fine-grained to aphanitic pseudotachylite veinlets formed during the earliest stages of the impact process are products of impact-related melting. A few large, brick-red breccia bodies and dikes with glassy-looking matrices, and a fine-grained, inclusion-rich, gray igneous rock tentatively had also been interpreted in the field as impact-melted rocks. They have been sampled for petrographic, chemical, and, most importantly, geochronologic investigations, because until recently the constraints on the age of the Slate Islands impact structure were weak. They were based on stratigraphic considerations (Sharpton and Dressler, 1996) and the structure's erosion level (Grieve et al., 1995). Sharpton et al. (1996), however, have shown that erosion at the Slate Islands structure was less severe than previously assumed. This interpretation is based on the previously unrecognized presence of allogenic breccias on the archipelago. Halls (1997) argued for ~1,000 m of erosion, and Grieve and Robertson (1997) for ~850–1,350 m, an interpretation refuted by Sharpton and Dressler (1997b) on the basis of a reevaluation of shatter-cone data. In the following section we report on the stratigraphic age considerations and on the results of our petrographic, chemical, and geochronologic studies.

Petrography and chemistry of melt rocks

Most terrestrial impact melt rocks are glassy or fine grained and commonly contain shocked and unshocked mineral and target rock inclusions. They are characterized by disequilibrium textures around inclusions and may exhibit clasts in various stages of melting and assimilation (e.g., Floran et al., 1978; Reimold, 1982; Andreoli et al., this volume; Reimold et al., this volume, and references therein). Geochemically they are surprisingly homogeneous, suggesting a very rapid and efficient melting and mixing process. Their bulk composition reflects a mixture of target rocks. At the Slate Islands structure no large body of a homogeneous melt is exposed. It may be present beneath the waters of Lake Superior, possibly forming dikes in the target rocks, and it may have a chemical composition similar to a mixture of target rocks. We computed the composition of a hypothetical Slate Islands structure melt (Table 1), based on a total of 64 target rock analyses reflecting the weighed aerial extent of corresponding rock units (geological map and chemical data from Sage, 1991). This hypothetical melt may be similar in composition to one or the other of the Slate Islands' rocks interpreted in the field as potential impact melt rocks and to the impact melt assumed to be present in the annular trough around the archipelago beneath the waters of Lake Superior. In the following we provide brief descriptions of two potential "melt rock" types and of pseudotachylite veinlets that were subjected to chemical and geochronologic investigations.

Medium-grained, inclusion-rich dark-gray impact melt-like rock. In one place at the western shore of Patterson Island, the main island of the archipelago, dark-gray, fine-grained, inclusion-bearing, igneous rocks overlie a breccia. Nearby they appear to wrap over Keweenawan target rocks. Macroscopically, they are similar to impact melts. Under the microscope, several features were recognized that have been described from impact melts, e.g., that of the Manicouagan structure (Floran et al., 1978). Among these features are skeletal plagioclase crystals indicative of rapid crystallization, quartz inclusions with disequilibrium reaction coronas, and checkerboard plagioclase inclusions (Dressler et al., 1995). A dendritic mesostasis fills the interstices between plagioclase laths and is suggestive of recrystallized glass. Some plagioclase crystals are filled with recrystallized glass; in microscopic appearance, they are similar to the mesostasis. A chemical analysis of this rock is listed in Table 1 (analysis C). One sample was radiometrically analyzed.

Pseudotachylite. We have noted pseudotachylites in only a few places on the Slate Islands (Dressler and Sharpton, 1997). They are not very conspicuous and can very easily be overlooked in the field. Some of the thin and anastomasing veinlets are relatively inclusion-free while others are not. However, no fragments are larger than a few millimeters across and mainly consists of quartz, which is more resistant to melting than other rock-forming minerals. A few of the quartz fragments exhibit PDFs. The gray to black matrix has flow lines. Mineral and rock clasts exhibit contacts and outlines suggestive of incipient melting. The veins have sharp contacts with the host rocks and, as stated above, we have observed pseudotachylites cut by polymict, clastic matrix breccia dikes and pseudotachylite-bearing target rock clasts in polymict breccias (Fig. 8B). On the basis of observations from other impact structures, pseudotachylites are geochemically similar to their host rocks (Speers, 1957; Dressler, 1984; Müller-Mohr, 1992; Thompson and Spray, 1994, and references therein), i.e., they were formed more or less in situ. At the Slate Islands structure, the investigated samples are somewhat altered, disallowing meaningful chemical comparison with their host rocks. Two small altered pseudotachylite specimens, one relatively inclusion free, the other inclusion rich, were subjected to isotopic analysis.

Dark brick-red impact melt-like rock. Several dark brick-red, inclusion-rich breccias with a commonly aphanitic, in places vesicular, matrix form several exposures up to several meters wide and a number of dikes at the western shore of Patterson Island (Fig. 9). Many of the inclusions are chert-hematite ironstones and are responsible for the conspicuous rock color. Other target rock inclusions, such as Keweenawan basalts, are less common. Under the microscope, mineral and rock clasts with PDFs are common, providing proof that the breccia was formed as the result of impact. However, optical mineralogy does not provide unequivocal evidence that the matrix represents an unaltered or recrystallized impact glass, because the matrix is strongly stained,

TABLE 1. AVERAGE COMPOSITION OF MAJOR TARGET ROCK UNITS AND OF VARIOUS "IMPACT MELTS"

Area (%)	Target Rocks*											
	Unit 1 50.6		Unit 2 26.2		Unit 4 5.6		Unit 5 13.8		Unit 7 0.7		Unit 8 3.1	
	N = 17	σ	N = 19	σ	N = 3	σ	N = 11	σ	N = 6	σ	N = 8	σ
SiO_2	52.81	4.84	65.01	7.02	48.83	0.67	83.48	2.25	48.45	1.03	46.39	3.16
TiO_2	0.99	0.24	0.50	0.15	0.72	0.28	0.57	0.45	2.18	0.35	1.46	0.86
Al_2O_3	14.95	1.34	15.58	1.58	15.43	0.80	15.45	0.80	11.90	2.49	13.47	2.49
Fe_2O_3	1.06	0.59	0.66	0.62	2.45	0.60	1.22	0.41	0.81	1.99	3.42	1.18
FeO	6.71	2.30	1.87	1.30	6.46	0.54	1.78	0.77	5.77	1.41	9.09	2.22
MnO	0.19	0.06	0.07	0.04	0.18	0.01	0.07	0.02	0.20	0.03	0.21	0.06
MgO	5.52	2.24	1.37	1.19	8.20	0.94	1.76	0.56	6.75	1.29	7.80	5.60
CaO	7.79	2.40	3.97	1.78	11.14	3.17	4.20	0.73	10.29	0.35	10.11	2.16
Na_2O	2.61	0.99	4.25	1.28	2.00	0.30	4.14	1.08	1.78	0.24	2.42	0.72
K_2O	0.60	0.52	1.45	0.63	0.14	0.03	1.60	0.50	0.34	0.14	0.86	0.55
P_2O_5	0.10	0.05	0.11	0.04	0.05	0.02	0.15	0.11	0.19	0.03	0.45	0.46
CO_2	2.91	2.01	3.12	2.48	1.10	0.49	3.15	1.67	3.08	1.40	0.78	0.38
S	0.07	0.15	0.04	0.04	0.08	0.03	0.02	0.01	0.02	0.00	0.27	0.37
H_2O^+	3.10	1.13	1.15	0.87	2.25	1.36	1.17	0.54	2.76	0.57	1.89	0.88
H_2O^-	0.57	0.20	0.53	0.13	0.60	0.03	0.38	0.09	1.84	0.39	1.19	0.32
Total	99.97		99.66		99.62		99.14		100.34		99.79	
(ppm)												
Ni	120	83	40	42	107	6	28	9	178	26	183	291
Cr	310	130	76	123	600	383	62	26	520	284	439	771
Co	38	14	11	7	37	3	11	5	56	5	56	19

	"Impact Melts"†				
	A	B	C N = 2	D N = 4 σ	E
SiO_2	56.99	80.80	47.10	55.48 11.64	58.36
TiO_2	0.81	0.86	1.53	0.28 0.07	0.29
Al_2O_3	15.12	16.08	13.84	10.09 8.91	10.61
Fe_2O_3	1.18	1.25	4.57	21.40 7.29	22.5
FeO	4.81	5.11	7.62	1.63 0.40	1.71
MnO	0.14	0.15	0.32	0.15 0.10	0.15
MgO	4.14	4.41	5.61	2.27 0.57	2.38
CaO	5.58	6.97	9.18	3.51 1.21	3.69
Na_2O	3.19	3.39	2.11	0.11 0.16	0.12
K_2O	0.94	1.00	0.49	0.18 0.03	0.19
P_2O_5	0.11	0.12	0.08	0.05 0.00	0.05
CO_2	2.82	n.a.	0.76	2.64 0.83	n.a.
S	0.06	0.06	0.06	0.04 0.01	0.04
H_2O^+	2.22	n.a.	2.36	1.79 0.47	n.a.
H_2O^-	0.55	n.a.	2.21	0.51 0.18	n.a.
Total	99.64	100.00	97.84	100.12	100.09
(ppm)					
Ni	87		98		
Cr	236				
Co	28		49		

*Unit 1 = Mafic to intermediate metavolcanic rocks. Unit 2 = Intermediate to felsic metavolcanic rocks. Unit 4 = Mafic to intermediate intrusive rocks. Unit 5 = Felsic to intermediate intrusive rocks. Unit 7 = Keweenawan mafic volcanic rocks. Unit 8 = Keweenawan diabase. Units 3, 6, and 9 = Very minor areal extent. Rock classifications and original geochemical data of units 1–8, Sarge, 1991.
N = Number of analyses; n.a. = not applicable
†A = Hypothetical impact melt. B = Hypothetical impact melt, LOI-free. C = Dark gray "impact melt." D = Brick-red impact melts. E = LOI-free. Weighted average of units 1–8; σ = standard deviation.
Analyses, mainly by XRF methods, were performed by the Geosciences Laboratories, Ontario Geological Survey, Sudbury, Ontario, Canada.

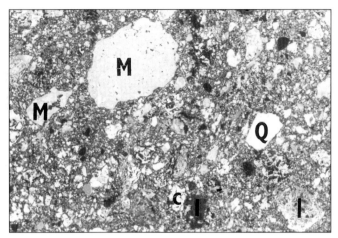

Figure 9. Brick-red "impact melt." Mineral and rock clasts in an aphanitic matrix, west shore of Patterson Island. Electronically scanned thin section. Scale, 3.5 mm. C = Chert; I = Hematite ironstone; M = metabasalt; Q = Vein quartz.

almost opaque, due to the presence of iron oxide minerals. Nevertheless, we subjected one dike specimen to isotopic analysis. Four samples of this impact rock type were investigated geochemically (Table 1, analyses D and E) for comparison with the hypothetical, computed impact melt.

Stratigraphic and other geologic age constraints

Grieve et al. (1995) proposed an age of ~350 Ma for the Slate Islands impact structure based on similarities of the structure's erosion level with that of the Charlevoix structure in Quebec. This structure in eastern Canada has been radiometrically dated at 357 ± 15 Ma (Rondot, 1971; Grieve et al., 1995). Somewhat more rigorous age constraints are provided by stratigraphic considerations (Sharpton et al., 1996, 1997): The youngest target rocks on the islands are either Keweenawan in age or belong to the Jacobsville Formation. According to Card et al. (1994), this formation is ~800 Ma old. Therefore, the structure has a maximum age of 800 or 1,100 Ma. Sage (1991) published K-Ar ages of 310 ± 18 Ma for antigorite and 282 ± 11 Ma for phlogopite of a lamprophyre dike that had been subjected to impact deformation. However, more recent and more reliable U-Pb age determinations of 1.1 Ga (Keweenawan) were obtained from perovskite of the same lamprophyre dike (L. Heaman, University of Alberta; personal communication, 1994). The archipelago lies between the Hudson Bay lowlands and Michigan basins. The wide region between these two Phanerozoic basins was almost certainly underlain by Ordovician to Devonian carbonate rocks (Norris and Sanford, 1968). Therefore, the absence of fragments of Phanerozoic carbonate rocks from Slate Islands impact breccias suggests an age of >350 Ma for the impact event. The nonisotopic age constraints are summarized in Table 2.

Isotopic age constraints

Various geochronologic methods have been applied in the study of impact structures; the K-Ar, Rb-Sr, U-Pb, and ^{40}Ar-^{39}Ar methods are the most common (Bottomley et al., 1990; Deutsch, 1990). Here we applied the ^{40}Ar-^{39}Ar method. During the impact

TABLE 2. STRATIGRAPHIC AGE CONSTRAINTS ON THE SLATE ISLANDS IMPACT

Observation	Age Constraint	Comment	Reference*
Brecciated Keweenawan	<1.1 Ga		1, 2
Shock-deformed and brecciated lamprophyre			
K-Ar (antigorite and phlogopite)	<282 to 310 Ma	U-Pb determinations more reliable	3
U-Pb (perovskite)	<1.1 Ga		4
Brecciated sandstone of the Jacobsville Formation	<800 Ma		5
Apparent absence of Ordovician/Devonian rock fragments in impact breccias	>350 Ma	Hudsons Bay Lowlands and Michigan basins were almost certainly connected during Ordovician/Devonian	6, 7
Similarity of Slate Islands erosion level with that of 357 Ma Charlevoix impact structure.	<350 Ma	Erosion levels are variable in various parts of the Canadian Shield	8

*1 = Halls and Grieve, 1976; 2 = Dressler and Sharpton, 1997; 3 = Sage, 1991; 4 = L. Heaman, U. of Alberta; 5 = Card et al., 1994; 6 = Norris and Sanford, 1968; 7 = Sharpton et al., 1996; 8 = Grieve et al., 1995.

process, target rocks melt instantaneously, allowing radiogenic argon to escape. On cooling, the melt rock begins to accumulate new radiogenic argon (Bottomley et al., 1990), and, if the radiometric clock has not been affected by postimpact tectono-metamorphic events, the K-Ar or ^{40}Ar-^{39}Ar methods should result in relatively precise determinations of the impact age. However, inclusions if present in the melt rocks may not be completely outgassed resulting in a mixed age, older than the age of the impact event. Other complications may result when, for example, radiogenic argon degasses from inclusions and becomes trapped elsewhere in the melt (Bottomley et al., 1990). We have isotopically analyzed a total of four rock specimens identified in the field as impact melts and pseudotachylites. Based on the geochronologic results, however, not all samples analyzed appear to be related to the impact event. The isotopic and geochemical results are interpreted as follows.

Geochronologic procedures, results, and discussion

Analytical procedures. ^{40}Ar/^{39}Ar analyses were conducted at the University of Houston's Thermochronology Laboratory. Samples were wrapped in Aluminum foil and stacked inside quartz tubing interspersed with fluence monitor biotite Fe-mica (Govindajaru, 1979) with an assumed age of 307.3 Ma. Included with unknowns and fluence monitors were a K-glass and optical grade CaF_2, used to monitor the magnitude of interfering nuclear reaction on K and Ca. Samples were sent to the Ford Reactor at the University of Michigan where they were irradiated for 120 hr. Measured correction factors were $(^{36}Ar/^{37}Ar)_{Ca} = 0.00026 \pm 2$, $(^{39}Ar/^{37}Ar)_{Ca} = 0.00070 \pm 5$, and $(^{40}Ar/^{39}Ar)_K = 0.022 \pm 2$. Gas was extracted from the fluence monitors K-glass and CaF_2, by heating individual crystals using a 10-W CO_2 laser that completely melted the crystals. Results from four to eight crystals were averaged to obtain the J factors. Heating of unknowns was done in a double-vacuum resistance furnace in up to 30 steps; heating times were generally 12 min. Prior to entery into the mass spectrometer, reactive gases were removed by a 50 l/s SAES getter over 5 min. The isotopic composition of the gas was measured using a Mass Analyser Products 215–50 rare-gas mass spectrometer. Relative abundances of the isotopes were measured using a Johnston electron multiplier by peak hopping through seven cycles; initial intensities were determined by regression to the time of gas admission. Mass spectrometer discrimination was measured by analysis of atmospheric argon from an online pipette system. Line blanks were determined at a variety of temperatures, and data were corrected using the values most appropriate for a given heating step.

Results and discussion. The medium-grained, dark-gray inclusion-rich "melt rock" of western Patterson Island has the macroscopic and microscopic appearance of an impact melt. However, it is silica-poor compared to the hypothetical impact melt composition (compare C with A and B in Table 1). Its composition is relatively close to that of Keweenawan mafic metavolcanics (Unit 7 in Table 1). Therefore, and based on its 1,074 Ma ^{40}Ar-^{39}Ar age (sample 95SL103, in Fig. 10), the rock is not an impact melt but represents a basal, inclusion-bearing basaltic flow similar to those observed at the base of the Keweenawan near Thunder Bay, ~150 km west of the Slate Islands structure (P. Lightfoot, Inco Ltd., Sudbury, Ontario; personal communication, 1996). Sample 95SL13:3e (Fig. 10) is from a similar inclusion-bearing Keweenawan basalt exposed at the eastern shore of Patterson Island. Both sample spectra show evidence of a post-Keweenawan reheating event.

The clast-poor pseudotachylite sample yields an age spectrum (sample 94 B1D2, in Fig. 10) consistent with an impact age of ~436 Ma. It shows a plateau-like pattern across most of its release spectrum with fluctuations between ~420 and ~470 Ma and an integrated age estimate of 436 ± 3 Ma. A clast-rich pseudotachylite sample yielded an early Keweenawan age of 1,240 ± 30 Ma, suggesting little degassing of clasts and possibly little or no melt generation in the pseudotachylite. The spectrum, not shown in Figure 10, as the analyzed Keweenawan specimens, shows evidence of a later reheating event. Because there is no evidence of volcanic or tectono-metamorphic activity in the northern Great Lakes region more recent than the ~1.1–1.2-Ga Keweenawan volcanism (Table 2), the spectrum of the clast-rich pseudotachylite most likely records a ~436-Ma impact event.

The brick-red rocks that form some spectacular large outcrops and several dikes intruding the target rocks on the western shore of Patterson Island are geochemically different from the hypothetical impact melt (compare analyses D and E with A and B, in Table 1). The high standard deviation of analyses D in Table 1 reflects a large and varied inclusion population. Occasional vesicles in the aphanitic matrix appears to support a melt-breccia classification for the rocks. Optical microscopy, however, is not conclusive. The ^{40}Ar-^{39}Ar release spectrum, not

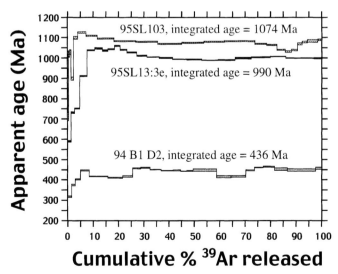

Figure 10. ^{40}Ar-^{39}Ar release spectra. Samples 95SL103 and 95SL13:3e: dark gray, inclusion-bearing "impact melts" (Keweenawan basalt). Sample 94B1D2: inclusion-poor pseudotachylite.

TABLE 3. FORMATION OF THE SLATE ISLANDS IMPACT STRUCTURE

Impact Stage	Contact and Compression	Excavation/Central Uplift	Modification	Long-term Readjustment
		Faulting		
	Pseudotachylite Shock Metamorphic Features Impact Melt Shatter Cones			Hydrothermal Alteration
		Allogenic Breccias		
		Polymict, Clastic Matrix Breccia ???		
		? ? ? ? Monomict, Autoclastic ?? Breccia		
Time (s)	0	10^0	10^1–60	$>10^3$

Breccia	Diagnostic Observations
In target rocks	
Pseudotachylite	Commonly thin dikes and anastomosing veins that have sharp contacts with the host rocks. They contain few clasts, some of which exhibit shock metamorphic features. Observed on Patterson and Mortimer Islands. In places, pseudotachylites are chloritized.
Polymict, clastic matrix breccia	These breccias form dikes and irregularly shaped bodies a few millimeters to more than 10 m in size. Contacts with the host rocks are sharp. They cut across pseudotachylite veins and contain pseudotachylite clasts and fragments cut by pseudotachylite veins. Rock and mineral fragments exhibit shock metamorphic features. Shatter-coned fragments are present. Altered glass fragments occur but are not common. As much as seven rock types have been observed in one breccia body forming clasts of various shapes and sizes. Common on Patterson Island, scare on outlying islands.
Monomict, autoclastic breccia	Angular fragments are set in a clastic matrix. Clasts and matrix have the same composition. Breccia bodies have transitional contacts with host rocks. Fragmented, polymict, clastic matrix breccia bodies and fragments of pseudotachylite veins have been noted in the autoclastic breccias. Mortimer and other outlying islands.
Allogenic breccias	
Suevite	Polymict, clastic matrix breccia containing 3–6 percent of altered, shard-like glass fragments and fragments exhibiting a variety of shock metamorphic features. Glass fragments have no aerodynamic shapes. Absence of these shapes are characteristic of "crater suevite." These allogenic breccias are relatively scarce and have been observed on south and east Patterson Island and Dupuis Island only.
Bunte breccia	This polymict, clastic matrix breccia does not contain glass fragments or strongly shock metamorphosed fragments. Fragments are mainly derived from Proterozoic rocks representing the upper stratigraphy of the target rocks. Relatively scarce; on south Patterson and Delaute Islands only.

Shock metamorphic features	Diagnostic Observations
Microscopic shock metamorphic features	Target rocks exhibit planar deformation features in quartz and feldspar and planar fractures in quartz. Non-shock-diagnostic kink bands in micaceous minerals are common. All these features have been noted in the parautochthonous target rocks and in breccia fragments. Minor maskelynite occurs in target rocks.
Impact melt	Altered glass fragments occur in suevite and to a minor extent in polymict, clastic matrix breccias. An impact melt sheet possibly exists in the annular trough around the Slate Islands beneath the water of Lake Superior.
Shatter cones	Shatter cones occur on all islands of the archipelago. In general, they point upward and range in size from about 2 cm to 10 m.
Hydrothermal alteration	All impact glass fragments in the various breccias are altered to smectite or chlorite. Shock metamorphic quartz, in places, is partially annealed.

shown in Figure 10, is that of a rock in isotopic disequilibrium. It does not reflect the age of the Slate Islands impact but is characteristic of a breccia containing Early Proterozoic and Archean clasts whose radiogenic clock was only partly reset by the impact.

Only the release spectrum obtained from the clast-poor pseudotachylite provides us with a reasonable radiometric age of about 436 Ma for the Slate Islands impact structure. This age supports the stratigraphic age constraints on the Slate Islands impact (Table 2).

CONCLUDING REMARKS

We have investigated various impact breccias and their distribution on the Slate Islands archipelago and correlated the frequency of low index planar deformation features in quartz of >100 specimens with results from shock experiments (Hörz, 1968; Müller and Déforneaux, 1968; Reimold and Hörz, 1989; Stöffler and Langenhorst, 1994). We were able to demonstrate that microscopic shock metamorphic features, shatter cones, impact glasses, and pseudotachylites were formed during the contact and compression phase of the impact process. Polymict, clastic matrix breccia dikes, suevite, and bunte breccia contain fragments exhibiting some or all of these earlier formed features (see above). They were formed during the excavation and central uplift stage of the impact process when target rocks were in a cohesionless state allowing long-range fragment mixing. Central uplift collapse and late stage crater modification resulted in the formation of autoclastic breccias on the outlying islands, while nearer to the center of the crater, earlier formed breccias accommodated late-stage seismic vibration and movement. The shock attenuation plan for the central peak is irregular. We believe this is mainly the result of two effects: the shock wave did not originate from a point or spherical source, and the central uplift, uplift collapse, and crater modification resulted in movement of large target rock blocks, destroying a shock attenuation plan that originally was more or less concentric. Table 3 summarizes various diagnostic observations and how they relate to the phases of the formation of the Slate Islands impact structure.

We collected several potential impact melt rocks on the islands. However, laboratory investigations revealed that not all of the rocks are impact melts. We were able to obtain ^{40}Ar-^{39}Ar release spectra on two pseudotachylite samples that suggest an age of ~436 Ma for the Slate Islands impact. This age is supported by stratigraphic age constraints. Continuous research on the Slate Islands impact structure will focus on additional geochronology, on a more in-depth study of allogenic breccias on the outlying islands of the archipelago, and possibly on an extension of our shock attenuation study. In this investigation we would concentrate on one rock type, namely, vein quartz, to eliminate effects rock types may have had on the formation of shock features.

ACKNOWLEDGMENTS

We thank John Wood, formerly director of the Ontario Geological Survey, for generous logistical support for our 1994 and 1995 field operations. R. R. Herrick, K. Klaus, B. Schnieders, B. C. Schuraytz, and J. Scott provided essential field assistance. The staff of the District Office, Ministry of Natural Resources, Terrace Bay, Ontario, provided us with all possible assistance, including radio communication services. S. Hokanson and D. Rueb assisted in preparations of graphics and photomicrographs. The manuscript benefited from the thorough reviews by W. von Engelhardt and C. Koeberl. Parts of this research was made possibly through research grants provided by the U.S. National Aeronautics and Space Administration's Planetary Geology and Geophysics Program. The Lunar and Planetary Institute is operated by the University Space Research Association under NASA Contract NASW-4574. This is Lunar and Planetary Institute Contribution 945.

REFERENCES CITED

Bottomley, R. J., York, D., and Grieve, R. A. F., 1990, ^{40}Argon-^{39}argon dating of impact craters: Proceedings, 20th, Lunar and Planetary Science Conference: Houston, Texas, Lunar and Planetary Institute, p. 421–431.

Card, K. D., Sanford, B. V., and Davidson, A., 1994, Bedrock geology, Lake Superior, Ontario, USA: Natural Resources Canada; Map NL-16/17-G.

Dence, M. R., 1971, Impact melts: Journal of Geophysical Research, v. 76, p. 5552–5565.

Deutsch, A., 1990, Die Datierung stoßwellenmetamorpher Gesteine und Minerale; Experiment-terrestrische Impaktkrater-lunare Proben: Münster, Germany, Habilitation Schrift, Fachbereich Geowissenschaften, University of Münster, 146 p.

Deutsch, A., Grieve, R. A. F., Avermann, M., Bischoff, L., Brockmeyer, P., Buhl, D., Lakomy, R., Müller-Mohr, V., Ostermann, M., and Stöffler, D., 1994, The Sudbury structure, Ontario, Canada: a tectonically deformed multi-ring impact basin: Geologische Rundschau, v. 84, p. 697–709.

Dressler, B. O., 1984, The effects of the Sudbury event and the intrusion of the Sudbury Igneous Complex on the footwall rocks of the Sudbury Structure, in Pye, E. G., Naldrett, A. J., and Giblin, P. E., eds., The geology and ore deposits of the Sudbury Structure: Ontario Geological Survey Special Volume 1, p. 97–136.

Dressler, B. O., and Sharpton, V. L., 1997, Breccia formation at a complex impact crater: Slate Islands, Lake Superior, Ontario, Canada: Tectonophysics, v. 275, p. 285–311.

Dressler, B. O., Sharpton, V. L., Schnieders, B., and Scott, J., 1995, New observations at the Slate Islands impact structure, Lake Superior: Ontario Geological Survey Miscellaneous Paper 164, p. 53–61.

Dressler, B. O., Crabtree, D., and Schuraytz, B. C., 1997, Incipient melt formation and devitrification at the Wanapitei impact structure, Ontario, Canada: Meteoritics and Planetary Science, v. 32, p. 249–258.

Dressler, B. O., Sharpton, V. L., and Schuraytz, B. C., 1998, Shock metamorphism and shock barometry at a complex impact structure: Slate Islands, Canada: Contribution to Mineralogy and Petrology, v. 130, p. 275–287.

Engelhardt, W. von., 1972, Shock produced rock glasses from the Ries crater: Contributions to Mineralogy and Petrology, v. 36, p. 265–292.

Engelhardt, W. von, 1997, Suevite breccia of the Ries impact crater, Germany: petrography, chemistry and shock metamorphism of crystalline rock clasts: Meteoritics and Planetary Science, v. 32, p. 545–554.

Engelhardt, W. von, and Bertsch, W., 1969, Shock induced planar deformation structures in quartz from the Ries crater, Germany: Contribution to Mineralogy and Petrology, v. 20, p. 203–234.

Floran, R. J., Grieve, R. A. F., Phinney, W. C., Warner, J. L., Simonds, C. H., Blanchard, D. P., and Dence, M. R., 1978, Manicouagan impact melt, Quebec, 1, Stratigraphy, petrology, and chemistry: Journal of Geophysical Research, v. 83, p. 2737–2759.

French, B. M., and Short, N. M., 1968, Shock metamorphism of natural materials: Baltimore, Maryland, Mono Book Corp., 644 p.

Govindajaru, K., 1979, Report (1968-1978) on two mica reference samples: biotite Fe-mica and phlogopite Mica-Mg: Geostandards Newsletter, v. 3, p. 3–24.

Grieve, R. A. F., 1987, Terrestrial impact structures: Annual Reviews of Earth and Planetary Science, v. 15, p. 245–270.

Grieve, R. A. F., and Ber, J., 1994, Shocked lithologies at the Wanapitei impact structure, Ontario, Canada: Meteoritics, v. 29, p. 621–631.

Grieve, R. A. F., and Robertson, P. B., 1976, Variations in shock deformation at the Slate Islands impact structure, Lake Superior, Canada: Contribution to Mineralogy and Petrology, v. 58, p. 37–49.

Grieve, R. A. F., and Robertson, P. B., 1997, New constraints on the Slate Islands impact structure: comments and reply: Geology, v. 25, p. 666–667.

Grieve, R. A. F., Dence, M. R., and Robertson, P. B., 1977, Cratering process: as interpreted from the occurrence of impact melts, in Roddy, D. J., Pepin, R. O., and Merrill, R. B., eds., Impact and explosion cratering: New York, Pergamon Press, p. 791–814.

Grieve, R. A. F., Robertson, P. B., and Dence, M. R., 1981, Constraints on the formation of ring impact structures based on terrestrial data, in Schultz, P. H., and Merill, R. B., eds., Multi-ring basins: New York, Pergamon, 295 p.

Grieve, R. A. F., Rupert, J., Smith, J., and Therriault, A., 1995, The record of terrestrial impact cratering: GSA Today, v. 5, p.189, 194–196.

Halls, H. C., 1997, New constraints on the Slate Islands impact structure: comments and reply: Geology, v. 25, p. 666.

Halls, H. C., and Grieve, R. A. F., 1976, The Slate Islands: a probable complex meteorite impact structure in Lake Superior: Canadian Journal of Earth Sciences, v. 13, p. 1301–1309.

Hildebrand, A. R., Pilkington, M., Connors, M., Ortiz-Aleman, C., and Chavez, R. E., 1995, Size and structure of the Chicxulub crater revealed by horizontal gravity gradients and cenotes: Nature, v. 376, p. 415–417.

Hörz, F., 1968, Statistical measurements of deformation structures and refractive indices in experimentally shock-loaded quartz, in French, B. M, and Short, N. M., eds., Shock metamorphism of natural materials: Baltimore, Maryland, Mono Books Corp., 644 p.

Langenhorst, F., 1994, Shock experiments on pre-heated a- and b-quartz: x-ray and TEM investigations: Earth and Planetary Science Letters, v. 128, p. 683–698.

Langenhorst, F., and Deutsch, A., 1994, Shock experiments on a- and b-quartz, I: Optical and density data: Earth and Planetary Science Letters, v. 125, p. 407–420.

Mariano, J., and Hinze, W. J., 1994, Structural interpretation of the Midcontinental Rift in eastern Lake Superior from seismic reflection and potential field studies: Canadian Journal of Earth Sciences, v. 31, p. 619–628.

Masaitis, V. L., 1994, Impactites from Popigai crater, in Dressler, B. O., Grieve, R. A. F., and Sharpton, V. L., eds., Large meteorite impacts and planetary evolution: Geological Society of America Special Paper 293, p. 153–162.

Melosh, H. J., 1989, Impact cratering —a geologic process: Oxford, United Kingdom, Oxford University Press, 245 p.

Melosh, H. J., 1997, Impact cratering, in Shirley, J. H., and Fairbridge, R. W., eds., Encyclopedia of planetary sciences: London, Chapman & Hall, p. 326–335.

Müller, W. F., and Défourneaux, W., 1968, Deformationsstrukturen in Quarz als Indikator für Stoßwellen: eine experimentelle Untersuchung an Quarzeinkristallen: Zeitschrift für Geophysik, v. 34, p. 483–504.

Müller-Mohr, V., 1992, Breccias in the basement of a deeply eroded impact structure, Sudbury, Canada: Tectonophysics, v. 216, p. 219–226.

Norris, A. W., and Sanford, B. V., 1969, Paleozoic and Mesozoic geology of the Hudson Bay lowlands, in Hood, P. J., ed., Earth science symposium on the Hudson Bay: Geological Survey of Canada Paper 68–53, 169–205.

Ontario Geological Survey, 1992, Chart A —Archean tectonic assemblages, plutonic suites and events in Ontario: Ontario Geological Survey, Map 2579.

Phinney, W. C., and Simonds, C. H., 1977, Dynamic implications of the petrology and distribution of impact melt rocks, in Roddy, D. J., Pepin, R. O., and Merrill, R. B., eds., Impact and explosion cratering: New York, Pergamon Press, p. 771–790.

Pierazzo, E., Vickery, A. M., and Melosh, H. J., 1997, A reevaluation of impact melt production: Icarus, v. 127, p. 408–423.

Reimold, W. U., 1982, The Lappajärvi meteorite crater, Finland: petrography, Rb-Sr, major and trace element geochemistry of the impact melt and basement rocks: Geochimica et Cosmochimica Acta, v. 46, p. 1203–1225.

Reimold, W. U., and Hörz, F., 1989, Textures of experimentally-shocked (5.1 GPa - 35.5 GPa) Witwatersrand quartzite, in Proceedings, 27th, Lunar and Planetary Science, Conference, Abstracts: Houston, Texas, Lunar and Planetary Institute, p. 703–704.

Roddy, D. J., and Davis, L. K., 1997, Shatter cones formed in large-scale experimental explosion craters, in Roddy, D. J., Pepin, R. O., and Merrill, R. B., eds., Impact and explosion cratering: New York, Pergamon Press, p. 715–750.

Rondot J., 1971, Impactite of the Charlevoix structure, Quebec, Canada: Journal of Geophysical Research, v. 76, p. 5414–5423.

Sage, R., 1991, Precambrian geology, Slate Islands: Ontario Geological Survey Report, 264, 111 p.

Sharpton, V. L., and Dressler, B. O., 1996, The Slate Islands impact structure: structural interpretations and age constraints, in Proceedings, 27th, Lunar and Planetary Science, Conference, Abstracts: Houston, Texas, Lunar and Planetary Institute, p. 1171–1178.

Sharpton, V. L., and Dressler, B. O., 1997a, Shock attenuation and breccia formation at a complex impact structure: Slate Islands, Northern Lake Superior, Canada (abstract), in Conference on Large Meteorite Impacts and Planetary Evolution, Abstracts: Houston, Texas, Lunar and Planetary Institute Contribution 922, p. 52.

Sharpton, V. L., and Dressler, B. O., 1997b, New constraints on the Slate Islands impact structure: reply, Geology, v. 25, p. 667–669.

Sharpton, V. L., Burke, K., Camargo-Zanoguera, A., Hall, S. A., Lee, D. S., Marin, L. E., Suarez-Reynoso, G., Quezada-Muñeton, J. M., Spudis, P. D., and Urrutia-Fucugauchi, J., 1993, Chicxulub multiring impact basin: size and other characteristics derived from gravity analysis: Science, v. 264, p. 1564–1566.

Sharpton, V. L., Dressler, B. O., Herrick, R. R., Schnieders, B., and Scott, J., 1996, New constraints on the Slate Islands impact structure, Ontario, Canada: Geology, v. 24, p. 851-854.

Sharpton, V. L., Copeland, P., Dressler, B. O., and Spell, T. L., 1997, New age constraints on the Slate Islands impact structure, Lake Superior, Canada, in Proceedings, 28th, Lunar and Planetary Science, Conference, Abstracts: Houston, Texas, Lunar and Planetary Institute, p. 1287–1288.

Sims, P. K., Card, K. D., Morey, G. B., and Peterman, Z. E., 1980, The Great Lake tectonic zone —a major crustal structure in central North America: Geological Society of America, Bulletin, v. 91, p. 690–698.

Speers, E. C., 1957, The age relationship and origin of common Sudbury Breccia: Journal of Geology, v. 65, p. 497–514.

Stesky, R. M., and Halls, H. C., 1983, Structural analysis of shatter cones from the Slate Islands, northern Lake Superior: Canadian Journal of Earth Sciences, v. 20, p. 1–18.

Stöffler, D., and Langenhorst, F., 1994, Shock metamorphism of quartz in nature and experiment: 1. Basic observations and theory: Meteoritics, v. 29, p. 155–181.

Therriault, A. M., Reid, A. M., and Reimold, W. U., 1993, Original size of the Vredefort structure, South Africa, in Proceedings, 23rd, Lunar and Planetary Science Conference, Abstracts: Houston, Texas, Lunar and Planetary Institute, p. 1419–1420.

Therriault, A. M., Grieve, R. A. F., and Reimold, W. U., 1995, How big is Vredefort?: Meteoritics, v. 30, 586–587.

Thompson, L. M., and Spray, J. G., 1994, Pseudotachylite rock distribution and genesis within the Sudbury impact structure, in Dressler, B. O., Grieve, R. A. F., and Sharpton, V. L., eds., Large meteorite impacts and planetary evolution: Geological Society of America Special Paper 293, p. 275–287.

MANUSCRIPT ACCEPTED BY THE SOCIETY DECEMBER 16, 1998

Lycksele structure in northern Sweden: Result of an impact?

H. Thunehed and S.-Å. Elming
Department of Applied Geophysics, Luleå University of Technology, 971 87 Luleå, Sweden
L. J. Pesonen
Laboratory for Palaeomagnetism, Geological Survey of Finland, P.O. Box 96, FIN-02151 Espoo, Finland

ABSTRACT

The Lycksele structure in northern Sweden is a large circular structure with a diameter of ~130 km. The structure has been defined from a combined analysis of topography, gravity, and magnetic data and is characterized by a circular system of faults, arc-shaped contacts between rocks, a quasi-circular system of downfaulted low-density granitic rocks, and an uplift of high-density rocks in the center. On the basis of ages of granite intrusions and a thrust zone that is cut by faults of the ring, together with paleomagnetic data, the age of the structure is between 1.80 and 1.26 Ga. The Bouguer gravity anomaly is similar to that associated with other known large impact structures on Earth. The rim of the structure, defined from the gravity gradients, magnetic data, and topography, is not restricted to specific rock types but cuts regional geologic structures as well as smaller intrusions, which may be expected for an impact structure in Precambrian deformed target rocks. Therefore, an impact origin seems most plausible for this structure, but to confirm this idea we need identifications of shock metamorphic features in the rocks.

INTRODUCTION

The current data base of impact structures in Fennoscandia includes 30 proven impact craters (Pesonen, 1996) of which the largest one is the Siljan Ring structure with a diameter of 55 km and an age of ~365 Ma. This situation is in contrast to the global database of impact structures, which reveals some 10 huge structures ($D > 100$ km) of which the largest ones, such as Sudbury in Canada ($D = 240$ km) and Vredefort in South Africa ($D = 300$ km) (Grieve and Pesonen, 1996; Henkel and Reimold, 1997), are found in the Precambrian shields. Witschard (1984) was the first researcher to look for such structures in Fennoscandia, using remote sensing data and to stress their economic importance. Therefore, we initiated a search program for large and old impact structures (Pesonen, 1996). They may, however, be difficult to recognize because of deformation and erosion.

The generally accepted criteria (Grieve and Pesonen, 1996) that would prove an impact origin for these old structures are not preserved in many Precambrian structures. Therefore we developed a new four-fold search strategy. First, we identify nearly circular structures on satellite, topographic, or geophysical images and maps. Second, after finding a suspect structure by remote sensing techniques, we try to find macroscopic evidence of impact in the rocks or minerals, such as suevites, impact breccias, pseudotachylite, or impact melt dikes and/or shatter cones. We also examine whether the petrophysical properties of the rocks show a radial variation away from the center of the structure, caused by the radial decay of the shock pressure (Grieve and Pesonen, 1992), as has been found in the case of several impact structures (e.g., Ries, Siljan, Lappajärvi, Iso-Naakkima, Kärdla) (Pilkington and Grieve, 1992; Plado et al., 1994; Pesonen, 1996). Third, although we are looking at a deep erosional level of an old structure, we search for such shock features that are resistant to postimpact metamorphism (e.g., Brazil twins in quartz) (Langenhorst and Deutsch, 1998). Fourth, paleomagnetic, isotopic, and geochemical analyses of rocks will be carried out to search for shock remanent magnetization, anomalously young ages, or geochemical fingerprints of an impact.

The Lycksele structure in northern Sweden is our first suspect structure (see, for example, Witschard, 1984; Nisca, 1995;

Pesonen, 1996), since it passes the first set of criteria for an impact origin to be described in this chapter. Later studies will show whether it fulfills the other three sets of criteria.

TOPOGRAPHY AND GEOPHYSICAL DATA

Figure 1 is a composite of the topography, aeromagnetic, gravity, and geologic maps of the Lycksele structure, showing that it fulfills the first set of criteria for an impact-generated structure (Nisca, 1995; Pesonen, 1996; Nisca et al., 1997).

The structure, with a diameter of ~130 km, appears as an arc-shaped possible rim on the topographic map, with a low land area inside it in the northeast (Fig. 1A). Analyses of satellite images (Witschard, 1984; Pesonen, 1996) and studies of drainage pattern (Nisca, 1995) also reveal a system of concentric circular structures interpreted as fault and fracture zones.

The aeromagnetic data (Fig. 1B) show a weak arc-shaped magnetic relief appearing in the northeastern and eastern part of the structure. Rock magnetic data suggest that the weak magnetic field inside the structure has its origin in a Proterozoic granite, and that the somewhat stronger field outside the structure may be related to migmatites. The magnetic features correlate very well with the topographic pattern. In the southeast there is a difference in magnetic pattern between migmatite inside the structure and migmatites outside it (Fig. 1B). This may indicate a difference in metamorphic grade and erosion level, with a more pronounced pattern of ductile deformation outside the structure. The good correlation between topographic lows and magnetic gradients indicates the contacts to be due to faulting.

The rocks in the center of the structure are mainly low-density (~2,670 kg m^{-3}) (Fig. 2A) Proterozoic granite, which is also the dominant rock type in most of the vicinity of the structure (Fig. 1E). Granodiorite, gabbro, and migmatite of higher density are found predominantly to the north and east outside the structure.

In the Bouguer gravity map, a prominent gravity high (130 µm s^{-2}) coincides with the central parts of the structure and gravity lows (–200 µm s^{-2}) characterize its edges (Fig. 1C). The gravity lows form almost a semicircle except for the southernmost part where a belt of gravity highs truncates the quasi-circular gravity pattern. The horizontal gradient map of the Bouguer data exhibits shallower structures. On the gradient map (Fig. 1D), the rim of the structure is very well defined and can also be traced through the gravity highs in the south. Topography and magnetic relief coincide with the structure as defined by gravity.

A noticeable feature in the area is the occurrence of various generations of dike swarms (Figs. 1B,E). A set of north-northwest–trending dikes of post-Jotnian age (~1.26 Ga) not clearly seen on the aeromagnetic map cuts the structure, whereas another set of dikes, striking northeast, approach the structure from the southwest but appear to be truncated by the structure. We do not yet have any radiometric ages or paleomagnetic data of the northeast-trending dikes.

The gravity data have been interpreted along two profiles running west-northwest (profile 1) and northeast (profile 2) through the structure (Figs. 1E and 2A,B). From the distribution of rock densities along profile 1 (Fig. 2A), it is clear that the maximum in the gravity field in the central part of the structure cannot be explained by density variations of the rocks at the surface. This gravity high is therefore interpreted to have its origin in high-density rocks beneath the low-density Proterozoic granites. The gravity lows that form a part of the ring structure can be interpreted as bodies of downfaulted granite with densities defined by the rocks at the surface. The uplift of the deeper parts of the crust in the southern end of profile 2 (Fig. 2B) is part of a regional thrusting along the Burträsk Shear Zone (Romer and Nisca, 1995) that truncates the structure. The high gradients in the gravity field indicate that the contacts at the edges of the structure are steep.

The high rock densities at ca. 50 km along the profile 1 (Fig. 2A) are related to a ~20 × 15 km^2 big block identified in the magnetic data some 25 km northwest of the center of the structure. The block is interpreted as a thin slab, and the sharp magnetic anomalies delimiting the block may indicate that the contacts to the surrounding rocks to be due to faulting.

DISCUSSION AND CONCLUSION

On the basis of a combined analysis of topography, gravity, and magnetic data, the Lycksele structure is seen to be characterized by a circular system of faults, arc-shaped contacts between rocks, and a quasi-circular distribution of downfaulted low-density granitic rocks.

The fault zone at the edge of the structure is inferred from high gradients in the gravity data, from arc-shaped magnetic anomalies, and from an arc-shaped topographic relief that is also reflected in the drainage pattern (Nisca, 1996). Based on an interpretation of gravity data, the faults are inferred to be vertical and they cut the granitic intrusions at the edge of the structure. The gravity, magnetic, and density data also indicate the rocks inside the structure to be downfaulted with respect to the surrounding rocks. The general feature of the gravity model can be explained by a structure related to an impact. In this model the granite at the edge of the structure has been downfaulted, and high-density

Figure 1. A–E, The Lycksele structure, northern Sweden, as seen in topographic and geophysical maps: A, topography; B, magnetic total field; C, Bouguer gravity field; D, northeast-southwest horizontal gradient of the Bouguer gravity field; E, geologic map of the area. Dark shading represent high values in the magnetic map and negative values in the gravity gradient map; dark values in the topography represent low altitudes. A second-order polynomial trend surface has been removed from the elevation data. Note the arc-shaped pattern in the topography in the northeastern part of (A). An arc-shaped pattern is also seen in the magnetic data. In the gravity data, the structure is evident as a quasi-circular system of gravity lows surrounding a gravity high in the center. The rim of the structure is best defined in the gravity gradient map, from which the diameter (~130 km) is estimated. In the geologic map, the rim is shown by the dotted circle. Also shown are two profiles, 1 and 2, along which the gravity data are interpreted. The Burträsk Shear Zone that truncates the structure in the south is marked with a dashed line.

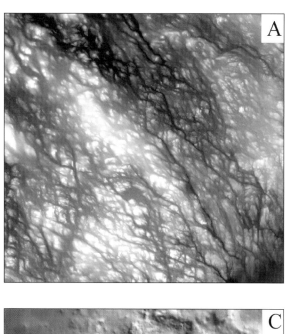
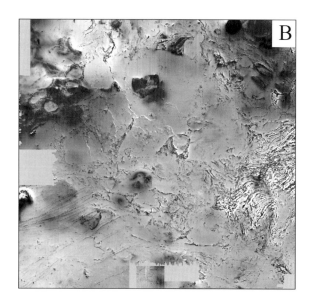
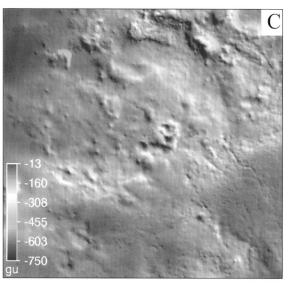
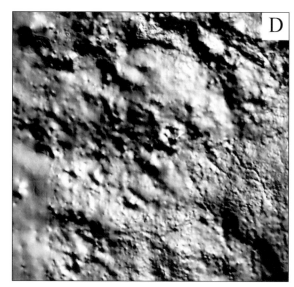
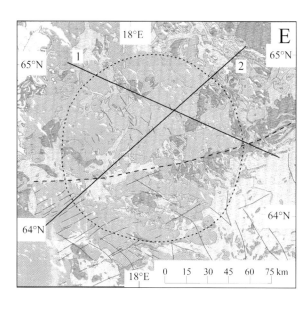
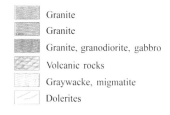

rocks from the deeper part of the crust have been uplifted in the center as may be expected in a large complex impact structure (Grieve and Pesonen, 1992).

The Bouguer anomaly pattern of this structure is similar to gravity anomalies associated with known impact structures (Fig. 3) like the ~2.0-Ga-old Vredefort structure in South Africa (Stepto, 1990), the younger Manicouagon structure in Canada (Sweeney, 1978; Pilkington and Grieve, 1992), and the Chicxulub (Hildebrand et al., 1991) structure in Mexico.

The ages of granites and associated pegmatites range from

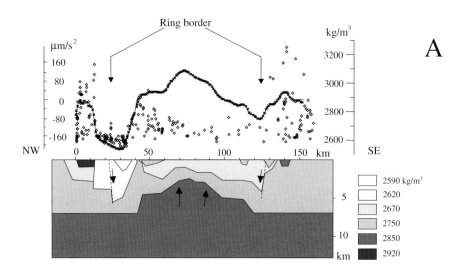

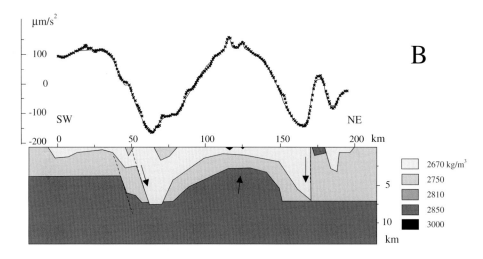

Figure 2. A and B, Residual Bouguer gravity data (crosses) along profiles across the Lycksele structure and interpreted models. A second-order polynomial for a regional field related to the Caledonian mountain range is subtracted from the original data. The location of the profile can be seen in Figure 1E. In (A), the density of rocks, sampled in a segment 6 km on both sides along the profile, is shown by diamonds. Note the absence of high-density rocks at the surface in the central area and the good correlation between rock densities and gravity anomaly at the edges of the structure. The gravity high in the center is interpreted as an uplift of high-density rocks from the lower part of the crust. The gravity low in the northwestern part of profile 1 (A) is related to a low-density granite intrusion that is cut by the ring structure. Note that the inner part of the intrusion is deeper than the granite outside the ring and that the density at the surface is also lower, probably indicating a different erosion level. Also in the second profile (B), the interpreted model is constrained by density data. In the southwest, outside the structure, the high-density rocks of the deeper crust are interpreted to be closer to the surface due to a prestructural uplift along the Burträsk Shear Zone.

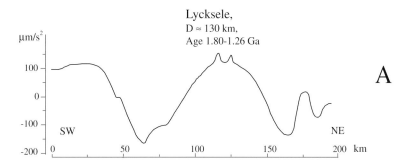
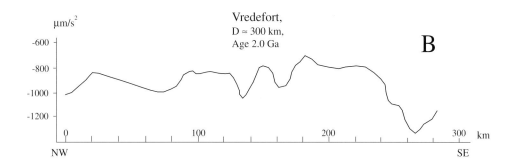
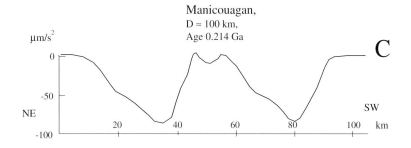
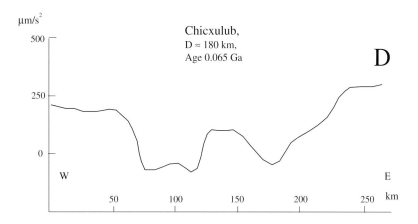

Figure 3. A–D, Comparison of Bouguer gravity anomalies along profiles crossing the Lycksele structure (1.8–1.2 Ga; this work) (A) and the Vredefort (~2.0 Ga) (B), the Manicouagan (0.214 Ga) (C), and the Chicxulub (0.065 Ga) (D) impact structures.

1.82 to 1.78 Ga (Romer and Smeds, 1994, and references therein), and the age of the ductile deformation along the Burträsk Shear Zone that is cut by the ring structure is 1.825 Ga. Based on these age data and on paleomagnetic results of north-northwest–trending dikes (Elming, unpublished data) the age of the structure is inferred to lie in the interval 1.80–1.26 Ga.

As erosion since 1.80–1.26 Ga may be of the order of 10–15 km, as indicated by paleotemperature, radiometric (Tullborg et al., 1996), and paleomagnetic data (Oveisy and Elming, unpublished data), the original topographic expression of a possible complex crater has disappeared. The arc-shaped pattern in the topography is probably due to differential erosion related to fracturing. However, on the basis of the diameter of the structure (~120 km), calculated from the gravity gradient map, remnants of the structural disturbance may still be expected at a depth corresponding to the interpreted model of uplift.

Interpretations of satellite images (Witschard, 1984; Nisca, 1995; Pesonen, 1996) suggest that the structure is a multi-ring structure. Sulfide ore deposits and other mineralizations of the Skellefte ore district skirt the northern area of the structure, and it is possible that they are controlled by the outer rims of the structure (Witschard, 1984).

The rocks have not been studied petrographically or geochemically to see if any traces of shock metamorphism can be found. Although four models for the origin of the structure are possible—namely, basement doming, meteorite impact, large granitic pluton, or fault-bounded block—we believe impact origin to be the most plausible for the following reasons. The gravity field is very similar to that associated with known impact structures, and the interpreted model agrees with a structure expected from a complex impact crater. The almost perfect circular structure, as defined from the gravity gradients, magnetic data, and topography, is not restricted to specific rock types but cuts regional geologic structures as well as smaller granite intrusions, which is also what can be expected for an impact structure in a Precambrian deformed target rock. These features are not likely to have come from basement doming, a large pluton, or a fault-bounded block.

ACKNOWLEDGMENTS

We thank the Swedish Geological Survey for permission to publish the gravity, magnetic, and petrophysical data, and the Swedish National Land Survey for permission to use the topographic data.

REFERENCES CITED

Grieve, R. A. F., and Pesonen, L. J., 1992, The terrestrial impact cratering record: Tectonophysics, v. 216, p. 1–30.

Grieve, R. A. F., and Pesonen, L. J., 1996, Terrestrial impact craters: their spatial and temporal distribution and impact bodies: Earth, Moon and Planets, v. 72, p. 357–376.

Henkel, H., and Reimold, W. U., 1997, Integrated gravity and magnetic modelling of the Vredefort impact structure—reinterpretation of the Witwatersrand basin as the erosional remnant of an impact basin: Royal Institute of Technology, Department of Geodesy and Photogrammetry, Sweden, technical report, p. 1–90.

Hildebrand, A. R., Penfield, G. T., Kring, D., Pilkington, M., Camargo, A., Jacobsen, S. B., and Boynton, W., 1991, Chicxulub Crater: a possible Cretaceous/Tertiary boundary impact crater on the Yucatán Peninsula, Mexico: Geology, v. 19, p. 867–871.

Langenhorst, F., and Deutsch, A., 1998, Mechanical Brazil twins. Mineralogy of astroblemes-terrestrial impact craters: (in press).

Nisca, D. H., 1995, Nya litologiska-tektoniska modeller för regionen Västerbotten-Södra Norrbotten [Ph.D. thesis]: Luleå, Sweden, Luleå University of Technology, 228 p.

Nisca, D. H., Thunehed, H., Pesonen, L. J., and Elming, S.-Å., 1997, The Lycksele structure, a huge ring formation in northern Sweden: result of an impact?, in Conference on Large Meteorite Impacts and Planetary Evolution (Sudbury '97): Houston, Texas, Lunar and Planetary Institute, Abstracts.

Pesonen, L. J., 1996, The impact cratering record of Fennoscandia: Earth, Moon and Planets, v. 72, p. 377–393.

Pilkington, M., and Grieve, R. A. F., 1992, The geophysical signature of terrestrial impact craters: Review of Geophysics, v. 30, p. 161–181.

Plado, J., Pesonen, L. J., Elo, S., Puura, V., and Suuroja, K., 1994, Geophysical research into the Kärdla meteorite crater, Hiiumaa, Estonia, in Törnberg, R., ed., The identification and characterization of impacts: 2nd International Workshop, Lockne-94, Impact Cratering and Evolution of Planet Earth, Abstracts.

Romer, R., and Nisca, D. H., 1995, Svecofennian crustal deformation of the Baltic Shield and U-Pb age of late-kinematic tonalitic intrusions in the Burträsk Shear Zone, northern Sweden: Precambrian Research, v. 75, p. 17–29.

Romer, R., and Smeds, S.-A., 1994, Implications of U-Pb ages of columbite-tantalites from granitic pegmatites for the Paleoproterozoic accretion of 1.90–1.85 Ga magmatic arcs to the Baltic Shield: Precambrian Research, v. 67, p. 141–158.

Stepto, D., 1990, The geology and gravity field in the central core of the Vredefort structure: Tectonophysics, v. 171, p. 75–103.

Sweeney, J. F., 1978, Gravity study of great impact: Journal of Geophysical Research, v. 83, no. B6, p. 2809–2815.

Tullborg, E.-L., Larsson, S.-Å., and Stiberg, J.-P., 1996, Subsidence and uplift of the present land surface in the southeastern part of the Fennoscandian Shield: Geol. Fören. Stockholm Förh., v. 118, p. 126–128.

Witschard, F., 1984, Large-magnitude ring structures on the Baltic Shield—metallogenic significance: Economic Geology, v. 79, p. 1400–1405.

MANUSCRIPT ACCEPTED BY THE SOCIETY DECEMBER 16, 1998

Lake Karikkoselkä impact structure, central Finland: New geophysical and petrographic results

Lauri J. Pesonen and Seppo Elo
Geological Survey of Finland, P.O. Box 96, FIN-02151 Espoo, Finland; lauri.pesonen@gsf.fi

Martti Lehtinen
Geological Museum, Finnish Museum of Natural History, P.O. Box 11, FIN-00014 Helsinki University, Finland

Tarmo Jokinen, Risto Puranen, and Liisa Kivekäs
Geological Survey of Finland, P.O. Box 96, FIN-02151 Espoo, Finland

ABSTRACT

The Lake Karikkoselkä meteorite impact structure is located at 62°13.3′N, 25°14.7′E within the Early Proterozoic granitic terrane in central Finland. This well-preserved simple impact structure has a diameter of ≤2.4 km. The structure owes its discovery to the circular bathymetry of Lake Karikkoselkä. The impact origin of the structure is indicated by extensive fracturing of the target rock, by shock metamorphic features in a breccia boulder, and by shatter cone–like features observed in lake shore outcrops of porphyritic granite. The microscopic shock metamorphic features include multiple sets of planar deformation features in quartz clasts, kink banding in biotite, mosaicism in quartz, and feldspars and incipient vesiculation in feldspars. Geophysical data show distinct anomalies related to the structure. In particular, the structure is associated with exceptionally strong airborne electromagnetic anomalies. The gravity data reveal a low of 3.8 mGal, consistent with the size of this impact structure.

Integrated geophysical data have been used to construct a five-layer model of the structure: (1) lake water, with (2) a thin veneer of mud at its bottom, (3) a sediment and/or impact breccia layer, (4) fractured granite below and beyond the lake, and (5) unfractured target granite (basement). The gravity and the electromagnetic models are mutually consistent and suggest a depth of ~120 m for the bottom of the sediment and/or breccia layer, thus implying an anomalously shallow structure (depth/diameter, <0.1). Paleomagnetic data suggest an age of 230–260 or 530–560 Ma for the impact, but older age interpretations are possible.

INTRODUCTION

Since the first Fennoscandian meteorite impact symposium in 1990 (Pesonen, 1991; Pesonen and Henkel, 1992), there has been an increased search activity for impact structures in Fennoscandia. This has resulted in the discovery of a considerable number of new impact structures. In 1995, we turned our attention to a small lake (Karikkoselkä) of the Petäjävesi water shed, and particularly to the lake's strikingly circular bathymetric contours. A fieldtrip made to the site resulted in the discovery of shatter-cone-like features in boulders and in outcrops around the lake (Lehtinen et al., 1996; Pesonen et al., 1997a). Paleomagnetic and petrophysical sampling of rocks was carried out at these outcrops, and a profile radially away from the lake to the presumably unshocked target rock outcrops was investigated and sampled in detail.

The purpose of these studies was fourfold: (1) to identify shock metamorphic features in the rocks, which, in addition to shatter cones, would prove the impact origin of the structure and would give an estimate for the degree of shock metamorphism (e.g., Grieve and Pesonen, 1992, 1996); (2) to measure petrophysical properties of the rocks to constrain geophysical models and to see if the properties show a gradual change with distance from the lake, as theoretically expected for impact structures (e.g., Pilkington and Grieve, 1992; Plado et al., 1997) (3) to carry out integrated geophysical modeling of the structure for calculation of its morphometric parameters to compare with terrestrial structures of similar size; and (4) to obtain an age for the impact event by applying paleomagnetic dating techniques (Pesonen, 1996b; Pesonen et al., 1996a,b).

In this chapter we summarize the major findings of these studies, which collectively demonstrate that the Lake Karikkoselkä is a small meteorite impact structure of presumably Late Permian–Early Triassic or Cambrian age. A comprehensive report of all the data and interpretations can be found in Pesonen et al. (1998a).

LOCATION AND GEOLOGIC SETTING

Location

The inset map in Fig. 1 shows the location of Lake Karikkoselkä, together with the other eight recognized impact structures in Finland. The structure is clearly visible in the Landsat-TM satellite image (Fig. 1) as a circular lake with a diameter of ~1.5 km. The coordinates of the lake center are 62°13.3′N, 25°14.7′E. The aerial photograph of the lake is shown in Figure 2. Note that there are three small islands (Levänsaaret) in the northeastern part of the lake, but no central island.

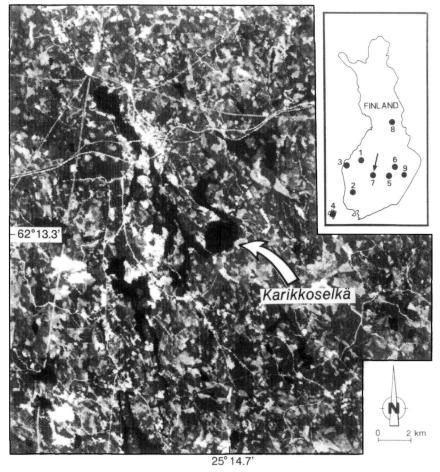

Figure 1. A Landsat-TM color composite image of bands 1, 2 and 3 of the Petäjävesi area, showing the circular Karikkoselkä structure at the center. Inset shows the locations of the recognized impact structures in Finland (modified after Pesonen, 1996a; Pesonen et al., 1997b, 1999a): 1, Lappajärvi; 2, Sääksjärvi; 3, Söderfjärden; 4, Lumparn; 5, Iso-Naakkima; 6, Suvasvesi N; 7, Lake Karikkoselkä (this work); 8, Saarijärvi; and 9, Paasselkä.

Figure 2. Aerial photograph of the Lake Karikkoselkä structure (oblique view from south). The sampling sites are shown with circled letters. Photo by Risto Aalto, Keskisuomalainen, Jyväskylä.

Geologic background and age data

The structure is situated in the southern part of a large granite batholith (Fig. 3A), which belongs to the Paleoproterozoic granite complex of central Finland, part of the 1,870–1,900-Ma Svecofennian orogenic domain (Rouhunkoski, 1959; Korsman et al., 1997). The target rock is a fine-grained, porphyritic, homogeneous granite consisting of microcline, oligoclase, quartz, biotite, traces of hornblende, and opaque oxides including magnetite and hematite (Rouhunkoski, 1959; Wetherill et al., 1962; this study). Two petrologically distinct types of porphyritic granites were found (Fig. 3A). A red porphyritic granite occurs on the southern, eastern, and northeastern sides of the lake (sites K, T and R of Fig. 3). The granite in the northwestern part (sites A, P, F and G in Fig. 3) is more grayish in color and contains more biotite and less hematite than the reddish type. The granite on the islands of Levänsaari (site L in Fig. 3) is intermediate between these two types. The two main granite types have distinct physical properties and are therefore treated separately in the analysis (see, for example, density values in Table 1). Outside the study area the basement consists of, in addition to various granites, granodiorites, tonalites, and quartz diorites, which are somewhat older than the granites and belong to the collisional phase of the Svecofennian Orogeny (Korsman et al., 1997).

Numerous U-Pb ages of the granites of central Finland show that they crystallized during the peak phase of the ~1,880–1,890-Ma Svecofennian Orogeny (Korsman et al., 1997). A U-Pb (zircon) age on the Petäjävesi granite, 5 km northwest of Lake Karikkoselkä, yielded an age of 1,882 ± 2 Ma, whereas an Rb-Sr determination (biotite) of the same rock gave 1,694 Ma (Wetherill et al., 1962; see also Pesonen et al., 1998a). These results are relevant in the interpretation of the paleomagnetic data described below.

The area has been tectonically stable since the Svecofennian Orogeny. However, the rocks show signs of mineral alterations (Rouhunkoski, 1959), which probably can be linked to late Svecofennian fracturing events.

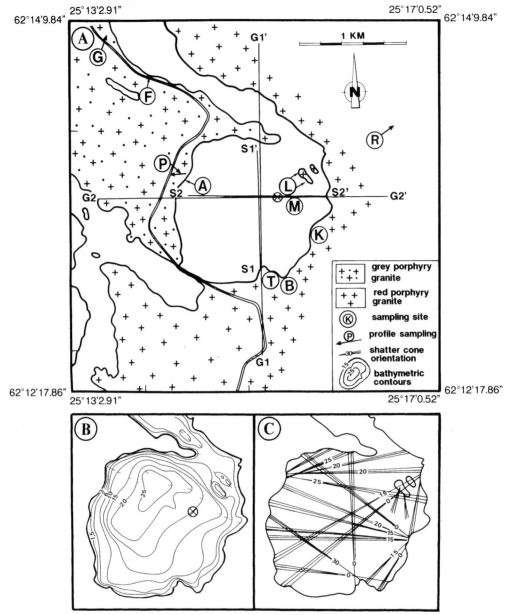

Figure 3. A, Simplified geologic map of the Petäjävesi area. The corner points (with coordinates) of this figure are the same as in subsequent figures. Rock types are shown in the legend. The locations of gravimetric (G–G′) and electromagnetic (S–S′) profiles are indicated. The sampling sites are denoted by letters. Site P denotes a profile (the arrow indicates its length) where rock sampling was done radially away from the shoreline, as explained in the text. This profile extends to the sites F and G farther away to west from the structure proper. Circled cross symbol denotes the site (M) where the Quaternary sediment coring took place. B, The bathymetric contours of the Lake Karikkoselkä (numbers refer to depth in meters). C, Orientation of shatter cone–like features as measured at the outcrops. The fans delineate roughly minimum (left) and maximum (right) directions of the striated conical surfaces projected on the horizontal plane. The numbers refer to upward (positive) or downward (negative) angles of the cone orientations, respectively.

MORPHOLOGIC CHARACTERISTICS

Bathymetry

The lake is anomalously deep (26 m) (Figs. 3B and 4) compared with typical Quaternary lakes in central Finland, which are generally less than 10 m deep. This feature can be considered as evidence for impact, since several other impact structures (but not all) associated with a central lake, have deep bathymetry (e.g., Deep Bay in Canada, Lake Suvasvesi in Finland, and Lake Bosumtwi in Ghana) (Innes et al., 1964; Pesonen et al., 1998b). Moreover, the bathymetric contours of Lake Karikkoselkä show

TABLE 1. PHYSICAL PROPERTIES OF ROCKS FROM THE KARIKKOSELKA IMPACT STRUCTURE

	Site	Material or Rock Type	d (m)	N	δ (kgm^{-3})	Φ (%)	ρ (Ωm)	χ (10^{-6}SI)	NRM (mAm^{-1})	J_{20}/J	Q
1.	L	Lake water	–	–	1000*	–	400	–	–	–	–
2.	M	Lake bottom mud	–	26	1500	–	50	50	–	–	–
3.	B	Impact breccia (boulder)	–	1	2490	7.20	1300	225	2.3	0.74	0.22
4.	A	Fractured *gray* porphyry granite	740	3	2691	1.04	–	249	0.95	0.86	0.30
5.	P	Less fractured *gray* porphyry granite	960	8	2710	1.03	–	297	0.76	0.70	0.08
6.	F, G	Unfractured *gray* porphyry granite	2300	2	2702	1.05	–	366	10.2	0.55	0.66
7.	K, T	Fractured *red* porphyry granite	600	13	2583	2.00	–	214	7.6	1.02	0.53
8.	R	Unfractured *red* porphyry granite	>3000	3	2603	1.51	21000	120	0.72	0.91	0.78

Site, rock types: see Figure 3a and text (*taken from literature, corresponding to 4°C)
d = approximate distance (meters) of the site from the lake center.
N = number of samples (see text)
δ, Φ, ρ = bulk density, porosity, resistivity.
χ, NRM, J_{20}/J = susceptibility, intensity of NRM, intensity of NRM.
Q = Koenigsberger ratio.

TOPOGRAPHY AND BATHYMETRY

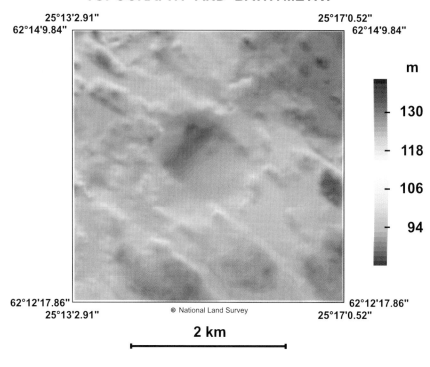

Figure 4. Combined topographic and bathymetric map of the Lake Karikkoselkä area. Colored shaded relief image. Scale bar shows elevation in meters above sea level. Coordinates of the map are the same as in Figure 3A. With permission of the National Land Survey of Finland (permit 13/MAR/98).

striking circularity toward the lake's center; again, this appears to be the case in many other impact structures (e.g., Lake Bosumtwi) (Pesonen et al., unpub. data). We interpret the unusual bathymetry to be a diagnostic feature of an impact structure. It is well-preserved despite several glacial episodes, erosion, and sedimentation.

Topography

The topographic map (Fig. 4) shows distinct northwest–southeast and northeast–southwest striking linear anomalies reflecting regional fractures, which postdate the granite intrusion and which probably have caused the mineral alterations in the granites. These anomalies appear to be truncated by the Lake Karikkoselkä structure. For example, a major northwest–southeast striking fracture at the southern part of the lake does not continue to the northwestern side of the lake, unlike a more southerly parallel fracture. However, these structural features possibly were modified during the slumping phase of the cratering event resulting in the formation of a more polygonal outline of the impact structure compared with a more circular crater originally.

The structure does not seem to have a well-developed topographic rim, or it may have been largely eroded. However, the weak topographic relief encircling the lake is probably the remnant of the rim, and it extends 50–100 m beyond the present lake perimeter. This topographic rim is faintly recognized in the western, northern, and eastern part of the lake (Fig. 4). Gamma radiation data, which have recently been shown to be useful in delineating rim features of impact structures (Pesonen et al., unpub. data), also suggest preservation of remnants of the rim within a few hundred meters inland from the lake perimeter (see below). There is no evidence of a central uplift or of multi-ring features associated with the Lake Karikkoselkä impact structure in the satellite image (Fig. 1) or in the topographic map (Fig. 4).

GEOPHYSICAL CHARACTERISTICS

Description of geophysical techniques

Various geophysical methods, including gravimetric, magnetic, electromagnetic (EM), acoustic, seismic, petrophysical, and paleomagnetic methods, were used to investigate the Lake Karikkoselkä structure. Both high-altitude and low-altitude airborne geophysical methods were applied. The geophysically surveyed area, the profile locations, and the paleomagnetic sampling sites are shown in Figure 3A. The seismic and acoustic profiles for the lake can be found in Pesonen et al. (1998a).

Lake bottom and subsurface soundings

To better constrain the gravity model, we carried out acoustic surveys in the lake in order to determine the topography of the lake bottom and to estimate the thickness of the postimpact Quaternary sediments. The survey was done on the lake with the Geological Survey of Finland (GSF) research boat using acoustic echo sounders, side-scan sonars, and a single-channel seismic reflection system described in Nuorteva (1994).

Deep sampling acquisition and analysis

One ~6-m-long core was taken from the Quaternary lake sediment by the GSF in winter 1996 (site M) (Fig. 3A). Unfortunately, a till layer did not allow sampling of deeper deposits and, consequently, basement was not reached. A petrographic investigation of till clasts did not exhibit any shock metamorphism features. The densities of the sediment were determined in the laboratory (Pesonen et al., 1998A) and these data were used in gravity modeling. We also attempted to locate outcrops or impactite samples from the lake bottom with the help of a scuba diver. Two boulders were recovered, but they turned out to be glacial erratics with no shock features. A local amateur geologist, however, found a breccia sample (boulder) from the southern shore of the lake (site B). This sample (Fig. 5B) turned out to be an impact breccia containing porphyritic granite clasts. Since it is the only sample of an indisputable impact lithology found to date, it is described in detail below.

Paleomagnetic and petrophysical sampling

Oriented samples were collected from all known outcrops around the lake and farther away inland for paleomagnetic and petrophysical studies. At sites A and P in the western part of the lake, we sampled rocks along a profile, about 200 m long, continuing farther to sites G and F located ~2.3 km outward from the center of the lake (Fig. 3A). The purpose of this profile sampling was to see if gradual changes in petrophysical or paleomagnetic properties occur in the rocks due to diminishing shock effects (e.g., Pesonen et al., 1996b; Plado et al., 1997). The paleomag-

Figure 5. A, Fragment of a shatter cone from the site K (boulder). B, Impact breccia sample B (boulder) found near site T, as discussed in the text.

netic measurements were carried out at the GSF laboratory using either the SQUID magnetometer or the Molspin magnetometer, as described in Järvelä et al. (1995) and Pesonen et al. (1998a). Samples were stepwise demagnetized with alternating fields (af) up to 100–150 mT, or thermally to 600°–700°C.

The majority of the petrophysical determinations, including bulk density, susceptibility, intensity of natural remanent magnetization (NRM), and Q-value, were carried out on the same cylindrical specimens studied paleomagnetically in the GSF paleomagnetic laboratory using instrumentation and techniques described in Pesonen et al. (1992). The porosity, electrical conductivity, wet and dry bulk density, and grain density of 20 selected samples in the GSF Petrophysics Laboratory were measured applying techniques described by Puranen (1991) and Kivekäs (1993). For the porphyritic granites located outside the area of Fig. 3A (site R), we used data from the GSF petrophysical database. Densities of the Quaternary lake sediment core were measured in the GSF Petrophysics Laboratory. A few magnetic hysteresis determinations were carried out at the University of Lund, Sweden, to identify the magnetic carriers of the remanent magnetization in the different rock types (see Pesonen et al., 1998a).

Thin-section studies

Several thin sections were prepared for petrographic study to search for shock metamorphic features. The techniques are the same as described in Lehtinen (1976) and Elo et al. (1993).

Airborne geophysical surveys

Older high-altitude airborne geophysical survey data over the Lake Karikkoselkä were available and are described in Pesonen et al. (1998a). The more recent high-resolution low-altitude airborne geophysical mapping includes magnetic, electromagnetic (AEM), and gamma radiation measurements carried out with a Twin Otter DHC-6/300 aircraft, the nominal flight altitude being ~30 m, flight directions north-south and east-west, and the line spacing 200 m. The magnetic and AEM measurements are registered four times per second, corresponding to about 12.5-m point spacing, whereas the radiation is recorded once in a second, corresponding to ~50-m point spacing. The magnetic recording system applies two cesium magnetometers. A multichannel spectrometer with NaI crystals is used for gamma radiation measurements. The AEM system applies a vertical coplanar broadside coil configuration, installed rigidly on wing tips and operating at two frequencies (3125 and 14,368 Hz). All data were processed at the GSF, the values corrected for aircraft disturbances, and regional trends removed.

Ground geophysical surveys

Gravity surveys. Two gravity profiles, G1–G1′ (south-north) and G2–G2′ (west-east) were measured in winter 1996. The standard error of gravity values measured on lake ice with a thermostated Worden Standard Master gravimeter was less than 0.1 mGal. The standard error of elevations measured with a hydrostatic chain level was less than 0.05 m. Water depths were measured at each observation point with a plumb line.

Wide-band electromagnetics. The same profiles as in the gravity survey, with somewhat shorter profile lengths (S1–S1′ and S2–S2′) (Fig. 3A), were measured using the electromagnetic apparatus (SAMPO) of the GSF. The SAMPO EM apparatus is a Finnish-made wide-band electromagnetic system in which the transmitter is a horizontal loop on the ground and the receiver measures the three perpendicular magnetic components of the electromagnetic field at 81 frequencies from 2 to 20,000 Hz (Soininen and Jokinen, 1991; Elo et al., 1992). Using the ratios of the vertical to horizontal field components, the measurements are transformed into apparent resistivity vs. depth curves (Aittoniemi et al., 1987). The SAMPO EM technique has been successfully used before in modeling impact structures, such as the Lappajärvi (Elo et al., 1992) and Iso-Naakkima structures (Elo et al., 1993).

RESULTS

Shock metamorphic features

Shatter cones. The first indication of shock metamorphism was the discovery of poorly developed shatter cones in boulders and also in the lake shore outcrops of the porphyritic granite. The size of the cones ranges from a few centimeters to about 1 m. The cones appear to be portions of fairly well developed shatter cones (Fig. 5A), but sometimes they appear only as conical, striated fan-shaped surfaces. The directions of the in situ shatter cones were measured: the mean directions point generally to the lake and generally upward (Fig. 3C).

There are, however, a number of points to be considered. First, when the shatter cones are broken with a hammer at the outcrops (or the same is done for boulders), smaller shatter cones become visible inside the samples, with cone orientations deviating from that of the major cone. Thus, shatter cone orientation is more random than Figure 3C indicates. We do not have any explanation for this observation. Secondly, the measurements of the cone directions were difficult to make, since the striated features are not fully developed cones, as has also been observed in some African impact structures (W.-U. Reimold, personal communication, 1997).

Rock fracturing. Another observation indicating shock effects is the intense fracturing of the rocks. In outcrops we noted that the rock fracturing is more intense at sites close to the shore line (particularly sites A, K, and T) than at sites farther from the lake (sites P, F and G). Multiple sets of rock-fracturing directions are seen at the outcrops. We interpret the tendency of rock fracturing to increase toward the lake to be caused by the shock and thus to be of impact origin. Consistent with this interpretation, the porosity data (Table 1) show slightly higher values for the fractured granites than for the unfractured ones. We also argue that the fracturing seen at the outcrops is distinct from the huge, linear fracture zones (seen, for example, in the topographic map of Fig. 4), which are presumably due to regional late Svecofennian tectonic events (i.e., preimpact).

Microscopic evidence of shock metamorphism. Twenty samples were studied petrographically to search for shock metamorphic features. These include samples from the shore line exhibiting shatter cones (Fig. 5A) and the impact breccia sample (Fig. 5B).

Several hundred quartz grains were petrographically studied. Up to four sets of planar deformation features (PDFs) per grain were observed. Orientations of the PDFs were estimated in tens of grains. Figure 6 of the breccia sample shows up to three sets of PDFs; the most obvious orientations are (0001) and $\{10\bar{1}1\}$ or $\{11\bar{2}2\}$. Many of the (0001) PDFs are clearly of decorated type, suggesting that the rocks have suffered some postimpact alteration. Other shock effects include kink bands in biotite and mosaicism of quartz and feldspars and incipient vesiculation of feldspars. The granite samples with shatter cones also revealed weakly developed PDFs in quartz and other shock effects. These are, however, much more weakly developed than those seen in the breccia sample. In summary, several lines of petrographic observations provide evidence that the Lake Karikkoselkä structure was formed as the result of hypervelocity meteorite impact.

Paleomagnetism

Four remanent magnetization components (PEF, A, B, and C) were isolated during the demagnetization of the rocks (Table 2; Fig. 7). They are superimposed so that in one specimen two to four components are present. In most cases their coercivities and unblocking temperatures do not overlap, allowing their separation with techniques of multicomponent analysis. Thermal demagnetization data reveal that both magnetite and hematite are carrying the remanences, as also seen in hysteresis data and in thin-section studies (Pesonen et al., 1998a).

The mean paleomagnetic poles of these components have been plotted in Figure 7, showing the Phanerozoic (Fig. 7A) and Proterozoic (Fig. 7B) apparent polar wander paths (APWPs) of Baltica (Elming et al., 1993; Torsvik et al., 1996). Because the curves have self-closing loops, the age interpretation obtained with the APW-path technique is not unique (see Järvelä et al., 1995).

The component with low coercivity is clearly due to the present Earth's magnetic field (PEF) and it is not further discussed. Poles A and B, both with a normal polarity, are of Svecofennian age. Pole A has a magnetization age about 1,880 Ma (Fig. 7A), in good agreement with the recent U-Pb (zircon) age of 1,882 ± 2 Ma from the unshocked Petäjävesi porphyritic granite (Pesonen et al., 1998a). Pole B, interpreted here as a late Svecofennian overprint magnetization at about 1,700 Ma ago (Fig. 7A), is similar to those recently found in Svecofennian and Archean rocks of Finland and Russian Karelia (Mertanen, 1995; Khramov et al., 1997). The Rb-Sr biotite age of 1,694 Ma (Wetherill et al., 1962; Pesonen et al., 1998a) of the same sample dated by the U-Pb technique (see above) is in good agreement with this interpretation. That both A and B are Svecofennian magnetizations is supported by the observations (Table 2; Fig. 8) that they are dominantly found in sites (P, F, and G), which are located farther away from the structure.

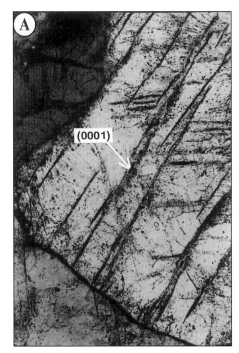 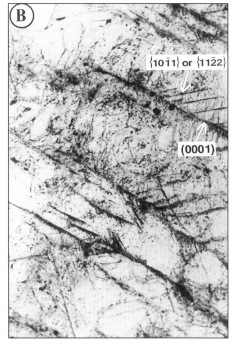

Figure 6. Two examples of quartz grains with PDFs. The PDFs have multiple sets of orientations (up to three) and at least directions parallel to (0001) and $\{10\bar{1}1\}$ and/or $\{11\bar{2}2\}$ have been identified as explained in text. Photomicrographs of thin sections, crossed polarizers. Impact breccia sample of Figure 5B. Width of figure is 0.4 mm.

TABLE 2. PALEOMAGNETIC RESULTS OF ROCKS FROM THE KARIKKOSELKA IMPACT STRUCTURE
(LAT. 62.22°N, LONG. 25.25°E)

Site	Name	Rock Type	d	B/N/n	PEF D, I	Component A D, I	Component B D, I	Component C D, I	Dominant Component
A	Vanha-Karikko	Fractured *gray* p.g.	740	1/4/17	349, 83	313, 58	021, 43	205, -35	C
P	Vanha-Karikko	Less fractured *gray* p.g.	960	1/13/37	55, 75	334, 45	035, 44	210, -47	A+C
F, G	Petäjälahti road	Unfractured *gray* p.g.	2300	1/2/11	73, 83	345, 58	27, 52	260, -35[1]	A
K	Kolu	Fractured *red* p.g.	600	1/5/11	107, 74	339, 38	55, 49	222, -42	C+A
T	Talasniemi	Fractured *red* p.g.	600	1/8/37	57, 75	333, 40	40, 43	231, -53	C
L	Levänsaari	Fractured *gray/red* p.g.	600	1/4/15	36, 38	354, 46	31, 38	226, -52	B+A
	Mean D, I			*6/36/128	51, 74	337, 48	35, 45	226, -46	
	k, α95				(18, 16)	(44, 10)	(72, 8)	(24, 14)	
Paleopole: Plat., Plon.				*6/36/128	66, 98	54, 240	48, 156	-44, 323	
	dp. dm (A95)					9, 13 (11)	6, 10 (8)	11, 18 (14)	

Site, name, rock type: see Table 1, Figure 3a, and text. Note: p.g. denotes porphyry granite.
d = approximate distance (meters) of the site from lake center (-impact center).
B/N/n = number of sites/samples/specimens (*denotes the level used in mean calculations).
PEF, A, B, C = components of NRM isolated by demagnetization treatments as explained in test. [1]faintly found in only one specimen.
Mean D, I = site mean NRM directions of each component.
k, α95 = precision parameter, 95% confidence circle of the mean direction (Fisher, 1953).
Paleopole = position (°N, °E) of the paleomagnetic pole in present-day coordinates calculated from the mean directions.
dp. dm (A95) = semi-axes of the oval of confidence (95% circle of confidence) of the pole.
Dominant component denotes the component(s) which is (are) most frequently isolated in specimens of a particular site (excluding PEF).

Unfortunately, no radiometric dates are so far available from either the shocked granite samples close to the shore or from the impact breccia sample.

Component C, with a reversed polarity, is the most characteristic magnetization. Based on three arguments, we interpret that this component is an impact-generated magnetization: (1) It is the dominant magnetization at sites K, T, and A along the lake shore, which show intense fracturing and shatter coning. In contrast, it is almost totally absent in the most distant sites from the lake (sites F and G). (2) Component C is most often found in those samples that show the reduced densities and susceptibilities, a feature we have previously interpreted to be caused by the impact (cf. Fig. 8). (3) The reversed polarity direction of C is atypical for Svecofennian rocks forming the basement of the area (e.g., Elming et al., 1993; Mertanen, 1995; Khramov et al., 1997).

If the C component proves to be of impact origin, it is not a shock remanent magnetization (SRM), however, as it is not of low coercivity type and does not have the small directional scatter expected for SRM (Halls, 1979). We propose that it is a chemical remanent magnetization (CRM) induced by hot circulating fluids during the impact process (e.g., Elming and Bylund, 1991). These chemical processes have also slightly reduced the susceptibilities in the rocks closer to the lake (Table 1). The reduced, negative aeromagnetic pattern of the structure (Figs. 9A,B and 10) is also consistent with this interpretation.

The age of the C magnetization is still unknown. The paleomagnetic pole position favors a Late Permian–Early Triassic age (230–260 Ma) (Fig. 7A). This interpretation is in agreement with the Mesozoic polarity time scale (Haag and Heller, 1991), which shows a dominance of reversed polarity during the Late Permian. Other possible age assignments for the C component are Early Cambrian (530–560 Ma) (Fig. 7A) or—but less likely—a late Svecofennian age (1,650–1,760 Ma) (Fig. 7B), as discussed previously in connection with component B and with the Rb-Sr age of 1,694 Ma of the porphyritic granite.

Petrophysics

Forty oriented samples from seven sites were collected for physical property determinations (Table 1). Figure 8 shows the susceptibility vs. density plot of the data. Three observations can be made. First, the gray porphyritic granite samples (sites A, P, F, and G) have higher densities and susceptibilities than the more reddish granites (sites K, T, and R) due to their higher contents of mafic minerals and opaque oxides. These results have implications for the gravity modeling. Second, there appears to be a trend that samples from lake shore outcrops (sites K, T, and A) (Table 1), with increased shatter coning and rock fracturing, have lower densities and susceptibilities than samples farther away from the lake (sites F, G, and R). Third, samples taken along a radial profile (sites A, P, F, and G) also follow this trend by showing a decrease of density toward the lake shore outcrops where the rock fracturing and shatter coning is most common. We interpret these trends to indicate a shock-enhanced increase in rock fracturing, which causes the

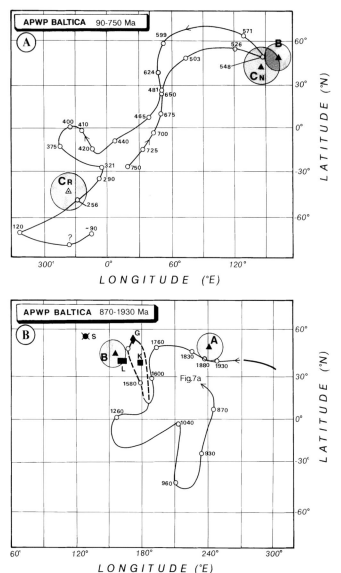

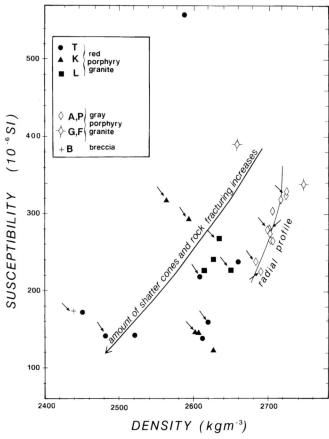

Figure 8. Susceptibility vs. density plot of all samples (mean values). Open (closed) symbols denote gray (red) porphyritic granite, respectively. Data from the radial profile (sites P, A, F, G) are indicated as explained in text. The increasing amount of shatter coning and rock fracturing toward the lake is marked as a trend. The arrows point to samples in which the C component (probably related to impact) of remanent magnetization is clearly identified.

Figure 7. Mean paleomagnetic poles of components A, B, and C plotted on the APWPs of Baltica. A, Phanerozoic APWP of 90–750 Ma by Torsvik et al. (1996), plotted as the south pole curve. Pole C_R denotes the reversed polarity C pole of Table 2, whereas pole C_N is its normal polarity counterpart. Pole B is the pole of component B in Table 2. B, The Proterozoic APWP of 870–1930 Ma by Elming et al. (1993), onto which the poles A and B of this work (Table 2) have been plotted, together with some additional B-type poles from Russian Karelia and from Finland (pole S is from Suoperä dikes; pole L is from layered intrusions; pole G is from Burakovskaja gabbro; and pole K is from the Kolvitsa porphyry (Mertanen, 1995; Khramov et al., 1997).

decrease in density and susceptibility (Pilkington and Grieve, 1992; Pesonen et al., 1992). The porosity values were available from only 22 specimens and are broadly consistent with this trend by showing somewhat higher porosities for the lake shore outcrops (sites K and T). There are not enough porosity determinations to study the correlation between porosity and density in the way described by Kivekäs (1993).

Investigation of the other petrophysical properties (NRM intensity, Q-value, magnetic hardness, J_{20}/J) also showed gradual changes with the radial distance from the lake but not so clearly as the density and susceptibility data (Pesonen et al., 1998a).

High-altitude airborne geophysical results

The high-altitude (flight altitude, 150 m) aeromagnetic map of the Petäjävesi area shows a generally weak magnetic relief with no distinct anomalies related to the Lake Karikkoselkä structure. However, the magnetic field contours are deflected systematically at the center of the lake, thus indicating a horizontal gradient in the aeromagnetics (see Pesonen et al., 1998a). The high-altitude airborne electromagnetic (AEM) data reveal a distinct phase-angle anomaly at the center of the lake (Pesonen et al., 1998a). As the low-altitude (flight altitude, 30 m) data, however, became available for us, we did not further interpret these high-altitude geophysical data.

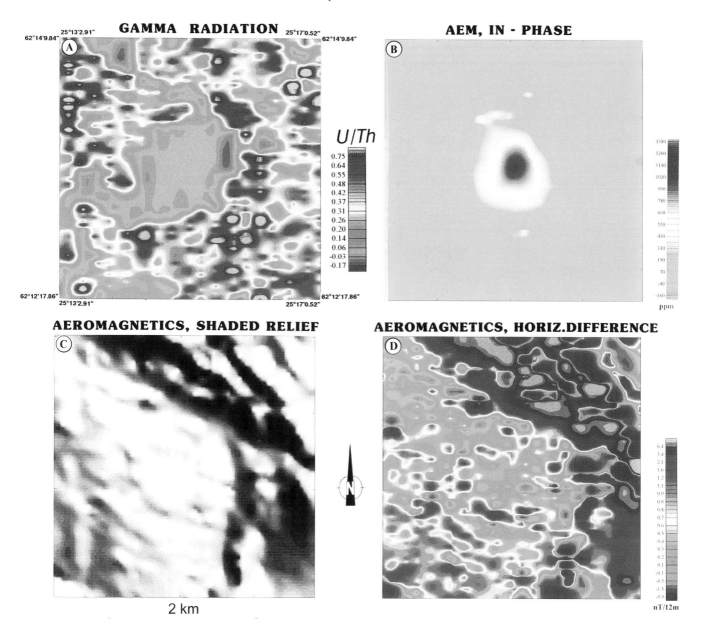

Figure 9. Examples of geophysical contour maps (colored) of the Lake Karikkoselkä structure. A, Gamma-radiation data showing U/Th-ratio. B, In-phase component of low-altitude AEM data. C, Gray-tone shaded relief (illumination from northeast) aeromagnetic map (note: data are not reduced to the pole). D, Horizontal differences (in N-S directions) of the aeromagnetic map of Figure 9C. Anomaly values are shown in the scale bars.

Low-altitude airborne radiation and magnetic results

Figure 9A shows the map of the ratio of uranium and thorium contents (U/Th) of the low-altitude airborne gamma radiation data. The map shows no anomalies in the water-covered areas, as expected. Note, however, possible traces of anomalies located slightly beyond the lake perimeter in the eastern side of the lake, possibly indicating relicts of the rim as observed at other impact structures such as the Lake Bosumtwi structure (Pesonen et al., unpub. data).

Two examples of high-resolution airborne magnetic maps of the Lake Karikkoselkä structure are shown in Figure 9, which is a shaded relief (illumination from the northeast) map of the airborne magnetic data. Although the northern and northeastern parts of the map are affected by strong regional magnetic anomalies, the structure can be seen due to its weak magnetic relief against the strong regional magnetic trends. Figure 9D outlines the horizontal gradient (in N-S direction) map of the airborne magnetic data. Both magnetic maps suggest that the structure extends slightly beyond the lake perimeter, particularly in the south in accordance with the

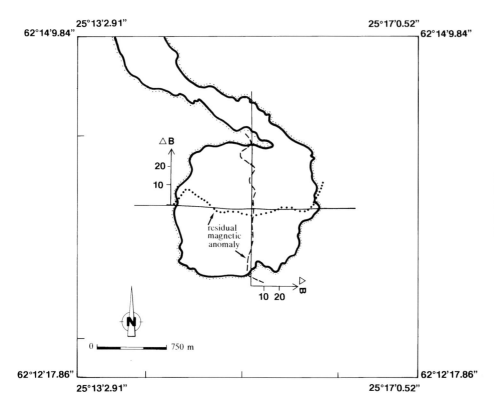

Figure 10. Two aeromagnetic profiles (south-north, west-east), from which the regional aeromagnetic trends have been removed. Coordinates of the corners are the same as in Figure 3A.

gravity (see below) and gamma radiation data (Fig. 9A). However, the presence of the northwest–southeast striking regional fractures in the south, as noted in the analysis of topographic map (Fig. 4), hampers this interpretation.

Two magnetic profiles, with the regional trends removed, are shown in Figure 10. The structure is clearly indicated in the east–west profile by a weak negative magnetic anomaly of ≤10 nT. The north–south profile does not show this anomaly, because the profile follows an anomaly deficient (fracture?) zone seen in the shaded relief map in Figure 9C. We have not carried out magnetic modeling of these profiles, but the negative anomaly can be explained with the paleomagnetic data showing a reversed polarity remanent magnetization at sites around the lake perimeter (component C in Table 2). This reversed remanent magnetization, combined with the present-day–induced magnetization, gives rise to the weak negative magnetic anomaly of the structure.

Low-altitude AEM results

Both the in-phase and out-of-phase AEM maps reveal strikingly circular anomalies related to the structure (Pesonen et al., 1998a). Figure 9B shows the AEM in-phase contour map of the structure with the positive anomaly roughly coinciding with the present lake. These AEM observations serve as diagnostics in searching for or identifying new impact structures in Precambrian terranes with electromagnetic methods (Pesonen et al., 1998a).

Since the Lake Karikkoselkä AEM anomaly covers an area of 200 × 200 m^2, it is sufficiently large for one-dimensional modeling with horizontal layer models. Detailed AEM interpretations were carried out using profile data. Figure 11 shows examples of in-phase (Fig. 11A) and out-of-phase (Fig. 11B) AEM profiles across the structure, which is characterized by strong AEM anomalies. The in-phase value (1,100 ppm) of the low-frequency (3125 Hz) anomaly is significantly higher than the typical value (<500 ppm) of inland lakes in Finland of non-impact origin. The ratio of the in-phase/out-of-phase AEM anomalies of the Lake Karikkoselkä structure is higher than 0.5, unlike in typical inland lakes (<<0.5).

We have carried out one-dimensional layer model interpretations of the AEM data (Pesonen et al., 1998A), and they are consistent with the interpretation resulting from the SAMPO wide-band EM ground survey (below). Another important aspect of the AEM anomalies is that they cannot be simply caused by the lake water and bottom sediments, but call for a deeper lying conductive body, such as a sediment or a breccia layer, as is shown below (see also Fig. 11). Therefore, strong AEM anomalies associated with circular lakes could prove to be diagnostic in the search for impact structures from the AEM maps, which cover more than 70% of Finland.

Wide-band ground electromagnetic results

Two SAMPO EM profiles, S1-S1′ and S2-S2′, were measured with a coil separation of 200 m. Each individual sounding

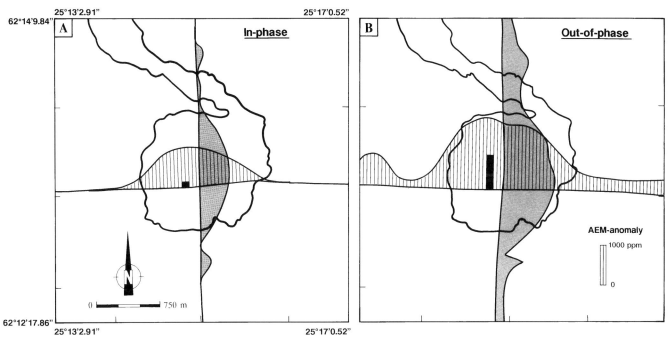

Figure 11. Two examples of high-resolution (low altitude) AEM profiles across the Lake Karikkoselkä structure. The coordinates of these maps match with those of Figure 3A. Low-frequency (3125 Hz) data measured at the altitudes of 26–32 m. Rastered (hatched) profiles are in south-north (west-east) directions, respectively. A, in-phase; B, out-of-phase (quadrature) component of the anomaly. The anomaly scale (in parts per million) is shown as a scale bar. The black bars near the lake center represent the theoretical AEM anomalies due to the water and mud bottom of the lake only, based on layer model computations.

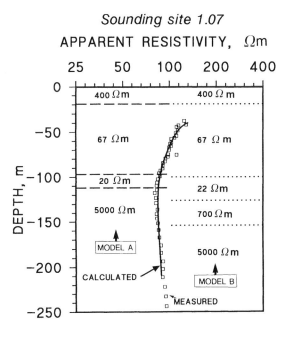

Figure 12. Example of the SAMPO wide-band ground electromagnetic sounding at site 1.07. Horizontal axis is the apparent resistivity and vertical axis is the depth. Site location is shown in Figure 13. Data points (solid line) show the measured (calculated) values, respectively. Two interpretation models (A and B) are shown and discussed in text. In both models, the topmost layer is water (400 Ωm) and the bottom layer is the unshocked target rock (resistivity >5,000 Ωm). In Model A, the middle layer consists of two sublayers (67 and 20 Ωm), whereas in Model B, it is divided into three sublayers (67, 22, and 700 Ωm). This double or triple layer represents either a sediment and/or an impact breccia layer with saline groundwater at its bottom, causing the low resistivities (20-22 Ωm).

curve was interpreted by means of horizontal layer model inversion using a GSF software package. In Figure 12 we show an example (site 1.07) of a typical apparent resistivity vs. depth sounding curve along with two interpretation models involving either four layers (Model A) or five layers (Model B). Both models gave equally good fits between the measured and calculated values. The uppermost layer corresponds to lake water, and its known depth and resistivity were used as fixed values in the parameter optimization. The resistivity of the lake water (400 Ωm in winter at 4°C) was determined from water samples. The interpretations of adjacent soundings were compared with each other to check for structural consistency. A consistency was noted between the two perpendicular soundings that cross each other near the lake center (Fig. 13).

The results of Figure 12 can be summarized as follows. The second layer (65–100 Ωm) beneath the lake water represents either a sedimentary layer (including mud) and/or an impact breccia layer. Below this layer, at the depth of about 100 m, there is a clearly more conductive layer (20–22 Ωm), which could contain saline ground water. Farther down there may be

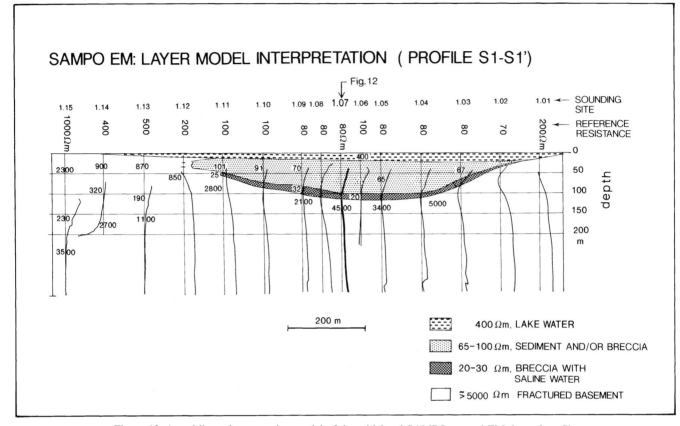

Figure 13. A multilayer interpretation model of the wideband SAMPO ground EM data of profile S1–S1' of Figure 3a. The horizontal scale is same as the vertical scale. The model used is a combination of model A and B of Figure 12, with four layers, from top to bottom: water (400 Ωm), sediment and/or impact breccia (65–100 Ωm), impact breccia with trapped saline water (20–30 Ωm) at its bottom, and fractured target rock (2,000–5,000 Ωm). The reference resistivities denote the values that form the basic resistivity level for each sounding curve.

another layer (700 Ωm), possibly representing a less conductive impact breccia as seen in the MODEL B. The resistivity of the breccia sample was 1,300 Ωm (Table 1), which is not far from the value (700 Ωm) used in Model B. In both models the lowermost layer, with resistivity >5,000 Ωm, represents the fractured basement.

Figure 13 shows the cross section with the four layers (Model B) for the whole profile. Since the model assumes horizontal layering, it works best in the central area of the structure where this assumption is approximately valid. Toward the margins of the structure the biasing effects due to the use of a two-dimensional model instead of a true three-dimensional one become increasingly significant and make the electromagnetic boundaries of the structure less certain. The applied coil separation sets the depth extent of modeling to about 200 m. We note that both models (A and B) give approximately equally good fits with measured data so that the breccia layer in model B could be a real feature even though the modeling does not require it. The estimated bottom of the sediment and/or breccia layer at the center of this model (depth, ~100 m) is compared with a the bottom estimate based on gravity data below.

Gravity results

Figure 14 shows the north-south Bouguer anomaly profile (G1–G1') across the structure. The amplitude of the residual anomaly is about –3.8 mGal, which is in good accordance with observations at other impact structures of this size within crystalline basement (Pilkington and Grieve, 1992; Pesonen et al., 1998a). A 2.5-dimensional interpretation model, with densities taken from petrophysical data in Table 1, consists of five layers, which are, from top to bottom, water (1,000 kgm^{-3}); Quaternary deposits (1,500 kgm^{-3}); sedimentary rocks and/or impact breccias (2,300 kgm^{-3}); slightly fractured bedrock (2,550 kgm^{-3}); and unfractured bedrock (2,650 kgm^{-3}).

The thickness of the Quaternary deposits (layers 1 and 2) was estimated from the results of the acoustic soundings. The third layer in the model is clearly required. It could be a sedimentary rock, as in the Iso-Naakkima structure (Elo et al., 1993), or an impact breccia lens, as in the Sääksjärvi or Lappajärvi structures (Elo et al., 1992), and should have a density of about 2,300 ± 100 kgm^{-3} (Elo, 1997). Only one breccia sample is so far available, and it yielded a slightly higher value of 2,490

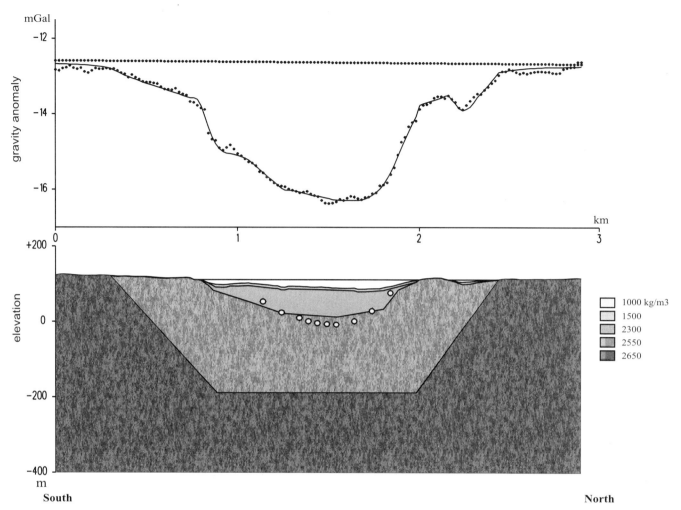

Figure 14. The south-north Bouguer anomaly profile G1–G1' across the lake with a 2.5–dimensional interpretation model consisting of five layers with appropriate densities, as shown in the legend. Dots denote the observed Bouguer anomaly values and also the interpreted regional background; solid line denotes the calculated anomaly. The five layers, from top to bottom, are water, Quaternary deposits (mud), sedimentary rock and/or impact breccia, slightly fractured basement of porphyritic granite, and unfractured basement. Open dots in the vertical section denote the bottom of the most conductive layer in the wideband electromagnetic model in Figure 13. Vertical scale is twice the horizontal scale.

kgm^{-3} (Table 1). Within the limits of all the errors, the bottom of the third layer in the gravity model coincides with the bottom of the most conductive layer in the wide-band electromagnetic model (see Fig. 14). Although the gravity profile shows that the most anomalous parts of the model are closely restricted to the present lake basin, the gravity anomaly tails still require, around and below the sedimentary unit or the impact breccia lens, an additional source with a bulk density of about 2,550 kgm^{-3}, slightly less than that of the unshocked target rock (2,650 kgm^{-3} on the average) (Table 1). This zone has a thickness of about 300–500 m thus increasing the diameter of the anomalous structure from 1.4 km to 2.4 km. The gravity model along the east-west profile (G2–G2') is essentially the same as that along the north-south profile.

DISCUSSION

The data presented in this chapter conclusively demonstrate that the Lake Karikkoselkä structure is a small, well-preserved simple one caused as the result of hypervelocity impact. The evidence includes petrographic observations of shock metamorphism, radial variations of petrophysical and paleomagnetic data, and several geophysical anomalies related to the structure. The data show that the structure has a diameter of ≤2.4 km and is likely to be of Permo-Triassic age (230–260 Ma), although older ages are possible. The Lake Karikkoselkä structure is another "small but old" impact structure in Fennoscandia (Pesonen, 1996a).

The most intriguing observation is the presence of shatter cones in such a small impact structure. According to our knowledge, this is so far the smallest impact structure found with shatter cones, which are generally found only in structures with diameters of more than 5 km. A possibility thus remains that we now observe the deeply eroded roots of a considerably older and larger structure. If this is the case, we might expect to see some relics of the original structure in the geophysical maps, as is the case of the Suvasvesi N structure in Finland (see Pesonen, 1996).

We studied in detail the geophysical data in order to find impact-related features beyond the present lake structure, but with negative results. There are no distinct circular shaped anomaly patterns in the airborne electromagnetic, magnetic, or gamma radiation data much beyond the periphery of the anomalies described in this chapter. We also analyzed digital elevation data in greater detail to find multi-ring features, but none were discovered (Pesonen et al., 1998a). Furthermore, as the petrophysical and paleomagnetic profiles indicate, and the shatter cones and intense rock fracturing show, the shock effects appear to be limited to a ~2.4-km-diameter structure. If the Lake Karikkoselkä shatter cones are true shatter cones, as we think, they must have been developed due to not well-understood impact conditions, possibly related to target rock characteristics. The only explanation we have is that the fine-grained porphyritic granite is somehow exceptionally susceptible to the development of shatter cones at the shock pressures prevailing at the erosional level now exposed, or that the environment conditions of the target rock (dry land, thin veneer of sedimentary material, or thin cover of surface weathering, etc.) were facilitating the development of shatter cones.

CONCLUSIONS

The following conclusions can be drawn from this study:

1. The Lake Karikkoselkä impact structure, located in a Paleoproterozoic granite area of the Fennoscandian Shield, was discovered due to its anomalous lake bathymetry. It is the seventh proven impact structure in Finland.

2. Several lines of evidence indicate that the structure has an impact origin. Among them are shock metamorphic features (particularly multiple sets of PDFs in quartz), presence of shatter cones, extensive fracturing of rocks around the lake, discovery of an impact breccia sample, and distinct geophysical anomalies related to the structure.

3. The paleomagnetic data suggest that the Lake Karikkoselkä impact took place in Late Permian–Early Triassic time (230–260 Ma ago) or in Cambrian time (530–560 Ma ago), but the results are complex and older estimates (1,650–1,760 Ma) are also possible.

4. Integrated geophysical data, constrained with petrophysical determinations of rock samples, reveal that the structure is very shallow (depth/diameter, <0.1). This can be due to a deep erosional level or it could be caused by an impact event of low-energy type.

5. There are no indications in geophysical, morphologic, or satellite data that the structure is larger than 2.4 km. Thus, the presence of shatter cones along the perimeter of the present lake is quite surprising. To our knowledge, the Lake Karikkoselkä structure is the smallest terrestrial impact structure with shatter cones.

6. Two useful criteria are highlighted in a search of new impact structures in Precambrian terranes rich in lakes. First, the AEM anomalies of impact-originated lakes seem to be much stronger than those of ordinary lakes. Second, the bathymetric data, including the depth of the lake and the circularity of the contours are commonly indicative of lakes that are associated with impact structures.

The model was checked by drilling during winter 1998. Eventually we hope to establish the presence of impact breccias within the lake basin and to obtain datable material to better constrain the age of the structure.

ACKNOWLEDGMENTS

A. Hamarus and T. Lahdelma called our attention to the anomalous bathymetry of Lake Karikkoselkä. M. Oksama and I. Suppala provided valuable assistance in geophysical modeling. M. Kurimo, A. Abels, and H. Arkimaa prepared some of the geophysical maps; A. Häkkinen assisted in interpreting the seismic and acoustic data; and T. Manninen helped in the digitizing work. O. Kouvo provided us with the new radiometric age data, and I. Snowball measured the magnetic hysteresis properties. S. Nässling drew the figures, and P. Jelkamäki and J. Väätäinen helped with photography. We appreciate constructive reviews of the manuscript by C. Koeberl and V. Sharpton. Many thanks to all these people.

REFERENCES CITED

Aittoniemi, K., Rajala, J., and Sarvas, J., 1987, Interactive inversion algorithm and apparent resistivity versus depth (ADP) plot in multifrequency depth soundings: Acta Polytechnica Scandinavica, Ph 157, 34 p.

Elming, S.-Å., and Bylund, G., 1991, Palaeomagnetism and the Siljan impact structure, central Sweden: Geophysical Journal International, v. 105, p. 757–770.

Elming, S.-Å., Pesonen, L. J., Leino, M., Khramov, A., Mikhailova, N., Krasnova, A., Mertanen, S., Bylund, G. and Terho, M., 1993, The drift of Fennoscandian and Ukrainian shields during the Precambrian: a paleomagnetic analysis: Tectonophysics, v. 223, p. 177–198.

Elo, S., 1997, Interpretations of the gravity anomaly map of Finland, in Pesonen, L. J., ed., The lithosphere of Finland–a geophysical perspective: Geophysica, v. 13, p. 51–80.

Elo, S., Jokinen, T., and Soininen, H., 1992, Geophysical studies of the Lappajärvi impact structure, western Finland: Tectonophysics, v. 216, p. 99–109.

Elo, S., Kuivasaari, T., Lehtinen, M., Sarapää, O., and Uutela, A., 1993, Iso-Naakkima, a circular structure filled with Neoproterozoic sediments, Pieksämäki, SE Finland: Bulletin of the Geological Society of Finland, v. 65, p. 3–30.

Fisher, R. A. F., 1953, Dispersion on a sphere: Proceedings of the Royal Society, London, v. A217, p. 293–305.

Grieve, R. A. F., and Pesonen, L. J., 1992, The terrestrial impact cratering record: Tectonophysics, v. 216, p. 1–30.

Grieve, R. A. F., and Pesonen, L. J., 1996, Terrestrial impact craters: their spatial and temporal distribution and impacting bodies: Earth, Moon, and Planets, v. 72, p. 357–376.

Haag, M., and Heller, J., 1991, Late Permian to Early Triassic magnetic stratigraphy: Earth and Planetary Science Letters, v. 107, p. 21–42.

Halls, H. C., 1979, The Slate Islands meteorite impact site: a study of shock remanent magnetization: Geophysical Journal of the Royal Astronomical Society, v. 59, p. 553–591.

Innes, M. J. S., Pearson, W. J., and Geuer, J. W., 1964, The Deep Bay crater: Publications of the Dominion Observatory (Ottawa), v. 31, p. 19–52.

Järvelä, J., Pesonen, L. J. and Pietarinen, H., 1995, On palaeomagnetism and petrophysics of the Iso-Naakkima impact structure, southeastern Finland: 2nd Laboratory for Paleomagnetism, Geological Survey of Finland, Espoo, Open File Report Q19/29.1/3232/95/1, 43 p.

Khramov, A. N., Fedotova, M. A., Pisakin, B. N., and Pryatkin, A. A., 1997, Paleomagnetism of lower Proterozoic intrusions and associated rocks in Karelia and the Kola Peninsula: a contribution to the model of the Precambrian evolution of the Russian-Baltic craton: Izvestiya, Physics of Solid Earth, v. 33, p. 447–463.

Kivekäs, L., 1993, Density and porosity measurements at the petrophysical laboratory of the Geological Survey of Finland, in Autio, S., ed., Current Research, 1991–1992, Geological Survey of Finland Special Paper 18, p. 119–127.

Korsman, K., Koistinen, T., Kohonen, J., Wennerström, M., Ekdahl, E., Honkamo, M., Idman, H., and Pekkala, Y., 1997, Bedrock map of Finland: Espoo, Geological Survey of Finland: Scale 1: 1 000 000

Lehtinen, M., 1976, Lake Lappajärvi, a meteorite impact site in western Finland: Geological Survey of Finland Bulletin 282, 92 p.

Lehtinen M., Pesonen, L. J., Puranen, R., and Deutsch, A., 1996, Karikkoselkä—a new impact structure in Finland, in Proceedings, 27th, Lunar and Planetary Science Conference, Abstracts: Houston, Texas, Lunar and Planetary Institute, p. 739–740.

Mertanen, S., 1995, Multicomponent remanent magnetizations reflecting the geological evolution of the Fennoscandian Shield—a paleomagnetic study with emphasis on the Svecofennian orogeny: Espoo, Geological Survey of Finland, 46 p.

Nuorteva, J., 1994, Topographically influenced sedimentation in Quaternary deposits—a detailed acoustic study from the western part of the Gulf of Finland: Espoo, Geological Survey of Finland, Report of Investigation 122, 88 p.

Pesonen, L.J., 1991, Probing impacts in Fennoscandia: EOS (Transactions of the American Geophysical Union), v. 72, p. 11–16.

Pesonen, L. J., 1996a, The impact cratering record of Fennoscandia: Earth, Moon, and Planets, v. 72, p. 377–393.

Pesonen, L. J., 1996b, The geophysical signatures of terrestrial impact craters, in International workshop on "The role of impact processes in the geological and biological evolution of planet Earth": Postojna, Slovenia, Ivan Rakovec Institute of Palaeontology, Scientific Research Centre SAZU, p. 61–62.

Pesonen, L. J., and Henkel, H., eds., 1992, Terrestrial impact craters and craterform structures with a special focus on Fennoscandia: Tectonophysics, v. 216, 234 p.

Pesonen, L. J., Marcos, N., and Pipping, F., 1992, Paleomagnetism of the Lake Lappajärvi impact structure, western Finland: Tectonophysics, v. 216, p. 123–142.

Pesonen, L. J., Lehtinen, M., Deutsch, A., Elo, S., and Lukkarinen, H., 1996a, New geophysical and petrographic results of the Suvasvesi N impact structure, Finland, proceedings, 27th, in Lunar and Planetary Science Conference, Abstract: Houston, Texas, Lunar and Planetary Institute, p. 1021–1022.

Pesonen, L. J., Järvelä, J., Sarapää, O., and Pietarinen, H., 1996b, The Iso-Naakkima metorite impact structure: physical properties and palaeomagnetism [abs.]: Meteoritics and Planetary Science, v. 31 supp., p. A105–A.106.

Pesonen, L. J., Elo, S., Puranen, R., Jokinen, T., Lehtinen, M., Kivekäs, L., and Suppala, I., 1997a, The Karikkoselkä impact structure, central Finland: new geophysical and petrographic results [abs.], in Conference on Large Meteorite Impacts and Planetary Evolution (Sudbury 97): Houston, Texas, Lunar and Planetary Institute, Contribution No. 922, p. 39.

Pesonen, L. J., Lehtinen, M., Tuukki, P., and Abels, A., 1997b, The Lake Saarijärvi—a new meteorite impact structure in northern Finland [abs.], in 9th, Lunar and Planetary Science Conference: Houston, Texas, Lunar and Planetary Institute, 2 p., CD-ROM.

Pesonen, L. J., Elo, S., Puranen, R., Jokinen, T., Lehtinen, M., Kivekäs, L., and Kouvo, O., 1998, The Karikkoselkä impact structure—integrated results and interpretation models: Geological Survey of Finland, Open File Report Q29.1/98/3, Laboratory for Paleomagnetism, 35 p.

Pesonen, L. J., Lehtinen, M., Kuivasaari, T., and Elo, S., 1999, Paasselkä—a new meteorite impact structure in eastern Finland: 62nd, Annual Meeting, Meteoritical Society, Abstracts, (in press).

Pilkington, M., and Grieve, R. A. F., 1992, The geophysical signature of terrestrial impact craters: Reviews of Geophysics, v. 30, p. 161–181.

Plado, J., Puura, V., and Pesonen, L. J., 1997, Effect of shock-induced stress on petrophysical properties of crater floor rocks, Kärdla impact structure, Estonia, in Proceedings, 3rd Nordic Symposium on Petrophysics: Gothenburg, Sweden, p. 108–110.

Puranen, R., 1991, Specifications of petrophysical sampling and measurements, in Autio, S., ed., Current Research 1989–1990, Special Paper 12, Geological Survey of Finland, p. 217–218.

Rouhunkoski, P. S., 1959, Petäjäveden alueen geologica [Phil. Lic. Dissertation]: University of Helsinki, Finland, 143 p.

Soininen, H., and Jokinen, T., 1991, SAMPO, a new wide-band electromagnetic system, in Technical Programme and Abstracts: 53rd, European Association of Exploration Geophysics Meeting and Technical Exhibition, Florence, Italy, p. 366–367.

Torsvik, T. H., Smethurst, M. A., Meert, J. G., Van der Voo, R., McKerrow, W. S., Brasier, M. D., Sturt, B. A., and Walderhaug, H. J., 1996, Continental break-up and collision in the Neoproterozoic and Palaeozoic—a tale of Baltica and Laurentia. Earth-Science Reviews, v. 40, p. 229–258.

Wetherill, G. W., Kouvo, O., Tilton, G. R., and Gast, P. W., 1962, Age measurements on rocks from the Finnish Precambrian: Journal of Geology, v. 70, p. 74–88.

Manuscript Accepted by the Society December 16, 1998

Seismic expression of the Chesapeake Bay impact crater: Structural and morphologic refinements based on new seismic data

C. Wylie Poag, Deborah R. Hutchinson, and Steven M. Colman
U.S. Geological Survey, 384 Woods Hole Road, Woods Hole, Massachusetts, 02543-1598, USA; wpoag@usgs.gov
Myung W. Lee
U.S. Geological Survey, Denver Federal Center, Denver, Colorado 80225-0046, USA

ABSTRACT

This work refines previous interpretations of the structure and morphology of the Chesapeake Bay impact crater on the basis of more than 1,200 km of multichannel and single-channel seismic reflection profiles collected in the bay and on the adjacent continental shelf. The outer rim, formed in sedimentary rocks, is irregularly circular, with an average diameter of ~85 km. A 20–25-km-wide annular trough separates the outer rim from an ovate, crystalline peak ring of ~200 m of maximum relief. The inner basin is 35–40 km in diameter, and at least 1.26 km deep. A crystalline(?) central peak, approximately 1 km high, is faintly imaged on three profiles, and also is indicated by a small positive Bouguer gravity anomaly. These features classify the crater as a complex peak-ring/central peak crater. Chesapeake Bay Crater is most comparable to the Ries and Popigai Craters on Earth; to protobasins on Mars, Mercury, and the Moon; and to type D craters on Venus.

INTRODUCTION

Poag et al. (1994) initially documented the structure and morphology of the late Eocene (~35 Ma) Chesapeake Bay impact crater on the basis of ~200 km of seismic reflection profiles. Subsequent studies by Poag (1996, 1997) provided ~150 km of additional profiles, and further confirmed its complex architecture and impact origin. The center of the crater is approximately beneath the town of Cape Charles on the Virginia segment of the Delmarva Peninsula (lat 37°16.5′N, long 76°0.7′W) (Fig. 1). The crater's circumference encloses the lower part of Chesapeake Bay, the lower reaches of the Rappahannock, York, and James Rivers, the lower part of the Delmarva Peninsula, and a section of inner continental shelf east of the Delmarva Peninsula. The outer rim of the crater, a steep escarpment of sedimentary rocks, is irregularly subcircular, and circumscribes an ~85-km-wide excavation filled with impact breccia (the Exmore breccia). The crater is buried under 300–1,000 m of postimpact late Eocene to Quaternary sedimentary strata. This crater is the largest impact feature yet documented in the United States; it is one of the two largest Cenozoic impact structures on the globe, and is one of four known submarine impact craters presently located on continental shelves. Its marine setting, relatively young geologic age, and location on a slowly subsiding passive continental margin account for nearly complete preservation of all impact-related features. As a result, it is probably the best preserved terrestrial impact crater in its size class (80–90 km diameter).

This chapter focuses on refinements of the structural and morphologic interpretations of the Chesapeake Bay Crater, as derived from an additional ~900 km of seismic reflection profiles.

SEISMIC DATA

Initial interpretations of crater structure and morphology at the Chesapeake Bay Crater relied on multichannel seismic reflection profiles donated to the U.S. Geological Survey (USGS) by Texaco, Inc. (Poag et al., 1994; Poag, 1996, 1997) (Fig. 2). The profiles were produced from 48-fold multichannel data collected

Poag, C. W., Hutchinson, D. R., Colman, S. M., and Lee, M. W., 1999, Seismic expression of the Chesapeake Bay impact crater: Structural and morphologic refinements based on new seismic data, *in* Dressler, B. O., and Sharpton, V. L., eds., Large Meteorite Impacts and Planetary Evolution II: Boulder, Colorado, Geological Society of America Special Paper 339.

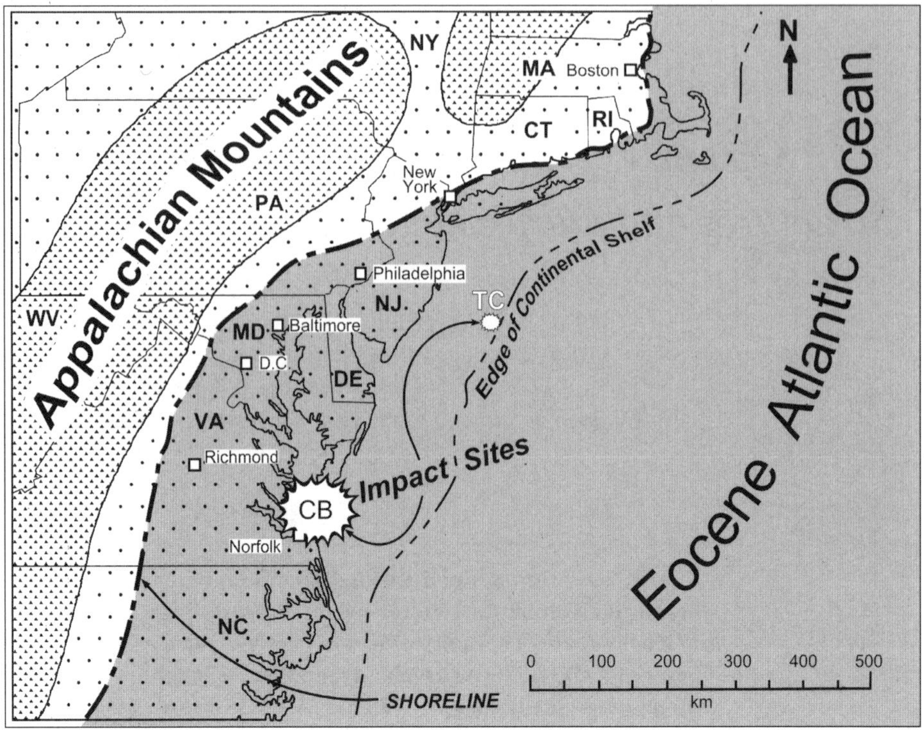

Figure 1. Location of Chesapeake Bay (CB) and Toms Canyon (TC) craters and general paleogeographic setting at time of impact, ~35 Ma.

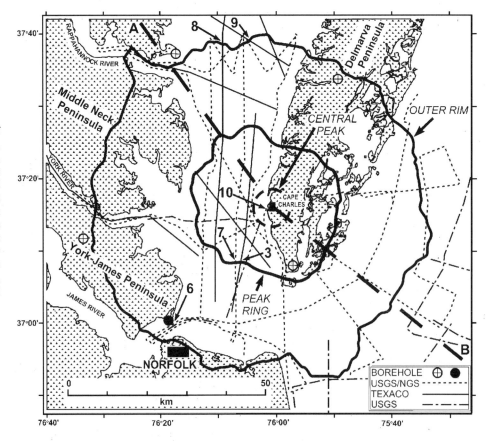

Figure 2. Distribution of ~1,200 km of intersecting seismic reflection profiles, which image the Chesapeake Bay Crater. Crossed circles indicate continuous core holes that sampled Exmore breccia; solid circle indicates a noncored bore hole that reached crystalline basement at 681-m drill depth; numbered arrows indicate locations of seismic profile segments illustrated in Figures 3 and 6–10; heavy dashed line A–B marks location of cross section shown in Figure 5; USGS = U.S. Geological Survey; NGS = National Geographic Society.

inside the bay in 1975 by Teledyne. The profiles clearly image the top of relatively undisturbed crystalline basement rocks, a series of megaslump blocks resting on the basement surface, a thick zone of chaotic reflections arising from the Exmore breccia (Poag, 1997), and most of the overlying postimpact sedimentary column (Fig. 3A, B). The Texaco profiles, however, do not image the upper ~80 m of sediment.

In 1996, the USGS collaborated with the National Geographic Society (NGS) to collect additional seismic data from the crater, both inside and outside the bay. The new data were collected in single-channel digital format, using a generator injector (GI) airgun as the seismic source. In total, more than 1,200 km of seismic reflection profiles form the basis of interpretations presented herein.

STRUCTURE AND MORPHOLOGY

Structural map and cross section

The structural and morphologic features of the Chesapeake Bay Crater indicate that it is a complex, peak-ring/central peak crater (Figs. 4 and 5). The surface mapped to determine the gross structural and morphologic features of the crater floor (Fig. 4) is a composite unconformity (heavy line in Fig. 5). Outside the crater, the unconformity is represented by the eroded surface of preimpact strata (unconsolidated middle Eocene to Lower Cretaceous sediments). At the crater's outer rim, the unconformity truncates preimpact sedimentary strata where they are cut by high-angle normal faults to form a steep outer rim escarpment. Within the annular trough, which lies between the outer rim and the peak ring, the unconformity defines the top of crystalline basement, a block-faulted erosional surface. Erosion of the basement surface took place at the termination of continental rifting (Late Triassic–Early Jurassic) prior to initiation of seafloor spreading between North America and Africa. Block faulting presumably was caused mainly by the Chesapeake Bay meteorite impact, because the upper surfaces of individual blocks displace the originally flat erosion surface. A few low-angle reverse faults also can be identified, some of which may predate the impact as part of the Appalachian orogen. On the peak ring and central peak and within the inner basin, the unconformity also is represented by the basement surface, which has been modified by direct shock, excavation, melting, and rebound from the meteorite strike.

Basement surface

The most prominent (highest amplitude) reflections on most seismic profiles used in this study are a pair of parallel reflections, the upper one of which represents the upper surface of crystalline basement (Fig 3). Boreholes that intersect the basement surface are rare in southeastern Virginia, but one of them is located on the northern bank of the mouth of the James River (Fig. 2), approximately 0.3 km north of one of our seismic lines (Fig. 6). This well was spudded approximately at sea level, and encountered granitic rocks at a depth of 681 m, equivalent to ~700 ms on the James River seismic profiles.

All the multichannel seismic profiles image the basement surface outside as well as inside the crater, with the exception of the inner basin, where deep seismic data are unavailable. The single-channel profiles, in contrast, display occasional gaps in basement reflections inside the bay. Furthermore, the single-channel seismic system we used did not resolve the basement surface in the eastern sector of the crater beneath the continental shelf, where the structural gradient of the basement steepens into the Baltimore Canyon trough (Poag, 1997). The structure of the basement surface is quite variable. It is disrupted by numerous high-angle normal faults and a few low-angle reverse faults, which are most evident on the single-channel profiles (Figs. 6 and 7). However, there are also broad areas where the basement surface is relatively flat and unfaulted. We refer to the basement surface within the circumference of the outer rim as the *structural floor* of the crater, as opposed to its *morphologic floor*, which is formed by the upper surface of the Exmore breccia; inferences and illustrations presented herein are based on this concept.

Outer rim

The outer rim of the Chesapeake Bay Crater is marked on seismic profiles by a steep fault scarp, which truncates the relatively continuous, horizontal reflections characteristic of sedimentary target rocks outside the crater (Figs. 5, 6, 8, and 9). Inside the crater, the fault scarp abuts a zone of chaotic, often hyperbolic, reflections, which arise from the Exmore breccia. On most seismic profiles, the main rim fault appears to be a listric surface, which flattens out along the top of crystalline basement (Figs. 8 and 9). However, on a few profiles the basement surface exhibits structural relief at the toe of the rim escarpment (Fig. 6). At these locations, the fault appears not to be listric, but has normal displacement, which offsets the basement as well as overlying sedimentary beds.

A variation of this structural/morphologic relationship is displayed on profiles along the northern rim of the crater. There, a group of concentric normal faults is present a few kilometers outside the crater rim. These concentric faults truncate the upper few hundred meters of sedimentary strata, which were excavated during impact to form a flat terrace (Fig. 9). The terrace is buried by a relatively thin layer (~80 m) of Exmore breccia.

The precise position of the crater's outer rim is known at 29 points where it intersects seismic profiles. Between intersections, the rim's position and shape must be interpolated. The closest lateral spacing of profile intersections with the crater rim occurs in the northern sector between the Rappahannock River and the Delmarva Peninsula (Fig. 2). There, the rim crosses a profile approximately every 2–4 km. The locations of the intersections clearly indicate that the trace of the outer rim is not a smooth arc, but is broken up into irregular blocks no more than 2–4 km on a side. We have interpolated that approximate spacing

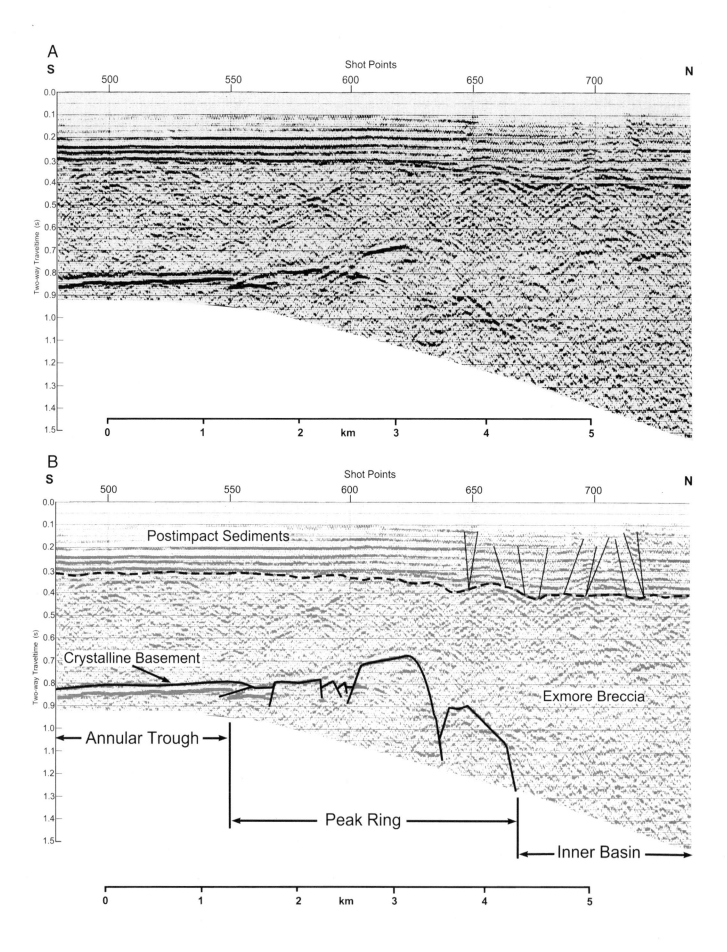

Figure 3. Segment of multichannel seismic reflection profile produced by Texaco, Inc., which crosses peak ring of Chesapeake Bay Crater (see Fig. 2 for location). A, Uninterpreted profile; B, geologic interpretation. Note clear differentiation between reflection signatures of crystalline basement (highest amplitude), Exmore breccia (chaotic, hyperbolic), and postimpact sediments (high-amplitude, continuous, parallel, nearly horizontal). This profile and all other Texaco profiles did not image upper 0.1 s (~80 m) of postimpact sediments. Compare with single-channel profile shown in Figure 7, which crosses peak ring ~1.5 km west of Figure 3 profile.

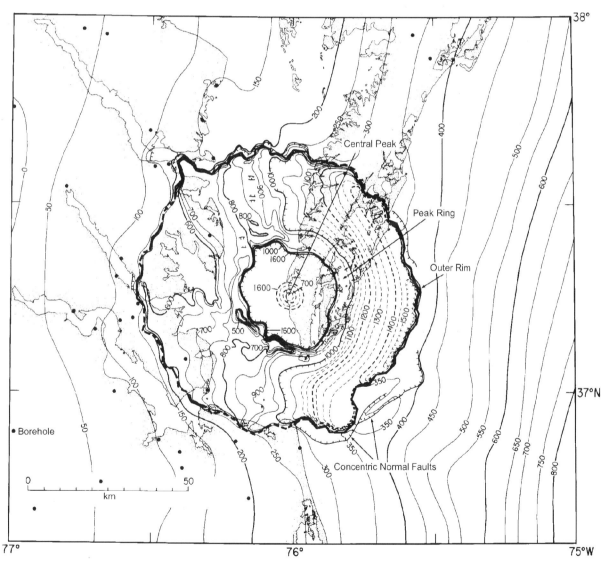

Figure 4. Structural map of structural floor of Chesapeake Bay Crater derived from interpretation of seismic reflection profiles. Note irregular geometry of outer rim and location of other structural and morphologic elements (compare with Fig. 5). Concentric normal faults (hachured) outside outer rim mark incipient slump blocks, which did not spall into crater. Dashed contours indicate poor structural control. Contour interval, 50 m.

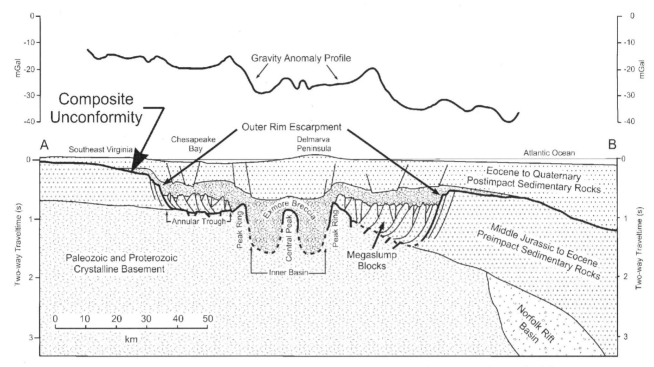

Figure 5. Geologic cross section A–B of Chesapeake Bay Crater (approximately to scale; see Fig. 2 for location). Lower heavy line traces structural floor of and its correlative surface outside the crater; together, these surfaces constitute the composite unconformable surface mapped in Figure 4. Upper surface of Exmore breccia forms morphologic floor of crater. Gravity anomaly profile derived from Figure 10.

along the entire extent of the crater rim to produce the irregular pattern shown in Figures 2 and 4. The orientation and the geometry of each small block, however, are speculative, and are intended only to show that the margin is irregular.

The diameter of the crater, as measured at the outer rim, varies from ~80 to 95 km, averaging ~85 km. Structural relief on the outer rim scarp varies from about 300 m on the northwest to 1,000 m on the southeast (Figs. 4 and 5). This differential relief reflects the trend of regional structural and depositional dip and attendant southeastward thickening of the sedimentary wedge.

Concentric normal faults

Two principal anomalies interrupt the general pattern of interpolated outer rim geometry discussed above. At two locations, the outer rim bulges markedly away from the crater center; one in the northwest quadrant, the other in the southeast quadrant (Figs. 2 and 4). The bulges are embayments caused by local wall failures along concentric normal faults that parallel the outer rim in several places (Figs. 4 and 9). Complete wall failure was limited to a few short segments of these ancillary faults, but incipient slumping can be seen on other segments. There may be additional bulges where seismic control is poor. For example, beneath the York-James and Middle Neck Peninsulas, only two (nearly superposed) profiles cross the rim in a distance of 70 km.

Annular trough

The block-faulted basement surface that constitutes the peripheral floor of the crater forms an annular trough ~20–25 km wide (Figs. 3–8). The trough separates the sedimentary outer rim escarpment from a low-relief crystalline peak ring. The structural floor of the trough is shallowest (less than –600 m) under the Middle Neck Peninsula, and dips eastward to more than –1,500 m (Figs. 4 and 5). The crater straddles a structural hinge line in the crystalline basement, which trends northeastward approximately under the eastern shoreline of the Delmarva Peninsula. At this hinge line, the structural gradient of the basement steepens noticeably toward the offshore depocenter of the Baltimore Canyon trough (Poag, 1996). Though the single-channel profiles do not image the basement surface directly beneath the crater on the continental shelf, four USGS offshore multichannel seismic reflection profiles (Fig. 2) document its depth a few kilometers southeast of the crater rim.

Peak ring

The peak ring of the Chesapeake Bay Crater is an irregular, low-relief, ovate ridge of crystalline basement rocks, averaging ~35–40 km in diameter, as measured along its crestline (Figs. 2–5). The peak ring is crossed at 17 locations by seismic reflec-

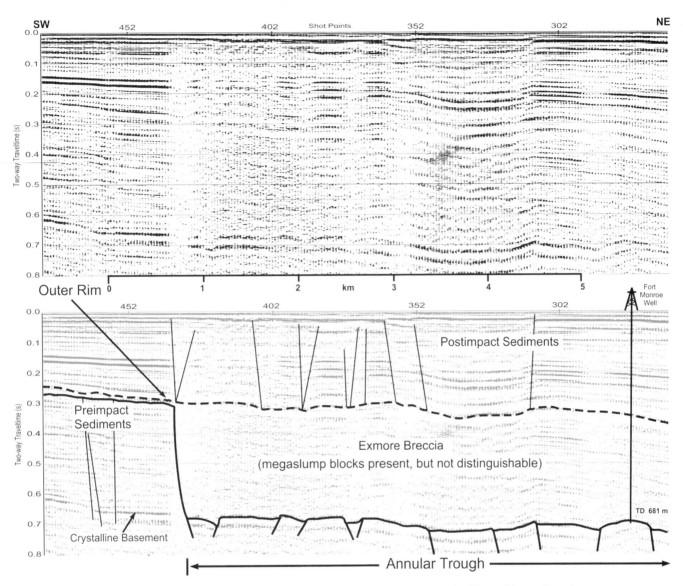

Figure 6. Segment of single-channel seismic reflection profile through mouth of James River, collected by USGS and NGS (see Fig. 2 for location). Above, uninterpreted profile; below, geologic interpretation. Note that this and other single-channel profiles image the upper ~80 m of postimpact sediments not discernable on Texaco profiles. Conversely, single-channel profiles display numerous internal multiples in Exmore breccia section, which mask reflections from megaslump blocks that rest on basement surface. Fort Monroe well was drilled (not cored) ~0.3 km north of seismic trackline.

tion profiles. Its highest elevation is on the southwest margin, where it rises ~250 m above the floor of the annular trough (~500–750 m below sea level) (Figs. 3 and 7). On most profiles, however, the structural relief is less than 100 m. In map view, the long axis of the peak ring trends northwest-southeast, nearly parallel to the primary bulge axis of the outer rim (Figs. 2 and 4). The upper surface of the peak ring is broken by normal and reverse faults (Figs. 3 and 7). Its inner flank slopes steeply and forms the wall of the inner basin.

Inner basin

Inside the peak ring, the inner basin approximates the location of the transient crater (Figs. 3–5 and 11). The basin outline is ovate, determined by the shape of the surrounding peak ring (Fig. 4). Its diameter ranges from 35 to 40 km, coincident with that of the crestline of the peak ring. The precise depth of the inner basin is not documented by the available seismic reflection data. Definitive multichannel records of deep basement seismic

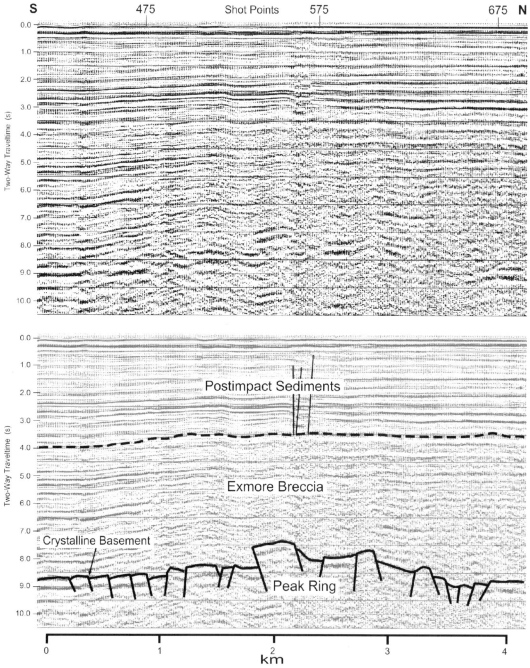

Figure. 7. Segment of single-channel seismic reflection profile across southwest sector of peak ring, collected by USGS and NGS (see Fig. 2 for location). Above, uninterpreted profile; below, geologic interpretation. Note distinct vertical offsets on faults that cut crystalline basement surface. Compare with multichannel profile shown in Figure 3, which crosses peak ring ~1.5 km east of Figure 7 profile.

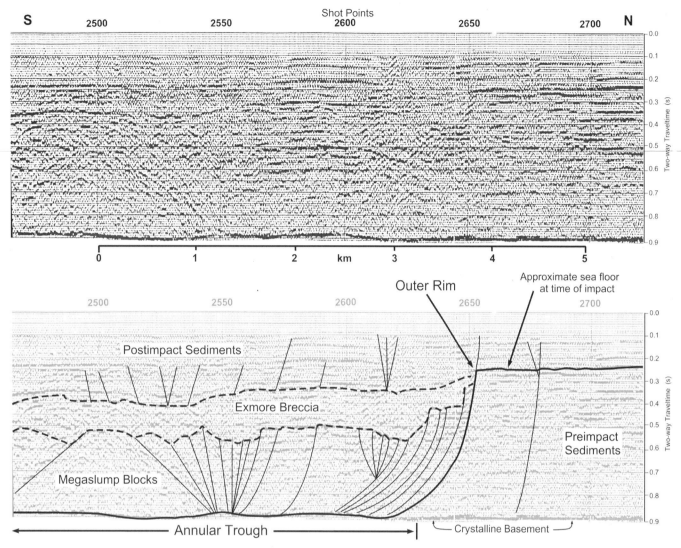

Figure 8. Segment of multichannel seismic profile produced by Texaco, which crosses northern rim of crater (see Fig. 2 for location). Top, uninterpreted profile; bottom, geologic interpretation. Note steep outer rim escarpment that truncates preimpact sediments, chaotic reflections in Exmore breccia, and megaslump blocks that rest on crystalline floor of crater.

data have not yet been released by Texaco, and the single-channel data of the USGS and NGS do not resolve structures that deep. If one uses Grieve and Robertson's (1979) scaling relationship ($d_t = 0.52D^{0.2}$, in which d_t is true depth, and D is the final diameter of the crater), the estimated inner basin depth is 1.26 km. This places the floor at ~1.6 s, which approximates the maximum depth of released multichannel data. Application of this equation is imprecise, however, because we cannot determine the original elevation of the outer rim, which is essential to estimate d_t.

Central peak

Evidence of a small central peak can be seen on three seismic reflection profiles. One single-channel profile, which terminates at the mouth of Cape Charles harbor, images slightly higher amplitude reflections at ~600 ms (~600 m deep). We tentatively interpret these to mark the top of the central peak (Fig. 10). Two adjacent profiles (not shown here) display anomalous, closely spaced clusters of inclined reflections, which may represent side echos from the flanks of the central peak. Though none of the three profiles is entirely definitive, supporting evidence can be derived from Bouguer gravity data (Fig. 11). A small positive gravity peak (–25 mGal) is present directly over Cape Charles, coincident with the location of the seismically inferred peak. Multichannel seismic data and/or drilling in the vicinity of Cape Charles are needed to confirm the presence of this peak.

158 C. W. Poag et al.

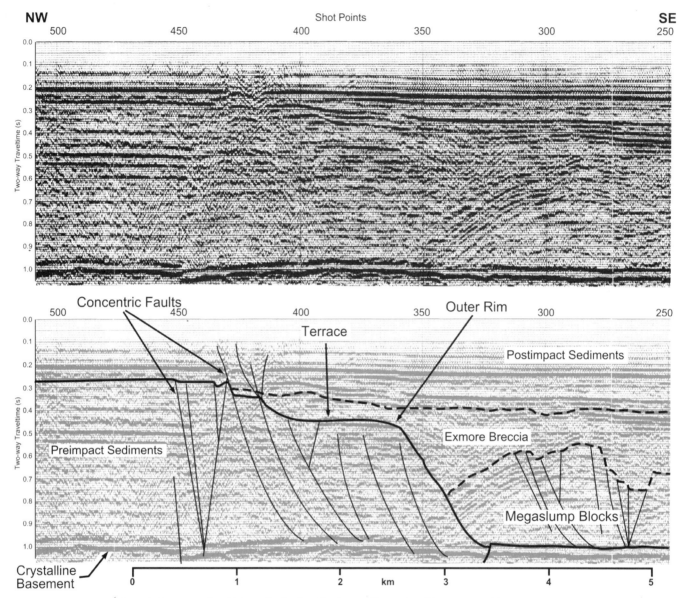

Figure 9. Segment of multichannel seismic reflection profile produced by Texaco, which crosses northern outer rim ~7 km east of profile shown in Figure 8 (see Fig. 2 for location). Top, uninterpreted profile; bottom, geologic interpretation. Note zone of concentric normal faults, which facilitated excavation of shallow terrace just outside outer rim escarpment. Rotated megaslump blocks are particularly well defined on this profile.

COMPARISONS

Submarine terrestrial craters

Chesapeake Bay Crater is one of four (possibly five) submarine impact structures found on modern continental shelves (Fig. 12). By far the largest of these is the Chicxulub Crater (Cretaceous-Tertiary boundary, Yucatán Peninsula, Mexico), whose huge diameter (200 km or more), great depth (4–5 km), and multi-ring morphology set it apart from the other four (Morgan et al., 1997a,b; Sharpton et al., 1997; Snyder et al., 1997; Warner et al., 1997). The Chesapeake Bay Crater is the next largest. The other three are the Montagnais Crater (45 km diameter, early Eocene, Scotian Shelf, Canada; Jansa et al., 1989), the Mjølnir Crater (40 km diameter, Late Jurassic, Barents Sea, Norway; Dypvik et al., 1996), and the Toms Canyon Crater, 20 km diameter, late Eocene, New Jersey shelf, US; Poag and Poppe, 1998).

From present evidence, it appears that the Chesapeake Bay Crater is the only unequivocal peak-ring/central peak crater

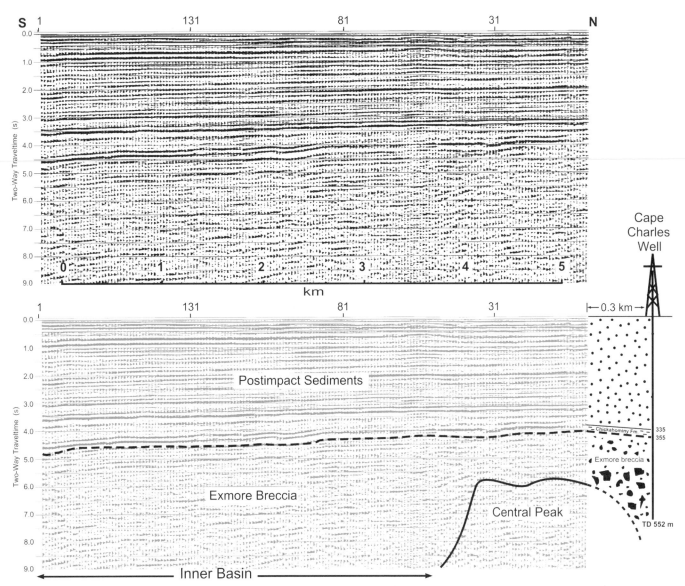

Figure 10. Segment of single-channel seismic reflection profile collected by USGS and NGS, which terminates (north end) at the mouth of Cape Charles Harbor (see Fig. 2 for location). Top, uninterpreted profile; bottom, geologic interpretation. The presence of a central peak is indicated by a slightly higher amplitude reflection, and edge-effect refractions. Cape Charles well was drilled (not cored) ~0.3 km east of the end of the profile; the well terminated in Exmore breccia at a total depth of 552 m.

among the submarine group. Chicxulub has a peak ring, but is considered by most researchers to be a multi-ring basin (Morgan et al., 1997b). Mjølnir has been described as having a peak ring, but the supposed ring is a subtle feature, formed entirely by sedimentary rocks; its precise nature is obscure. Definition of the Mjølnir Crater's morphology is complicated by the fact that it has undergone significant postimpact erosion, which may have altered the original morphology of a peak ring. The Toms Canyon structure also formed entirely within sedimentary strata and gives no hint of a peak ring.

Subaerial terrestrial craters

Among the few well-documented and well-preserved subaerial impact craters, the closest analogues for the Chesapeake Bay peak-ring/central peak structure appear to be the Ries Crater (25 km diameter, late Miocene, Nördlingen, Germany; Pohl et al., 1977; Newsom et al., 1990) and the Popigai Crater (85 km diameter, late Eocene, Anabar Shield, Siberia, Russia; Masaitis et al., 1994) (Fig. 12). Both of these craters formed in mixed sedimentary-crystalline target rocks and have distinct peak rings of

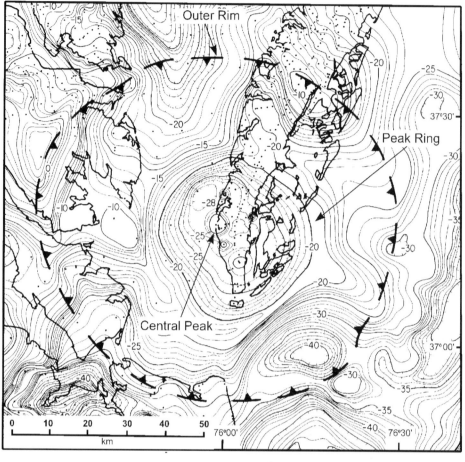

Figure 11. Simple Bouguer gravity map of lower Chesapeake Bay region. A negative bull's-eye anomaly (–20 to –28 mGal) marks inner basin of crater. Small relative positive anomaly (–25 mGal) over town of Cape Charles coincides with location of central peak shown in Figure 10. Small solid dots indicate locations of gravity stations onshore and in bay. Gravity data courtesy of John Costain, Virginia Polytechnic Institute and State University.

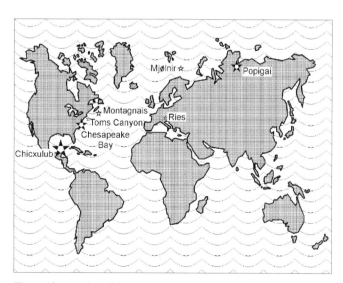

Figure 12. Location of five buried submarine impact craters and two subaerial craters discussed in text.

upraised crystalline basement and possibly central peaks. Cross sections of the Ries Crater based on drilling and geophysical surveys (seismic reflection, gravity, geomagnetic) show comparable structure and morphology, except for its smaller size and questionable presence of a central peak (Fig. 13).

Popigai, on the other hand, is practically a twin of Chesapeake Bay Crater. Their ages are indistinguishable (Chesapeake Bay = 35.2–35.5 m.y., Poag, 1997; Popigai = 35.7 m.y., Bottomley et al., 1997). They are also virtually identical in size. Though the diameter of Popigai is commonly given as ~100 km (e.g., Grieve et al., 1997; Masaitis et al., 1997), the geologic map of Masaitis (1994) indicates otherwise; both craters average ~85 km in diameter. Aside from minor irregularities in the peripheral outlines, the outer rims of the two craters can hardly be distinguished from one another when superposed (Fig. 14). The other structural and morphologic features are quite similar. The principal difference appears to be a slight asymmetry to the Popigai structure; the peak ring and central peak are offset to the west relative to the center of the outer rim. The diameters of the peak ring and central peak also may be slightly larger at Popigai.

The deeper parts of the inner basin at Popigai, as at Chesapeake Bay and Ries, have not yet been thoroughly studied. However, Masaitis et al. (1997) reported a low-relief cen-

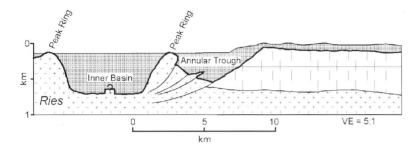
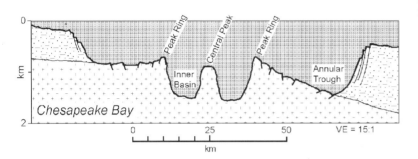

Figure 13. Comparative structural and morphologic cross sections of Ries Craters and Chesapeake Bay. Ries section modified from Newsom et al. (1990).

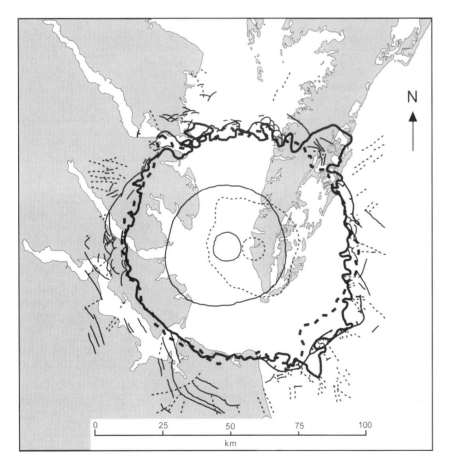

Figure 14. Superposed outlines of outer rim, peak ring, and central peak of Popigai Crater (solid lines) and Chesapeake Bay Crater (dashed lines). Outer rims (average diameter ~85 km) are nearly indistinguishable, aside from modest irregularities in outline. Peak ring and central peak of Chesapeake Bay Crater are nearly symmetrically centered within outer rim, whereas equivalent features at Popigai are offset to the west. Popigai geometries taken from geologic map published by Masaitis (1994). Concentric faults (solid lines) and radial faults (dashed lines) shown outside outer rims are mainly from the Popigai map. Diameter of Popigai central peak estimated (not shown on geologic map, but reported to be present by Masaitis et al., 1997).

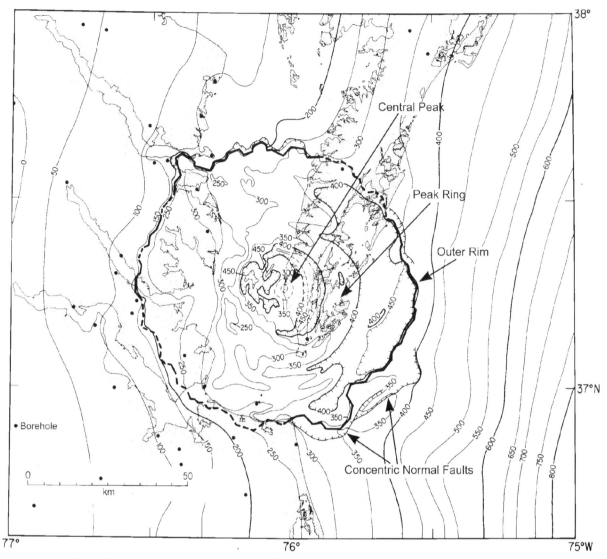

Figure 15. Structural map of top surface of Exmore breccia used for morphologic comparison with extraterrestrial craters. Morphology is similar to that of structural crater floor (Fig. 4), but with reduced structural relief. Contours derived from seismic reflection profiles; contour interval, 50 m.

tral peak at Popigai on the basis of three-dimensional computer modeling.

Extraterrestrial craters

Geometric analyses (morphology, structure) of impact craters on other planetary bodies are limited by our inability to image the structural floor of the craters. That is because the craters are partly filled with deposits such as impact breccia, basalt flows, and melt sheets, whose upper surfaces produce the observable morphologic floor. Peak rings and central peaks can not be recognized on other planets unless they protrude above the crater-fill deposits. In the case of Chesapeake Bay Crater, on the other hand, the seismic profiles image the peak ring and central peak even though they are deeply buried by the thick Exmore breccia. A datum comparable to the morphologic floor of a typical planetary crater would be at a level within the Exmore breccia above which the peak ring and central peak would protrude. Such a surface, however, is a conceptual feature, which cannot be traced within the chaotic reflections of the breccia. The best available proxy for that surface is the upper surface of the Exmore breccia. Because of differential breccia compaction, the morphology of this breccia surface mimics the morphology of the underlying basement surface. Thus, a structural map of the Exmore breccia surface (i.e., the morphologic floor of the crater) (Fig. 15) displays the same principal features (outer rim, annular trough, peak ring, central peak) as the buried structural floor (Fig. 4), though the elevation relief is much reduced.

Morphometric studies of peak-ring and central peak craters on the Moon (Hale and Head, 1979, 1980), Mercury (Hale and

Head, 1980; Pike, 1988), Mars (Wood, 1980), and Venus (Alexopolous and McKinnon, 1992, 1994; Wood and Tam, 1993) indicate that peak-ring craters that also have central peaks (protobasins of Pike, 1988; type D craters of Alexopolous and McKinnon, 1994), are comparatively rare in the solar system. Melosh's (1982) hydrodynamic model for ring and peak formation explains this by treating peak rings as late-stage products of peak collapse. Thus peak-ring craters are envisioned as evolutionary transitions between craters with only central peaks and those with only peak rings. Many field data support this model (Hale and Grieve, 1982; Alexopolous and McKinnon, 1994), but agreement is not unanimous (Hale and Head, 1980). Variability may depend on such factors as the preimpact geology of target rocks, the velocity, size, and incidence angle of the impactor, and the degree to which crater-fill deposits bury the structural crater floor.

The morphology of Chesapeake Bay Crater agrees well with that of both Martian protobasins and Venusian type D craters in most respects. The Chesapeake Bay Crater differs, however, in having a larger diameter than most type D craters. Alexopolous and McKinnon (1994) showed that type D craters cluster between diameters of 40 and 50 km, though one example was 70 km wide. In contrast, the diameters of protobasins on Mars cluster at 100–150 km with maximum diameters of 215–285 km (Wood, 1980). Alexopolous and McKinnon (1994) suggested that the diameter differences may be related to geologic differences. This would fit the increasing body of evidence that Earth's geologic history has been more comparable to that of Mars than of Venus.

SUMMARY

On the basis of more than 1,200 km of seismic reflection profiles, the principal structural and morphologic features of the Chesapeake Bay Crater are well established, though the presence of a central peak requires further substantiation. This impact structure can be classified as a complex peak-ring/central peak crater, with an average outer rim diameter of ~85 km. The structure and the morphology of the Chesapeake Bay Crater are closely analogous to those of the smaller Ries Crater of southern Germany, and of the equal-sized Popigai crater of northern Siberia. Chesapeake Bay Crater also is closely comparable morphologically to the type D peak-ring craters on Venus and to protobasins on Mars.

ACKNOWLEDGMENTS

We thank Michael Rampino, Chuck Pillmore, Jo Morgan, and an anonymous reviewer for helping to improve an earlier version of the manuscript.

REFERENCES CITED

Alexopolous, J. S., and McKinnon, W. B., 1992, Multiringed impact craters on Venus: An overview from Arecibo and Venera images and initial Magellan data: Icarus, v. 100, p. 347–363.

Alexopolous, J. S., and McKinnon, W. B., 1994, Large impact craters and basins on Venus, with implications for ring mechanics on the terrestrial planets, in Dressler, B. O., Grieve, R. A. F., and Sharpton, V. L., eds., Large Meteorite Impacts and Planetary Evolution: Geological Society of America Special Paper 293, p. 29–50.

Bottomley, R., Grieve, R., York, D., and Masaitis, V., 1997, The age of the Popigai impact crater event and its relation to events at the Eocene/Oligocene boundary: Nature, v. 388, p. 365–368.

Dypvik, H., Gudlaugsson, S. T., Tsikalas, F., Attrep, M., Jr., Ferrell, R. E., Jr., Krinsley, D. H., Mørk, A., Faleide, J. I., and Nagy, J., 1996, Mjølnir structure: an impact crater in the Barents Sea: Geology, v. 24, p. 779–782.

Grieve, R. A. F., and Robertson, P. B., 1979, The terrestrial cratering record. 1, Current status of observations: Icarus, v. 38, p. 212–229.

Grieve, R. A. F., Rupert, J., Smith, J., and Therriault, A., 1997, The record of terrestrial impact cratering: GSA Today, v. 5, p. 189, 194–196.

Hale, W. S., and Grieve, R. A. F., 1982, Volumetric analysis of complex lunar craters: Implications for basin ring formation: Journal of Geophysical Research, v. 87 (suppl.), p. A65–A76.

Hale, W. S., and Head, J. W., 1979, Central peaks in lenar craters: Morphology and morphometry; in Proceedings, 10th Lunar and Planetary Science Conference: Houston, Texas, Lunar and Planetary Institute, p. 2623–2633.

Hale, W. S., and Head, J. W., 1980, Central peaks in Mercurian craters: comparisons to the moon, in Proceedings, 11th Lunar and Planetary Science Conference: Houston, Texas, Lunar and Planetary Institute, p. 2191–2205.

Jansa, L. F., Pe-Piper, G., Robertson, B. P., and Friedenreich, O., 1989, A submarine impact structure on the Scotian Shelf, eastern Canada: Geological Society of America Bulletin, v. 101, p. 450–463.

Masaitis, V. L., 1994, Impactites from Popigai crater, in Dressler, B. O., Grieve, R. A. F., and Sharpton, V. L., eds., Large meteorite impacts and planetary evolution: Geological Society of America Special Paper 293, p. 153–162.

Masaitis, V. L., Naumov, M. V., and Mashchak, M. S., 1997, Three-dimensional modeling of impactite bodies of the Popigai impact crater, Russia; in Conference on Large Meteorite Impacts and Planetary Evolution; Houston, Texas, Lunar and Planetary Institute, p. 32–33.

Melosh, H. J., 1982, A schematic model of crater modification by gravity: Journal of Geophysical Research, v. 87, p. 371–380.

Morgan, J., and the Chicxulub Working Group, 1997a, The Chicxulub seismic experiment: crater morphology; in Conference on Large Meteorite Impacts and Planetary Evolution; Houston, Texas, Lunar and Planetary Institute, p. 35–36.

Morgan, J., Warner, M., and the Chicxulub Working Group, 1997b, Size and morphology of the Chicxulub impact crater: Nature, v. 390, p. 472–476.

Newsom, H. E., Graup, G., Iseri, D. A., Geissman, J. W., and Keil, K., 1990, The formation of the Ries crater, West Germany; evidence of atmospheric interactions during large cratering event, in Sharpton, V. L., and Ward, P. D., eds., Global catastrophes in Earth history: Geological Society of America Special Paper 247, p. 195–206.

Pike, R. J., 1988, Geomorphology of impact craters on Mercury, in Vilas, F., Chapman, C. R., and Matthews, M. S., eds., Mercury: Tucson, University of Arizona Press, p. 165–273.

Poag, C. W., 1996, Structural outer rim of Chesapeake Bay impact crater (late Eocene): seismic and bore hole evidence: Meteoritics and Planetary Science, v. 31, p. 218–226.

Poag, C. W., 1997, The Chesapeake Bay bolide impact: a convulsive event in Atlantic Coastal Plain evolution, in Segall, M. P., Colquhoun, D. J., and Siron, D., eds., Evolution of the Atlantic Coastal Plain—sedimentology, stratigraphy, and hydrogeology: Sedimentary Geology Special Issue, v. 108, p. 45–90

Poag, C. W., and Poppe, L. J., 1998, The Toms Canyon structure, New Jersey outer continental shelf: a possible late Eocene impact crater: Marine Geology, v. 145, p. 23–60.

Poag, C. W., Powars, D. S., Poppe, L. J., and Mixon, R. B., 1994, Meteoroid mayhem in ole Virginny: source of the North American tektite strewn field: Geology, v. 22, p. 691–694

Pohl, J., Stöffler, D., Gall, H., and Ernston, K., 1977, The Ries impact crater, *in* Roddy, D. J., Pepin, R. O., and Merrill, R. B., eds., Impact and explosion cratering: New York, Pergamon Press, p. 343–404.

Sharpton, V. L., Marin, L. E., Corrigan, C. M., and Dressler, B. O., 1997, Impact deposits from the southern inner flank of the Chicxulub impact basin; *in* Conference on Large Meteorite Impacts and Planetary Evolution: Houston, Texas, Lunar and Planetary Institute, p. 53–54.

Snyder, D. B., Hobbs, R. W., and the Chicxulub Working Group, 1997, Crustal-scale structural geometries of the Chicxulub impact from British Institutions Reflection Profiling Syndicate seismic reflection profiles; *in* Conference on Large Meteorite Impacts and Planetary Evolution: Houston, Texas, Lunar and Planetary Institute, p. 57.

Warner, M., and the Chicxulub Working Group, 1997, Chicxulub seismic experiment: overview and size of the transient cavity; *in* Conference on Large Meteorite Impacts and Planetary Evolution: Houston, Texas, Lunar and Planetary Institute, p. 61–62.

Wood, C. A., 1980, Martian double ring basins: new observations, *in* Proceedings, 11th, Lunar and Planetary Science Conference: Houston, Texas, Lunar and Planetary Institute, p. 2221–2241.

Wood, C. A., and Tam, W., 1993, Morphologic classes of impact basins on Venus, *in* Proceedings, 24th, Lunar and Planetary Science Conference: Houston, Texas, Lunar and Planetary Institute, p. 1535–1536.

MANUSCRIPT ACCEPTED BY THE SOCIETY DECEMBER 16, 1998

Gravity signature of the Teague Ring impact structure, Western Australia

Jeffrey B. Plescia*
California Institute of Technology, Jet Propulsion Laboratory, MS 183-501, 4800 Oak Grove Drive, Pasadena, California 91109, USA

ABSTRACT

Teague Ring is an impact structure in Western Australia (25° 50′ S; 120° 55′ E) having a diameter of about 31 km and an age of approximately 1,630 Ma. A gravity survey over the feature defines a significant negative Bouguer anomaly of about –12 mGal. This gravity anomaly is modeled as a mass of material, corresponding to the central crystalline core of the structure, having a negative density contrast, with respect to the adjacent rocks of –0.13 g cm^{-3} and extending to a depth of ~5 km. At Teague Ring, the extensive allochthonous and autochthonous breccia deposits (typically producing low-density material) have been removed by several kilometers of erosion. Therefore, the low-density material is interpreted to be caused by the fracturing of the crystalline basement complex below such brecciated material. The magnitude of the Teague Ring gravity anomaly is consistent with those of other similar-size terrestrial impact structures.

BACKGROUND

Teague Ring is a large ancient impact structure in Western Australia, located at 25° 50′ S and 120° 55′ E (Figs. 1 and 2). The objective of this chapter is to define the crustal aspects of the structure based on gravity data; it is accepted here, based on geologic data, that Teague Ring is of impact origin. Recently, the Geological Survey of Western Australia (Pirajno, 1998) has renamed the structure the "Shoemaker Impact Structure" after the late Eugene Shoemaker, who conducted considerable research on the feature. For consistency with other relevant publications, the structure is referred to here by its original name, Teague Ring.

Teague Ring has a diameter of 31 km and an age of ~1,630 Ma (Shoemaker and Shoemaker, 1996). Dimensions of the structure are based on the diameter of the disturbed Early Proterozoic Earaheedy Group sedimentary rocks. The age is only loosely constrained on the basis of ages from the crystalline rocks exposed within the structure. Teague Ring was first recognized as a structural anomaly by Butler (1974), and various models of origin have been proposed by several investigators over the years. Published studies of the structure have been conducted only at the reconnaissance scale. More detailed work was undertaken by Eugene and Carolyn Shoemaker, but it has been published only in limited form (Shoemaker and Shoemaker, 1996); currently, the area is being mapped as part of a broader regional study of the Nabberu 1:100,000 sheet (F. Pirajno, personal communication, 1998).

As part of a multidisciplinary effort to better define the nature of the Teague Ring impact structure and to understand specifics about the subsurface crustal structure, a gravity survey was undertaken in austral winter of 1996. The objective of the survey was to define the gravity field over the Teague Ring region, identify any anomaly associated with the structure, and interpret that anomaly in terms of crustal structure.

---

*Present address: Astrogeology Program, U.S. Geological Survey, 2255 North Gemini Drive, Flagstaff, Arizona 86001; e-mail: jplescia@flagmail.wr.usgs.gov.

Plescia, J. B., 1999, Gravity signature of the Teague Ring impact structure, Western Australia, *in* Dressler, B. O., and Sharpton, V. L., eds., Large Meteorite Impacts and Planetary Evolution II: Boulder, Colorado, Geological Society of America Special Paper 339.

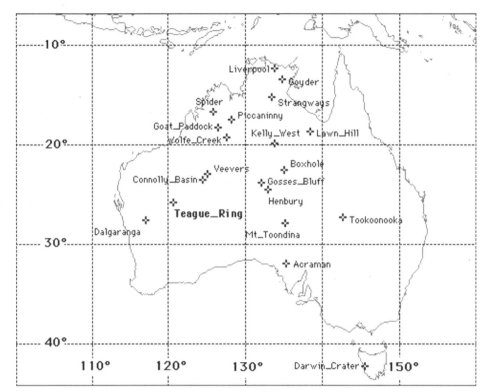

Figure 1. Map of Australia showing location of Teague Ring and other recognized impact structures.

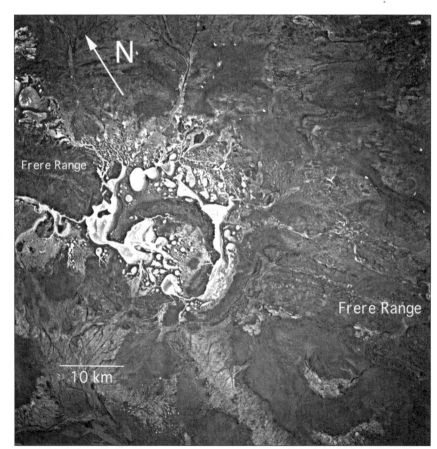

Figure 2. Oblique Space Shuttle image (STS41D-42-39) of Teague Ring structure. White areas are playa surfaces of the Lake Teague/Lake Nabberu system. Dark band extending from lower right to upper left is the Frere Range, dominated by the dark banded iron of the Frere Formations.

GEOLOGIC SETTING

The Teague Ring structure occurs near the unconformable contact between the gently dipping Early Proterozoic sediments of the Nabberu Basin and the underlying crystalline basement of the Archean Yilgarn Block. The Archean crystalline basement consists of granitic rocks intruded into remnants of greenstone belts. The granitic rocks in the Teague Ring region are medium- to coarse-grained adamellites exhibiting a strong gneissic fabric (Bunting et al., 1980). Exposures of basement rock are quite limited as the region is characterized by widespread laterite soil and lake deposits and wind-blown material associated with the Lake Nabberu–Lake Teague drainage system.

The Early Proterozoic sediments are part of the Earaheedy Group of the Nabberu Basin (Hall et al., 1977; Bunting et al., 1977, 1982). The Earaheedy Group consists of approximately 6,000 m of shallow-water marine sediments and includes eight formations. Of this section, only a few units are relevant to the Teague Ring structure (Fig. 3); from oldest to youngest, these include the Yelma Formation, Frere Formation, Windidda Formation, Wandiwarra Formation, Princess Ranges Quartzite, and Wongawol Formation.

The Yelma Formation is a clastic unit (medium- to coarse-grained quartz arenite, shale, phyllite, and chert with minor stromatolitic carbonate) unconformably overlying the basement rocks. The Frere Formation, conformable on the Yelma, consists of ferruginous sediments and clastics and is dominated by interbedded banded iron units. The Windidda Formation, a carbonate and clastic unit, is not exposed in the immediate vicinity of Teague Ring but occurs to the east in scattered outcrops. The Wandiwarra Formation is separated from the lower units by a disconformity and is composed of quartz sandstone and shale. The Princess Ranges Quartzite consists of interbedded quartz arenite. The uppermost relevant unit the Wongawol Formation, is a fine-grained arkosic sandstone.

The Earaheedy Group is considered to be Lower Proterozoic in age, as it overlies the Archean basement, which is dated at 2.7–2.4 Ga (Compston and Arriens, 1968; Roddick et al., 1976; Cooper et al., 1978; Stuckless et al., 1981) and is unconformably overlain by the Bangemall Group, which is dated at 1.0–1.1 Ga (Compston and Arriens, 1968; Gee et al., 1976). Dates on units within the Earaheedy Group (Horwitz, 1975a; Preiss et al., 1975; Goode et al., 1983) include an age of >1,685 Ma for the Wandiwarra Formation (K/Ar on glauconite) and 1,590 to >1,700 Ma (K/Ar on glauconite; whole-rock Rb/Sr; Pb-Pb on galena) for the Yelma Formation.

GEOLOGY AND STRUCTURE OF TEAGUE RING

The Teague Ring structure (Figs. 2 and 3) is defined at the surface by a collar of deformed Frere Formation rocks and cored by Archean granites (Bunting et al., 1982). In map view,

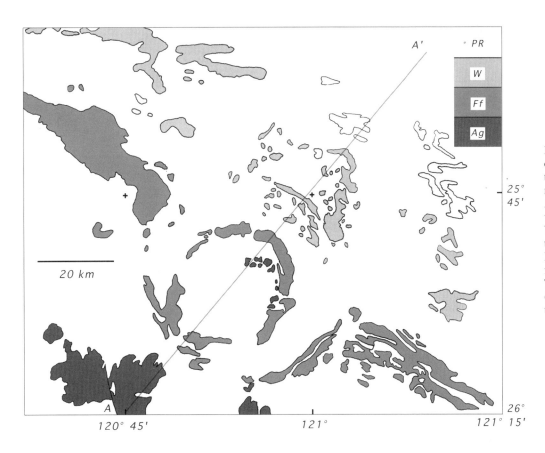

Figure 3. Basement geology of Teague Ring region; Quaternary surface deposits are not illustrated. Ag denotes Archean granites (undivided); Ff is Frere Formation; W is the Windiwarra Formation; and PR is the Princess Range Quartzite. Adapted from Bunting et al. (1982). The northeast-trending line (A–A′) indicates the orientation of the gravity model.

the northwest trend of the Frere rocks (and the Frere Range itself) are interrupted and deformed at Teague Ring. The overall structure is one of a circular core of crystalline basement about 13 km across surrounded by a ring syncline whose axis has a diameter of about 20 km.

Frere Formation rocks form the collar immediately surrounding the crystalline core and a second, outer arc of rocks extending from the northwest through the south to southeast side. On the northeastern margin, additional units are involved in the deformation including the Wongawol Formation, Princess Ranges Quartzite, and Wandiwarra Formation. Outside the structure, these units have gentle northeast dips (5°–10°) and northwest strikes, whereas within the structure the dips become steep (25°–75°). Along the northeast margin of the structure dips are subvertical to locally overturned, whereas dips are shallower and the folding more open on the southwest margin. In detail, deformation can be quite complex with numerous steeply dipping tight folds and faults and a general pattern of circumferential shortening and thrusting (Shoemaker and Shoemaker, 1996).

Crystalline rocks (quartz syenite and leucogranite) are exposed only in the northeast corner of the core inside the collar of Frere Formation rocks (Bunting et al., 1980). Most of the interior is covered by quartz sand and gypsum deposits. Additional limited exposures of hornblende-quartz monzonite and granite to adamelite occur a short distance outside the structure to the southwest and south. The intrusive relations between the granitic rock types and the overlying Yelma Formation are not observed. However, regional relations and geochronologic data suggest these granitic rocks are part of the crystalline basement and that the Yelma Formation unconformably overlies them.

ORIGIN OF TEAGUE RING

Various modes of origin have been suggested to explain the Teague Ring structure. An evolution of ideas regarding the origin has been fostered by the recognition over the last decade that impact structures are indeed important features of the terrestrial geologic record. Based on the geologic relations observed at Teague Ring, within the context of a better understanding of the nature of terrestrial impact structures, an impact origin seems the most likely explanation, which is accepted here.

Originally, Butler (1974), in his first description of the structure, suggested Teague Ring could be either an intrusion of the granites or an impact, although he did not reach a specific conclusion in the short paper. Horwitz (1975b) mentioned Teague Ring only briefly in a summary of the geology of the Yilgarn Block and suggested that it resulted from the interference of mild folds typical of the region. In a discussion of the Lower Proterozoic stratigraphy and structure of the area, Bunting et al. (1977), proposed that Teague Ring was formed by the cold reemplacement of a syenite plug at high strain rates driven by compressive stress associated with the northwest-trending regional folds. Bunting et al. (1980) later suggested that Teague Ring was formed by the explosion of volatiles related to alkaline magmatism. In each of these discussions, only brief mention is made of the structure, and details are not provided regarding the suggested mode of origin.

Shoemaker and Shoemaker (1996) have argued for an impact origin on the basis of the structural style and deformational features observed. They have noted a few shatter cones within the Frere Formation granular iron strata and a single shatter cone within the central quartz syenite. Planar deformation lamellae in quartz grains from the core granitic rocks and, locally, pseudotachylite veins in the granite and syenite have been reported by Bunting et al. (1980). The structural style of tightly folded strata and inward directed thrusting and shortening are consistent with the deformational style observed in the central uplifts of impact structures (Shoemaker and Shoemaker, 1996).

AGE OF TEAGUE RING

Constraints on the age of the Teague Ring structure are provided by geologic relations of the units involved and from geochronology of samples. A maximum age is provided by the age of the Earaheedy Group rocks (~1,700 Ma) that were deformed by the impact event. Bunting et al. (1980) conducted Rb-Sr dating on the crystalline rocks from inside and outside the core of Teague Ring; material from nine sites within the structure and two outside the structure for comparison (5 km south of the structure and 185 km south) were analyzed. The southernmost sample, outside the structure, gave an age of 2,367 Ma for the crystalline basement. The nine samples from within the structure exhibit isochrons indicative of ages of 1,630 and 1,260 Ma.

Bunting et al. (1980) considered the 1,630-Ma age to be either the original emplacement age of the granitic rocks or a regional metamorphic event; Shoemaker and Shoemaker (1996) interpreted the 1,630 Ma age as a resetting age due to the impact. Shoemaker and Shoemaker (1996) argued this point on the basis that the granitic rocks of the core show shock features (shatter cones and planar deformation lamellae); that the age is younger than the overlying Yelma rocks; and that samples of similar rocks from immediately outside the structure and elsewhere have radiometric ages of 2,367 Ma that appear unaffected by younger metamorphic events. Both Bunting et al. (1980) and Shoemaker and Shoemaker (1996) have speculated that the 1,260-Ma age is the result of weathering-induced increase in the Rb / Sr ratio. Thus, Teague Ring was already deeply eroded by the mid-Proterozoic time.

COLLECTION AND REDUCTION OF GRAVITY DATA

In order to define the gravity signature of Teague Ring and in an attempt to understand the crustal structure of the feature, a gravity survey was undertaken in August 1996. An earlier reconnaissance survey conducted in 1986 (Shoemaker and Shoemaker, 1996) suggested a significant negative anomaly over the feature. Because of the scale of the structure and the low precision of the topographic maps of the region (Natmap, 1988), positions of the gravity stations were established by a Global Positioning System (GPS)–controlled survey. The GPS survey was tied to the only accessible nearby bench mark, NMF-429, located at 26°0′9.828″S, 120°59′9.122″E. Horizontal and vertical control information for this bench mark was provided by the Department of Land Administration, Mapping and Survey Division, Geodetic Survey Services. Because the bench mark was fairly inaccessible, the GPS base station was set up at camp and tied to the bench mark by occupying the bench mark as a rover station.

Locations for the gravity stations were computed by collecting positional information at each station for 10 min at 1-s intervals (600 readings) and differentially correcting the locations after the fact using the camp base station as a reference. The typical standard deviations of the station locations were 0.2 m horizontal north, 0.2 m horizontal east, 0.6 m vertical. The datum for reduction of the GPS data was Australia Geodetic Datum 1966 to conform with the 1:250,000 Nabberu topographic map sheet.

Approximately 140 stations were acquired over the structure using a Lacoste Romberg gravity meter. Additional stations were planned over the western portion of the structure but a malfunction of the meter precluded completion of the survey. Gravity readings were collected along roads and fence lines and were spaced approximately 500 m apart (station spacing was estimated in the field using the vehicle odometer). About 10 data points on the western edge of the structure across the Frere Range, used in this analysis, were collected in 1986 by Eugene Shoemaker.

Due to the absence of an absolute gravity base within the region, gravity values were tied to a base station value determined using the theoretical value for gravity at this latitude (979015.76 mGal). Base station readings were made approximately every 4 h; observed meter drift was of the order 0.007 dial divisions (0.01 mGal) per hour. Reductions were made using the U.S. Geological Survey Bouguer Gravity Program and include latitude, free air, Bouguer, and earth curvature. Terrain corrections were not applied because relief was quite limited (station elevations ranged from 534.1–569.1 m: total relief of only 35 m) and topographic maps of the region at sufficient precision to make terrain corrections are lacking. A density of 2.67 g cm^{-3} was assumed for the Bouguer correction. The typical 0.6-m standard deviation mean in the vertical coordinate would result in a corresponding uncertainty of ± 0.12 mGal in the reduced Bouguer gravity value.

GRAVITY DATA ANALYSIS

The reduced Bouguer gravity values for the stations were contoured using a 35- × 35-node grid (resulting in grid spacing of approximately 1 km) and applying a first-order polynomial to the data for smoothing to produce the Bouguer gravity anomaly map (Fig. 4). The gridding program uses a weighting function based on distance to adjacent data points to calculate the value of the grid point (the farther the data point from the grid node, the less weight given; the more dense the data points, the better control on the contours). The result is a contoured surface over the entire grid area regardless of whether there are any data points within a particular area. Thus, in the southeast and southwest quadrants, where data are missing, contours are shown. These contours represent the field averaged across the area and do not necessarily accurately portray the field. Small anomalies or changes in those areas would not be depicted by this method. The contours in these areas are not intended as an accurate representation of the field.

Variable order polynomial surfaces were applied to the data and a set of residuals calculated to better isolate the anomaly associated with the structure. Polynomial surfaces of first to sixth order were examined. The first-order surface (a simple planar surface), accounting for 46% of the signal, provides a reasonable illustration of the residual anomaly associated with Teague Ring. Because of the nonuniform distribution of the data points and the heterogeneous density of the basement rocks, it is difficult to resolve whether individual short-wavelength features observed in the contoured gravity field are associated with the structure or the underlying crystalline basement. Subtraction of higher order polynomial surfaces results in the creation of numerous short-wavelength anomalies that may be artifacts and not geologic features.

A simple Bouguer gravity map (including all standard corrections except terrain) of the Teague Ring structure is illustrated in Figure 5A. A three-dimensional view of the same data is presented in Figure 5B. The data show a gravity field decreasing to the northeast with a pronounced superimposed circular negative anomaly. Typical regional gradients are just under 1 mGal km^{-1}. The general gravity decrease from southwest to northeast is consistent with a northward thickening wedge of the Earaheedy Group rocks within the Nabberu Basin overlying the crystalline basement of the Archean Yilgarn Block. The close contour low is centered over the Teague Ring structure.

The prominent closed contour low over the structure has approximately 12 mGal of negative relief and closely parallels the collar of Frere Formation surrounding the crystalline core. Isogals are deflected around the structure out a diameter of 31 km, consistent with the dimension of the gap of the northwest-trending Frere Formation (Fig. 3). Locally, the contours suggest that the Frere Formation exhibits a local gravity high. The contours define an anomaly elongate to the northwest; this may be an artifact of the data distribution and gridding procedure or it may represent the effects of the northwest strike of the regional geology. Data are insufficient to resolve which is the case.

Residual gravity maps enhance the negative anomaly associated with the structure. Figure 5A,B shows a contour map and a three-dimensional view of the first-order residual gravity field. The residual data indicates typical gravity relief of about

170 J. B. Plescia

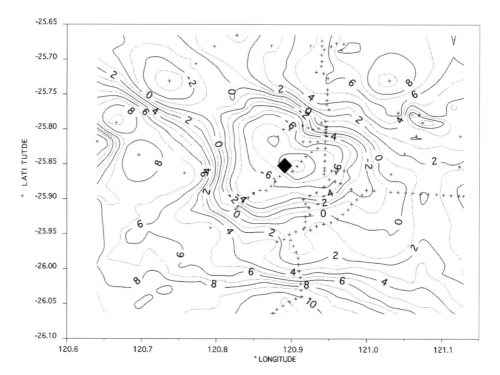

Figure 4. A, Simple Bouguer gravity map of the Teague Ring structure. Crosses indicate station locations; large diamond indicates center of the structure. Contour interval, 2 mGal. B, Three-dimensional view of the Bouguer gravity field.

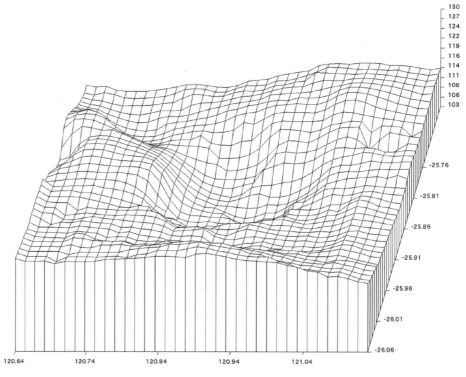

12 mGal between the center of the structure and the surroundings, not significantly different from the Bouguer gravity. Gravity relief appears to be greater with respect to the area north of the structure than to the south. This difference in relative gravity relief correlates with the surface geology. To the north lies exposures of the Princess Range Quartzite and the Wandiwarra Formation; to the south lies exposures of granite. These data suggest that the crystalline rocks to the south have a relatively higher bulk density.

Deflection of the isogals in the residual map over the northern margin of the crystalline core indicates that the quartz syenite exhibits ~1-mGal higher gravity than adjacent areas of

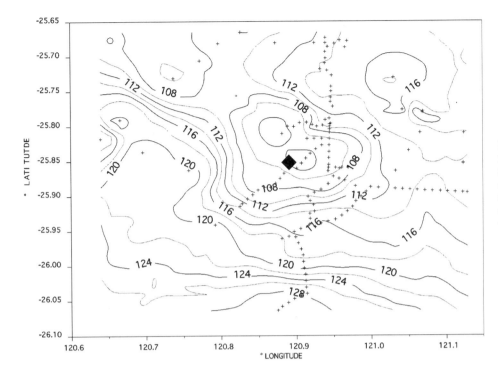

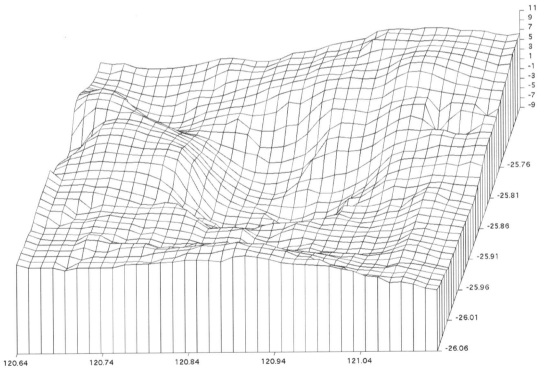

Figure 5. A, First order residual gravity map of the Teague Ring structure. Crosses indicate station locations; large diamond indicates center of the structure. Contour interval is 2 mGal. B, Three-dimensional view of the first-order residual gravity field.

the crystalline core, implying the syenite has a slightly higher bulk density relative to the other crystalline rock.

GRAVITY MODEL

In an attempt to characterize the subsurface structure of Teague Ring, a series of two and half dimensional gravity models were constructed. The model is oriented in a northeast-southeast direction, depicted by the line in Figure 3, along the regional dip of the Earaheedy Basin and lying along the densest distribution of data. Zero kilometers in the profile corresponds the contact between the Earaheedy Group rocks of the ring syncline and the crystalline basement.

Detailed modeling is inhibited by the lack of subsurface

information and the general lack of physical property data for the rock units in the area. In addition, the widespread cover of Quaternary material hides the details of the basement lithology. Thus, only generalized model can be constructed. However limited, these models do provide broad constraints on the crustal structure around Teague Ring.

The basic elements of the regional geology incorporated into the gravity model include the Proterozoic crystalline basement (Yilgarn Block); the shallow structural trough of Earaheedy Group rocks along the southern edge of the core, the deeper structural trough of Earaheedy Group rocks along the northern edge of the core and the northeast thickening wedge into the Nabberu Basin, and a central core of crystalline rock. Using just these four basic elements reproduces the general amplitude and shape of the long-wavelength anomaly associated with Teague Ring. However, such a simple model misses some of the short-wavelength attributes. Specifically, this simple model does not completely match the observed data along the southern portion of the profile. This discrepancy indicates that the structure along the southern side of Teague Ring is more complicated than a simple uniform basement.

In order to better match the long- and short-wavelength attributes of the observed gravity, a more complex model is presented in Figure 6. This model includes the four elements noted above plus two additional bodies. Density contrasts of all of the modeled bodies with respect to the deeper crystalline basement are Earaheedy Group rocks (horizontally lined bodies) northeast and southeast of the core is -0.07 g cm^{-3}; the crystalline core material (stippled body) is -0.13 g cm^{-3}; and small bodies representing inhomogeneities in the crystalline basement (shaded bodies) -0.10 g cm^{-3} (body at 0 km) and $+0.05$ g cm^{-3} (body near -10 km).

The lack of seismic reflection or refraction data in the region precludes better constraints on the model. The central zone of relatively low-density material can be modeled as either a larger body with a relatively small density contrast or a smaller body with a larger density contrast. The model illustrates a body extending to a depth of 5 km with a density contrast of -0.13 cm^{-3}. There are no data to indicate the specific nature of smaller bodies on the southwest end of the profile; they may simply represent higher and lower density zones of the crust with respect to the large-scale average density or individual plutons or other compositionally distinct zones.

The model displays the regional geology as reflected in the gravity data. Because of the level of erosion, most of the structure typically associated with a complex impact crater (e.g., central peak, terraced rim) have been eroded. Hence this model does not depict the crustal cross section typically associated with an impact. Rather, the only aspects of the model directly associated with the structure are the central core area of low-density material and the adjacent troughs of Earaheedy rocks. The central core of low-density represents the area associated with the lowermost portions of the central uplift. The low density suggests that the material has been fractured, reducing its bulk density; large-scale brecciation is not observed. The synclinal troughs of Earaheedy rocks represent the ring syncline typi-

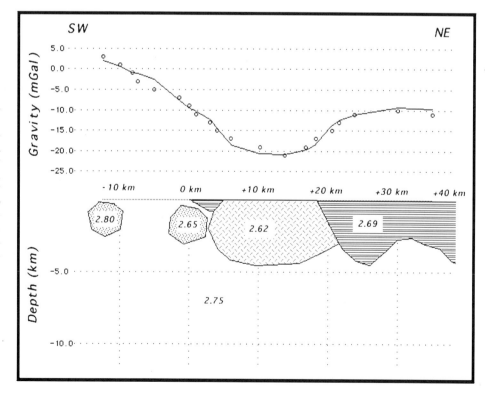

Figure 6. Two and half dimensional gravity model of the Teague Ring structure. Profile is oriented in a northeast-southwest direction. Top, Relative observed gravity (open circle) and modeled gravity (continous line). Vertical exaggeration, 2×. Bottom, Model bodies: horizontally striped bodies indicate Earaheedy Group rocks; stippled body represents the low-density core material; shaded bodies represent density anomalies within the crystalline crust. Numbers denote the densities of the bodies in g cm^{-3}.

cally observed around the margin of the central uplift in complex craters. These synclines do not exhibit pronounced individual anomalies; rather, they merely influence the slope of the overall anomaly.

DISCUSSION

The gravity data for Teague Ring clearly define a significant negative anomaly (–12 mGal) associated with the structure. Of the approximately 150 known impact structures (Grieve et al., 1995), only about one-third have been studied to determine their gravity signature (summarized in Pilkington and Grieve, 1992). Gravity anomalies are associated with all size structures; structures as small as 1–2 km have significant anomalies. Figure 7 illustrates the gravity anomaly for Teague Ring and for other features (>1 km) whose anomalies have been defined. The anomaly observed for Teague Ring is consistent with the anomalies associated with other impact features. Teague Ring lacks the paired central positive anomaly and annular negative anomaly typical of many complex impact structures because of the erosional level at Teague.

Impact structures of similar diameters to Teague Ring have comparable gravity anomalies. For example, the Mjolnir Structure in the Barents Sea north of Scandinavia is ~39 km in diameter and exhibits a negative annular anomaly of –3 mGal and a central positive anomaly +4 mGal above the annular low (Gudlaugsson, 1993), resulting in total gravity relief of 7 mGal. The West Clearwater Lake structure in Quebec has a diameter of 32 km and exhibits a –16-mGal anomaly (Plante et al., 1990). Finally, the Manson structure in Iowa is about 38 km in diameter and exhibits a central positive anomaly of +4 mGal over the central peak surrounded by an annular negative anomaly of –2 to –4 mGal corresponding to the structural moat and terrace zone, producing about 6–8 mGal of gravity relief (Plescia, 1996).

The gravity model data (Fig. 7) suggest that the low-density material associated with the crystalline core extends to a depth of several kilometers. The density and the vertical extent of the core density contrast region can be varied to produce similar anomalies. The model uses a vertical extent of 5 km and has a negative density contrast of –0.12 cm^{-3}; unfortunately, there are no other data types (e.g., seismic refraction) for Teague Ring to constrain this aspect. These dimensions and densities are, however, consistent with data for a few other structures.

Pilkington and Grieve (1992) summarized density contrasts between fractured and unfractured target rocks for several impact structures. In crystalline terrain, the relatively contrast between the unfractured and fractured basement rocks is between –0.13 and –0.17 g cm^{-3}. If the effects of sediments and breccias are included, the density contrast can reach –0.34 g cm^{-3}. Dyrelius (1988) presented data for the Siljan impact structure in Sweden (~40 km diameter), where a density contrast of –0.04 to –0.09 g cm^{-3} extends to depths of >5 km.

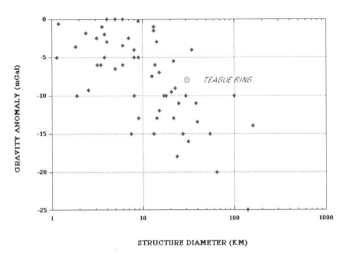

Figure 7. Gravity anomalies of impact structures plotted as a function of the structure diameter. Anomaly for Teague Ring is denoted by the crossed circle; other structures are indicated by diamond shapes.

Pohl et al. (1977) presented data for the Ries Crater in Germany (22-km diameter). There, velocity and density values are reduced below normal to depths of as much as 6 km beneath the present structural level. Thus, the modeled density contrast of –0.13 g cm^{-3} extending to a depth of 5 km for Teague Ring is consistent with data for similar-sized impact structures elsewhere in the world.

The diameter of Teague Ring as deduced from either the gravity or the currently exposed geology does not necessarily reflect the original diameter of the structure. Teague Ring is clearly a deeply eroded impact structure and the 31-km diameter represents the extent of the deformation at a structural level below the original surface. Complex impact structures are characterized by a central uplift surrounded by a structurally depressed zone filled with allochthonous and autochthonous breccia, in turn surrounded by a faulted terrace zone that marks the edge of the structure. Only the central uplift seems to be preserved at Teague Ring; there is no evidence of a breccia layer nor terraced ring fault blocks. This degree of erosion might suggest that the original diameter of Teague Ring was greater.

The amount of erosion at Teague Ring can be estimated from two methods, stratigraphy and morphometry. The first is a simple stratigraphic assessment of the thickness of the units involved in the deformation. This is a minimum estimate, as it includes only preserved units; higher stratigraphic levels may have been involved in the deformation and completely removed. The highest unit observed within the ring syncline is the Wongawol Formation (observed in a small exposure in the northeast quadrant of the ring syncline). Based on average regional thicknesses of the units below the Wongawol Formation, about 2–3 km of erosion would have taken place to reach the crystalline basement.

The morphometric approach involves the use of a relation presented by Grieve et al. (1995) and Grieve and Pilkington

(1996), which indicates the degree of structural uplift in the central peak of a complex crater to be:

$$h_{SU} = 0.086 D^{1.03}, \quad (1)$$

where h_{SU} represents the structural uplift of the central peak material, and D is the diameter of the crater in kilometers. At Teague Ring, the present 31 km diameter would imply an uplift of 3.0 km. This value must be combined with the topographic expression of the crater which, based on data from Grieve et al. (1995) and Grieve and Pilkington (1996), would amount to 300–700 m. As a deformed core of crystalline basement is still exposed, it suggests that the erosion is less than the 4.0 km. Both methods provide similar approximate bounds on the amount of erosion, probably a few kilometers. In addition, the reported presence of shatter cones (Shoemaker and Shoemaker, 1996) suggests that a deep structural level has been reached as shatter cones occur only at relatively low shock pressures (2–30 GPa).

In an attempt to assess the original diameter, several approaches can be taken using morphometric and structural comparisons to other structures. One limit on the diameter can be inferred from the morphometry of the structure; knowing the diameter of the central peak, the postcollapse rim-to-rim diameter can be estimated based on using the relation (Pike, 1985):

$$D_{cp} = 0.199 D^{1.058}, \quad (2)$$

where D_{cp} is the diameter of the central peak, and D is the rim-to-rim diameter. If the assumption is made that the core of crystalline rocks at Teague Ring represents the diameter of the central peak (10 km), the original diameter would have been 42 km.

An alternate approach is the use of structural comparisons with other large terrestrial structures, in this case a comparison of the relative position of the ring syncline to the total diameter. At Upheaval Dome, Utah (Kriens et al., 1997), the ring syncline axis occurs at a distance of $0.69 R$ (where R is the radius of the structure). At the Carswell structure in Canada, the axis of the ring syncline occurs at about $0.93 R$ (Currie, 1969). Large-scale explosive craters also show ring synclines. At the Snowball and Prairie Flat tests, each of which was a 500-ton TNT explosion (Roddy, 1977), the ring syncline axis occurs at distances of 0.6–$0.75 R$. Such a structural analogy predicts an original diameter of 22–33 km. The low end of the range is clearly less than the current observed diameter; the high end would suggest a structure slightly larger than the present diameter. Although there is considerable uncertainty in these estimates, it would suggest that, although there has been considerable erosion, the original diameter of Teague Ring was probably not significantly greater than its present diameter.

CONCLUSIONS

Teague Ring is a 31-km diameter impact structure in Western Australia that is geologically defined by an annulus of deformed Earaheedy Group rocks (principally Frere Formation) and a core of fractured crystalline granitic rock. The structure interrupts the northwest-trending strike of the Frere Range and deforms the gentle northeast dips of the sedimentary rocks of the Nabberu Basin.

A gravity survey reveals a significant negative gravity anomaly of –12 mGal and a diameter of approximately 31 km. The anomaly can be modeled as a mass of low-density material centered over the crystalline core, having a relative density contrast of about –0.13 g cm^{-3}, extending to a depth of ~5 km. The low density of the central core is interpreted to result from fracturing of the crystalline basement complex to a depth of several kilometers and is consistent with density contrasts observed at other similar size impact structures. Teague Ring is heavily eroded, with several kilometers of erosion having occurred; thus, the exposures represent the deepest structural levels of a complex impact crater.

ACKNOWLEDGMENTS

This research is dedicated to the memory of Eugene M. Shoemaker, who sparked my interest in the study of terrestrial impact craters and whose unbounded enthusiasm for the subject played a major role in this research and that of many others. It was also that enthusiasm that kept me going despite the flies. The assistance in the field of Donald Peipgras and Carolyn Shoemaker are also greatly appreciated. Reviews by Buck Sharpton and Lauri Pesonen helped improve this manuscript. This work was supported by the National Aeronentics and Space Administration Planetary Geology and Geophysics Program.

REFERENCES CITED

Bunting, J. A., Commander, D. P., and Gee, R. D., 1977, Preliminary synthesis of Lower Proterozoic stratigraphy and structure adjacent to the northern margin of the Yilgarn Block: Perth, Western Australia Geological Survey Annual Report 1976, p. 43–48.

Bunting, J. A., de Laeter, J. R., and Libby, W. G., 1980, Evidence for the age and cryptoexplosive origin of the Teague Ring Structure, Western Australia: Perth, Report of the Department of Mines, Western Australia, for the Year 1979 Annual Report, p. 125–129.

Bunting, J. A., Brakel, A. T., and Commander, D. P., 1982, Nabberu Western Australia, geologic series map: Perth, Geological Survey of Western Australia, sheet SG/51–5, scale 1:250,000, 27 p.

Butler, H., 1974, The Lake Teague Ring Structure, Western Australia: an astrobleme? Search, v. 5, p. 533–534.

Compston, W., and Arriens, P. A, 1968, The Precambrian geochronology of Australia: Canadian Journal of Earth Science, v. 5, p. 561–583.

Cooper, J. A., Nesbitt, R. W., Platt, J. P., and Mortimer, G. E., 1978, Crustal development in the Agnew region, Western Australia, as shown by Rb-Sr isotopic and geochemical studies: Precambrian Research, v. 7, p. 31–59.

Currie, K. L., 1969, Geological notes on the Carswell circular structure, Saskatchewan: Geological Survey of Canada Paper 67–32, 60 p.

Dyrelius, D., 1988, The gravity field of the Siljan ring structure, in Boden, A., and Eriksson, K. G., eds., Deep drilling in crystalline bedrock, Vol. 1, The deep gas drilling in the Siljan Impact Structure, Sweden and astroblemes: New York, Springer-Verlag, p. 85–94.

Gee, R. D., de Laeter, J. R., and Drake, J. R., 1976, Geology and geochronology of altered rhyolite from the lower part of the Bangemall Group near Tangadee, Western Australia: Perth, Western Australia Geological Survey Annual Report 1975, p. 112–117.

Goode, A. D. R., Hall, W. D. M., and Bunting, J. A., 1983, The Nabberu Basin of Western Australia, in Trendall, A. F., and Morris, R. C., eds., Iron-formations: facts and problems: Amsterdam, Netherlands, Elsevier Scientific, p. 295–325.

Grieve, R. A. F., and Pilkington, M., 1996, The signature of terrestrial impacts: AGSO Journal of Australian Geology and Geophysics, v. 14, p. 399–420.

Grieve, R. A. F., Rupert, J., Smith, J., and Therriault, A., 1995, The record of terrestrial cratering: GSA Today, v. 5, p.189, 194–196.

Gudlaugsson, S. T., 1993, Large impact crater in the Barents Sea: Geology, v. 21, p. 291–294.

Hall, W. D. M., and Goode, A. D. T., 1978, The Early Proterozoic Nabberu Basin and associated Iron Formations of Western Australia: Precambrian Research, v. 7, p. 129–184.

Hall, W. D. M., Goode, A. D. T., Bunting, J. A., and Commander, D. P., 1977, Stratigraphic terminology of the Earaheedy Group, Nabberu Basin: Perth, Western Australia Geological Survey Annual Report 1976, p. 40–43.

Horwitz, R. C., 1975a, The southern boundaries of the Hamersley and Bangemall Basins of sedimentation, in Australian Geological Convention,1st, Proterozoic Geology: Adelaide, Geological Society of Australia, p. 91.

Horwitz, R. C., 1975b, Provisional geology map at 1:250,000 of the north-east margin of the Yilgarn Block, Western Australia: Wembley, Western Australia, CSIRO Mineral Research Laboratory Report F, P. 10.

Kriens, B. J., Herkenhoff, K. E., and Shoemaker, E. M, 1997, Structure and kinematics of a complex crater: Upheaval Dome, southeast Utah, in Conference on Large Meteorite Impact and Planetary Evolution: Houston, Texas, Lunar and Planetary Institute, LPI Contribution 922, p. 29–30.

Natmap, 1988, Nabberu Sheet SG 51-1, National Topographic Map Series: Belconnen, Australia, NATMAP, scale 1:250,000

Pike, R. J., 1985, Some morphologic systematics of complex impact structures: Meteoritics, v. 20, p. 49–68.

Pilkington, M., and Grieve, R. A. F., 1992, The geophysical signature of terrestrial impact craters: Reviews of Geophysics, v. 30, p. 161–181.

Pirajno, F., 1998, Shoemaker Impact Structure—GSWA renames Teague Ring Structure after Eugene Shoemaker: Fieldnotes, Geological Survey Newsletter for the Mineral and Petroleum Industries, February 1998, no. 9, p. 1–2.

Plante, L., Seguin, M. K., and Rondot, J., 1990, Etude gravimetrique des astroblems du Lac a l'Eau Clari, Nouveua Quebec: Geoexploration, v. 26, p. 303–323.

Plescia, J. B., 1996, Gravity investigation of the Manson impact structure, Iowa, in Koeberl, C., and Anderson, R. R., eds., The Manson Impact Structure, Iowa: anatomy of an impact crater: Geological Society of America Special Paper 302, p. 89–104.

Pohl, J., Stoffler, D., Gall, H., and Ernstson, K., 1977, The Ries impact crater, in Roddy, D. J., Pepin, R. O., and Merrill, R. B., eds., Impact and explosion cratering: New York, Pergamon Press, p. 343–404.

Preiss, W. V., Jackson, M. J., Page, R. W., and Compston, W., 1975, Regional geology, stromatolite biostratigraphy and isotopic data bearing on the age of a Precambrian sequence near Lake Carnegie, Western Australia, Australian Geological Convention, 1st, Proterozoic Geology: Adelaide, Australian Geological Convention, p. 92–93.

Roddick, J. C., Compston, W., and Durney, D. W., 1976, The radiometric age of the Mount Keith Granodiorite, a maximum age estimate for an Archean greenstone sequence in the Yilgarn Block, Western Australia: Precambrian Research, v. 3, p. 55–78.

Roddy, D. J., 1977, Large-scale impact and explosion craters: comparisons of morphological and structural analogs, in Roddy, D. J., Pepin, R. O., and Merrill, R. B., eds., Impact and explosion cratering: New York, Pergamon Press, p. 185–246.

Shoemaker, E. M., and Shoemaker, C. S., 1996, The Proterozoic impact record of Australia: AGSO Journal of Australian Geology and Geophysics, v. 16, p. 379–398.

Stuckless, J. S., Bunting, J. A., and Nkomo, I. T., 1981, U-Th-Pb systematics of some granitoids from the northeastern Yilgarn Block, Western Australia, and implications for uranium source potential: Journal of the Geological Society of Australia, v. 28, p. 365–375.

MANUSCRIPT ACCEPTED BY THE SOCIETY DECEMBER 16, 1998

BP and Oasis impact structures, Libya, and their relation to Libyan Desert Glass

Begosew Abate* and Christian Koeberl
Institute of Geochemistry, University of Vienna, Althanstrasse 14, A-1090 Vienna, Austria;
e-mail: christian.koeberl@univie.ac.at

F. Johan Kruger
Hugh Allsopp Laboratory, Bernard Price Institute, University of the Witwatersrand, Johannesburg 2050, South Africa

James R. Underwood, Jr.*
Department of Geology, Kansas State University, Manhattan, Kansas 66506-3201, USA

ABSTRACT

We have conducted petrographic, geochemical, and isotopic studies on a suite of target rock samples from two impact structures, BP and Oasis, in southeastern Libya. Both structures occur in Lower Cretaceous sandstone of the Nubia Group and are deeply eroded. Earlier microscopic thin-section studies provided evidence for the impact origin of the structures by demonstrating the presence of shock-characteristic planar deformation features in quartz. No absolute ages have yet been determined for these structures. From the proximity of these structures to the occurrence of the enigmatic Libyan Desert Glass (LDG), previously interpreted as impact glasses formed from a mature sandstone, and from the absence of disturbed strata of the sandstone of the Nubia Group in the area of the occurrence of the LDG, a possible relation was suggested between BP and Oasis impact structures and the LDG.

Most of the target rocks at both structures have somewhat lower SiO_2 and higher contents of other major oxides than LDG, but this distinction disappears when the composition of the sandstones is recalculated to a water-free basis. Also, there is a good correlation between the major and trace element compositions of samples from the target rocks of BP and Oasis structures and samples of LDG, but the refractory trace element content of the LDG is generally somewhat higher than that of the target rocks of the two Libyan craters. Rare earth elements (REE) from the BP and Oasis impact structures and from LDG have similar abundances and display a similar chondrite-normalized pattern. The isotopic ratios of Nd and Sr for samples of the two structures and for the LDG are in a similar range. They are characterized by negative ε_{Nd} values and positive ε_{Sr} values, which are characteristic of upper continental crustal rocks. In an ε_{Nd} vs. ε_{Sr} diagram, LDG values plot within the field defined by the BP and Oasis rocks, and in a $1/Sr$ vs. ε_{Sr} plot LDG is within the range defined by BP and Oasis target rocks.

From the available petrographic, geochemical, and isotopic data, we conclude that the target rocks of the Libyan structures could represent the parent material for LDG. However, without further age information it is not possible to conclude unambiguously that the BP or Oasis structures are the source craters for the LDG.

---

* Present addresses: (Abate) Ethiopian Institute of Geological Surveys, P.O. Box 30389, Addis Ababa, Ethiopia; (Underwood) 9518 Topridge Dr. #3, Balcones Place, Austin, TX 78750-3500, USA.

Abate, B., Koeberl, C., Kruger, F. J., and Underwood, J. R., Jr., 1999, BP and Oasis impact structures, Libya, and their relation to Libyan Desert Glass, *in* Dressler, B. O., and Sharpton, V. L., eds., Large Meteorite Impacts and Planetary Evolution II: Boulder, Colorado, Geological Society of America Special Paper 339.

INTRODUCTION

Currently, only 18 impact structures are known to occur in the whole African continent; for details, see the review by Koeberl (1994a), who listed 15 African impact structures. Since then the Gweni Fada (Chad; Vincent and Beauvilain, 1996), Sinamwenda (Zimbabwe; Master et al., 1996), and Morokweng (South Africa; Corner et al., 1997; Koeberl et al., 1997) structures have been added. Compared to other continents, this represents a number that is very small, indicating that many other impact structures remain to be discovered in Africa. In addition, the majority of these impact structures are not only small (a few kilometers in diameter, with the exception of Vredefort and Morokweng), but most of them are also not well studied. Impact structures are not only of considerable scientific interest, they may also have economic importance (e.g., Masaitis and Grieve, 1994). For these and other reasons, it is important to detect impact structures and to undertake detailed geologic, petrographic, and geochemical studies.

The BP and the Oasis structures in southeast Libya are deeply eroded and appear as concentric ridges of deformed rocks that rise above the surrounding desert plain. They were first mentioned by Kohman et al. (1967), who described their location based on space and aerial photography. Martin (1969) provided further descriptions of BP, the smaller of the two structures. The BP structure is located at 25°19′ N and 24°20′ E, and the Oasis structure is located at 24°35′ N and 24°24′ E (Fig. 1), close to the Libyan-Egyptian border. The names resulted from the fact that the Oasis structure was first visited in the 1960s by a field party of the Oasis Oil Company of Libya, and the other structure was visited around that time by a team from the B.P. Exploration Company. The impact origin of the structures was confirmed in the early 1970s by French et al. (1972, 1974). It has been suggested that they were formed simultaneously by a double impact (Underwood and Fisk, 1980). This is based on rather weak evidence: they are close to each other, about 80 km apart, and the degree of erosion of the two structures is similar. So far it has not been possible to obtain an absolute age for any of these structures; the ages are inferred to be post-Nubian (i.e., postdating the Lower Cretaceous target rocks), thus providing no evidence for or against a simultaneous origin.

Based on the proximity of these structures to the area of the occurrence of Libyan Desert Glass (LDG) about 150 km to the east, and the lack of disturbed sandstone strata of the Nubia Group in the area of the occurrence of the glass, several workers (e.g., Martin, 1969; Underwood and Fisk, 1980; Murali et al., 1988; cf. Koeberl, 1997) have suggested that the LDG might be related to the two Libyan impact structures.

Libyan Desert Glass is an enigmatic natural glass found in an area of about 6,500 km² between sand dunes of the southwestern corner of the Great Sand Sea in western Egypt, near the Libyan border (Fig. 1). The glass occurs as centimeter- to decimeter-size irregular and strongly wind-eroded pieces. The total quantity of the glass present has been estimated at 1.4×10^9 g,

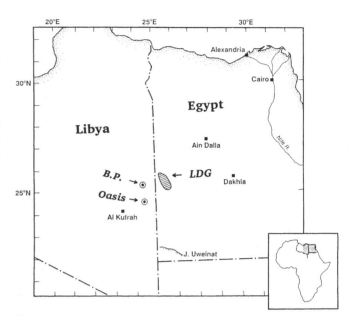

Figure 1. Location map, showing parts of Egypt and Libya. The position of the LDG field (in western Egypt) is shown in relation to the BP and Oasis impact structures in neighboring Libya. In general, no LDG is found in Libya. The name derives from the classical name for the northeast part of the Sahara, known since the time of Herodotus as the Libyan Desert.

with a much larger original mass assumed (Barnes and Underwood, 1976; Diemer, 1997). The age of the LDG was determined by K-Ar and fission-track methods. While it is possible to calculate a K-Ar age for LDG (58.3 ± 16.4 Ma; Matsubara et al., 1991), these values suffer from the very low K content of the glass, and other methodologic problems (Horn et al., 1997). Better suited for an age determination of the LDG is the fission-track method, which gave ages ranging from 28.5 ± 2.3 Ma (Storzer and Wagner, 1971) to 29.4 ± 0.5 Ma (plateau age; Storzer and Wagner, 1977), which was recently confirmed by Bigazzi and de Michele (28.5 ± 0.8 Ma; 1996); see also Bigazzi and de Michele (1997) and Horn et al. (1997).

The origin of LDG has been the subject of much debate since it was discovered (cf. Clayton and Spencer, 1934; Weeks et al., 1984; Diemer, 1997). Many workers were of the opinion that LDG is an impact glass (e.g., Kleinmann, 1969; Jessberger and Gentner, 1972; Barnes and Underwood, 1976), but were deterred by the lack of a suitable impact crater. To avoid this problem, Seebaugh and Strauss (1984) suggested that the glass formed in a cometary collision, where shock-melting occurred without crater formation. Another suggestion for the origin of LDG involved a sol-gel process, based on the asserted occurrence of organic material in LDG samples (Jux, 1983). In a similar model, Feller (1997) suggested a sedimentary origin, and Futrell and O'Keefe (1997) hypothesized—ignoring geochemical data—that the glass formed from welded lunar glassy spherules.

However, most researchers have now accepted the geochemical and geologic evidence for an impact origin of LDG (see, for example, Diemer, 1997; Storzer and Koeberl, 1991; Koeberl, 1997; Murali et al., 1987, 1988, 1989, 1997; Rocchia et al., 1996, 1997; Abate et al., 1997). Evidence for an impact origin includes the presence of schlieren and partly digested mineral phases, such as lechatelierite (a high temperature melt of quartz); baddeleyite, a high-temperature breakdown product of zircon (Kleinmann, 1969; Storzer and Koeberl, 1991; Horn et al., 1997); and the likely existence of a meteoritic component (Murali et al., 1987, 1988, 1989, 1997; Barrat et al., 1997). Although in the last few years the evidence for an impact origin of the LDG has grown, the source crater, if it exists, remained undiscovered, leading Horn et al. (1997) to revive the suggestion of in situ formation of the glass from desert sand and sandstone.

As mentioned above, the geographic proximity of the LDG site to the BP and Oasis structures has previously led to speculation that one of these two structures might be the LDG source. However, Diemer (1997) speculated that the BP and Oasis structures, which occur in Nubian sandstone of Early Cretaceous age, are older than the LDG. Giegengack and Underwood (1997) also thought that the target rocks that are exposed at the BP and Oasis structures at the present time were covered at the time of the crater formation under about 400 m of younger sediment. However, recent geochemical investigations related to the origin of Libyan Desert Glass (Barrat et., 1997; Horn et al., 1997; Koeberl, 1997; Abate et al., 1997) have suggested that the target rocks of the BP and Oasis structures are good candidates for the source material of LDG. In particular, Horn et al. (1997) pointed out the importance of Sr and Nd isotope analysis of the target rocks from the BP and Oasis structures in inferring any direct possible relation of LDG with these structures. In the present study we have conducted petrographic studies and, for the first time, major element, trace element, and isotopic analyses to characterize the rocks from these two structures and to compare their geochemical characteristics with those of LDG.

GENERAL GEOLOGY OF OASIS AND BP IMPACT STRUCTURES

At the present erosion level, the rocks that currently crop out at the both structures are quartz sandstones with minor conglomerates and siltstones. These rocks are sandstone strata of the Nubia Group probably of Early Cretaceous age (Goudarzi, 1970), which is a ferruginous fine- to medium-grained quartz sandstone containing abundant cross-beds, mud cracks, ripple marks and silicified wood. It is characterized by buttes and mesas of horizontal beds and northwest-trending sandstone ridges. The Nubian sandstone in the general area of the structures reaches a thickness of about 1,700 m and has experienced low-grade metamorphism and regional deformation (Underwood, 1975, 1976; Underwood and Fisk, 1980). Both structures are visible from space and air (cf. NASA Space Shuttle image of Oasis structure: Fig. 7 in Koeberl, 1994a). Figure 2A shows aerial photographs of Oasis and BP (Fig. 2B). Figure 3a–d shows some landforms of two crater

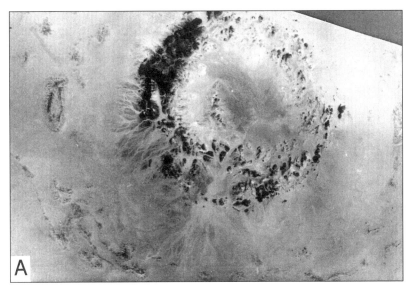

Figure 2. A, Aerial photograph of the Oasis impact structure. The inner ring of hills (diameter, 5.1 km) is clearly visible. The outer set of disturbed strata (11.5-km diameter) is partly visible on the bottom part of the image. B, Aerial photograph of BP impact structure, showing the two rings of hills and the central peak. North is up in both images.

Figure 3. A, BP structure; view of central peak (central block; cf. Fig. 2b) from the middle ring; view toward north-northwest. B, center of BP structure, with strata dipping in a variety of directions (location of samples 15-1 to 22-2). C, View across center Oasis structure from high point on southwest rim; view toward northeast (cf. Fig. 2A). Location of samples 41-1 /-2 in center of picture (low-lying hills). D, Prominent ring of hills (5.1-km diameter) of Oasis structure, west quadrant of structure.

structures. The following description is largely summarized after Underwood and Fisk (1980).

The Oasis structure is a ring-shaped feature with a prominent circular range of hills approximately 100 m high, which forms an inner ring with a diameter of 5.1 km (Fig. 2A). The Oasis structure lacks a central peak (unlike the BP crater). Most of the rocks within the central ring of hills are intensely folded, and beds may be vertical or overturned. Previously, a diameter of 11.5 km was assigned to the Oasis structure, based on an outer zone of disturbed strata that is visible in the field (Underwood and Fisk, 1980) and from the air (Fig. 3A). However, disturbed strata, which are covered by a thin veneer of sand, extend to a diameter of 18 km, as determined from recent NASA Space Shuttle radar studies (McHone et al., 1995a,b).

The BP structure was first described by Martin (1969), who suggested that the structure might be of meteorite impact origin. It was earlier occasionally referred to as the Jebel Dalma structure (cf. Underwood and Fisk, 1980) and is a relatively small structure defined by two discontinuous rings of hills and a central peak (Figs. 2B and 3A,B). The inner ring has a diameter of 2 km, with an average height of 30 m, and the outer ring, which is characterized by strata that dip inward at fairly low angles (3°–15°), has a diameter of 2.8 km and a maximum relief of only 20 m. The rocks at the center of the structure show intense jointing and bedding is difficult to discern. Rocks exposed are the Cretaceous Nubia Group and include quartz sandstone, siltstone, and conglomerate. Although the outer ring has a diameter of 2.8 km, recent NASA Space Shuttle radar studies have shown that disturbed beds, which are covered by a thin veneer of sand, extend to a diameter of 3.2 km (McHone et al., 1995a,b).

SAMPLES AND ANALYTICAL TECHNIQUES

Initial field work and sampling were carried out by Underwood and Fisk (French et al., 1974). From their collection, we used 15 samples from the BP structure, 11 from the Oasis structure, and 3 samples from the East site, to the east of the two impact structures, about 160 km east-northeast of the Kufra Oasis, representing undeformed sandstone of the Nubia Group that is not related to either of the two impact structures (cf. French et al., 1974). Because no detailed descriptions of rocks from the structures are available, we have characterized the sample locations in as much detail as possible, from 1970s field notes. Furthermore, Figure 3A–D gives field impressions of the sample sites. Sample numbers were originally preceded by U; for our study, we indicated the location by adding the letters BP, O (Oasis), and E (East).

The samples were from the following locations: BP-structure: 11-1: from middle (or intermediate), outward dipping ring of strata just east of water gap, allowing drainage of structure to the south; several gently plunging folds are present, with axes tangential to the structure; 15-1: from high-standing north half of the center block (center of Fig. 3A); 16-1: from highest point of high-standing center block; 17-1: reddish-purple sample from high north flank of central block, overlooking deeply eroded south half of block, just beyond saddle west of the highest point (Fig. 3B); 18-1: from northwest segment of high-standing central block, overlooking the eroded southern half toward the southeast; 19-1: sample of light-colored Nubia sandstone from the south margin of low-lying central block (center foreground of Fig. 3A); 22-1A: sample from white, complexly folded Nubia sandstone from low-lying part of center block (center foreground of Fig. 3A); 22-2A–C: random samples of light-colored sandstone from severely eroded, low-lying southern half of center block, complexly folded, with fold axes gently and randomly plunging, should be the oldest rocks exposed in the structure; 24-1: sandstone that caps butte of outermost, inward dipping ring of strata; 4-1: sandstone from unusual circular outcrops, 100–200 m in diameter, several 100 m south-southeast of the BP structure; 6-1: sandstone from southwest segment of outermost, inward dipping ring of strata; 9-1: from east of isolated block in west flank, from outer, inward-dipping rim.

Oasis structure: 32-1A: from restricted, low-lying tongue of outcrops extending northward from southern interior of structure; these beds are closest to the center of Oasis; some beds strike N75–80°W, dip 77–80°SW; 32-1B: as 32-1A; 32-2A: as 32-1A; 33-1B: just east of north entrance to structure, northnortheast segment of prominent 5.1-km-diameter ring of disturbed strata (background, left side of Fig. 3C); 34-1: northeast segment of rock exposed in prominent 5.1-km-diameter ring of disturbed strata of the Nubia Group (background center of Fig. 3C); 39-1: very complexly folded segment of 5.1-km-diameter ring on east flank of structure (background right side of Fig. 3C); 39-1-2: other fragment from same location as 39-1; 41-1: southwest quadrant of interior of Oasis, from base of three small hills at end of ridge extending generally north (possibly dikes?) (center foreground of Fig. 3C,D); 41-1-2: other fragment from same location as 41-1; 43-1: chalky, white rock from northwest segment of low-lying outcrops in central part of structure (inside 5.1-km-diameter prominent ring of disturbed strata); coarse grains (quartz?) scattered throughout, beds dip steeply; could be a clastic dike; 43-1-2: other fragment from same location as 43-1. Samples 51-1 and 55-1 are from the east rim of the East structure, and sample 48-1 is from near the center of the feature opposite a breach in the east rim.

From most of these samples, polished thin sections were made and studied with the petrographic microscope. Unfortunately, the amount of sample available was very limited (mostly <5 g), so it was not possible to obtain thin sections from all samples. Also, great care had to be taken to have enough material available for the chemical analyses. Major elements and selected trace elements were analyzed by X-ray fluorescence (XRF) (see Reimold et al., 1994, for details), and trace elements were determined using instrumental neutron activation analysis (INAA), following methods described by Koeberl (1993).

Three samples were selected from the BP and four from the Oasis structure for isotopic analysis in the Hugh Allsopp Laboratory of the Bernard Prince Institute at the University of the Witwatersrand, South Africa. Natural isotopic compositions of $^{143}Nd/^{144}Nd$ and $^{87}Sr/^{86}Sr$ were determined from samples digested under clean room conditions, and chemical separations were made using HDP resin cation exchange columns. Methods utilized are similar to those described by Smith et al. (1985) and Brown et al. (1989). A single-collector Micromass 30 mass spectrometer and a multiple collector VG 354 mass spectrometer were used to determine the abundances of Rb, Sr, Sm, and Nd by isotope dilution and the isotopic ratios of Sr and Nd. Due to the low elemental contents, some of the isotope ratio measurements were repeated without spiking.

PETROGRAPHY RESULTS

In hand specimen, most of the samples appear very friable. Microscopic study shows that they are fine- to medium-grained quartz sandstones with little variation between samples. Rocks from all three locations (BP, Oasis, and East) are fairly similar to each other. The cement in the samples consists of quartz, iron oxides, and carbonates (Fig. 4A). Quartz grains are poorly to moderately sorted, i.e., texturally submature. Though they are compositionally mature, being composed of quartz, fracturing of the grains contributes to the large variations in grain size, introducing textural immaturity by altering the original roundness and sorting of the grains. Iron staining is present in some samples, in which case iron oxide-rich cement (Fig. 4B) gives a dark brown color to the sandstone. In addition, some samples from both structures display, in rare grains, one to four sets of planar fractures (PFs) in quartz (Fig. 5A). Many grains show intense nonplanar fracturing (Fig. 5B) and undulatory extinction (Fig. 6). Shattered and fractured quartz grains are often surrounded by a matrix of even finer, angular quartz grains, which may represent either a monomict breccia or shattered sandstone. A specimen from the Oasis structure contains fragments of microcrystalline quartz. A total of 1,844 quartz grains in 10 thin sections were counted to determine the proportion of grains showing undula-

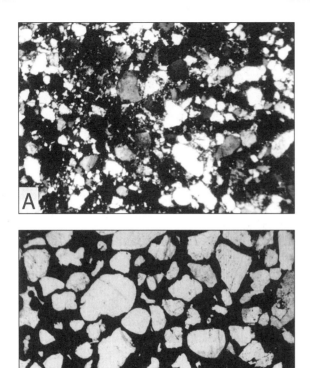

Figure 4. A, Photomicrograph showing unsorted angular fragments of quartz cemented with iron oxide and carbonate from the Oasis structure (sample O-34-1). Crossed polars, 2.5 mm across. B, Moderately rounded quartz grains in iron oxide-rich matrix in sample BP-4-1 (BP structure), indicating some sediment transport. Plane polarized light, width of image: 2.5 mm.

tory extinction, fracturing, and PFs. The results (Fig. 7) show that most grains show undulatory extinction, but the abundance of fractured grains or those containing PFs varies widely. Not enough material was available to prepare thin sections of the unaffected sandstones from the East site.

Apart from the occurrence of planar fracturing, which does not constitute unique evidence of shock metamorphism, no clear signs of shock metamorphism were found in our samples. Two quartz grains (one in a rock from BP, one from Oasis) show closely spaced PFs with two orientations, which are difficult to distinguish from true planar deformation features (PDFs), but the scarcity of these features makes a detailed study difficult. Some samples show distinct evidence of brecciation (e.g., O-41-1), which may or may not be impact-related. Thus, the shock metamorphic effects described by French et al. (1974) appear to be rare. Detailed petrographic characterizations of the samples are given in Table 1.

GEOCHEMICAL AND ISOTOPIC DATA

The major and trace element compositions of rocks from the BP structure are given in Table 2. Table 3 reports these data for

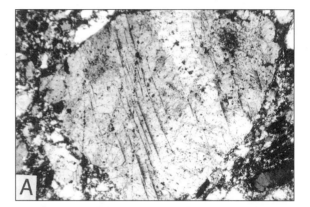

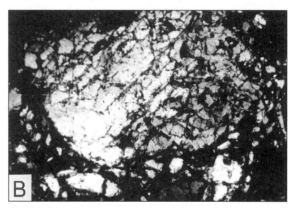

Figure 5. A, Photomicrograph showing a shocked (?) quartz grain from the Oasis structure (sample 32-1B), clearly showing two distinct sets of planar fractures. Two more sets are barely visible. Crossed polarizers, width of image 2.5 mm. B, Photomicrograph of a brecciated quartz grain (sample BP-22-2B) surrounded by iron oxide cement and angular fragments of quartz from the BP structure. Crossed polarizers, width of image = 1.0 mm.

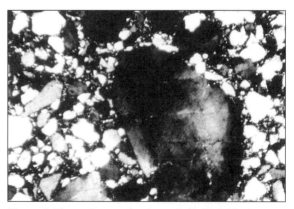

Figure 6. Photomicrograph of sample O-39-1 (Oasis structure), showing undulatory extinction in a quartz grain, surrounded by small angular quartz fragments with minor feldspar and opaque minerals. Quartz shows this effect in several sections. Crossed polarizers, width of image = 2.5 mm.

TABLE 1. PETROGRAPHIC DESCRIPTIONS OF THE TARGET ROCKS FROM THE BP AND OASIS IMPACT STRUCTURES AND FROM THE EAST SITE

Sample	Rock Type	Description
BP crater quartz sandstone		
BP-11-1	Quartz sandstone	Submature, moderately sorted, medium- to fine-grained quartz sandstone. Strong intragranular fracturing and brecciation are common in several grains. No PDFs, but one grain shows one set of poorly developed planar fractures and undulatory extinction.
BP-17-1	Quartz sandstone	Submature, poorly sorted, medium- to fine-grained quartz sandstone. The matrix consists of clay and monomict quartzitic breccia. No evidence of shock.
BP-22-1A	Quartz sandstone	Poorly sorted and brecciated quartz sandstone. The matrix contains clay; rich in iron oxide. Two sets of planar fractures, undulatory extinction, and mosaicism are observed in several grains.
BP-22-2B	Quartz sandstone	Immature, poorly sorted, fine-grained quartz sandstone. Many quartz grains in this section show intense intragranular fracturing and brecciation. Quartz displays undulatory extinction and mosaicism. No PDFs.
BP-22-2C	Quartz sandstone	Moderately sorted, brecciated quartz sandstone. Fracturing and undulatory extinction are common. One grain displays well-developed set of planar fractures.
BP-4-1	Quartz sandstone	Well-rounded and moderately sorted, iron oxide-cemented quartz sandstone. No evidence of shock.
BP-6-1	Quartz sandstone	Submature, moderately sorted, quartz sandstone, unshocked.
BP-9-1	Quartz sandstone	Immature, poorly sorted, medium- to fine-grained quartz sandstone, unshocked.
East site quartz sandstone		
E-48-1	Quartz sandstone	Submature, poorly sorted, medium- to fine-grained quartz sandstone. Quartz displays undulatory extinction and mosaicism, unshocked
Oasis crater quartz sandstone		
O-32-1A	Quartz sandstone	Unsorted, fine- to medium-grained quartz sandstone. Quartz displays fracturing, undulatory extinction, and mosaicism. Few quartz grains display poorly developed set of planar fractures.
O-32-2A	Quartz sandstone	Moderately sorted carbonate- and iron oxide-cemented quartz sandstone. No fracturing or PDFs.
O-33-1B	Quartz sandstone	Submature, poorly sorted, fine-grained quartz sandstone. Quartz displays undulatory extinctions and minor brecciation. No PDFs.
O-34-1	Quartz sandstone	Poorly sorted, fine- to medium-grained carbonate cemented quartz sandstone. Texturally and compositionally immature. Quartz shows undulatory extinction and mosaicism. No PDFs. The section also contains a grain of amphibole.
O-39-1	Quartz sandstone	Fine- to medium-grained moderately sorted quartz sandstone. Quartz displays minor fracturing and brecciation. Unshocked.
O-41-1	Quartz sandstone	Submature, moderately sorted, medium- to fine-grained, hematite-cemented quartz breccia. Quartz displays undulatory extinction and intragranular fracturing. No PDFs.
O-43-1	Quartz sandstone	Poorly sorted, fine- to medium-grained, monomict quartz breccia. Several quartz grains show two sets of planar fractures.

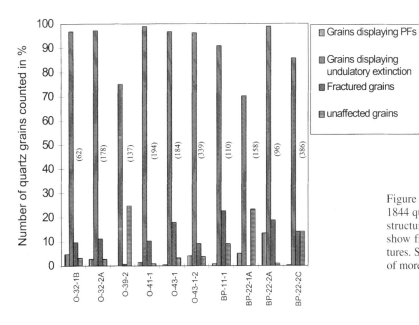

Figure 7. Histogram showing the results of point counting of 1844 quartz grains in 10 samples from the BP and Oasis impact structures. Most grains show undulatory extinction, and some show fracturing of (possibly shock-characteristic) planar fractures. Some grains show more than one effect, resulting in totals of more than 100% for each thin section. PFs = planar fractures.

TABLE 2. CHEMICAL COMPOSITION OF BP CRATER TARGET ROCKS

	BP-11-1	BP-15-1	BP-16-1	BP-17-1	BP-18-1	BP-19-1	BP-22-1A	BP-22-2A
(wt. %)								
SiO_2	92.96	94.12	93.6	92.87	94.42	93.55	95.85	90.95
TiO_2	0.08	0.05	0.07	0.07	0.09	0.07	0.07	0.11
Al_2O_3	0.48	0.33	1.97	3.71	0.47	1.84	0.92	2.83
Fe_2O_3*	0.27	0.11	0.19	0.15	0.17	0.36	0.04	0.07
MnO	0.02	0.02	0.02	0.02	0.02	0.02	0.02	0.07
MgO	<0.02	<0.02	<0.02	<0.02	<0.02	<0.02	<0.02	0.04
CaO	0.06	0.08	0.11	0.14	0.07	0.09	0.05	0.07
P_2O_5	<0.02	0.02	0.05	0.03	0.03	0.03	0.03	0.04
LOI	5.15	0.33	3.54	3.09	4.41	3.79	2.84	2.64
Total	99.02	100.06	99.55	100.08	99.68	99.75	99.82	96.82
(ppm, except where noted)								
Na	25	28	47	40	50	25	50	75
K	24	44	129	68	100	60	2.5	170
Sc	0.56	0.63	0.46	0.86	0.94	0.78	0.92	1.61
V	<20	<20	<20	<20	<20	<20	<20	<20
Cr	19.7	175	159	633	15.1	24.7	11.7	40.6
Fe	0.15	0.05	0.13	0.13	0.14	0.25	0.03	0.17
Co	0.2	0.2	0.25	0.24	0.24	0.23	0.18	0.37
Ni	<0.8	<10	2	9.52	<10	3	4.8	9
Cu	<3	<3	<3	<3	<3	<3	<3	<3
Zn	3.5	3	2.7	<2	4.95	2.2	<3	3.6
Ga	0.2	0.82	1.1	1.62	1.61	1.1	1	1.7
As	0.33	0.16	0.29	0.18	0.26	0.53	0.16	0.43
Se	0.002	0.012	0.1	0.014	0.048	0.2	0.008	0.3
Br	2.1	1.26	3.1	0.66	1.04	3.4	0.96	0.5
Rb	0.7	0.56	0.3	0.3	<1	0.4	<0.15	0.8
Sr	17	12.1	11	17.2	18.3	9.3	49	41
Y	<8	<8	<8	<8	<8	<8	<8	8
Zr	32.8	42	51	40	95	91	80	151
Nb	<6	4	7	10	9	<6	<6	6
Ag	<0.002	0.04	n.d.	<0.08	<0.04	n.d.	<0.008	n.d.
Sb	0.04	0.007	0.052	0.05	0.03	0.13	0.05	0.11
Cs	0.021	0.06	0.031	0.02	<0.12	0.011	0.024	0.018
Ba	14.4	13	72	15.7	27.7	32	17.6	56
La	5.62	5.64	4.56	5.41	6.36	5.07	26.1	24.3
Ce	9.9	10.6	8.67	11.1	13.4	10.2	54.9	52.3
Nd	4.6	5.08	3.9	7.51	7.58	5.1	28.8	25.9
Sm	0.83	0.88	0.76	1.48	1.30	0.95	4.98	4.64
Eu	0.34	1.18	0.13	0.62	0.52	0.17	1.48	0.79
Gd	1.0	0.7	0.69	0.9	0.91	0.93	2.9	3.05
Tb	0.15	0.11	0.095	0.12	0.16	0.12	0.53	0.36
Tm	0.05	0.06	0.055	0.06	0.08	0.063	0.086	0.14
Yb	0.26	0.35	0.31	0.35	0.45	0.39	0.51	0.84
Lu	0.037	0.05	0.048	0.05	0.06	0.068	0.055	0.11
Hf	0.98	1.27	1.38	1.15	2.84	2.55	1.99	3.61
Ta	0.07	0.11	0.058	0.06	0.19	0.11	0.13	0.32
W	0.08	<1	0.36	0.02	0.13	0.39	0.06	0.36
Ir (ppb)	<1	<1	<0.2	0.15	<1	<0.4	<1	<0.2
Au (ppb)	0.14	0.38	0.2	0.27	0.063	0.6	0.07	0.4
Hg	<0.006	<0.8	<0.1	<0.021	<0.003	<0.05	0.003	<0.05
Th	0.88	1.27	1.03	1.3	1.62	1.44	2.14	3.33
U	0.27	0.21	0.26	0.34	0.54	0.44	0.33	0.54
K/U	88.8	209.5	469.1	200	185.1	136.3	7.57	314.8
Zr/Hf	33.47	33.07	36.96	34.78	33.45	35.69	40.20	41.83
Zr/U	121.4	200.0	196.1	117.6	175.9	206.8	242.4	279.6
Co/Ni	0.20	0.20	0.13	0.03	0.24	0.08	0.04	0.04
La/Sc	10.04	8.95	9.91	6.29	6.77	6.50	28.37	15.09
La/Th	6.39	4.44	4.43	4.16	3.93	3.52	12.20	7.30
Th/U	3.26	6.05	3.96	3.82	3.00	3.27	6.48	6.17
La_N/Sm_N	4.26	4.03	3.78	2.30	3.08	3.36	3.31	3.29
La_N/Yb_N	14.6	10.9	9.94	10.4	9.55	8.78	34.6	19.5
Eu/Eu*	1.13	0.70	0.55	1.64	1.46	0.55	1.19	0.64

TABLE 2. CHEMICAL COMPOSITION OF BP CRATER TARGET ROCKS (continued)

	BP-22-2B	BP-22-2C	BP-24-1	BP-4-1	BP-6-1	BP-8-1	BP-9-1
(wt. %)							
SiO_2	95.04	95.82	88.53	52.4	90.45	88.6	90.23
TiO_2	0.07	0.08	0.12	0.07	0.84	0.31	0.31
Al_2O_3	0.82	0.21	2.8	1.5	3.4	2.52	4.08
Fe_2O_3*	0.09	0.21	0.97	39.63	0.88	0.34	0.85
MnO	0.02	0.02	0.02	0.01	0.04	0.03	0.03
MgO	<0.02	<0.02	<0.02	<0.02	<0.02	0.02	<0.02
CaO	0.02	0.1	0.14	0.67	0.61	0.16	0.1
P_2O_5	0.02	0.04	0.03	0.12	0.04	0.02	0.02
LOI	3.78	2.31	6.93	5.47	3.33	7.78	3.92
Total	99.86	98.79	99.54	99.87	99.59	99.78	99.54
(ppm, except where noted)							
Na	43	42	95	122	156	58	54
K	25	36	105	150	127	99	61
Sc	0.98	0.38	1.19	2.04	5.01	1.5	4.11
V	<20	<20	<20	27	25	<20	<20
Cr	8.3	19.4	35	74	110	14.7	22.2
Fe	0.045	0.14	0.66	24.4	0.51	0.23	0.53
Co	0.12	0.16	0.96	12.6	1.09	1.07	3.38
Ni	4	3.02	20	11.5	<9.6	<11	<16
Cu	4	<3	6	37	5	<3	<3
Zn	2.5	3.88	<3	169	<3	11.8	<4
Ga	0.9	0.26	0.63	5.09	4.7	3.32	7.06
As	0.24	0.14	0.6	4.13	0.34	0.6	0.49
Se	0.2	0.45	0.044	0.10	0.02	0.02	0.33
Br	2.8	0.22	1.32	0.08	0.88	1.32	0.54
Rb	0.3	0.39	0.82	1.98	<2	1.27	<2
Sr	25	9.6	8.4	44	18	8	36
Y	<8	<8	14	<8	25	9	<8
Zr	105	39	80	76	500	367	163
Nb	<6	<6	<6	<6	24	12	7
Ag	n.d.	<0.02	<0.06	<0.07	<0.012	<0.03	0.012
Sb	0.064	0.01	0.05	0.11	0.12	0.09	0.08
Cs	0.039	0.02	0.12	0.15	<0.03	0.12	0.06
Ba	38	20.1	37.2	29.4	48	22.7	48
La	9.07	7.35	64.8	2.97	19.6	16.1	19.6
Ce	21.3	12.8	187	7.05	35.9	21	35
Nd	12.6	5.87	81.5	4.82	18	15.2	14.1
Sm	2.44	1.02	16.2	1.14	3.32	2.8	2.45
Eu	0.42	0.18	2.68	0.18	1.22	0.40	0.38
Gd	1.65	0.57	9.99	0.49	4.8	3.08	2.58
Tb	0.18	0.07	1.18	0.09	0.79	0.38	0.38
Tm	0.095	0.03	<0.05	0.05	0.46	0.18	0.24
Yb	0.55	0.19	1.73	0.38	3.11	1.09	1.35
Lu	0.079	0.03	0.09	0.07	0.39	0.14	0.18
Hf	2.79	0.98	1.49	3.27	10.3	9.64	4.57
Ta	0.14	0.05	0.15	0.41	1.72	0.44	1.19
W	0.28	0.1	<0.2	0.39	0.85	0.18	0.34
Ir (ppb)	<0.2	<1	<1	<2	<1	<1	<1
Au (ppb)	0.5	0.16	1.62	0.34	0.28	0.21	0.4
Hg	<0.05	0.009	<0.003	0.024	0.006	<0.1	<0.3
Th	1.61	0.81	2.05	0.98	12.4	5.25	4.77
U	0.43	0.27	1.18	2.64	1.94	1.02	1.08
K/U	58.1	133.3	88.9	56.82	65.46	97.06	56.48
Zr/Hf	37.63	39.80	53.69	23.24	48.54	38.07	35.67
Zr/U	244.1	144.4	67.80	28.79	257.7	359.8	150.9
Co/Ni	0.03	0.05	0.05	1.10	1.09	1.07	3.38
La/Sc	9.26	19.34	54.45	1.46	3.91	10.73	4.77
La/Th	5.63	9.07	31.61	3.03	1.58	3.07	4.11
Th/U	3.74	3.00	1.74	0.37	6.39	5.15	4.42
La_N/Sm_N	2.34	4.54	2.52	1.64	3.72	3.62	5.04
La_N/Yb_N	11.1	26.1	25.3	5.28	4.26	9.98	9.81
Eu/Eu*	0.64	0.72	0.64	0.74	0.93	0.42	0.46

*All Fe as Fe_2O_3. n.d. = no data.

TABLE 3. CHEMICAL COMPOSITION OF THE ROCKS FROM THE EAST SITE AND OASIS CRATER TARGET ROCKS

	E-51-1	E-55-1	E-48-1	O-32-1A	O-32-1B	O-32-2A	O-33-1B	O-34-1	O-39-1	O-41-1-1	O-41-1.2	O-43-1	O-42-1-2	O-39-1-2
(wt. %)														
SiO_2	62.56		93.95	95.40	94.97	86.63	93.19	88.72	93.56		90.81	96.81	98.07	97.11
TiO_2	0.08		0.67	0.12	0.02	0.58	0.17	0.07	0.08		0.13	0.12	0.14	0.10
Al_2O_3	2.54	0.34	1.74	1.23	0.83	1.64	3.96	2.44	2.86	5.14	0.62	0.15	0.17	1.17
Fe_2O_3*	29.20		0.40	0.07	0.06	5.68	0.17	1.95	0.57		5.16	0.12	0.14	0.63
MnO	0.03		<0.02	0.02	0.04	0.09	0.02	0.04	0.08		0.05	0.02	0.03	0.10
MgO	<0.02		<0.02	<0.02	0.37	0.15	<0.02	<0.02	<0.02		<0.02	<0.02	<0.02	<0.02
CaO	0.33		0.26	0.78	1.56	0.56	0.07	1.45	0.10		0.50	0.14	0.09	0.14
P_2O_5	0.21		0.04	0.02	0.04	0.15	0.03	0.02	0.03		0.02	0.03	0.07	0.03
LOI	5.02		1.53	2.27	1.88	3.78	2.35	4.35	2.76		2.73	3.38	0.34	0.75
Total	99.97		98.59	99.91	99.77	99.26	99.96	99.04	100.04		100.02	100.77	100.05	100.03
(ppm, except where noted)														
Na	100	78	237	54	120	58	247	54	39	85	60	33	32	74
K	520	198	52	<322	46	98	150	87	100	79	80	179	30	169
Sc	4.38	3.31	2.79	1.67	0.22	3.78	2.54	1.2	1.29	0.68	0.66	0.4	0.53	1.74
V	<20	24	n.d.	<20	<20	<20	<20	<20	<20	<20	<20	<20	<20	<20
Cr	53.3	39.6	49.9	20.8	27.2	32.5	32.5	16.7	68.6	16.5	23.5	22.4	9.7	16.8
Fe	10.10	0.24	0.29	0.05	0.04	3.74	0.16	1.54	0.26	3.6	3.14	0.11	0.12	0.49
Co	106	4.62	0.91	0.80	0.74	16.6	0.88	5.90	2.15	8.96	9.52	0.62	0.06	4.82
Ni	107	10	19	4.8	9.8	14.2	11	10.1	3.16	<22	13.8	<4	2.33	10.2
Cu	<3	<3	n.d.	30	<3	8	<3	<3	6	<3	8	4	5	3
Zn	891	15.2	<3	8	1.99	153	9.4	17	11	77.7	78	4	7	22
Ga	2.94	2.36	2.07	1.55	0.35	2.42	2.10	2.11	4.55	1.22	1.06	1.91	0.56	1.47
As	18.6	1.35	5.04	0.10	0.41	0.26	0.35	3.70	1.74	5.06	3.59	0.48	0.61	2.46
Se	0.23	0.3	0.11	0.06	0.13	0.116	0.20	0.19	0.22	0.04	0.08	0.21	0.018	0.10
Br	0.20	0.4	0.10	0.002	0.20	0.06	0.90	0.40	0.54	0.04	0.54	0.02	0.38	0.02
Rb	1.27	1.18	22.50	1.18	<3	7.70	0.50	0.98	1.86	1.27	1.00	0.44	1.05	2.72
Sr	117	14	12	11	97	96	5	3	6	16	11	109	10	5
Y	31	37	n.d.	<8	<8	27	18	<8	<8	8	<8	<8	<8	10
Zr	100	934	232	140	33	1080	186	58	76	187	198	100	95	140
Nb	<6	16	n.d.	8	<8	11	12	<6	<6	<6	<6	<6	<6	9
Ag	0.088	<0.05	<0.8	0.024	<0.04	0.030	n.d.	<0.03	0.012	<0.07	0.012	0.012	0.026	0.024
Sb	0.32	0.28	0.07	0.07	0.05	0.02	0.091	0.20	0.14	0.06	0.52	0.03	0.07	0.21
Cs	0.69	0.08	0.09	0.045	0.06	0.19	0.053	0.08	<0.045	0.05	0.045	<0.02	0.02	0.045
Ba	473	84.8	35.9	45.7	19.8	516	48.0	57.3	60.0	73.0	47.3	20.9	11.7	76.2
La	14.9	29.7	9.09	8.38	3.38	35.4	16.2	6.97	4.26	9.52	9.62	2.88	2.42	6.02
Ce	27.3	88.4	12.5	14.9	6.96	71.7	25.1	8.27	8.01	18.3	18.40	4.59	4.38	14.10
Nd	11.4	24.8	9.20	8.13	3.77	35.5	20.9	7.08	5.04	8.1	8.83	4.51	2.50	6.90
Sm	3.76	5.78	1.57	1.53	0.74	5.95	5.99	1.26	1.16	1.48	1.50	0.47	0.52	1.57
Eu	0.9	0.76	0.28	0.28	0.34	0.85	1.38	0.24	0.26	0.24	0.26	0.1	0.1	0.32
Gd	6.1	6.65	2.32	1.81	0.55	6.60	5.33	0.80	1.50	1.2	0.99	1.00	0.58	1.51
Tb	1.13	1.02	0.39	0.23	0.07	0.83	0.68	0.13	0.28	0.19	0.16	0.13	0.09	0.27
Tm	0.65	0.51	0.73	0.13	0.03	0.46	0.31	0.08	0.18	0.1	0.10	0.05	0.05	0.18
Yb	1.86	3.78	1.35	0.81	0.14	3.25	1.71	0.48	1.14	0.59	0.61	0.25	0.32	1.40
Lu	0.29	0.53	0.2	0.11	0.02	0.50	0.23	0.06	0.15	0.09	0.09	0.03	0.05	0.19
Hf	1.54	29.2	6.00	3.73	0.93	32.5	4.33	1.70	1.87	5.49	5.48	1.79	3.34	4.25
Ta	0.27	1.1	1.48	0.25	0.04	1.00	0.92	0.22	0.34	0.22	0.21	0.05	0.09	0.51
W	0.21	0.19	1.20	0.45	0.04	0.21	1.11	0.09	0.08	<0.2	0.11	0.15	0.04	0.20
Ir (ppb)	<2	<1	<1.6	<0.8	<1	<2	<0.3	<1	<1	<1	<1	<1	<1	<1
Au (ppb)	0.20	0.67	0.29	0.98	0.47	1.03	1.10	0.34	0.39	0.93	0.41	0.46	0.42	0.39
Hg	<0.09	<0.04	0.024	0.018	<0.01	<0.06	<0.05	0.006	<0.018	0.01	<0.1	<0.05	0.075	<0.09
Th	3.84	11.9	5.46	2.30	0.85	17.1	3.34	1.56	1.23	2.45	2.43	0.66	0.78	1.71
U	0.66	3.86	1.6	1.04	0.32	1.03	0.99	0.55	0.51	0.72	0.79	0.21	0.29	0.66
K/U	787.8	51.3	32.5	n.d.	143.7	95.15	151.5	158.1	196.0	109	101.2	868.9	103.4	256.0
Zr/Hf	64.94	31.9	38.67	37.53	35.48	33.23	42.9	34.1	40.64	34.1	36.13	55.87	28.7	32.9
Zr/U	151.5	242	145.0	134.0	103.1	1048	187.8	105.4	149.0	259	250.6	485.4	327.5	212.1
Co/Ni	0.99	0.46	0.05	0.17	0.08	1.17	0.08	0.58	0.68	n.d.	0.69	0.62	0.026	0.47
La/Sc	3.40	8.97	3.26	5.02	15.36	9.37	6.38	5.81	3.30	14	14.6	7.2	4.57	3.46
La/Th	3.88	2.49	1.66	3.64	3.98	2.07	4.85	4.47	3.46	3.88	3.96	4.36	3.10	3.52
Th/U	5.82	3.08	3.41	2.21	2.66	16.60	3.37	2.84	2.41	3.4	3.08	3.20	2.69	2.59
La_N/Sm_N	2.49	3.23	3.64	3.45	2.87	3.74	1.70	3.48	2.31	4.05	4.04	3.86	2.93	2.41
La_N/Yb_N	5.41	5.32	4.55	6.99	16.3	7.36	6.40	9.81	2.53	11.1	10.66	7.78	5.11	2.91
Eu/Eu*	0.57	0.37	0.45	0.51	1.63	0.41	0.75	0.72	0.60	0.55	0.66	0.45	0.55	0.63

*All Fe as Fe_2O_3. n.d. = no data. For samples E-55-1 and O-41-1-1 not enough material was available for XRF analysis.

rocks from the East site and from the Oasis impact structure. The SiO₂ content of the samples from both structures shows a wide variation from 99 wt%, the composition of a pure quartzite, to 52 wt% for a hematite-cemented sandstone. Most rocks contain about 90–95 wt% SiO₂. The other major oxide contents are very low in the rocks from all three sites, with the exception of the hematite-rich specimens (Tables 2 and 3). The generally high silica contents of the samples show that the rocks are compositionally mature. The high hematite contents displayed by a few porous specimens show the extent of the clay mineral content in the matrix. Silica shows inverse correlation against most major oxides, whereas there is a positive correlation between the other major element oxides. Some examples are shown in Harker correlation diagrams in Figure 8A–C (silica vs. K and Al gives a negative correlation, and CaO vs. Fe₂O₃ shows a poorly defined positive correlation). It is interesting to note the low alkali contents (Na, K, Rb, Cs) for almost all of the rocks.

Regarding trace element contents, it is obvious that the contents are very low. Some of the elemental contents are positively correlated with each other, as shown in Figure 9A–C. The distinct

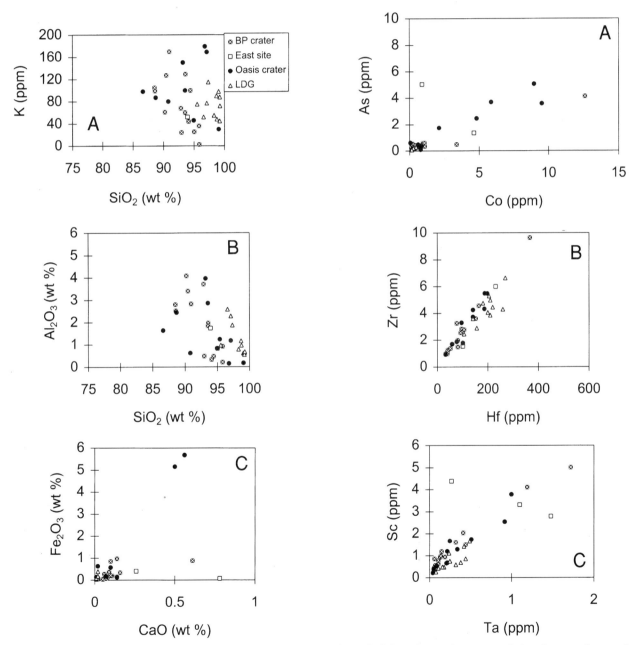

Figure 8. Harker correlation diagrams from the BP and Oasis structures, compared to LDG samples (triangles): A, Silica vs. K. B, Silica vs. Al₂O₃. C, CaO vs. Fe₂O₃.

Figure 9. Selected trace element correlation diagrams for samples BP and Oasis target rocks, and LDG (legends as in Fig. 8A): A, As vs. Co. B, Hf vs. Zr. C, Ta vs. Sc.

positive correlation between Zr and Hf (Fig. 9B) is the result of Hf being contained only in the accessory mineral zircon. The rare earth element contents show a limited variation in most samples from all three locations, with most samples having La contents between about 5 and 20 ppm. There are some exceptions, with La contents ranging up to 65 ppm (sample BP-24-1), which is from the outermost ring of disturbed strata. All samples from the BP and Oasis structures and the sandstone from the East site are enriched in the light rare earth elements (LREE), with chondrite-normalized ratios of La_N/Sm_N of about 4 and relatively flat HREE patterns (Fig. 10). The Eu anomalies are mostly negative (average Eu/Eu* = 0.7), but a few of the rocks have positive anomalies. The chondrite-normalized REE patterns are similar to those expected of post-Archean upper crustal rocks (Taylor and McLennan, 1985).

Isotopic data (Sr and Nd isotopic ratios, and the concentrations of Rb, Sr, Sm, and Nd as determined by isotope dilution) are reported in Table 4. The natural isotopic ratios of both Sr and Nd show a relatively small range, with $^{87}Sr/^{86}Sr$ ratios of 0.710557–0.715702, and $^{143}Nd/^{144}Nd$ ratios of 0.511739–0.512064. Despite the low abundances of all elements, we attempted to keep errors low by measuring isotope ratios in a second unspiked aliquot. Table 4 also gives ε_{Sr} and ε_{Nd} values, which range from 86 to 159, and –11.2 to –17.3, respectively. The mean $^{87}Sr/^{86}Sr$ ratios for the BP and Oasis rocks are close to the average continental crustal value (0.72), and the ε_{Sr} and ε_{Nd} values indicate that the Nubian quartz sandstone is of typical crustal composition and was derived from old crustal rocks (Faure, 1986, p. 211).

DISCUSSION

Evidence for impact

As the structures are deeply eroded and only the deeper, older rocks of the structures are exposed, crater fill material and local continuous ejecta deposits are most likely completely absent. Crater fill usually contains breccias and impact melts with abundant evidence of shock metamorphism. Due to the significant erosion of BP and Oasis, there is an absence of major macroscopic and microscopic evidence for impact processes, such as shatter cones, impact melt rock, high-pressure polymorphs of quartz, and possible meteoritic fragments, as noted by French et al. (1974). In our samples, the available evidence in support of shock metamorphism includes two to three sets of planar fractures in quartz (Fig. 4), but no unambiguous occurrences of PDFs (as reported by French et al., 1972, 1974) were found.

Frequency and degree of development of planar fractures is identical in specimens of both the BP and Oasis target rocks. Open fractures, similar to those observed in Coconino Sandstone from Meteor Crater, which have been interpreted as the result of sudden grain motion and contact during passage of a shock wave (Kieffer, 1971; French et al., 1974), are common features in both

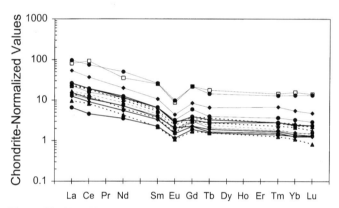

Figure 10. Chondrite-normalized (Taylor and McLennan, 1985) REE patterns of samples from the BP and Oasis impact structures, the East site, and LDG samples (legends as in Fig. 8A). The samples shown in this figure represent the variation that is present among the sample suite. LDG samples are plotted with dashed lines for better distinction.

TABLE 4. ISOTOPIC COMPOSITIONS OF Sr AND Nd, AND CONCENTRATIONS OF Sr, Rb, Nd, AND Sm IN SAMPLES FROM BP AND OASIS IMPACT STRUCTURES

Sample	Rb (ppm)	Sr (ppm)	$^{87}Sr/^{86}Sr$ (± 1σ)	Sm (ppm)	Nd (ppm)	$^{143}Nd/^{144}Nd$ (± 1σ)	ε_{Sr}	ε_{Nd}
BP-U-16-1	0.3188	13.78	0.710557 (49)	0.6915	3.720	0.511943 (17)	86	–17.3
BP-U-19-1	0.1825	9.80	0.710858 (16)	0.7230	3.229	0.511807 (126)	90	–16.2
BP-U-22-2B	0.1463	25.03	0.710870 (40)	2.45	n.d.	0.512064 (40)	90	–11.2
OU-32-2A	5.8445	89.40	0.715702 (12)	5.72	74.221	n.d.	159	n.d.
OU-39-1	0.5123	8.81	0.712830 (10)	1.36	6.512	0.511923 (29)	118	–13.9
OU-41-1	0.0965	31.23	0.713830 (40)	n.d.	9.018	0.511841 (24)	132	–15.5
OU-43-1	0.1356	5.23	0.710980 (30)	0.4656	4.553	0.512043 (31)	92	–11.6

Note: The Sr and Nd isotopic compositions are given for the present time. ε_{Sr} and ε_{Nd} are the deviations in the $^{87}Sr/^{86}Sr$ and $^{143}Nd/^{144}Nd$ ratios in parts per 10^4, from the present values in a uniform reservoir of $^{87}Sr/^{86}Sr$ = 0.7045 and $^{143}Nd/^{144}Nd$ = 0.512638 (Faure, 1986). n.d. = no data.

BP and Oasis rocks. The intense fracturing and undulatory extinctions displayed by several grains can be taken as supportive evidence of shock effects on the target rocks. According to Grieve et al. (1996), quartz starts to melt under shock compression of >50 GPa. Between shock pressures of 5 and 50 GPa, mosaicism, planar microstructures (PDFs and PFs), partial transformation to stishovite and coesite, and melting and quenching to lechatelierite can be observed. None of the latter have yet been found at BP and Oasis (French et al., 1974). While the microcrystalline quartz in a sample from the Oasis structure could be devitrified silica glass, it is more likely that it simply represents chert.

In the BP and Oasis structures French et al. (1974) observed a so-called cored inclusion, consisting of a quartz grain surrounded by a glassy and recrystallized matrix, typical of impact-derived breccia, but poorly developed planar microstructures. The dominant orientations of the poles of the PDFs found by French et al. (1974) to the c-axis of quartz the (0001), $\{11\bar{2}2\}$, $\{10\bar{1}1\}$, and $\{51\bar{6}1\}$ forms (c, ξ, r,z, and x, respectively; see also Fig. 22 in Grieve et al., 1996), and not the $\{10\bar{1}3\}$ and $\{10\bar{1}2\}$ (π and ω), which are common in shocked crystalline rocks (cf. French and Short, 1968; Stöffler and Langenhorst, 1994). The presence of devitrified glass, abundant planar fractures, and the higher angle orientations of PDFs has been ascribed to the large difference in shock impedance between solid grains and pores in the target rocks, which resulted in an irregular distribution of the effects of shock metamorphism, and to the fact that these particular orientations occur at relatively higher shock pressures (Grieve et al., 1996; French et al., 1974).

Such a behavior is typical of porous (sedimentary) rocks; compare, for example, Wabar (Chao et al., 1961; Bunch and Cohen, 1964; Short 1966) and Kamensk (Grieve et al., 1996). In porous target rocks, as at BP and Oasis, PDFs are rare even if the samples contain diaplectic silica glass (Kieffer, 1971). According to some workers (Robertson and Grieve, 1977; Robertson, 1980; Stöffler and Langenhorst, 1994), the lower orientation PDFs (π and ω) do not develop, because PDFs are produced only when the rocks are sufficiently compacted, and at lower pressure, volume work on the rocks is absorbed in closing existing pores and voids. Hence, only evidence of higher shock pressures are likely to be present. While our limited set of thin sections did not yield any unambiguous PDFs, lechatelierite, or devitrified glass, we did observe multiple sets of PFs and abundant fractured grains.

RELATION OF BP AND OASIS ROCKS TO LIBYAN DESERT GLASS

As mentioned above, there is abundant evidence for an impact origin of LDG, and little or no evidence for a low-temperature origin. Besides a low water content of the LDG (Beran and Koeberl, 1997), a very important observation is the existence of a meteoritic component in LDG (Murali et al., 1989, 1997; Rocchia et al., 1996; Barrat et al., 1997; Koeberl, 1997). These authors have found that the contents of siderophile elements, such as Co, Ni, and Ir, are significantly enriched in some rare dark bands that occur in some LDG samples. Together with Re-Os isotopic data of dark bands in LDG (C. Koeberl, unpublished data), these observations provide an unambiguous fingerprint of a (rare) cosmic component and, by inference, of an origin by impact.

Our geochemical results (Tables 2–4) provide data for comparison of the BP and Oasis structures and the LDG (see Koeberl, 1997, for the description and a detailed table of composition of LDG). The similarities in the major element compositions between rocks from Oasis and BP and LDG is obvious. Average LDG contains (in wt%): SiO_2, 98.4; Al_2O_3, 1.19, Fe_2O_3, 0.13; CaO, 0.01; MgO, 0.01. Although the silica contents of the rocks from BP and Oasis at first appear to be slightly lower than those of the LDG, recalculation of the target rock abundances to volatile-free compositions yields comparable contents. For example, BP-11-1 and O-32-1A yield 97.7 and 97.6 wt% SiO_2, respectively. The alumina contents of the Oasis and BP rocks range from about 0.15 to 4.1 wt% and include the average LDG values (Fig. 8B). The target rocks from both craters and LDG have low alkali element contents and similar interelement ratios (Tables 2 and 3). Selected plots of major and trace element abundances (Harker diagrams; Figs. 8 and 9) show the presence of similar interelement correlations among samples of the BP, Oasis target rocks, and the LDG. On the basis of major element compositions of LDG, Fudali (1981, p. 247), concluded that the glass had formed from "a sand or sandstone composed of quartz grains coated with a mixture of kaolinite, hematite and anatase," which closely describes the BP and Oasis rocks.

Chondrite-normalized REE plots also show the similarity in absolute REE abundance and shape of the pattern between rocks from Oasis/BP and LDG (Fig. 10). While the patterns of samples from the BP and Oasis rocks and from the East site show considerable variation in abundance (Fig. 10), the range includes the LDG values. From the point of view of trace element composition, the low alkali contents of the LDG (Na: 34 ppm, K: 74 ppm) are mirrored by equally low contents in most of the BP and Oasis rocks (Tables 2 and 3). This is in contrast to surface sands from the LDG site, for which Koeberl (1997) reported Na and K values that are one to two orders of magnitude higher than the LDG values. Similar observations can be made for some interelement ratios. For example, in average LDG, K/U = 89, La/Sc = 9.9, Zr/Hf = 46, and Th/U = 2.7 (Koeberl, 1997). These values compare very favorably with the respective data for BP and Oasis rocks (Tables 2 and 3). Thus, a comparison of major and trace element abundances and interelement ratios between target rocks from the BP and Oasis structures and LDG does not contradict a connection. The fit is much better than for the comparison between LDG and surface sands from the LDG area (for analytical data of surface sands, see Koeberl, 1997).

We hoped that the isotopic data would provide evidence that would allow us to confirm or reject a connection between LDG and the BP/Oasis structures. The similarity in isotopic composition between tektites (impact glasses) and target rocks has been used as evidence for a common origin of the tektites and to determine the type of source rocks (e.g., Shaw and Wasserburg, 1982;

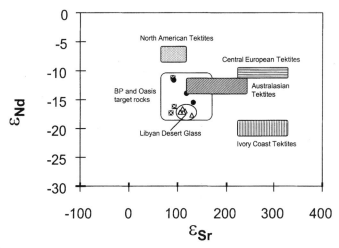
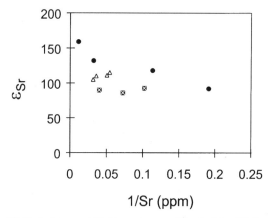

Figure 11. Plot of present day ε_{Nd} vs. ε_{Sr} values for rocks from the BP and Oasis impact structures as compared to Libyan Desert Glass samples (legends as in Fig. 8A), contrasted with data from the four tektite-strewn fields. LDG data from C. Koeberl (unpublished data) and Barrat et al. (1997). Fields of tektite data from Blum et al. (1992) and Koeberl et al. (1998).

Figure 12. Plot of ε_{Sr} vs. 1/Sr (in parts per million) of the BP, Oasis, and LDG samples (legends as in Fig. 8A), indicating that the LDG data can be represented by a mixture of rocks with isotopic characteristics similar to those of the BP and Oasis target rocks.

Blum et al., 1992, 1996; Koeberl, 1994b). Figure 11 shows that the present day ε_{Nd} vs. ε_{Sr} values for the BP and Oasis target rocks and those of LDG plot in the same general area. In the ε_{Sr} vs. 1/Sr plots the LDG plot, the relationship between the LDG and the BP/Oasis data could represent a mixing line (Fig. 12). These observations indicate that the LDG could have been produced from target rocks of BP and Oasis craters. This allows us to reach the conclusion that our data show that a connection between LDG and the BP/Oasis structures is possible. However, because rock compositions in the area (representing Nubia Group sandstones) show relatively little variation, the reverse conclusion cannot be made without ambiguity.

SUMMARY AND CONCLUSIONS

The 3.2-km-diameter BP impact structure in Libya consists of two eroded and discontinuous rings of hills surrounding a central block, the south half of which is deeply eroded. The inner ring, which is about 2 km in diameter with an average relief of 30 m, is surrounded by another ring (2.8 km in diameter; maximum relief of about 20 m), and by an outer disturbed zone 3.2 km in diameter. Rocks exposed are quartz sandstone, siltstone, and conglomerate from the Cretaceous Nubia Group. The Oasis impact structure, also in Libya, has a diameter of about 18 km, but the most prominent part is a central ring of hills, about 5.1 km in diameter and 100 m high. The structure exposes the same rocks as the BP structure (ca. 85 km north-northwest of Oasis). The age of both structures is only constrained as being younger than the Cretaceous target rocks.

The proximity of the BP and Oasis structures to the Libyan Desert Glass–strewn field (in Egypt) led to the speculation that they might be the source of the LDG. Our geochemical studies provide the first data for such a comparison. We have studied the petrographic characteristics and geochemical composition of 29 samples from the BP and Oasis sites. Petrographic studies of thin sections of the samples show that they represent mostly submature, moderately to poorly sorted, medium- to fine-grained quartz sandstone, or quartzitic breccia. Most of the studied samples do not show evidence of shock, but in a few sections some quartz grains with up to four sets of planar fractures (which are indicative of shock metamorphism, but do not provide proof) were found.

Major element compositions of the samples were determined by X-ray fluorescence and trace element compositions by neutron activation analysis. The results available to date indicate a limited range in composition of all analyzed samples. The average LDG composition is very similar to the composition of some of the BP and Oasis sample compositions, indicating a possible relation. In a plot of ε_{Nd} vs. ε_{Sr} for BP and Oasis target rocks, the LDG values plot within the field defined by the BP and Oasis rocks, providing further evidence in favor of a relation between LDG and those rock types.

Thus, the available data indicate that the most suitable source rock is sandstone from the Nubia Group, which is exposed in both the BP and Oasis impact structures. The chemical similarity between rocks from Oasis/BP and LDG is much closer than that between surface sands at the LDG–strewn field and LDG. There is no geochemical evidence to suggest that the source rocks of the LDG are significantly different from those currently exposed at the BP and Oasis structures. This agrees with the estimate by McHone et al. (1995b), who suggested, from orbital radar observations, that only about 100 m of erosion has taken place since the formation of the Oasis structure. On the other hand, the Nubia Group rocks in the area could be very thick, as 100 m seems to be a minimum estimate for the erosion considering the current appearance of the structure.

While the BP and/or Oasis impact structures are currently the best candidates as source craters for the LDG, more field work is clearly necessary. This includes detailed field work at both impact structures (which have not been studied geologically since the 1970s), especially with the aim of obtaining datable material to allow constraints on the age of these impact structures. Furthermore, collection of more (and representative) samples for additional petrologic, geochemical, and isotopic work is desirable.

ACKNOWLEDGMENTS

We acknowledge the Director, Council for Geoscience, Pretoria, South Africa, for permission to use the XRF laboratory, and we are grateful to H. C. Cloete, Council of Geoscience, Pretoria, for performing XRF analyses. We are indebted to Dona Jalufka, Vienna, for drafting assistance. Paul Buchanan, University of Vienna, provided helpful comments on the manuscript. We are grateful to Ray Anderson and Mike Zolensky for constructive reviews and to Burkhard Dressler for editorial comments. The research reported here was supported by the Austrian Academic Exchange Service (for a Ph.D. stipend and travel funds to B.A.), and by the Austrian Fonds zur Förderung der wissenschaftlichen Forschung project Y58-GEO (to C.K.).

REFERENCES CITED

Abate, B., Koeberl, C., Underwood, J. R., Jr., Fisk, E. P., and Giegengack, R. F., 1997, BP and Oasis impact structures, Libya, and their relation to Libyan Desert Glass: petrography, geochemistry, and geochronology [abs.]: Large Meteorite Impacts and Planetary Evolution (Sudbury 97): Houston, Texas, Lunar and Planetary Institute, Contribution 922, p. 1.

Barnes, V. E., and Underwood, J. R., Jr., 1976, New investigations of the strewn field of Libyan Desert Glass and its petrography: Earth and Planetary Science Letters, v. 30, p. 117–122.

Barrat, J. A., Jahn, B. M., Amosse, J., Rocchia, R., Keller, F., Poupeau G., and Diemer, E., 1997, Geochemistry and origin of Libyan Desert Glasses: Geochimica et Cosmochimica Acta, v. 61, p. 1953–1959.

Beran, A., and Koeberl, C., 1997, Water in tektites and impact glasses by Fourier-transformed infrared spectrometry: Meteoritics and Planetary Science, v. 32, p. 211–216.

Bigazzi, G., and de Michele, V., 1996, New fission-track age determinations on impact glasses: Meteoritics and Planetary Science, v. 31, p. 234–236.

Bigazzi, G., and de Michele, V., 1997, New fission-track ages of Libyan Desert Glass, in Proceedings, Silica 96, Meeting on Libyan Desert Glass and related desert events: Milan, Italy, Pyramids, p. 49–58.

Blum, J. D., Papanastassiou, D. A., Koeberl, C., and Wasserburg, G. J., 1992, Neodymium and strontium isotopic study of Australasian tektites: new constraints on the provenance and age of target materials: Geochimica et Cosmochimica Acta, v. 56, p. 483–492.

Blum, J. D., Chamberlain, C. P., and Hingston M. P., 1996, Neodymium, strontium, and oxygen isotope investigation of the target stratigraphy and impact melt rock from the Manson impact structure, in Koeberl, C., and Anderson, R. R., eds., The Manson impact structure, Iowa: anatomy of an impact crater: Geological Society of America Special Paper 302, p. 317–324.

Brown, R. W., Allsopp, H. L., Bristow, J. W., and Smith, C. B., 1989, Improved precision of Rb-Sr dating of kimberlitic micas: an assessment of a leaching technique: Chemical Geology, v. 79, p. 125–136.

Bunch, T. F., and Cohen, A. J., 1964, Shock deformation of quartz from two meteorite craters: Geological Society of America Bulletin, v. 75, p. 1263–1266.

Chao, E. C. T., Fahey, J. J., and Littler, J., 1961, Coesite from Wabar Crater, near Al Hadida, Arabia: Science, v. 133, p. 882–883.

Clayton, P. A., and Spencer, L. J., 1934, Silica glasses from the Libyan Desert: Mineralogical Magazine, v. 23, p. 501–508.

Corner, B., Reimold, W. U., Brandt, D., and Koeberl, C., 1997, Morokweng impact structure, Northwest Province, South Africa: geophysical imaging and some preliminary shock petrographic studies: Earth and Planetary Science Letters, v. 146, p. 351–364.

Diemer, E., 1997, Libyan Desert Glass: an impactite; State of the art in July 1996, in Proceedings, Silica 96, Meeting on Libyan Desert Glass and related desert events: Milan, Italy, Pyramids, p. 95–110.

Faure, G., 1986, Principles of isotope geology: New York, John Wiley, 589 p.

Feller, M., 1997, Vitreous silica from the Sahara, in Proceedings, Silica 96, Meeting on Libyan Desert Glass and related desert events: Milan, Italy, Pyramids, p. 111–114.

French, B. M., and Short, N. M., eds., 1968, Shock metamorphism of natural materials: Baltimore, Maryland, Mono Book Corp., 644 p.

French, B. M., Underwood, J. R., Jr., and Fisk, E. P., 1972, Shock metamorphic effects in two new Libyan impact structures: Geological Society of America Abstracts with Programs, v. 4, no. 7, p. A510–A511.

French, B. M., Underwood, J. R., Jr., and Fisk, E. P., 1974, Shock metamorphic features in two meteorite impact structures, southeastern Libya: Geological Society of America Bulletin, v. 85, p. 1425–1428.

Fudali, R. F., 1981, The major element chemistry of Libyan Desert Glass and the mineralogy of its precursor: Meteoritics, v. 16, p. 247–259.

Futrell, D. S., and O'Keefe, J. A., 1997, A brief discussion of the petrogenesis of Libyan Desert Glass, in Proceedings, Silica 96, Meeting on Libyan Desert Glass and related desert events: Milan, Italy, Pyramids, p. 115–120.

Giegengack, R., and Underwood, J. R., Jr., 1997, Origin of Libyan Desert Glass: some stratigraphic considerations, in Proceedings, Silica 96, Meeting on Libyan Desert Glass and related desert events: Milan, Italy, Pyramids, p. 37–40.

Goudarzi, G. H., 1970, Geology and mineral resources of Libya; a reconnaissance: U.S. Geological Survey Professional Paper 660, 104 p.

Grieve, R. A. F., Langenhorst, F., and Stöffler, D., 1996, Shock metamorphism of quartz in nature and experiment: II. Significance in geoscience: Meteoritics and Planetary Science, v. 31, p. 6–35.

Horn, P., Müller-Sohnius, D., Schaaf, P., Kleinmann, B., and Storzer, D., 1997, Potassium-argon and fission-track dating of Libyan Desert Glass, and strontium and neodymium isotope constraints in its source rocks, in Proceedings, Silica 96, Meeting on Libyan Desert Glass and related desert events: Milan, Italy, Pyramids, p. 59–76.

Jessberger, E., and Gentner, W., 1972, Mass spectrometric analysis of gas inclusions in Muong Nong glass and Libyan Desert Glass: Earth and Planetary Science Letters, v. 14, p. 221–225.

Jux, U., 1983, Zusammensetzung und Ursprung von Wüstengläsern aus der Großen Sandsee Ägyptens: Zeitschrift der deutschen geologischen Gesellschaft, v. 134, p. 521–553.

Kieffer, S. W., 1971, Shock metamorphism of the Coconino Sandstone at Meteor Crater, Arizona: Journal of Geophysical Research, v. 76, p. 5449–5473.

Kleinmann, B., 1969, The breakdown of zircon observed in the Libyan Desert Glass as evidence of its impact origin: Earth and Planetary Science Letters, v. 5, p. 497–501.

Koeberl, C., 1985, Trace element chemistry of Libyan Desert Glass: Meteoritics, v. 20, p. 686.

Koeberl, C., 1993, Instrumental neutron activation analysis of geological and cosmochemical samples: a fast and reliable method for small sample analysis: Journal of Radioanalytical and Nuclear Chemistry, v. 112, p. 481–487.

Koeberl, C., 1994a, African meteorite impact craters: characteristics and geological importance: Journal of African Earth Sciences, v. 18, p. 263–295.

Koeberl, C., 1994b, Tektite origin by hypervelocity asteroidal or cometary impact: target rocks, source craters, and mechanisms, in Dressler, B. O., Grieve, R. A. F., and Sharpton, V. L., eds., Large meteorite impacts and planetary evolution: Geological Society of America Special Paper 293, p. 133–152.

Koeberl, C., 1997, Libyan Desert Glass: geochemical composition and origin, in Proceedings, Silica 96, Meeting on Libyan Desert Glass and related desert events: Milan, Italy, Pyramids, p. 121–132.

Koeberl, C., Armstrong, R. A., and Reimold, W. U., 1997, Morokweng, South Africa: a large impact structure of Jurassic-Cretaceous boundary age: Geology, v. 25, p. 731–734.

Koeberl, C., Reimold, W. U., Blum, J. D., and Chamberlain, C. P., 1998, Petrology and geochemistry of target rocks from the Bosumtwi impact structure, Ghana, and comparison with Ivory Coast tektites: Geochimica et Cosmochimica Acta, v. 62, p. 217–2196.

Kohman, T. P., Lowman, P. D., Jr., and Abdelkhalek, M. L., 1967, Space and aerial photography of the Libyan Desert glass area: 30th Annual Meteoritical Society Meeting, Moffett Field, California, Abstracts.

Martin, A. J., 1969, Possible impact structure in southern Cyrenacia, Libya: Nature, v. 223, p. 940–941.

Masaitis, V. L., and Grieve, R. A. F., 1994, The economic potential of terrestrial impact craters: International Geology Reviews, v. 36, p. 105–151.

Master S., Reimold W. U., and Brandt D., 1996, Evidence for shock metamorphic origin of multiply-striated joint surfaces (MSJS) in sandstones of the Sinamwenda meteorite impact structure, Zimbabwe [abs.]: Lunar and Planetary Science, v. 27, p. 827–828.

Matsubara, K., Matsuda J., and Koeberl, C., 1991, Noble gases and K-Ar ages in Aouelloul, Zhamanshin, and Libyan Desert impact glasses: Geochimica et Cosmochimica Acta, v. 55, p. 2951–2955.

McHone, J. F., Blumberg, D. G., Greeley, R., and Underwood, J. R., Jr., 1995a, Space Shuttle radar images of terrestrial impact structures; SIR-C/X-SAR [abs.]: Meteoritics and Planetary Science, v. 30, p. 543.

McHone, J. F., Blumberg, D. G., Greeley, R., and Underwood, J. R., Jr., 1995b, Orbital radar images of Libyan impact structures: Geological Society of America Abstracts with Programs, v. 27, p. A209.

Murali, A. V., Zolensky, M. E., Carr, R., Underwood, J. R., Jr., and Giegengack, R. F., 1987, Libyan Desert Glass: trace elements and gas inclusions: Geological Society of America Abstracts with Programs, v. 19, p. A782.

Murali, A. V., Zolensky, M. E., Underwood, J. R., Jr., and Giegengack, R. F., 1988, Formation of Libyan Desert Glass: Lunar and Planetary Science, v. 19, p. 817–818.

Murali, A. V., Linstrom, E. J., Zolensky, M. E., Underwood, J. R., Jr., and Giegengack, R. F., 1989, Evidence of extraterrestrial component in the Libyan Desert Glass: EOS (Transactions, American Geophysical Union), v. 70, p. 1178.

Murali, A. V., Zolensky, M. E., Underwood, J. R., Jr., and Giegengack, R. F., 1997, Chondritic debris in Libyan Desert Glass, in Proceedings, Silica 96, Meeting on Libyan Desert Glass and related desert events: Milan, Italy, Pyramids, p. 133–142.

Reimold, W. U., Koeberl, C., and Bishop, J., 1994, Roter Kamm impact crater, Namibia: Geochemistry of basement rocks and breccias: Geochimica et Cosmochimica Acta, v. 58, p. 2689–2710.

Robertson, P. B., 1980, Anomalous development of planar deformation features in shocked quartz of porous lithologies [abs.]: Lunar and Planetary Science, v. 11, p. 938–940.

Robertson, P. B., and Grieve, R. A. F., 1977, Shock attenuation at terrestrial impact structures, in Roddy, D. J., Pepin, R. O., and Merrill, R. B., eds., Impact and explosion cratering: New York, Pergamon Press, p. 687–702.

Rocchia, R., Robin, E., Fröhlich, F., Meon, H., Frogent, L., and Diemer, E., 1996, L'origine des verres du désert libyque: un impact météorique: Comptes Rendus Academie de Science (Paris), v. 322 (Ser. IIa), p. 839–845.

Rocchia, R., Robin, E., Fröhlich, F., Ammosse, J., Barrat, J. A., Meon, H., Froget, L., and Diemer, E., 1997, The impact origin of Libyan Desert Glass, in Proceedings, Silica 96, Meeting on Libyan Desert Glass and related desert events: Milan, Italy, Pyramids, p. 143–158.

Seebaugh, W. R., and Strauss, A. M., 1984, A cometary impact model for the source of Libyan Desert Glass: Journal of Non-crystalline Solids, v. 67, p. 511–519.

Shaw, H. F., and Wasserburg, G. J., 1982, Age and provenance of the target materials for tektites and possible impactites as inferred from Sm-Nd and Rb-Sm systematic: Earth and Planetary Science Letters, v. 60, p. 155–177.

Short, N. M., 1966, Shock-lithification of unconsolidated rock materials: Science, v. 154, p. 382–384.

Smith, C. B., Allsopp, H. L., Kramers, J. D., Hutchinson, G., and Roddick, J. C., 1985, Emplacement ages of Jurassic-Cretaceous South African kimberlites by the Rb-Sr method on phlogopite and whole-rock samples: Transactions of the Geological Society of South Africa, v. 88, p. 249–266.

Stöffler, D., and Langenhorst, F., 1994, Shock metamorphism of quartz in nature and experiment I: basic observation and theory: Meteoritics, v. 29, p. 155–181.

Storzer, D., and Koeberl, C., 1991, Uranium and zirconium enrichments in Libyan Desert Glass: Lunar and Planetary Science, v. 22, p. 1345–1346.

Storzer, D., and Wagner G. A., 1971, Fission-track ages of North American tektites: Earth and Planetary Science Letters, v. 10, p. 435–440.

Storzer, D., and Wagner, G. A., 1977, Fission track dating of meteorite impacts: Meteoritics, v. 12, p. 368–369.

Taylor, S. R., and McLennan, S. M., 1985, The continental crust: its composition and evolution: Oxford, Blackwell Scientific, 312 p.

Underwood, J. R., 1975, Reconnaissance geology of meteorite impact structures in SE Libya: Geological Society of America Abstracts with Programs, v. 7, p. A242.

Underwood, J. R., Jr., 1976, Impact structures in the Libyan Sahara: some comparisons with Mars, in Proceedings, International Colloquium of Planetary Geology, Geologica Romana (Rome), v. 15, p. 337–340.

Underwood, J. R., Jr., and Fisk, E. P., 1980, Meteorite impact structures, southeast Libya, in Salem, M. J., and Busrewil, M. T., eds., Proceedings, The geology of Libya, Symposium, 1978: London, Academic Press, p. 893–900.

Vincent P., and Beauvilain A., 1996, Découverte d'un nouveau cratère d'impact météoritique en Afrique: l'astrobleme de Gweni-Fada (Ennedi, Sahara du Tchad): Comptes Rendus de l'Academie des Sciences, v. 323 (II), p. 987–997.

Weeks, R. A., Underwood, J. R., Jr., and Giegengack, R., 1984, Libyan Desert Glass: a review: Journal of Non-crystalline Solids, v. 67, p. 593–619.

Manuscript Accepted by the Society December 16, 1998

Mjølnir Structure, Barents Sea: A marine impact crater laboratory

Filippos Tsikalas, Steinar Thor Gudlaugsson*, Jan Inge Faleide, and Olav Eldholm
Department of Geology, University of Oslo, P.O. Box 1047 Blindern, N-0316 Oslo, Norway

ABSTRACT

Integrated geophysical, mineralogic and geochemical analyses substantiate that the 40-km-diameter Mjølnir Structure in the Barents Sea is a buried marine impact crater of Volgian-Berriasian age, 141–149 Ma. The discovery of a proximal ejecta layer that correlates with the seismically defined deformation event at Mjølnir proves the impact origin of the structure. The ejecta layer contains unequivocal impact indicators, such as shocked quartz grains and a strong enrichment in iridium. Furthermore, detailed geophysical analysis of the structure shows that Mjølnir is one of the best preserved terrestrial impact craters, and provides new and improved constraints on the poorly understood aspects of meteorite impacts into marine areas and sedimentary basins. In particular, the low-strength sedimentary target rocks and the presence of water have led to increased gravitational collapse and infilling, resulting in a shallow apparent crater depth. The observed geophysical anomalies closely reflect the structural crater expression and the laterally varying physical properties induced by an impact in a shallow marine sedimentary basin. Sedimentary loading of the original impact relief has caused substantial postimpact deformation and structural modification. The estimated magnitude of the Mjølnir impact with an energy release of the order of 10^{21} J was not large enough to have caused significant mass extinction. It may, however, be associated with considerable short-term environmental disturbance on a regional scale.

INTRODUCTION

Large impacts are considered to have affected the evolution of the Earth's lithosphere (e.g., Grieve and Parmentier, 1984), atmosphere (e.g., Zahnle, 1990), and biosphere (e.g., McLaren and Goodfellow, 1990) on short- and long-time scales. Knowledge of impact parameters and the physical and chemical processes associated with large impacts may lead to improved insight into the syn- and postimpact effects. Discovery of new impact structures is, therefore, particularly important. However, almost all of the approximately 150 currently known impact structures are located on land and are commonly modified and partly eroded by several geologic processes (Grieve et al., 1995). Detailed geophysical studies are not available for the majority of these structures. Alternatively, postimpact sedimentation may preserve craters in a marine environment. Nonetheless, only a few underwater craters are yet identified due to the difficulty in recognizing buried submarine craters.

The few known marine craters, only about 10% of the total number of terrestrial impacts (Grieve et al., 1995) make the study of any large marine crater important because the outcome will provide new data on the poorly understood aspects of meteorite impacts into marine areas and sedimentary basins. Marine geophysical measurements are the major tools during the initial recognition and study of such features, and Pilkington and Grieve (1992) have established a set of general criteria to evaluate whether the observed geophysical anomalies may be caused by impact. However, geophysical studies alone do not confirm an impact origin, and the need for geologic, mineralogic and/or geochemical evidence is indispensable (Sharpton and Grieve, 1990).

The 40-km-diameter Mjølnir Structure on the Bjarmeland

---

*Present address: National Energy Directorate, Reykjavik, Iceland; e-mail: stg@os.is.

Tsikalas, F., Gudlaugsson, S. T., Faleide, J. I., and Eldholm, O., 1999, Mjølnir Structure, Barents Sea: A marine impact crater laboratory, *in* Dressler, B. O., and Sharpton, V. L., eds., Large Meteorite Impacts and Planetary Evolution II: Boulder, Colorado, Geological Society of America Special Paper 339.

Platform in the Barents Sea (Figs. 1 and 2) was first interpreted as an impact structure by Gudlaugsson (1993) from its geophysical signature and overall geologic setting. This inference was derived from a limited amount of multichannel seismic profiles, and regional gravity, and magnetic profiles. The impact hypothesis prompted the acquisition of high-resolution seismic, gravity and magnetic profiles by the Norwegian Defense Research Establishment in 1992 and 1993. Together with the previously acquired shallow and conventional multichannel seismic profiles, these data comprise an extensive and unique geophysical database (Fig. 2; Table 1). In addition, stratigraphic and sedimentologic information exist from a drillhole 30 km north-northeast of Mjølnir's periphery (well 7430/10-U-01) (Fig. 2). The existing regional grid of seismic profiles on the Bjarmeland Platform has made it possible to correlate the main seismic sequence boundaries at Mjølnir to the established stratigraphic framework of the Barents Sea (Worsley et al., 1988; Gabrielsen et al., 1990; Richardsen et al., 1993).

The extensive geophysical and geologic database has the potential of verifying or rejecting the proposed impact structure interpretation. It also provides an excellent opportunity to study the structure and its geophysical characteristics in detail, and to search for an ejecta layer. Most results from our detailed studies of the Mjølnir Structure have been presented in a series of papers by Dypvik et al. (1996), Tsikalas et al. (1998a–c). Here, we integrate and summarize the main findings and outline an avenue for continued research.

SOURCE-CRATER AND EJECTA-LAYER PAIR

For an impact origin to be unambiguously confirmed, documented occurrences of meteoritic material and/or shock metamorphic features are needed (Sharpton and Grieve, 1990). Hence, detailed studies of ejecta layers and corresponding source craters are a key to understanding the complex physical and chemical phenomena caused by collision of large meteorites with the Earth. However, onshore impact exposures suffer degradation of both the crater proper and the proximal ejecta by erosion, tectonic deformation, and subduction. Of the 24 known craters ≥30 km in diameter, excluding the Mjølnir Structure, only 8 have preserved proximal ejecta deposits, and several are eroded beneath the crater floor (Grieve et al., 1995). Well-preserved crater-ejecta pairs are therefore extremely rare

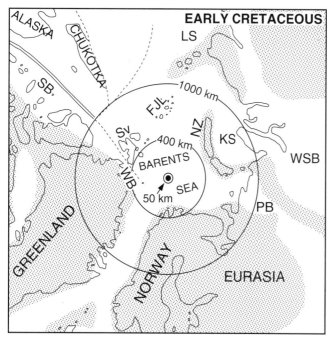

Figure 1. Location of Mjølnir Structure (black dot) superimposed on an Early Cretaceous paleogeographic reconstruction based on Ziegler (1988) and Faleide et al. (1993). Radii of various tsunami wave heights are described in the text. Dotted areas represent positive areas with no Early Cretaceous sedimentation. FJL = Franz Josef Land; KS = Kara Sea; LS = Lapten Sea; NZ = Novaya Zemlya; PB = Pechora Basin; SB = Sverdrup Basin; Sv = Svalbard; WB = Wandel Sea Basin; WSB = West Siberian Basin.

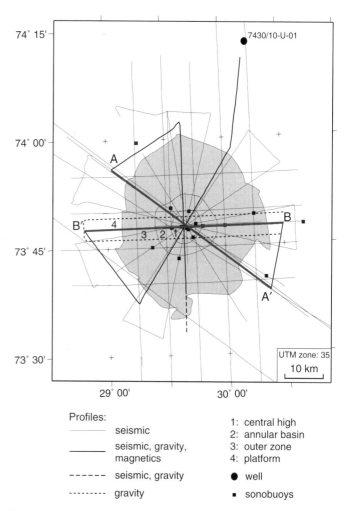

Figure 2. Geophysical profiles in the Mjølnir area superimposed on the structure defined by its radial zonation boundaries.

in the terrestrial impact cratering record. Therefore, many aspects of the impact process remain poorly constrained. In addition, only one-fourth of the approximately 100 pre-Cenozoic craters have been dated with a precision better than ±10 Ma (Grieve et al., 1995).

The seismic correlation from Mjølnir to well 7430/10-U-01 by continuous seismic reflection profiles (Fig. 2) show that the well penetrates the stratigraphic interval spanning the postulated impact event (Dypvik et al., 1996; Tsikalas et al., 1998b). The seismic tie to the well, the regional seismic stratigraphy, and the core analysis suggest that the stratigraphic position of the impact event is bounded by a prominent lower Barremian limestone unit and the base reflector of the Upper Jurassic sequence corresponding to an Upper Callovian unconformity (Fig. 3). The top of the seismic disturbance at Mjølnir intersects the well at a stratigraphic level estimated to be of Volgian-Berriasian age, 141–149 Ma (Dypvik et al., 1996; Tsikalas et al., 1998b).

The search for impact indicators in the well revealed that a 0.8-m-thick core section contained shocked quartz grains and a strong enrichment in iridium (Dypvik et al., 1996). In particular, the shocked quartz grains showed the presence of planar fractures and recrystallized planar deformation features, both common diagnostics for the high shock pressures caused by impact. An iridium peak of about 1,000 ppt, about 15 times the background value, was recorded approximately at the same level. These findings are strongly suggestive of extraterrestrial material (Dypvik et al., 1996; Dypvik and Ferrell, 1998).

STRUCTURAL AND MORPHOLOGIC CHARACTERISTICS

The seismic reflection method is the most powerful geophysical method in sedimentary targets, where the pre-impact stratification provides a series of reference horizons for the large-scale impact-induced structures to be identified and mapped, e.g., the Montagnais Structure (Jansa et al., 1989), the Chesapeake Bay Structure (Poag et al., 1994), and the Manson Structure (Keiswetter et al., 1996). Most craters presently under water have never been subaerially exposed and may be well-preserved.

The seismic investigation of the Mjølnir Structure includes a deep and a shallow part, indirectly reflecting differentiation into primary impact-induced and secondary crater-influenced deformation. This distinction also reflects the resolution of the three seismic datasets comprising single-channel, shallow multichannel, and deep multichannel profiles (Table 1).

The seismic data clearly show that the major structural features of Mjølnir are typical of large complex impact structures (Tsikalas et al., 1998b,c). The seismic profiles image the distinct radial zonation pattern, comprising the following structures: (1) a 12-km-wide complex outer zone, including a marginal fault zone and a modestly elevated ring; (2) a 4-km-wide annular depression; and (3) an uplifted 8-km-diameter central high (Fig. 4). In addition, we observe distinct boundary faults forming a ~150-m-high near-circular rim wall separating highly deformed strata within the crater from intact platform strata, and a 45–180-m-thick disturbed and incoherent seismic reflectivity unit caused by the impact and confined by prominent fault-blocks and postimpact strata (Figs. 4 and 5).

Tsikalas et al. (1998c) have shown that the observed broad-brimmed bowl-shaped disturbance (Fig. 3) is characterized by a systematic loss of coherence toward its central and upper parts caused by a progression of seismic facies from disrupted layering and diffractions to chaotic and reflection-free zones. The seismic disturbance is divided into an intensely disturbed and a less disturbed transitional zone (Fig. 3). Moreover, analysis of structural features within 850–1,400 km³ of disturbed volume provides insight into major cratering processes, such as brecciation and excavation, melting, gravitational collapse of the transient crater, and structural uplift. The seismic sections also differentiate between autochthonous target rocks and up to 1.3-km-thick allogenic breccia units (Tsikalas et al., 1998c, their Figs. 11 and 12), provide estimates of the amount of structural uplift (1.5–2 km) that approximate the lower limits of what is expected for Mjølnir based on empirical relations between crater dimensions and the amount of uplift, and finally document absence of compact melt-generated bodies (Tsikalas et al., 1998c). A transient crater of 16 km in diameter and 4.5 km in depth has been determined. The estimated collapse factor for Mjølnir, i.e., the ratio of final crater diameter to transient crater diameter, is of the order of 2.5

TABLE 1. MJØLNIR STRUCTURE: GEOPHYSICAL DATA

Data Type*	Profiles	Profile Length (km)	Velocity Analyses
Conventional multichannel seismic reflection profiles (NPD, IKU, BGR)	20	1081	361
Shallow multichannel reflection profiles (IKU)	4	174	128
Shallow high-resolution single-channel reflection profiles (NDRE)	23	872	
Shallow refraction profiles (sonobuoys) (NDRE)	16		
Side-scan sonar profiles (IKU)	4	174	
Marine gravity measurements along single-channel profiles (NDRE)	9	397	
Marine magnetic measurements along single-channel profiles (NDRE)	7	292	
Base station magnetic measurements at Hopen, Bjørnøya, and Tromsø (UT)			
Aeromagnetic profiles (GSN)			

*BGR = Bunderanstalt für Geowissenschaften und Rohstoffe, Hannover; GSN = Geological Survey of Norway, Trondheim; IKU = IKU Petroleum Research, Trondheim; NDRE = Norwegian Defense Research Establishment, Horten; NPD = Norwegian Petroleum Directorate, Stavanger; UT = University of Tromsø.

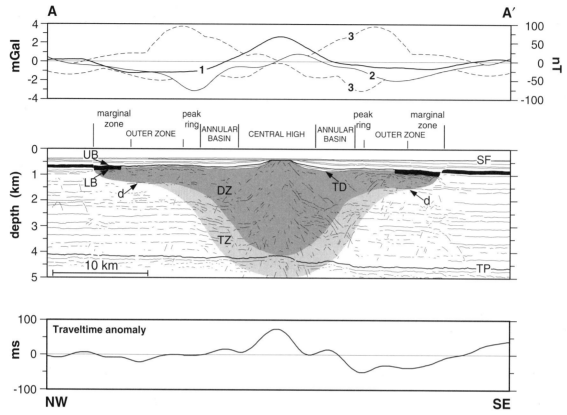

Figure 3. Geophysical type-section of the Mjølnir Structure (Fig. 2) (Tsikalas et al., 1998a). The observed gravity, magnetic and travel-time data were Gaussian-filtered with filter lengths of 7, 10, and 10 km, respectively. Top; 1 = Free-air gravity anomaly; 2 = residual magnetic anomaly; 3 = "envelope"-magnetic anomaly curves, corresponding to the minimum and maximum magnetic anomalies of all profiles crossing through the center. Middle; Interpreted seismic section based on both a high-resolution single-channel profile and a multichannel profile recorded along the same line. DZ = area of intense disturbance; TZ = transitional area of less disturbance. The stratigraphic interval bounding the time of impact is shown in black. SF = seafloor; UB = lower Barremian; LB = base Upper Jurassic; TD = first continuous reflector above the seismic disturbance; TP = Top Permian; d = low-angle décollement. The profile was depth converted using smoothed interval velocities derived from stacking velocities (Tsikalas et al., 1998c). Bottom; Two-way travel-time anomaly with reference to a planar Top Permian reflector.

(Tsikalas et al., 1998c), considerably larger than the expected values for typical terrestrial craters (1.4–2.0; average 1.6) (Melosh, 1989). The general structural, geometric, and volumetric comparison of the Mjølnir Structure with similar-size terrestrial craters exhibits several similarities, strengthening the impact interpretation.

IMPACT-INDUCED CHANGES IN PHYSICAL PROPERTIES

Terrestrial craters have geophysical characteristics that are largely associated with the passage of a shock wave and the initiation of subsequent crater-forming processes. The most conspicuous geophysical signature is a residual negative gravity anomaly (Pilkington and Grieve, 1992) caused by low-density material resulting from both lithologic and physical changes associated with the cratering process (Grieve and Pesonen, 1992). The gravity low commonly extends to or slightly beyond the crater rim. In addition, there is a tendency for the gravity signature of complex craters >30-km-diameter to exhibit a central positive high, ascribed to central structural uplift (e.g., Pilkington et al., 1994; Plescia, 1996). However, it is the counteracting processes of uplift and brecciation that determine the final density distribution within the impact-affected rock volume.

The Mjølnir gravity data are of high quality and analysis of cross-over errors indicates minimal deviations (Tsikalas et al., 1998a). The residual free-air gravity field exhibits a circular symmetric anomaly over the structure. The anomaly is divided into an annular low, with an outer diameter of 45 km attaining minimum values of –1.5 mGal over the periphery, and a central 14-km-wide gravity high, with a maximum value of +2.5 mGal (Fig. 3) (Tsikalas et al., 1998a).

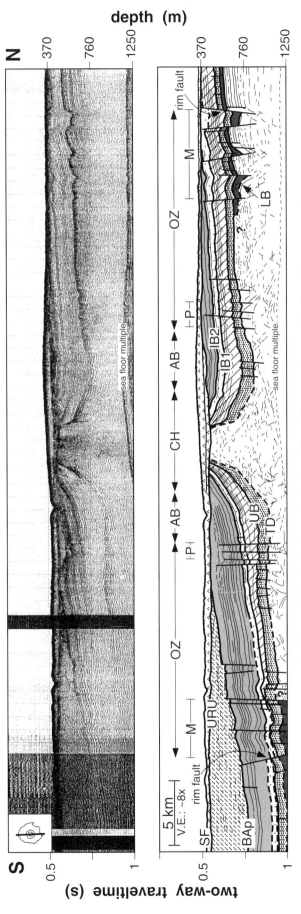

Figure 4. High-resolution single-channel seismic profile, and interpretation, crossing the entire structure through the center. Annotations as in Figure 3. IB1, IB2 = Intra-Barremian reflectors; BAp = Base Aptian; URU = Late Cenozoic upper regional unconformity. CH = central high; AB = annular basin; OZ = outer zone; M = marginal fault zone; P = peak ring.

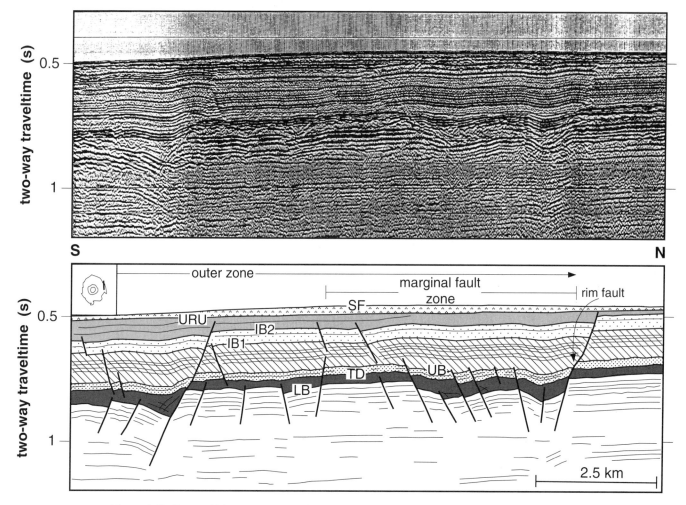

Figure 5. Shallow multichannel seismic profile, and interpretation, across and parallel to the crater rim. Annotations as in Figures 3 and 4.

The magnetic signature of impact craters is variable. The most common impact effect is a broad magnetic low, primarily observed in crystalline target impacts, resulting from disturbance of the regional magnetic trends. In addition, the presence of local magnetic anomalies can be ascribed to uplift of magnetized rocks or impact-generated melt bodies (Pilkington and Grieve, 1992). The amount of melt generated by impacts into sedimentary targets is expected to be similar to that generated in crystalline target rocks, but with a more dispersed and scattered rather than coherent character (Kieffer and Simonds, 1980; Cintala and Grieve, 1994). The Mjølnir residual magnetic field anomalies, corrected for temporal variations, are quite complex, but there is a clear correspondence to the structure proper. We note in particular an absence of any magnetic anomaly over its center and several moderate amplitude anomalies toward the periphery (Fig. 3) (Tsikalas et al., 1998a).

Fracturing and brecciation are expected to induce changes in the seismic velocity (Pilkington and Grieve, 1992). The lateral changes may, in turn, produce pull-up or pull-down effects on continuous reflectors below the impact structure. Extensive analysis of stacking velocities and sonobuoy profiles (Fig. 2; Table 1) has not resolved major lateral changes with reference to the Bjarmeland Platform strata. However, even small lateral velocity anomalies may induce robust travel-time anomalies in planar reflectors. Indeed, the seismic profiles reveal a small pull-up of the high-amplitude, originally planar Top Permian reflector beneath the structure. The travel-time anomaly is 16 km in diameter and rises to +80 ms beneath the central crater (Fig. 3) (Tsikalas et al., 1998a).

Integrated geophysical modeling results support the lateral differentiation of the Mjølnir seismic disturbance into a central uplift and a peripheral region (Fig. 6). A qualitative model to explain the modeled distribution of physical properties postulates interaction of several cratering processes including formation of transient cavity, brecciation, gravitational collapse, and structural uplift (Tsikalas et al., 1998a). The primary effect of the impact event is an impact-induced porosity increase due to extensive fracturing and brecciation. Subsequent modification of the density field takes place as a result of mass transport during gravitational collapse and structural uplift of the crater floor displacing

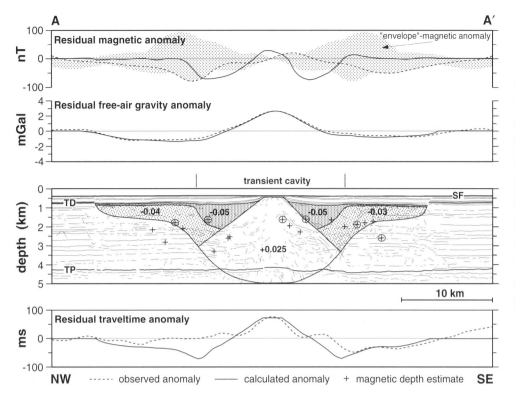

Figure 6. Impact structure model with differentiated brecciation and observed and calculated gravity, magnetic, and travel-time anomalies (Tsikalas et al., 1998a). Darker raster denotes allogenic breccia, while lighter raster denotes autochthonous target rock breccia. Magnetic source depth (slope and half-slope) estimates refer to all profiles in Figure 2. In view of the circular structure, the depth estimates have been plotted according to distance from the center. Circled depth estimates from profile AA′. Density contrasts are given in grams per cubic centimeter. Annotations as in Figure 3.

deep, denser strata to shallower levels beneath the central structure. Furthermore, differential compaction of postimpact sediments results in long-term alteration of the impact-induced density field.

The absence of high-amplitude seismic reflectivity (Tsikalas et al., 1998b,c) precludes the presence of impact-generated compact melt bodies or sheets of reasonable dimensions. In fact, the modeling (Tsikalas et al., 1998a) demonstrates that the observed low-amplitude magnetic anomalies can be interpreted in terms of dislocation of weakly magnetized platform strata, perhaps associated with local concentrations of dispersed melts or minor melt dikes in the peripheral region (Fig. 6). Moreover, the geophysical data are not compatible with alternative origins for the structure, such as salt or clay diapirs and igneous complex (Tsikalas et al., 1998a). In contrast, the Mjølnir geophysical signatures are consistent with those produced by similar-size craters elsewhere (Tsikalas et al., 1998a). Thus, the spatial correspondence of structural features and geophysical anomalies substantiates the impact interpretation.

IMPACT INTO A MARINE SEDIMENTARY BASIN

Although the principles of cratering mechanics have been largely established (e.g., Melosh, 1989), the nature of the cratering processes will, to some extent, depend on whether the target is crystalline or sedimentary, and on the presence of water (e.g., McKinnon, 1982; Ahrens and O'Keefe, 1983; Sonett et al., 1991). However, the global cratering record will tend to introduce a bias toward crystalline, water poor targets in models of typical impact structures. Nonetheless, integrated studies of craters believed to have formed in a marine environment attribute several features associated both with the final crater and the related deposits to the presence of water (e.g., Jansa, 1993; Poag, 1996). Similarly, there is both empirical and experimental evidence that impact craters in fluid-rich sediments collapse to a larger degree than those in crystalline targets (Roddy, 1977; Melosh, 1989). In particular, laboratory experiments have shown that meteorite impacts in unconsolidated, water-covered sedimentary targets may result in a more modulated crater topography than inferred from statistics of known terrestrial impact craters (Gault and Sonett, 1982). In addition, marine impacts have greater chances to be preserved because they are immediately covered by postimpact sediments, which in some cases can reach considerable thicknesses. Finally, sediment loading above the primary impact relief may result in substantial postimpact deformation and structural modification. However, the quantification of such postimpact effects is almost entirely absent from the terrestrial impact record.

In this context, the seismic mapping of the Mjølnir Structure profiles has revealed some unusual features (Tsikalas et al., 1998b; Tsikalas et al., 1998c). In particular, the expected step-like terraces at the periphery are replaced by prominent fault blocks floored by apparent low-angle décollement surfaces (Fig. 3). Continuous reflectors beneath the fault blocks are best imaged by the shallow multichannel seismic profiles (Tsikalas et al., 1998b, their Fig. 8). Furthermore, reconstruction of the original crater, accounting for sediment compaction and fault restoration, brings out a shallower than predicted structure without a raised crater rim

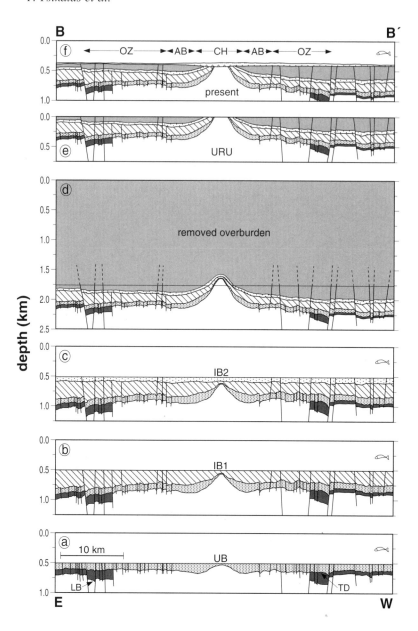

Figure 7. Reconstruction of the original crater relief along the east-west–trending profile BB′ (Fig. 2) by decompaction and fault restoration of the entire postimpact sedimentary succession (after Tsikalas et al., 1998b). Time-steps (a-f) correspond to the main unit boundaries. Annotations and raster codes as in Figures 3 and 4.

(Fig. 7). In addition, the seismic profiles image extensive postimpact deformation expressed by structural reactivation and differential subsidence (Figs. 4 and 5). It is postulated that the shallow expression of the initial crater is related to a large degree of collapse and the coeval unusually extensive infilling. The outward structural expansion resulted from the inward collapse of the initial crater rim along faults floored by apparent low-angle décollement surfaces at the periphery (Tsikalas et al., 1998b). Toward the center, the collapse is characterized by debris mass-flows. The massive collapse was probably caused by the low strength of the siliciclastic sedimentary target. Tsikalas et al. (1998b) have suggested that the collapse of the impact-induced water cavity (Fig. 8) and the subsequent rapid surge of sea water into the excavated crater transported large amounts of ejecta and crater wall material back into the crater, accounting for the extensive infilling.

Hydraulic excavation by back-rushing turbulent sea water may explain the lack of a raised crater rim. Subsequently, the extensive postimpact deformation is triggered by prograding postimpact sediments (Fig. 4) and governed by the instability and radially varying changes in physical properties within the impact-affected rock volume (Fig. 6). It is also evident that the postimpact deformation has considerably enhanced the structural expression of the original, subtle crater (Fig. 7). Thus, the present distinct expression of Mjølnir is largely a postimpact burial phenomenon.

ENVIRONMENTAL CONSEQUENCES

The Mjølnir Structure is one of the 20 largest impact structures discovered on Earth, ranking eighth among those presently not exposed at the surface (Grieve et al., 1995), and its structure

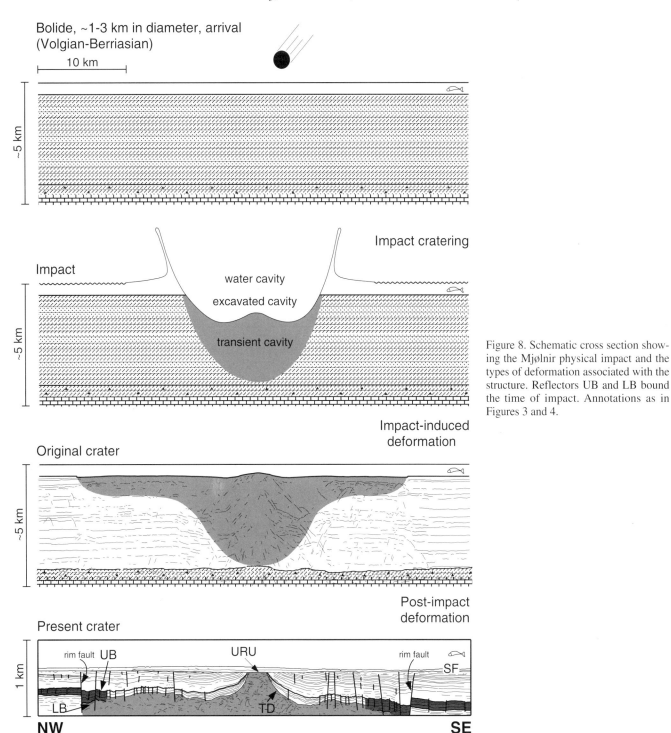

Figure 8. Schematic cross section showing the Mjølnir physical impact and the types of deformation associated with the structure. Reflectors UB and LB bound the time of impact. Annotations as in Figures 3 and 4.

and associated exterior ejecta are among the best-preserved source-crater/ejecta-layer pairs in the terrestrial impact record (Dypvik et al., 1996). In order to access the possible consequences of the impact, Tsikalas et al. (1998c) have attempted to constrain the magnitude of the impact event, i.e., the energy release, impactor size, and mass. They estimated that the energy release was in the order of 16×10^{20} J (range, $2.4–53 \times 10^{20}$ J), and the impactor's size and mass were 1.8 km (range, 0.9–3 km) and 10×10^{12} kg (range, $1.5–33 \times 10^{12}$ kg), respectively. These estimates are based on various scaling laws (e.g., Schmidt and Holsapple, 1982; Melosh, 1989) and on reasonably well-documented average values for average impact velocities, impactor angles, and densities (e.g., Rahe et al., 1994; Shoemaker et al., 1994).

Studies of the Montagnais impact crater, in Canada, an

appropriate marine impact analog to Mjølnir, have shown that the threshold of global extinction initiation is placed at a 3.5–km impactor diameter, which translates into a 45-km final crater (Aubry et al., 1990; Jansa, 1993). The smaller Mjølnir Structure is inferred to have been caused by the impact of a projectile ~1.8 km in diameter that produced a 40-km-diameter final structure, decreasing to 26 km after correction for the extensive collapse (Tsikalas et al., 1998c). In addition, the time interval between impacts of Mjølnir's magnitude (energy release, 10^{21} J) is approximately 0.7–1 Ma (e.g., Grieve and Shoemaker, 1994), far shorter than the typical 5-Ma average stage duration in the geologic time scale. Although the Mjølnir impact is not expected to have led to a major biologic extinction, it is predicted to have been large enough to cause serious environmental disturbance on a short-term regional scale.

Tsikalas et al. (1998c) have estimated that the Mjølnir impact may have led to a potential 8.3-magnitude earthquake, within a magnitude range of 7.7–8.7. An earthquake of such magnitude may trigger slumping on nearby sedimentary slopes. These authors also estimated the volume of the displaced material during the Mjølnir impact to have been on the order of 140–180 km^3. Most of the material was redeposited in the vicinity of the crater and was probably associated with a major disturbance of the water column. In fact, the impact of the Mjølnir projectile into a shallow shelf environment of 300–500-m water depth (Dypvik et al., 1996), together with the subsequent collapse of the impact-generated water cavity (Fig. 8), is expected to have induced large-amplitude tsunami waves. According to the impact-wave height relationships for shallow water depths (e.g., Hills et al., 1994; Toon et al., 1994) the Mjølnir impact resulted in a 30–60-m high wave at a distance of 50 km, decreasing to 5–10 m at 400 km and to 2–3 m at 1,000 km. The tsunami waves may have caused seafloor erosion and sediment redeposition both in the proximity of the crater and in the adjacent coastal regions (Fig. 1). In fact, impact ejecta signatures from the Mjølnir Structure have been found just above a mud-flake conglomerate zone in well 7430/10-U-01 (Fig. 2), 50 km away from the impact (Dypvik et al., 1996).

FUTURE AVENUES OF MJØLNIR RESEARCH

Plate reconstructions to the Early Cretaceous time of impact and the estimated wave amplitudes suggest that impact-generated tsunami waves may have led to coastal erosion accompanied by deposition of turbidites into nearby basins in northeast Greenland, northern Fennoscandia, and Novaya Zemlya (Fig. 1). The waves probably also spread, with greatly reduced amplitudes, into the North Atlantic rift basins between Norway and Greenland, the Pechora Basin, and the Sverdrup Basin in the Canadian Arctic. The presence of land masses along Novaya Zemlya and in the Urals to the east most likely prevented the wave propagation into the southern Kara Sea and West Siberian Basin (Fig. 1) (Tsikalas et al., 1998c). Therefore, a search should be initiated to locate traces of ejecta layer and associated impact-wave deposits, as well as other regional short-term perturbations in the appropriate sedimentary successions recorded by outcrops and cores in the Arctic realm.

The Mjølnir Structure is an obvious candidate for relatively shallow scientific drilling. It would unambiguously prove the impact origin and provide additional material for detailed sedimentologic, mineralogic, and geochemic analyses. In addition, the total volume of 850–1,400 km^3 of impact-deformed platform strata at Mjølnir (Tsikalas et al., 1998c), located in a region of hydrocarbon-generating rocks (e.g., Larsen et al., 1992), makes it a potential reservoir unit. However, aside from its economic potentials, the Mjølnir Structure provides a well-confined and easily accessible laboratory for the study of both impact and postimpact deformational processes. We strongly recommend a local three-dimensional high-resolution seismic survey followed by a transect of continuously cored and logged wells across the structure.

SUMMARY

Integrated geophysical and geologic analyses have been used to study the 40-km-diameter Mjølnir Structure of Volgian-Berriasian age, 141–149 Ma. The results support the impact origin and have led to the acceptance of the Mjølnir Structure as a terrestrial impact crater. In fact, Mjølnir provides one of the best records of marine impacts yet available. Furthermore the analysis has yielded the following conclusions:

1. The seismically defined deformation event at Mjølnir correlates with a proximal ejecta layer that contains shocked quartz grains and iridium enrichment.

2. The observed geophysical anomalies correspond closely to the radially varying distribution of structural and morphologic units, and to the physical property distribution induced by impact in a shallow marine sedimentary basin modified by postimpact compaction caused by sediment loading.

3. Integrated geophysical modeling is a powerful tool to discriminate between impact craters and similar features of different origin, particularly in a marine environment.

4. The stratigraphy, morphology, and structure of the 40-km-diameter impact crater is similar to the majority of large complex craters. However, the shallow crater depth of both the original and the present craters, and the extensive postimpact deformation, are features specific for Mjølnir.

5. Impact in a shallow marine sedimentary basin led to greater gravitational collapse and considerable infilling, accounting for the shallow crater depth. The extensive postimpact deformation is closely related to the instability and lateral changes in physical properties within the impact-affected rock volume, triggered by prograding postimpact sediments.

6. Reconstruction of the crater relief immediately after impact demonstrates that the deformation associated with the extensive postimpact overburden has considerably enhanced the structural expression of an originally subtle crater. Thus, the present distinct expression of Mjølnir is largely a postimpact burial phenomenon.

7. The characteristic broad-brimmed and bowl-shaped disturbance results from collapse on fault blocks floored by apparent low-angle décollement at the periphery and chaotic debris–mass flows toward the center.

8. Although the magnitude of the Mjølnir impact is not expected to be associated with significant mass extinction, it may have caused environmental disturbance on a short-term regional scale.

ACKNOWLEDGMENTS

We gratefully acknowledge the Norwegian Defense Research Establishment, the Norwegian Petroleum Directorate, and IKU Petroleum Research for providing seismic, gravity, and magnetic data, and information from well 7430/10-U-01. We also thank J. Pohl and B. M. French for helpful comments and suggestions as referees. The study was supported through grants from the Research Council, Norway, and Athens Academy, Greece.

REFERENCES CITED

Ahrens, T. J., and O'Keefe, J. D., 1983, Impact of an asteroid or comet in the ocean and the extinction of terrestrial life, *in* 13th, Proceedings, Lunar and Planetary Science Conference: Journal of Geophysical Research, v. 88 (suppl.), p. A799–A806.

Aubry, M. P., Gradstein, F. M., and Jansa, L. F., 1990, The late early Eocene Montagnais bolide: no impact on biotic diversity: Micropaleontology, v. 36, p. 164–172.

Cintala, M. J., and Grieve, R. A. F., 1994, The effects of differential scaling of impact melt and crater dimensions on lunar and terrestrial craters: some brief examples, *in* Dressler, B. O., Grieve, R. A. F., and Sharpton, V. L., eds., Large meteorite impacts and planetary evolution: Geological Society of America Special Paper 293, p. 51–59.

Dypvik, H., Gudlaugsson, S. T., Tsikalas, F., Attrep, M., Jr., Ferrell, R. E., Jr., Krinsley, D. H., Mørk, A., Faleide, J. I., and Nagy, J., 1996, Mjølnir structure: an impact crater in the Barents Sea: Geology, v. 24, p. 779–782.

Dypvik, H., and Ferrell, R. E., Jr., 1998, Clay mineral alteration associated with a meteorite impact in the marine environment (Barents Sea): Clay Minerals, v. 33, p. 51–64.

Faleide, J. I., Vågnes, E., and Gudlaugsson, S. T., 1993, Late Mesozoic–Cenozoic evolution of the south-western Barents Sea in a regional rift-shear tectonic setting: Marine and Petroleum Geology, v. 10, p. 186–214.

Gabrielsen, R. H., Færseth, R. B., Jensen, L. N., Kalheim, J. E., and Riis, F., 1990, Structural elements of the Norwegian continental shelf. Pt. I: The Barents Sea region: Norwegian Petroleum Directorate Bulletin 6, 33 p.

Gault, D. E., and Sonett, C. P., 1982, Laboratory simulation of pelagic asteroidal impact: atmospheric injection, benthic topography, and surface wave radiation field, *in* Silver, L. T., and Schultz, P. H., eds., Geological implications of impacts of large asteroids and comets on Earth: Geological Society of America Special Paper 190, p. 69–92.

Grieve, R. A. F., and Parmentier, E. M., 1984, Impact phenomena as factors in the evolution of the Earth, *in* 27th International Geological Congress, Moscow: Utrecht, Netherlands, VNU Science, p. 99–114.

Grieve, R. A. F., and Pesonen, L. J., 1992, The terrestrial impact record: Tectonophysics, v. 216, p. 1–30.

Grieve, R. A. F., and Shoemaker, E. M., 1994, The record of past impacts on Earth, *in* Gehrels, T., ed., Hazards due to comets and asteroids: Tucson, University of Arizona Press, p. 417–462.

Grieve, R. A. F., Rupert, J., Smith J., and Therriault, A., 1995, The record of terrestrial impact cratering: GSA Today, v. 5, p. 193–196.

Gudlaugsson, S. T., 1993, Large impact crater in the Barents Sea: Geology, v. 21, p. 291–294.

Hills, J. G., Nemchinov, I. V., Popov, S. P., and Teterev, A. V., 1994, Tsunami generated by small asteroid impacts, *in* Gehrels, T., ed., Hazards due to comets and asteroids: Tucson, University of Arizona Press, p. 779–789.

Jansa, L. F., 1993, Cometary impacts into ocean: their recognition and the threshold constraint for biological extinctions: Palaeogeography, Palaeoclimatology, Palaeoecology, v. 104, p. 271–286.

Jansa, L. F., Pe-Piper, G., Robertson, B. P., and Friedenreich, O., 1989, Montagnais: a submarine impact structure on the Scotian shelf, eastern Canada: Geological Society of America Bulletin, v. 101, p. 450–463.

Keiswetter, D., Black, R., and Steeples, D., 1996, Seismic reflection analysis of the Manson Impact Structure, Iowa: Journal of Geophysical Research, v. 101, p. 5823–5834.

Kieffer, S. W., and Simonds, C. H., 1980, The role of volatiles and lithology in the impact cratering process: Reviews of Geophysics and Space Physics, v. 18, p. 143–181.

Larsen, R. M., Fjæran, T., and Skarpnes, O., 1992, Hydrocarbon potential of the Norwegian Barents Sea based on recent well results, *in* Vorren, T. O., Bergsager, E., Dochl-Stamnes, Ø. A., Holter, E., Johansen, B., Lie, E., and Lund, T. B., eds., Arctic geology and petroleum potential: Amsterdam, Netherlands, Elsevier Scientific, Norwegian Petroleum Society Special Publication 2, p. 321–331.

McKinnon, W. B., 1982, Impact into the Earth's ocean floor: preliminary experiments, a planetary model, and possibilities for detection, *in* Silver, L. T., and Schultz, P. H., eds., Geological implications of impacts of large asteroids and comets on Earth: Geological Society of America Special Paper 190, p. 129–142.

McLaren, D. J., and Goodfellow, W. D., 1990, Geological and biological consequences of giant impact: Annual Review of Earth and Planetary Sciences, v. 18, p. 123–171.

Melosh, H. J., 1989, Impact cratering—a geologic process: New York, Oxford University Press, 245 p.

Pilkington, M., and Grieve, R. A. F., 1992, The geophysical signature of terrestrial impact craters: Reviews of Geophysics, v. 30, p. 161–181.

Pilkington, M., Hildebrand, A. R., and Ortiz-Aleman, C., 1994, Gravity and magnetic field modeling and structure of the Chicxulub Crater, Mexico: Journal of Geophysical Research, v. 99, p. 13147–13162.

Plescia, J. B., 1996, Gravity investigation of the Manson impact structure, Iowa, *in* Koeberl, C., and Anderson, R. R., eds., The Manson Impact Structure, Iowa: Anatomy of an impact crater: Geological Society of America Special Paper 302, p. 89–104.

Poag, C. W., 1996, Structural outer rim of Chesapeake Bay impact crater: seismic and bore hole evidence: Meteoritics and Planetary Science, v. 31, p. 218–226.

Poag, C. W., Powars, D. S., Poppe, L. J., and Mixon, R. B., 1994, Meteoroid mayhem in the Ole Virginny: Source of the North American tektite strewn field: Geology, v. 22, p. 691–694.

Rahe, J., Vanysek, V., and Weissman, P. R., 1994, Properties of cometary nuclei, *in* Gehrels, T., ed., Hazards due to comets and asteroids: Tucson, University of Arizona Press, p. 597–634.

Richardsen, G., Vorren, T. O., and Tørudbakken, B. O., 1993, Post–Early Cretaceous uplift and erosion in the southern Barents Sea: a discussion based on analysis of seismic interval velocities: Norsk Geologisk Tidsskrift, v. 73, p. 3–20.

Roddy, D. J., 1977, Large-scale impact and explosion craters: Comparisons of morphological and structural analogs, *in* Roddy, D. J., Pepin, R. O., and Merrill, R. B., eds., Impact and explosion cratering: New York, Pergamon Press, p. 185–246.

Schmidt, R. M., and Holsapple, K. A., 1982, Estimates of crater size for large-body impact: gravity-scaling results, *in* Silver, L. T, and Schultz, P. H., eds., Geological implications of impacts of large asteroids and comets on Earth: Geological Society of America Special Paper 190, p. 93–102.

Sharpton, V. L., and Grieve, R. A. F., 1990, Meteorite impact, cryptoexplosion, and shock metamorphism; A perspective on the evidence at the K/T boundary, *in* Sharpton, V. L., and Ward, P. D., eds., Global catastrophies in Earth history; an interdisciplinary conference on impacts, volcanism, and mass mortality: Geological Society of America Special Paper 247, p. 301–318.

Shoemaker, E. M., Weissman, P. R., and Shoemaker, C. S., 1994, The flux of periodic comets near Earth, *in* Gehrels, T., ed., Hazards due to comets and asteroids: Tucson, University of Arizona Press, p. 313–335.

Sonett, C. P., Pearce, S. J., and Gault, D. E., 1991, The oceanic impact of large objects: Advances in Space Research, v. 11, p. 77–86.

Toon, O. B., Zahnle, K., Turco, R. P., and Covey, C., 1994, Environmental perturbations caused by asteroid impacts, *in* Gehrels, T., ed., Hazards due to comets and asteroids: Tucson, University of Arizona Press, p. 791–826.

Tsikalas, F., Gudlaugsson, S. T., Eldholm, O., and Faleide, J. I., 1998a, Integrated geophysical analysis supporting the impact origin of the Mjølnir Structure, Barents Sea: Tectonophysics, v. 289, p. 257–280.

Tsikalas, F., Gudlaugsson, S. T., and Faleide, J. I., 1998b, Collapse, infilling, and postimpact deformation at the Mjølnir impact structure, Barents Sea: Geological Society of America Bulletin, v. 110, p. 537–552.

Tsikalas, F., Gudlaugsson, S. T., and Faleide, J. I., 1998c, The anatomy of a buried complex impact structure: the Mjølnir Structure, Barents Sea: Journal of Geophysical Research, v. 103, p.30, 369–30, 483.

Worsley, D., Johansen, R., and Kristensen, S. E., 1988, The Mesozoic and Cenozoic succession of Tromsøflaket, *in* Dalland, A., Worsley, D., and Ofstad, K., eds., A lithostratigraphic scheme for the Mesozoic and Cenozoic succession offshore mid- and northern Norway: Norwegian Petroleum Directorate Bulletin 4, p. 42–65.

Zahnle, K., 1990, Atmospheric chemistry by large impacts, *in* Sharpton, V. L., and Ward, P. D., eds., Global catastrophies in Earth history; an interdisciplinary conference on impacts, volcanism, and mass mortality: Geological Society of America Special Paper 247, p. 271–288.

Ziegler, P. A., 1988, Evolution of the Arctic–North Atlantic and the Western Tethys: American Association of Petroleum Geologists Memoir, 43, 198 p. (30 plates).

MANUSCRIPT ACCEPTED BY THE SOCIETY DECEMBER 16, 1998

Geological Society of America
Special Paper 339
1999

Shock-induced effects in natural calcite-rich targets as revealed by X-ray powder diffraction

Roman Skála
Czech Geological Survey, Geologická 6, CZ-15200 Praha 5, Czech Republic; e-mail: skala@cgu.cz
Petr Jakeš
*Institute of Geochemistry, Mineralogy and Raw Materials, Faculty of Science, Charles University, Albertov 6,
CZ-12843 Praha 2, Czech Republic; e-mail: jakes@prfdec.natur.cuni.cz*

ABSTRACT

Four shocked limestone samples from the Kara Structure in Russia and 13 limestone and marlstone specimens from the Steinheim Crater in Germany have been characterized by X-ray powder diffraction. Two different approaches were applied during the study. Single-peak profile fitting has provided unit-cell dimensions, peak intensities, and widths. Of these parameters, only the peak half-widths are a reliable indicator of the degree of shock metamorphism. The Rietveld crystal structure refinements substantiated this observation. The unit-cell parameter ratios and the volumes of the unit cell or the coordination polyhedron around calcium atoms, respectively, can also be used to estimate the magnitude of shock stress that the samples were subjected to. When compared, the latter of the two applied data reduction methods provides a better estimate of the shock level. In the future, after proper calibration on experimentally shocked materials, it will probably be used as a shock barometer for carbonate-rich target rocks.

INTRODUCTION

In the past, impact researchers paid little attention to carbonate-rich target materials, although a substantial number of terrestrial impact structures exist in carbonate rocks. This is surprising since impacted carbonates could be a possible source of greenhouse gases. Lange and Ahrens (1986) speculated on the role of carbon dioxide released during impact events in early Earth's atmosphere.

Several papers report on the results of shock experiments in calcite, limestones, marbles, and dolomites. Ahrens and Gregson (1964) presented Hugoniot data for calcite and calcite-rich rocks. Boslough et al. (1982), Kotra et al. (1983), Lange and Ahrens (1986), Martinez et al. (1993, 1995), and Tyburczy and Ahrens (1986) have pointed out that shock compression of carbonates results in degassing. On the basis of experiments with limestones and marbles, Miura and Okamoto (1997) suggested that carbon is formed at the expense of calcite during impact.

Conversely, Martinez et al. (1994) and Bell (1997) showed that, even under shock loads of up to 60–65 and 40 GPa, respectively, carbonate-rich materials do not decompose, and carbon dioxide is not released.

Miura (1995) determined the unit-cell dimensions of calcite from powder data for limestone samples from the Ries Impact Crater in Germany. He distinguished two types of calcite—one formed under reducing conditions of the vapor plume, and the other under oxidizing conditions outside it. Unfortunately, he did not provide experimental details. Moreover, his cell dimensions are systematically lower than those published in the literature (Reeder, 1990).

The aim of this chapter is to inspect the effects of shock metamorphism in calcite-rich materials from two impact structures, the Kara Crater in Russia and the Steinheim Basin in Germany.

The optical microscopic study of these did not exhibit any obvious and pervasive features that could be indicative of high strain rates. Subtle changes indicating mechanical deformation, however, have been observed, such as mosaic-like specific bent fractures cutting across grains, common twin lamellae, and darkening of grains. Here we report on an X-ray diffraction study. The objective of our research is to characterize crystallographic changes in calcite due to strong shock pressures and to lay the foundation for future research in establishing a calcite shock barometry.

SAMPLES, SAMPLING, AND DATA COLLECTION

Four samples composed dominantly of calcite from the Kara Impact Structure in Russia and 13 specimens from the Steinheim Impact Crater in Germany were investigated. For an unshocked standard we used a clear, flattened rhombohedron of optical quality calcite from Iceland (so-called Iceland spar), which is characterized by very sharp diffraction maxima over the whole range of 2θ values studied. In addition to Iceland spar, shocked limestone specimens from the Steinheim Basin were compared with unshocked Solnhofen limestone—the Malmian limestone, which is stratigraphically equivalent to the shocked materials from the Steinheim structure.

Samples from the Kara Structure represent a sequence of shock-induced phenomena from almost unshocked clasts in suevite (NSKA02) or a weakly shocked specimen from the crater basement (NSKA06) to moderately shocked carbonates from shatter cones (NSKA09) and clasts in suevite (NSKA05). Limestone and marlstone samples from the Steinheim Crater can be divided into several distinct groups according to the pressure load they were subjected to. Solnhofen (RS01) and Heidenheim (ST02) samples coming from localities outside the crater were not shocked at all. Very low levels of shock metamorphism we expected in materials from the crater rim (ST03, ST12, ST13) and from a monomict limestone breccia, the so-called "Gries" (ST01). Carbonate clasts from the fall-back breccia (ST04) and shatter cones (ST05, ST06, ST07) represented the lithology with moderate shock-induced strain, the highest shock level available for our study.

The Kara Structure lies on the Kara River almost on the shore of the Kara Sea on the slopes of the Pai-Khoi Range in Russia, centered roughly at latitude N69°12′ and longitude E65°0′. There are two interpretative concepts about this impact structure. Traditionally, it was assumed (Masaitis, 1980; Sazonova et al., 1981) that the structure consists of the main Kara Crater 60–65 km wide on land and the neighboring Ust-Kara Crater with diameter of 25 km located mostly below sea level. Recently, Nazarov et al. (1993) reevaluated field observations, geophysical, topographic, petrologic, and geochemical data and postulated that the Ust-Kara Crater is not a separate structure. Instead, it is a part of large complex crater with the former Kara Crater corresponding to a central feature, and the so-called Ust-Kara Structure representing parts of this large crater cavity north of the central feature. The latter authors estimated that the total diameter of this crater is at least 120 km. Nazarov et al. (1993) also critically summarized all available geochronologic data of the Kara impact event. Based on these data, the structure is 64.57 ± 1.56 (1σ) Ma old. More recently, however, Trieloff et al. (1998) defined the age of the Kara Structure from three impact meltrock samples using ^{40}Ar-^{39}Ar plateau ages as 70.3 ± 2.2 Ma (2σ).

The Kara Structure is located in an intensively deformed complex of Paleozoic sedimentary rocks of Ordovician (Cambrian?) to Permian age overlying a complex of metasedimentary and volcaniclastic rocks of Proterozoic age. At the moment of impact, the Paleozoic complex was overlain by Mesozoic (mainly Cretaceous) quartz-clayey sediments with a thickness less than 500 m. This complex was later eroded, but fragments of Cretaceous rocks are preserved as clasts in suevites, in other ejecta deposits, and in clastic dikes. The largest portions of limestones occur in the Ordovician to Carboniferous target sequence.

The Steinheim Basin is located in Baden-Württemberg in southern Germany in the upland region of the Swabian Alb at latitude N42°02′ and longitude E10°04′. Its diameter varies between 3.1 km and 3.8 km (3.4 being an average) and its present-day depth reaches 90 m. The central uplift of this crater is 900 m across and 50 m high (Reiff, 1977).

Target lithologies at Steinheim consist dominantly of limestones and marls of the Upper and Lower Malmian, and clays, limestones, and sandstones of Dogger. The impact origin of this crater is, according to Reiff (1977), supported by the presence of the following: (1) shatter cones in the central uplift and annular breccia lens; (2) a large, annular breccia lens within the crater cavity; (3) faulted, folded, and locally brecciated rocks at the crater rim; (4) faulted, folded, and locally brecciated rocks below the breccia lens; and (5) planar deformation features in detrital quartz from the breccia.

An age of the impact event forming this structure has never been determined. Since it is accepted that both the adjacent Ries Crater and the Steinheim Basin were formed during a single impact event, the age of the Steinheim Crater is traditionally cited as that of the Ries, which is 15.21 ± 0.15 (1σ) old (Staudacher et al., 1982). Characteristics and numbering of the samples studied are given in Table 1.

We found that the method of sample preparation prior to diffraction experiments is critical. Because of the mineralogic heterogeneity encountered in some specimens, an electrical hand microdrill was used to obtain specific materials. Extremely high revolution rates of this microdrill led to a loss of the carbonate structure and the loss of shock effects. An excessive increase in peak broadening was noted. The high-speed microdrilling resulted in temperatures sufficient to decompose part of the carbonate. Eventually, all materials were resampled using a chisel and hammer. Rock chips obtained by this method were then ground in an agate mortar under alcohol to reduce the possibility that grinding again would modify the sample structure. Powdered samples with grain sizes of between 20 and 50 μm prepared in this way were used for all X-ray powder diffraction runs reported here.

TABLE 1. CHARACTERISTICS OF STUDIED CALCITE-RICH SAMPLES

Sample	Description
NSKA02	Bedded limestone clast from suevite. Individual beds differ by their color; grayish material (batch G) is intercalated with black (batch B), probably bitumen-rich one. Drilled out at the Kara River 20 km SE of the crater center. MGU (Moscow State University) no. KA1-053,4.
NSKA05	Black carbonaceous limestone clast from suevite. Drilled out at the Saa-Yaga River 25 km NW of the crater center. MGU no. SA1-327,0.
NSKA06	Limestone from the shocked crater basement. Drilled out at the Khonde-Yaga River 25 km S of the crater center. MGU no. KH0-059,0.
NSKA09	Shatter coned light gray massive compact limestone without bitumen. Batch 2 taken from the depth about 2 cm below the surface of shatter cone, while batch 4 represents thin surface and subsurface layer of the shatter cone. Collected at the Kara River.
RS01	Unshocked thin-bedded lithographic Solnhofen limestone of Upper Malmian age.
ST01	Tectonic monomict breccia ("Gries") consisting of centimeter-size angular to subangular limestone clasts (batch A) cemented by heavily fractured and brecciated very fine grained carbonate-rich matrix (batch B).
ST02	Unshocked bedded massive limestone collected at Heidenheim 7 km W of the crater center.
ST03	Fractured massive bedded limestone. Collected at the crater rim at Burgstall.
ST04	Fallback breccia consisting of claystones, clayey limestones, and marlstones. Sampled from the drill core SE of the crater center.
ST05	Shatter-coned massive limestone. Batch A corresponds to the surface layer as much as 3 mm thick; batch B comes from the depth about 2–3 cm below shatter cone surface. Collected at the central uplift.
ST06	Shatter-coned massive limestone. Surface 1–1.5-mm-thick layer of the shatter cone samples. Collected at the central uplift.
ST07	Shatter-coned massive limestone. Batch A represents thin film of the shatter cone surface; batch B corresponds to clayey limestone sampled at the depth about 5 mm below it. Collected at the central uplift.
ST12	Easily disintegratable fractured limestone. Collected at the crater rim at Burgstall.
ST13	Fractured massive bedded limestone. Collected at the crater rim at Burgstall.

Step-scanned powder diffraction data were collected using a Philips X'Pert PW-3710 MPD powder diffraction system. The system was equipped with a sealed copper tube operated at 40 kV and 40 mA. A bent graphite secondary monochromator removed the β-lines in the characteristic spectrum. The primary and secondary Soller slits lowered the effect of asymmetry of low-angle diffractions. Wafers cut from single-crystal silicon were used as sample holders to minimize the effects of complicated background shape commonly observed when normal glass holders are used. Powder patterns were collected in the range of 15°–140° 2θ with a step width of 0.02° 2θ; count-time was varied from 4 to 10 s per step.

RESULTS

Examples of raw scans in the range of 55.5°–62.5° 2θ CuK_α are shown in Figure 1. These scans reveal considerable line broadening in the case of samples for which higher magnitude of shock was expected. Therefore, we focused our attention mainly on peak widths. The obtained powder data were examined by two different data reduction methods—single-peak profile fitting refinement and the Rietveld method for crystal structure refinement.

Single-peak profile fitting

Reflection positions, widths, and intensities were refined using the ZDS program of Ondruš (1995), applying the approximation of reflection shapes by split Pearson VII function corrected for asymmetry. The obtained reflection positions were indexed using theoretical powder data generated from crystal structure data applying the LAZY PULVERIX program (Yvon et al., 1977). From such indexed patterns the unit-cell parameters were calculated by the appropriate routine in the program package ZDS (Ondruš, 1995), containing corrections for sample eccentricity. Unit-cell dimensions with the number of observed diffraction and the number of diffraction lines used for the refinement of lattice parameters are summarized in Table 2. Differences in unit cell volumes among all samples do not exceed 0.2% in the case of the Kara Structure and 0.3% in the case of the Steinheim Basin samples. Moreover, these differences can be caused by slightly varying chemistry of the carbonates studied. We believe therefore that the knowledge of the size of the unit cell is of little importance when studying the effects of shock metamorphism. Valuable information, however, is obtained when the ratio of unit-cell parameters is plotted against cell volume. This approach, however, does not eliminate the above problem of chemical composition.

Another result of single-peak profile fitting is a parameter describing the width of individual reflections—so-called full width in half-maximum (FWHM or half-width). In contrast to the unit-cell dimensions, the half-widths do not change with chemical composition. Therefore, this parameter was eventually used to distinguish samples with weak shock-induced deformations from those shocked more intensively. Caglioti et al. (1958) showed that the value of FWHM reveals an angular dependency on the diffraction angle, and they postulated the following equation describing this variance:

$$(FWHM)^2 = U\tan^2\theta + V\tan\theta + W$$

where θ is half the diffraction angle; U, V, and W are refinable parameters; and FWHM corresponds to half-width. When we applied this formula to our measurements, otherwise dispersed

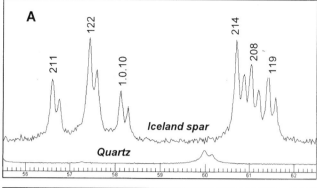
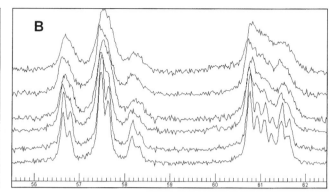
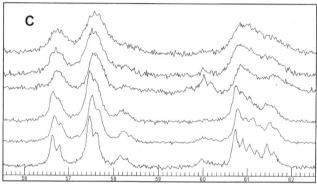

Figure 1. Raw θ/2θ scans in the region from 55.5° to 62.5° 2θ CuK$_\alpha$. A, Iceland spar (upper curve) represents an unshocked standard, whereas quartz (lower curve) is a common admixture in the limestone samples studied; diffraction indexes corresponding to calcite are shown in the pattern of Iceland spar. B, Limestones from the Kara Crater; sequence of scans, from bottom to top, is NSKA02G, NSKA02G, NSKA06, NSKA09,2, NSKA09,4, and NSKA05. C, Limestone from Solnhofen (RS01) and limestone and marlstone samples from the Steinheim Crater; the sequence of scans, from bottom to top, is ST01,B, ST13, RS01, ST04, ST06, and ST05,A. Note an increase in line widths with higher expected shock-induced strain.

data points produced curves, which can be used for estimating the intensity of shock metamorphism to which a certain sample was subjected. We have plotted data in graphs depicting directly FWHM and diffraction angle instead of squared value of half-width and tangent of halved diffraction angle. Angular dependence in this coordinate system, however, became more complicated; a polynomial function of the fourth order was finally used to fit the data properly.

From plots in Figure 2 it can be inferred that the limestone samples NSKA05 and NSKA09,2 from the basement of the Kara Structure and from a shatter cone, respectively, and the samples ST05, ST06, and ST07 representing shatter-coned limestones from the Steinheim Basin were subjected to the shock of the highest degree in the respective impact structures. Another feature, which can be easily seen in these plots, is a limited reliability of FWHM data from single-peak profile fitting at higher diffraction angles. Explanation of this effect is simple: high-angle reflections are generally of low intensities and when an effect of poor crystallinity induced by impact metamorphism influences the phase studied, those reflection almost disappear from the powder pattern. Therefore, high-angle reflections are poorly approximated by profile shape function, which results in biased values of intensities and FWHM parameters, whereas peak positions are usually correctly calculated (so they can be used for unit-cell parameters refinement). This is the reason why only FWHM values below 80° (or 100°) 2θ were considered to be reliable for estimation of the degree of shock in the specimens studied.

Miura (1995) used intensity of (104) reflection for characterization of samples from the Ries Structure. We were unable to determine whether his numbers correspond to peak integral intensities or to peak heights. Nevertheless, both quantities reflect (similarly as unit-cell parameters) the degree of shock-induced comminution and the variation in chemical composition. Moreover, they are very sensitive to the admixture of any phase—not only crystalline, but glassy one as well. Therefore, the determination of unbiased intensity of any reflection either in the form of integral or height is very problematic in the case of mixtures, and its use in an uncorrected form is obviously meaningless.

RIETVELD CRYSTAL STRUCTURE REFINEMENTS

The Rietveld method is a procedure for crystal structure refinement from powder diffraction data of phases with a known crystal structure model. The method seeks the best fit between the data set obtained during the diffraction experiment and that calculated on the basis of refined crystal structure. The quantity minimized, usually employing least-square methods, is the residual S_y:

$$S_y = \sum_i w_i(y_i - y_{ci})^2$$

where $w_i = 1/y_i$, y_i is observed (gross) intensity at ith step; y_{ci} is calculated intensity at the ith step; and the sum is over all data points. For details see Young (1993) and references therein.

The program FullProf (Rodriguez-Carvajal, 1997) was used

TABLE 2. UNIT-CELL DIMENSIONS OF STUDIED CARBONATE SAMPLES AS REFINED BY SINGLE-PEAK PROFILE FITTING PROCEDURE

	a	σ_a	c	σ_c	V	σ_V	c/a	M	N
Iceland	4.98941	0.00007	17.0616	0.0005	367.83	0.01	3.4196	60	60
NSKA02B	4.98583	0.00016	17.0456	0.0010	366.96	0.03	3.4188	60	60
NSKA02G	4.98537	0.00015	17.0429	0.0009	366.83	0.03	3.4186	60	60
NSKA05	4.98456	0.00049	17.0506	0.0033	366.88	0.10	3.4207	57	56
NSKA06	4.98629	0.00029	17.0513	0.0018	367.15	0.06	3.4196	61	58
NSKA09,2	4.98384	0.00024	17.0325	0.0017	366.38	0.05	3.4175	59	59
NSKA09,4	4.98661	0.00037	17.0464	0.0027	367.09	0.08	3.4184	59	59
RS01	4.98672	0.00033	17.0507	0.0022	367.20	0.07	3.4192	60	60
ST01,A	4.98680	0.00025	17.0531	0.0016	367.26	0.05	3.4196	60	60
ST01,B	4.98749	0.00012	17.0558	0.0008	367.42	0.03	3.4197	60	60
ST02	4.98734	0.00018	17.0527	0.0011	367.33	0.04	3.4192	61	61
ST03	4.98569	0.00037	17.0498	0.0022	367.03	0.07	3.4197	60	60
ST04,A	4.98178	0.00105	17.0457	0.0060	366.37	0.20	3.4216	33	33
ST04,B	4.98306	0.00087	17.0349	0.0053	366.32	0.17	3.4186	33	31
ST05,A	4.98555	0.00074	17.0538	0.0046	367.09	0.14	3.4206	61	56
ST05,B	4.98648	0.00083	17.0495	0.0053	367.14	0.17	3.4191	61	61
ST06	4.98628	0.00057	17.0531	0.0036	367.19	0.11	3.4200	61	58
ST07,A	4.98525	0.00053	17.0470	0.0034	366.90	0.11	3.4195	60	60
ST07,B	4.98425	0.00071	17.0539	0.0046	366.91	0.14	3.4216	61	61
ST12	4.98666	0.00032	17.0522	0.0020	367.22	0.06	3.4196	61	61
ST13	4.98574	0.00037	17.0476	0.0024	366.99	0.08	3.4193	61	61

Note: a, c, and V indicate unit cell edges and volume; c/a denotes ratio of unit cell parameters; M and N are numbers of reflections measured and used in refinement, respectively.

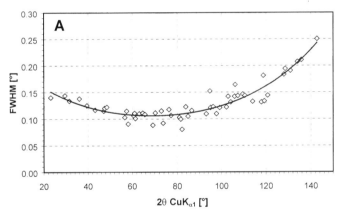
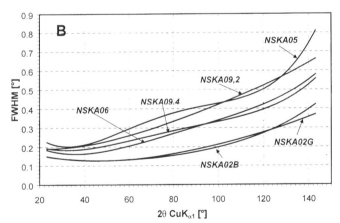
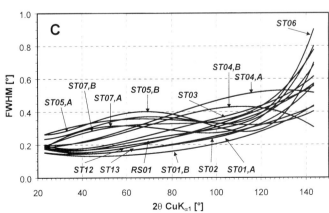

Figure 2. Dependence of peak half-widths FWHM from single-peak profile fitting on diffraction angle 2θ. A, Iceland spar, both regression curve and experimental data are shown. B, Limestones from the Kara Structure. C, Limestones from the Steinheim Crater and Solnhofen (RS01). The higher the expected shock stress, the broader the reflections. Note substantially different scales in panels A–C.

TABLE 3. BASIC CRYSTAL STRUCTURE PARAMETERS OF STUDIED CARBONATE SAMPLES OBTAINED BY RIETVELD REFINEMENT

	a	σ_a	c	σ_c	V	c/a	x_o	σ_x
Iceland	4.9886	0.0001	17.0613	0.0004	367.70	3.4201	0.2659	0.0005
NSKA02B	4.9856	0.0002	17.0476	0.0005	366.97	3.4193	0.2673	0.0005
NSKA02G	4.9851	0.0002	17.0450	0.0005	366.84	3.4192	0.2674	0.0005
NSKA05	4.9831	0.0004	17.0435	0.0012	366.51	3.4203	0.2670	0.0004
NSKA06	4.9849	0.0003	17.0512	0.0008	366.95	3.4206	0.2665	0.0004
NSKA09.2	4.9837	0.0004	17.0406	0.0010	366.53	3.4193	0.2675	0.0004
NSKA09,4	4.9858	0.0004	17.0494	0.0010	367.04	3.4196	0.2667	0.0004
RS01	4.9863	0.0004	17.0504	0.0009	367.12	3.4195	0.2662	0.0005
ST01,A	4.9860	0.0002	17.0530	0.0006	367.14	3.4202	0.2660	0.0005
ST01,B	4.9876	0.0002	17.0557	0.0005	367.43	3.4196	0.2657	0.0005
ST02	4.9866	0.0002	17.0536	0.0006	367.25	3.4199	0.2661	0.0004
ST03	4.9831	0.0003	17.0434	0.0008	366.50	3.4203	0.2656	0.0004
ST05,A	4.9795	0.0006	17.0243	0.0016	365.56	3.4189	0.2663	0.0005
ST05,B	4.9809	0.0005	17.0330	0.0014	365.96	3.4197	0.2660	0.0004
ST06	4.9842	0.0005	17.0386	0.0013	366.57	3.4185	0.2663	0.0005
ST07,A	4.9838	0.0004	17.0358	0.0011	366.46	3.4182	0.2661	0.0005
ST07,B	4.9821	0.0004	17.0391	0.0012	366.26	3.4201	0.2661	0.0005
ST12	4.9857	0.0003	17.0495	0.0007	367.03	3.4197	0.2658	0.0005
ST13	4.9849	0.0002	17.0443	0.0006	366.80	3.4192	0.2663	0.0004

Note: a, c, and V indicate unit cell edges and volume; c/a denotes ratio of unit cell parameters; x_o indicates the fractional coordinate of oxygen atom in the structure. Quantities above and their respective estimated standard deviations refined from measured powder data using the program FullProf (Rodriguez-Carvajal, 1997). Starting structure model for calcite was that adopted from Table 2a in Reeder, 1990.

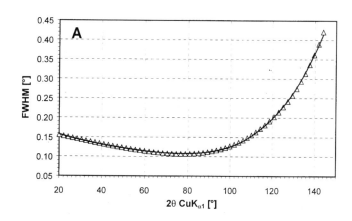

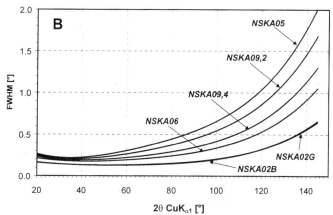

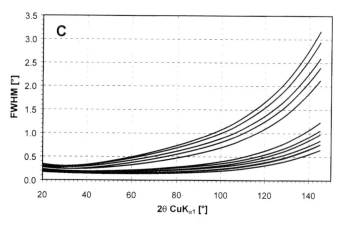

Figure 3. Dependence of peak half-widths FWHM from the Rietveld crystal structure refinement on diffraction angle 2θ. A, Iceland spar; regression curve representing the polynomial of fourth order through equidistantly spaced data points given by the formula of Caglioti et al. (1958) from the Rietveld refinement. B, Limestones from the Kara Structure. C, Limestones from the Steinheim Crater and Solnhofen. The sequence of curves for samples from the Steinheim Structure, from the bottom to the top, is as follows: ST01,B, ST02 overlapping ST01,A, ST13, ST12, RS01, ST03, ST07,A, ST07,B, ST06, ST05,B, and ST05,A. Note substantially different scales in individual panels and an increase in FWHM with increased expected shock level.

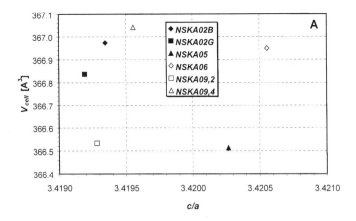

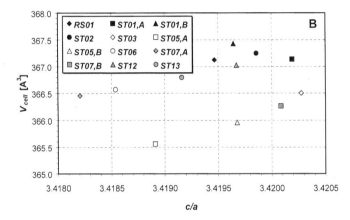

Figure 4. Variations in unit-cell volume with c/a ratio for studied carbonates. A, Samples from the Kara Structure. B, Samples from the Steinheim Crater. Cell volume of samples subjected to the higher magnitude of the shock metamorphism is reduced, compared to samples experiencing low or no shock.

TABLE 4. CHARACTERISTICS OF [CaO$_6$] OCTAHEDRA IN CALCITE STRUCTURES THROUGH OCTAHEDRAL VOLUME, QUADRATIC ELONGATION, AND BOND ANGLE VARIANCE

	V_o	σ_V	QE	σ_{QE}	σ^2
Iceland	16.92	0.02	1.0030	0.0003	11.011
NSKA02B	16.82	0.01	1.0031	0.0002	11.590
NSKA02G	16.80	0.02	1.0032	0.0003	11.657
NSKA05	16.81	0.01	1.0031	0.0002	11.557
NSKA06	16.86	0.01	1.0031	0.0002	11.335
NSKA09,2	16.78	0.01	1.0032	0.0002	11.695
NSKA09,4	16.85	0.01	1.0031	0.0002	11.335
RS01	16.89	0.01	1.0030	0.0003	11.084
ST01,A	16.90	0.01	1.0030	0.0003	11.042
ST01,B	16.93	0.02	1.0029	0.0003	10.868
ST02	16.90	0.01	1.0030	0.0002	11.074
ST03	16.89	0.01	1.0030	0.0002	10.897
ST05,A	16.81	0.01	1.0030	0.0002	11.098
ST05,B	16.85	0.01	1.0030	0.0002	10.998
ST06	16.86	0.01	1.0030	0.0002	11.058
ST07,A	16.86	0.01	1.0030	0.0002	10.961
ST07,B	16.85	0.01	1.0030	0.0003	11.098
ST12	16.91	0.01	1.0030	0.0002	10.912
ST13	16.89	0.01	1.0030	0.0003	11.051

Note: Quantities above and their respective estimated standard deviations were calculated by the program VOLCAL (Finger, 1996).

for the Rietveld refinements of crystal structures of the studied calcite-rich target rocks and unshocked standards. Refined parameters were scale factor, sample displacement correction, background polynomial parameters, half-width parameters, profile shape function parameters, preferred orientation parameter, unit-cell dimensions, and fractional coordinate of oxygen atom. During refinements, a standard turning-on sequence of refined parameters was kept. Background was modeled by a polynomial function of the fourth- or fifth-order implemented in the program. Sample displacement was corrected by cosine term. Pseudo-Voigt profile shape function corrected for asymmetry approximated measured reflection profiles. Preferred orientation due to pronounced calcite (104) cleavage was corrected by the March-Dollase correction. A convergence criterion was set to 0.1. Data from Table 2a in Reeder (1990) were adopted as a starting model of the calcite structure for the Rietveld structure refinements.

The advantage of the Rietveld method is the resolution of overlapping reflections, which could not be separated in any way during single-peak profile fitting. Therefore this method provides relatively very reliable unit-cell dimensions and peak half-widths. Moreover, the separation of otherwise overlapping reflections allows also to refine the positions of atoms within the crystal structure. On the other hand, admixtures of clay minerals or phases with unknown or poorly characterized structures complicate this procedure, which is why samples ST04,A and ST04,B were not used for the refinement—because they contain significant amounts of kaolinite-like phases. Crystal structure parameters characterizing the samples studied are summarized in Table 3.

Values of FWHM derived by the Rietveld refinements were recalculated from coefficients of the Caglioti et al. (1958) formula to equidistant data points in the coordinate system diffraction angle vs. FWHM. These data were then approximated by polynomial functions of fourth order using a least-squares routine. Obtained regression curves are shown in Figure 3. These data not only confirm the data from single-peak profile fitting (Fig. 2), they also seem to be of higher precision. That means that the Rietveld refinements provide more reliable data for the distinction of samples subjected to different shock pressures.

Dimensions of unit-cell, which were products of the whole powder pattern fitting procedure, can also finally distinguish samples with different pressure histories. The results obtained from these parameters agree well with those provided by half-widths. It seems to be very useful to plot unit-cell parameter ratios against the cell volume (see Fig. 4).

Quadratic elongations and bond angle variances (Robinson et al., 1971), together with volumes of coordination polyhedra around calcium atoms, were calculated from crystal structure

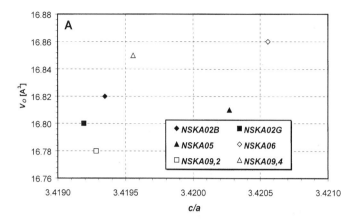

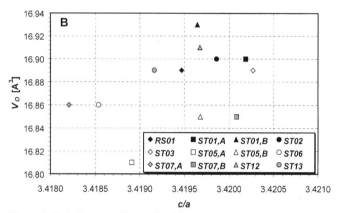

Figure 5. Variations in volume of polyhedron around calcium atom with c/a ratio for studied carbonates. A, Samples from the Kara Structure. B, Samples from the Steinheim Crater.

parameters by the program VOLCAL (Finger, 1996) and are listed in Table 4. Volumes of octahedra are plotted against unit-cell parameter ratios (Fig. 5), indicating that the higher the shock load, the smaller the volume of the coordination polyhedron.

Table 5 summarizes principal atomic distances characterizing groups [CO_3] and octahedra [CaO_6]. They were calculated from refined crystal structure data using the BONDSTR program (Rodriguez-Carvajal, 1995). Unfortunately, variability of particular distances among individual samples from an impact structure does not exceed standard deviations of respective distances. Therefore, these distances are of only limited relevance when assigning a degree of shock metamorphism.

DISCUSSION

During the study of the system $CaCO_3$ under elevated static pressures, high-pressure metastable polymorphs $CaCO_3$ (II) and $CaCO_3$ (III) were produced and characterized (e.g., Carlson, 1990; Biellmann et al., 1993). Ahrens and Gregson (1964) speculated that in the region below 10 GPa there are several phase transitions to these metastable $CaCO_3$ polymorphs. Pressures under which shatter cones are formed are assumed to be on the order of 2–10 GPa (Marakushev, 1981) and planar deformation features (PDFs) in quartz grains in suevites from the Kara Structure (Sazonova et al., 1981) indicate minimum pressures of 3–12 GPa. However, the present study did not detect any other polymorphs except calcite in these naturally shocked limestone samples.

Although Boslough et al. (1982), Kotra et al. (1983), Lange and Ahrens (1986), and Tyburczy and Ahrens (1986) observed

TABLE 5. PRINCIPAL ATOMIC DISTANCES WITHIN CRYSTAL STRUCTURES OF STUDIED CARBONATE SAMPLES

	C–O	σ	Ca–O	σ	O1–O2	σ	O1–O6	σ
Iceland	1.3265	0.0011	2.3364	0.0006	3.2113	0.0035	3.3946	0.0020
NSKA02B	1.3324	0.0010	2.3315	0.0005	3.2021	0.0031	3.3898	0.0018
NSKA02G	1.3331	0.0010	2.3308	0.0006	3.2009	0.0034	3.3890	0.0019
NSKA05	1.3306	0.0009	2.3312	0.0005	3.2018	0.0028	3.3891	0.0016
NSKA06	1.3285	0.0009	2.3334	0.0005	3.2057	0.0028	3.3914	0.0016
NSKA09,2	1.3330	0.0009	2.3300	0.0005	3.1996	0.0030	3.3880	0.0017
NSKA09,4	1.3297	0.0009	2.3331	0.0005	3.2053	0.0028	3.3910	0.0016
RS01	1.3272	0.0010	2.3346	0.0005	3.2084	0.0032	3.3922	0.0018
ST01,A	1.3260	0.0010	2.3351	0.0005	3.2094	0.0032	3.3928	0.0018
ST01,B	1.3251	0.0010	2.3364	0.0006	3.2119	0.0034	3.3940	0.0019
ST02	1.3268	0.0009	2.3350	0.0005	3.2091	0.0029	3.3928	0.0017
ST03	1.3237	0.0009	2.3346	0.0005	3.2092	0.0029	3.3914	0.0017
ST05,A	1.3260	0.0010	2.3309	0.0005	3.2033	0.0031	3.3869	0.0018
ST05,B	1.3247	0.0009	2.3326	0.0005	3.2061	0.0029	3.3890	0.0016
ST06	1.3272	0.0010	2.3331	0.0005	3.2065	0.0031	3.3899	0.0018
ST07,A	1.3263	0.0009	2.3332	0.0005	3.2071	0.0030	3.3897	0.0017
ST07,B	1.3257	0.0010	2.3329	0.0006	3.2061	0.0032	3.3899	0.0019
ST12	1.3251	0.0009	2.3354	0.0005	3.2102	0.0030	3.3926	0.0018
ST13	1.3272	0.0009	2.3337	0.0005	3.2071	0.0028	3.3909	0.0016

Note: Quantities above and their respective estimated standard deviations were calculated by the program BONDSTR (Rodriguez-Carvajal, 1995).

decomposition of carbonates, our research of shocked limestones does not show decarbonization under pressures of less than 10 GPa. This pressure estimate is based on the occurrence of shatter cones and PDFs in quartz. Similarly, no indication of carbonate breakdown as expected by Miura (1995) and Miura and Okamoto (1997) was observed in the samples studied. Conversely, previous observations by Martinez et al. (1994) and Bell (1997), indicating substantial stability of calcite up to 65 GPa shock pressure, are fully compatible with our X-ray diffraction data. The explanation of this behavior is not too difficult: the crystal structure of calcite consists of layers built up by triangular [CO_3] groups forming octahedral interstices filled by calcium atoms. This structure reacts to shear compression under conditions of moderate shock metamorphism by shifting of individual layers and formation of twins. Perhaps this mechanism is able to accommodate pressures up to at least 20 GPa. Therefore, we believe that some theories describing major biotic impact catastrophes taking into account large volumes of CO_2 released from weakly to moderately shocked carbonates have to be reevaluated. Apparently the process of carbonate degassing is effective only at pressures significantly higher than 20 GPa, and the amount of carbonate target material subjected to stresses of such magnitude is limited, even in very large impacts.

Based on our data we were able to qualitatively estimate the degree of shock to which the individual samples studied were subjected. The most valuable feature extracted from the X-ray powder diffraction patterns used for these estimates is the width of the reflections. More reliable reflection width values were obtained when the Rietveld crystal structure refinement was employed as a data reduction method than in the case of a common single-peak profile fitting procedure. On the other hand, lattice parameters depending significantly on the chemical composition of the carbonates studied seem to be of limited importance when speculating on the pressure history of the materials studied.

The results obtained in this research are rather qualitative and require detailed calibration using experimentally shocked calcite samples. Natural material does not provide a reliable record of actual pressure load and duration.

ACKNOWLEDGMENTS

We thank Winfried Reiff (Stuttgart), Mikhail A. Nazarov (Moscow), and Victor L. Masaitis (St. Petersburg), who kindly provided samples of shocked carbonate samples for the study. The manuscript benefited from thoughtful and constructive reviews by Fred Hörz and an anonymous reviewer. This research was conducted under Project 205/95/0980 from the Grant Agency of the Czech Republic.

REFERENCES CITED

Ahrens, T. J., and Gregson, V. G., Jr., 1964, Shock compression of crustal rocks: data for quartz, calcite, and plagioclase rocks: Journal of Geophysical Research, v. 69, p. 4839–4874.

Bell, M. S., 1997, Experimental shock effects in calcite, gypsum, and quartz [abs.]: Meteoritics and Planetary Science, v. 32, suppl., p. A11.

Biellmann, C., Guyot, F., Gillet, Ph., and Reynard, B., 1993, High-pressure stability of carbonates: quenching of calcite-II, high-pressure polymorph of $CaCO_3$: European Journal of Mineralogy, v. 5, p. 503–510.

Boslough, M. B., Ahrens, T. J., Vizgirda, J., Becker, R. H., and Epstein, S., 1982, Shock-induced devolatilization of calcite: Earth and Planetary Science Letters, v. 77, p. 409–418.

Caglioti, G., Paoletti, A., and Ricci, F. P., 1958, Choice of collimators for a crystal spectrometer for neutron diffraction: Nuclear Instruments and Methods, v. 3, p. 223–228.

Carlson, W. D., 1990, The polymorphs of $CaCO_3$ and the aragonite-calcite transformation, in Reeder, R. J., ed., Carbonates: mineralogy and chemistry: Reviews in Mineralogy, v. 11, p. 191–225.

Finger, L. W., 1996, VOLCAL—Program to calculate polyhedral volumes and distortion parameters: Washington, D.C., Carnegie Institution of Washington.

Kotra, R. K., See, T. H., Gibson, E. K., Hörz, F., Cintala, M. J., and Schmidt, R. G., 1983, Carbon dioxide loss in experimentally shocked calcite and limestone, in Proceedings, 14th, Lunar and Planetary Science Conference, Abstracts: Houston, Texas, Lunar and Planetary Institute, p. 401–402.

Lange, M. A., and Ahrens, T. J., 1986, Shock-induced CO_2 loss from $CaCO_3$: implications for early planetary atmospheres: Earth and Planetary Science Letters, v. 77, p. 409–418.

Marakushev, A. A., ed., 1981, Impactites: Moscow, Publishing House of the Moscow State University, 240 p. [in Russian].

Martinez, I., Schärer, U., and Guyot, F., 1993, Impact-induced phase transformations at 50–60 GPa in continental crust: an EPMA and ATEM study: Earth and Planetary Science Letters, v. 119, p. 207–223.

Martinez, I., Agrinier, P., Schärer, U., and Javoy, M., 1994, CO_2-production by impact in carbonates? An SEM-ATEM and stable isotope (C,O) study of carbonates from the Haughton impact crater: Earth and Planetary Science Letters, v. 121, p. 559–574.

Martinez, I., Deutsch, A., Schärer, U., Ildefonse, Ph., Guyot, F., and Agrinier, P., 1993, Shock recovery experiments on dolomite and thermodynamical modelling of impact induced decarbonization: Journal of Geophysical Research, v. 100, B8, p. 15465–15476.

Masaitis, V. L., Danilin, A. N., Mashak, M. S., Raikhlin, A. I., Selivanovskaya, T. V., and Shadenkov, E. M., 1980, Geology of astroblemes: Leningrad, Russia, Nedra, 231 p. [in Russian].

Miura, Y., 1995, New shocked calcite and iron grains from Nördlinger Ries impact crater [abs.]: Meteoritics, v. 30, p. A550–A551.

Miura, Y., and Okamoto, M., 1997, Change of limestone by impacts—source of impact induced graphite from target; IGCP Project 384, Symposium Impacts and Extraterrestrial Spherules: New Tools for Global Correlation: Tallinn, Estonia, Excursion Guide and Abstracts, p. 38–40.

Nazarov, M. A., Badyukov, D. D., Alekseev, A. S., Kolesnikov, E. M., Kashkarov, L. L., Barsukova, L. D., Suponeva, I.V., and Kolesov, G. M., 1993, The Kara impact structure and its relation to the Cretaceous/Paleogene event: Byuletin Moskovskogo Ova Ispytatelei Prirody, Otdel Geologii, v. 68, p. 13–32 [in Russian].

Ondruš, P., 1995, ZDS—software for analysis of X-ray powder diffraction patterns: Prague, Czech Republic, Czech Geological Survey, X-ray Diffraction Laboratory.

Reeder, R. J., 1990, Crystal chemistry of the rhombohedral carbonates, in Reeder, R. J., ed., Carbonates: mineralogy and chemistry: Reviews in Mineralogy, v. 11, p. 1–47.

Reiff, W., 1977, The Steinheim Basin—an impact structure, in Roddy, D. J., Pepin, R. O., and Merrill, R. B., eds., Impact and explosion cratering: New York, Pergamon Press, p. 309–320.

Robinson, K., Gibbs, G. V., and Ribbe, P. H., 1971, Quadratic elongation: a quantitative measure of distortion in coordination polyhedra: Science, v. 172, p. 567–570.

Rodriguez-Carvajal, J., 1995, BONDSTR—distance, angle and bond-strength calculations: Gif-sur-Yvette Cedex, France, Laboratoire Léon Brillouin, CEA-CNRS, Saclay.

Rodriguez-Carvajal, J., 1997, FullProf version 3.3. Rietveld, profile matching and integrated intensity refinement of X-ray and/or neutron data: Gif-sur-Yvette Cedex, France, Laboratoire Léon Brillouin, CEA-CNRS, Saclay.

Sazonova, L. V., Karotaeva, N. N., Ponomarev, G. Ya., and Dabizha, A. I., 1981, The Kara meteorite crater, in Marakushev, A. A., ed., Impactites: Moscow, Publishing House of the Moscow State University, p. 93–170 [in Russian].

Staudacher, Th., Jessberger, E. K., Dominik, B., Kirsten, T., and Schaeffer, O. A., 1982, ^{40}Ar-^{39}Ar ages of rocks and glasses from the Nördlinger Ries crater and the temperature history of impact breccias: Journal of Geophysics, v. 51, p. 1–11.

Trieloff, M., Deutsch, A., and Jessberger, E. K., 1998, The age of the Kara impact structure: Meteoritics and Planetary Science, v. 33, p. 361–372.

Tyburczy, J. A., and Ahrens, T. J., 1986, Dynamic compression and volatile release of carbonates: Journal of Geophysical Research, v. 91, p. 4730–4744.

Young, R. A., 1993, Introduction to the Rietveld method, in Young, R. A., ed., The Rietveld method: Oxford, United Kingdom, Oxford University Press, p. 1–38.

Yvon, K., Jeitschko, W., and Parthé, E., 1977, LAZY PULVERIX, a computer program for calculating X-ray and neutron diffraction powder patterns: Journal of Applied Crystallography, v. 10, p. 73–74.

MANUSCRIPT ACCEPTED BY THE SOCIETY DECEMBER 16, 1998

Carbon isotope study of impact diamonds in Chicxulub ejecta at Cretaceous-Tertiary boundary sites in Mexico and the Western Interior of the United States

R. M. Hough, I. Gilmour, and C. T. Pillinger
Planetary Sciences Research Institute, The Open University, Walton Hall, Milton Keynes MK7 6AA, United Kingdom

ABSTRACT

Diamonds have been found in the ejecta of the Chicxulub Crater in northeastern Mexico and in the Western Interior of the United States. They are located in both the ejecta claystone and the fireball layer that form the impact doublet at the Cretaceous-Tertiary boundary and range in size from 5 nm to 30 µm. Stepped heating experiments consistently reveal carbonaceous components in the acid-resistant residues, and cubic diamond was the only form observed using transmission electron microscopy. The carbon isotopic profiles for combustion of the diamond indicate multiple components with a range of compositions from –11 to –30‰ ($\delta^{13}C$). The isotopic compositions are representative of a mixture of carbon sources in which a meteoritic component is at most very minor along a mixing trend with carbon from target lithologies. The question remains whether diamond is also present within crater lithologies at Chicxulub in the same way as at other craters such as Popigai in Russia and Ries in Germany.

INTRODUCTION

Impact-produced diamonds have previously been reported at the 100-km Popigai Impact Crater in Russia (Masaitis et al., 1972, 1995; Masaitis 1993; Koeberl et al., 1997) and at the 24-km Ries Crater in Germany (Rost et al., 1977; Hough et al., 1995). Evidence is present in each case for two different formation processes for diamond: Masaitis et al. (1993, 1995) proposed a shock origin for the large Popigai diamonds, which is supported by the presence of the hexagonal polymorph of diamond, lonsdaleite, and other mineralogic similarities between the diamonds and precursor graphite, whereas Hough et al. (1995) proposed a process similar to chemical vapor deposition (CVD) for fine-scale aggregates from the Ries Crater as supported by silicon carbide and diamond intergrowths.

The Cretaceous-Tertiary (K-T) boundary comprises an impact doublet with the upper layer formed by vaporization of the projectile and target rocks with a global distribution and called the fireball layer, and a lower claystone, which also contains impact ejecta but is more geographically restricted and is called the ejecta layer (Hildebrand and Boynton, 1990). Diamonds at the K-T boundary were first reported by Carlisle and Braman (1991), who had selected a sample of the K-T boundary fireball layer from the Knudsen's Farm locality, Alberta, Canada; they used acid treatments to form a resistant residue to look for diamonds. Within this residue they reported cubic diamonds 3–5 nm in size and, with a single carbon isotopic analysis ($\delta^{13}C$ of –48‰), together with the small grain size, they proposed that the diamonds were meteoritic in origin, similar to those found in carbonaceous chondrites (Carlisle, 1992).

Diamonds were subsequently found in the fireball layer from the Brownie Butte, Montana, and Berwind Canyon, Colorado, K-T localities described by Gilmour et al. (1992). They were cubic and averaged 6 nm in size, but carbon and nitrogen isotope studies yielded a terrestrial isotopic signature with no known meteoritic (interstellar) component such as isotopically light

Hough, R. M., Gilmour, I., and Pillinger, C. T., 1999, Carbon isotope study of impact diamonds in Chicxulub ejecta at Cretaceous-Tertiary boundary sites in Mexico and the Western Interior of the United States, *in* Dressler, B. O., and Sharpton, V. L., eds., Large Meteorite Impacts and Planetary Evolution II: Boulder, Colorado, Geological Society of America Special Paper 339.

nitrogen detected (Gilmour et al., 1992). Gilmour et al. (1992) concluded that the diamonds were impact products rather than meteoritic. Hough et al. (1997) came to the same conclusion for diamonds extracted from the K-T boundary fireball layer at Arroyo El Mimbral, Mexico.

Presented here are the results of a study of the K-T boundary at the Western Interior sites of Raton Pass (RP), New Mexico, and Clear Creek North (CCN), Colorado, as well as data from Berwind Canyon (BC), Colorado, Brownie Butte, Montana, and Mimbral, Mexico.

TRANSMISSION ELECTRON MICROSCOPY

All samples were treated with acids to produce an acid-resistant residue in the same way as has been used to study diamonds as a minor phase in meteorites (e.g., Lewis et al., 1987; Russell, 1992) and at other impact sites (Hough et al., 1995), including the work of Gilmour et al. (1992) on K-T samples. This process removes carbonates, silicates, organic carbon, and graphite from the samples with only the most resistant phases known to survive. To study the acid-resistant residues by microscopy and ultimately to identify individual mineral phases requires the use of a transmission electron microscope (TEM) with elemental analysis (EDS) and selected area electron diffraction (SAED) pattern crystallographic analysis capabilities. A detailed study of the acid residues from the fireball and ejecta layers of each site revealed the presence of a relatively small number of minerals, and using energy dispersive scan for elemental analyses (EDS), it was possible to identify areas rich in Ti, Zr, and C elements.

Titanium

Those areas giving Ti peaks also gave oxygen peaks presumably due to the presence of rutile grains (TiO_2). The rutile grains form prismatic lath shapes up to 2 μm in size (Fig. 1) and also finer crystals either by the breakup of larger crystals or as a separate population. In the Clear Creek North fireball layer, a crystal 300 nm wide was identified, which had an etched appearance and was unlike any other phase present in the residue. EDS analyses gave peaks for Ti, O, and Fe, suggesting that the phase is ilmenite (Fig. 1). Although rutile is a common detrital mineral in sedimentary rocks, it is also known to exist in impact melts (von Engelhardt et al., 1995). The presence in impact melts of zircon transformed to baddeleyite and ilmenite decomposed to rutile indicate that original melt temperatures were in excess of 2,000°C (El Goresy, 1968; von Engelhardt et al., 1995). Thus, although the rutile is not shock deformed, it may well be an altered condensation product produced in the impact fireball.

Zircon

Particles that gave Zr peaks also gave signals for Si and O, presumably indicating zircon crystals ($ZrSiO_4$). Previous work has shown zircons to be present in the K-T impact layers, and some were found to display shock features (Bohor et al., 1993). TEM analyses revealed zircon to be present in all samples including the ejecta layers, occurring as prismatic laths and finer crystals.

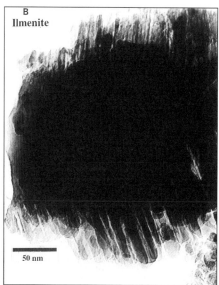
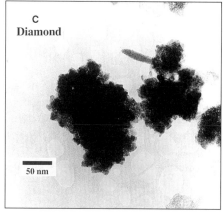

Figure 1. TEM micrographs. A, Rutile lath; B, ilmenite; C, diamond in acid-resistant residues from K-T boundary samples. The diamond cluster in C, is a cubic diamond (individual crystals up to 10 nm), determined from selected area electron diffraction patterns.

Carbon

Several areas were observed that gave pure carbon peaks in the EDS; on analysis at high magnification, small crystals up to 10 nm in size were identified (Fig. 1) and an SAED pattern gave d-spacings of 2.06, 1.26, and 1.08 Å, corresponding closely to those of cubic diamond. Locating the diamonds proved extremely difficult due to the high abundance of both rutile and zircon in the residues and also because of the very small size of the individual diamond crystals themselves. However, clusters similar to those described by Gilmour et al. (1992) were found in the residue of the fireball layer from Berwind Canyon (Fig. 1).

There was no indication on the diffraction patterns of any lonsdaleite component within the cubic diamond clusters. Diamond was not only located in the fireball layers of Berwind Canyon and Brownie Butte (Gilmour et al., 1992; this study), it was also in the ejecta layer of Clear Creek North (this study). There was also an indication of diamond in the Raton Pass fireball layer. A carbon peak was observed on the EDS, but due to the very fine grain size and the limitations of the microscope, imaging proved too difficult to obtain SAED patterns. No other carbon phases were observed in these acid-resistant residues, and therefore the phase is assumed to be diamond. The lack of a positive confirmation of diamond in some residues necessitated the use of a different technique, i.e., stepped combustion.

CARBON ISOTOPES AND CARBON CONTENT

Stepped combustion combined with static mass spectrometry was used to determine the overall carbon distribution and stable isotopic compositions of the acid residues. Gilmour et al. (1992) used the same techniques and satisfactorily resolved diamond in preliminary experiments performed on the residues from Brownie Butte and Berwind Canyon. Diamonds have also been resolved from other carbonaceous phases in studies on meteorite residues using the same experiments (Russell, 1992; Wright and Pillinger, 1989). In this study, residues of the fireball and ejecta layers from Raton Pass, Clear Creek North, and Berwind Canyon, and also the fireball layer from Brownie Butte and Mimbral, were analyzed. A background sample 6–8 cm below the K-T boundary from Brownie Butte was also treated and analyzed in the same way to give a control sample. The experiments were from room temperature to 1,200°C. In each case some carbon was released below 400°C, but this is typical of organic contamination from sample handling (Wright and Pillinger, 1989). The carbon release profile of the background sample indicates that there were no carbonaceous components present in the residue apart from typical organic contamination (<400°C) (Fig. 2).

Fireball and ejecta layers

The stepped combustion experiments reveal the presence of carbonaceous components in all of the residues analyzed (Figs. 3–5). Carbon was consistently released within the temperature range of 400°–600°C, with the exception of the Mimbral fireball layer residue (650°–900°C) and the ejecta layer from Clear Creek North (600°–800°C).

Fireball layers. The Western Interior fireball layer residues from Raton Pass, Clear Creek North, Berwind Canyon, and Brownie Butte are remarkably homogeneous in terms of their carbon release temperature ranges (Fig. 3; Table 1) and consistently reveal a carbonaceous component combusting between 400° and 575°C. Peak yields, which represent the maximum carbon release at an individual temperature step, occur from 500° to 550°C apart from for Brownie Butte, where the peak is at 450°C. The Berwind Canyon, Brownie Butte, and Mimbral fireball layers are known to contain diamond (Gilmour et al., 1992; Hough et al., 1997), and stepped combustion data indicate that the Raton Pass and Clear Creek North fireball layers are also diamondiferous (Fig. 3). This supports observations of these residues under the TEM where carbon-rich (EDS) areas were observed but could not be positively identified as diamond. The combustion temperature of diamond is known to be controlled by its grain size (Ash, 1990); therefore, the homogeneous major carbon release temperatures obtained for the Western Interior fireball layer residues suggest that the diamond grain size in those residues is the same. TEM observations gave a grain size of 6 nm for the diamond in both the Berwind Canyon and Brownie Butte fireball layers, which corresponds well to that expected for the temperature range of combustion obtained by stepped heating experiments (Ash, 1990). The Mimbral fireball layer contains higher concentrations and much larger diamonds (up to 30 μm) (Hough et al., 1997); hence, the combustion temperature is higher, with a major release of carbon from 650° to 900°C and peak release at 850°C (Fig. 4).

Ejecta layers. Study of the ejecta layers has revealed the possible presence of diamond as well as in the fireball layer

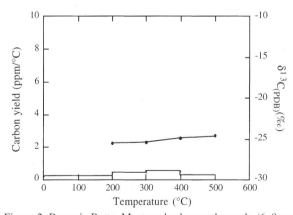

Figure 2. Brownie Butte, Montana background sample (6–8 cm below K-T boundary). Background sample stepped combustion with no major release of carbon; any carbon present is probably organic contamination from sample handling. The histogram represents the carbon release due to combustion, and the line graph with filled dots represents the corresponding carbon isotopic composition.

Figure 3. Stepped combustion plots of acid-resistant residues. A, Fireball layer from Raton Pass; B, fireball layer from Clear Creek North. Both show a major carbon release from 400° to 600°C.

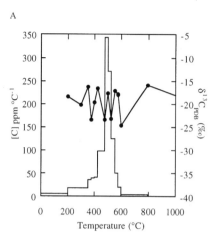 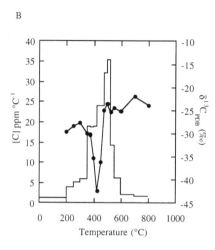

Figure 4. Stepped combustion plots of acid-resistant residues. A, Fireball layer from Brownie Butte (Gilmour et al., 1992); B, fireball layer from Mimbral (Hough et al., 1997).

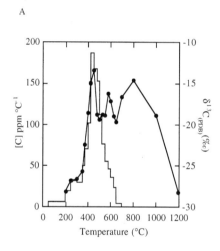 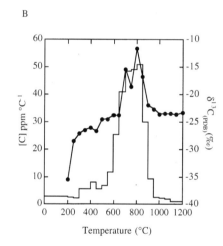

Figure 5. Stepped combustion plots of acid-resistant residues. A, Raton Pass ejecta layer; B, Clear Creek North ejecta layer.

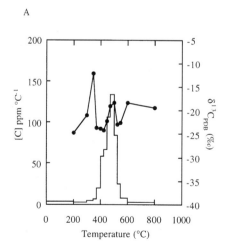 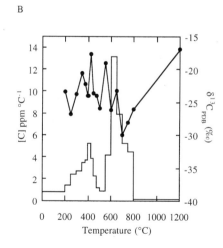

TABLE 1. WESTERN INTERIOR (USA) AND MIMBRAL (MEXICO) FIREBALL LAYERS: RESULTS OF STEPPED COMBUSTION OF DIAMOND IN THE FIREBALL LAYER RESIDUES

Site	C (wt%)	[C] Total (ng)	[C] Major (ng)	Σ δ¹³C Major (‰)	Temp Range Major (°C)
CCN	0.63	321.3	189	-23	425 – 550
RP	3.3	1094	844	-23 to -25.5	425 – 575
BC1	3.5	1356.5	1019.9	-14.7 to -17.4	425 – 575
BC2	4.9	3571.9	1500	-11.1 to -17.6	480 – 580
BC3	12.2	3284.8	693	-14.5 to -19.5	475 – 600
BB	3.0	32163.3	1375.1	-13.4 to -19.4	400 – 575
MIM	1.6	2560.5	1752	-11 to -15	650 – 900

Note: BC1 and BB from Gilmour et al., 1992. MIM data from Hough et al., 1997. Major refers to major carbon component released; C wt% refers to the residue.

TABLE 2. EJECTA LAYERS: RESULTS OF MAJOR CARBON RELEASE OBTAINED BY STEPPED COMBUSTION

Site	C (wt%)	[C] Total (ng)	[C] Major (ng)	Σ δ¹³C Major (‰)	Temp Range Major (°C)
CCN	0.28	643	398	-23 to -30	600–800
RP	1.48	1649.6	1376	-18.4 to -22	400–550
BC	1.87	563.2	305	-25	500–650

Major refers to major carbon component released; C (wt%) refers to the residue.

(Table 2; Fig. 5). TEM observations may have identified diamond in one of the ejecta layer residues (in CCN) and also in the ejecta layer equivalent (cf. spherule bed from El Penon, Mexico). Stepped combustion data (Table 2; Fig. 5) revealed carbonaceous components in the three Western Interior ejecta layers analyzed, namely, from Raton Pass, Clear Creek North, and Berwind Canyon, which presumably indicate that they are also diamondiferous. It should be noted that the major release temperatures are different from those of the fireball layers, which can be accounted for in one of two ways: (1) The diamond may be of a coarser grain size, resulting in higher combustion temperatures (e.g., CCN, 600°–800°C); or (2), if the diamonds are occluded by other mineral matter such as zircon or rutile (abundant in the residues), then this may effectively shield the diamonds from combustion until much higher temperatures than expected for their grain size. The second explanation is quite likely, as the residues contain an abundance of other minerals (though not carbonaceous) that could easily shield the diamond, either by surrounding or enclosing the diamond crystals (the same minerals are present in the fireball layers as well, although with no apparent effect). The possibility of coarser grained diamond in those residues with higher combustion temperatures is unresolved; no evidence for a substantially coarser grain size was obtained when the residues were observed under the TEM. If larger crystals are present, they may be missed under the TEM, due to heterogeneities in sampling. It is apparent from the stepped combustion experiments that all of the ejecta residues analyzed contain carbonaceous components, presumably diamond, and that the probable grain sizes (from release temperatures) of those diamonds is in nanometers rather than anything substantially larger.

Carbon isotopic composition of the K-T diamonds

The peak combustion of diamond (between 400° and 600°C in most residues) shows a U-shaped profile for δ¹³C values that is typical of more than one component (Ash, 1990). A more easily burned portion (lower temperature) often has a δ¹³C value that is isotopically heavier than the higher temperature component. Figure 6 shows the peak carbon isotopic profiles of the fireball layer residues over the major carbon release range (400°–600°C). From 500° to 550°C the U-shaped profile is clearly visible for several residues (CCN, RP, BC); this matches closely to the peak release of carbon in those residues. The same follows for the Mimbral residue, where the isotope profile reveals more than one component (–11 to –15‰) (Fig. 4).

The peak release of carbon for the Brownie Butte fireball layer occurs at 400°–425°C, which again yields δ¹³C values indicating more than one component (Fig. 4).

There is also a variation in the carbon isotopic composition of the components in the diamonds between the fireball layers, with a maximum δ¹³C of –10‰ down to a minimum of –25‰. This is highlighted by a plot of the median temperature of release (major release) against the δ¹³C of the carbon (Fig. 7), and even for the three (replicate) samples from Berwind Canyon (BC 1, BC 2, and BC 3) there is a difference in median temperature and δ¹³C. Taking the component being released to be diamond then the heterogeneity may be accounted for by grain size variations even within the same residue (BC).

The ejecta layers show variability in the carbon isotopic composition of the diamonds. The δ¹³C values, ranging from –18 to –30‰, are isotopically lighter than the values obtained for the fireball layers but with a similar difference between the heavy and light end-members, as with the fireball layers. The median temperatures of release are also highly variable, with a range of 475°–700°C, which is much greater than that obtained for the fireball layers.

"Light" components

Isotopically light components are present in many of the residues but the carbon isotopic composition of the diamond in the Clear Creek North ejecta layer (Fig. 5) is isotopically lighter than in any of the other residues. The light components tend to be observed at approximately 400°C; this is typically at lower temperatures than for the major diamond combustion range (500°C).

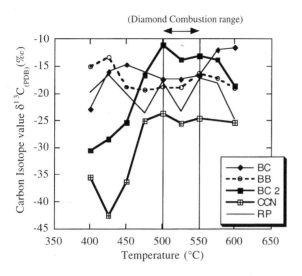

Figure 6. Carbon isotope data for the peak combustion range: fireball layers. Graph shows the carbon isotopic profile for diamond combustion in the fireball layer residues with multiple isotopically different components.

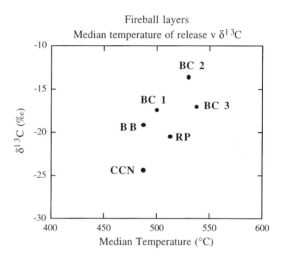

Figure 7. Median release temperatures vs. $\delta^{13}C$: fireball layers. Diagram highlights the variation in isotopic compositions of the carbonaceous components (inferred to be diamond) within the residues of the fireball layers.

The component is present in both experiments on the BC fireball layer (–30‰), as well as in the BC ejecta layer (–35‰) and in the CCN fireball layer (–42‰). The nature of this component is unresolved; unlike the higher temperature components (attributed to diamond), it is seemingly not present in all residues. It is, however, accompanied by a release of carbon, which suggests that it comes from a carbonaceous source that may not be present in all residues. The only identified form of carbon present in the residues is diamond. It is conceivable that there is more than one population in the residues that has different carbon isotopic compositions. For this to be the case, the isotopically light diamond must be very fine grained to combust at low temperatures (Ash, 1990). It should be noted that Koeberl et al. (1997) reported CO_2 inclusions within impact diamonds. If such inclusions were present in the K-T nanodiamonds, this could account for inconsistencies during stepped combustion. Unfortunately, the nanodiamonds are so small that testing such a theory is outside the scope of our present study. Carlisle (1992) reported a $\delta^{13}C$ for the K-T diamonds of –48‰ and attributed this to a meteoritic diamond signature. The data presented here support an isotopically light component, possibly diamond in the residues, which could be the same component as analyzed by Carlisle (1992). However, the light isotopic compositions obtained in the experiments presented here are purely components; the residues themselves have a bulk isotopic composition that is considerably heavier.

FORMATION OF THE DIAMOND

Shock formation

During impact, shock pressures are extremely high; this has been shown to form high-pressure polymorphs of minerals: carbon is transformed to diamond by a martensitic-type transformation (Erskine and Nellis, 1991). TEM analyses of the K-T diamonds yield SAED patterns of purely cubic diamond with no indication of lonsdaleite (thought to be a shock indicator). The diamonds are, however, very small, and the diffraction patterns obtained are for the clumps of crystals and not individual crystals. It is possible that any lonsdaleite present is in such small quantities that it fails to register on the SAED pattern. As the sensitivity of the mass spectrometer used in stepped combustion experiments is so high, it can detect minute amounts of carbon (1 ng is routine), this may be a better indicator than first thought to detect the presence of minor phases such as lonsdaleite. Even without lonsdaleite, the diamonds may still be shock produced, and small (nanometer-size up to approximately 10 μm) cubic diamonds have been produced in laboratory shock experiments at very high pressures (De Carli, 1995).

Chemical vapor deposition

An alternate mode of formation for the diamond is by a process similar to CVD in a plasma, either within the impact fireball or in vapors generated within shocked target rocks. Shock waves passing through such vapors have been shown to produce diamond (Kleiman et al., 1984). Nanometer-sized diamonds have been formed industrially using CVD when they grow on a silicon substrate via an interfacial layer of silicon carbide (Sato and Kamo, 1992; J. Hutchison, personal communication, 1995). Such a process was proposed for the formation of diamond and silicon carbide found in suevite from the Ries Crater in Germany (Hough et al., 1995). Silicon carbide is also present in diamondiferous acid-residues from the Popigai Impact Crater in Russia (Hough et al., 1997). The K-T diamonds are cubic and could well have grown within a plasma, which would contain the necessary species for diamond growth (H, Si, and C ions). Silicon carbide has also been reported in these K-T boundary claystones (B. F. Bohor, personal communication, 1995). Confirmation of its presence would further support a CVD-type process within the impact-produced fireball at the K-T boundary as silicon carbide cannot form by direct shock transformation. The stepped combustion experiments show no significant peak carbon releases at high temperatures (>1,000°C) to suggest the combustion of silicon carbide in the residues used in this study; however, we cannot entirely rule out its presence, since it typically combusts at higher temperatures than our system currently allows, and more carbon was released at the higher temperatures than in typical blanks. Further TEM analyses and higher temperature stepped combustion experiments should help to resolve this issue.

CARBON SOURCES

The carbon isotopic compositions are a useful indicator of possible carbon sources for the diamonds. The values obtained in this study and by Gilmour et al. (1992) do not match any single major carbon reservoir; they can, however, easily be obtained by a mixing of several sources or by heterogeneity of a single carbon source. The $\delta^{13}C$ of the soot (–25‰) in K-T boundary samples can directly be attributed to a source of biomass (also approximately –25‰); the soot is notably homogeneous in terms of its isotopic composition, unlike the impact diamonds. This serves to further support the fire origin for the soot (Wolbach et al., 1990), rather than from carbon in target lithologies at the crater. Other features have indicated that the diamonds are probably not of interstellar origin and any meteoritic component is minor in a mixture.

In this study the most likely source of carbon for the diamonds is from the target rocks at the impact site. The occurrence of the diamonds within both the fireball and ejecta layers, together with shocked minerals and the iridium anomaly, suggests a direct association with the ejecta of the impact. The majority of material associated with the impact originates from the target lithologies. The fireball would have contained vaporized target rocks, shocked target rock materials, and vaporized meteorite (represented by the Ir anomaly). At Chicxulub the target rocks include carbonates and evaporites at the surface, but components in the melt rocks and breccias of the crater itself suggest sandstones, granites, granitic gneisses, and schists that were also shocked, melted, and vaporized (Hildebrand et al., 1991; Sharpton et al., 1996). Carbon to form the diamond may have been derived from the sediments (as organic matter), from the carbonates (by dissociation of $CaCO_3$), or from the crystalline rocks that are known to contain graphite (V. Sharpton, personal communication, 1997). With $\delta^{13}C$ for the K-T diamonds ranging from –11 to –30‰, any variation in the mixture of the carbon forms could account for the different isotopic compositions obtained for the diamonds. The nature of the fireball is obviously unknown; inevitably there would be a mixing of the various carbon forms within this fireball, and heterogeneities in the fireball itself would be reflected in the $\delta^{13}C$ of the diamonds, if that is where they grew. If, however, they were purely shock-produced, then they might be expected to have more homogeneous isotopic compositions representative of their original carbon source.

GEOGRAPHIC DISTRIBUTION AND CONCLUSIONS

At all the localities studied, a carbonaceous component was observed in acid-resistant residues of both the ejecta and fireball layers. The nature of the acid treatment, TEM observations, and the characteristics of the carbon release profiles point to the component being diamond. The direct association between the diamond in the K-T boundary claystones and impact features, such as shock-deformed minerals and iridium anomalies, suggests that they, too, were impact-produced. Stable isotope studies have shown that any meteoritic component is minor, and if present, is probably an end-member in a mixing trend. The most likely source of the carbon to form the diamond is from target rocks at the impact site, namely, Chicxulub, Mexico, and the formation process could either be by shock transformation of the carbon in the target rocks (in the same way as stishovite and coesite occur in quartz), or by a process similar to chemical vapor deposition within a plasma in the fireball, or in shock-produced vapors in the target rock. The study by Hough et al. (1995) of the Ries Crater in Germany found diamond and silicon carbide intergrown (previously never observed naturally), which is typical of that observed during CVD growth of diamond. Silicon carbide has been reported in K-T boundary samples; if confirmed, this would support the CVD theory.

One question that remains unresolved is the geographic distribution of the K-T diamonds. They are present at K-T boundary sites throughout the Western Interior of the United States and are also present in Canada. The K-T boundary from Petriccio, Italy (marine limestones), was studied by Gilmour et al. (1992), but neither stepped combustion nor TEM analyses revealed any carbonaceous components. Carlisle (1995) believed that the extremely small grain size of the diamond would exclude their presence from marine localities, as they would not settle through the water column. If this were true, this would account for their absence in Italian sections, as these are marine localities. Mimbral is a proximal marine locality, and the diamonds are preserved there in the fireball layer probably due to their coarser grain size, which enabled them to settle through the water column.

It should be noted that the diamond in the Brownie Butte fireball layer combusted at lower temperatures than in the other residues (400°–425°C); this may represent finer grained diamond in the residue. Even though TEM analyses identified very similar grain sizes in the Brownie Butte and Berwind Canyon samples, it may be that a finer grained portion predominates in the Brownie Butte fireball layer. This would suggest that there is a grading up through the United States and away from the impact crater (Chicxulub, Mexico) from relatively coarse to finer grained diamond and from higher to lower concentrations in the same way as for shocked quartz grains (Alvarez et al., 1995). If this was to continue farther away from the impact, ultimately, at more distal sites, the diamond may either be undetectable (too small) or not present at all. A wider search of K-T boundary samples from around the world is needed to resolve this problem.

ACKNOWLEDGMENTS

We thank Naomi Williams, The Open University, for assistance with microscopy, and Bruce Bohor, Art Sweet, and Jack Lerbekmo for assistance with fieldwork. Appreciation is extended to Ted Bunch and Dieter Heymann for their constructive reviews. The Particle Physics and Astronomy Research Council and the Natural Environment Research Council are thanked for funding.

REFERENCES CITED

Alvarez, W., Claeys, P., and Kieffer, S. W., 1995, Emplacement of Cretaceous-Tertiary boundary shocked quartz from Chicxulub crater: Science, v. 269, p. 930–935.

Ash, R. D., 1990, Interstellar dust from primitive meteorites: a carbon and nitrogen isotope study [Ph.D. thesis]: Milton Keynes, United Kingdom, The Open University, 240 p.

Carlisle, D. B., 1992, Extraterrestrial diamond: Nature, v. 357, p. 119–120.

Carlisle, D. B., 1995, Dinosaurs, diamonds and things from outer space, the great extinction: Stanford, California, Stanford University Press, 241 p.

Carlisle, D. B., and Braman, D. R., 1991, Diamonds at the K/T boundary: Nature, v. 352, p. 709.

De Carli, P. S., 1995, Shock wave synthesis of diamond and other phases: Proceedings, Materials Research Society Symposium, v. 383, p. 21–31.

El Goresy, A., 1968, The opaque minerals in impactite glass, in French, B. M., and Short, N. M., eds., Shock metamorphism of natural materials: Baltimore, Maryland, Mono Book Corp., p. 531–533.

Erskine, D. J., and Nellis, W. L., 1991, Shock-induced martensitic phase transformation of oriented graphite to diamond: Nature, v. 349, p. 317–349.

Gilmour, I., Russell, S. S., Arden, J. W., Lee, M. R., Franchi, I. A., and Pillinger, C. T., 1992, Terrestrial carbon and nitrogen isotopic ratios from Cretaceous-Tertiary boundary nanodiamonds: Science, v. 258, p. 1624–1626.

Hildebrand, A. R., and Boynton, W. V., 1990, Proximal Cretaceous-Tertiary boundary impact deposits in the Caribbean: Science, v. 248, p. 843–847.

Hildebrand, A. R., Penfield, G. T., Kring, D. A., Pilkington, M., Camargo, Z., Jacobsen, S. B., and Boynton, W. V., 1991, Chicxulub crater: a possible Cretaceous/Tertiary boundary impact crater on the Yucatan peninsula, Mexico: Geology, v. 19, p. 867–871.

Hough, R. M., Gilmour, I., Pillinger, C. T., Arden, J. W., Gilkes, K. W. R., Yuan, J., and Milledge, H. J., 1995, Diamond and silicon carbide in suevite from the Nördlinger Ries impact crater: Nature, v. 378, p. 41–44.

Hough, R. M., Langenhorst, F., Montanari, A., Pillinger, C. T., and Gilmour, I., 1997, Diamonds from the iridium-rich K-T boundary layer at Arroyo el Mimbral, Tamaulipas, Mexico: Geology, v. 25, p. 1019–1022.

Kleiman, J., Heimann, R. B., Hawken, D., and Salansky, N. M., 1984, Shock compression and flash heating of graphite/metal mixtures at temperatures up to 3200 K and pressures up to 25 Gpa: Journal of Applied Physics, v. 56, p. 1440–1454.

Koeberl, C., Masaitis, V. L., Shafronovsky, G. I., Gilmour, I., Langenhorst, F., and Schrauder, M., 1997, Diamonds from the Popigai impact structure, Russia: Geology, v. 25, p. 967–970.

Lewis, R., Ming, T., Wacker, J. F., Anders, E., and Steel, E., 1987, Interstellar diamond in meteorites: Nature, v. 326, p. 160–162.

Masaitis, V. L., 1993, Impactites from the Popigai impact structure: Regionalaia Geologia i Metallogenia, v. 1, p. 121–134. [in Russian].

Masaitis, V. L., Futergendler, S. I., and Gnevyshev, M. A., 1972, The diamonds in the impactites of the Popigai meteoritic crater: Zapiski Vsesouznogo Mineralogicheskogo Obstchestva, v. 101, p. 108–113 [in Russian].

Masaitis, V. L., Shafranovsky, G. I., and Fedorova, I. G., 1995, The apographic impact diamonds from astroblemes Ries and Popigai: Proceedings of the Russian Mineralogy Society, v. 124, p. 12–19 [in Russian].

Rost, R., Dolgov, Y. A., and Vishnevsky, S. A., 1977, Gases in inclusions of impact glass in the Ries crater, West Germany, and finds of high pressure carbon polymorphs: Doklady Akademii Nauk, USSR, v. 241, p. 165–168 [in Russian].

Russell, S. S., 1992, A carbon and nitrogen isotope study of chondritic diamond and silicon carbide [Ph.D. thesis]: Milton Keynes, United Kingdom, The Open University, 228 p.

Sato, Y., and Kamo, M., 1992, Synthesis of diamond from the vapour phase, in Field, J. E., ed., The properties of natural and synthetic diamond: London, London Academic Press, p. 423–469.

Sharpton, V. L., Marin, L. E., Carney, J. L., Lee, S., Ryder, G., Schuraytz, B. C., Sikora, P., and Spudis, P. D., 1996, A model of the Chicxulub impact basin based on evaluation of geophysical data, well logs, and drill core samples: Geological Society of America Special Paper 307, p. 55–74.

von Engelhardt, W., Arndt, J., Fecker, B., and Pankay, H. G., 1995, Suevite breccia from the Ries crater, Germany: origin, cooling history and devitrification of impact glasses: Meteoritics, v. 30, p. 279–293.

Wolbach W. S., Gilmour I., and Anders E., 1990, Major wildfires at the Cretaceous/Tertiary boundary, in Sharpton, V. L., and Ward, P. D., eds., Global catastrophes in Earth history: Geological Society of America Special Paper 247, p. 391–400.

Wright, I. P., and Pillinger, C. T., 1989, Carbon isotopic analysis of small samples by use of stepped-heating extraction and static mass spectrometry: U.S. Geological Survey Bulletin 1890, p. 9–34.

MANUSCRIPT ACCEPTED BY THE SOCIETY DECEMBER 16, 1998

Formation of a flattened subsurface fracture zone around meteorite craters

Yevgeny Zenchenko and Vsevolod Tsvetkov
Institute for Dynamics of Geospheres, 38 Leninsky Prospect, Building 6, 117979 Moscow, Russia

ABSTRACT

Studies of well-preserved impact craters on the Earth's surface demonstrate the presence of an approximately hemispherical fracture zone under the crater floor. In addition, there is a flattened subsurface fracture zone that becomes shallower with increasing distance from the crater's center. High-explosive laboratory modeling allowed us to observe all features of this fracturing process, including subsurface fracture zone formation. Pine rosin was used as a model material. Target fragmentation was recorded by optical techniques facilitated by the relative transparency of rosin. The experiments showed that both subsurface and hemispherical fracture zone formation takes place simultaneously and precedes crater formation. Furthermore, the ratio of horizontal to vertical dimensions of the fracture zone is the same 0.5 as that for the Earth's craters.

INTRODUCTION

Rock fracturing is one of the basic parts of the cratering process. The presence of a fracture zone that surrounds a crater is well known in different geophysical fields. It may be detected using different geophysical techniques. Early investigations of several terrestrial meteorite craters (e.g., Sander et al., 1964; Robertson and Grieve, 1975; Innes, 1964; Aaloe et al., 1976) demonstrated a complex fracture zone surrounding these craters. The results of seismic investigations of Meteor Crater (Ackerman et al., 1975) are presented in Figure 1, which clearly demonstrates the peculiarities of the fracture zone shape.

Dabizha and Ivanov (1978) generalized the results of fracture zone investigations in different terrestrial craters and offered a geophysical model of meteorite craters. A fracture zone in this model is described as consisting of two regions: a hemispherical inner region and a shallow outer subsurface region. The radius of the hemispherical region equals to a crater radius, whereas the radius of the shallow zone is two times greater than the crater radius (Bazilevsky et al., 1983). It was suggested (Dabizha and Ivanov, 1978) that the complex form of the fracture zone is a result of secondary fragmentation during crater excavation. The cratering flow lines that terminate at the free surface are considered to be responsible for subsurface fracture zone formation in this model. This suggestion has a serious drawback because secondary fragmentation takes place in the inner part of the fracture zone rather than at its periphery (Tsvetkov et al., 1977).

Target fragmentation was the subject of a broad variety of high-explosive as well as projectile impact laboratory cratering experiments. Most of them were conducted using natural rock samples. However, the use of such materials has some substantial disadvantages. First, it is impossible to observe the fracture pro-

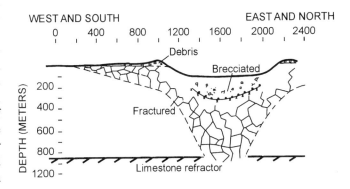

Figure 1. Schematic geologic cross section of Meteor Crater, Arizona (after Ackerman et al., 1975).

Zenchenko, Y., and Tsvetkov, V., 1999, Formation of a flattened subsurface fracture zone around meteorite craters, *in* Dressler, B. O., and Sharpton, V. L., eds., Large Meteorite Impacts and Planetary Evolution II: Boulder, Colorado, Geological Society of America Special Paper 339.

cess develop in time; we see only the final result of target fragmentation. Second, the elastic and strength moduli of rocks exceed the initial shock pressure, resulting in only a small number of cracks fragmenting a rock sample.

Using silicate glasses as a model material allows us to observe the fracturing process by optical techniques. Unfortunately, even in this case problems arise due to the high values of elastic and strength parameters in comparison with initial shock pressure. Transparent polymeric materials are also used as a model medium in cratering experiments, but sometimes their utilization results in unpredictable effects.

Thus the choice of model material is an important step in small-scale cratering experiments from the point of view of the target fragmentation process. We used pine rosin as a model material in our experiments. Rosin was previously used for the modeling of a contained explosion (Tsvetkov et al., 1977), and results of that experiment were in good agreement with data from large-scale underground explosions. Mindful of this experience, we believe that our results will be fruitful for understanding the process of target fragmentation caused by meteorite impact.

EXPERIMENTAL PROCEDURE

Pine rosin is a brittle material having a fracture behavior similar to the majority of rocks. Rosin has the following mechanical properties: density $\rho = 1.08$ g/cm³; longitudinal wave velocity $v_l = 2,470$ m/s; transverse wave velocity $v_t = 1,040$ m/s; unconfined compressive strength = 20 MPa. Spherical charges of pentaerythrol tetranitrate (PETN), with radii $R = 4$ mm and mass = 0.38 g, were used as an explosive. We examined half-buried as well as full-buried charge configurations. The initial pressure in this explosive source is over 10 GPa and exceeds the value $\rho v_l^2 = 6.5$ GPa, characterizing the elastic module of a rosin, and hence the experimental conditions are similar to the case of hypervelocity meteorite impact.

Experimental samples were made by casting melted rosin into an organic glass mold that also served as an acoustic extension of a rosin block in order to delay stress wave reflections from free boundaries. The bottom of the sample was glued onto an organic glass slab for the same purpose. These measures enabled us to observe target fragmentation during the necessary period of time. Figure 2 illustrates the experimental arrangement. Due to the relative transparency of rosin, we could observe the crushing front propagation using a high-speed frame camera at $5 \cdot 10^5$ frames/s.

EXPERIMENTAL RESULTS

The photos in Figure 3 illustrate serial frames of the fracture process development in a rosin block. The successive projections of these frames are presented in Figure 4. These projections were used for further treatment of experimental results. The zone of fracture process development in Figure 4 can be divided into

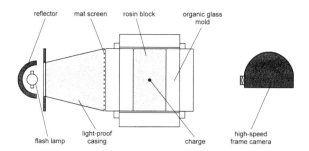

Figure 2. Experimental arrangement for optical recording of fragmentation process.

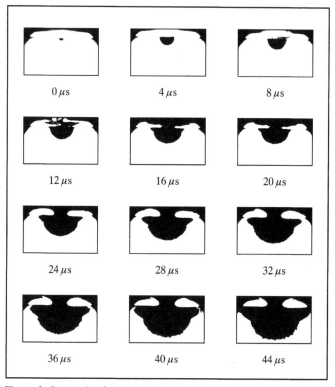

Figure 3. Successive frames of the fragmentation process development in pine rosin. The explosive charge is full-buried. Numbers under the frames express the time intervals after detonation in microseconds. The frame width = 180 mm.

three regions. Central region 1 is confined by a cone whose apex position and apex angle are dependent on charge location. The fracture front in this region has an approximately spherical form. Its velocity, v_f, equals 1,200 m/s at distances less than $10R$ and 960 m/s in the range of $10–15R$, where R is charge radius. At larger distances the fracture front decays.

In region 2, the fracture front possesses a conical surface with a constant angle between its element and free surface. The value of this angle also depends on charge position. The velocity of the conical front is $v_f = 960$ m/s in the entire region 2, regardless of charge position. The elements of the conical front

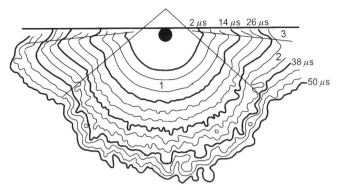

Figure 4. Successive positions of the fracture front. The time interval between adjacent contours equals 4 µs.

TABLE 1. GEOMETRIC CHARACTERISTICS OF THE FRACTURE FRONT FOR DIFFERENT CHARGE POSITIONS

Charge Position	l/R*	Θ (°)	β (°)
Half-buried[†]	4.6	98	29
Full-buried[§]	4.9	110	36

*R = charge radius.
[†]"Half-buried" means the center of a charge is at the target surface.
[§]"Full-buried" means the center of a charge is under the target surface by the value of its radius.

terminate at some depth under the free surface at the lower boundary of zone 3.

In zone 3, which borders with the free surface, fragmentation is caused by spall phenomena. The spall region formation was previously described analytically by Melosh (1984) and was observed in the hypervelocity impact experiments (Polanskey and Ahrens, 1990) The geometric characteristics of these zones for different charge positions are presented in Table 1 in accordance with Figure 5.

The photos in Figure 3 also show ejecta plume formation and growth. On the basis of these observations it should be noted that crater excavation takes place substantially later than target fragmentation. Therefore we can conclude that the existence of a conical crushing front is an independent process that is not affected by crater excavation. The final fracture zone shape that is observed on the cross section of the experimental sample (Fig. 6) repeats the fracture front configuration. Its horizontal radius equals four crater radii, its depth, two radii. In other words, the ratio of the horizontal to vertical fracture zone dimension is the same as that for meteorite craters.

DISCUSSION

The velocity of the conical crushing front is equal to the spherical crushing front velocity at the periphery of the spherical part of the fracture zone. It was shown for the case of a contained explosion in rosin (Tsvetkov et al., 1977) that the value v_f = 960 m/s corresponds to the maximum tension crack velocity, which is equal to the Rayleigh wave velocity in rosin. The crushing front in the outer part of the fracture zone is a front of tension cracks propagating with maximum velocity. Hence we conclude that the conical crushing front in the entire region of its existence is also a front of tension cracks propagating with maximum velocity.

We note that the pattern of fracture process development and a pattern of waves propagating in a half-space after a half-buried explosion on its surface (Fig. 7) are very similar. First, this resemblance consists of the existence of a conical transverse wave K generated by longitudinal wave P during its interaction with the free surface. The angle α between the K-wave and the free surface is given by

$$\alpha = \frac{v_t}{v_l};$$

for pine rosin, $\alpha = 26°$ The value of α depends on the distance from the charge epicenter when the charge is under the free surface and is defined by the crossing point velocity of the P-wave and the surface. The dependence of this velocity on epicentral distance can be readily obtained from simple geometric consideration. For the case of a full-buried explosion the velocity of the crossing point v_c is defined as follows:

$$v_c = v_l \sqrt{\frac{r/R+1}{r/R-1}},$$

where r is the epicentral distance, R is the charge radius, and v_l is the longitudinal wave velocity. The angle α increases from zero at the epicenter to an asymptotic value equals that of the case of a half-buried charge.

The velocity of the crossing point of the conical fracture front extension and a free surface v_{fc} is defined as

$$v_{fc} = \frac{v_f}{\sin \beta},$$

where v_f is the conical fracture front velocity, and β is the angle between them, according to Figure 5. In the case of a half-buried explosion v_{fc} = 1,980 m/s, whereas for full-buried explosion, v_{fc} = 1,630 m/s. In both cases this velocity is less than the longitudinal wave velocity and the velocity of the crossing point of longitudinal wave and the free surface. This means that the conical fracture front is always behind conical wave front K. That is, the stress-strain state due to the transverse conical wave affects the fracture process on the conical crushing front. Therefore, we would expect that both the wave propagation pattern and the fracture process development change together as a function of the explosive source depth. But even in the case of a half-buried explosion, which is the simplest from the point of view of geometry, there is a statistically significant difference between the slope angles of the conical crushing front and the K-wave front. In the

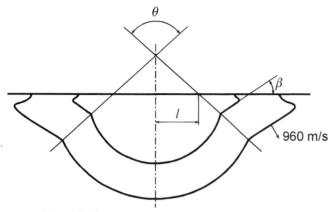

Figure 5. The scheme explaining notations in Table 1.

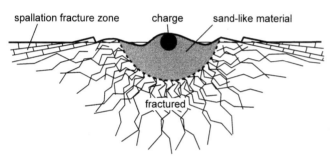

Figure 6. Schematic cross section of the rosin block after the explosion.

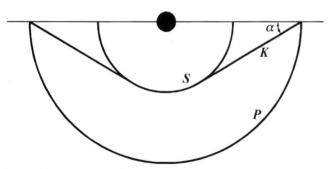

Figure 7. The scheme of elastic wave fronts produced by an explosion of a half-buried charge on the surface of an elastic half-space. P = longitudinal wave; S = transverse wave; K = conical transverse wave.

case of a full-buried charge, angle α changes as shown above, whereas the slope angle of the conical crushing front β remains constant; moreover, its value becomes greater than in the case of half-buried explosion.

Thus the connection between the crushing front shape and the elastic waves front geometry is not so obvious as it first seems. An exact knowledge of the medium stress-strain state in the area of a conical crushing front is very important for a detailed description of the target surface influence on the fracture mechanism. These kinds of data may be obtained only from numeric calculations that are beyond the scope of this work.

It should be noted in comparing the results of laboratory cratering experiments with data on meteorite craters that gravity substantially affects the cratering process including target fracturing. Lithostatic pressure may serve as a factor preventing rock fracture under the action of tensile stresses. In the terrestrial craters ranging from 10 to 20 km in diameter, a sharp reduction of gravity anomaly growth vs. crater diameter takes place. This fact is connected with the suppression of rock bulking at depths of 5–10 km (Bazilevsky et al., 1983). The subsurface flattened fracture zone formation in meteorite craters is also affected by gravity. But the subsurface fracture zone depth does not exceed approximately one-third of the depth of the fracture zone in the central region for several kilometer-diameter craters (Fig. 1).

Thus, one may expect that the proposed mechanism for subsurface fracture zone formation is valid for large craters up to about 50 km in diameter. Moreover, even in large craters the influence of gravity on target fracturing will vanish at the subsurface fracture zone periphery. Rock fragmentation under the action of tensile stresses may differ from that of modeling materials. Rocks often fragment at discontinuities of their internal structure vs. brittle homogeneous materials those fragmentation is accompanied by the formation of new surfaces. But from the point of view of the final result (that is, the alteration of rock properties such as density or sound velocity), there is no major difference between these two modes of fragmentation.

CONCLUSIONS

The experiments conducted displayed a series of interesting features characteristic of target fragmentation caused by an explosion on the surface of a half-space. It was shown that the fracture zone shape observed in laboratory experiments resembles that of well-preserved meteorite craters on the Earth's surface. The ratios of horizontal to vertical dimensions of the fracture zone are practically the same for both laboratory and meteorite craters, whereas fracture zone dimensions with respect to crater diameter are substantially different. This discrepancy may be caused by the relatively low strength of rosin vs. rock, as well as by the alteration of the fractured rock properties due to geologic processes that affect crater appearance in geophysical fields. The experiments directly demonstrate that fragmentation in both subsurface and hemispherical fracture zones takes place simultaneously and precedes crater formation and is controlled by the same fracture mechanism. Thus the subsurface fracture zone formation is not a result of the cratering flow.

ACKNOWLEDGMENTS

We express particular appreciation to the Organizing Committee of the Sudbury '97 conference, which permitted our participation in the meeting. This work was supported by Grant 99-05-64002 from the Russian Foundation for Basic Research.

REFERENCES CITED

Aaloe, A., Dabizha, A., Karnaukh, B., and Starodubtsev, V., 1976, Geophysical investigations at the main Kaali Crater: Izvestiya Eytonian Academy of Science: Chemistry, Geology, v. 25, p. 58–65.

Ackerman, H. D., Godson, R. H., and Watkins, J. S., 1975, A seismic refraction technique used for subsurface investigations at Meteor Crater, Arizona: Journal of Geophysical Research, v. 80, p. 765–775.

Bazilevsky, A. T., Ivanov, B. A., Florensky, K. P., Yakovlev, O. I., Feldman, V. I., and Granovsky, L. V., 1983, Impact craters on the Moon and planets: Moscow, Nauka, 200 p.

Dabizha, A. I., and Ivanov, B. A., 1978, Geophysical model for meteorite craters construction and particular questions of cratering mechanics, *in* Krinov, Y. L., ed., Meteoritika: Moscow, Nauka, v. 37, p. 160–167.

Innes, M. J. S., 1964, Recent advances in meteorite crater research at the Dominion Observatory: Meteoritics, v. 2, p. 219–241.

Melosh, H. J., 1984, Impact ejection, spallation, and the origin of meteorites: Icarus, v. 59, p. 234–260.

Polanskey, C. A., and Ahrens, T. J, 1990, Impact spallation experiments: fracture patterns and spall velocities: Icarus, v. 87, p. 140–155.

Robertson, R. B., and Grieve, R. A. F., 1975, Houghton dome and Slate Islands: recently discovered Canadian impact structures: Meteoritics, v. 10, p. 480–481.

Sander, G. W., Overton, A., and Bataille, R. D., 1964, Seismic and magnetic investigation of the Deep Bay Crater: Journal of the Royal Astronomical Society of Canada, v. 58, p. 16–30.

Tsvetkov, V. M., Sisov, I. A., and Syrnikov, N. M., 1977, On fracture mechanism of rocks by explosion, *in* Roddy, D. J., Pepin, R. O., and Merrill, R. B., eds., Impact and explosion cratering: New York, Pergamon Press, p. 669–685.

MANUSCRIPT ACCEPTED BY THE SOCIETY DECEMBER 16, 1998

Effect of erosion on gravity and magnetic signatures of complex impact structures: Geophysical modeling and applications

J. Plado
Institute of Geology, University of Tartu, 51014 Tartu, Estonia
L. J. Pesonen
Laboratory for Palaeomagnetism, Geological Survey of Finland, FIN-02151 Espoo, Finland
V. Puura
Institute of Geology, University of Tartu, 51014 Tartu, Estonia

ABSTRACT

The changes in the gravity and magnetic anomalies of meteorite impact structures as a function of erosion have been investigated. The model structure represents a typical midsize, complex impact crater in Precambrian target rocks, with a diameter of 30 km and a height of the central uplift of 1.5 km. We used a three-dimensional forward modeling technique. Six erosional levels from 1 to 6 km which successively followed the crater formation time, were modeled from the time of primary erosional leveling of the surface to the time when the structure was completely eroded. In the gravity field, the major effect of erosion is a pronounced decrease in the amplitude of the negative anomaly, with only minor change in its diameter (or half-width), making the gravity anomaly appear progressively more flat. The amplitude of the central positive anomaly due to the structural uplift also decreases with erosion but not as rapidly as the main anomaly. The diameter of the central gravity anomaly is unaffected by erosion. The model agrees with observations of gravity amplitudes and erosion levels of 13 impact structures with diameter ranges of 20–40 km. The magnetic anomalies also change during erosion but in a more complex way than the gravity anomalies. Moreover, the shape and amplitudes of magnetic anomalies and their changes due to erosion are latitude-dependent. Therefore, the magnetic data and modeling results presented in this chapter are valid only for Fennoscandia.

INTRODUCTION

In comparison with other terrestrial planets, the relatively small number (~170) of meteorite impact structures on Earth is due to the active geologic processes reshaping the Earth's surface. A great number of impact structures have been completely destroyed during the convergent plate tectonic processes of subduction and crustal collision. In addition, geologic processes such as volcanism, sedimentation, deformation, and erosion either deform, hide or remove the morphologic features of terrestrial impact structures (Grieve and Pesonen, 1992, 1996). Therefore, indirect methods, particularly high-resolution geophysical techniques, have become important in tracing new impact structures beneath cover sediments and in identifying the remnants of those impact structures that have been severely eroded or deformed by tectonism (Pilkington and Grieve, 1992; Elo et al., 1992; Ormö and Blomqvist, 1996; Scott et al., 1997; Plado et al., 1997; Pesonen et al., this volume). Moreover, the geophysical data are useful in calculating the morphometric parameters

---

Plado, J., Pesonen, L. J., and Puura, V., 1999, Effect of erosion on gravity and magnetic signatures of complex impact structures: Geophysical modeling and applications, *in* Dressler, B. O., and Sharpton, V. L., eds., Large Meteorite Impacts and Planetary Evolution II: Boulder, Colorado, Geological Society of America Special Paper 339.

(e.g., the melt volumes) of buried impact craters (Pilkington and Grieve, 1992). These applications of geophysics in impact cratering research are, however, hampered by the fact that the geophysical anomalies of impact structures depend strongly on the state of erosion and on the degree of deformation of the structures.

Very little has been done to estimate the changes in the geophysical anomalies of impact structures as a function of erosion and deformation (e.g., Pesonen et al., 1993). Pilkington and Grieve (1992) were the first to point out a trend (a decrease) in the amplitudes of the negative gravity anomalies of impact structures caused by progressive erosion. Plado et al. (1997) attempted to model the changes in gravity and magnetic anomalies of impact structures due to erosion and deformation. However, they used a 2.5-dimensional modeling technique, which turned out to be insufficient to accurately describe the anomalies of the truly three-dimensional structures.

In this chapter we present novel geophysical modeling results whereby the effect of erosion on gravity and magnetic anomalies was investigated using a three-dimensional approach as applied to a hypothetical, midsize complex impact structure in a Precambrian shield area. We restricted the modeling to consider only the effect of erosion on gravity and magnetic anomalies of impact structures; the effect of deformation is presented elsewhere (Plado et al., 1997). The main emphasis was on analysis of the gravity data, as the magnetic anomalies are more complex and depend on several parameters including the geographic site (latitude) of the structure.

In the first portion of this study we show that the erosion produces distinct changes in the amplitudes of gravity and magnetic anomalies and less pronounced changes in their widths. Based primarily on the gravity modeling, we demonstrate that some parameters derived from the gravity anomalies are practical useful measures of the erosion level of an impact structure. In the second portion we test our model on 13 real impact structures for which gravity and erosion level data are available (Pilkington and Grieve, 1992).

THE MODEL

Figure 1 shows the cross section of a hypothetical impact structure and its morphometric parameters following Croft (1985) and Melosh (1989). The original diameter (D) (rim to rim) is 30 km, characterizing the model structure as a typical midsize, complex crater in the global data base of impact structures (Grieve and Pesonen, 1996). The height (h_{CU}) and the diameter (D_{CU}) of the central structural uplift are calculated from D using equations (1) and (2) (see Melosh, 1989):

$$h_{CU} = 0.06 D^{1.1} \approx 2.5 \text{ km} \quad (1)$$

and

$$D_{CU} = 0.22 D \approx 6.6 \text{ km}. \quad (2)$$

Croft (1985) has given an expression for the diameter of the transient cavity (D_{TC}) for complex terrestrial impact structures (Eq. 3),

$$D_{TC} \cong D_Q^{0.15 \pm 0.4} D^{0.85 \pm 0.04} \approx 22 \text{ km}, \quad (3)$$

where D_Q is the transition diameter for simple-to-complex crater (=4 km for crystalline targets on Earth). The depth (h_{TC}) of the transient cavity has been estimated to be roughly one-third or one-fourth of its diameter D_{TC} (Melosh, 1989). The maximum rim height of the final crater (with $D = 30$ km) lies between 0.5 and 1 km (Fig. 1).

To simplify the computations, we conventionally leveled the surface (dotted line in Fig. 1A). Thus, we assume that the structural rim and the uppermost 1 km of the central uplift have been eroded away. The depression is filled by 0.2-km-thick impact

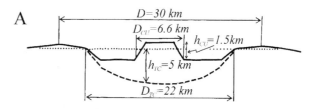

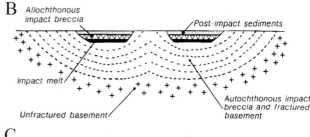

Figure 1. A, Schematic cross section of a complex impact structure with its dimensions. Dashed line indicates the shape of the transient cavity (TC); dotted line indicates the pre-impact target rock level. Symbols are explained in text. B, An idealized distribution of various impact produced/influenced lithologies as it could be in the complex structure of A. Dashed lines describe the artificial layers to count for the gradual changes of density and magnetic properties within the structure in the autochthonous breccias and fractured bedrock. C, A simplified geophysical model for B, consisting of several vertical prisms that have 16 corner points on a plan view (see Fig. 2). The thickness of each prism is 500 m, except the two prisms describing the allochthonous impact breccia with thicknesses of 500 and 300 m, and the prism of the impact melt, which has a thickness of 200 m. The model is directly derived from Figure 1B. Letters (*a* through *e*) indicate five subparallel layers to describe the autochthonous breccias (*a–c*) and fractured basement (*d–e*) within which radial changes in density and magnetic properties take place progressively (see Table 1). Arrows mark erosional levels at 1, 2, 3, 4, 5, and 6 km. No vertical exaggeration.

melt layer (volume, $V = 48$ km^3) and a 0.8-km-thick allochthonous impact breccia ($V = 221$ km^3) that is covered by 0.5-km-thick postimpact sediments ($V = 180$ km^3). Below the impact melt layer, filling the bottom of the transient cavity, there is a ~6-km-thick bowl-shaped unit consisting of autochthonous breccias and fractured basement (Fig. 1B). The shape of these layers follows that of the primarily flattened final crater of Figure 1A with the structural uplift at the center. However, since the modeling does not allow a continuous parameterized change in the breccias and fractured target rocks to take place, we have split them artificially into five successive layers with equal thicknesses of 1 km (Fig. 1C, Table 1) where layers *a* through *c* correspond to the autochthonous *breccias* and layers *d* through *e* to fractured basement, respectively. Below layer *e* the basement is virtually unaffected by the shock.

Geometry of the model

The final model is shown in Figure 1C and consists of several vertical prism-like bodies with 16 corner points on a plan view (Fig. 2) and with a diameter decreasing stepwise downward. The thickness of the prisms in both the autochthonous breccia layer and the fractured bedrock layer is 500 m. The two prisms in the allochthonous impact breccia have thicknesses of 500 and 300 m, respectively, whereas the impact melt prism is 200 m thick.

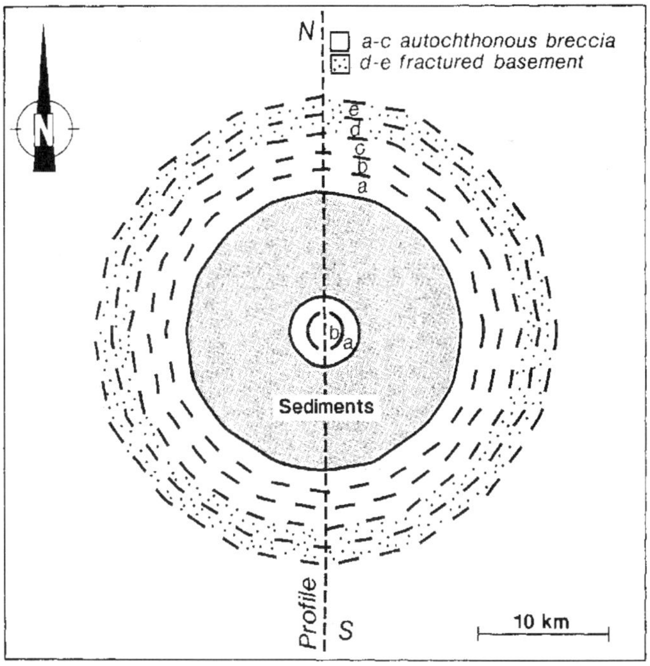

Modeling density variations

In the model, the density is increasing radially away from the point of impact (Figs. 1 and 2; Table 1), as is the case in many

Figure 2. Plan view of the modeled area (50 × 50 km), including distribution of the uppermost layers of the vertical prisms. The impact is supposed to take place at the center of the area. Location of the central profile (north-south), described in Figures 6 and 7, is shown. Letters *a* through *e* as in Figure 1.

TABLE 1. PHYSICAL PROPERTIES OF ROCK LAYERS

Rock Type	ρ (kgm^{-3})	χ (10^{-6} SI)	NRM (Am^{-1})	Q	D (°)	I (°)	Age (Ma)
Surrounding (mica schist)							
	2689	300			304	73	2680
Autochthonous breccia and fractured basement							
Layer 1	2480	200	0.04	4.9	328	41	1930
Layer 2	2530	220	0.043	4.8	328	41	1930
Layer 3	2580	240	0.046	4.7	328	41	1930
Layer 4	2620	260	0.049	4.6	321	57	2680 + 1930
Layer 5	2660	280	0.052	4.5	321	57	2680 + 1930
Impact melt							
	2500	2000	0.82	10	328	41	1930
Allochthonous (impact) breccia							
	2380	50	0.01	5	328	41	1930
Post-impact sediments							
	2350	100	0.01	2.44	328	41	1930

ρ = density.
χ = magnetic susceptibility.
NRM = intensity of natural remanant magnetization.
Q = Koenigsberger ratio.
D and I = declination and inclination of NRM, respectively.

known complex impact structures. This increase of density with depth is caused by the decrease of porosity and fracturing in impact rocks and also in the upper part of the fractured target, as observed, for example, in Clearwater West, Canada (Plante et al., 1990), in Siljan ring in Sweden (Dyrelius, 1988), in Lappajärvi and Iso Naakkima in Finland (Kukkonen et al., 1992; Pesonen, 1996), and in Kärdla, Estonia (Plado et al., 1996). The model density values for the various layers are taken from the data of impact structures of Lappajärvi (Kukkonen et al., 1992) and Kärdla (Plado et al., 1996) and are listed in Table 1. The density is increasing downward from the postimpact sediments (2,350 kgm^{-3}) to allochthonous impact breccia (2,380 kgm^{-3}) to the five layers of autochthonous breccia and fractured basement (2,480–2,660 kgm^{-3}). The impact melt has a density of 2,500 kgm^{-3} while the surrounding target rock (mica gneiss) has a density of 2,689 kgm^{-3}.

Modeling magnetic variations

Considering investigations of terrestrial impact structures and laboratory experiments, the effects of the transient shock on magnetic properties of different rock types of the impact structures are more variable than those in density. Generally, shock produces a drop in the magnetic susceptibility and often (but not always) also in the natural remanent magnetization (NRM), thus causing the weak magnetic relief associated with many impact structures (Hargraves and Perkins, 1969; Pohl et al., 1975; Cisowski and Fuller, 1978; Pilkington and Grieve; 1992; Pesonen, 1996; Scott et al., 1997). However, in some cases the impact influenced rocks may acquire a new remanence by transient stresses, the shock remanent magnetization (SRM), along the direction of the Earth's magnetic field at the time of impact (e.g., Halls, 1979).

Slowly cooled crystalline impact melt rocks may acquire a thermoremanent magnetization (TRM) in the direction of the magnetic field at the time of impact, e.g. Manicouagan, Canada (Larochelle and Currie, 1967) and Lappajärvi, Finland (Pesonen et al., 1992). The volume and magnetic contrast of melt, and therefore the magnetic anomaly, is largely controlled by the composition and properties of target rocks.

To describe the direction of the NRM in the model, we used the remanent magnetization directions of the Fennoscandian paleomagnetic data base (Pesonen et al., 1989, 1991). The use of this data base requires the knowledge of the ages of the impactites and target rocks. In our model, the age of the surroundings (=unshocked target rocks) was ~2,680 Ma (i.e., Archean) and the impact occurred at 1,930 Ma. The age of the postimpact sediments was also assumed to be 1,930 Ma. The polarity of the magnetic field was normal. The remanent magnetization directions in Table 1 were taken from the NRM data of rocks having ages of 2,680 and 1,930 Ma in the paleomagnetic data base of Fennoscandia. For the uppermost three layers of the autochthonous breccia and fractured basement (layers a–c in Fig. 1C), we used the same NRM direction as for the impact melt, assuming that these layers have an SRM. For the two lowermost layers (layers d and e in Fig. 1C), we used the vectorial sum of the pre-impact and impact NRM directions, respectively.

The shape and amplitudes of magnetic anomalies depend on latitude, in addition to the previously discussed geometrical and petrophysical properties of the rock units constituting the structure. Since the magnetic data and modeling results are not reduced to the pole, the results presented in this chapter are valid only for Fennoscandian latitudes (~60°–70°).

The values for the magnetic properties in the model used for different layers of the structure were assigned according to literature values of known Precambrian and impact rocks described by Puranen (1989), Pesonen et al. (1989), Pilkington and Grieve (1992), and Järvelä et al. (1995), and are listed in Table 1. The density and magnetic properties for the various layers of our model structure were stated to be conforming. However, physical and chemical processes taking place during the impact and later on may alter the petrophysical properties of these rocks. Postimpact thermal and chemical processes (Pilkington and Grieve, 1992) may cause considerable changes in the impact-generated rocks independent of geologic boundaries. In our simple model, we did not consider all these effects. The direction, amplitude, and range of NRM produced by postimpact thermochemical processes are different for every particular impact case. Therefore, the magnetic properties of each structure should be studied individually and separately from the gravity model.

To confirm density and magnetic layers, we assumed the following conditions: (1) that the cooling of the structure took place rapidly, (2) that the crater rim was eroded away and the crater depression was rapidly filled by postimpact sediments, and (3) that postimpact physical-chemical processes affecting physical properties of the rocks in the structure were not taking place.

The magnetic properties of the present Earth's magnetic field (intensity, 41 Am^{-1}; declination, 6°; inclination, 73.5°) used in calculations correspond to the values for the central part of the Fennoscandian Shield (latitude, ~62°) and were same for all models.

Modeling Methods

The main objective of the modeling was to study the progressive changes in the gravity and magnetic anomalies of the structure through six successive erosional levels (h_E) at 1-km intervals. The first level (=0 km, Fig. 1C) corresponds to the early postimpact phase when the surface became flattened at the target level. The lowermost erosional level (=6 km) corresponds to the level where the main units of the structure (i.e., the postimpact sediments; the allochthonous breccia; the impact melt; the autochthonous breccia, layers a–c; and the upper part of fractured basement layer d) have been eroded, and only the lowermost part of the fractured basement, layer e, has been preserved. The other erosional levels (corresponding to h_E values of 1, 2, 3, 4, and 5 km in Fig. 1C) lie between levels 0 and 6 km.

The gravity and magnetic anomaly values were calculated over a 50 × 50 km area (Figs. 3 and 4) centered on the impact point

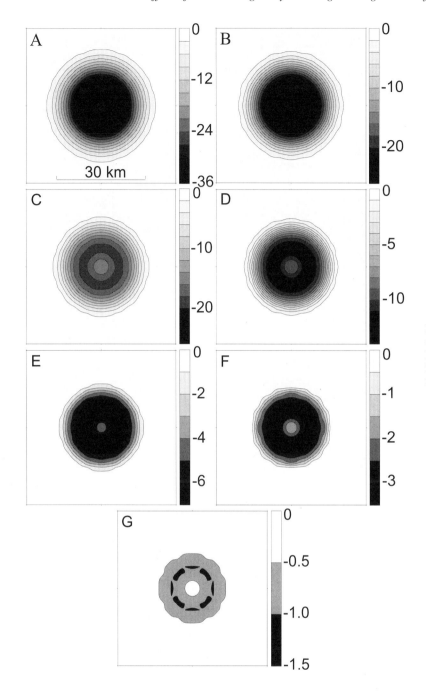

Figure 3. Plan views (50 × 50 km) of the gravity anomalies (mGal) of: A, early postimpact, at various erosion levels: B, 1 km; C, 2 km; D, 3 km; E, 4 km; F, 5 km; G, 6 km. The impact is taking place at the center of the area. Note that the amplitude scale varies.

using the ModelVision Software Package by Encom Technology Pty, Ltd., Australia (1995). Using this three-dimensional program, the Bouguer gravity and the total magnetic field intensities for observation points with a grid of 2 × 2 km were calculated. The models are simplified from the real geologic situation with no background variations in gravity and magnetism and with no regional trends. The final maps of the gravity (Fig. 3) and total field magnetic anomalies (Fig. 4) are produced with kriging for the same grid size as the calculations were done above, and shown at seven successive erosional levels. The north-south profile data of the gravity and magnetic anomalies and the effects of erosion on these profiles and their derivatives (horizontal and vertical gradients) are shown in Figures 6 and 7 at the various erosional levels.

To numerically study the progressive effect of erosion on the gravity anomaly, we used the following parameters to describe the shape of the gravity anomaly (see Parasnis, 1979): diameter (D_G), the half-width ($W^{1/2}$), maximum amplitude (A) of the main negative gravity anomaly, and corresponding values (d_{CU}, a_{CU}, and $w^{1/2}{}_{CU}$) for the central positive anomaly (due to structural uplift) (Fig. 5).

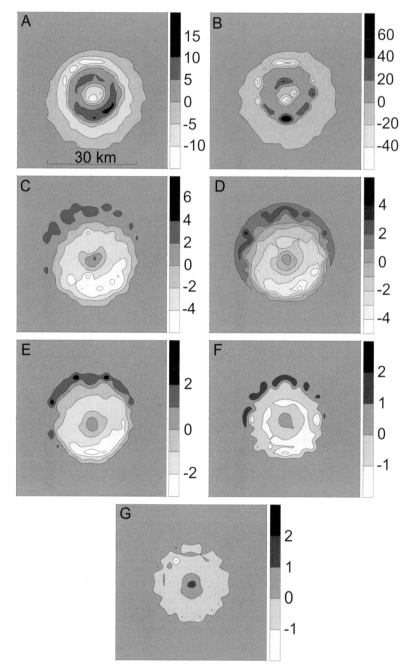

Figure 4. Plan view of the magnetic anomalies (total intensity, nT) of: A, early postimpact, at various erosion levels: B, 1 km; C, 2 km; D, 3 km; E, 4 km; F, 5 km; and G, 6 km. The impact is taking place at the center of the area. Note that the amplitude scale varies.

RESULTS

Anomaly maps

The gravity anomaly of the model (Fig. 3A) is perfectly circular and has a diameter of ~36 km, thus slightly exceeding the original diameter of the structure. It is due to the radial distribution of the density layers, extending farther than the original rim. The half-width of the anomaly is 25.2 km. The minimum value of the gravity anomaly is –36 mGal, which is consistent with a complex impact structure of this size unaffected by erosion (Pilkington and Grieve, 1992) (Fig. 6). The central uplift produces a positive gravity high of 6.2 mGal at the center of the main negative gravity anomaly (Figs. 3A and 6A) as is often observed in many complex impact structures, e.g., Vredefort, South Africa (Henkel and Reimold, 1997) and Lappajärvi, Finland (Elo et al., 1992). The horizontal gradient of the gravity anomaly (Fig. 6B) shows two peaks, one minima and one maxima symmetrically on the sides of the center of the impact structure. The vertical gradient (Fig. 6C) shows two maxima and two minima and a central maximum due to structural uplift.

Effect of erosion on gravity and magnetic signatures of complex impact structures 235

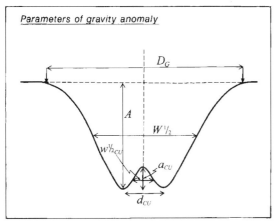

Figure 5. Schematic cross section of a gravity anomaly of a complex impact structure. The numerical parameters to describe the geometry of the anomaly are indicated. D_G, A, and $W\frac{1}{2}$ denotes the diameter, amplitude, and half-width, respectively, of the major negative anomaly; d_{CU}, a_{CU} and $w\frac{1}{2}_{CU}$ are corresponding values for the central positive anomaly, respectively.

The initial magnetic anomaly map (Fig. 4A) reflects two features. First, the strongly magnetic melt body produces a circular positive anomaly with maximum amplitude of 13.1 nT. It is surrounded by the negative anomaly, which is most intensive at the northwestern edge of the structure. Second, in the central part, the positive anomaly is distorted by the negative anomaly (–14.1 nT). This kind of magnetic anomaly is expected for a body in the Northern Hemisphere (Parasnis, 1979). Correspondingly, two magnetic highs and three lows are visible on the profile (Fig. 7A).

Effects of erosion on gravity anomalies

Erosion progressively removes the impact-produced gravity signatures. During the erosion of the structure, both the amplitude of impact gravity anomaly and its diameter decrease (Table 2; Figs. 3, 6A and 8A). Since the decrease is much stronger in amplitude, erosion progressively flattens the gravity

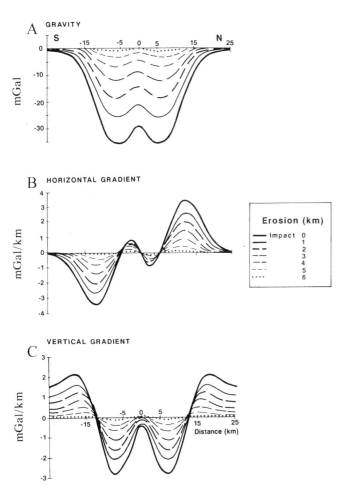

Figure 6. North-south profiles of the Bouguer gravity across the model as a function of erosion. A, Bouguer gravity; B, its horizontal gradient; and C, its vertical gradient. See Figure 2 for location. Curves are calculated for the impact time and at six erosional levels, as indicated in the index figure and shown in Figure 1C.

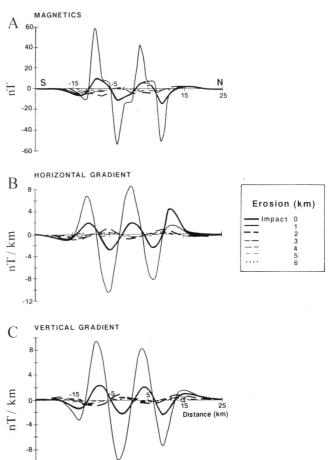

Figure 7. North-south profiles of the total magnetic field intensity across the model as a function of erosion. A, Magnetic (total component, nT); B, its horizontal gradient; and C, vertical gradient profiles across the hypothetical impact structure. See Figure 2 for location. Curves are calculated for the impact time and at six erosional levels, as indicated in the index figure and shown in Figure 1C.

TABLE 2. PROPERTIES OF IMPACT GRAVITY ANOMALIES IN DIFFERENT SITUATIONS

	h_E/D	Impact Gravity Anomaly			Anomaly of Central Uplift		$|A/a_{CU}|$
		A (mGal)	$W^{1/2}$ (km)	$|A/W^{1/2}|$ (mGal/km)	a_{CU} (mGal)	$w^{1/2}{}_{CU}$ (km)	
Impact		-35.7	25.2	1.4	6.2	4.6	5.8
Erosion (km)							
1	0.03	-25.7	24.8	1.0	4.3	5.2	6.0
2	0.07	-18.4	24.2	0.8	3.7	5.8	5.0
3	0.10	-11.9	23.8	0.5	2.8	5.2	4.3
4	0.13	-6.8	23.2	0.3	1.9	5.4	3.6
5	0.17	-3.2	22.8	0.1	1.5	5.2	2.1
6	0.20	-0.9	22.4	0.0	0.7	5.0	1.3

h_E = depth of erosion.
D = the original impact rim diameter.
A = the maximum amplitude of impact gravity anomaly.
$W^{1/2}$ = half-width of the impact gravity anomaly.
a_{CU} = the maximum amplitude of gravity anomaly corresponding to the central uplift.
$w^{1/2}{}_{CU}$ = half-width of the gravity anomaly corresponding to the central uplift.

anomalies of the impact structure. The flattening is clearly seen in the profile data (Fig. 6A) and can be numerically expressed by a parameter F, which is obtained by dividing the maximum gravity amplitude with its half-width ($A/W^{1/2}$) at each erosional level (Table 2; Fig. 8C). At all erosional levels the presence of the central uplift is seen as the positive peak within the central negative anomaly, although it decreases progressively with erosion (Fig. 8A). In spite of the significant decrease of the positive gravity anomaly of the central uplift during erosion, it remains relatively more pronounced, as compared with the corresponding negative impact anomaly.

The amplitudes of the horizontal and vertical derivatives also diminish with erosion (Figs. 6B, C). The locations of the maximum horizontal gradients do not shift considerably with erosion. Their position on the profiles approximately corresponds to the diameter of 24 km, which is $0.8 \times D$. This is the diameter where the vertical derivative curves intersect at 0 mGal/km. However, the central uplift produces significant shifts in the horizontal and vertical derivative curves during progressive erosion.

Effects of erosion on magnetic anomalies

Due to the highly magnetic (Table 1) impact melt layer, the first erosional model ($h_E/D \approx 0.03$) produces intensive magnetic anomalies up to 60 and –60 nT (Figs. 4B and 7A). The position of two positive and three negative anomalies conforms with those of the starting model. After the removal of impact melt layer, at the erosional level of 2 km and also in further erosional levels, the amplitude of the magnetic anomalies decreases and the configuration alters so that the position of negative and positive anomalies changes (Figs. 4 and 7). These five circular anomalies at different erosional levels are mainly negative, followed by positive anomalies at the northwest. These are due to weaker

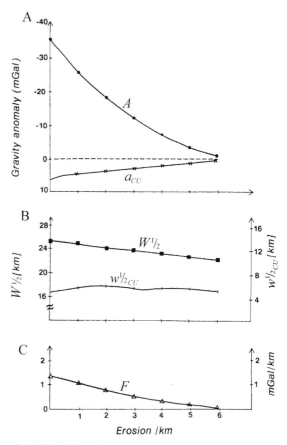

Figure 8. A, The effect of erosion on the amplitudes of gravity anomalies. Black circles indicate amplitude, A, of the main anomaly; crosses indicate amplitude (a_{CU}), respectively, of the central positive anomaly. B, Effect of erosion on the half-widths of the main anomaly ($W^{1/2}$, black squares) and of the central uplift anomaly ($w^{1/2}{}_{CU}$, vertical bars). C, Effect of erosion on the parameter $F = A/W^{1/2}$ (amplitude of the main anomaly divided by its half-width).

magnetization of autochthonous impact breccia and fractured basement as compared with surroundings and main field direction. The central uplift produces a distinct magnetic high. The horizontal and vertical derivatives (Fig. 7B, C) also diminish after the removal of the melt layer. The amplitudes of the erosional magnetic anomalies are so weak, that, in real geologic situations (presence of regional field), they will go unnoticed.

Testing the model

Our erosional model shows clearly that the erosion decreases the amplitudes of the gravity and magnetic anomalies more effectively than their widths. The magnetic amplitudes depend not only on the shape of the structure and rock types and their petrophysical properties but also on the latitude of the structure. Therefore the magnetic data of various structures are not directly comparable unless transformed into the magnetic pole that is not done here. This is the prime reason why we restrict our model testing to the gravity data.

In Figure 9 we have tested our model by plotting the gravity amplitude and its change due to erosion (solid curve), together with the gravity data of 13 terrestrial impact structures for which gravity amplitudes and erosional levels have been measured or estimated. The data are listed in Pilkington and Grieve (1992) and include only structures with diameters ranging from 20 to 40 km. The gravity anomalies of these 13 test structures have been corrected to correspond with that of a nominal 30-km diameter, which was done by applying a linear fit (Eq. 4) to the gravity anomaly vs. D:

$$A = -0.068\,D - 11.34 \qquad (4)$$

This correction increases slightly the negative gravity amplitude of structures with $D < 30$ km, and decreases the amplitude of structures with $D > 30$ km. The effect of the above correction is 0.68 mGal (maximum).

Note that the erosion level index of Pilkington and Grieve (1992) runs from 1 (uneroded) to 7 (almost totally eroded) and does not correspond to the erosion values of 0–6 km of our model structure. The terrestrial impact data (Fig. 9) show clear trends in decrease of negative gravity anomalies due to progressive erosion in rough agreement with our model (solid curve). Figures 8 and 9 show that it could be possible to use the amplitude and half-width of the impact gravity anomalies as diagnostic criteria for estimating of the regional erosional level and the original diameter of the structure. The trend in Figure 9, described with the linear regression, allows us to calculate the erosional level (h_E) from the negative gravity anomaly (A) for a structure of ~30 km in original diameter:

$$h_E = (A + 31.8)/5.8 \text{ (km).} \qquad (5)$$

Including the original impact diameter, D, gives:

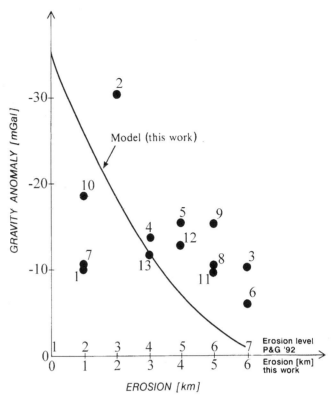

Figure 9. Testing the erosion model with gravity data of known impact structures. Vertical axis shows the negative gravity anomaly. Horizontal axis shows the erosion in kilometers for the model (solid curve) as redrawn from Figure 8A. The data points (black circles) denote data of 13 such complex impact structures for which the gravity data and also an estimate of the erosion level are available from Table 2 of Pilkington and Grieve (1992) after minor correction due to their departure from the nominal diameter of 30 km (see text). Note that the erosion level index of Pilkington and Grieve (1992) runs from 1 (uneroded) to 7 (almost totally eroded) and does not directly correspond to the erosion values of 0–6 km of this modeling structure. 1, Azuara, Spain; 2, Boltysh, Ukraine; 3, Carswell, Canada; 4, Clearwater East, Canada; 5, Clearwater West, Canada; 6, Gosses Bluff, Australia; 7, Haughton, Canada; 8, Lappajärvi, Finland; 9, Mistastin, Canada; 10, Ries, Germany; 11, Rouchouart, France; 12, Saint Martin, Canada; 13, Steen River, Canada.

$$h_E = D\,(A + 31.8)/171.0 \text{ (km)} \qquad (6)$$

These theoretical equations results in overestimating h_E (~3.5 km for Gosses Bluff, Australia; ~2.9 km for Lappajärvi, Finland; ~0.3 km for Boltysh, Ukraine), especially for structures with low A. This is probably due to the heterogeneity of the target and subsequent geologic processes, postimpact thermal and geochemical processes and tectonic modification, which are able to reduce the negative gravity amplitude. These effects, however, are not considered in our model. Nevertheless, in general gravity data allow us to estimate the erosion level of impact structures, and, as a consequence, also the regional erosional level.

CONCLUSIONS

Impact events generate various types of rocks, differing in their physical properties from rocks unaffected by impact. Allochthonous impact breccias and fractured target rocks usually have lower density than their source rocks, and produce most of the negative gravity anomaly associated with impact. The amplitude of the gravity anomaly is dependent on the volume, shape, density contrast, and thickness of the rocks affected by the impact beneath the crater. In the case of a young impact crater, all these properties are largely controlled by the crater's diameter, i.e., by the energy of impact, and by the properties and state of the target rocks. However, erosion is able to change the volume of the structure as well as the distance to the source. Therefore, this allows us to use gravity anomalies to estimate the erosion level of impact structures. With the present modeling we found that the amplitude and diameter of the negative impact gravity anomaly significantly decrease due to erosion. Since the decrease is much stronger in amplitude, the erosion progressively flattens the gravity anomalies of impact structures.

Erosion may magnify the gravity response of the central uplift in comparison with the total anomaly. This conclusion is consistent with real situations: some deeply eroded complex impact structures, e.g., Vredefort, South Africa (Henkel and Reimold, 1997), and Lappajärvi, Finland (Elo et al., 1992), show positive gravity anomalies in their central parts. This is in good agreement with the fundamental conclusion, that central parts of craters represent blocks of less crushed rocks uplifted during the modification stage of crater formation (Melosh, 1989).

We have shown the model with the prevailing effect of a strongly remanent magnetized impact melt body, e.g., Dellen in Sweden (Henkel, 1992), and with a concentric region with decreased magnetization of target (e.g., Slate Islands, Canada) (Halls, 1979), located in Fennoscandia. The calculated magnetic anomaly reflects a circular positive anomaly surrounded by a negative, produced mainly by the melt layer. In the central part, the positive anomaly is distorted by the negative anomaly, corresponding to the central uplift. The removal of impact melt by erosion significantly decreases the amplitudes of the anomalies.

ACKNOWLEDGMENTS

We thank Seppo Elo, Geological Survey of Finland, and Markku Peltoniemi, Helsinki University of Technology, for helpful comments. The figures were drawn by Salme Nässling, Geologic Survey of Finland. We also appreciate the thoughtful comments of Gary L. Kinsland, and W. A. Morris, and an anonymous reviewer. The earlier stage of this research was supported by the Center for International Mobility of Finland and by Grant 2063 from the Estonian Science Foundation.

REFERENCES CITED

Cisowski, S. M., and Fuller, M., 1978, The effect of shock on the magnetism of terrestrial rocks: Journal of Geophysical Research, v. 83, p. 3441–3458.

Croft, S. K., 1985, The scaling of complex craters: Proceedings, 15th Lunar and Planetary Science Conference: Houston, Texas, Lunar and Planetary Institute, p. 828–842.

Dyrelius, D., 1988, The gravity field of the Siljan ring structure: Deep Drilling in Crystalline Bedrock, v. 1, p. 85–94.

Elo, S., Jokinen, T., and Soininen, H., 1992, Geophysical investigations of the Lake Lappajärvi impact structure, western Finland: Tectonophysics, v. 216, p. 99–109.

Encom Technology Pty Ltd., 1995, ModelVision, Geophysical data display, analysis and modeling, Version 1.20: Milson's Point, Australia, p. 1–212.

Grieve, R. A. F., and Pesonen, L. J., 1992, The terrestrial impact cratering record: Tectonophysics, v. 171, p. 1–30.

Grieve, R. A. F., and Pesonen, L. J., 1996, Terrestrial impact craters: their spatial and temporal distribution and impacting bodies: Earth, Moon and Planets, v. 72, p. 357–376.

Halls, H. C., 1979, The Slate Islands meteorite impact site: a study of shock remanent magnetization: Geophysical Journal of the Royal Astronomical Society, v. 59, p. 553–591.

Hargraves, R. B., and Perkins, W. E., 1969, Investigations of the effect of shock on natural remanent magnetism: Journal of Geophysical Research, v. 74, p. 2576–2589.

Henkel, H., 1992, Geophysical aspects of meteorite impact craters in eroded shield environment, with special emphasis on electric resistivity: Tectonophysics, v. 216, p. 93–90.

Henkel, H., and Reimold, W. U., 1997, Integrated gravity and magnetic modeling of the Vredefort impact structure—reinterpretation of the Witwatersrand basin as the erosional remnant of an impact basin: Stockholm, Royal Institute of Technology, Department of Geodesy and Photogrammetry, 90 p.

Järvelä, J., Pesonen, L. J., and Pietarinen, H., 1995, On palaeomagnetism and petrophysics of the Iso-Naakkima impact structure, southeastern Finland: Espoo, Geological Survey of Finland, Internal Report Q19/29.1/3232/95/1, 53 p.

Kukkonen, I. T., Kivekäs, L., and Paananen, M., 1992, Physical properties of kärnäite (impact melt), suevite and impact breccia in the Lappajärvi meteorite crater, Finland: Tectonophysics, v. 216 (1/2), p. 111–122.

Larochelle, A., and Currie, K. L., 1967, Paleomagnetic study of igneous rocks from the Manicouagan structure, Quebec: Journal of Geophysical Research, v. 72, p. 4163–4169.

Melosh, H. J., 1989, Impact cratering: a geologic process: New York, Oxford University Press, 245 p.

Ormö, J., and Blomqvist, G., 1996, Magnetic modelling as a tool in the evaluation of impact structures, with special reference to the Tväeren Bay impact crater, SE Sweden: Tectonophysics, v. 262, p. 291–300.

Parasnis, D., 1979, Principles of applied geophysics: London, Chapman & Hall, 275 p.

Pesonen, L. J., 1996, The Iso-Naakkima meteorite impact structure: physical properties and paleomagnetism of a drill core: Meteoritics and Planetary Science, v. 31 (suppl.), p. A105–A106.

Pesonen, L. J., Torsvik, T. H., Elming, S.-Å., and Bylund, G., 1989, Crustal evolution of Fennoscandia—paleomagnetic constraints: Tectonophysics, v. 162, p. 27–49.

Pesonen, L. J., Bylund, G., Torsvik, T. H., Elming, S.-Å., and Mertanen S., 1991, Catalogue of paleomagnetic directions and poles from Fennoscandia—Archean to Tertiary: Tectonophysics, v. 195, p. 151–207.

Pesonen, L. J., Marcos, N., and Pipping, F., 1992, Palaeomagnetism of the Lappajärvi impact structure, western Finland: Tectonophysics, v. 216 (1/2), p. 123–142.

Pesonen, L. J., Masaitis, V., and Lindström, M., 1993, Report on topic 4: Terrestrial craters, geophysics, economics and formations, in Montanari, A., and Smit, J., eds., Post-Nördlingen Newsletter, Scientific Network of the European Science Foundation: Strasbourg, France, p. 8–11.

Pilkington, M., and Grieve, R. A. F., 1992, The geophysical signature of terrestrial impact craters: Reviews of Geophysics, v. 30, p. 161–181.

Plado, J., Pesonen, L. J., Elo, S., Puura, V., and Suuroja, K., 1996, Geophysical research on the Kärdla impact structure, Hiiumaa Island, Estonia: Meteoritics and Planetary Science, v. 31, p. 289–298.

Plado, J., Pesonen, L. J., and Puura, V., 1997, Gravity and magnetic modeling of a complex impact structure: effect of deformation and erosion, in Proceedings, 27th, Lunar and Planetary Science Conference, Abstracts: Houston, Texas, Lunar and Planetary Institute, p. 42.

Plante, L., Seguin, M.-K., and Rondot, J., 1990, Etude gravimétrique des astroblêmes du Lac à l'Eau Claire, Nouveau-Quebec: Geoexploration, v. 26, p. 303–323.

Pilkington, M., and Grieve, R. A. F., 1992, The geophysical signature of terrestrial impact craters: Reviews of Geophysics, v. 30, p. 161–181.

Pohl, J., Bleil, U., and Hornemann, U., 1975, Shock magnetization and demagnetization of basalt by transient stress up to 10 kbar: Journal of Geophysics, v. 41, p. 23–41.

Puranen, R., 1989, Susceptibilities, iron and magnetite content of Precambrian rocks in Finland: Geological Survey of Finland Report of Investigation, v. 90, p. 45.

Scott, R. G., Pilkington, M., and Tanczyk, E. I., 1997, Magnetic investigations of the West Hawk, Deep Bay and Clearwater impact structures, Canada: Meteoritics and Planetary Science, v. 32, p. 293–308.

MANUSCRIPT ACCEPTED BY THE SOCIETY DECEMBER 16, 1998

Impact crises, mass extinctions, and galactic dynamics: The case for a unified theory

Michael R. Rampino
NASA, Goddard Institute for Space Studies, New York, New York 10025, USA; Earth and Environmental Science Program, New York University, New York, New York 10003, USA

ABSTRACT

A general model in which mass extinctions of life on Earth are related to impacts of asteroids and comets whose flux is partly modulated by the dynamics of the Milky Way Galaxy is supported by several lines of astronomical and geologic evidence. The flux of Earth-crossing objects, and calculations of the environmental effects of impacts, predict that collisions with large bodies (mostly comets) ≥5 km in diameter ($\geq 10^7$ Mt TNT equivalent) could be sufficient to explain the record of ~25 extinction pulses in the last 545 m.y. The same evidence implies that the five major mass extinctions are related to impacts of the largest bodies (≥10 km in diameter, $\geq 10^8$ Mt events). Tests of "kill curve" relationships for impact-induced extinctions suggest that a possible threshold for catastrophic global extinction pulses occurs with impacts that produce craters with diameters between ~100 and 150 km. Seven of the recognized extinction peaks have thus far been correlated with concurrent (in some cases, multiple) stratigraphic impact markers and/or large impact craters.

Recent statistical analyses support previous findings that the record of extinction events exhibits a periodic component of ~30 m.y. Additionally, spectral peaks of 30 ± 0.5 m.y. and 35 ± 2 m.y. have been extracted from recently revised sets of well-dated large impact craters. These results could be explained by periodic or quasi-periodic showers of Oort cloud comets. The pacemaker for such comet showers may involve the Sun's vertical oscillation through the galactic disk, with a similar cycle time between crossings of the galactic plane. Further tests of the model will involve the identification and quantification of the dark matter component in the galactic disk and improved resolution of the putative periodic signature in the record of large craters and extinction events.

INTRODUCTION: THE GALACTIC COMET SHOWER MODELS

In the early 1980's, two important scientific papers focused attention on the question of extraterrestrial causes of mass extinctions of life of Earth Alvarez et al. (1980) provided hard physical evidence for a link between the end-Cretaceous mass extinctions and a large-body impact. Raup and Sepkoski (1984) presented data suggesting that extinctions were periodic, and thus might have a common cause. In response to that report of an approximately 26–30 m.y. periodic component in the record of mass extinctions, Rampino and Stothers (1984a) proposed a galactic model in which the periodic extinctions were related to comet showers modulated by the periodic passage of the solar system through the central plane of the Milky Way Galaxy.

In this model, the probability of encounters with molecular clouds that could perturb the Oort comet cloud and cause comet showers is modulated by the Sun's oscillation about the galactic disk. The rather flat distribution of clouds in the galactic disk suggests that an encounter would be more likely as the Sun passes through the plane region, and hence the encounters would be quasi-periodic, with a period equal to the time between plane crossings (Rampino and Stothers, 1986). Although the model was criticized on the grounds that any quasi-periodic modulation of comets would be too weak to show up over the background of nonperiodic impacts (Thaddeus and Chanan, 1985), numerical simulations, using the best available astronomical data, suggested that this effect should be detectable in the terrestrial record of impact cratering with at least a 50% a priori probability (Stothers, 1985).

Another problem involved the phase of the modulation, as the most recent extinction event identified in the initial Raup and Sepkoski (1984) study was in the mid-Miocene at ~11 Ma, whereas the Sun's last crossing of the galactic plane occurred in the last million years or so. Rampino and Stothers (1984a) originally argued that there could be enough scatter in the cloud encounters to explain the apparent offset between the most recent extinction and the plane crossing. Subsequent work, however, has identified a significant extinction event in the Pliocene (~2.3 Ma), although both the mid-Miocene and Pliocene extinctions were minor events compared with the Late Cretaceous (65 Ma) or even the late Eocene (35 Ma) extinctions (Raup and Sepkoski, 1986; Sepkoski, 1995). It is thus quite possible that the pulse of extinction that would be related to our most recent plane crossing has not yet reached a peak.

Perturbations of cometary orbits can also be induced by the tidal forces produced by the overall gravitational field of the Galaxy. Since these forces vary with the changing position of the solar system in the Galaxy, they provide a mechanism for the periodic variation in the flux of Oort cloud comets into the inner Solar System (for a full discussion, see Bailey et al., 1990). Matese et al. (1995, 1996) calculated the time modulation of the comet flux into the inner Solar System from gravitational perturbations induced by the adiabatically varying galactic tides during the in-and-out of plane oscillation. The cycle time and the degree of modulation depend critically on the mass distribution in the galactic disk, particularly the distribution of dark matter in the disk. Using a galactic model, Matese et al. (1995) showed that: (1) if there is no dark matter in the disk, then the mean plane crossing period is ~44 m.y., and the peak-to-trough ratio in the comet flux is about 2.5 to 1; (2) if the then-current best estimate of local disk density is used, then the mean plane crossing period is ~33 m.y., and the peak-to-trough flux ratio is about 4 to 1; and (3) if the extreme range of the various recent estimates in the disk matter density is considered (e.g., in agreement with values of Bahcall et al. [1992] at the 1-σ level), then the period could be as short as about 28 m.y., with a comet flux ratio of 4 to 1. The standard deviations of the flux peaks were 4–5 m.y. in most model cases. At present, the issue of dark disk matter remains an open question (e.g., Gould et al., 1996; Crezé et al., 1998), although the lastest study suggests about 30% dark matter, implying a half period of oscillation of 34 ± 3 m.y. (Stothers, 1998). The most likely remaining source of dark matter in the galactic disk may be very cold molecular clouds (Stothers, 1984; Lequeux et al., 1993).

In a simulation utilizing the then-current best estimate of disk mass density and the Sun's present position close to, but just above, the galactic plane (see Reed, 1997), Matese et al. (1995) found that the most recent times of peak comet flux were ~1 m.y. in the future, and also about 31, 65, and 98 Ma, and that the cycle interval varied from 29.5 to 34.2 m.y. over a run of their galactic model of several hundred million years. The amplitude of the comet pulses and the length of the cycle interval are modulated somewhat by the epicyclic motion of the Sun about the galactic center, with a period of about 180 m.y. The times of peak comet flux lag the times of galactic plane crossing by about 1 m.y. (Valtonen et al., 1995).

As there are good reasons for believing that the largest impactors are mostly comets (Shoemaker et al., 1990), the largest craters should preferentially show the galactic modulation of comet flux. On the other hand, if comets commonly break up, then small impacts may also show periodic clustering. Pulses of increased comet flux could explain the stepped nature of some extinction events, clusters of similar-age craters, and impact layers seen in the geologic record (Hut et al., 1987; Shoemaker and Wolfe, 1986). Recently, Farley et al. (1998) reported evidence of enhanced extraterrestrial ^{3}He in sediments bracketing the time of two known large impacts in the late Eocene (~35 Ma), suggesting increased dust from a pulse of comets or comet breakup in the inner Solar System over an interval of ~2.5 m.y. Both galactic models must assume that shower comets come from the Oort cloud, in numbers moderately greater than background rates for other comets and asteroids (especially comets from the Kuiper belt).

Rampino and Haggerty (1996a) and Rampino (1998) applied the name Shiva hypothesis to the galactic comet shower hypothesis in reference to the Hindu deity of cyclical destruction and renewal, and this hypothesis could result in a conjoining of Earth history with our wider galactic environment. The galactic models of mass extinction may be broken down into a series of testable hypotheses: (1) that there is a general cause-and-effect relationship between large impacts and mass extinctions; (2) that the record of mass extinctions shows a periodic component of about 30 m.y.; (3) that the record of large impacts on Earth shows a similar periodicity; and (4) that astronomical parameters such as galactic disk mass and Oort comet cloud structure could produce a significant modulation of comet impacts on the Earth, with a periodicity and phase matching those found in the impact/mass extinction records. This chapter updates the status of these aspects of the galactic models, and reports on some recent tests that use newer impact crater and mass extinction records.

IMPACT-EXTINCTION HYPOTHESIS

The predictions of Alvarez et al. (1980) have been confirmed to a remarkable degree: there is now considerable evidence that the mass extinction of life marking the Cretaceous/Tertiary (K/T) boundary (65 Ma) coincided with the impact of an asteroid or comet ~10 km in diameter, which created the ~180-km diameter Chicxulub impact structure in the Yucatán region of Mexico (e.g., Hildebrand et al., 1995). Such a large impact (releasing ~10^{24} J, or ~10^8 Mt TNT equivalent) is calculated to have caused a severe global catastrophe, primarily related to dense clouds of fine ejecta, production of nitric oxides and acid rain, and smoke clouds from fires triggered by atmospheric reentry of ejecta (Toon et al., 1997). Other effects, which depend on the geology and paleoenvironment at the site of impact include tsunami, enhanced greenhouse effect from atmospheric water vapor derived from an ocean impact, CO_2 released by impact into carbonate rocks, and cooling and acid rain from large amounts of sulfuric acid aerosols derived from calcium sulfate target rocks (Pope et al., 1997; Toon et al., 1997).

We can estimate the expected times between collisions of bodies of various sizes with the Earth based on observations of Earth-crossing asteroids and comets and the cratering record of the inner planets (Shoemaker et al., 1990). On the large-diameter end of the size spectrum (where cometary bodies should dominate), these data predict that, in an ~100 m.y. period, the earth should be hit by several bodies larger than a few kilometers in diameter (these release ~10^{23} J, or 10^7 Mt TNT equivalent) and perhaps one ~10-km diameter (~10^{24} J, or 10^8 Mt) object (most likely cometary) (Shoemaker et al., 1990).

Modeling of the effects of impacts has shown that a $\geq 10^8$-Mt event would produce a more severe and widespread environmental disaster than an ~10^7 Mt event (Melosh et al., 1990; Toon et al., 1997). These results lead to the prediction of approximately five major mass extinctions, and about 25 ± 5 less severe pulses of extinction during the Phanerozoic eon (the last 545 m.y.) resulting from impacts. The independent paleontologic record of extinctions for that interval clearly shows 5 major mass extinctions and ~20 less severe extinction pulses, in agreement with the estimates for impact-induced extinctions (Fig. 1; note that overall extinction rates were probably higher during the period prior to ~500 Ma, and the lack of resolution of distinct peaks in that interval is a result of poor knowledge of the fossil record, e.g., Sepkoski, 1995).

A first-order "kill curve" relationship between mass extinctions and impacts of various magnitudes (Fig. 2) was first proposed by Raup (1992). Raup's proposed kill curve can be compared with data representing the known largest (>80 km diameter) impact craters with well-defined ages that overlap the ages of mass extinction boundaries (when the full dating uncertainties in both are taken into consideration) (Table 1). The observed points in Figure 2 agree with the predicted curves within the errors permitted by the geologic data, supporting at least a first-order relationship between large impacts and mass extinctions. Subsequent work by Poag (1997) suggests that craters ~100 km in diameter (~10^7-Mt impact events) are not related to significant immediate global extinction events.

Recent work on the Triassic/Jurassic boundary (Rhaetian/Hettangian), as apparently evidenced by a palynologic break in the Newark Basin (Fowell et al., 1994), suggests a revision in the age of that boundary from ~205.7 ± 4.0 Ma (Gradstein and Ogg, 1996) to a younger age of ~201 Ma. Thus, the ~100-km diameter Manicouagan impact structure, dated by U-Pb methods at 214 ± 1 Ma, may be too old to mark the end of the Rhaetian (as had earlier

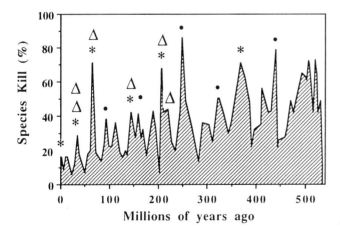

Figure 1. Percentage extinction of marine species per geologic stage (or substage) during the Phanerozoic (species data derived from genus-level data of Sepkoski [1995], treated with the reverse rarefaction estimates of Raup, 1979), and compiled record of proven and possible impact evidence. Triangles indicate large, dated impact structures; asterisks indicate impact ejecta; dots indicate iridium spikes above background; see text.

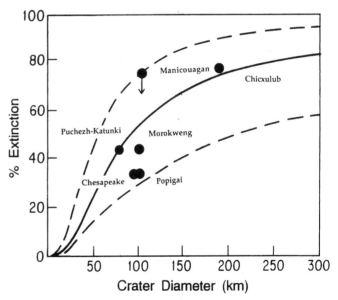

Figure 2. Proposed kill curve for Phanerozoic marine species plotted against estimated size of impact craters associated with extinctions of various magnitudes (assuming that the two are related) (solid line with dashed estimated error lines) (after Raup, 1992). Largest well-dated impact craters with ages overlapping mass extinction times are plotted for comparison (see Table 1). New dating of the extinction record suggests that the Manicouagan Crater may be associated with the end-Carnian extinction event (~40% species kill). The kill curve of Poag (1997) suggests a step function in the curve, with a rapid rise in global species kill from very low values, at a crater size somewhat greater than 100 km in diameter.

TABLE 1. LARGE DATED IMPACT CRATERS
(Grieve and Personen, 1996, with updates)
WITH ASSOCIATED MARINE EXTINCTIONS
(Sepkoski, 1995)

Name	Diameter (km)	Age (Ma)	Extinction	Species (%)
Popigai	100	35.7 ± 0.8	Late Eocene	30
Chesapeake	90	35.2 ± 0.3	Late Eocene	30
Chicxulub	180	65.2 ± 0.4	K/T	76
Morokweng	100?	145 ± 2	J/K	42
Manicouagan	100	214 ± 1	Late Triassic	
			end-Rhaetian	76
			or end-Norian	40
			or end-Carnian	42
Puchezh-Katunki	80	220 ± 10	end-Carnian?	42

TABLE 2. STRATIGRAPHIC EVIDENCE OF IMPACT DEBRIS
AT OR NEAR EXTINCTION EVENTS
(For references, see text)

Age	Evidence
Pliocene (2.3 Ma)	Impact debris
Late Eocene (35 Ma)	Microtektites (multiple), tektites, microspherules, shocked quartz
Cretaceous/Tertiary (65 Ma)	Microtektites, tektites, shocked minerals, stishovite, Ni-rich spinels, iridium
Jurassic/Cretaceous (~142 Ma)	Shocked quartz
Late Turassic (~201[?] Ma)	Shocked quartz (multiple)
Late Devonian (~368–365 Ma)	Microtektites (multiple)

been proposed, with 76% marine species extinction), but could in fact be related to the older Norian/Rhaetian boundary (~40% species extinction) dated at ~209.6 ± 4.1 Ma, or the Carnian/Norian boundary at 220.7 ± 4.4 Ma (~42% species extinction) (Fig. 1). This interpretation supports modification of the original kill curve, with a possible threshold level for significant global mass extinctions at impacts that create craters somewhat larger than 100 km in diameter (~10^7-Mt events) (Poag, 1997).

IMPACTS AND EXTINCTION EVENTS: EVIDENCE

The impact hypothesis now represents the most plausible theoretical framework within which to carry out research on mass extinctions. One direction of research is the search for impact signatures at geologic boundaries marked by faunal changes. Among the materials considered diagnostic of impact are shocked minerals (including shocked quartz, stishovite, and shocked zircons), impact glass (microtektites/tektites), microspherules with structures indicating high-temperature origin, and Ni-rich spinels. Shocked minerals and tektite glass are quite rare in the geologic record, and yet these materials have been reported in stratigraphic horizons close to the times of six recorded extinction pulses (Table 2) (Rampino et al., 1997; Grieve, 1997). A recent preliminary report of shocked quartz from the Permian/Triassic boundary (~248 Ma) in Antarctica and Australia (Retallack et al., 1998) is still unconfirmed.

Thus far, 6 of the approximately 25 extinction peaks in Figure 1 (Pliocene ~2.3 Ma; late Eocene, ~35 Ma; Cretaceous/Tertiary, 65 Ma; Jurassic/Cretaceous, ~144 Ma; Late Triassic, ~201 Ma; and Late Devonian, ~365 Ma) are associated with large impact craters and/or some form of stratigraphic evidence of impacts (Tables 1 and 2). Several other extinction events are associated with "possible" evidence of impact consisting of iridium concentrations somewhat elevated with respect to background values (see tables in Rampino et al., 1997, and Grieve, 1997). The impact model predicts that further correlations between extinction intervals and impact evidence will be discovered.

Significant statistical correlation between impacts and extinctions has been demonstrated for the major biotic events of the last 250 m.y. (Matsumoto and Kubotani, 1996), and for some of the stratigraphic stage boundaries that are defined by lesser faunal changes during the Cenozoic era (the last 65 m.y.) (Stothers, 1993). Thus, there may be a general correlation between impacts and extinctions of various magnitudes, ranging from significant mass extinctions (related to >10^7-Mt impacts) to the less severe faunal turnover events that mark many stratigraphic stage boundaries (related to ~10^6–10^7-Mt impacts).

Clusters of impacts could explain the difficulty in global correlation of some boundaries. For example, the Jurassic/Cretaceous boundary does not yet have an internationally accepted definition, stemming in part from the difficulty of correlating the Tethyan and Boreal paleogeographic realms. In the Tethyan realm, the boundary is placed between the Tithonian and Berriasian stages (144.2 ± 2.6 Ma), whereas in the Boreal realm, the boundary is apparently slightly younger, corresponding to the Lower/Middle Berriasian boundary (142.0 ± 2.6 Ma) (Gradstein and Ogg, 1996). Dypvik et al. (1996) recently reported the ejecta layer from the 40-km diameter Mjølnir impact structure in the Barents Sea close to the Boreal J/K boundary, and a large Ir anomaly is reported from the same level in a section on the Siberian Platform (Zhakarov et al., 1993). The large Morokweng impact structure in South Africa (145 ± 2 Ma), and the smaller Gosses Bluff impact in Australia (142.5 ± 0.5 Ma) are dated in the same J/K boundary interval (Table 3). Thus, it would not be surprising if the faunal turnover events that have been used to mark the J/K boundary in marine and terrestrial sequences in various parts of the world differ somewhat in age.

PERIODICITY IN THE MASS EXTINCTION RECORD

Raup and Sepkoski (1984, 1986) reported a statistically significant 26.4-m.y. periodicity in extinction time series for the last 250 Ma, and a secondary periodicity of 30 m.y. Periods of ~26–31 m.y. have been derived using various subsets of extinction events (family and genus levels), different geologic time scales, and various methods of time-series analysis. Rampino and Stothers (1986) reported detection of a similar ~29-m.y. component in the record of extinctions of nonmarine vertebrates. The interpretation of these results has been a subject of considerable debate (e.g., see Rampino and Haggerty, 1996a). The detection of a periodic component in extinctions does not imply either that the record contains a strict periodicity, or that every extinction event follows a regular schedule; the extinction record is likely to be a mixture of periodic and random events.

Fourier analysis of an extended record of 21 extinctions going back to about 515 Ma, more than doubling the length of the original time series, resulted in a spectrum with the highest peak at 27.3 m.y. (Rampino and Haggerty, 1996a,b). When additional Fourier analyses were performed on a series of truncated extinction time series starting from 0 to 545 Ma, and subtracting one extinction at a time back to 250 Ma, the result was a stable peak between 26.5 and 27.3 m.y., which remained the dominant feature in the spectra (Rampino and Haggerty, 1996b). To extend that study, the extinction record of the last 250 m.y. was truncated in the same way, and Fourier analysis revealed that all records longer than 144 m.y. exhibited a major spectral peak at 26–27 m.y. and a subsidiary peak at ~30–32 m.y. (Fig. 3).

To further test the consistency of the extinction data with the period *and* phase of the galactic models, Rampino and Stothers (1998) recently performed a series of new linear spectral analyses on these data. Using the 11 mass extinctions over the last 250 m.y., they performed the time-series analyses either allowing the phase to be a free parameter, or with fixed phasing, in which case they only used trial starting epochs in the range 0 ± 1 Ma (the time of most recent galactic plane crossing). When the phase is allowed to vary, only one high spectral peak at 27 m.y. is apparent. When the phase is fixed at the present time, however, the highest peak shifts to 28 m.y. But perhaps more significantly, two somewhat smaller peaks at 32 and 35 m.y. become considerably more prominent. Thus, although the average time interval between these mass extinctions is 25 m.y., the spacing is actually somewhat irregular. Considering the problems involved in determining statistical significance of spectral peaks that are not specified a priori, it is diffficult to choose which period should be considered the "best" fit to the data.

A longer cycle may also be present in the extinction time series. The three most severe mass extinctions (Late Ordovician, ~435 Ma; Late Permian, ~250 Ma; and Late Cretaceous, 65 Ma) are separated by ~180 Ma. The Solar System undergoes a perigalactic revolution cycle, with an estimated period of ~170 ± 10 m.y. (Matese et al., 1995); this cycle might also modulate the flux of Oort cloud comets (Bailey et al., 1990). Matese and Whitmire (1996) have found evidence for such a perturbation of the present Oort comet cloud by the galactic radial tide, due to the entire distribution of matter interior to the solar orbit.

TABLE 3. IMPACT CRATERS
(D ≥ 5 km and age estimates ≤ 250 ± 10 Ma)

Crater	Location	Diameter (km)	Age (Ma)	Method
Zhamanshin	Kazakhstan	13.5	0.90 ± 0.10	Ar-Ar
Bosumtwi	Ghana	10.5	1.03 ± 0.02	K-Ar
El'gygytgyn	Russia	18	3.5 ± 0.5	K-Ar
Bigach	Kazakhstan	7	6 ± 3	Strat.
Karla	Russia	12	10 ± 10	Strat
Ries	Germany	24	15.1 ± 1	Ar-Ar
Haughton	Canada	24	23.4 ± 1	Ar-Ar
Chesapeake	USA	90	35.2 ± 0.3	Biostrat.
Popigai	Russia	100	35.7 ± 0.8	Ar-Ar
Wanapitei	Canada	7.5	37 ± 2	Ar-Ar
Mistastin	Canada	28	38 ± 4	Ar-Ar
Logoisk	Belarus	17	40 ± 5	Strat.
Montagnais	Canada	45	50.5 ± 0.8	Ar-Ar
Ragozinka	Russia	9	55 ± 5	Strat.
Marquez	USA	22	58 ± 2	Strat.
Chicxulub	Mexico	180	64.98 ± 0.05	Ar-Ar
Kamensk	Russia	20	65 ± 2	Strat.
Kara, Ust-Kara	Russia	65	70.3 ± 2.2	Ar-Ar
Manson	USA	35	73.8 ± 0.3	Ar-Ar
Lappajarvi	Finland	23	77 ± 0.4	Ar-Ar
Boltysh	Ukraine	24	88 ± 3	K-Ar
Dellen	Sweden	15	89 ± 2.7	Rb-Sr
Steen River	Canada	25	95 ± 7	K-Ar
Avak	USA	12	100 ± 5	Strat.
Carswell	Canada	39	115 ± 10	Ar-Ar
Mien	Sweden	9	121 ± 2.3	Ar-Ar
Tookoonooka	Australia	55	128 ± 5	Strat.
Gosses Bluff	Australia	22	142.5 ± 0.5	Ar-Ar
Mjølnir	Barents Sea	40	142.6 ± 2.6	Strat.
Morokweng	South Africa	100	145 ± 2	Ar-Ar
Rochechouart	France	23	214 ± 8	Ar-Ar
Manicouagan	Canada	100	214 ± 1	U-Pb
Puchezh-Katunki	Russia	80	220 ± 10	Strat.
Araguainha	Brazil	40	247 ± 5.5	Rb-Sr

EVIDENCE FOR PERIODIC IMPACT PULSES

The original report of periodicity in extinction events by Raup and Sepkoski (1984) prompted immediate tests of the impact record for evidence of periodicity. The initial time-series analyses (Rampino and Stothers, 1984a,b; Alvarez and Muller, 1984), and some subsequent studies (Shoemaker and Wolfe, 1986; Rampino and Stothers, 1986; Yabushita 1991, 1992, 1996a,b) found evidence for a possible 28–32 m.y. period in impact crater ages. Others have argued, however, that the number of well-dated craters (~35) is still too small to extract a consistent statistically significant periodicity, should one exist (e.g., Grieve and Shoemaker, 1994; Grieve and Pesonen, 1996).

The differences in the formal periods derived from analyses of extinctions and cratering might at first seem problematic. However, several studies have concluded that the observed differences in the formal periodicity are to be expected, taking into consideration problems in dating and the likelihood that both

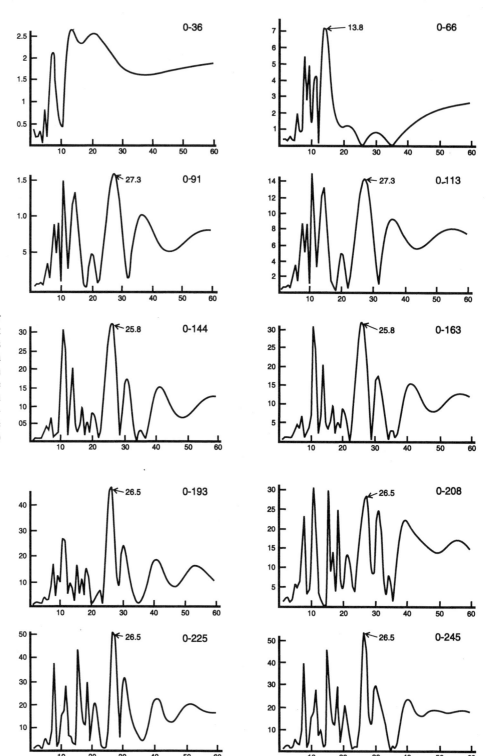

Figure 3. Fourier power spectrum for extinction events from Figure 1 (subset from 0 to 250 Ma) computed as described in Rampino and Caldeira (1993). X-axes indicate period in million years; y-axes indicate spectral power. The time series was truncated, one extinction at a time, from 250 to 36 Ma. The dominant stable peaks in the spectrum (aside from noise in the 0–20 m.y. period range) are situated at ~26–27 and ~30 m.y.

records would be a mixture of periodic and random events (Stothers, 1988, 1989; Trefil and Raup, 1987; Fogg, 1989).

Considering that most Earth-crossing asteroids fall in the ≤1 km size range (Shoemaker et al., 1990), we might expect that for impact craters on Earth ≤20 km in diameter, the presumably random asteroid flux would dominate over any signal from periodic comets in that size range. If this is true, then the cratering period should show up preferentially in the largest craters. Using only the nine largest well-dated impact craters, Matese et al. (1998) recently found a best-fitting period of 36 ± 2 m.y. when the phase was fixed to the time of most recent galactic plane crossing. To further test the consistency of the crater record with the galactic

TABLE 4. RESULTS OF SPECTRAL ANALYSES OF IMPACT CRATER AGES

Diameter (km)	No	Phase	Highest Peak (m.y.)	Second Highest Peak (m.y.)
5	31	Free	30	35
		Fixed	30	35
35	11	Free	35	None
		Fixed	35	None
90	5	Free	36	29
		Fixed	36	30

models, Rampino and Stothers (1998) performed linear time-series analyses on sets of various sized craters using the revised list of 34 impact craters shown in Table 3 (Grieve and Pesonen, 1996; with updates from Grieve, 1996, 1997, and other sources).

As Table 4 shows, two peaks were detected, a narrow peak at 30 ± 0.5 m.y. and a broader peak at 35 ± 2 m.y. In the cases in which only the largest craters were utilized, the highest peak in the period spectrum is located at ~36 m.y., in agreement with the results of Matese et al. (1998). In an earlier study designed to test the effects of the errors in the ages of the craters on the periods detected, Stothers (1988) showed that the dating errors alone were capable of shifting the dominant period between 30 and 35 m.y. We note, however, that the periods detected in the mass extinction and cratering records are statistically significant (at the 5% level) only when they are treated as periods (and phases) known a priori. Thus, it is still difficult to say whether the "best" period in cratering is closer to 30 or 36 m.y. The width of the comet flux pulses (from ~2 to ~8 m.y., depending on the model in which large comet impacts become most probable) might preclude assignment of high significance in any case.

CONCLUSIONS

Only recently have we begun to appreciate the importance of comet and asteroid impacts in Earth history. A further link between catastrophic impact events on the Earth and the cycles of the Milky Way Galaxy would represent a significant connection of terrestrial biologic evolution with the larger astrophysical environment. The galactic hypothesis for mass extinctions can be evaluated according to the four qualities that, according to Wilson (1998), scientists seek in theories generally: (1) parsimony, (2) generality, (3) consilience among disciplines, and (4) predictiveness. The galactic hypothesis meets the criterion of parsimony by connecting extinction events on Earth with the simple gravitational dynamics of the galaxy; it is general through time and among planets circling stars within the galactic disk; it unexpectedly brings consilience among the fields of astrophysics, geology, and evolutionary biology; and it makes a number of testable predictions across many phenomena in diverse areas of research.

The geologic data on mass extinctions and evidence of large impacts, although still sparse, are thus far consistent with a quasi-periodic modulation of the flux of Oort cloud comets with a mean period of about 30–36 m.y., as predicted by galactic models utilizing the current range of estimates of the total mass of the galactic disk (including dark matter). Discrepancies in the periodicities detected in the extinctions and cratering may be the result of a combination of dating errors, mixtures of periodic and nonperiodic events, and true irregularities in the underlying cycle. Further tests of the agreement between the galactic models and the geologic record will depend partly on the identification and improved quantification of the dark matter component in the galactic disk. Further astronomical and geologic studies (including the discovery and age-dating of large impact craters) should help to clarify and refine both the expected astronomical cycle time and the periodicities detectable in the geologic record.

ACKNOWLEDGMENTS

Thanks to K. Caldeira, B. Haggerty, J. Matese, R. B. Stothers, M. Valtonen, and S. Yabushita for discussions and criticism. C. W. Poag and J. Sepkoski critically reviewed the manuscript, and Sepkoski provided an updated version of his extinction data set.

REFERENCES CITED

Alvarez, L. W., Alvarez, W., Asaro, F., and Michel, H. V., 1980, Extraterrestrial cause of the Cretaceous/Tertiary extinction: Science, v. 208, p. 1095–1108.

Alvarez, W., and Muller, R. A., 1984, Evidence from crater ages for periodic impacts on the Earth: Nature, v. 308, p. 718–720.

Bahcall, J. N., Flynn, C., and Gould, A., 1992, Local dark matter from a carefully selected sample: Astrophysical Journal, v. 389, p. 234–250.

Bailey, M. E., Clube, S. V. M., and Napier, W. M., 1990, The Origin of comets: Oxford, United Kingdom, Pergamon Press, 577 p.

Crezé, M., Chereul, E., Bienaymé, O., and Pichon, C., 1998, The distribution of nearby stars in phase space mapped by Hipparcos: I. The potential well and local dynamical mass: Astronomy and Astrophysics, v. 329, p. 920–936.

Dypvik, H., Gudlaugsson, S. T., Tsikalas, F., Attrep, M., Jr., Ferrell, R. E., Jr., Krinsley, D. H., Mørk, A., Faleide, J. I., and Nagy, J., 1996, Mjølnir structure: an impact crater in the Barents Sea: Geology, v. 24, p. 779–782.

Farley, K. A., Montanari, A., Shoemaker, E. M., and Shoemaker, C. S., 1998, Geochemical evidence for a comet shower in the late Eocene: Science, v. 280, p. 1250–1253.

Fogg, M. J., 1989, The relevance of the background impact flux to cyclic impact/mass extinction hypotheses: Icarus, v. 79, p. 382–395.

Fowell, S. J., Cornet, B., and Olsen, P. E., 1994, Geologically rapid Late Triassic extinctions: palynological evidence from the Newark Supergroup, in Klein, G. D., ed., Pangea: paleoclimate, tectonics, and sedimentation during accretion, zenith, and breakup of a supercontinent: Geological Society of America Special Paper 288, p. 197–206.

Gould, A., Bahcall, J. N., and Flynn, C., 1996, Disk M dwarf luminosity function from Hubble Space Telescope star counts: Astrophysical Journal, v. 465, p. 759–768.

Gradstein, F. M., and Ogg, J. G., 1996, A Phanerozoic time scale: Episodes, v. 19, nos. 1 & 2, insert.

Grieve, R. A. F., 1996, Chesapeake Bay and other terminal Eocene impacts: Meteoritics and Planetary Science, v. 31, p. 166–167.

Grieve, R. A. F., 1997, Extraterrestrial impact events: the record in the rocks and the stratigraphic column: Palaeoclimatology, Palaeogeography, Palaeoecology, v. 132, p. 5–23.

Grieve, R. A. F., and Pesonen, L. J., 1996, Terrestrial impact craters: their spatial and temporal distribution and impacting bodies: Earth, Moon, and Planets, v. 72, p. 357–376.

Grieve, R. A. F., and Shoemaker, E. M., 1994, The record of past impacts on Earth, in Gehrels, T., ed., Hazards due to comets & asteroids: Tucson, University of Arizona Press, p. 417–462.

Hildebrand, A. R., Pilkington, M., Connors, M., Ortiz-Aleman, C., and Chavez, R. E., 1995, Size and structure of the Chicxulub Crater revealed by horizontal gravity gradients and cenotes: Nature, v. 376, p. 415–417.

Hut, P., Alvarez, W., Elder, W. P., Hansen, T., Kauffman, E. G., Keller, G., Shoemaker, E. M., and Weissman, P. R., 1987, Comet showers as a cause of mass extinctions: Nature, v. 329, p. 118–126.

Lequeux, J., Allen, R. J., and Guilloteau, S., 1993, CO absorption in the outer Galaxy, abundant cold molecular gas: Astronomy and Astrophysics, v. 280, p. L23–L26.

Matese, J., and Whitmire, D., 1996, Tidal imprint of distant galactic matter on the Oort comet cloud: Astrophysical Journal, v. 472, p. L41–L43.

Matese, J. J., Whitman, P. G., Innanen, K. A., and Valtonen, M. J., 1995, Periodic modulation of the Oort Cloud comet flux by the adiabatically changing galactic tide: Icarus, v. 116, p. 255–268.

Matese, J. J., Whitman, P. G., Innanen, K. A., and Valtonen, M. J., 1996, Why we study the geological record for evidence of the solar oscillation about the galactic midplane: Earth, Moon and Planets, v. 72, p. 7–12.

Matese, J. J., Whitman, P. G., Innanen, K. A., and Valtonen, M. J., 1998, Variability of the Oort comet flux: Can it be manifest in the cratering record? Highlights in Astronomy, v. 11A, p. 252–256.

Matsumoto, M., and Kubotani, H., 1996, A statistical test for correlation between crater formation rate and mass extinctions: Monthly Notices of the Royal Astronomical Society, v. 282, p. 1407–1412.

Melosh, H. J., Schneider, N. M., Zahnle, K. J., and Latham, D., 1990, Ignition of global wildfires at the Cretaceous/Tertiary boundary: Nature, v. 343, p. 251–254.

Poag, C. W., 1997, Roadblocks on the kill curve: testing the Raup hypothesis: Palaios, v. 12, p. 582–590.

Pope, K. O., Baines, K. H., Ocampo, A. C., and Ivanov, B. A., 1997, Energy, volatile production, and climate effects of the Chicxulub Cretaceous/Tertiary impact: Journal of Geophysical Research, v.

Rampino, M. R., 1998, The Shiva hypothesis: impacts, mass extinctions, and the Galaxy: The Planetary Report, v. 18, p. 6–11.

Rampino, M. R., and Caldeira, K., 1993, Major episodes of geologic change: correlations, time structure and possible causes: Earth and Planetary Science Letters, v. 114, p. 215–227.

Rampino, M. R., and Haggerty, B. M., 1996a, The "Shiva Hypothesis": impacts, mass extinctions, and the Galaxy: Earth, Moon and Planets, v. 72, p. 441–460.

Rampino, M. R., and Haggerty. B. M., 1996b, Impact crises and mass extinctions: a working hypothesis, in Ryder, G., Fastovsky, D., and Gantner, S., eds., The Cretaceous-Tertiary event and other catastrophes in Earth history: Geological Society of America Special Paper 307, p. 11–30.

Rampino, M. R., and Stothers, R. B., 1984a, Terrestrial mass extinctions, cometary impacts and the Sun's motion perpendicular to the Galactic plane: Nature, v. 308, p. 709–712.

Rampino, M. R., and Stothers, R. B., 1984b, Geological rhythms and cometary impacts: Science, v. 226, p. 1427–1431.

Rampino, M. R., and Stothers, R. B., 1986, Geologic periodicities and the Galaxy, in Smoluchowski, R., Bahcall, J. N., and Matthews, M., eds., The Galaxy and the Solar System: Tucson, University of Arizona Press, p. 241–259.

Rampino, M. R., and Stothers, R. B., 1998, Mass extinctions, comet impacts, and the Galaxy: Highlights in Astronomy v. 11A, p. 246–251.

Rampino, M. R., Haggerty, B. M., and Pagano, T. C., 1997, A unified theory of impact crises and mass extinctions: quantitative tests: Annals of the New York Academy of Sciences, v. 822, p. 403–431.

Raup, D. M., 1979, Size of the Permian-Triassic bottleneck and its evolutionary implications: Science, v. 206, p. 217–218.

Raup, D. M., 1992, Large-body impact and extinction in the Phanerozoic: Paleobiology, v. 18, p. 80–88.

Raup, D. M., and Sepkoski, J. J., Jr., 1984, Periodicity of extinctions in the geologic past: Proceedings of the National Academy of Sciences USA, v. 81, p. 801–805.

Raup, D. M., and Sepkoski, J. J., Jr., 1986, Periodic extinctions of families and genera: Science, v. 231, p. 833–836.

Reed, B. C., 1997, The Sun's displacement from the galactic plane: limits from the distribution of OB-star latitudes: Publications of the Astronomical Society of the Pacific, v. 109, p. 1145–1148.

Retallack, G. J., Seyedolali, A., Holser, W. T., Krull, E. S., Amber, C. P., and Kyte, F. T., 1998, Search for evidence of impact at the Permian-Triassic boundary in Antarctica and Australia and Antarctica: Geology, v. 26, p. 979–982.

Sepkoski, J. J., Jr., 1995, Patterns of Phanerozoic extinction: a perspective from global data bases, in Walliser, O. H., ed., Global events and event stratigraphy in the Phanerozoic: Berlin, Springer, p. 35–51.

Shoemaker, E. M., and Wolfe, R. F., 1986, Mass extinctions, crater ages, and comet showers, in Smoluchowski, R., et al., eds., The Galaxy and the Solar System: Tucson, University of Arizona Press, p. 338–386.

Shoemaker, E. M., Wolfe, R. F., and Shoemaker, C. S., 1990, Asteroid and comet flux in the neighborhood of Earth: in Sarpton, V. L., and Ward, P. D., Global catastrophes in Earth history: Geological Society of America Special Paper 247, p. 155–170.

Stothers, R. B., 1984, Mass extinctions and missing matter: Nature, v. 311, p. 17.

Stothers, R. B., 1985, Terrestrial record of the solar system's oscillation about the galactic plane: Nature, v. 317, p. 338–341.

Stothers, R. B., 1988, Structure of Oort's comet cloud inferred from terrestrial impact craters: Observatory, v. 108, p. 1–9.

Stothers, R. B., 1989, Structure and dating errors in the geologic time scale and periodicity in mass extinctions: Geophysical Research Letters, v. 16, p. 119–122.

Stothers, R. B., 1993, Impact cratering at geologic stage boundaries: Geophysical Research Letters, v. 20, p. 887–890.

Stothers, R. B., 1998, Galactic disk dark matter, terrestrial impact cratering and the law of large numbers: Monthly Notices of the Royal Astromical Society, v. 300, p. 1098–1104.

Thaddeus, P., and Chanan, G., 1985, Cometary impacts, molecular clouds, and the motion of the Sun perpendicular to the galactic plane: Nature, v. 314, p. 73–75.

Toon, O. B., Zahnle, K., Morrison, D., Turco, R. P., and Covey, C., 1997, Environmental perturbations caused by the impacts of asteroids and comets: Reviews of Geophysics, v. 35, p. 41–78.

Trefil, J. S., and Raup, D. M., 1987, Numerical simulations and the problem of periodicity in the cratering record: Earth and Planetary Science Letters, v. 82, p. 159–164.

Valtonen, M. J., Zheng, J. Q., Matese, J. J., and Whitman, P. G., 1995, Near-Earth populations of bodies coming from the Oort cloud and their impacts with planets: Earth, Moon, and Planets, v. 71, p. 219–223.

Wilson, E. O., 1998, Consilience: New York, Alfred A. Knopf, 322 p.

Yabushita, S., 1991, A statistical test for periodicity hypothesis in the crater formation rate: Monthly Notices of the Royal Astronomical Society, v. 250, p. 481–485.

Yabushita, S., 1992, Periodicity in the crater formation rate and implications for astronomical modeling, in Clube, S. V. M. et al., eds., Dynamics and evolution of minor bodies with galactic and geological implications: Dordrecht, Netherlands, Kluwer, p. 161–178.

Yabushita, S., 1996a, Are cratering and probably related geological records periodic? Earth, Moon, and Planets, v. 72, p. 343–356.

Yabushita, S., 1996b, Statistical tests of a periodicity hypothesis for crater formation rate-II: Monthly Notices of the Royal Astronomical Society, v. 279, p. 727–732.

Zhakarov, V. A., Lapukhov, A. S., and Shenfil, O. V., 1993, Iridium anomaly at the Jurassic-Cretaceous boundary in northern Siberia: Russian Journal of Geology and Geophysics, v. 34, p. 83–90.

MANUSCRIPT ACCEPTED BY THE SOCIETY DECEMBER 16, 1998

Printed in U.S.A.

Late Archean impact spherule layer in South Africa that may correlate with a Western Australian layer

Bruce M. Simonson
Department of Geology, Oberlin College, Oberlin, Ohio 44074-1044, USA
Scott W. Hassler
Department of Geological Sciences, California State University, Hayward, California 94542, USA
Nicolas J. Beukes
Department of Geology, Rand Afrikaans University, Auckland Park 2006, Johannesburg, South Africa

ABSTRACT

The Late Archean Monteville Formation in the Griqualand West basin (South Africa) contains a single layer rich in distinctive millimeter-size spherules of former silicate melt that have external shapes and/or internal textures similar to well-documented impact spherules such as microtektites in Cenozoic strewn fields and microkrystites in the Cretaceous/Tertiary boundary layer. Most of the Monteville Formation was deposited below storm wave base in a deep shelf environment, yet the spherule layer has wave-formed sedimentary structures and contains intraclasts up to 1.8 m long that could be the product of waves generated by an oceanic impact. The Monteville spherule layer may be part of the same ejecta strewn field as one of two similar spherule layers in the Hamersley basin of Western Australia. Ejecta layers from large impacts therefore have the potential to establish precise time-equivalence between Precambrian successions on different continents and to constrain paleogeographic reconstructions.

INTRODUCTION

Impacts by extraterrestrial bodies have played an important role in the geologic and biologic history of the Earth, yet our picture of the effects of these impacts may be very biased. More than 150 impact structures have been recognized (Grieve et al., 1995), most of which occur in the subaerially exposed parts of continents. A few impact structures have been recognized in shallow seas on continental margins, such as the Mjølnir (Gudlaugsson, 1993; Dypvik et al., 1996) and Montagnais (Jansa and Pe-Piper, 1987; Jansa et al., 1989; Jansa, 1993) structures. In addition, one impact has been documented solely from evidence in the pelagic realm, the late Pliocene ejecta of the "Eltanin asteroid" (Kyte and Brownlee, 1985; Margolis et al., 1991; Gersonde et al., 1997).

Since ~60% of the Earth's surface is underlain by oceanic crust, most incoming extraterrestrial bodies have probably landed in the open ocean, so this record is clearly skewed.

Tectonic recycling of oceanic lithosphere has long since eliminated any possibility of finding pre-Mesozoic oceanic craters in situ, so our best hope of elucidating the record of open ocean impacts lies in ejecta layers preserved in continental margin successions. They are most likely to be preserved as recognizable entities in low-energy environments below wave base where they can avoid physical reworking, yet even on the deep sea floor, bioturbation is generally vigorous enough to mix impact ejecta with ambient sediment after deposition (Glass, 1969). Perhaps for these reasons, well-documented ejecta layers are a rarity in the stratigraphic record. In this chapter, we describe

Simonson, B. M., Hassler, S. W., and Beukes, N. J., 1999, Late Archean impact spherule layer in South Africa that may correlate with a Western Australian layer, *in* Dressler, B. O., and Sharpton, V. L., eds., Large Meteorite Impacts and Planetary Evolution II: Boulder, Colorado, Geological Society of America Special Paper 339.

the sedimentology of a newly discovered ejecta layer in South Africa related to a Late Archean impact (Simonson et al., 1997). Precambrian sediments deposited in low-energy environments are particularly well suited for the preservation of thin, distal impact ejecta layers, as biogenic reworking was not a factor.

GEOLOGIC BACKGROUND

Two of the best-preserved early Precambrian volcanic and sedimentary successions on Earth are the Mount Bruce Supergroup of Western Australia and the Transvaal Supergroup of South Africa. The area in which the Mount Bruce Supergroup was deposited has been referred to as the Hamersley basin (Trendall, 1983). In contrast, the area occupied by the Transvaal Supergroup has historically been subdivided into two parts known as the Transvaal and Griqualand West basins because there is a narrow gap between them where the strata have been completely eroded (Fig. 1). The strata in all of these basins are strikingly similar at levels ranging from regional stratigraphic successions (Button 1976; Cheney, 1996) (Fig. 2) to the types of stratification and sedimentary structures they display at the scale of hand samples (compare, for example, the carbonates described by Beukes, 1987, and Simonson et al., 1993). Moreover, recent radiometric age dates (Jahn et al., 1990; Arndt et al., 1991; Barton et al., 1994; Jahn and Simonson, 1995; Walraven and Martini, 1995; Sumner and Bowring, 1996; Woodhead et al., 1998; Trendall et al., 1998; Altermann and Nelson, 1998) indicate that the two successions are largely contemporaneous (Fig. 2).

At least three distinct layers in the Hamersley basin are rich in spherules that have been interpreted as impact ejecta by Simonson (1992) and Simonson and Hassler (1997). The impact interpretation of the middle layer is supported by the presence of anomalously high concentrations of iridium and certain other platinum group elements in association with the spherules (Simonson et al., 1998). Given the many similarities with the Mount Bruce Supergroup, we initiated a search for possible equivalents of these spherule layers in the Transvaal Supergroup. Before presenting our results from South Africa, more description of the Hamersley spherules is necessary.

As with microtektites far from their source (Glass et al., 1997), the vast majority of the Hamersley spherules are spheroids and ovoids, although smaller numbers of irregular grains representing fragments and more exotic splash forms are present (Simonson, 1992, Table 1). The spherules in the Hamersley layers average roughly 0.75 mm in diameter and consist largely of K-feldspar, with lesser amounts of carbonate (primarily calcite and ankerite), quartz, muscovite, and stilpnomelane. We interpret these as replacive phases formed during diagenesis. K-feldspar also replaces some of the K/T boundary layer spherules (Montanari, 1991; Smit et al., 1992), as well as basaltic pyroclasts in many tuffs in iron formation basins (Hassler and Simonson, 1989), including the Hamersley. Most of the Hamersley spherules contain spherulitic arrays of fibrous crystals that radiate inward from the edges (Fig. 3A) (Simonson, 1992, Figs. 4 and 9A) and

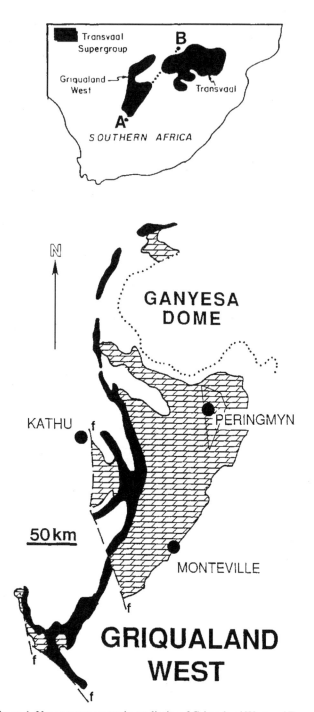

Figure 1. Upper map: approximate limits of Griqualand West and Transvaal basins within South Africa. Dotted line between points A and B shows approximate line of section in Figure 4. Lower map: three study locations of Monteville spherule layer (solid circles) and general geology of the Griqualand West basin. Black areas indicate iron formation, primarily the Kuruman; slanted brick pattern, carbonate, primarily the Campbellrand Subgroup; white areas, other rock units; dotted line, eroded edge of Transvaal Supergroup strata around Ganyesa Dome; dashed "f" lines, major faults. Ganyesa Dome is also known as the Morokweng impact structure, which formed ca. 145 Ma (Corner et al., 1997; Koeberl et al., 1997). Light polygon around Pering Mine (Peringmyn) is same as one shown in Figure 8. Both maps after Beukes, 1983a, Fig. 4-1.

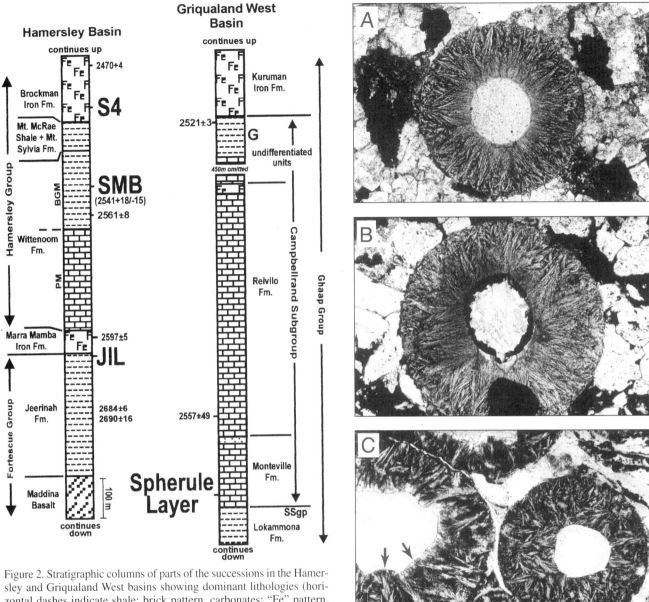

Figure 2. Stratigraphic columns of parts of the successions in the Hamersley and Griqualand West basins showing dominant lithologies (horizontal dashes indicate shale; brick pattern, carbonates; "Fe" pattern, banded iron formation; and diagonal lines, basaltic lavas). Tic marks indicate stratigraphic levels of isotopically dated samples (ages in Ma; see text for references), Monteville spherule layer, and three spherule layers in Hamersley basin (see text for further details). PM = Paraburdoo Member; BGM = Bee Gorge Member, both the Wittenoom Formation; SSgp = Schmidtsdrif Subgroup; G = Gamohaan Formation. Scale bar in lower left applies to both columns, but note omission of 450 m of section from upper part of Campbellrand Subgroup.

Figure 3. Photomicrographs in plane polarized light of spherules from Hamersley and Griqualand West basins. A, Spherule from layer in Billy-goat Bore core in the Hamersley basin (see Simonson and Hassler, 1997, Fig. 3B). Spherule consists of an outer shell of inward-radiating K-feldspar and a central spot of finely crystalline muscovite; it is surrounded by dolomite with opaque pyritic and carbonaceous intraclasts. Circular cross section of central spot suggests it is a filled-in gas bubble. Diameter of spherule, 0.72 mm. B, Spherule from Monteville layer in core SF1 in Pering area (Fig. 8) with same constituents as (A), plus some opaque pyrite replacing parts of the spherule. Diameter of spherule, 0.91 mm. C, Several spherules from Monteville layer in core V1 from Pering area (Fig. 8), which also consist mainly of radial-fibrous K-feldspar crystals with central spots of finely felted muscovite, yet have a somewhat different appearance. Mammillary contact between outer shell and central spot in spherule on left (arrows) suggests it is a former devitrification front. Sutured contact between two adjacent spherules in upper left is attributed to pressure solution. Diameter of spherule on right, 0.65 mm.

are comparable to devitrification textures in impact spherules from the moon (Lofgren, 1971, Figs. 1 and 2) and the K/T boundary layer (Bohor and Glass, 1995). In contrast, a minority of the spherules display either radiating to dendritic crystals similar to those in the Eocene "cpx spherules" (Glass et al., 1985; Glass and Burns, 1988) and microkrystites from the K/T boundary layer (Montanari, 1991; Smit and Klaver, 1982; Smit et al., 1992); or randomly oriented laths typical of plagioclase in partially crystallized basaltic melts (Lofgren, 1977, 1983). Where muscovite and quartz are present, they generally occupy central spots that we interpret as both filled-in gas bubbles and replaced glass cores that survived partial crystallization (similar to Fig. 5H in Glass et al., 1985) or spherulitic devitrification, only to be replaced later in diagenesis. By definition (Glass and Burns, 1988), microkrystites contain crystals formed during cooling whereas microtektites lack crystals and consist entirely of glass, so we believe both types are represented in the Hamersley spherules.

Other types of millimeter-scale spheroidal clasts are locally present in the stratigraphic units of the Hamersley basin, particularly volcanic accretionary lapilli and carbonate ooids, but they can be reliably distinguished from the impact spherules by petrographic means. The ooids consist of dolomite, display both concentric laminae and radial structures internally, and have sprays that diverge outward from centers rather than inward from edges (Simonson and Jarvis, 1993). In contrast, impact spherules that are replaced by carbonate consist of a mosaic of coarse crystals that are generally calcite and rarely show any fibrous textures. Many of the accretionary lapilli have external shapes and mineralogic compositions similar to the impact spherules, but their internal textures are totally different. The volcanic accretionary lapilli in both the Hamersley Group (Hassler, 1993) and the underlying Fortescue Group (Boulter, 1987) consist of fine-grained ash that is massive or displays crude concentric zoning rather than inwardly radiating crystallites.

Our search of the Transvaal Supergroup in South Africa resulted in the discovery of a single layer rich in impact spherules in the Monteville Formation. This formation is located at the base of the Campbellrand Subgroup in the Griqualand West basin (Fig. 4) and was the last unit deposited in deeper water before the onset of platformal carbonate deposition for an extended period of time (Beukes, 1987; Grotzinger, 1989; Altermann and Siegfried, 1997). The strata surrounding the Monteville spherule layer belong to what Beukes (1987) called the Prieska facies and consist primarily of fine-grained, thin-bedded black shales and carbonate lutites. The lack of wave-formed sedimentary structures and the presence of thin, normally graded beds of volcanic tuff in the enclosing strata indicate the impact spherule layer was deposited below storm wave base in a deep shelf to slope environment, even though the upper Monteville Formation contains peritidal carbonates.

DESCRIPTION OF THE MONTEVILLE IMPACT SPHERULE LAYER

We recognized the impact spherule layer at three distinct locations in the Griqualand West basin (Fig. 1). At all three locations, the layer occurs at a comparable stratigraphic level, based on its position relative to the upper and lower contacts of the Monteville Formation. The nature of the layer differs significantly in character from one site to the next (Fig. 5), so we describe them individually, but the spherules themselves look very similar and closely resem-

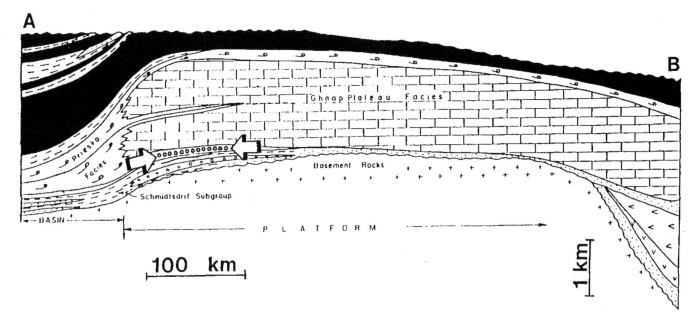

Figure 4. Schematic stratigraphic cross section of the Ghaap Group and equivalent units along the line connecting points A and B in upper part of Figure 1. Black areas indicate banded to granular iron formation; brick pattern, platformal carbonates (Ghaap Plateau facies); comma pattern, basinal carbonates (Prieska facies); horizontal dash pattern, shale; stipple pattern, quartzose sandstone; V pattern, volcanic rocks; cross pattern crystalline basement rocks. Row of open circles between two large arrows indicates the approximate lateral extent (known to date) and stratigraphic level of the Monteville spherule layer. After Beukes, 1983a, Fig. 4-4.

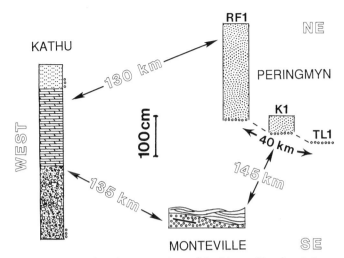

Figure 5. Schematic columnar sections of the Monteville spherule layer at all three study sites. Open circles indicate spherules; stipple pattern, quartzose sandstone; long dashes, gravel-size intraclasts; wavy lines, cross-stratified sand; gravel pattern, matrix-rich layer with gravel-size intraclasts; double horizontal stipple pattern, thinly laminated sandy layer; thin brick pattern, basinal carbonate and black shale; and open circles to right of Kathu column, stratigraphic levels of spherule occurrences. Vertical scale bar applies to all columns, and open numbers indicate horizontal distances between sites (see Fig. 1 for study site locations and Fig. 8 for locations of cores within Pering area).

ble the Hamersley spherules at all three locations in terms of size, external shapes, and internal textures. Preliminary measurements on the spherules in the Pering mine area yielded an average diameter of 0.65 mm with a maximum of 1.6 mm, and they consist largely of K-feldspar with lesser amounts of carbonate, quartz, and muscovite. Likewise, many spherules contain spherulitic sprays of K-feldspar crystals that radiate inward from their edges, and where muscovite and quartz are present, they again occupy central spots (Fig. 3B,C). These similarities suggest altered microtektites and microkrystites are both present in the Monteville layer, and further studies are in progress to see if there are any systematic differences among the three sites. The chief differences we did note in the relative abundances of the constituent minerals appear to reflect the fact that our Australian material is mainly from weathered surface exposures, whereas our South African material is mostly from fresh drill core. As in the Hamersley basin, volcanic accretionary lapilli (Altermann, 1996) and carbonate ooids (Beukes, 1983b, 1987) are present in the Monteville and associated formations, but can be reliably distinguished from the impact spherules petrographically. An impact origin for the spherules is likewise supported by a significant siderophile enrichment, especially in iridium (up to 6.4 ppb), in samples from the Monteville layer (Koeberl et al., 1999).

Monteville farm location

The only place we have seen the spherule layer in surface exposures is in the type area of the Monteville Formation on the Monteville farm (Fig. 1). Here the layer ranges from 29 to 80 cm in thickness and is clearly subdivided into two distinct parts (Figs. 5–7), described below. The spherule layer is underlain by the same series of carbonate and shale layers in most or all of the exposures where we examined it. The contact between the spherule layer and the underlying strata is usually either concordant or discordant at a low angle (Fig. 6), but locally the underlying strata are truncated along steep-walled scours up to 21 cm deep filled with coarse sandy detritus and gravel-size intraclasts.

The lower part of the spherule layer consists of a combination of tabular, gravel-size intraclasts and coarse to very coarse sand-size material (Fig. 7A). The gravel-size intraclasts are not always present, but where they are, they appear to be supported in a matrix of the sandy material. The gravel-size intraclasts consist of carbonate, shale, and early diagenetic pyrite with sedimentary textures and structures comparable to those in the underlying strata. The carbonate intraclasts range up to 180 cm in length and 13 cm in thickness, whereas the shale intraclasts are smaller, ranging up to 13 × 8 cm in cross section. The pyritic intraclasts are smaller still, rarely more than several centimeters long (Fig. 7A). Although they develop a chaotic, swirly texture in places, the intraclasts are typically inclined at relatively low angles, which explains how a layer that is only half a meter thick on average can contain clasts that are nearly 2 m in length. The matrix sand is rich in spherules, the remainder consisting of intraclasts of pyrite, carbonate, and shale that range down to at least fine sand size. It is difficult to determine if any finer grained detritus is present because of pervasive replacement by coarsely crystalline carbonate, but the sand grains usually appear to be forming clast-supported frameworks, which suggests little if any mud was originally present. Locally, tongues of spherule-rich sand up to 6 cm long are injected along bedding planes into the tips of some of the carbonate intraclasts, which indicates they were still soft at the time they were redeposited.

In contrast, the upper part of the spherule layer on the Monteville farm consists almost entirely of fine- to medium sand-size detritus; gravel-size intraclasts are present but quite rare. The sand in the upper part of the layer consists largely of carbonate with lesser numbers of spherules. The carbonate is coarsely crystalline dolomite of obvious diagenetic origin. The spherules show the same textures as those in the lower part of the layer, but they are finer on average and most have irregular outlines instead of the circular to oval cross sections that are so prevalent among the larger spherules. The upper part of the spherule layer also differs from the lower part in having ubiquitous thin laminations with low to moderate angle dips. These laminations are gently curved on a broad scale with both convex-up and convex-down geometries and low-angle internal truncations (Fig. 7B). Although we did not measure the dips systematically, this cross-stratification appears to be of the hummocky variety. In addition, the top of the layer is capped by bedforms with symmetrical cross sections (Fig. 7B).

Both the upper and lower parts of the spherule layer are about 25 cm thick on average, but like the layer as a whole, each one

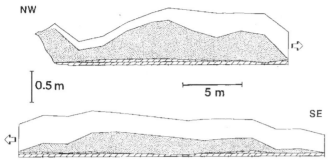

Figure 6. Fence diagram showing lateral variability of spherule layer along ca. 45 m of nearly continuous exposure on Monteville farm. The two segments are both part of one continuous crop, and they join together where indicated by open arrows. Vertical sections through the layer and adjacent strata were measured every 2 m horizontally on average (indicated by tic marks along bottom). No measurements were made farther to northwest due to lack of exposure, and they were arbitrarily terminated at southeast end of the diagram. Bottom line of diagram is base of an individual carbonate layer that persists along the entire crop. Slanted brick indicate underlying layers of carbonate and shale; stipple marks, lower, spherule-rich part of layer with tabular, gravel-size intraclasts; and white area, upper, sandy, thinly laminated to cross-stratified part of spherule layer. Note 10-fold vertical exaggeration.

varies dramatically in thickness as it is traced laterally (Fig. 6). The lower part of the layer, for example, can be up to 64 cm thick in one place, yet dwindle to less than 1 cm in another. While we did not detect any clear correlation between the thickness variations of the upper and lower parts of the layer, their composite variation resembles cross sections through dunelike bedforms. We describe these bedforms as "dunelike" because they have relatively planar bases and gently convex tops and are slightly asymmetric (Fig. 6). If they truly are duneforms, they appear to lack internal cross-stratification to match their external shapes, although the gravel-size intraclasts are concentrated in piles in the thickest part of the duneforms. Where the layer is less than 4 cm thick, gravel-size intraclasts are generally scarce and the layer displays wavy laminations.

Pering Mine area

We also identified the Monteville spherule layer in cores drilled ca. 145 km north of the Monteville farm in connection with a breccia-hosted zinc deposit at Pering Mine (Wheatley et al., 1986). We examined nine of the Pering cores that intersected the appropriate stratigraphic level within the Monteville Formation and located the spherule layer in eight of these. We also located the layer in core SF1 (Fig. 9A), which was drilled independently and is archived at the Gold Fields Geological Centre (Ries and Jolly, 1988). Taken together, these nine cores define a trapezoidal area that is ca. 65 km north to south and 20 km east to west (Fig. 8), and the spherule layer changes dramatically in character across this area. The surrounding strata are identical in character to those at the Monteville farm, the only difference being that the pyrite and carbonaceous shales are fresh since they come from cores rather than surface exposures.

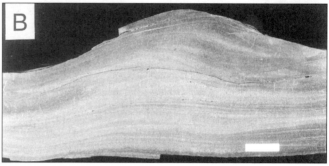

Figure 7. Spherule layer in type area of Monteville Formation on Monteville farm. A, Surface exposure in lower part of spherule layer. Immediately to right of coin is tabular intraclast of carbonate with rounded ends and cross section that measures about 15 × 3.5 cm. Also visible are long, thin, weathered intraclasts of formerly black (now reddish) shale (indicated by arrows); many smaller, coarse sand- to fine gravel-size intraclasts of early diagenetic pyrite weathered to iron oxyhydroxides (black pieces, for example, above the large carbonate intraclast); and many sand-size impact spherules (fine white spots). Diameter of coin, 2 cm. B, Sample from upper part of spherule layer; sample consists of coarsely crystalline dolomite with finer spherule-type debris and shows thin, wavy lamination with possible hummocky cross-stratification. At top is cross section through crest of a symmetrical bedform that is linear in plan; sample was cut at right angles to the crestline of the bedform. Scale bar in lower right, 2 cm long.

Throughout the Pering Mine area, the spherule layer has a sharp basal contact overlain by ca. 5 cm of nearly pure spherules with minor amounts of intraclastic pyrite and black shale (Fig. 9). What rests on top of this basal layer varies considerably. In most cores, the basal zone is overlain by a thin layer of quartzose sandstone with a few scattered spherules (Figs. 9B, and 10), then one or more layers that are again richer in spherules, but generally spherules that are finer and less abundant than those in the basal zone. The thickness of the quartzose sandstone layer decreases progressively across the Pering area (Figs. 5, and 8) from a maximum of 2 m in core RF1 (the northernmost) to no more than 5 mm in core SF1 (the southernmost); the quartz in the latter is incorporated into thin lenses in the black shale directly above the pure spherule layer (Fig. 9A). The maximum grain size of the

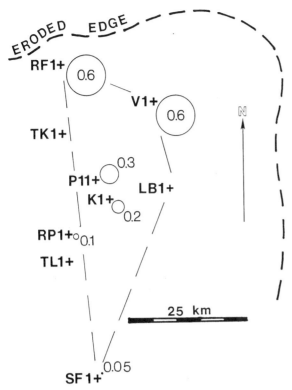

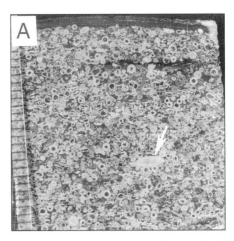

Figure 8. Map of Pering area; straight dashed lines outline same area as light polygon in Figure 1. Cross symbols with names immediately to left indicate locations of all cores known to contain Monteville spherule layer. Numbers indicate average diameter (in millimeters) of coarsest quartz grains in spherule layer for selected cores; and circles are drawn at same scale to illustrate the proportionate decrease in size of quartz grains south across Pering area. Curved dashed line labeled "eroded edge" is the erosional limit of the Monteville Formation; to immediate north and east, only older units of the Ghaap Group are present. Note: core P11 was drilled directly on the Pering Mine site.

Figure 9. Polished slabs of Monteville spherule layer from cores in Pering area (see Fig. 8 for locations). A, Spherule-rich layer from core SF1. Largest clast visible (arrow) is ovoid pyrite concretion that has been detritally reworked. Layer shown here is underlain by 50-cm-thick zone of gravel-size carbonate intraclasts with interstitial matrix rich in spherules (including the one shown in Fig. 3B). Note thinly laminated black shale that caps the layer; it contains a few thin lenses with additional spherules and coarse quartz silt. Tic marks along left side of photo are in millimeters. B, Most of Monteville spherule layer from two cores drilled ca. 15 km apart (see Fig. 8 for locations): TK1 (on right) and P11 (on left). Both cores have light basal layer consisting almost entirely of spherules overlain by thin, dark gray bed of quartzose sandstone with scattered spherules concentrated in lower part. Sandstone bed is overlain by another light gray spherule-rich layer along a sharp contact in P11, whereas it is overlain by a darker bed of quartzose dolomite with relatively few spherules in TK1. Scale bar, 2 cm long.

coarsest quartz grains in the sandstone layer likewise decreases progressively from 0.6 to 0.05 mm from north to south across the area (Figs. 8, and 10). In contrast, the size of the spherules does not appear to change significantly across the Pering area (Fig. 10).

Petrographically, the quartz grains associated with the spherules in the Pering area have much in common with those described by Simonson et al. (1998) from spherule layers in the Wittenoom Formation and Carawine Dolomite of the Hamersley basin. Like the quartz in the Hamersley layers, most of the quartz grains in the Pering area are rounded and have quasiplanar arrays of fluid (commonly two-phase) inclusions that appear to be healed microfractures (Fig. 10). A few also contain hair-like mineral inclusions (possibly rutile crystals) that are bent and/or segmented. However, the quartz in the Pering area is dominated by monocrystalline grains with relatively straight extinction, whereas much of the quartz in the Hamersley layers is polycrystalline or has undulose extinction. In fact, the monocrystalline nature of the quartz in the spherule layer more closely matches

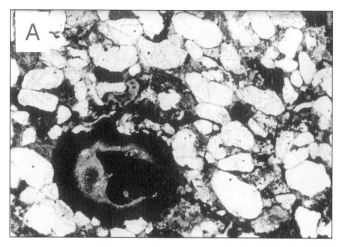

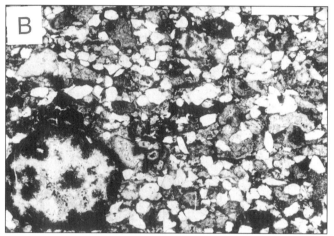

Figure 10. Photomicrographs from quartzose sandstone bed within Monteville spherule layer from two cores in Pering area. A, Some of coarsest quartz grains in core RF1. Note their well-sorted, well-rounded nature, and variably oriented, quasi-linear traces of fluid inclusions inside many grains. Outline of the impact spherule in lower left corner has been penetrated by several quartz grains, presumably via pressure solution. Interstitial material is probably a combination of finer pieces of spherule-type debris, flattened shale intraclasts, and finely crystalline authigenic material. B, Some of coarsest quartz grains in core K1, which was drilled about 35 km south of RF1 (see Fig. 8). Quartz grains in this core are finer than those in RF1, but impact spherules like one in lower left corner are about the same size. Scale bar in upper right, 0.5 mm long, and applies to both (A) and (B).

the textures shown by detrital quartz grains in sandstones of the Motiton Member, which caps the Monteville Formation (Beukes, 1987), than it does the Hamersley quartz grains. In any event, the quartz grains in all of these layers show textures that are commonplace in the granitoids and gneisses of Precambrian shields. We did not spot any good candidates for planar deformation features in our preliminary reconnaissance of the quartz grains associated with the spherules, but we have yet to search for them systematically.

One other difference we noted within the Pering area is the presence of what appears to be a layer of gravel-size carbonate intraclasts beneath the 5 cm of pure spherules in SF1. This intraclast-rich zone is 50 cm thick and very similar to the lower part of the spherule layer on the Monteville farm. We did not observe an intraclast-rich zone of this sort in any of the other cores from the Pering area. Given that the layer fluctuates dramatically along strike and effectively thins to nothing in some places on the Monteville farm, intersecting it would be hit or miss in a core, but we believe the intraclast-rich zone is generally absent from the Pering area in view of the fact that it is present in only one of nine cores.

Kathu drill core

Finally, we also identified spherules in a deep research core drilled near Kathu on the western edge of the Griqualand West basin (Fig. 1) (Altermann and Siegfried, 1997). Unlike either of the occurrences described above, spherules are present in two layers in the Kathu core rather than one. The upper layer occurs between drilling depths of 3,142.70 and 3,143.35 m and is continuously present through a measured stratigraphic thickness of 60 cm. The lower layer occurs between drilling depths of 3,145.30 and 3,147.23 m and probably corresponds to the "flat pebble conglomerate in shale matrix" noted in Figure 3 of Altermann and Siegfried (1997). The true thickness of this lower layer is uncertain; the drilling depths that were listed on the core boxes and samples suggest a thickness of 193 cm, yet this interval was represented by six discontinuous segments of core with an aggregate thickness of 134 cm. There are other signs of problems with the coring in this vicinity, e.g., 2 m of core are missing a few meters lower (Altermann and Siegfried, 1997, Fig. 3). This casts some uncertainty on the interpretation of these layers, but given how different they are (as described below), it is clear that they are separate layers formed by different processes. The strata in between these two layers, as well as those directly above and below, consist of black shale and carbonate typical of Beuke's (1987) Prieska facies. They also contain distinctive soft-sediment deformation structures known as roll-up structures and interpreted as diagnostic of deeper water or basinal sedimentation (Simonson and Carney, 1999).

The lower of the two spherule-bearing layers in the Kathu core consists of a mixture of gravel-size intraclasts of carbonate, coarse sand- to fine gravel-size intraclasts of early diagenetic pyrite, and a dark matrix of carbonaceous mud rich in quartzose silt (Fig. 11A). Spherules are quite rare in this layer; they were observed only within a few decimeters of the presumed basal contact (Fig. 5). These few spherules are almost completely replaced by carbonate, yet some of them retain pseudomorphic traces of the characteristic inwardly radiating fibrous crystal sprays. The largest intraclasts we observed are pebbles of laminated calcitic carbonate that were 4 × 1.5 cm in cross section, but given the fact that the core is only 4 cm wide, larger ones could certainly be present. Most of the larger intraclasts in this layer appear to be matrix-supported. The size and abundance of the gravel-size intraclasts vary within the layer, but we did not detect

Figure 11. Sawn slabs from both spherule-bearing layers in Kathu core (see Fig. 5). A, Wetted slab from poorly sorted lower layer with gravel-size carbonate intraclasts (elongated light to dark gray bodies) and coarse sand- to fine gravel-size intraclasts of early diagenetic pyrite (white bodies) in matrix of carbonaceous mud rich in quartzose silt. No spherules are present on this surface, and faint diagonal lines from upper right toward lower left are saw marks. B, Polished slab from sandy, well sorted upper layer consisting of combination of intraclastic pyrite (white bodies) plus carbonate sand and impact spherules (black to medium gray). Arrows indicate several of the spherules present on this surface, but many others are present. Black layer visible to right of scale bar is carbonaceous shale, which is sharply overlain by base of spherule layer. Note laminations within spherule layer that were originally horizontal, but are now intersected at angle of 25° by core axis (indicated by vertical boundaries on right and left sides of sample). Scale bars in both (A) and (B) 2 cm long.

any preferred orientation or systematic packaging of the clasts, nor any distinct stratification within the layer.

The upper spherule-bearing layer in the Kathu core, which is 60 cm thick, is different from the lower one in almost every respect. It consists in large part of well-sorted, fine to medium sand grains, many of which are spherules. The rest of the sand consists mainly of calcitic carbonate grains (including some concentrically laminated ooids) and intraclastic pyrite grains (Fig. 11B). The upper layer contains little or no gravel-size detritus, but mud is present in the form of discrete, centimeter-scale beds of black shale. These shale beds are thinly laminated, contain abundant quartzose silt, and are located at stratigraphic distances of 18, 30, 34.5, 42.5, 50, and 56 cm above the base of the upper layer. The sandy zones bounded by these black shale beds are generally thinly laminated, normally graded (Fig. 11B), and tend to have basal contacts that are sharper than their upper ones. The laminations in these sandy zones appear to be nearly planar and horizontal in core, but some are gently curved and there are a few internal discordances. Given the narrowness of the core, we cannot rule out the possibility that some or all of these laminations are involved in larger scale cross-stratification similar to that seen in the upper part of the spherule layer on Monteville farm.

INTERPRETATION OF THE MONTEVILLE SPHERULE LAYER

In each of the three areas where we identified it, the spherule layer represents the coarsest grained and highest energy deposit within a stratigraphic interval that is at least tens of meters thick. In all three areas, the layer is composite in nature, i.e., spherules are present in two or more subunits that differ in grain size and display different sedimentary structures. This indicates the layer was not deposited by a single, continuous event such as gravitational settling in still water, but rather the product of a series of related depositional events. The variation in sedimentary structures we observed from site to site indicates that different processes were at work when the spherules were deposited in different parts of the Griqualand West basin. We believe the most plausible explanations of what those processes were and how they related to one another in space and time are detailed below.

The deep scours and abundance of gravel-size intraclasts eroded from the substrate in the lower part of the layer, both at Monteville farm and in core SF1, all indicate the spherules were emplaced by an anomalously high-energy event that caused extensive bottom erosion. The symmetrical bedforms and the nature of the related cross-stratification in the upper part of the layer (Fig. 7B) further indicate that this event must have involved oscillatory or wave movement, whether or not unidirectional currents were also involved. The absence of wave-formed structures in surrounding strata, even sandy tuffs, indicates the strata that host the spherule layer were deposited below wave base. We therefore suggest the waves active during deposition of the Monteville spherules layer were produced either by impact-generated tsunamis or by unusually large waves perhaps caused by the intense winds generated during a large impact (Schultz, 1992) or an impact-induced "hypercane" (Emanuel et al., 1995). Comparable interpretations have been proposed for similar structures in the K/T boundary layer (Bourgeois et al., 1988; Smit et al., 1996; Yancey, 1996; Albertão and Martins, 1996).

One unusual feature of the Monteville spherule layer in the Pering area is the local abundance of quartzose sand. As noted above, some of the spherule layers in the Hamersley basin contain similar detrital grains of quartz derived from continental basement rocks (Simonson, 1992; Simonson et al., 1998), but it is finer and much scarcer than the quartz in the Monteville layer in the Pering area. The striking gradients in both thickness and grain size of the sandstone bed across the Pering area (Fig. 8) clearly indicate that the source of this quartz sand lay to the north, probably in the Ganyesa Dome area (Fig. 1). Paleogeographic reconstructions based on lithologic gradients indicate that land lay in this direction during deposition of the Ghaap Group (Beukes, 1987, Figs. 14–17). We therefore suggest this sand was transported offshore by bottom return flow caused by storm set-up and/or the waves responsible for making the symmetric bedforms in the Monteville farm exposures.

As noted above, coarse quartzose sand also occurs in the Motiton Member at the top of the Monteville Formation (Beukes, 1987, p. 17–18), so quartz was clearly dispersed across the Griqualand West basin at other times as well. However, we believe the sandstone in the Motiton Member has a much different origin, as they display normal wave ripples and mudcracks (Beukes, 1987, Fig. 7C). This indicates they were deposited under much shallower water conditions, and given the fact that the Motiton Member persists well south of the Pering area (it is present, for example, on the Monteville farm), these sands were probably transported southward via eolian and/or fluvial processes during a time of general regression.

Finally, there is the occurrence of spherules in two layers separated by nearly 2 m of basinal shale and carbonate (Fig. 5) instead of a single layer in the Kathu core. We interpret the matrix-rich lower spherule layer with gravel-size intraclasts as the product of a mass movement such as a debris flow. Perhaps this layer represents a mass flow triggered by impact related phenomena, as suggested for certain layers associated with the Chicxulub (Bralower et al., 1998) and Acraman (Wallace et al., 1996) impact structures. The well-sorted, thinly laminated upper spherule layer is very different in character and is similar to the upper part of the spherule layer in the exposures on the Monteville farm in some ways. However, the upper spherule layer in the Kathu core could not have been deposited during the same event that produced the spherules, provided the two spherule layers in the Kathu core and the basinal strata between them (Fig. 5) are all in proper stratigraphic order. We therefore interpret the upper spherule layer in the Kathu core as the product of a turbidity current that swept across the basin floor much later. These interpretations must remain tentative, however, given the state of the core through the interval in which the two spherule layers occur. Unfortunately, it is unlikely that any additional data will be available from this site in the foreseeable future, given the great depth of the Kathu core.

AGE AND COMPARISON WITH THE HAMERSLEY BASIN

Although the spherule layer has not been directly dated, one layer in the Monteville Formation and others in several associated units have been dated isotopically. Zircons from a tuff in the Vryburg Formation ~150 m beneath the Monteville Formation yielded a date of 2,642.2 ± 2.3 Ma (Walraven and Martini, 1995), whereas others from tuffs in the Gamohaan Formation ~2,000 m above the Monteville Formation yielded ages of 2,521 ± 3 Ma (Sumner and Bowring, 1996) and 2,516 ± 4 Ma (Altermann and Nelson, 1998). Carbonates from the Reivilo Formation, which rests directly on the Monteville Formation, have been directly dated at 2,557 ± 49 Ma using Pb-Pb (Jahn et al., 1990), and zircons from a tuff in the upper part of the Oaktree Formation, which is the correlative of the Monteville Formation in the Transvaal basin, yielded a date of 2,550 ± 3 Ma (Walraven and Martini, 1995). Finally, dates from zircons in a reworked tuffaceous layer near the top of the Monteville Formation fell into four distinguishable age groups, among which Altermann and Nelson (1998) believed 2,555 ± 19 Ma is most likely to be the date of deposition.

Impact spherules have been identified in four formations in the Hamersley basin of Western Australia. In ascending stratigraphic order, the host units and informal names for these layers are: Jeerinah Formation (JIL for Jeerinah Impact Layer); Wittenoom Formation (SMB, for Spherule Marker Bed); Carawine Dolomite, the dolomixtite; and Dales Gorge Member of the Brockman Iron Formation–S4 (occurs in the fourth shale macroband) (Fig. 2). The Wittenoom Formation and Carawine Dolomite occur in mutually exclusive parts of the Hamersley basin, and based on sedimentologic arguments, Simonson (1992) and Simonson and Hassler (1997) argued that the SMB and dolomixtite are both part of a single, reworked strewn field of spherules. The subsequent discovery of JIL (Simonson et al., 1998; Hassler and Simonson, 1998) raises the possibility this correlation is incorrect, although it is supported by isotopic ages recently obtained on carbonates from the SMB and Carawine Dolomite. An age of 2,541 +18/–15 Ma was obtained for the SMB, whereas carbonates of the Carawine Dolomite yielded an age of 2,548 +26/–29 Ma (Woodhead et al., 1998) Interpolating between zircon-bearing tuffs that have been dated isotopically (shown in Fig. 2), we estimate JIL was probably deposited around 2.63 Ga and the S4 spherule layer around 2.49 Ga.

Given these age constraints, the Monteville spherule layer could be contemporaneous with either the lower (JIL) or middle (SMB) spherule layer in the Hamersley basin, but not the stratigraphically highest one (S4) (Fig. 2). Relatively little is known about JIL at present, but at most of the sites where we have studied it, the Monteville layer has much in common with the SMB. Like the Monteville layer, the spherules in the Wittenoom layer are locally concentrated into symmetrical bedforms that appear to be large wave-formed sedimentary structures (Simonson, 1992; Hassler et al., 1996). Moreover, the Monteville and Wittenoom Formations were both deposited largely in deep shelf environments and have very similar sedimentary features (Beukes, 1987; Simonson et al., 1993). The SMB and dolomixtite also contain gravel-size intraclasts of carbonate, shale, and early diagenetic pyrite, and the largest of the carbonate intraclasts are significantly bigger than the largest shale and pyrite intraclasts at a given location. Since we did not recognize the pyritic intraclasts as such until we observed them in core (they are uniformly weathered to iron oxides in surface exposures), Simonson (1992) referred to them as "metallic flat pebbles" of uncertain origin. Finally, the main constituents of the SMB, other than spherules, are sand-size intraclasts of carbonate and shale, just like the layer on the Monteville farm and the upper layer in the Kathu core.

In contrast, the lower layer in the Kathu core is unlike the SMB, but bears a strong resemblance to the basal 85 cm of the

24.7 m of dolomixtite intersected by drill core RHDH2A in the Ripon Hills area (Simonson and Hassler, 1997, Fig. 4). Almost all of the dolomixtite layer in this core consists of decimeter- to meter-scale dolomite intraclasts with a matrix of sandy dolomite grains (Simonson, 1992). In the aforementioned basal zone, however, it bears a striking resemblance to the lower spherule-bearing layer in the Kathu core because it has a dark carbonaceous matrix and finer gravel-size carbonate intraclasts. While the lower layer in the Kathu core and the dolomixtite both contain spherules and originated as carbonate debris flows, they may differ in terms of when they formed relative to the time of the actual spherule-producing impact. Simonson and Hassler (1997) suggested the dolomixtite is from a debris flow that took place long after the impact, whereas available evidence suggests the lower layer in the Kathu core may actually be a more direct result of an impact. However, neither of these relationships is well constrained at present.

CONCLUSIONS

In summary, we interpret the Monteville spherule layer as a reworked strewn field of impact ejecta that may correlate with one of two spherules layers in the Hamersley basin, based on numerous petrographic and sedimentologic similarities and published isotopic age dates. The prospect that different parts of a single Late Archean strewn field can be recognized on different parts of the globe offers great promise for precise time-stratigraphic correlations of Precambrian successions on two different continents. Even the most precise isotopic dating techniques still involve uncertainties on the order of several million years, but impact ejecta layers are emplaced essentially instantaneously. Given their areally extensive nature, the potential exists to construct an intercontinental time-stratigraphic framework for Precambrian strata using compositionally similar and isotopically dated impact–strewn fields. Our discovery of a previously undetected impact spherule layer in a well-studied succession such as the Transvaal Supergroup suggests that more spherule layers are likely to be found in well-preserved, low-energy early Precambrian sedimentary successions. These in turn may help to constrain paleogeographic reconstructions in the Precambrian and point the way to Precambrian impact structures in addition to those that have already been found (Shoemaker and Shoemaker, 1996).

ACKNOWLEDGEMENTS

Eugene Siepker and J. Brouwer (Gold Fields of South Africa, Ltd., Oberholzer), Frans Dooge (Pering Mine Pty., Ltd., Reivilo), and H. Peter Siegfried (Geological Survey of South Africa, Upington) arranged access to cores. Derrick and Anthea Shaw permitted access to the outcrops on the Monteville farm. Darian Davies and Brooke Wilkerson assisted in the field. The field work in South Africa was supported by the National Geographic Society (Washington, D.C.), Oberlin College (Oberlin, Ohio) and Rand Afrikaans University (Johannesburg).

REFERENCES CITED

Albertão, G. A., and Martins, P. P. Jr., 1996, A possible tsunami deposit at the Cretaceous-Tertiary boundary in Pernambuco, northeastern Brazil: Sedimentary Geology, v. 104, p. 189–201.

Altermann, W., 1996, Sedimentology, geochemistry and palaeogeographic implications of volcanic rocks in the Upper Archaean Campbell Group, western Kaapvaal craton, South Africa: Precambrian Research, v. 79, p. 73–100.

Altermann, W., and Nelson, D. R., 1998, Sedimentation rates, basin analysis and regional correlations of three Neoarchaean and Paleoproterozoic subbasins of the Kaapvaal Craton as inferred from precise U-Pb zircon ages from volcaniclastic sediments: Sedimentary Geology, v. 120, p. 255–256.

Altermann, W., and Siegfried, H. P., 1997, Sedimentology and facies development of an Archaean shelf: carbonate platform transition in the Kaapvaal Craton, as deduced from a deep borehole at Kathu, South Africa: Journal of African Earth Sciences, v. 24, p. 391–410.

Arndt, N. T., Nelson, D. R., Compston, W., Trendall, A. F., and Thorne, A. M., 1991, The age of the Fortescue Group, Hamersley Basin, Western Australia, from ion microprobe zircon U-Pb results: Australian Journal of Earth Sciences, v. 38, p. 261–281.

Barton, E. S., Altermann, W., Williams, I. S., and Smith, C. B., 1994, U-Pb zircon age for a tuff in the Campbell Group, Griqualand West Sequence, South Africa: implications for Early Proterozoic rock accumulation rates: Geology, v. 22, p. 343–346.

Beukes, N. J., 1983a, Palaeoenvironmental setting of iron-formations in the depositional basin of the Transvaal Supergroup, South Africa, in Trendall, A. F., and Morris, R. C., eds., Iron-formations: facts and problems: Amsterdam, Netherlands Elsevier, p. 131–209.

Beukes, N. J., 1983b, Ooids and oolites of the Proterophytic Boomplaas Formation, Transvaal Supergroup, Griqualand West, South Africa, in Peryt, T. M., ed., Coated grains: Berlin, Springer-Verlag, p. 199–214.

Beukes, N. J., 1987, Facies relations, depositional environments and diagenesis in a major early Proterozoic stromatolitic carbonate platform to basinal sequence, Campbellrand Subgroup, Transvaal Supergroup, southern Africa: Sedimentary Geology, v. 54, p. 1–46.

Bohor, B. F., and Glass, B. P., 1995, Origin and diagenesis of K/T impact spheules—From Haiti to Wyoming and beyond: Meteoritics, v. 30, p. 182–198.

Boulter, C. A., 1987, Subaqueous deposition of accretionary lapilli: significance for palaeoenvironmental interpretations in Archaean greenstone belts: Precambrian Research, v. 34, p. 231–246.

Bourgeois, J., Hansen, T. A., Wiberg, P. L., and Kauffman, E. G., 1988, A tsunami deposit at the Cretaceous-Tertiary boundary in Texas: Science, v. 241, p. 567–570.

Bralower, T. J., Paull, C. K., and Leckie, R. M., 1998, The Cretaceous-Tertiary boundary cocktail: Chicxulub impact triggers margin collapse and extensive sediment gravity flows: Geology, v. 26, p. 331–334.

Button, A., 1976, Transvaal and Hamersley Basins—review of basin development and mineral deposits: Minerals Science and Engineering, v. 8, p. 262–293.

Cheney, E. S., 1996, Sequence stratigraphy and plate tectonic significance of the Transvaal succession of southern Africa and its equivalent in Western Australia: Precambrian Research, v. 79, p. 3–24.

Corner, B., Reimold, W. U., Brandt, D., and Koeberl, C., 1997, Morokweng impact structure, Northwest Province, South Africa: geophysical imaging and shock petrographic studies: Earth and Planetary Science Letters, v. 146, p. 351–364.

Dypvik, H., Gudlaugsson, S. T., Tsikalas, F., Attrep, M. Jr., Ferrell, R. E. Jr., Krinsley, D. H., Mørk, A., Faleide, J. I., and Nagy, J., 1996, Mjølnir structure: an impact crater in the Barents Sea: Geology, v. 24, p. 779–782.

Emanuel, K. A., Speer, K., Rotunno, R., Srivastava, R., and Molina, M., 1995, Hypercanes: a possible link in local extinction scenarios. Journal of Geophysical Research, v. 100, p. 13755–13765.

Gersonde, R., Kyte, F. T., Bleil, U., Diekmann, B., Flores, J. A., Gohl, K., Grahl,

G., Hagen, R., Kuhn, G., Sierro, F. J., Völker, D., Abelmann, A., and Bostwick, J. A., 1997, Geological record and reconstruction of the late Pliocene impact of the Eltanin asteroid in the Southern Ocean: Nature, v. 390, p. 357–363.

Glass, B. P., 1969, Reworking of deep-sea sediments as indicated by the vertical dispersion of the Australasian and Ivory Coast microtektite horizons: Earth and Planetary Science Letters, v. 6, p. 409–415.

Glass, B. P., and Burns, C. A., 1988, Microkrystites: A new term for impact-produced glassy spherules containing primary crystallites, in Proceedings, 18th, Lunar and Planetary Science Conference, Abstracts: Houston, Texas, Lunar and Planetary Institute, p. 455–458.

Glass, B. P., Burns, C. A., Crosbie, J. R., and DuBois, D. L., 1985, Late Eocene North American microtektites and crystal-bearing spherules, Proceedings, 16th Lunar and Planetary Science Conference, Pt. 1: Journal of Geophysical Research, v. 90, (suppl.), p. D175–D196.

Glass, B. P., Muenow, D. W., Bohor, B. F., and Meeker, G. P., 1997, Fragmentation and hydration of tektites and microtektites: Meteoritics and Planetary Science, v. 32, p. 333–341.

Grieve, R., Rupert, J., Smith, J., and Therriault, A., 1995, The record of terrestrial impact cratering: GSA Today, v. 5, p. 189–196.

Grotzinger, J. P., 1989, Facies and evolution of Precambrian carbonate depositional systems: emergence of the modern platform archetype, in Crevallo, P. D., Wilson, J. L., Sarg, J. F., and Read, J. F., eds., Controls on carbonate platform and basin development: Society of Economic Paleontologists and Mineralogists Special Publication 44, p. 79–106.

Gudlaugsson, S. T., 1993, Large impact crater in the Barents Sea: Geology, v. 21, p. 291–294.

Hassler, S. W., 1993, Depositional history of the Main Tuff Interval of the Wittenoom Formation, Late Archean–Early Proterozoic Hamersley Group, Western Australia: Precambrian Research, v. 60, p. 337–359.

Hassler, S. W., and Simonson, B. M., 1989, Deposition and alteration of volcaniclastic strata in two large, early Proterozoic iron-formations in Canada: Canadian Journal of Earth Sciences, v. 26, p. 1574–1585.

Hassler, S. W., and Simonson, B. M., 1998, Are wave-formed sedimentary structures and large rip-up clasts in early Precambrian microkrystite horizons diagnostic of oceanic impacts?, in Proceedings, 29th, Lunar and Planetary Sceince Conference, Abstracts: Houston, Texas, Lunar and Planetary Science Institute, CD-ROM 1086.

Hassler, S. W., Robey, H. F., Davies, D., and Simonson, B., 1996, Modeling of sedimentary bedforms produced by impact-induced tsunami in the ~2.6 Hamersley Basin, in Western Australia, in Proceedings, 27th, Lunar and Planetary Science Conference, Abstracts: Houston, Texas, Lunar Planetary Institute, p. 503–504.

Jahn, B.-M., Bertrand-Sarfati, J., Morin, N., and Macé, J., 1990, Direct dating of stromatolitic carbonates from the Schmidtsdrif Formation (Transvaal Dolomite), South Africa, with implications on the age of the Ventersdorp Supergroup: Geology, v. 18, p. 1211–1214.

Jahn, B.-M., and Simonson, B. M., 1995, Carbonate Pb-Pb ages of the Wittenoom Formation and Carawine Dolomite, Hamersley Basin, Western Australia (with implications for their correlation with the Transvaal Dolomite of South Africa): Precambrian Research, v. 72, 247–261.

Jansa, L. F., 1993, Cometary impacts into ocean: their recognition and the threshold constraint for biological extinctions: Palaeogeography, Palaeoclimatology, Palaeoecology, v. 104, p. 271–286.

Jansa, L. F., and Pe-Piper, G., 1987, Identification of an underwater extraterrestrial impact crater: Nature, v. 327, p. 612–614.

Jansa, L. F., Pe-Piper, G., Robertson, B. P., and Friedenreich, O., 1989, Montagnais: a submarine impact structure on the Scotian shelf, eastern Canada: Geological Society of American Bulletin, v. 101, p. 450–463.

Koeberl, C., Armstrong, R. A., and Reimold, W. U., 1997, Morokweng, South Africa: a large impact structure of Jurassic-Cretaceous boundary age: Geology, v. 25, p. 731–734.

Koeberl, C., Simonson, B. M., and Reimold, W. U., 1999, Geochemistry and petrography of the late Archean spherule layer in Griqualand West basin, South Africa, in Proceedings, 30th, Lunar and Planetary Science Conference, Abstracts: Houston, Texas, Lunar and Planetary Institute, CD-ROM 1755.

Kyte, F. T., and Brownlee, D. E., 1985, Unmelted meteoritic debris in the Late Pliocene iridium anomaly: evidence for the ocean impact of a nonchondritic asteroid: Geochimica et Cosmochimica Acta, v. 49, p. 1095–1108.

Lofgren, G. E., 1971, Devitrified glass fragments from Apollo 11 and Apollo 12 lunar samples in Proceedings, 2nd, Lunar and Planetary Science Conference: Houston, Texas, Lunar and Planetary Institute, v. 1, p. 949–955.

Lofgren, G. E., 1977, Dynamic crystallization experiments bearing on the origin of textures in impact-generated liquids, in Proceedings, 8th Lunar and Planetary Science Conference: Houston, Texas, Lunar and Planetary Institute, p. 2079–2095.

Lofgren, G. E., 1983, Effect of heterogeneous nucleation on basaltic textures: A dynamic crystallization study: Journal of Petrology, v. 24, p. 229–255.

Margolis, S. V., Claeys, P., and Kyte, F. T., 1991, Microtektites, microkrystites, and spinels from a Late Pliocene asteroid impact in the southern Ocean: Science, v. 251, p. 1594–1597.

Montanari, A., 1991, Authigenesis of impact spheroids in the K/T boundary clay from Italy: new constraints for high-resolution stratigraphy of terminal Cretaceous events: Journal of Sedimentary Petrology, v. 61, p. 315–339.

Ries, W., and Jolly, M., 1988, Geological borehole log of SF1: Oberholzer, Goldfields of South Africa Limited, unpublished report.

Schultz, P. H., 1992, Atmospheric effects on ejecta emplacement and crater formation on Venus: Journal of Geophysical Research, v. 97, p. 16183–16248.

Shoemaker, E. M., and Shoemaker, C. S., 1996, The Proterozoic impact record of Australia: AGSO Journal of Australian Geology and Geophysics, v. 16, p. 379–398.

Simonson, B. M., 1992, Geological evidence for a strewn field of impact spherules in the early Precambrian Hamersley Basin of Western Australia: Geological Society of America Bulletin, v. 104, p. 829–839.

Simonson, B. M., and Carney, K. E., 1999, Roll-up structures: evidence of in-situ microbial mats in Late Archean deep shelf environments: Palaios, v.14, p.13–24.

Simonson, B. M., and Hassler, S. W., 1997, Revised correlations in the Early Precambrian Hamersley Basin based on a horizon of resedimented impact spherules: Australian Journal of Earth Sciences, v. 44, p. 37–48.

Simonson, B. M., and Jarvis, D. G., 1993, Microfabrics of oolites and pisolites in the early Precambrian Carawine Dolomite of Western Australia, in Rezak, R., and Lavoie, D., eds., Carbonate microfabrics: Heidelberg, Germany Springer-Verlag, p. 227–237.

Simonson, B. M., Schubel, K. A., and Hassler, S. W., 1993, Carbonate sedimentology of the early Precambrian Hamersley Group of Western Australia: Precambrian Research, v. 60, p. 287–335.

Simonson, B. M., Beukes, N. J., and Hassler, S. W., 1997, Discovery of a Neoarchean impact spherule horizon in the Transvaal Supergroup of South Africa and possible correlations to the Hamersley Basin of Western Australia, in Proceedings, 28th, Lunar and Planetary Science Conference, Abstracts: Houston, Texas, Lunar Planetary Institute, p. 1323–1324.

Simonson, B. M., Davies, D., Wallace, M., Reeves, S., and Hassler, S. W., 1998, Iridium anomaly but no shocked quartz from Late Archean microkrystite layer: oceanic impact ejecta?: Geology, v. 26, p. 195–198.

Smit, J., and Klaver, G., 1982, Sanidine spherules at the Cretaceous-Tertiary boundary indicate a large impact event: Nature, v. 292, p. 47–49.

Smit, J., Alvarez, W., Montanari, A., Swinburne, N., Van Kempen, T. M., Klaver, G. T., and Lustenhouwer, W. J., 1992, "Tektites" and microkrystites at the Cretaceous Tertiary boundary: two strewn fields, one crater?, in Proceedings, 23rd, of Lunar and Planetary Science Conference: Houston, Texas, Lunar and Planetary Institute, v. 22, p. 87–100.

Smit, J., Roep, Th. B., Alvarez, W., Montanari, A., Claeys, P., Grajales-Nishimura, J. M., and Bermudez, J., 1996, Coarse-grained, clastic sandstone complex at the K/T boundary around the Gulf of Mexico: Deposition by tsunami waves induced by the Chicxulub impact?, in Ryder, G., Fastovsky, D., and Gartner, S., eds., The Cretaceous-Tertiary event and other catastrophes in Earth history: Geological Society of America Special Paper 307, p. 151–182.

Sumner, D. Y., and Bowring, S. A., 1996, U-Pb geochronologic constraints on deposition of the Campbellrand Subgroup, Transvaal Supergroup, South Africa: Precambrian Research, v. 79, p. 25–35.

Trendall, A. F., 1983, The Hamersley Basin, in Trendall, A. F., and Morris, R. C., eds., Iron-formations: facts and problems: Amsterdam, Netherlands, Elsevier, p. 69–129.

Trendall, A. F., Nelson, D. R., de Laeter, J. R., and Hassler, S. W., 1998, Precise zircon U-Pb ages from the Marra Mamba Iron Formation and Wittenoom Formation, Hamersley Group, Western Australia: Australian Journal of Earth Sciences, v. 45, p. 137–142.

Wallace, M. W., Gostin, V. A., and Keays, R. R., 1996, Sedimentology of the Neoproterozoic Acraman impact-ejecta horizon, South Australia: AGSO Journal of Australian Geology and Geophysics, v. 16, n. 4, p. 443–451.

Walraven, F., and Martini, J., 1995, Zircon Pb-evaporation age determinations of the Oak Tree Formation, Chuniespoort Group, Transvaal Sequence: implications for Transvaal-Griqualand West basin: South African Journal of Geology, v. 98, p. 58–67.

Wheatley, C. J. V., Whitfield, G. G., Kenny, K. J., and Birch, A., 1986, The Pering carbonate-hosted zinc-lead deposit, Griqualand West, in Anhaeusser, C. R., and Maske, S., eds., Mineral deposits of Southern Africa, vol. I: Johannesburg, Geological Society of South Africa, p. 867–874.

Woodhead, J. D., Hergt, J. M., and Simonson, B. M., 1998, Isotopic dating of an Archean bolide impact horizon, Hamersley Basin, Western Australia: Geology, v. 26, p. 47–50.

Yancey, T. E., 1996, Stratigraphy and depositional environments of the Cretaceous-Tertiary boundary complex and basal Paleocene section, Brazos River, Texas: Transactions, Gulf Coast Association of Geological Societies, v. 46, p. 433–442.

MANUSCRIPT ACCEPTED BY THE SOCIETY DECEMBER 16, 1998

Deep seismic reflection profiles across the Chicxulub crater

D. B. Snyder* and R. W. Hobbs
British Institutions Reflections Profiling Syndicate, Bullard Laboratories, Madingley Rise, Madingley Road, Cambridge CB3 0EZ, United Kingdom

ABSTRACT

The deep seismic reflection profiles acquired during the British Institutions Reflection Profiling Syndicate (BIRPS) survey off the north coast of Yucatán, during September–October 1996, have clarified the geometries of several major features of the Chicxulub impact structure. The most prominent reflectors observed on each of the four radial profiles are related to a Jurassic-Cretaceous stratum at 2–3-km depths and to the Moho at 30–35-km depths. The prominent Mesozoic reflector provides a key marker of near-surface deformation within the preimpact strata at depths great enough to avoid immediate erosion after the impact, and it indicates three major "rings" of impact-related deformation to the upper crust. Prominent reflectors within the lower crust are generally continuous for 1–10 km. The Moho reflectors indicate, in conjunction with the velocity models, variations of only a few kilometers in the thickness of crust beneath the crater. Several crustal reflectors project upward at dips of ca. 30° from Moho depths to the primary deformation zones defined by disruptions to the prominent Mesozoic reflector, suggesting that these reflectors represent shear zones whose reflectivity was enhanced by the rapid injection of impact melt.

INTRODUCTION

Since the time that Alvarez et al. (1980) first linked the Cretaceous-Tertiary boundary with a large cometary impact 65.0 Ma ago, numerous field studies worldwide have traced the impact site to the Yucatán peninsula of Mexico (e.g., Hildebrand et al., 1991; Sharpton et al., 1992). The Chicxulub structure provides the best known opportunity to study in detail the preserved structures of a large complex impact crater because of its relatively recent formation and its rapid burial after formation (Lopez-Ramos, 1975; Hildebrand, 1997). The Tertiary sediments that completely buried the impact crater soon after its formation have prohibited direct study of deformed rocks except by drill core, but these sediments do provide an ideal environment for study of the crater structure by seismic profiling methods.

The only clear expression of the onshore half of the impact crater is a semi-circular ring of cenotes (sinkholes) with a diameter of about 160 km (Pope et al., 1993; Connors et al., 1996). Drill cores show that postimpact Tertiary sediments thicken from a few hundred meters outside the cenote ring to more than a kilometer in the interior of the basin (Ward et al., 1995; Sharpton et al., 1996). Overall, the crater has a distinct circular gravity and magnetic signature, with a particularly distinctive gradient at a diameter of about 180 km (Pilkington et al., 1994; Hildebrand et al., 1995).

Herein we show that, offshore, the new seismic data reveal a peak-ring and three other rings of deformation with some asymmetric structures. This makes Chicxulub the first documented candidate for a terrestrial multi-ring basin (Morgan et al., this volume). The crater's apparent transient cavity during impact and excavation had an estimated diameter of 90–105 km, whereas the final (collapsed) crater, with heavily eroded rims, is estimated to lie just outside several large slump block terraces and to have a diameter of 180–210 km (Morgan et al., 1997). The cenote ring and distinct gravity gradients coincide most closely with this second ring of structures. A third, outermost ring, defined here for the first time, has a diameter of 240–260 km, as marked by

---

*Present address: Geological Survey of Canada, 615 Booth St., Rm. 204, Ottawa, Ontario K1A 0E9, Canada; snyder@cg.nrcan.gc.ca.

Snyder, D. B., and Hobbs, R. W., 1999, Deep seismic reflection profiles across the Chicxulub crater, *in* Dressler, B. O., and Sharpton, V. L., eds., Large Meteorite Impacts and Planetary Evolution II: Boulder, Colorado, Geological Society of America Special Paper 339.

gentle monoclines or disrupted strata within the Cretaceous-Jurassic stratigraphic section.

SEISMIC DATA ACQUISITION

Between 25 September and 3 October 1996, the seismic vessel *Geco Sigma* acquired 647.6 km of seismic reflection profiles off the northern coast of the Yucatán peninsula in Mexico (Fig. 1) for the British Institutions Reflection Profiling Syndicate (BIRPS). The weather was excellent and the surveying delayed only once for 4 h due to an airhose failure. Timing of the survey was critical because the airgun array served as the seismic source for the streamer hydrophones as well as for 34 ocean bottom seismometers and 99 REFTEK land station recorders (Fig. 1) (Christeson et al., this volume).

The shallow, gently dipping ocean shelf in the area required that the seismic vessel steam no closer than 20 km from the shore; thus, the reflection profiles do not image structures beneath the proposed ground zero of the impact, a point on the coast defined by the concentric patterns in gravity anomalies (Hildebrand et al., 1995). The seismic profiles provide four separate radial transects of the impact crater to reveal possible azimuthal variations in rock structures over the half the impact structure that lies offshore. Two transects (B and C) are oblique to the coastline (Fig. 1) and image the lithosphere at radial distances of 20–180 km from ground zero. The other two transects are joined in a single profile (A–A1) that forms a chord across the impact structure, offset about 20 km from its center and parallel to the coastline (Fig. 1).

Both the deep crustal-scale structures and near-surface morphology that resulted from the impact 65.0 Ma ago were targets of the reflection profiling. To achieve both adequate spatial resolution and trace redundancy (fold) in the interpretative stacks, we chose a composite configuration of the analogue recording streamer: 162 hydrophone groups near the ship at 12.5-m group spacing, and 78 groups at 50.0-m spacing (Table 1). The crustal-scale survey thus used an active streamer 5,925 m in length with shot and receiver intervals of 50 m. The higher resolution survey of the uppermost 8 km utilized a 2,025-m active streamer with the same 50-m shot intervals and four times the receiver density, a 12.5-m interval, and it is not discussed further here. The 36-element airgun array was designed to have a relatively flat spectral response between 8 and 50 Hz to serve well both the high- and low-resolution surveys. It produced a peak-to-peak pressure wave of 171 bar-m over a bandpass window of 3–125 Hz, very high by BIRPS standards (Hobbs and Snyder, 1992).

SEISMIC DATA PROCESSING

The composite streamer configuration allowed processing of the seismic data within two independent processing streams: one with low-resolution coverage of the entire crust and one with high-resolution coverage of the sedimentary section and uppermost 8–10 km. The entire streamer length was used in the low-resolu-

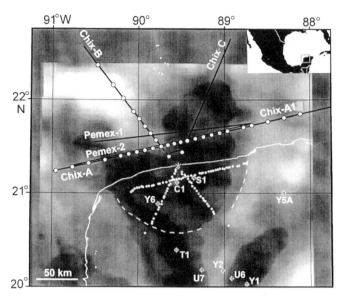

Figure 1. Location map showing the Yucatán peninsula of Mexico, the seismic reflection profile survey lines of the 1996 BIRPS experiment, two older profiles acquired by the Pemex and locations of stationary seismometers (white dots), onshore and on the seafloor, that recorded the survey shots in order to provide velocity estimates.

TABLE 1. ACQUISITION PARAMETERS

Source volume	150 liters	(9162 in^3)
Source pressure	13.8 MPa	(2000 psi)
Source array	36 airguns	75 m wide by 18.5 m
Source power	171 bar-m	(in 3–125 Hz bandwidth)
Source depth		6 ± 1 m
Shot interval		50 m
Streamer length		5925 m
Receiver groups	240	Configured 162 x 12.5 m and 78 x 50.0 m
Streamer depth		10 ± 1.5 m
Near offset		152 m
Sample rate		2 ms
Record length		18 s
Filters	Low cut	3 Hz @ 18 dB/octave
	High cut	128 Hz @ 72 dB/octave
Navigation		Multifix DGPS

tion study but the near-source receiver group spacing was expanded to 50 m by summing the recorded response of four adjacent groups (Table 2). The sampling rate was reduced to 8 ms. The 12.5-m receiver spacing used in the high-resolution study retained the original 2-ms sample rate. Usual methods of estimating spatial resolution indicate that the high-resolution study can distinguish structures 25 m apart vertically and 50 m apart horizontally.

The main sources of noise on these sections were "mudroll" and peg-leg multiples. The mudroll is interpreted as Stoneley waves, seismic waves travelling along the interface between the sea water and the seafloor sediments. These waves were particularly strong on the near-shore lines (A and A1) because the shallow water depth of ~20 m allowed efficient coupling of seismic

TABLE 2. PROCESSING PARAMETERS

1. Adjacent trace sum to uniform group spacing and appropriate gain balancing.	120 channels @ 50 m
2. Resample and bandpass filter	8 ms; 3 Hz @ 18 dB/octave–48 Hz @ 72 dB/octave
3. Geometric divergence correction	$1/(t\,v^2) + 1$ dB/s
4. FK filter	Mute group velocities <1400 m/s, 50% taper
5. Receiver array simulation	1:3:1 mix ratio
5. Source array simulation	1:3:1 mix ratio
7. Deconvolution (lag, operator length, window applied)	48 ms, 464 ms, 0.5–4.5 s; 72 ms, 640 ms, 4.0–9.0 s; 98 ms, 720 ms, 8.0–18.0 s
8. Velocity analysis	5 km spacing average
9. Front mute picks	e.g., 200 ms @ 200 m, 3500 ms @ 6000 m
10. NMO correction and stack	Nominal 60 fold
11. Revised gain correction (see step 3) + inelastic attenuation compensation to 3 s	$1/(t\,v^2)$ to 8 s
12. Running mix	0.5, 1.0, 0.5 weighted mix
13. FK filter	± 1500 m/s, 70% taper
14. Deconvolution (as above)	48 ms, 464 ms, 0.5–7.0 s; 72 ms, 640 ms, 5.0–16.0 s
15. 15° finite difference time migration	90% refraction velocities
16. Running mix	7 trace gaussian weighted
17. Trace sum	4 trace sum
18. Bandpass filters (Hz @ dB/octave)	8 @ 18 to 24 @ 48
19. Depth conversion	100% refraction velocities
20. Gain balancing	30 km depth window

Note: Residual statics were tested prior to velocity analysis (8) but were not found to be effective.

energy from the airgun array at 6-m depth into the soft seafloor sediments and then into the hydrophones at 10-m depth. Trials indicated that an FK filter applied to the shot gathers (step 4 in Table 2) was most effective at removing this unwanted energy.

The peg-leg multiples arise from the strong impedance contrasts between strata within the Cretaceous, and, to a lesser extent, the Tertiary, sedimentary sequences. Reverberations within the shallow water layer are predictable and can be suppressed using predictive deconvolution. Reverberations within sedimentary strata are traveling in media with variable velocities that are very similar to the primary velocities of the deeper target rocks that the multiples obscure. It is therefore impossible to remove these peg-leg multiple reflections with complete confidence (Hardy and Hobbs, 1991), but the two deconvolution processors and the inherent velocity filter within the stacking process (steps 7, 10, and 14 in Table 2) did greatly reduce the amplitudes of the multiples.

Stacking velocities used throughout the processing were picked using semblance, mini-stack panels, and trial midpoint gathers (e.g., McBride et al., 1993) to maximize the coherency of primary reflections. These stacking velocities were kept consistent with coincident interval velocity models estimated from refracted seismic P-wave arrivals recorded on the streamer, ocean bottom seismometers, and landstations (Brittan et al., this volume; Christeson et al., this volume). The stacking velocities were also used to perform finite-difference time migrations and subsequent time-to-depth conversions of the stack sections for interpretation (steps 15 and 19 in Table 2). Thus, the resulting seismic sections (Plate 1, inside back cover) have correct vertical and horizontal scales with little exaggeration, as best as we can determine from the existing models of the rock velocities beneath the Chicxulub area.

INTERPRETATIONS

Seismic reflection profiles produce pseudo cross sections of the lithosphere that provide the highest available resolution information about the geometries of prominent, large structures within the crust and uppermost mantle. These profiles provide little constraint on the rock types or their physical properties such as P- or S-wave velocity or density, and must rely on other geophysical techniques and drill cores for this information. Many such sources of independent information now exist for Chicxulub, some of which are reported in this volume (Brittan et al., Christeson et al., Sharpton et al.) and in Sharpton et al. (1996).

This chapter focuses on two major reflective horizons within the crust beneath the Chicxulub impact structure: the prominent Jurassic(?)–Cretaceous reflector sequence at a few kilometers' depth and the Moho. Numerous prominent dipping reflectors have been revealed by these profiles and define the proposed multi-ring structure of the Chicxulub impact, but these are tentatively related to and discussed in the context of the two horizontal ones (Table 3).

The prominent group of reflectors at depths of 3–5 km on all the seismic sections is identified as Jurassic(?)–Cretaceous strata via correlation with the onshore drill cores (Ward et al., 1995) and the knowledge that interlayered anhydrites and carbonates typically produce strong reflections. This Mesozoic reflector sequence appears relatively undisturbed throughout most of the Yucatán continental shelf, but becomes gently domed or uplifted slightly on all four seismic profiles at about 120–150-km radial distance from the hypothesized "ground zero" of the impact (F in Plate 1, Figs. 2 and 3). Reflections dipping downward and inward from this annular distance are best observed below a monocline on line A1 (7-km depth at SP 6300 to 20-km depth at SP 5750), but also beneath a subtler monocline on line A (9-km depth at SP 1800 to 11-km depth at SP 2000 to 12.5-km depth at SP 2300), and on line C where the reflector is partly obscured by peg-leg multiples and by its ramp and flat geometry (7-km depth at SP 3000 to 18-km depth at SP 2000). On line B, these dipping reflections are very weak but appear between 4-km depth at SP 3650 and 6-km depth at SP 3550 (Plate 1). The local nature of the folding and doming suggest that the Mesozoic strata are uplifted above crustal-scale blind thrusts possibly related to outward

TABLE 3. CHICXULUB IMPACT FEATURES BASED PRIMARILY ON STRUCTURES IN CRETACEOUS STRATA

CHIX Survey Line	B	A	A1	C	Ring Diameter at the Surface
P: Peak ring	35–43	35–48	(35–53)	27–42	~80 km
S: End strata	40	38	42	42	80 km = I
A: Outer edge of collapsed terraces	(65)	61–64	60–65	55–58	130 km = II
Cenotes	....	75	80	....	
R: Restored "rim"	88	88 or 98	85–96	(75)	195 km = III
N: Normal fault		(110)	(97)	97	
F: "Ring fracture"	(135)	119	126	127	~250 km = IV
	Small folds	monoclines		apex of broad fold	

Note: Unlabeled numbers are radial distances in kilometers. Roman numerals indicate the four ringed zones of deformational structures described in the text; these have no direct correlation to multi-ringed impact structures on other planets. Letters in CHIX survey line column further described in Figure 2 and 3 captions.

displacement of the upper crust by the initial shock wave from the impact.

At a distance of 75–90 km, the Mesozoic reflectors are strongly folded, contorted, and faulted (R in plate 1, Figs. 2 and 3). The structural geometries are complex on each profile, they differ greatly from one profile to another, and these features occur at a range of radial distances (Table 3) (also see Morgan et al., this volume). On line A the prominent reflector sequence is thinner and ramps down at SPs 2250–2450 (Plate 1), with locally anomalous higher velocities near SP 2220 (Brittan et al., this volume). Strata at 2-km depth are disrupted near SP 2080. On line A1 the reflectors at 3–5-km depths are locally folded at SPs 5500–5820, and possibly out to SP 6000 at shallower depths (Plate 1). On line B a prominent horizontal gap occurs in the reflector sequence between SPs 2330 and 2460 (Plate 1). On line C the prominent reflectors at 5-km depth are strongly disrupted from SPs 1780–1880 and a local high (1-km depth) occurs on the Cretaceous–Tertiary boundary at SP 1730 (Plate 1). Shallower reflectors become relatively continuous at SP 1940. Collectively, these features suggest that these radial distances represent the highly eroded rim along the outermost edge of the collapsed crater. The complex reflector geometries may represent the contorted and back-folded strata often associated with the crater rim (e.g., Melosh, 1989, and references therein). Because the Chicxulub impact likely occurred in shallow marine conditions (Ward et al., 1995), the original topography of the crater, and particularly its rim, would have been greatly modified by collapse of saturated sediments and erosion by tsunami shortly after the impact. Little topographic expression would therefore be expected, but deeper strata might preserve such indicators of the former rim as are reported here.

At 55–65-km radial distance, the Jurassic(?)–Cretaceous strata are either downdropped 4 km (lines A, A1, and C) or begin to ramp into the crater interior at dips of 10°–15° (line B) (see feature labeled A in plate 1, Figs. 2 and 3). At 40–45 km, the prominent reflector sequence disappears completely (feature S in Plate 1, Figs. 2 and 3). The former feature (A) is interpreted as the outer edge of the collapsed terraced zone (also the inner ring of Morgan et al., 1997). The latter (S) represents the outermost edge of the collapsed transient cavity, the boundary within which the Mesozoic strata were pulverized, vaporized and ejected into the atmosphere by the impact processes (Melosh, 1989). By reconstructing the overlying sedimentary section, Morgan et al. (1997) estimated the apparent transient cavity to have a diameter of 90–105 km and the original collapsed crater to have a diameter of 180–210 km at the surface (their outer ring). These represent the second and third rings of a candidate multi-ring basin structure (Morgan et al., this volume).

The innermost ring structure is the peak ring (P in Plate 1, Figs. 2 and 3). Competing hypotheses of its nature suggested either uplifted and fractured basement (Sharpton et al., 1993) or impact breccia (Pilkington et al., 1994). The seismic profiles provide no evidence of uplifted basement here. A layer with more chaotic reflectivity, interpreted as impact breccia that contains variable amounts of melt throughout, separates the Cretaceous and Tertiary sedimentary sequences out to radial distances of 100–125 km. Nearer the central region of the impact structure, this more chaotic seismic layer is at least 5 km thick, and probably includes significant amounts of melt. It appears to make up the ca. 400 m topography associated with the peak ring (Morgan et al., 1997; Brittan et al., this volume). Near its upper boundary it contains several blocks, presumed to represent secondary impactors, with diameters of several kilometers and Tertiary sediments draped over their top surfaces (SPs 3890, 4000, and 4070 on line A, Plate 1).

The lower crust is generally reflective throughout the survey area at depths greater than 20 km. The base of prominent reflectivity occurs at 32–38 km, with no regionally consistent variation. Individual reflections or bands of reflectivity are continuous for several to tens of kilometers, and these deepen or shallow grad-

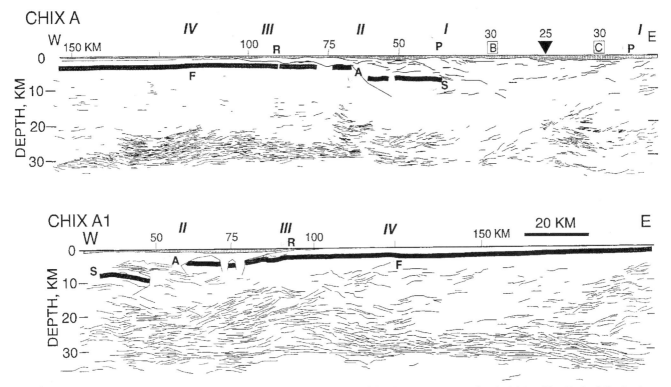

Figure 2. Line drawing made from migrated, depth-converted seismic sections of the along-coast survey line A–A1 (see Fig. 1); the full seismic sections are illustrated in Plate 1 (inside back cover). Migration and depth conversion velocities used to "correct" the geometries of the reflectors illustrated here were made using the stacking velocities used to process the data (Table 2 and text). P = central peak ring; S = edge of reflective strata, outermost edge of the collapsed transient cavity; A = crater edge or the outermost limit of downdropped megaterrace blocks; R = crater rim structures; F = ring fracture or blind thrust; see text for discussion of these interpretative labels. Roman numerals indicate the ringed zones as defined in Table 3 and described in the text. Letters in squares indicate line intersection locations.

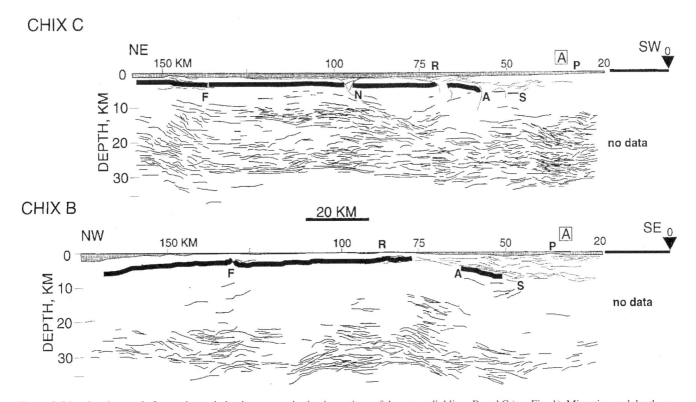

Figure 3. Line drawing made from migrated, depth-converted seismic sections of the two radial lines B and C (see Fig. 1). Migration and depth conversion were done as in Figure 2. Annotation as in Figure 2, plus N = major normal fault.

ually, suggestive of folding or flow rather than of brittle faulting. Within the resolution of our various seismic methods, the base of prominent reflectivity coincides with the Moho defined from wide-angle reflections and velocity models (cf. Plate 1 with Fig. 5 of Christeson et al., this volume).

The upper crust is generally less reflective except for a few prominent reflectors that dip at 15°–25° toward the center of the impact structure. One reflector dips at ~30° and projects to a radial distance of ~97 km where it offsets the Mesozoic strata by ~250 m with a normal sense of offset (N on Line C, Plate 1, Fig. 3). Another reflector projects to the near-surface at 126 km where the doming of Mesozoic strata described above suggests a reverse "blind thrust" sense of displacement (F on Line A1, Plate 1). Similar, albeit weaker, reflectors are observed on the other three lines (Figs. 2 and 3).

A third set of dipping reflectors occurs beneath the interior of the impact structure and these project up from 25-km depth toward the downdropped blocks of Mesozoic strata. These dipping reflectors within the crystalline basement probably represent shear zones with their reflectivity enhanced by intrusions. If these shears were active during the impact or during the collapse phase, these intrusions could be related to melts generated deep in the crust by the impact (Wu et al., 1995; Spray, 1997).

Taken as a whole, the four seismic lines reveal a pattern of disruption to near-surface, preimpact strata that suggests a zoned or multi-ring geometry to the impact structure. Deeper reflections indicate that the sources of these rings of disruption are crustal-scale shears or fault zones that dip toward the center of the crater at 15°–25°, and in one case (N), at about 30°. Interestingly, each ring structure appears to have a different characteristic deformation, varying from a lag deposit (ridge) of fall-back breccia (P) to slump block margins (A), an isolated normal fault (N) to blind thrusts (F). Some of these structures were suspected before the acquisition of the deep seismic data, but no clear picture of the details was previously available. These seismic profiles thus represent a major contribution not only to the understanding of the Chicxulub impact, but also to impact structures throughout the inner solar system.

ACKNOWLEDGMENTS

The seismic reflection profiles referred to in this chapter were acquired by Geco-Prakla for the British Institutions Reflection Profiling Syndicate (BIRPS) and were processed by Bedford Interactive Processing Services and the authors. They are available for the cost of reproduction from the British Geological Survey (Marine Geophysics Programme Manager, West Mains Road, Edinburgh EH9 3LA, UK). This research was done under the auspices of the Universidad Nacional Autónoma de México and the Natural Environment Research Council of the UK. The survey was funded by the British Natural Environment Research Council and BIRPS Industrial Associates Program (Amerada Hess, Ltd., British Gas, PLC, BP Exploration Operating Company, Ltd., Chevron UK, Ltd., Conoco (UK) Ltd., Lasmo North Sea, PLC, Mobil North Sea, Ltd., Shell UK Exploration & Production, Ltd, and Statoil (UK), Ltd.). This is Cambridge Earth Sciences Contribution 5211.

REFERENCES CITED

Alvarez, L. W., Alvarez, W., Asaro, F., and Michel, H. V., 1980, Extraterrestrial cause for the Cretaceous-Tertiary extinction: Science, v. 208, p. 1095–1108.

Connors, M., Hildebrand, A. R., Pilkington, M., Ortiz-Aleman, C., Chavez, R. E., Urrutia-Fucugauchi, J., Graniel-Castro, E., Camara-Zi, A., Vasquez, J., and Halpenny, J. F., 1996, Yucatán karst features and the size of Chicxulub crater: Geophysical Journal International, v. 127, p. F11–F14.

Hardy, R. J. J., and Hobbs, R. W., 1991, A strategy for multiple suppression: First Break, v. 9, p. 139–144.

Hildebrand, A. R., 1997, Contrasting Chicxulub crater structural models: what can seismic velocity studies differentiate? Journal of Conference Proceedings, v. 1, p. 37–46.

Hildebrand, A. R., Penfield, G. T., Kring, D. A., Pilkington, M., Camargo, Z. A., Jacobsen, S., and Boynton, W. V., 1991, Chicxulub crater: a possible Cretaceous-Tertiary boundary impact crater on the Yucatán peninsula, Mexico: Geology, v. 19, p. 867–871.

Hildebrand, A. R., Pilkington, M., Connors, M., Ortiz-Aleman, C., and Chavez, R. E., 1995, Size and sructure of the Chicxulub crater revealed by horizontal gravity gradients: Nature, v. 376, p. 415–417.

Hobbs, R. W., and Snyder, D. B., 1992, Marine seismic sources used for deep seismic reflection profiling: First Break, v. 10, p. 417–426.

Lopez-Ramos, E., 1975, Geological summary of the Yucatán Peninsula, in Nairn, A. E. M., and Stehli, F. G., eds., The ocean basins and margins, Vol. 3—The Gulf of Mexico and the Caribbean: New York, Plenum Press, p. 257–282.

McBride, J. H., Lindsey, G., Snyder, D. B., Hobbs, R. W., and Totterdell, I. J., 1993, Some problems in velocity analysis for marine deep seismic profiles: First Break, v. 11, p. 345–356.

Melosh, H. J., 1989, Impact cratering: a geologic process: New York, Oxford University Press, 245 p.

Morgan, J., Warner, M., and the Chicxulub Working Group, 1997, Size and morphology of the Chicxulub impact crater: Nature, v. 390, p. 472–476.

Pilkington, M., Hildebrand, A. R., and Ortiz-Aleman, C., 1994, Gravity and magnetic field modeling and structure of the Chicxulub Crater, Mexico: Journal of Geophysical Research, v. 99, p. 13147–13162.

Pope, K. O., Ocampo, A. C., and Duller, C. E., 1993, Surficial geology of the Chicxulub impact crater, Yucatán, Mexico, Earth, Moon, and Planets, v. 63, p. 93–104.

Sharpton, V. L., Burke, K., Camargo-Zanoguera, A., Hall, S., Lee, D. S., Marin, L. E., Suarez-Reynoso, G., Quezada-Muneton, J. M., Spudis, P. D., and Urrutia-Fucugauchi, J., 1993, Chicxulub multi-ring impact basin: Size and other characteristics derived from gravity analysis: Science, v. 261, p. 1564–1567.

Sharpton, V. L., Dalrymple, G. B., Marín, L. E., Ryder, G., Schuraytz, B. C., and Urrutia-Fucugauchi, J., 1992, New links between the Chicxulub impact structure and the Cretaceous/Tertiary boundary: Nature, v. 359, p. 819–821.

Sharpton, V. L., Marín, L. E., Carney, C., Lee, S., Ryder, G., Schuratyz, B. C., Sikora, P., and Spudis, P. D., 1996, A model of the Chicxulub impact basin based on evaluation of geophysical data, well logs and drill core samples, in Sharpton, V. L., and Ward, P. D., eds., Global catastrophes in Earth history: an interdisciplinary conference on impacts, volcanism, and mass mortality: Geological Society of America Special Paper 247, p. 55–74.

Spray, J. G., 1997, Superfaults: Geology, v. 25, p. 579–582.

Ward, W. C., Keller, G., Stinnesbeck, W., and Adatte, T., 1995, Yucatan subsurface stratigraphy: implications and constraints for the Chicxulub impact: Geology, v. 23, p. 873–876.

Wu, J., Milkereit, B., and Boerner, D. E., 1995, Seismic imaging of the enigmatic Sudbury structure: Journal of Geophysical Research, v. 100, p. 4117–4130.

MANUSCRIPT ACCEPTED BY THE SOCIETY DECEMBER 16, 1998

Near-surface seismic expression of the Chicxulub impact crater

John Brittan*, Joanna Morgan, Mike Warner
T. H. Huxley School of Environment, Earth Sciences, and Engineering, Imperial College, London SW7 2BP, United Kingdom
Luis Marin
Instituto de Geofisica, Universidad Nacional Autonoma de Mexico, Mexico DF 04510

ABSTRACT

We show seismic velocity and reflectivity images of the shallow offshore structure of the Chicxulub impact crater, derived from short-offset reflected and long-offset refracted seismic waves. These data, generated by airgun shots and recorded on a 6-km multichannel streamer, were acquired as part of the whole-crustal Chicxulub seismic experiment. The P-wave velocity structure was derived from inversion of the travel-times of the first-arriving waves at each hydrophone. The velocity data clearly image the ~1-km-thick impact basin, which is identified by its low-velocity (2–3.5 km/s) Tertiary fill. The impact breccias have average velocities of between 3 and 4.5 km/s; the pre-impact Cretaceous carbonates have average velocities >5 km/s. Radial asymmetry in the offshore crater structure, which is apparent in the gravity data, is confirmed by analysis of both the reflection and velocity images. A thick (>1.5 km) low-velocity Tertiary basin to the northeast of the crater center is coincident with a region of low gravity. On the seismic reflection images, the now-buried topographic peak ring is clear. It has an average diameter of 80 km, and stands several hundred meters above the otherwise relatively flat crater basin floor. Seismic and gravity data indicate that the peak ring does not have a strong velocity or density signature. On several seismic profiles the innermost edge of the inwardly slumped target rocks lie directly beneath the peak ring. Seismic reflections, which dip toward the crater center at ~30° and run from the outer edge of the peak ring to the inner edge of the slumped blocks, indicate a generic relationship between these two features. We suggest that these dipping reflections represent the boundary between the inwardly collapsed uplifted rim of the transient crater and the outwardly collapsed central uplift, and that the peak ring at Chicxulub was formed as a result of the interaction of these two collapse regimes.

INTRODUCTION

The Chicxulub impact structure, which lies partly off the Yucatan Peninsula in Mexico, is one of the most important geologic structures to have come to prominence in the last 20 years. From its initial recognition as an impact-related structure (Penfield and Camargo-Zanoguera, 1981; Hildebrand et al., 1991; Pope et al., 1993; Sharpton et al., 1992), and its subsequent dating (Sharpton et al., 1992; Swisher et al., 1992), the Chicxulub structure has become widely accepted as the remains of a catastrophic extraterrestrial Cretaceous-Tertiary (K-T) impact proposed by Alvarez et al. (1980).

The crater floor lies buried beneath approximately 1 km of Tertiary sediments and has been largely unaffected by erosion and tectonic forces. The main constraints on the composition and structure of the crater have been provided by potential field studies and from stratigraphic analysis of onshore well data (e.g.,

---

*Present address: PGS Tensor (UK) Ltd., Walton-Upon-Thames KT12 5PL, United Kingdom.

Brittan, J., Morgan, J., Warner, M., and Marin, L., 1999, Near-surface seismic expression of the Chicxulub impact crater, *in* Dressler, B. O., and Sharpton, V. L., eds., Large Meteorite Impacts and Planetary Evolution II: Boulder, Colorado, Geological Society of America Special Paper 339.

Pilkington et al., 1994; Sharpton et al., 1996). In an attempt to further elucidate the dimensions of the crater and the effects of the impact on the morphology of the deep crust, a large seismic data set was acquired across the structure by an international group (Morgan et al., 1997; Snyder et al., this volume). In this chapter we present wide-angle velocity analyses of data from four marine multichannel seismic lines, providing further structural and compositional controls on the Chicxulub Crater. In addition, we address the spatial location, seismic velocity, morphologic character, and gravity signature of the peak ring.

The shallow seismic velocity structure has been studied at a number of terrestrial impact craters and has provided information on craters and the cratering process (Pilkington and Grieve, 1992). For example, at the Barringer Crater in Arizona, fracture-related low velocities in the country rock extend to at least a crater diameter beyond the crater rim (Ackermann et al., 1975), with the magnitude of the negative velocity anomaly decreasing with distance from the crater center. In addition, the low-velocity zone penetrates at least a crater diameter below the breccia lens. At the complex Ries Crater in Germany, a low velocity zone extends to a depth of nearly 6 km below the crater surface and correlates closely with a zone of low density (Pohl et al., 1977). Near-surface high velocities are observed at the very center of some large craters (e.g., Vredefort) (Green and Chetty, 1990) and are thought to represent uplifted deep crustal material.

Seismic velocities have been previously measured across the Campeche Bank in two-ship shallow refraction studies (Ewing et al., 1960; Antoine and Ewing, 1963). The location of some of these profiles (labeled A15–A18 and E16) are shown in Figure 1 and the velocity/depth function along each profile is illustrated in Figure 2. The results from these studies are comparable to velocities obtained from shots fired within well C-1 (Cué, 1953) and recorded on receivers at surface. The velocity/depth profile derived from this experiment is easily related to the geologic profile obtained from the drill core.

SHALLOW SEISMIC VELOCITY ANALYSES

The British Institutions Reflection Profiling Syndicate (BIRPS) acquired four deep-crustal, multichannel seismic profiles (Morgan et al., 1997; Snyder et al., this volume) across the impact structure in fall 1996 (Fig. 1). The stacked seismic reflection sections resulting from these profiles illustrate many important structural features of the impact structure and lead to estimates of the transient cavity diameter, and the extent of both shallow and deep deformation (Morgan et al., 1997; Morgan and

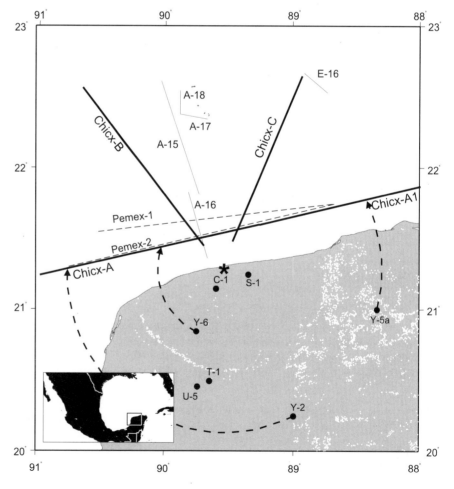

Figure 1. The thick offshore lines show the position of the multichannel seismic reflection profiles presented in this chapter; the dashed lines indicate two Pemex reflection profiles for which data are available. The thin lines mark the positions of previous two-ship refraction surveys in the area: the profiles marked E16 Ewing et al. (1960) and A15–A18 (Antoine and Ewing, 1963). The solid black dots indicate onshore wells referred to in the text; dashed lines from Y-6, Y-5a, and Y-2 indicate the projection of well data onto the seismic data shown in Figure 5. The white dots locate the positions of water-filled surface sinkholes (cenotes) and show a clear ring distribution. The asterisk marks the nominal center of the Chicxulub impact structure, chosen as it lies between the centers of the inner and outer gravity structure, the magnetic anomaly, and the cenote ring. All radial distances have been calculated using this center at ~89.54°W, 21.3°N.

Warner, 1999). In this chapter we show shallow seismic velocity models along two radial lines, Chicx-B and C, and along Chicx-A and A1, which cross the crater along a chord (Fig. 1).

Shallow seismic velocities are usually derived from coherency analysis of pre-critical reflections on short-offset, pre-stack, common-midpoint gathers, that is, from stacking velocities. While this technique is straightforward, the necessarily limited pathlength of wave propagation in each component stratigraphic unit means that the resultant velocity information is imprecise and not spatially well resolved. With the trend toward increasing streamer length in marine geophysical surveys, it is now possible to obtain accurate, shallow velocity information using postcritical refracted arrivals at long offset. In the Chicxulub seismic experiment, the distance from shot to end hydrophone was ~6.1 km, enabling seismic velocities to be determined from wide-angle refracted arrivals to depths of about 2 km.

In Figure 3 we show two typical shot gathers from within and outside the impact basin along Chicx-A (locations 4 and 1 in Fig. 4A). These data show clear differences in both reflection character and apparent velocities. Inside the postimpact basin we observed several reflections within the top 1-s two-way traveltime (TWTT), and the apparent velocities of the first arrivals increase from ~2.3 km/s between 0 and 2 km, to >5 km/s at offsets greater than 5 km. Outside the basin there are no clear near-surface reflections, there are two deep reflections at 0.9 and 1.6 s, and the first arrivals have a more constant apparent velocity. The arrivals at offsets of >5 km have traveled to a maximum depth of about 2.1 km within the basin and about 1.3 km outside the basin.

To determine the shallow subsurface velocity structure, we modeled the first arrivals recorded on the hydrophone array. We modeled data from every 40th shot (every 2 km) along the full extent of each profile. Travel-times of the first arrivals were picked and a velocity model was derived using a ray-based, two-dimensional inversion scheme (Zelt and Smith, 1992). Our aim in using this scheme was to determine the simplest and smoothest tomographic velocity model capable of fitting the travel-times to within the error inherent in picking the arrival times. Any model that is less smooth or more complex than our model, and that

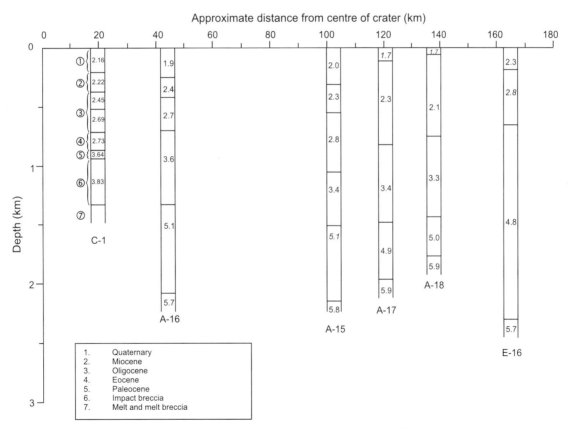

Figure 2. Summary of preexisting seismic velocity information from the Chicxulub impact structure. Velocity-depth functions are shown for the two-ship refraction experiments: profiles E16 (Ewing et al., 1960) and A15–A18 (Antoine and Ewing, 1963). The velocity-depth function illustrated is an average velocity at about the center of the profile, and is plotted at a radial distance measured from the crater center to the center of the seismic profile. Velocities given in italics are from unreversed profiles or represent assumed velocities. The velocity-depth function given for well Chicxulub-1 (C-1) was derived from a downhole seismic experiment (Cué, 1953). The stratigraphic correlation for well C-1 is derived from interpretations of the borehole logs (Cué, 1953; Ward et al., 1995).

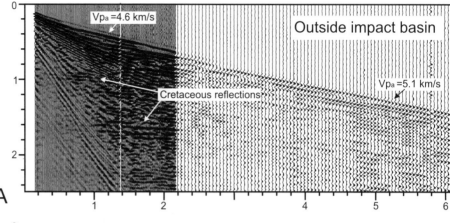
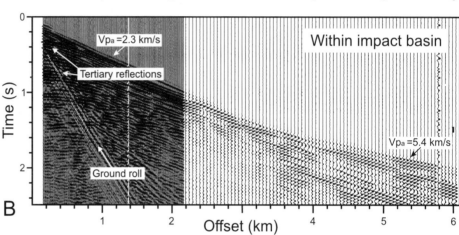

Figure 3. Two typical shot gathers of data collected outside (A) and within (B) the impact basin along Chicx-A. The location of these gathers is marked by arrows 1 and 4, respectively in Figure 4A. Vp_a represents apparent P-wave velocity of the first arrivals. The first 162 traces are 12.5 m apart; traces 163–240 are 50 m apart.

averages (to within error) to the same smooth model as ours, will also be an acceptable fit to the data. However, the apparently increased resolution of such a model will not be justified by the first arrival times in the seismic data. In order to improve resolution in a way that is meaningful, it is necessary to incorporate amplitude and/or waveform information into the inversion. We are actively applying two-dimensional full-wavefield inversion (Pratt et al., 1996) to the entire Chicxulub dataset, but such methods are computationally expensive. Brittan et al. (1997) have applied this technique to a short section of the Pemex-2 line showing the potential of this high-resolution technique.

In the case of the Chicxulub data, the simplest initial model that could satisfactorily explain the first-arrival travel-times was a fixed water layer underlain by a two-layer model of the Earth. Both layers had a positive, vertical velocity gradient that, along with the position of the interface, was allowed to vary during the inversion. Within each layer, lateral velocity variations are smooth. Using only two subseabed layers, we were able to model the observed data with a mean mismatch of ~20 ms and an acceptable resolution factor of >0.5 for the vast majority of nodes (Zelt and Smith, 1992). The close fit between observed and synthetic travel-times implies that our model velocities are accurate; however, these velocities are local averages, smoothed vertically over distances of the order of the seismic wavelength (~150 m)

and laterally over distances of the order of the Fresnel zone (~1,000 m). In this model, we will therefore not be able to resolve velocity changes whose spatial wavelength is <150 m. Since Chicx-B and C both cross Chicx-A, we can compare coincident but independently derived velocity functions and check our velocity models. There is negligible misfit where the lines cross.

A composite of the upper part of the stacked reflection data and the velocity model along Chicx-A and A1 is shown in Figure 4. The combined data clearly image a number of important compositional and structural features of the crater. The deep impact basin is observed in the reflection data as a ~1-km layer of subhorizontal reflectors, and extends from a radial distance of around 65 km on Chicx-A and 82 km on Chicx-A1. The base of this reflective zone is likely to represent or lie close to the Cretaceous/Tertiary boundary, as interpreted by Camargo-Zanoguera and Suarez Reynoso (1994) on the basis of correlation with onshore wells C-1, Y-6, and S-1. The velocity data mirror the reflection data: the thickening of the Tertiary section is paralleled by a thickening of a low velocity zone. The velocities within the impact basin increase from 2.0 km/s near surface to ~3.5 km/s near the base of the Tertiary, and are comparable to velocities observed near C-1 (Cué, 1953), and velocities determined along the nearby refraction line A-16 (Antoine and Ewing, 1963); see Figure 2. The low velocities measured within the impact are com-

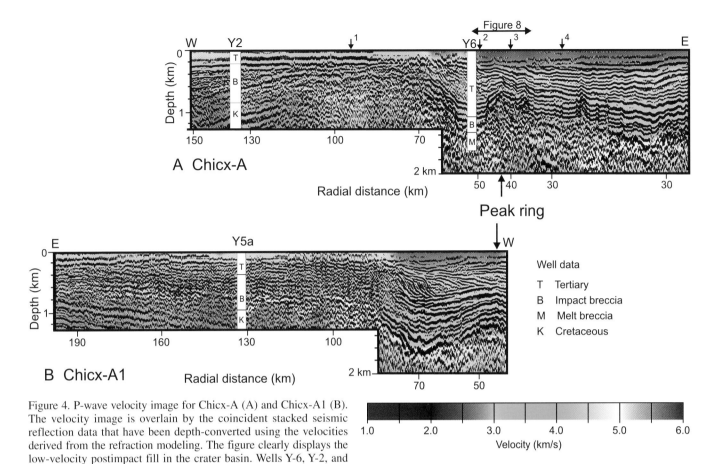

Figure 4. P-wave velocity image for Chicx-A (A) and Chicx-A1 (B). The velocity image is overlain by the coincident stacked seismic reflection data that have been depth-converted using the velocities derived from the refraction modeling. The figure clearly displays the low-velocity postimpact fill in the crater basin. Wells Y-6, Y-2, and Y-5a have been projected onto the seismic profiles at the appropriate radial distances (see Fig. 1 for projection). T = Tertiary rocks; B = impact breccia; M = melt breccia; K = Cretaceous stratigraphy. Arrows 1 through 4 above Chicx-A indicate the location of shot gathers in Figures 3 and 7.

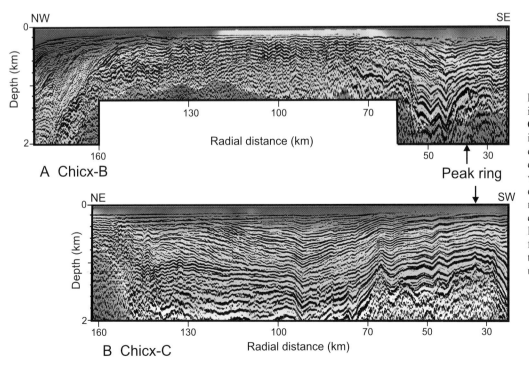

Figure 5. P-wave velocity image for Chicx-B (A) and Chicx-C (B). The velocity image is overlain by the coincident stacked seismic reflection data that have been depth-converted using the velocities derived from the refraction modeling. On Chicx-C, the edge of the Tertiary basin clearly lies at much greater offsets from the crater center than on the other three profiles. See Figure 4 for color key.

patible with stacking velocities of 2.8 km/s for near-normal-incidence reflections near the base of the Tertiary. Tomographic modeling by Christeson et al. (this volume) using arrivals at the sea-bottom stations also shows a low-velocity impact basin.

The unreflective material below the impact basin is likely to represent the allogenic and melt breccia observed below the Tertiary in Y6, C-1, and S-1 (Hildebrand et al., 1991; Ward et al., 1995; Sharpton et al., 1996). Immediately beneath the impact basin the velocities do not appear to show a dramatic change in average velocity. The fastest arrivals (>5 km/s) seen at >5 km offset on data within the basin (Fig. 3B) are modeled as returning from depths a few hundred meters beneath the Tertiary section. The increase in velocities from <4 km/s to >5 km/s appears to occur at a depth close to the expected boundary between the allogenic impact breccia and melt breccia, assuming the lithology in wells Y6, S1, and C1 (Sharpton et al., 1996) can be projected offshore. Therefore, we are probably imaging the transition from allogenic impact breccia to melt breccia with the seismic velocity, but we emphasize that the resolution of these travel-time-only models is limited.

Outside the deep basin, the near-surface velocities start at 3–3.5 km/s and gradually increase to ~4.5 km/s at ~800 m. From the stratigraphy in the onshore wells, we would expect 200–500 m of Tertiary sediments to be underlain by impact breccia; in wells Y5a, Y1, and Y2, the depth to the base of the impact breccia is between 850 and 950 m (Sharpton et al., 1996). Near-surface velocities with values around 3–3.5 km/s probably represent the Tertiary sequence, whereas the deeper material with velocities of up to 4.5 km/s is likely to be impact breccia. Below ~800 m velocities increase to >5.0 km/s, and coincide with reflective target stratigraphy previously interpreted as Cretaceous in age (Camargo-Zanoguera and Suarez-Reynoso, 1994). Stacking velocities obtained on Cretaceous reflections outside the basin confirm that average velocities are much higher outside than inside the basin: the reflectors at 0.9 and 1.6 s in Figure 3A have stacking velocities of 3.6 and 5.2 km/s, respectively, whereas within the basin the stacking velocities are ~2.8 and ~3.8 km/s at comparable travel times.

Figure 5 shows a composite velocity image/stacked section for the two radial lines Chicx-B and Chicx-C. On Chicx-B the deep Tertiary basin extends from around 75 km in radial distance to the southeastern end of the section illustrated. To the northwestern end of the line at radial distances >150 km, we observe a thickening in the sedimentary sequence as we approach the continental margin. At radii of 100–150 km, the velocity of material below ~1-km depth appears to be slightly higher (by ~0.5 km/s) than that along Chicx-A and A1. Using conventional velocity-density relationships (Gardner et al., 1974), and assuming a 4-km-thick Mesozoic section, these rocks would produce a gravity high of ~10 mGals along this profile. The observed gravity is 5–20 mGals higher along this line than in adjacent areas outside the basin.

The velocity/reflection image along Chicx-C is distinct. The postimpact basin has no clear edge at between 65- and 82.5-km radius, as seen on other lines. There is a topographic high at ~67 km radius (labeled inner ring in Fig. 5), which appears to correlate with the basin edge observed on the other lines. The shallow Tertiary fill has similar velocities along Chicx-C to those seen on the other lines; however, the basin thickens away from the center to reach a thickness of >1.5 km between 75- and 140-km radial distance. The deepest Tertiary sediments have velocities of 4.0–4.5 km/s. The extension of a low-velocity deep Tertiary basin out to a crater radius of ~140 km along Chicx-C is in agreement with the gravity signature in this region: there is a gravity low with only weak evidence for the ring structure that is seen across much of the crater (Hildebrand, 1997). The velocities along Chicx-C are in agreement with velocities obtained along refraction lines A16 and A17 (Antoine and Ewing, 1963); these profiles also show the low velocity zone thickening to the north (Fig. 2).

In summary, the velocity models, when combined with coincident stacked reflection sections, show a good correlation between velocity (compositional) features and reflective (structural and stratigraphic) features, and are compatible with the lithologies drilled onshore. The velocities obtained are in agreement with the preexisting onshore and offshore seismic velocity data and stacking velocities obtained from the reflection data. The deep Tertiary marls and carbonates that filled the impact basin have a relatively low velocity (2–3.5 km/s) compared to the adjacent shallow water Tertiary carbonates and impact breccias (3–4.5 km/s) at equivalent depths outside the impact basin. Beneath the impact basin, we observe an increase in velocity from <4 km/s to >5 km/s close to the expected boundary between impact and melt breccias. The Cretaceous stratigraphy correlates with velocities of >5.0 km/s.

PEAK RING: LOCATION AND GEOPHYSICAL SIGNATURE

Within the sediment-filled crater basin there is a series of topographic highs interpreted as the peak ring of the crater (Camargo-Zanoguera and Suarez-Reynoso, 1994; Morgan et al., 1997). In the west and northwest (e.g. Chicx-A and B), the increase in elevation from the annular trough to peak ring is 700 m (Figs. 4 and 5). To the northeast the peak ring has slightly less relief, rising 400–500 m above the annular trough (e.g., Chicx-A1 and C). Table 1 gives the location of the peak ring on the seismic reflection lines shot during the Chicxulub Seismic Experiment (Chicx-A, A1, B, and C) and on two seismic lines shot by Pemex (Camargo-Zanoguera and Suarez-Reynoso, 1994). The locations of the peak ring are displayed in Figure 6. The average peak ring diameter is 80 ± 5 km when measured from peak to peak. Pemex-1 has been excluded from these averages because the line runs at a tangent to the peak ring.

The peak ring at Chicxulub appears analogous to those seen in peak-ring craters on other planets. The peak ring is observable as a series of irregular, blocky massifs, it is only approximately circular, and it has significant lateral variability. Similar charac-

TABLE 1. LOCATION OF PEAK RING WITHIN THE IMPACT BASIN

Seismic Line	Approximate Radius (km)	Location of Peak (km)
A	35–48	42.5
A1	35–55	45
B	35–43	37
C	27–42	35
Pemex-1	38–39	38.5
Pemex-2	44–53	51

teristics are seen in the peak ring of the 220-km-diameter crater Galle on Mars (cf. Fig. 8.8c of Melosh, 1989) and in that of the 95-km-diameter Potanina crater on Venus (cf. Fig. 11b of Alexopoulos and McKinnon, 1994). While a number of crater basins show peak rings that are continuous around the impact center, many show a much more discordant arrangement of highs in the crater floor (Alexopoulos and McKinnon, 1994).

Figure 6 shows the position of the peak ring relative to the gravity anomaly across the crater, and illustrates the spatial relationship between the observable peak ring on each seismic line. It is clear from Figure 6 is that there is no simple relationship between the position of the peak ring and the underlying gravity anomaly. For example, the peak ring on Chicx-A correlates reasonably well with a local ring of low gravity, while, on the nearby lines Chicx-A1 and C, the peak ring appears to lie between a local gravity high and gravity low. We conclude that the peak ring does not have a clear expression in the gravity data. The inner edge of the collapsed transient cavity (Morgan et al., 1997) partially underlies the peak ring, and it seems likely that the gravitational anomaly produced by this edge interferes with that produced directly by the peak ring to produce the complicated composite anomaly observed.

On all four lines our shallow velocity models suggest that the material that composes the peak ring has a P-wave velocity that is comparable to that of the sediments in the lowest part of the basin and to the material immediately beneath the impact basin (Figs. 4 and 5). In Figure 7 we show a shot gather from within the annular trough, above the peak ring, and within the central basin along Chicx-A (see Fig. 4 for precise location). For offsets 0–2 km the first-arrivals above the peak ring are within ±10 ms of the arrivals within the annular trough and central basin. These arrivals have traveled to depths of ~600 m; there is correspondingly minimal lateral change in velocity in the top 600 m above the peak ring on Chicx-A. Between 2 and 4 km offset, the first-arrivals are different phases on the different gathers. Across the peak ring at least some first arrivals are postcritical reflections from the top of the peak ring (normal incidence TWTT, 0.55 s); on the other two gathers, the first arrivals are simple turning waves. Between 4 and 6 km offset the first arrivals have traveled beneath the Tertiary section. The comparable travel-times at offsets >4 km suggest there are only small changes in average velocity between the peak ring and adjacent areas; the deepest arrivals from beneath

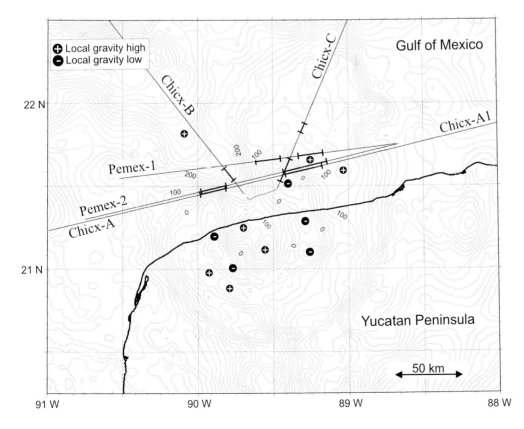

Figure 6. Spatial location of topographic highs interpreted as the peak ring. The spatial extent of these highs has been picked from Chicx-A, A1, B, and C, and Pemex-1 and 2, and is marked by a bold line. The dotted lines represent the Bouguer gravity field across the crater contoured at 20-gravity unit intervals.

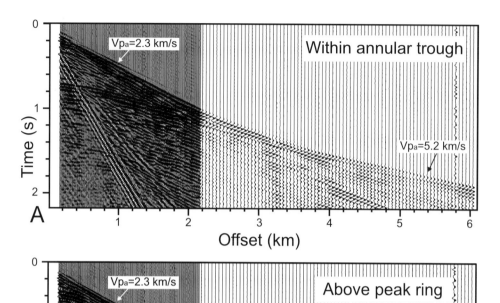
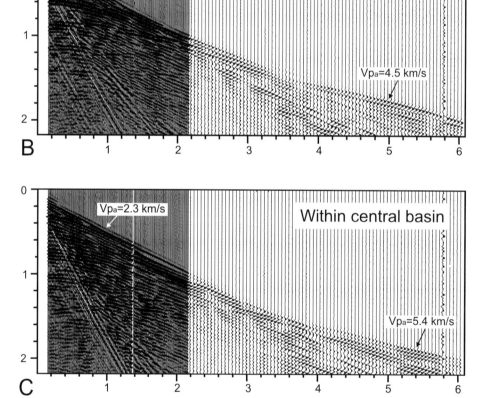

Figure 7. Three typical shot gathers of data collected within the annular trough (A), above the peak ring (B), and within the central basin along Chicx-A (C). The location of these gathers is marked by arrows 2, 3, and 4, respectively, in Figure 4A. The first 162 traces are 12.5 m apart; traces 163–240 are 50 m apart.

the peak ring have slightly lower apparent velocities than within the annular trough and central basin. We conclude that the traveltimes, and consequently the velocity models, indicate that there is no large lateral change in average velocity associated with or immediately beneath the peak ring. There may, however, be local changes in velocity not detectable with our smooth velocity models. The longest-offset travel-times suggest a small decrease in velocity beneath the peak ring, perhaps indicating that this region is more brecciated.

PEAK RING: FORMATION

On the seismic reflection data, the peak ring is located above or just inboard of reflections that have been interpreted as the inner edge of slumped blocks (Camargo-Zanoguera and Suarez Reynoso, 1994) that locate the collapsed transient cavity (Morgan et al., 1997). On most lines, prominent reflectors with a dip of ~30° toward the crater center run between the outer edge of the peak ring and the inner edge of the last slumped block. Examples of these reflectors along Chicx-A can be seen in Figure 8A. The location of slumped blocks immediately beneath the peak ring, and the presence of craterward-dipping reflectors between these two features, indicates a possible generic relationship between the peak ring and the collapsed transient cavity.

Prior to the seismic experiment, there were contradictory models for the location, composition, and formation of the peak ring at Chicxulub. Sharpton et al. (1993) associated the peak ring with a gravity high at ~50-km radius, and proposed that the peak ring consisted of fractured, uplifted, deep, basement material. In their model the density of the peak-ring material is 60–80 kg/m^3 higher than the surrounding allogenic breccia. In the simplest version of this model, the basement uplift was thought to have involved purely vertical (as opposed to radial) motion. In contrast, in a less conventional model, Pilkington et al., (1994) associated the peak ring with a prominent low in the gravity field that surrounds the central gravity high at a radius from the crater center of around 35 km. This low was modeled with a ring of low-density breccia (a local density contrast of –320 kg/m^3), located at a depth of 1–2 km below surface and with a maximum vertical extent of less than 1 km. The breccia was thought to be above, and originally floating on, a coherent melt sheet, and to be composed of material sloughed off from the rebounding central uplift. Density contrasts of –320 and +60–80 kg/m^3 correspond to velocity contrasts of about –2000 and +350 m/s, respectively (Gardner et al., 1974). Such large velocity contrasts would be readily detectable in the seismic data using the analysis methods we have employed. We do not observe them, and it is clear that the density models of the peak ring at Chicxulub (Sharpton et al., 1993; Pilkington et al., 1994) are untenable in their original forms.

The new seismic data locate the peak ring accurately, and it is now clear that the peak ring cannot be the principal cause of the 35-km-radius gravity low (Hildebrand, 1997). Thus, the seismic data, while not directly disproving the idea that peak rings are formed by accumulations of breccia above a melt sheet, do remove the main observational support for such a model. Rings of uplifted basement are observed at some terrestrial craters—for example, Popigai, Siberia (Masaitis, 1994) and Ries, Germany (Pohl et al., 1977)—and have been identified as eroded peak rings (e.g., Sharpton et al., 1996). In addition, spectral analyses of craters on the Moon have been used to support the model that peak rings are formed from uplifted target rocks (Bussey and Spudis, 1997). However, a simple model in which the peak ring at Chicxulub, formed as an antiform of vertically uplifted basement, is not consistent with the observation that slumped blocks lie directly below of the peak ring as seen on the seismic data. If we assume that the peak ring at Chicxulub is formed from uplifted basement material, then the question arises as to how the peak ring can lie directly above the collapsed transient cavity. One possibility is that the material that formed the peak ring became fluidized during the impact.

Baldwin (1981) proposed a tsunami model of peak-ring basin formation in which a large proportion of the subsurface is

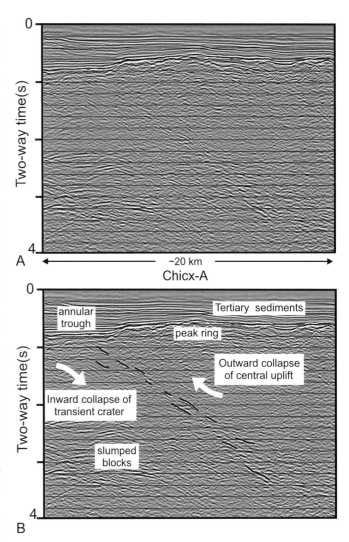

Figure 8. Close-up of the reflection data across the peak ring on Chicx-A: (A) raw data; (B) interpreted data. The existence of slumped blocks beneath the peak ring and the prominent dipping reflectors (highlighted in black) that lie between the peak ring and the slumped blocks suggest a generic relationship between these two features.

liquidized by an impact event and waves in this material radiate outward from ground-zero until their energy density decays sufficiently to freeze them into topographic highs in the surface. In this model, the peak ring is formed by the collapse and outward radiation of the liquidized material forming the central uplift. The nature of a hydrodynamical collapse of a central uplift and the subsequent "freezing" of the collapsing material is given a more physical basis in Melosh (1989). He suggested that the stress wave generated by the impact could cause acoustic fluidization of much of the target material. Solid material that is acoustically fluidized can flow without the presence of a fluid or vapor, thus enabling the central uplift to behave temporally as a low-viscosity fluid.

The seismic data are consistent with a model in which the peak ring is formed as a result of the collapse of the central uplift in which this collapse interacts with the inwardly collapsing inner edge of the transient cavity. There are two end members in this model. One is a variation of the proposal of Sharpton et al. (1996), and is consistent with the observations of uplifted basement rings at eroded craters. In this version, the rising central basement uplift is large. On reaching a transient and unstable position of maximum uplift, the central uplift recollapses downward and outward. If the outward collapse is sufficient, it will meet and interact with the inward-collapsing rim of the transient cavity that is moving inward to form the slumped blocks. The peak ring is thus formed by the occlusion of the outward collapse of the central uplift. It is composed of brecciated uplifted basement rocks, but underlain at depth by inwardly slumped blocks of Cretaceous sediment. At the other extreme, a process more closely akin to Baldwin's (1981) proposal is possible. In this version, a smaller central uplift collapses downward into a region of acoustically fluidized rheology, causing a wave to propagate outward. This wave then interacts with the inward collapsing rim, and occlusion of the wave produces the peak ring. In this scheme, there is no actual material transport of the central uplift outward. Hybrid models involving different degrees of both wave and material transport outward are also possible.

It seems likely that the prominent dipping reflectors between the peak ring and inner edge of the slumped blocks are indicative of the meeting between outwardly moving material and/or wave of the collapsing central uplift, and the inwardly collapsing transient crater (Fig. 8B). In the description of crater evolution by Croft (1981), this boundary was designated the "inflection line" and marked the boundary between uplifted material (inboard) and down-dropped material (outboard).

SUMMARY

Using seismic waves refracted within the shallow subsurface, we have constructed preliminary seismic velocity images of the top 1.3–2 km of the Chicxulub impact structure. These images show clear differences between the seismic velocity within the postimpact basin, the impact-related breccias, and the preexisting Cretaceous stratigraphy. A velocity boundary within the unreflective breccia beneath the impact basin appears to represent the boundary between allogenic impact breccia and melt breccia.

The peak ring has no consistent gravity or seismic velocity signature. The position of the peak ring above the inner edge of the last slumped blocks indicates that the peak ring is not vertically uplifted basement material that is emergent through impact breccia. The prominent craterward-dipping reflectors that lie between the outer edge of the peak ring and the inner edge of the slumped blocks suggest a generic relationship. We hypothesize that the peak ring is formed by an agglomeration of material around the point where the outward movement of weak or fluidized material from the central uplift is rescinded by inward movement of the collapsing transient cavity.

ACKNOWLEDGMENTS

The seismic reflection data from lines D92-RP001 and D92-RP002 were kindly provided by Petroleos Méxicanos. A fellowship from the Royal Commission for the Exhibition of 1851 (to J.B.) and a fellowship from the Leverhulme Trust (to J.M.) helped to support this work. The Chicxulub Seismic Experiment was funded by the Natural Environmental Research Council, The U.S. National Science Foundation, The British Institutions Reflection Profiling Syndicate Industrial Associates and the Royal Society. The seismic experiment was planned and executed by the Chicxulub Working Group.

REFERENCES CITED

Ackermann, H. D., Godson, R. H., and Watkins, J. S., 1975, A seismic refraction technique used for subsurface investigations at Meteor Crater, Arizona: Journal of Geophysical Research, v. 80, p. 765–775.

Alexopoulos, J. S., and McKinnon, W. B, 1994, Large impact craters and basins on Venus, with implications for ring mechanics on the terrestrial planets, *in* Dressler, B. O., Grieve, R. A. F., and Sharpton, V. L., eds., Large meteorite impacts and planetary evolution: Geological Society of America Special Paper 293, p. 29–50.

Alvarez, L. W., Alvarez, W., Asaro, F., and Michel, H. V., 1980, Extraterrestrial cause of the Cretaceous-Tertiary extinction: Science, v. 208, p. 1095–1108.

Antoine, J., and Ewing, J., 1963, Seismic refraction measurements on the margins of the Gulf of Mexico: Journal of Geophysical Research, v. 68, p. 1975–1996.

Baldwin, R. B., 1981, On the tsunami theory of the origin of multi-ring basins, *in* Schultz, P. H., and Merrill, R. B., eds., Multi-ring basins: New York, Pergamon Press, p. 275–288.

Brittan, J., Forgues, E., Pratt, G. R., Morgan, J., Warner, M., Macintyre, H., and Marin, L., 1997, Wavefield inversion across the edge of the Chicxulub impact structure: Journal of Conference Proceedings, v. 1, p. 65–72.

Bussey, D. B. J., and Spudis, P. D., 1997, Compositional analysis of the Orientale basin using full resolution Clementine data: some preliminary results: Geophysical Research Letters, v. 24, p. 445–448.

Camargo-Zanoguera, A., and Suarez-Reynoso, G., 1994, Evidencia sismica del crater impacto de Chicxulub [Seismic evidence of the Chicxulub impact crater]: Boletín de la Asociación Mexicana de Geofisicos de Exploración, v. 34, p. 1–28 (in Spanish).

Croft, S. K., 1981, The modification stage of basin formation: conditions of ring formation, *in* Schultz P. H., and Merrill, R. B., eds., Multi-ring basins: New York, Pergamon Press, p. 227–257.

Cué, A. V., 1953, Determinacion de velocidades sismicas en el pozo Chicxulub [Determination of seismic velocity in the well Chicxulub-1]: Boletín de la Asociación Mexicana de Geólogos Petroles, v. 5, p. 285–290 (in Spanish).

Ewing J., Antoine, J., and Ewing, M., 1960, Geophysical measurements in the western Caribbean Sea and in the Gulf of Mexico: Journal of Geophysical Research, v. 65, p. 4087–4126.

Gardner, G. H. F., Gardner, L. W., and Gregory, A. R., 1974, Formation velocity and density—the diagnostic basics for formation traps: Geophysics, v. 39, p. 770–780.

Green, R. W., and Chetty, P., 1990, Seismic refraction studies in the basement of the Vredefort structure: Tectonophysics, v. 171, p. 105–113.

Hildebrand, A. R., 1997, Contrasting Chicxulub crater structural models: what can seismic velocities differentiate? Journal of Conference Proceedings, v. 1, p. 37–46.

Hildebrand, A. R., Penfield, G. T., Kring, D. A., Pilkington, M., Camargo., A. Z., Jacobsen, S. B., and Boynton, W. V., 1991, A possible Cretaceous-Terti-

ary boundary impact crater on the Yucatan peninsula, Mexico: Geology, v. 19, p. 867–871.
Masaitis, V. L., 1994, Impactites from Popigai crater, *in* Dressler, B. O., Grieve, R. A. F., and Sharpton, V. L., eds., Large meteorite impacts and planetary evolution: Geological Society of America Special Paper 293, p. 153–162.
Melosh, H. J., 1989, Impact cratering: a geologic process: New York, Oxford University Press, 245 p.
Morgan J. V., Warner, M. R., and the Chicxulub Working Group, 1997, Size and morphology of the Chicxulub impact crater: Nature, v. 390, p. 472–476.
Morgan J., and Warner, M., 1999, The third dimension of a multi-ring impact basin: Geology, v. 27, p. 407–410.
Penfield, G., and Camargo-Zanoguera, A., 1981, Definition of a major igneous zone in the central Yucatán platform with aeromagnetics and gravity, *in* Technical Program, Abstracts and Bibliographies, Society of Exploration Geophysicists, 51st Annual General Meeting: Tulsa, Oklahoma, Society of Exploration Geophysicists, p. 37.
Pilkington, M., and Grieve, R. A. F., 1992, Gravity and magnetic field modelling and structure of the Chicxulub Crater, Mexico: Reviews of Geophysics, v. 30, p. 161–181.
Pilkington, M., Hildebrand, A. R., and Ortiz-Aleman, C., 1994, Gravity and magnetic field modeling and structure of the Chicxulub crater, Mexico: Journal of Geophysical Research, v. 99, p. 13147–13162.
Pohl, J., Stöffler, D., Gall, H., and Ernstson, K, 1977, The Ries impact crater, *in* Roddy, D. J., Pepin, R. O., and Merrill, R. B., eds., Impact and explosion cratering: New York, Pergamon Press, p. 343–404.
Pope, K. O., Ocampo, A. C., and Duller, C. E., 1993, Surficial geology of the Chicxulub impact crater, Yucatan, Mexico: Earth, Moon and Planets, v. 63, p. 93–104.
Pratt, R. G., Song, Z.-M., Williamson, P., and Warner, M., 1996, Two-dimensional velocity models from wide-angle seismic data by wavefield inversion: Geophysical Journal International, v. 124, p. 323–341.
Sharpton, V. L., Dalrymple, G. B., Marin, L. E., Ryder, G., Schuraytz, B. C., and Urrutia-Fucugauchi, J., 1992, New links between the Chicxulub impact structure and the Cretaceous/Tertiary boundary: Nature, v. 359, p. 819–821.
Sharpton, V. L., Burke, K., Camargo-Z., A., Hall, S. A., Lee, S., Marin, L. E., Suarez-R., G., Quezada-M., J. M., Spudis, P. D., and Urrutia-F., J., 1993, Chicxulub multi-ring impact basin: size and other characteristics derived from gravity analysis: Science, v. 261, p. 1564–1567.
Sharpton, V. L., Marin, L. E., Carney, J. L., Lee, S., Ryder, G., Schuraytz, B. C., Sikora, P., and Spudis, P. D., 1996, Model of the Chicxulub impact basin, *in* Ryder, G., Fastovsky, D., and Gartner, S., eds., The Cretaceous-Tertiary event and other catastrophes in Earth history: Geological Society of America Special Paper 307, p. 55–74.
Swisher III, C. C., Grahales-Nishimura, J. M., Montanari, A., Margolis, S. V., Claeys, P., Alvarez, W., Renne, P., Cedillo-Pardo, E., Florentin, J.-M., Maurrasse, R., Curtis, G. H., Smit, J., and McWilliams, M. O., 1992, Coeval $^{40}Ar/^{39}Ar$ ages of 65.0 million years ago from Chicxulub crater melt rock and Cretaceous-Tertiary boundary tektites: Science, v. 257, p. 954–958.
Ward, W. C., Keller, G., Stinnesbeck, W., and Adatte, T., 1995, Yucatan subsurface stratigraphy: implications and constraints for the Chicxulub impact: Geology, v. 23, p. 873–876.
Zelt, C. A., and Smith, R. B., 1992, Seismic travel time inversion for 2-D crustal velocity structure: Geophysical Journal International, v. 108, p. 16–34.

MANUSCRIPT ACCEPTED BY THE SOCIETY DECEMBER 16, 1998

Morphology of the Chicxulub impact: Peak-ring crater or multi-ring basin?

Joanna Morgan and Mike Warner
T. H. Huxley School of Environment, Earth Sciences, and Engineering, Imperial College, London SW7 2BP, United Kingdom

ABSTRACT

We address the issue of whether the Chicxulub Crater has a peak-ring or multi-ring-basin morphology. To date, models of the crater morphology have been based on potential field data, a small number of onshore wells, the surface expression of the crater, and two marine seismic reflection profiles shot by Pemex along chords across the crater. Here we present four new seismic profiles, acquired in October 1996 as part of the Chicxulub seismic experiment. The advantage of these profiles is that they all extend to >150 km from the crater center, that they are deep profiles recorded to 18 s, and that two of the profiles are radial. These new data, along with some reprocessed Pemex data, provide the most comprehensive image to date of crater structure. In particular, they image the preimpact, Mesozoic stratigraphy as a series of bright subhorizontal reflectors, enabling us to track deformation across the crater. We use offsets in the Mesozoic stratigraphy to infer the surface topography across the crater immediately after the impact. The marine seismic data strongly suggest that Chicxulub has a multi-ring basin morphology; it contains a peak-ring and at least two widely separated rings showing inward-facing asymmetric scarps. The new seismic data provide the first direct evidence that Chicxulub Crater is a multi-ring impact basin, and provide the first direct evidence that such basins occur on Earth.

INTRODUCTION

The structure at Chicxulub is now widely accepted as the long-sought Cretaceous-Tertiary (K-T) impact site (Hildebrand et al., 1991; Pope et al., 1993; Sharpton et al., 1993; Camargo-Zanoguera and Suarez-Reynoso, 1994). The complexity of the potential field data and the small number of wells cored at isolated intervals have led to some ambiguity in determining both the crater size and morphology. The crater has been interpreted as an ~180-km peak-ring crater (Hildebrand et al., 1991; Pilkington et al., 1994; Espindola et al., 1995; Kring, 1995), an ~240-km peak-ring crater (Pope et al., 1993 & 1996), and a ~300-km multi-ring basin (Sharpton et al., 1993, 1996; Urrutia-Fucugauchi et al., 1996). The range of models highlights the problem of nonuniqueness inherent in modeling these data.

The mechanism of collapse in large craters is not well understood, but it is generally accepted that, in peak-ring craters, the peak-ring is formed by collapse of a gravitationally unstable central uplift (see Melosh, 1989, Fig. 8.14). In addition, structural rim uplift produced during the compressive stage of impact collapses to form a broad final crater, and, by definition, the outermost topographic high forms the crater rim. In multi-ring basins, at least one further inward-facing asymmetric scarp (a ring) forms either inside or outside the crater rim. Early studies of the moon and satellites of Jupiter revealed a number of large, old, multi-ring basins with several rings, and a plethora of models of ring formation were proposed (Hartman and Yale, 1969; Head, 1977; Hodges and Wilhelms, 1978; Melosh and McKinnon, 1978). It is unlikely that very large multi-ring basins of this nature exist on planets with a relatively high gravitational field such as

Morgan, J., and Warner, M., 1999, Morphology of the Chicxulub impact: Peak-ring crater or multi-ring basin? *in* Dressler, B. O., and Sharpton, V. L., eds., Large Meteorite Impacts and Planetary Evolution II: Boulder, Colorado, Geological Society of America Special Paper 339.

Earth. The high gravity causes reduced topography across the crater, and, for large craters, the melt volume becomes close to, or exceeds, the volume of the transient cavity (Grieve and Cintala, 1992). Grieve and Cintala (1992) have suggested that large impacts would produce palimpsest-like structures filled with impact melt, rather than a ringed crater topography.

Recent *Magellan* images have revealed four relatively young (<1 Ga) multi-ring basins on Venus (Mead, Isabella, Klenova, and Meitner), each with two rings; Isabella and Klenova also have clearly observable peak rings (Alexopoulos and McKinnon, 1994). The discovery of small (140–280-km diameter) multi-ring basins on Venus, and the fact that Earth and Venus have comparable gravitational fields, suggests that similar-size multi-ring basins might exist on Earth. On Venus the transition from peak-ring crater to multi-ring basin occurs between 110- and 140-km diameter (Alexopoulos and McKinnon, 1994). Hence, if the transition from peak-ring-crater to multi-ring-basin morphology were purely gravity driven, we would predict that craters on Earth of >140 km in diameter would be multi-ring basin. However, the Venusian near-surface is significantly hotter than Earth's, and it might be easier to form rings on Venus, assuming a viscosity-driven model for ring formation, e.g., the ring tectonic model of Melosh and McKinnon (1978).

There are three known craters on Earth with diameters greater than 140 km: Vredefort in South Africa, Sudbury in Canada, and Chicxulub in Mexico. The first two are both around 2 Ga old; Vredefort has been eroded to several kilometers beneath the crater floor, whereas Sudbury is severely tectonically deformed. In both cases, reconstructing the original crater structure is difficult (Therriault et al., 1993; Stöffler et al., 1994; Deutsch et al., 1995; Spray and Thompson, 1996; Henkel and Reimold, 1996), and their crater morphology is equivocal. The Chicxulub crater has been buried by Tertiary marls and carbonates, and lies on a tectonically quiet carbonate platform (Hildebrand et al., 1991; Camargo-Zanoguera and Suarez-Reynoso, 1994; Sharpton et al., 1996). It thus remains relatively pristine and is the best-preserved large crater known on Earth.

The existing K-T boundary at Chicxulub is unlikely to represent the original surface topography immediately after the impact. Before burial by Tertiary sediments, there would have been extensive erosion of any emergent topographic highs (the crater rim and any ring structure), and perhaps significant reworking of ejecta deposits. Clues to the original surface topography therefore must be sought in the deeper Mesozoic stratigraphy. In this chapter we present a detailed study of impact-related offsets in the Mesozoic stratigraphy, as revealed by high-resolution seismic reflection data, and relate these offsets to original surface topography. Once we have reconstructed the original topography, we can ascertain the crater morphology. If Chicxulub is a peak-ring crater it will possess a peak ring and a crater rim. If it is a multi-ring basin, it must meet three criteria: possess at least two distinct rings visible as asymmetric scarps at the topographic surface, the rings must be separated by a distance that is greater than that expected in terraces in peak-ring craters, and the rings must extend around a large proportion of the crater.

SEISMIC DATA

Burial of the Chicxulub crater, its location partly offshore, and the reflective preimpact stratigraphy make Chicxulub an ideal target for seismic investigation. As part of the 1996 British Institutions Reflection Profiling Syndicate (BIRPS) Chicxulub seismic experiment, we acquired ~650 km of marine reflection profile (Morgan et al., 1997; Snyder et al., this volume), recorded to 18 s along four lines (Fig. 1), together with a variety of onshore and offshore wide-angle seismic recordings (Brittan et al., this volume; Christeson et al., this volume). These data reveal the crater structure from the crater fill to the base of the crust. In addition we reprocessed Pemex seismic data that were acquired in 1992 (Camargo-Zanoguera and Suarez-Reynoso, 1994), and were able to improve the image in the shallow section at <2.0-s two-way travel-time (TWTT), which is critical for assessing crater morphology.

In Figures 2 through 5 we present the shallow seismic data; the crater center is always shown to the right of the section. We can track the Mesozoic stratigraphy as it becomes increasingly disturbed toward the crater center, and is downthrown to form a

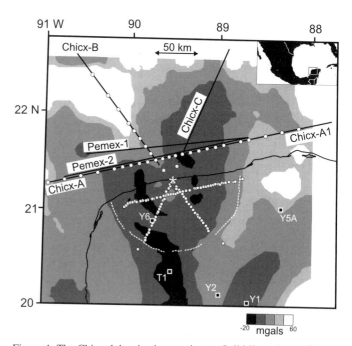

Figure 1. The Chicxulub seismic experiment. Solid lines show offshore reflection lines; white dots show wide-angle receivers. Shading shows Bouguer gravity anomaly; the crater is marked by a ~30-mgal circular gravity low. The dashed white line marks the position of the cenote ring. Squares show well locations. We calculate all radii using a nominal center at 89.54°W, 21.3°N, located by an asterisk. There is ambiguity in defining an exact center as the inner and outer gravity structures, the magnetic data, and the cenote ring all have slightly different centers (Pilkington et al., 1994; Sharpton et al., 1996; Pope et al., 1996).

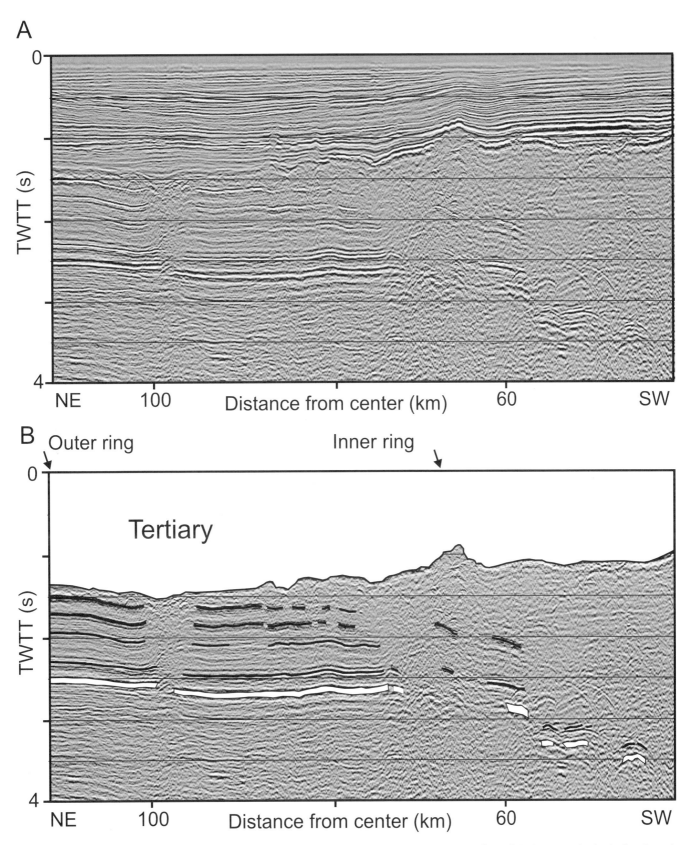

Figure 2. A, Seismic data along part of line Chicx-C. The data are conventionally processed, but not migrated. B, Interpreted seismic data from A, showing the Tertiary section, a line drawing highlighting a selection of Mesozoic reflectors, and a strong reflector at the base of the reflective section shaded white. Arrows indicate the location of the outer and inner rings.

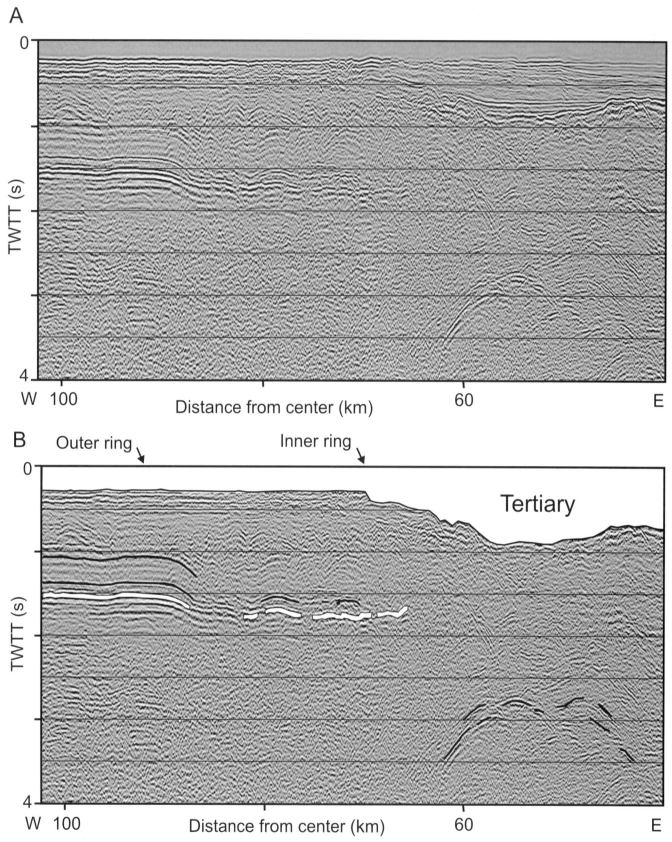

Figure 3. A, Seismic data along part of line Chicx-A. B, Interpreted seismic data from A, showing the Tertiary section, a line drawing highlighting a selection of Mesozoic reflectors, and a strong reflector at the base of the reflective section shaded white. Arrows indicate the location of the outer and inner rings.

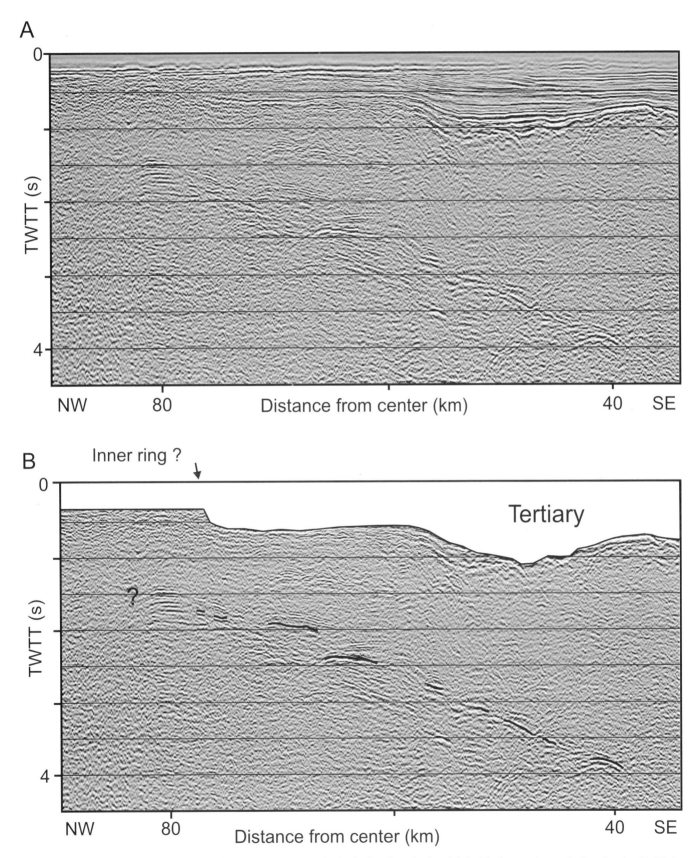

Figure 4. A, Seismic data along part of line Chicx-B. B, Interpreted seismic data from A, showing the Tertiary section, and a line drawing highlighting a selection of Mesozoic reflectors. Arrow indicates the location of the inner ring.

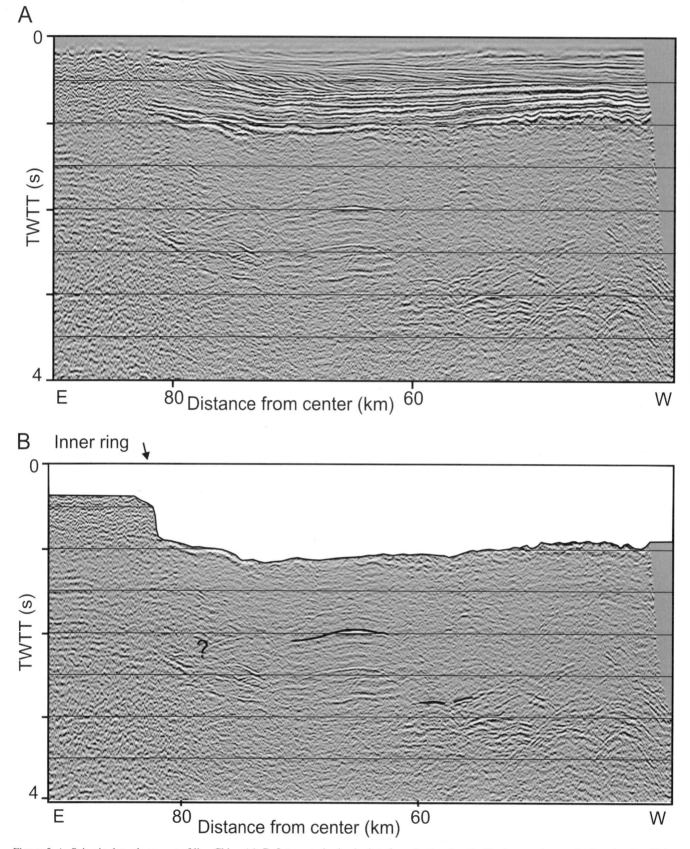

Figure 5. A, Seismic data along part of line Chicx-A1. B, Interpreted seismic data from A, showing the Tertiary section, and a line drawing highlighting a selection of Mesozoic reflectors. Arrow indicates the location of the inner ring.

series of slumped blocks. On several profiles we have identified a particularly bright reflection that occurs near the base of the reflective Mesozoic stratigraphy. This reflector lies at between 1.5- and 2.5-s TWTT at large radii, and at 3–4-s TWTT when it eventually disappears as it approaches a 40-km radius. Preimpact strata inside this diameter are interpreted as having been excavated (Morgan et al., 1997).

Tertiary section

In the central region of the crater, the Tertiary section is observable as a sequence of subhorizontal reflections from surface to ~1-s TWTT (Figs. 2–6). On all profiles except Chicx-C, the Tertiary sequence thins away from the crater center, suggesting a deep postimpact Tertiary basin with a radius of 65–83 km. This is in agreement with Tertiary thicknesses observed in onshore wells at comparable radii (Hildebrand et al., 1991; Camargo-Zanoguera and Suarez-Reynoso, 1994; Ward et al., 1995; Sharpton et al., 1996). In contrast, on the Chicx-C profile, the Tertiary basin thickens outward; on this line the edge of the deep basin is ~150 km from the crater center. Within the central, deep postimpact basin there are topographic highs that form a peak-ring with a diameter of ~80 km (Morgan et al., 1997; Snyder et al., this volume; Brittan et al., this volume).

Preimpact stratigraphy

On the Chicx-C profile, the outermost significant disturbance of the Mesozoic stratigraphy occurs at about 100-km radius, where we observe a fault-bounded asymmetric graben (Fig. 2A). This graben is linked to a fault or shear zone that dips at 30°–40° toward the crater center, and penetrates the whole crust (Morgan et al., 1997). The deformation of the Mesozoic stratigraphy occurs partly within the graben and partly across a monocline, producing a structural high in the uppermost Mesozoic stratigraphy at around 110-km radius. The Mesozoic stratigraphy remains relatively flat, between 100- and 75-km radius, the uppermost stratigraphy is initially around 150–200 ms (equivalent to 400–500 m) lower than the same stratigraphy at 110- km radius. Between 70- and 45-km radius, the Mesozoic rocks are downthrown to form a series of slumped blocks. The reflectivity of the Mesozoic stratigraphy, and, in particular, the high-amplitude low-frequency reflector at the base of the section (coloured white in Fig. 2B), allow us to track the stratigraphy across the faults relatively easily.

On Chicx-A/Pemex-2 the Mesozoic sequence is also reflective. The outermost significant disturbance of the Mesozoic stratigraphy occurs at around 122-km radius (Morgan et al., 1997; Snyder et al., this volume), and appears to be produced by outwardly directed thrusting. The next major offset in the Mesozoic stratigraphy occurs at around 87-km radius, where we observed a monocline that downdrops the stratigraphy toward the crater center (Fig. 3A). As in Chicx-C, the Mesozoic stratigraphy drops by about 200 ms (400–500 m) across the monocline, and

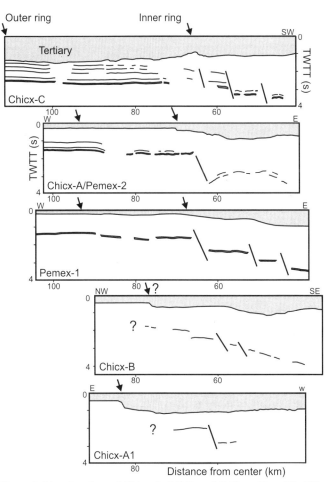

Figure 6. Line drawings of Chicx-A, A1, B, and C (from Figs. 2B–5B) and Pemex-1. Pemex-1 has not been reprocessed; our interpretation is based on the images shown in Camargo-Zanoguera and Suarez-Reynoso (1994). Arrows indicate the location of the outer and inner rings; straight lines are faults.

remains relatively flat for around 20 km. There is then a major offset in the stratigraphy at between 60- and 65-km radius, where the Mesozoic rocks are downthrown again toward the crater center. As seen on Chicx-C, there is a bright low-frequency reflector near the base of the section, which is easy to track from 150- to 65-km radius (shaded white in Fig. 3B).

The stratigraphy along Chicx-B is reflective between 40- and 82-km radius, but is only weakly reflective beyond this (Fig. 4A). Although it is clear in a general sense that the Mesozoic stratigraphy deepens from around 1.5-s TWTT at 80-km radius to 4-s TWTT at 40-km radius, it is difficult to track unequivocally a single reflector between these two points. We have attempted to pick these data (Fig 4B), but the lateral correlations remain uncertain. Similarly, the Mesozoic stratigraphy along Chicx-A1 is weakly reflective between 85 and 100 km (Fig. 5A), and it is difficult to track the stratigraphy across this seismically transparent zone. We can be certain, however, that there is a major offset in the Mesozoic rocks at around 62-km radius on this line (Fig. 5B).

We also observed a thrust in the Mesozoic stratigraphy similar to that observed on Pemex-2/Chicx-A, at around 128-km radius.

DISCUSSION

In Figure 6, we have plotted line drawings of the four lines shown in Figures 2B through 5B, together with Pemex-1. For Pemex-1 we have picked the same bright reflector as was picked in Chicx-A/Pemex-2 (see Fig. 3B). The offsets in the Mesozoic stratigraphy along Pemex-1 are comparable to those observed on Chicx-A/Pemex 2, although the innermost inward slumping occurs over several faults, as it does on Chicx-B and C.

In the top three profiles we observed a large separation between an outer fault and an inner zone of intense slumping. On Chicx-A1 we observed a major offset at 62-km radius that correlates well with the onset of intense slumping observed on the other profiles; on Chicx-B the zone of intense slumping appears to start at a larger radius. The offsets we observed in the Mesozoic rocks would have been observable in the crater topography (immediately following the impact), provided that the offsets are large enough to not be entirely obscured by ballistic ejecta deposits, and that the top of the footwall remained above the crater fill.

In Figure 7 we have plotted a cartoon of the postimpact crater topography along profiles Chicx-C, Chicx-A/Pemex-2, and Pemex-1. The reflective Mesozoic stratigraphy (colored dark gray) represents the bright reflector near the base of the Mesozoic rocks on these three profiles. On each profile there is an outer isolated fault with an offset of 400–500 m. Primary ejecta thicknesses are difficult to determine accurately, but with an apparent transient cavity diameter of ~100 km (Morgan et al., 1997), they should be less than 400 m at these ranges. Thus, these outer offsets would not have been obscured by ejecta deposits, even if the faulting occurred entirely before any ejecta deposition, and would have produced an inward-facing asymmetric scarp at the surface (hereafter called the outer ring).

As we move inward, the Mesozoic stratigraphy is relatively flat, although there is some indication on all three lines that the Mesozoic rocks shallow as they approach the zone of intense slumping. Using velocity models determined by Brittan et al. and Christeson et al. (this volume), the basal reflector on Chicx-A lies at ~4 km at 90-km radius, drops ~0.5 km just inboard of the monocline, and rises to ~4.2 km at 65-km radius. If this impact had occurred on a dry planet, this gentle increase in elevation of the Mesozoic rocks between 85- and 65-km radius would have been accentuated by ejecta that thicken toward the crater center. Thus, just outboard of the zone of intense slumping, the Mesozoic stratigraphy would have been <200 m lower than the same stratigraphy immediately outboard of the outer ring.

At present, the flat crater floor at Chicxulub lies at about 600–900-m beneath the K-T boundary outside the crater (Hildebrand et al., 1991; Camargo-Zanoguera and Suarez-Reynoso, 1994; Ward et al., 1995; Sharpton et al., 1996). The deepest slumped blocks would therefore have been covered by the impact

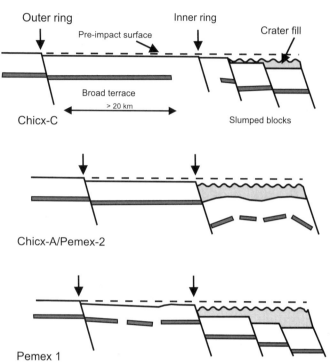

Figure 7. Cartoon of topography across the crater immediately following the impact (excluding primary and secondary ejecta) along Chicx-C, Chicx-A and Pemex-1. The peak-ring is not shown. The crater fill material would normally form a flat floor; it is drawn with an undulating surface to indicate that we are unsure of its precise topographic position.

deposits (labeled crater fill in Fig. 7) that form the crater floor, but the head scarp of the outermost fault would have been visible as a major offset at the surface. On all three profiles, this head scarp would have appeared as an inward-facing asymmetric scarp at surface, an inner ring (see Fig. 7). The separation between the inner and outer rings is much larger than that of individual terraces in peak-ring craters. These three profiles show multi-ring basin topography; they satisfy the first two criteria outlined in the introduction. The third condition that must be satisfied, to determine if Chicxulub is truly a multi-ring basin, is that the rings must extend around a significant portion of the crater.

In Figure 8 we have plotted the location of the rings over the Bouguer gravity data. The rings have been picked from seismic data discussed here, and from a subset of shallow proprietary seismic data acquired by Pemex across the crater. The peak-ring in Figure 8 is picked on the topographic high. The inner and outer rings have been picked from offsets in the bright reflector near the base of the Mesozoic stratigraphy, and thus mark the location of the rings at depth. To determine their location at the surface, immediately following the impact, we must extrapolate up the scarps (see direction of arrows in Fig. 6). The original location of the inner and outer rings at the surface would thus have been several kilometers outboard of that shown in Figure. 8 Our best estimates of the original ring diameters place the inner ring at ~145 km, and the outer ring at ~200 km.

Because of the Mesozoic stratigraphy along Chicx-A1 and

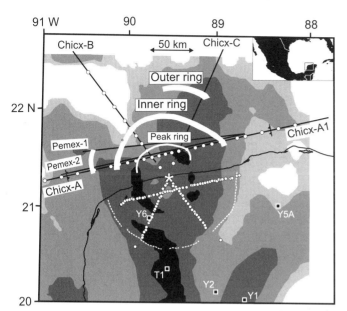

Figure 8. Location of rings (in white) over the Bouguer gravity anomaly. The innermost ring marks the topographic high of the peak ring. The middle and outermost rings are the inner and outer rings shown in Figure 6. The rings on this map represent their location at depth and have been picked directly from the seismic data. The location of the rings on the postimpact surface would be several kilometers outboard of this. Outside of the outer ring we observe some disruption of the Mesozoic stratigraphy. On Chicx-A and A1 we see two outwardly directed thrusts, shown by black lines with triangles, which appear to penetrate the entire crust and offset the Mesozoic stratigraphy (Morgan et al., 1997; Snyder et al., this volume) at 122 and 128 km, respectively. Assuming these features were not covered by ejecta, they would have produced outward-facing asymmetric scarps at surface, and are thus not directly analogous to the rings observed on the moon and Venus. They may, however, be analogous to the rings seen on Ganymede and Callisto.

B being poorly imaged outside the deep Tertiary basin and the lack of onshore seismic data, we cannot be certain at present that the rings at Chicxulub do indeed extend right around the crater. We are certain, however, that all the seismic lines available that do show a well-imaged Mesozoic section (i.e., all those capable of revealing the shallow structure) also show multi-ring-basin morphology. There are no well-defined craters observed on other terrestrial planets that show an obvious multi-ring morphology in some directions but not in others. The existing seismic evidence therefore directly and strongly supports the hypothesis that Chicxulub is a multi-ringed impact basin, and we propose that Chicxulub should be regarded as the first well-founded example of a multi-ring impact basin on Earth. To test this definitively, however, will require additional data, especially seismic and/or drill core, from those parts of the crater that are at present unexplored at high resolution.

ACKNOWLEDGEMENTS

The seismic reflection data from lines D92-RP001 and D92-RP002 were provided by Petroleos Méxicanos via the Universidad Nacional Autónoma de México. Morgan was supported by a fellowship from the Leverhulme Trust. The Chicxulub Seismic Experiment was funded by the Natural Environment Research Council, the U.S. National Science Foundation, the British Institutions Reflection Profiling Syndicate Industrial Associate Programme, the Royal Commission for the Exhibition of 1851, the Royal Society, and Leicester and London Universities. The Chicxulub seismic reflection data were acquired by Geco-Prakla and processed by Bedford Interactive Processing Services.

REFERENCES CITED

Alexopoulos, J. S., and McKinnon, W. B., 1994, Large impact craters and basins on Venus, with implications for ring mechanics on the terrestrial planets, in Dressler, B. O., Grieve, R. A. F., and Sharpton, V. L., eds., Large meteorite impacts and planetary evolution: Geological Society of America Special Paper 293, p. 29–50.

Camargo-Zanoguera, A., and Suarez-Reynoso, G., 1994, Evidencia Sismica del crater impacto de Chicxulub [Seismic evidence of the Chicxulub impact crater]: Boletín de la Asociación Mexicana de Geofisicos de Exploracion, v. 34, p. 1–28 (in Spanish).

Deutsch, A., Grieve, R. A. F., Avermann, M., Bischoff, L., Brockmeyer, P., Buhl, D., Lakomy, R., Müller-Mohr, V., Ostermann, M., and Stöffler, D., 1995, The Sudbury structure (Ontario, Canada): a tectonically deformed multiring impact basin: Geologische Rundschau, v. 84, p. 697–709.

Espindola, J. M., Mena, M., de la Fuente, M., and Campos-Enriquez, J. O., 1995, A model of the Chicxulub impact structure (Yucatan, Mexico) based on gravity and magnetic signatures: Physics of the Earth and Planetary Interiors, v. 92, p. 271–278.

Grieve, R. A. F., and Cintala, M. J., 1992, An analysis of differential impact melt crater scaling and implications for the terrestrial impact record: Meteorics, v. 27, p. 526–538.

Hartmann, W. K., and Yale, F. G., 1969, Mare Orientale and its intriguing basin: Sky and Telescope, v. 37, p. 4–7.

Head, J. W., 1977, Origin of outer rings in lunar multi-ringed basins: evidence from morphology and ring spacing, in Roddy, D. J., Pepin, R. O., and Merrill, R. B., eds., Impact and explosion cratering: New York, Pergamon Press, p. 563–573.

Henkel, H., and Reimold, W. U., 1997, Integrated gravity and magnetic modelling of the Vredefort impact structure—reinterpretation of the Witwatersrand basin as the erosional remnant of an impact basin: Stockholm, Sweden, The Royal Institute of Technology Information Circular 100, 90 p.

Hildebrand, A. R., Penfield, G. T., Kring, D. A., Pilkington, M., Camargo, A. Z., Jacobsen, S. B., and Boynton, W. V., 1991, A possible Cretaceous-Tertiary boundary impact crater on the Yucatan peninsula, Mexico: Geology, v. 19, p. 867–871.

Hodges, C. A., and Wilhelms, D. E.,1978, Formation of rings in lunar multi-ringed basins: Icarus, v. 34, p. 294–323.

Kring, D. A., 1995, The dimensions of the Chicxulub impact crater and impact melt sheet: Journal of Geophysical Research, v. 100, p. 16979–16986.

Melosh, H. J., 1989, Impact cratering: a geologic process: New York, Oxford University Press, 245 p.

Melosh, H. J., and McKinnon, W. B., 1978, The mechanics of ringed basin formation: Geophysical Research Letters, v. 5, p. 985-988.

Morgan, J. V., Warner, M. R., and the Chicxulub Working Group, 1997, Size and morphology of the Chicxulub impact crater: Nature, v. 390, p. 472–476.

Pilkington, M., Hildebrand, A. R., and Ortiz-Aleman, C., 1994, Gravity and magnetic field modelling and structure of the Chicxulub crater, Mexico: Journal of Geophysical Research, v. 99, p. 13147–13162.

Pope, K. O., Ocampo, A. C., and Duller, C. E., 1993, Surficial geology of the Chicxulub impact crater, Yucatan, Mexico: Earth, Moon and Planets, v. 63, p. 93–104.

Pope, K. O., Ocampo, A. C., Kinsland, G. L., and Smith, R., 1996, Surface expression of the Chicxulub crater: Geology, v. 24, p. 527–530.

Sharpton, V. L., Burke, K., Camargo-Z., A., Hall, S. A., Lee, S., Marin, L. E., Suarez-R., G., Quezada-M., J. M., Spudis, P. D., and Urrutia-F., J., 1993, Chicxulub multi-ring impact basin: size and other characteristics derived from gravity analysis: Science, v. 261, p. 1564–1567.

Sharpton, V. L., Marin, L. E., Carney, J. L., Lee, S., Ryder, G., Schuraytz, B.C., Sikora, P., and Spudis, P. D., 1996, Model of the Chicxulub impact basin, in Ryder, G., Fastovsky, D., and Gartner, S., eds., The Cretaceous Tertiary event and other Catastrophes in Earth History: Geological Society of America Special Paper 307, p. 55–74.

Spray, J. G., and Thompson, L. M., 1996, Friction melt distribution in a multi-ring impact basin: Nature, v. 373, p. 130–132.

Stöffler, D., Deutsch, A., Avermann, M., Bischoff, L., Brockmeyer, P., Buhl, D., Lakomy, R., and Müller-Mohr, V., 1994, The formation of the Sudbury structure, Canada: toward a unified impact model, in Dressler, B. O., Grieve, R. A. F., and Sharpton, V. L., eds., Large meteorite impacts and planetary evolution: Geological Society of America Special Paper 293, p. 303–318.

Therriault, A. M., Reid, A. M., and Reimold, W. U., 1993, Original size of the Vredefort structure, South Africa [abs.]: Lunar and Planetary Science, v. 24, p. 1419–1420.

Urrutia-Fucugauchi, J., Marín, L., and Trejo-Garcia, A., 1996, UNAM scientific drilling program of the Chicxulub impact structure—evidence for a 300 kilometre crater diameter: Geophysical Research Letters, v. 23, p. 1565–1568.

Ward, W. C., Keller, G., Stinnesbeck, W., and Adatte, T., 1995, Yucatan subsurface stratigraphy: implications and constraints for the Chicxulub impact: Geology, v. 23, p. 873–876.

MANUSCRIPT ACCEPTED BY THE SOCIETY DECEMBER 16, 1998

Upper crustal structure of the Chicxulub impact crater from wide-angle ocean bottom seismograph data

G. L. Christeson, R. T. Buffler, Y. Nakamura
Institute for Geophysics, 4412 Spicewood Springs Road, Building 600, University of Texas at Austin, Austin, Texas 78759, USA

ABSTRACT

We used travel times recorded by 23 ocean bottom seismograph (OBS) receivers to model crustal structure of the Chicxulub impact crater. The OBSs were located along Chicx-A/A1, a deep reflection profile extending as a chord across the offshore portion of the Chicxulub structure. A tomographic inversion of crustal refractions was used to produce the upper crustal structure. The prominent feature in the shallow velocity structure is the low velocities associated with the Tertiary basin. The low velocity section is thickest near the center of the crater, and thins on the flanks. High velocities observed near the surface along several portions of the crater flanks may represent the widespread presence of reefal buildups at the margins of the Tertiary basin. Negative velocity anomalies observed beneath the Tertiary basin within the collapsed transient cavity are interpreted as melt rocks. The largest negative anomaly is located near the center of Chicx-A/A1 (~26 km from the interpreted center of the crater), and extends laterally 50 km. The depth to the top of this anomaly is about 1 km; its average thickness is 1.3 km. Two smaller anomalies are located near the eastern and western margins of the collapsed transient cavity. The average thickness of these three negative velocity anomalies, when averaged over the width of the collapsed transient cavity, is 1 km.

INTRODUCTION

The Chicxulub structure, located along the northern Yucatán Peninsula in Mexico, is now widely believed to represent a large Cretaceous-Tertiary boundary impact (e.g., Hildebrand et al., 1991; Sharpton et al., 1992; Swisher et al., 1992; Pope et al., 1993). The structure was first identified in the late 1960s as a circular feature on gravity (Fig. 1) and magnetic data, and was subsequently drilled at several locations to depths of 1,500–3,000 m by PEMEX (Lopez-Ramos, 1975; Ward et al., 1995). The lithologies observed in these exploration wells (bunte breccia, suevite breccia, melt, and melt-matrix breccia) (Hildebrand et al., 1991; Ward et al., 1995; Sharpton et al., 1996) are clearly consistent with an impact origin for the Chicxulub structure.

The size and internal geometry of this structure are still relatively poorly constrained by potential field and drill-hole data. Some researchers advocate a 180-km peak-ring structure (Hildebrand et al., 1991, 1995; Pilkington et al., 1994; Connors et al., 1996; Hildebrand, 1997); some, a 240-km peak ring structure (Pope et al., 1993, 1996); and others, a 300-km multi-ring basin (Sharpton et al., 1993, 1994, 1996). The 180-km peak-ring model of Hildebrand (1997) includes a 90-km diameter collapsed transient cavity filled with a 3-km-thick central melt pool. Underlying the melt pool is a thick (up to 13 km) region of megabreccia

---

Christeson, G. L., Buffler, R. T., and Nakamura, Y., 1999, Upper crustal structure of the Chicxulub impact crater from wide-angle ocean bottom seismograph data, *in* Dressler, B. O., and Sharpton, V. L., eds., Large Meteorite Impacts and Planetary Evolution II: Boulder, Colorado, Geological Society of America Special Paper 339

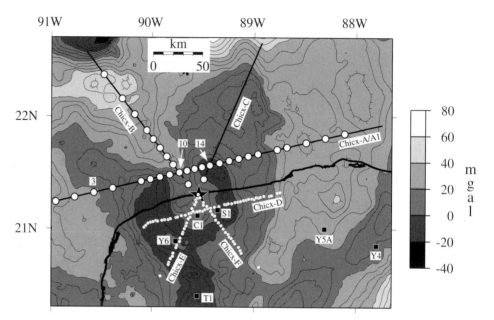

Figure 1. Location of the Chicxulub seismic experiment. Deep reflection profiles are shown by the solid lines. Large white circles mark positions of offshore OBS receivers, and smaller white circles mark positions of onshore PASSCAL receivers. Record sections for OBSs 3, 10, and 14 (labeled on figure) are shown in Figure 2. Star at coastline marks approximate center of crater. Squares show deep (>1.5 km) well locations. Gravity data courtesy of A. Hildebrand and M. Pilkington, Geological Survey of Canada.

surrounding a 40-km diameter central uplift with an uplift magnitude of 18 km. In contrast, the 300-km multi-ring basin model of Sharpton et al. (1994, 1996) has a 170-km diameter collapsed transient cavity with a 120-km diameter central structural uplift. The central uplift forms a peak ring instead of the central peak of the Hildebrand (1997) model, and has an uplift magnitude of approximately 30 km. The melt sheet has a width similar to that of Hildebrand (1997), but a thickness of <1 km. The transient cavity is filled with impact breccias that thicken to ~2.5 km in a trough surrounding the central uplift.

The large density contrasts forming the prominent gravity anomalies of the Chicxulub structure (e.g., Pilkington et al., 1994; Sharpton et al., 1996) should also be associated with large changes in velocity (Pilkington and Grieve, 1992; Hildebrand, 1997). Thus, velocity information can be used to distinguish among various features in the potential field models and to provide new constraints on the Chicxulub impact structure. This chapter presents new velocity results from a recent seismic experiment carried out by an international group of collaborators over the offshore portion of the Chicxulub impact structure (Morgan et al., 1997; Brittan et al., this volume; Morgan and Wagner, this volume; Snyder and Hobbs, this volume). Another chapter in this volume (Brittan et al., this volume) uses arrivals recorded by a multichannel streamer to constrain the shallow structure across the impact crater; this chapter complements that work by using data recorded by an array of ocean bottom seismograph (OBS) receivers to constrain the deeper crustal structure along one profile across the Chicxulub crater.

DATA AND INTERPRETATION

The Chicxulub seismic experiment consisted of ~650 km of deep reflection profiles acquired by the British Institutions Reflection Profiling Syndicate (Morgan et al., 1997; Snyder and Hobbs, this volume), with simultaneous wide angle data recording by OBS and land receivers (Fig. 1). This chapter presents analysis of OBS data recorded by 23 instruments along Line A/A1, a profile that extends as a chord across the offshore portion of the Chicxulub structure. The recording instrument, a University of Texas OBS, has three-component gimbaled geophones with 126-dB dynamic range. The seismic lines were shot with a 36-element airgun array, source volume of 9,162 in^3, at a shot spacing of 50 m (Snyder and Hobbs, this volume). Processing of OBS data included applying a correction for clock drift during deployment and inverting the water wave and upper sediment arrivals for instrument location and orientation. Average errors in instrument location are estimated to be <50 m, and timing errors after clock drift corrections are <10 ms.

Data quality for all instruments was excellent. Sample record sections for three OBS receivers are shown in Figure 2. Differences in structure beneath the three instruments are evident from the travel times of the observed arrivals. For example, the crustal refraction for OBS 10, located within the crater, reaches reduced travel times >1 s at offsets <15 km. In contrast, that for OBS 3, located on the western flank of the crater, reaches reduced travel times >1 s at an offset of ~60 km. These differences in crustal refraction travel times indicate that a greater volume of low-velocity materials is present beneath OBS 10 than beneath OBS 3. In the next section we use the crustal refraction travel times to constrain crustal structure beneath the Chicx-A/A1 profile.

Upper crustal structure

We used the tomographic code of Hole (1992) to invert for upper crustal structure. This code is an iterative nonlinear inversion that uses a finite difference travel-time scheme (Vidale,

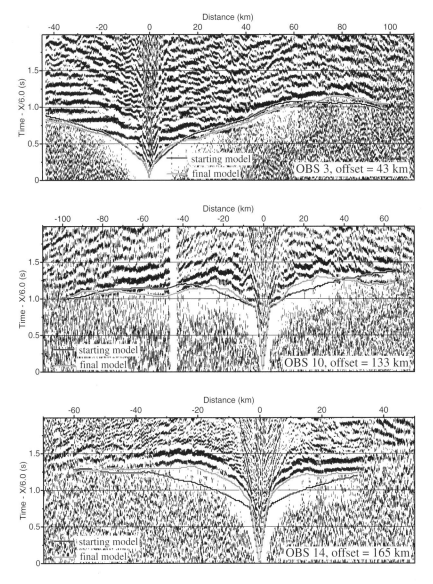

Figure 2. Sample record sections for three sample OBS receivers (see Fig. 1 for instrument location). Record sections are plotted with a reduction velocity of 6 km/s; data have been passed through a Butterworth filter (low-cut 3 Hz, high-cut 15 Hz, 48 dB/octave rolloff) and have a 1-s automatic gain control applied. Travel time curves for the two models shown in Figure 3 are plotted for the offsets over which observed arrivals were picked. Note the differences in crustal refraction travel times at near offsets for the three instruments.

1990; Hole and Zelt, 1995) for the forward step and simple back-projection (Hole, 1992) for the inversion step. Approximately 24,000 first-arrival travel times were picked from the 23 OBS record sections; picks were made for every trace for offsets <12 km and every third trace for offsets >12 km. We estimate the average picking error to be approximately 25 ms.

The starting model for the tomographic inversion was constructed using the following steps: (1) Travel-time picks from the coincident MCS shot gathers (Brittan et al., this volume) were merged with picks from OBS record sections at offsets <12 km and used to invert for the shallow velocity structure to a depth of 3 km using the Zelt and Smith (1992) two-dimensional modeling code in a manner similar to that described by Brittan et al. (this volume). However, velocities in the lowermost layer (velocities >5 km/s) were not allowed to vary laterally. (2) The velocity model was extended to a depth of 15 km. Velocity gradients from depths of 3–15 km were chosen to minimize the residual in the travel-time picks. (3) The water layer (thickness <30 m) was removed, and the velocity model was gridded using a 100-m spacing to create the starting model for tomographic inversion (Fig. 3A).

The iterative tomographic inversion of Hole (1992) was carried out until the root-mean-square residual between observed and calculated travel times decreased from a starting value of 85 ms to the approximate picking error of 25 ms; this took six iterations. The primary feature of the resulting velocity model (Fig. 3B) is lowered velocities within the crater as compared to velocities at the same depths for the crater flanks. These lowered velocities are interpreted as the Tertiary impact basin (Brittan et al., this volume). The velocity anomaly between the final and starting models is plotted in Figure 3C. The largest anomaly is located near the center of the crater, below the base of the Tertiary basin; velocities within this anomaly are as much as 0.7 km/s lower than that of the starting model.

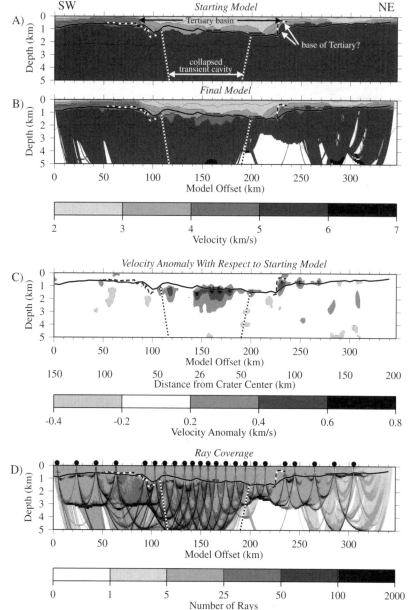

Figure 3. A, Starting model for the tomographic inversion. Depth to the base of the Tertiary basin, as interpreted from the reflection profile, is shown by the black line; an alternate interpretation along the crater flanks is indicated by the dashed lines and represents mapping a slightly shallower horizon on the flanks of the basin. The dotted lines mark the collapsed transient cavity, which is interpreted from the coincident seismic reflection profile (Morgan et al., 1997; Brittan et al., this volume). B, Velocity model for Chicx A/A1 obtained from tomographic inversion. The model is shown only where constrained by wide-angle data. C, Velocity anomalies, calculated between the starting and final models of the tomographic inversion. Note the negative velocity anomalies located beneath the Tertiary basin that we interpret as melt bodies. D, Ray coverage for the tomographic model shown in Figure 3B. (For comparison purposes with the Chicx-A/A1 reflection data presented by Snyder and Hobbs [this volume], shotpoint = 987 + model_offset/0.05.)

A qualitative measure of the need for the anomalies plotted in Figure 3C can be obtained by comparing the travel-time curves for the starting and final models (Fig. 2). For an instrument located on the western crater flank, there are only small differences between the travel-time curves of the two models. For OBS 10, located at model offset 133 km, both travel-time curves agree at near offsets, but the travel-time curve for the starting model is too early for shots at distances 10–30 km both east and west of the instrument (with the discrepancy greater to the east). These travel-time residuals help construct the negative velocity anomalies both east and west of the receiver. OBS 14 is located directly over the large negative anomaly; travel-time curves for the starting model at this instrument are too early by several hundred milliseconds at most offsets.

What is the resolution of the tomographic modeling? One measure of resolution is ray coverage, which is plotted in Figure 3D. The ray coverage indicates that structure to a depth of 5 km might be resolved in the central portion of the model, but only the upper 3 km will be resolved on the flanks. Another measure of resolution can be made by using test anomalies. In Figure 4 we show the results of two test inversions. For one test we added three 0.75-km-thick, −0.5-km/s anomalies to the starting model at similar positions to the anomalies observed in our tomographic model. Travel times were then calculated through this model using the identical source-receiver geometry as that used in our experiment. Random noise (10–30 ms, with larger noise at longer offsets) was added to the calculated travel times, which were then used with the starting model to invert for velocity structure

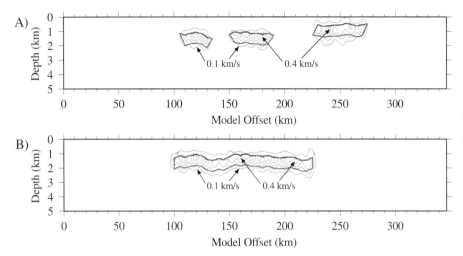

Figure 4. Resolution tests. Gray lines outline –0.5 km/s anomalies added to the starting model shown in Figure 3A. Black lines show results of the tomographic inversions contoured at 0.1-km/s intervals (starting model–inversion model). Note that the inversions underestimate the magnitude of the anomalies, but that the contours >0.2 km/s resolve the dimensions of the test anomalies.

(Fig. 4A). As is typical with tomography, the resulting anomalies show smearing, with anomalies of larger area but lesser magnitude recovered in comparison to the actual model. However, if velocity anomalies <0.2 km/s are disregarded, the dimensions of the original anomalies are recovered. A similar test was done for an anomaly extending the width of the Tertiary basin (Fig. 4B), and again velocity anomalies ≥0.2 km/s recover the dimensions of the original anomaly. These resolution tests indicate that the negative velocity anomalies beneath the Tertiary basin and on its eastern flank are real. However, the results in Figure 4 do indicate that the lateral variability within the large negative anomaly located at the center of the model (alternating 0.4–0.6- and 0.6–0.8-km/s anomalies) is an artifact of the experimental geometry.

Central crater structure

Gravity (Fig. 5C) was computed over the velocity model using a standard velocity to density conversion (Ludwig et al., 1970). This density conversion is somewhat arbitrary, but it does allow for an indication of expected gravity over the profile. To the east, the velocity contrast at model offset 225 km produces a large gravity gradient that fits well with the observed gravity values. To the west the expected gravity gradient at model offset 90 km does not fit the observed gravity data; however, the gravity anomalies in this region appear to be affected by a large northwest-trending gravity high (Fig. 1). Within the central portion of the basin, no velocity contrasts are great enough to provide a good fit to the central gravity high and surrounding gravity low observed in the gravity data. However, there are some small

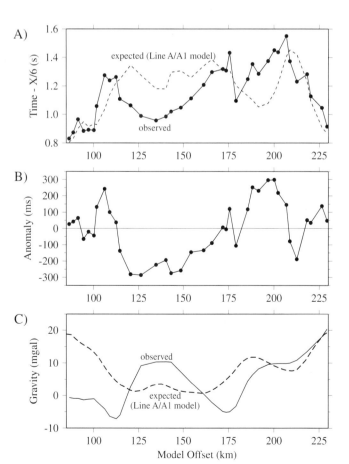

Figure 5. A, Expected (dashed line) and observed (solid line, with circles) travel times for receivers along Chicx-D from shots along Chicx-A/A1; only travel times at minimum distance are shown. Expected travel times are calculated by extending the Chicx-A/A1 velocity model through the crater center to Chicx-D. Travel times are plotted with a reduction velocity of 6 km/s, which largely removes the effect of differing shot to receiver ranges for the Chicx-D receivers. B, Travel time anomaly (observed-expected) of Figure 5A. C. Average gravity value (solid line) between Chicx-D and Chicx-A, and calculated (dashed line) gravity along Chicx A/A1. Calculated gravity values are for the velocity model shown in Figure 3B using a density conversion discussed in the text. Note the general correlation between gravity and travel time anomalies between 100 and 200 km model offset – travel times are later than expected over gravity lows and earlier than expected over the central gravity high.

anomalies (0.25–0.5 km/s) in the 1.5–2.5-km depth range that generally correlate with the central high and surrounding lows. We suspect that the primary structure producing the large central gravity anomalies lies shoreward of Chicx-A/A1.

Unfortunately, the location of the crater center approximately on the coastline (Fig. 1) precluded the possibility of a marine crater center profile. Chicx-A/A1 was positioned as close to the crater center as possible, in 20 m of water, but still remains ~26 km from the crater center at its closest approach. For this reason Chicx-D, a land profile parallel to Chicx-A/A1, was located such that the midpoints of shots from Chicx-A/A1 recorded by Chicx-D form a profile across the center of the crater. A full analysis of the Chicx-D data is beyond the scope of this chapter; however, we can obtain some hints about central crater structure by comparing Chicx-D travel times with estimated travel times calculated by extending the Chicx-A/A1 velocity model across the crater to Chicx-D. These travel times are plotted in Figure 5A for shots at minimum distance (42–45 km) from Chicx-A/A1 to Chicx-D. The observed travel times have a general pattern of fast-slow-fast-slow, fast arrivals across the array, with a spread of 620 ms in reduced travel times. Shown in Figure 5C is the average gravity value between Chicx-D and Chicx-A, calculated along the path between Chicx-D receiver and closest Chicx-A/A1 shot. Although not perfect, there is a general correlation between gravity and travel-time anomalies (Fig. 5B) between 100- and 200-km model offset, with travel times later than expected over gravity lows and earlier than expected over the central gravity high. This suggests that much of the structure responsible for the large central gravity anomalies is located between Chicx-A/A1 and Chicx-D.

DISCUSSION

Shallow velocity structure

The velocity model for the shallow part of the Chicxulub crater presented here along Chicx-A/A1 based on the OBS data fits well with what is known about the shallow geology of the region based on both drill holes and preliminary interpretations of seismic reflection profiles. Following the KT impact, the crater and surrounding area were progressively covered by a layer of Tertiary sediments. The center and deeper part of the crater is filled with approximately 1,000 to 1,500 m of sediments. Several wells drilled through this section (C1, S1, and Y6 in Fig. 1) suggest that these sediments are relatively deep-water marls and limestones filling a deeper water basin (Lopez-Ramos, 1975). This area presumably was a crater or depression that had been formed by the impact and subsequent modifications (e.g., slumping, backwash). This interpretation is supported by the new seismic reflection data (Morgan et al., 1997; Morgan and Warner, this volume; Snyder and Hobbs, this volume), as well as by two older PEMEX seismic reflection lines (Camargo-Zanoguera and Suarez-Reynoso, 1994). The seismic facies consists of generally uniform, continuous reflections with a prograding geometry that indicates a progressive filling of the basin from all sides.

This geometry suggests that shallow-water shelfal environments progressively migrated from the flanks of the basin to the center during the Tertiary. The base of the Tertiary section is interpreted on the reflection data as the base of the uniform reflections (black line in Fig. 3B) and forms the boundary with the underlying crater materials. This boundary is best constrained within the crater between model offsets 85–225 km, and is more speculative on the crater flanks. The base of the Tertiary section also has been identified in several of the deep wells drilled onshore in the center (~1,000–1100 m in Y6, C1) and on the flanks (~200–500 m in T1, YSA, Y4) of the basin (Ward et al., 1995; Sharpton et al., 1996). This supports the general geometry of the basin as interpreted on the seismic lines.

The wells drilled onshore also support the idea that the thinner (300–500 m) Tertiary section on the flanks of the crater are equivalent shallow-water carbonate shelfal deposits (Lopez-Ramos, 1975), and probably includes reefs similar to the modern-day reef buildups on the present shelf (i.e., Arrecife Alacrañ). This interpretation is supported by the seismic reflection data, which show a corresponding change in seismic facies at the basin margin (clinoform breakpoint) from the continuous prograding reflections (basinal sediments) to a more discontinuous facies typical of a platform setting with mixed environments. This change occurs near model offsets 85 and 225 km on the west and east flanks, respectively.

This geologic setting can be used to explain the observed shallow-velocity structure (Fig. 3). In the center of the basin (model offsets, 85–225 km), the velocities at the base of the Tertiary section range from 3 to 4 km/s, which would fit well with the buried and more consolidated older Tertiary section drilled. The overlying slower velocities (2–3 km/s) would correspond to younger (late Tertiary?) deeper water basin fill (more unconsolidated limestones and marls). The slightly lower velocities in the shallowest part of the basinal section (2.2–2.3 km/s) would correspond to the youngest and most unconsolidated basin fill.

Along the crater flanks (model offsets <85 and >225 km), the shallow velocity structure is more variable than observed in the center of the crater, with velocities >3 km/s reaching close to the seafloor. We speculate that these shallow high velocities represent the widespread presence of reefal buildups within the Tertiary section at the margins of the basin (and perhaps also on the eastern end of the profile). The lower velocities observed elsewhere on the flanks could represent lateral facies changes to lower velocity backreef deposits. The thin, low-velocity zone at the seafloor (2–3 km/s) probably corresponds to the youngest, more unconsolidated Tertiary section, which in places could be cavernous or contain cenotes (as observed onshore) formed during late Cenozoic drops in sea level.

Melt layer

Three deep onshore wells located within 50 km of the crater center (wells C1, S1, and Y6 in Fig. 1) have encountered a suevite breccia (mixed melt glass and shocked lithic fragments) at depths

~1 km below sea level (Lopez-Ramos, 1975; Ward et al., 1995; Sharpton et al., 1996). This unit is approximately 300 m thick, and is underlain by impact melt and melt-matrix breccia of undetermined thickness at wells C1 and S1; well Y6 intersected 380 m of andesitic melt rock before bottoming in 6–8 m of laminated anhydrite (Lopez-Ramos, 1975; Schuraytz et al., 1994; Sharpton et al., 1996). Sharpton et al. (1996) have argued that these melt rocks may represent local pods of melt, possibly within a thick suevite unit, rather than a thick coherent melt sheet. The melt matrix is fine-grained, unlike the coarse-grained, igneously layered impact melt rocks observed at the 2.5–4 km thick Sudbury Igneous Complex, which could suggest that the total yield of impact melt was much less than observed at the Sudbury Crater (Warren et al., 1994; Sharpton et al., 1996). In contrast, Pilkington et al. (1994) argued that the origin of the high-amplitude magnetic field anomalies that outline an area over the Chicxulub structure with an average radius of 45 km is due to the presence of a central melt pool with a thickness of approximately 3 km. In their model, the Y6 well is located where the impact melt has flowed outward over slumped blocks and the outer breccia.

Melt rock densities measured at several craters show negative density contrasts with the surrounding target rocks (Pilkington et al., 1994). Since density and velocity anomalies are usually correlated, we would expect melt rocks to be represented by negative velocity anomalies in Figure 3. Indeed, several large (>0.25 km/s) negative velocity anomalies are observed beneath the base of the Tertiary basin, extending to depths of 2–3 km (Fig. 3C). The depth to the top of these anomalies is 1,000–1,250 m, similar to the modeled depth of 1,100 m to the top of the magnetic source bodies (Penfield and Camargo Z., 1981). The largest velocity anomaly extends approximately 50 km from model offsets 140 to 190 km, with an average thickness of 1.3 km. It is located near the center of the collapsed transient cavity. Smaller velocity anomalies, with lateral extents of 15–20 km, are located near the western and eastern margins of the collapsed transient cavity. There are several negative velocity anomalies located on the eastern flank of the crater, between model offsets 225–270 km. These anomalies are located within the interpreted Tertiary section, and may be related to structure within the reefal buildup on this crater flank. Alternatively, if the dashed line in Figure 3 is the base of the Tertiary and the solid line the top of Cretaceous (an interpretation some of our colleagues favor), then the velocity anomalies might be related to lateral variability in the breccia unit.

If the negative velocity anomalies between model offsets 100–200 km represent melt rock, then it appears that a coherent melt sheet extending the width of the transient cavity is not present. Instead, the melt is concentrated near the center of the collapsed transient cavity, with smaller pockets located along the eastern and western walls of the transient cavity. The average thickness of the negative velocity anomalies along Chicx-A/A1, if spread out over the entire transient cavity, would be about 1 km. The velocity model does not preclude the existence of a thicker coherent melt pool located closer to the crater center, and indeed the observed travel-time arrivals along Chicx-D (Fig. 5) indicate a lateral variability in crustal structure from Chicx-A/A1 to Chicx-D.

Basement uplift?

Geologic studies of terrestrial complex craters show that the central peak uplift is composed of deformed and fractured rocks that originally underlay the transient crater, and that the amount of uplift is comparable to the initial crater depth (Wilshire and Howard, 1968; Pohl et al., 1977; Melosh, 1989). Peak-ring craters are thought to form when the central uplift collapses; basement uplift may remain beneath the crater center (e.g., Melosh, 1989). Although the amount and lateral extent differ, basement uplift is a component of most deep structural models for Chicxulub (e.g., Pilkington et al., 1994; Sharpton et al., 1996; Hildebrand, 1997), and is envoked to explain the prominent central gravity high. Our velocity model shows no indication of basement uplift (Fig. 3); however, it is not apparent that basement uplift should produce an observable velocity contrast. No velocity contrast is observed at depths of 2,000–4,000 m on the crater flanks (Fig. 3) where basement is expected according to the onshore wells (Lopez-Ramos, 1975; Ward et al., 1995). It is probable that basement velocities do not significantly differ from the high-velocity Mesozoic carbonate platform rocks, and thus basement uplift cannot be constrained from seismic refraction data (or at least not with an analysis of first-arrival travel times alone).

Alternatively, deep central uplift may be present but similar to the model of Pilkington et al. (1994), have a radius <26 km, and thus not extend to Chicx-A/A1. This could be the explanation for the early arrivals observed from Chicx-A/A1 shots to Chicx-D receivers located near the center of the crater (Fig. 5), and future analysis of these data may provide additional constraints on central uplift.

SUMMARY OF CONCLUSIONS

1. The prominent feature in the shallow-velocity structure is low velocities associated with the Tertiary basin. The low-velocity section is thickest at the center of the crater, and thins on the flanks. High velocities observed near the surface along several portions of the crater flanks may represent the widespread presence of reefal buildups at the margins of the Tertiary basin. The large velocity contrast between the Tertiary basin and a reefal buildup along its eastern margin is associated with a prominent gravity gradient along the profile.

2. Negative velocity anomalies observed beneath the Tertiary basin within the collapsed transient cavity are interpreted as melt rocks. The largest negative anomaly is located near the center of Chicx-A/A1 (~26 km from the interpreted center of the crater), and extends laterally 50 km. The depth to the top of this anomaly is about 1 km, its average thickness is 1.3 km. Two smaller anomalies are located near the eastern and western margins of the collapsed transient cavity. The average thickness of the three negative velocity anomalies, when averaged over the width of the collapsed transient cavity, is 1 km.

3. There is no evidence for large central uplift under the Chicx-A/A1 profile. This is either because central uplift has a radius <25 km and does not reach Chicx-A/A1, or because no velocity contrast is associated with the deeper uplifted rocks.

ACKNOWLEDGMENTS

We would like to thank the Chicxulub Working Group (J. Brittan, R. Buffler, A. Camargo, G. Christeson, P. Denton, A. Hildebrand, R. Hobbs, H. Macintyre, G. Mackenzie, P. Maguire, L. Marin, J. Morgan, Y. Nakamura, M. Pilkington, V. Sharpton, D. Snyder, G. Suarez, A. Trejo, and M. Warner) for their assistance in carrying out this project and their discussions about the significance of the velocity results. We thank the captain, crew, and science party of the R/V Longhorn for their expert aid in collecting these data. John Hole kindly provided his tomographic inversion code. Dale Sawyer and an anonymous reviewer provided constructive criticism that improved this manuscript. This research was supported by National Science Foundation Grant OCE-9415716. This is University of Texas Institute for Geophysics Contribution 1393.

REFERENCES CITED

Camargo-Zanoguera, A., and Suarez-Reynoso, G., 1994, Evidencia sismica del crater impacto de Chicxulub [Seismic evidence of the Chicxulub impact crater]: Boletín de la Asociación Mexicana de Geofisicos de Exploracion, v. 34, p. 1–28.

Connors, M., Hildebrand, A. R., Pilkington, M., Ortiz-Aleman, C., Chavez, R. E., Urrutia-Fucugauchi, J., Graniel-Castro, E., Camara-Zi, A., Vasquez, J., and Halpenny, J. F., 1996, Yucatán karst features and the size of Chicxulub crater: Geophysical Journal International, v. 127, p. F11–F14.

Hildebrand, A., 1997, Contrasting Chicxulub crater structural models: what can seismic velocity studies differentiate?: Journal of Conference Proceedings, v. 1, p. 37–46.

Hildebrand, A. R., Penfield, G. T., Kring, D. A., Pilkington, M., Camargo, Z. A., Jacobsen, S. B., and Boynton, W. V., 1991, Chicxulub Crater: a possible Cretaceous/Tertiary boundary impact crater on the Yucatán Peninsula, Mexico: Geology, v. 19, p. 867–871.

Hildebrand, A. R., Pilkington, M., Connors, M., Ortiz-Aleman, C., and Chavez, R. E., 1995, Size and structure of the Chicxulub crater revealed by horizontal gravity gradients and cenotes: Nature, v. 376, p. 415–417.

Hole, J. A., 1992, Nonlinear high-resolution three-dimensional seismic travel time tomography: Journal of Geophysical Research, v. 97, p. 6553–6562.

Hole, J. A., and Zelt, B. C., 1995, Three-dimensional finite-difference reflection travel times: Geophysical Journal International, v. 121, p. 427–434.

Lopez-Ramos, E., 1975, Geological summary of the Yucatan Peninsula, in Nairn, A. E. M., and Stehli, F. G., eds., The ocean basins and margins, Vol. 3, The Gulf of Mexico and the Caribbean: New York and London, Plenum Press, p. 257–282.

Ludwig, W. J., Nafe, J. E., and Drake, C. L., 1970, Seismic refraction, in Maxwell, A. E., ed., The sea, vol. 4, pt. 1: New York, Wiley-Interscience, p. 53–84.

Melosh, H. J., 1989, Impact cratering: a geological process: New York, Oxford University Press, 245 p.

Morgan, J., Warner, M., and Chicxulub Working Group, 1997, Size and morphology of the Chicxulub impact crater: Nature, v. 390, p. 472–476.

Penfield, G. T., and Camargo Z., A., 1981, Definition of a major igneous zone in the central Yucatán platform with aeromagnetics and gravity: Society of Exploration Geophysicists, 51st Annual International Meeting, Los Angeles, Abstracts, p. 37.

Pilkington, M., and Grieve, R. A. F., 1992, The geophysical signature of terrestrial impact craters: Reviews of Geophysics, v. 30, p. 161–181.

Pilkington, M., Hildebrand, A. R., and Ortiz-Aleman, C., 1994, Gravity and magnetic field modeling and structure of the Chicxulub Crater, Mexico: Journal of Geophysical Research, v. 99, p. 13147–13162.

Pohl, J., Stöffler, D., Gall, H., and Ernstson, K., 1977, The Ries impact crater, in Roddy, D. J., Pepin, R. O., and Merrill, R. B., eds., Impact and explosion cratering: New York, Pergamon Press, p. 343–404.

Pope, K. O., Ocampo, A. C., and Duller, C. E., 1993, Surficial geology of the Chicxulub impact crater, Yucatan, Mexico: Earth, Moon, and Planets, v. 63, p. 93–104.

Pope, K. O., Ocampo, A. C., Kinsland, G. L., and Smith, R., 1996, Surface expression of the Chicxulub crater: Geology, v. 24, p. 527–530.

Schuraytz, B. C., Sharpton, V. L., and Marín, L. E., 1994, Petrology of impact-melt rocks at the Chicxulub multiring basin, Yucatán, Mexico: Geology, v. 22, p. 868–872.

Sharpton, V. L., Dalyrymple, G. B., Marin, L. E., Ryder, G., Schuraytz, B. C., and Urrutia-Fucugauchi, J., 1992, New links between the Chicxulub impact structure and the Cretaceous/Tertiary boundary: Nature, v. 359, p. 819–821.

Sharpton, V. L., Burke, K., Camargo-Zanoguera, A., Hall, S. A., Lee, D. S., Marín, L. E., Suárez-Reynoso, G., Quezada-Muñeton, J. M., Spudis, P. D., and Urrutia-Fucugauchi, J., 1993, Chicxulub multiring impact basin: size and other characteristics derived from gravity analysis: Science, v. 261, p. 1564–1567.

Sharpton, V. L., Marín, L. E., and Schuraytz, B. C., 1994, The Chicxulub multiring impact basin: evaluation of geophysical data, well logs, and drill core samples [abs.] in Conference on New Developments Regarding the KT Event and Other Catastrophes in Earth History: Houston, Texas, Lunar and Planetary Institute, Contribution 825, p. 108–112.

Sharpton, V. L., Marín, L. E., Carney, C., Lee, S., Ryder, G., Schuraytz, B. C., Sikora, P., and Spudis, P. D., 1996, A model of the Chicxulub impact basin based on evaluation of geophysical data, well logs, and drill core samples, in Ryder, G., and Fastovsky, D., and Schultz, P., eds., The Cretaceous-Tertiary event and other catastrophes in Earth history: Geological Society of America Special Paper 307, p. 55–74.

Swisher, C. C., III, Grajales-Nishimura, J. M., Montanari, A., Margolis, S. V., Claeys, P., Alvarez, W., Renne, P., Cedillo-Pardo, E., Maurrasse, F. J.-M. R., Curtis, G. H., Smit, J., and McWilliams, M. O., 1992, Coeval $^{40}Ar/^{39}Ar$ ages of 65.0 million years ago from Chicxulub crater melt rock and Cretaceous-Tertiary boundary tektites: Science, v. 257, p. 954–958.

Vidale, J. E., 1990, Finite-difference calculation of traveltimes in three dimensions: Geophysics, v. 55, p. 521–526.

Ward, W. C., Keller, G., Stinnesbeck, W., and Adatte, T., 1995, Yucatán subsurface stratigraphy: implications and constraints for the Chicxulub impact: Geology, v. 23, p. 873–876.

Warren, P. H., Claeys, P., and Cedillo-Pardo, E., 1994, Where are the Chicxulub coarse-grained, igneously layered impact melt rocks analogous to those at Sudbury? [abs.], in Conference on New Developments Regarding the KT Event and Other Catastrophes in Earth History: Houston, Texas, Lunar and Planetary Institute, Contribution 825 p. 128–130.

Wilshire, H. G., and Howard, K. A., 1968, Structural pattern in central uplifts of cryptoexplosion structures as typified by Sierra Madera: Science, v. 162, p. 258–261.

Zelt, C. A., and Smith, R. B., 1992, Seismic traveltime inversion for 2-D crustal velocity structure: Geophysical Journal International, v. 108, p. 16–34.

MANUSCRIPT ACCEPTED BY THE SOCIETY DECEMBER 16, 1998

Sudbury Structure 1997: A persistent enigma

B. O. Dressler and V. L. Sharpton
Lunar and Planetary Institute, Houston, Texas 77058, USA

ABSTRACT

It took almost 30 years for the origin of the Sudbury Structure as the result of asteroid or comet impact to gain wide acceptance in the geosciences community. Most field and laboratory observations can be reconciled with an impact origin. However, there is disagreement among proponents of the impact hypothesis as to the interpretations of several field observations and laboratory results. And there is no agreement on such issues as the size of the structure; the origin of the Sudbury Igneous Complex, including the Sublayer and mineral deposits; the origin and distribution of Sudbury pseudotachylites (Sudbury Breccia); the provenance of carbon in the rocks filling the Sudbury Basin; the presence or absence of fullerenes in these basin rocks; and the meaning of high-precision U-Pb geochronological dates of the Sudbury Igneous Complex.

INTRODUCTION

Geologic research on the Sudbury Structure, consisting of the Sudbury Basin, the Sudbury Igneous Complex, and its brecciated and shock metamorphosed footwall rocks, has been going on for more than a century; a voluminous literature exists that deals with the various rock units of the structure. Interest in the origin of the Sudbury Igneous Complex and the world-renowned nickel-copper deposits at its base, of the breccias within the Sudbury Basin, of the carbon in these breccias and the mudstones overlying them, and of high-pressure deformational features and widespread pseudotachylites in the footwall rocks of the Sudbury Igneous Complex is as vivid as it was almost 100 yr ago. Sudbury has attracted the interest of several generations of geoscientists, and there is probably no living geologist who is not aware of the existence of this geologically enigmatic structure.

Prior to the advances made in the understanding of planetary impact processes (French and Short, 1968; Melosh, 1989, and references therein), the various features of the Sudbury Structure were interpreted to be the result of endogenic, i.e., magmatic and volcanic, processes (e.g., Coleman, 1905; Williams, 1957; Stevenson, 1972; Naldrett and Hewins, 1984). Nearly 35 yr ago, however, a major new interpretation was made. In a landmark paper, R. S. Dietz (1964) suggested that the Sudbury Structure is the result of a meteorite impact. At this time this hypothesis was revolutionary and, not surprisingly, took almost 30 yr to gain wide acceptance in the geoscience community. Shatter cones in the rocks around the Sudbury Igneous Complex (Dietz, 1964), shock metamorphic features in the footwall rocks and in the clasts of the Onaping Formation within the Sudbury Basin (French, 1968; Dence, 1972; Dressler, 1984), and the formation of pseudotachylite breccias in the rocks around and up to a distance of ~80 km from the Sudbury Igneous Complex all support an asteroid or comet impact origin for the structure.

However, there are features that, until a few years ago, were seen as evidence for an endogenic origin. Among them are the noncircular shape of the Sudbury Igneous Complex (most impact structures are approximately circular); the occurrence of breccia-in-breccia clasts in the heterolithic breccias of the Onaping Formation, suggesting a prolonged sequence of different brecciation processes; and sharp contacts between various breccia units of the Onaping Formation, an observation also thought to be in conflict with a single event such as meteorite impact (Muir, 1984). Research over the last 15 yr, however, has shown that the Sudbury Structure probably was originally circular (Shanks and Schwerdtner, 1990) and that the presence of breccia-in-breccia clasts in breccia deposits (Dressler and Sharpton, 1997) do not represent unequivocal evidence against an impact origin of the structure. In general, it has been shown over the last two decades

Dressler, B. O., and Sharpton, V. L., 1999, Sudbury Structure 1997: A persistent enigma, *in* Dressler, B. O., and Sharpton, V. L., eds., Large Meteorite Impacts and Planetary Evolution II: Boulder, Colorado, Geological Society of America Special Paper 339.

that Sudbury field and laboratory observations can be reconciled with an impact origin.

However, this does not mean that all features of Sudbury geology are now well understood. Even among Sudbury researchers who favor the impact origin, there is disagreement as to the interpretation of a number of observations. There is still no consensus on (1) the size of the Sudbury Structure; (2) the origin of the Sudbury Igneous Complex and the associated Sublayer and ore deposits; (3) the origin and distribution of pseudotachylites, i.e., the Sudbury Breccias; (4) the provenance of the carbonaceous matter, including fullerenes, in the rocks of the Whitewater Group within the Sudbury Basin; and (5) the meaning of high precision U-Pb geochronological dates obtained from the various units of the Sudbury Igneous Complex. Prior to the Sudbury 1997 Conference on Large Meteorite Impacts and Planetary Evolution, several of these issues were being discussed in the literature. A few of them are briefly addressed in this introductory note. A summary account on the Sudbury papers included in this Special Paper and on the papers presented at the conference and suggestions for future research are given by Naldrett in the final chapter of this volume.

SOME 1997 UNSOLVED PROBLEMS IN SUDBURY GEOLOGY

Size of the Sudbury Structure

One of the fundamental parameters of an impact structure is its original rim diameter. Knowing a structure's size allows a rough estimate of the energy responsible for the formation of the structure. Several attempts to estimate the size of the Sudbury Structure have been made in the past. The original rim of the structure is no longer preserved and when the impact origin was first proposed, the size of the impact structure was considered to be the size of the Sudbury Igneous Complex, which has a long axis of ~60 km, and a short axis of ~25 km. Since then, with the increased acceptance of the impact origin and a better understanding of the cratering process, it has become generally accepted that the shock metamorphosed and brecciated footwall rocks around the Sudbury Igneous Complex form an integral part of the original Sudbury Structure, thus enlarging the structure considerably. There is now general agreement that the original structure was much larger than 60 km. The size estimates in Table 1 are based on modern cratering models, that is, on shatter cone distribution, shock attenuation studies, and the distribution of breccias up to ~80 km away from the lower contact of the Sudbury Igneous Complex. These estimates are rough approximations at best and do not take into account deformation and erosion since the formation of the structure. Based on paleomagnetic studies, Schwarz and Buchan (1982) estimated that erosion since the intrusion of the 1.2-Ga-old Sudbury olivine diabase dike swarm amounted to ~9 km. Therefore, erosion possibly was ~15 km since the Sudbury Impact 1.85-Ga ago, assuming that the rate of erosion was similar between 1.2 and 1.85 Ga. The Sudbury Structure is strongly deformed and, based on interpreta-

TABLE 1. SIZE OF THE SUDBURY STRUCTURE

Transient Crater Diameter (km)	Rim Diameter (km)	References
70	190	Peredery and Morrison, 1994
(100)	190	Dressler et al., 1987
100	180 – 200	Lakomy, 1990
100	150 – 200	Grieve et al., 1991
100 – 140	200 – 280	Deutsch et al., 1995

tion of vibroseismic data (Milkereit et al., 1992; Wu et al., 1994), it may be considerably larger than presently thought. Overall, estimates of the rim diameter of the Sudbury Structure are presently little constrained.

Origin of the Sudbury Igneous Complex and its ore deposits

One of the issues that has attracted much interest ever since the discovery of the Sudbury Structure is the origin of the Sudbury Igneous Complex and associated sulfide ore deposits. This elliptical body consists, from bottom to top, of inclusion-rich, in places ore-bearing, quartz diorite (Sublayer), norite, quartz gabbro, and granophyre layers, and, within the target rocks surrounding the Sudbury Igneous Complex, the quartz dioritic Offset dikes. The complex has been variously considered as a differentiated, crust-contaminated magma (Naldrett and Hewins, 1984), a differentiated impact melt sheet (Faggart et al., 1985; Grieve et al., 1991; Deutsch, 1994), or possibly a combination of these two alternatives (Dence, 1972; Shanks et al., 1990; Chai and Eckstrand, 1993, 1994; Johns and Dressler, 1995; Dressler et al., 1996). It has been long known that the complex is geochemically and mineralogically different from other large terrestrial mafic bodies (Naldrett and Hewins, 1984, and references therein).

Isotopic investigations provide evidence that the entire Sudbury Igneous Complex, including the Sublayer and its ores, represent a single unit of impact melt (Faggart et al., 1985; Grieve et al., 1991; Dickin et al., 1996, and references therein). However, there are field observations that suggest a longer sequence of igneous activity, that the Sublayer consists of several postnorite intrusive phases, and that the norite and quartz gabbro were emplaced after the granophyre. There are several locations where the granophyre and footwall rocks beneath the Sudbury Igneous Complex exhibit solid state deformation, where the entire Sudbury Igneous Complex is cut by major deformation zones, and where the norite and quartz gabbro (between the undeformed granophyre and the footwall) are virtually undeformed (Dressler, 1987; Shanks et al., 1990; Johns and Dressler, 1995). Therefore, the entire Sudbury Igneous Complex is possibly not a simple, single differentiated crustal impact melt sheet. Based on the petrogenetic significance of chromium spinels from the inclusions in the Sublayer, Zhou et al. (1997) have suggested that, subsequent to the formation of a crustal melt, mantle-derived high-Mg magmas mixed with this crustal melt. On the basis of structural observations, Cowan (1996) and Riller (1996) have questioned the impact melt sheet origin of the Sudbury Igneous Complex (see Cowan et al., this volume). Geochemical considerations and models, some structural observations, and the interpretations of several intrusive relationships are presently not reconcilable within a one-impact–sheet model for the origin of the Sudbury Igneous Complex.

Similarly, there is no agreement as to the position and origin of the Sublayer and its associated ores within an impact model. In the early years of Sudbury investigations, Bell (1891a,b) and Barlow (1904, 1906) suggested that the ores formed as a result of differentiation from a magma. A hydrothermal origin for the ores was forwarded by Knight (1917) and Yates (1948), while Howe (1914) and Bateman (1917) described the ores as sulfide melt injections. Several similar scenarios for the genesis of the Sudbury ore deposits were presented in later years before Dietz (1972) suggested that the ores were possibly cosmogenic and related to a splash-emplaced Sublayer. There is, however, isotopic evidence that the Sublayer and its ores are of crustal origin (Dickin et al., 1996, and references therein) and that the Sublayer is not necessarily part of a simple differentiated impact melt sheet. Dressler (1982) and Naldrett et al. (1984) have suggested, as has Yates (1948), that the Sublayer and its associated sulfide deposits were emplaced after the main body of the Sudbury Igneous Complex. This is believed to be supported by specific field relationships, e.g., the presence of Sublayer dikes intruding the norite. Again, chemical considerations and models are presently not reconcilable with the interpretations of a number of field observations. A new look at present models for the origin of both the Main Mass (norite, quartz gabbro, and granophyre) of the Sudbury Igneous Complex and of the Sublayer coupled with new field investigations is needed to further a better understanding of the Sudbury Structure.

Origin and distribution of Sudbury pseudotachylite

The footwall rocks around the Sudbury Igneous Complex contain the pseudotachylites long-known as Sudbury Breccias. They are very similar to the pseudotachylites of the Vredefort impact structure, in South Africa, the pseudotachylite type location (Reimold and Colliston, 1994, and references therein). They have been the subject of numerous investigations (e.g., Fairbairn and Robson, 1942; Speers, 1957; Dupuis et al., 1982; Dressler, 1984; Müller-Mohr, 1992a,b; Spray and Thompson, 1995). There is now general agreement that the formation of the Sudbury pseudotachylites is related to the impact origin of the Sudbury Structure, a conclusion supported by the spatial relationship of the breccia occurrences around the Sudbury Igneous Complex. Dressler (1984) has shown that the pseudotachylites are abundant within a distance of ~5–10 km from the Igneous Complex. Field evidence also exists for the presence of zones of increased pseudotachylite development ~20–25 km and ~80 km away (Dressler, 1984; Peredery and Morrison, 1984). Between these zones, breccia bodies are less abundant and smaller. It has been suggested that these zones of stronger brecciation may be related to multiple ring systems of "multi-ring basins" where brecciation should be particularly intense (F. Hörz, Johnson Space Center, Houston, Texas, personal communication, 1983; Brockmeyer, 1990; Spray and Thompson, 1995).

Because of the lack of continuous outcrop, however, it is not clear that these zones of stronger brecciation are indeed continuous about the Sudbury Igneous Complex, as has been proposed by Spray and Thompson (1995). Sudbury Breccia dikes strike randomly and dip vertically or steeply (Dressler, 1984). The only known large continuous breccia body, the >14-km-long Frood-Stobie zone (Speers, 1957; Dressler, 1982; Dressler et al., 1992; Müller-Mohr, 1992a,b; Spray and Thompson, 1995) is not concentric with the Sudbury Igneous Complex, but may be related to the formation of terraces around the transient crater (Spray, 1997). It is generally accepted that Sudbury Breccias originated by brittle fracture and comminution during the Sudbury impact event leading, in places, to partial or complete frictional melting of crushed rocks. However, further field research is required to test old and new ideas on the formation mechanism of large pseudotachylite breccia bodies, including their distribution, orientation, and relationship to multi-ring basin formation.

Carbonaceous matter and fullerenes in the rocks of the Sudbury Basin

Another major enigma at Sudbury is the large amount of carbon in the impact-produced breccias and overlying mudstones of the Whitewater Group within the Sudbury Basin. This situation has been difficult to interpret even before the impact origin was considered and still contrasts strongly with the carbon-poor or carbon-free character of virtually all other known terrestrial impact breccia deposits. As much as 1% elemental carbon has been found in the heterolithic impact breccias of the Black Member of the Onaping Formation. The mudstones of the Onwatin Formation contain considerably more, i.e., up to 7% carbon. All this amounts to an enormous ~10^{14} kg of elemental carbon present in the two formations, which raises the question of the origin of the carbon. Several potential sources are being proposed in two of the papers included in this Special Paper (Bunch et al. and Heymann et al.). Among them are the projectile, the Precambrian target rocks, the atmosphere above ground zero, fumarolic/hydrothermal activity, and biogenic activity.

In this context, the recent discovery by Becker et al. (1994) of fullerenes in the heterolithic breccias of the Black Member of the Onaping Formation is of considerable significance. An equally epoch-making discovery of extraterrestrial helium in the fullerenes (Becker et al., 1996) further raised the interest of geoscientists and planetologists in the carbonaceous matter found in the rocks of the Sudbury Basin. However, not all researchers looking for fullerenes were successful finding these large carbon molecules in the breccias of the Black Member (D. Heymann, Rice University, Houston, Texas, personal communications, 1996, 1997), but if they are present, they were either derived from the impacting projectile or were synthesized within the impact plume from carbon originally contained in the projectile, the atmosphere, or the target rocks. In either case, they must have survived postimpact thermal overprint and low-grade greenschist facies metamorphism. The presence of extraterrestrial helium would support a cosmogenic origin for the Sudbury fullerenes. Independent third-party analytical investigations are needed to establish the presence and abundance or the absence of fullerenes in the rocks of the Sudbury Basin.

Geochronology

The Sudbury Igneous Complex is among the best-dated terrestrial igneous bodies. Most of its units have been dated by precise U-Pb methods on zircon and baddeleyite at about 1.85 Ga (Krogh et al., 1984). Corfu and Lightfoot (1996) have recently documented that the mafic/ultramafic inclusions in the Sublayer also have this 1.85-Ga age. Furthermore, Early Proterozoic, pre-Sudbury–impact anorthosites at the lower contact of the Sudbury Igneous Complex (S. Prevec, Laurentian University, Sudbury, Ontario, Canada, personal communication, 1996) and Archean granitic rocks at the contact with the Foy Offset (A. Deutsch, University of Münster, Germany, personal communication, 1994) have been reset to this 1.85-Ga U-Pb age. The precise and uniform dates of the Sudbury Igneous Complex appear to support a one-time, one-magma origin for all phases of the Sudbury Igneous Complex. However, the mafic/ultramafic rocks found as inclusions in the Sublayer must have formed prior to their incorporation in the Sublayer magma and the emplacement of the Sublayer at the base of the norite of the Sudbury Igneous Complex. Although some contact features suggest a postnorite emplacement of the Sublayer (Dressler, 1982), the length of the time gaps between the formation of the mafic/ultramafic rocks, their incorporation as inclusions in the Sublayer, and the emplacement of the Sublayer is unknown. It is also conceivable that the mafic/ultramafic inclusions acquired their 1.85-Ga age within the Sublayer magma and that they are xenoliths derived from rocks older than the Igneous Complex. Isotopic evidence, however, demonstrates that the mafic/ultramafic inclusions in the Sublayer are not derived from target rocks but from primitive, moderately Mg-rich Sudbury Igneous Complex magmas (Lightfoot et al., 1997). The mafic/ultramafic inclusions are virtually unaltered and lack shock metamorphic features, also suggesting that the inclusions did not originate from target rocks.

The minerals dated in the Sudbury Igneous Complex are zircon and baddeleyite, zircon in the norite and baddeleyite in the granophyre (Krogh et al., 1984). As refractory minerals they formed early in the differentiation process. In the opinion of some researchers, some field evidence (see above) exists for lithification and solid-state deformation of the granophyre before emplacement of the lower phases of the Sudbury Igneous Complex. The length of time for the granophyre to solidify is not known. However, if the interpretation of the field observations is correct and a substantial time gap occurred, allowing solidification of the granophyre before emplacement of the norite, one has to accept that norite and Sublayer magmas containing 1.85-Ga-old refractory zircon crystals intruded at the base of the lithified granophyre containing baddeleyite, also 1.85-Ga old. Such a sequence of events is in conflict with present interpretations of geochemical data (e.g., Faggart et al., 1985; Deutsch, 1994; Dickin et al., 1996). However, a somewhat revised impact-melt model can perhaps accommodate the conflicting data and observations. In it, all the Sudbury Igneous Complex was formed due to impact melting. The lower phases of the Sudbury Igneous Complex may represent an impact melt formed through melting of mainly deep-crustal mafic target rocks or an endogenic, impact-triggered, strongly crustally contaminated magma. The granophyre may represent an impact melt of upper-crustal, silica-rich rocks (Dressler et al., 1996). It did not mix with the deep-crustal, more mafic melt that resided at depth, possibly beneath the excavation cavity, before it intruded underneath the mostly lithified granophyre during a late phase of the Penokean orogeny. This mafic melt may contain components derived from mantle-derived high-Mg magmas (Zhou et al., 1997). As stated above, some field observations appear to be in conflict with laboratory results, geochemical models, and single-impact melt models. The new model, somewhat modified from Dressler et al. (1996) and briefly described above, satisfies most data and observations. Additional laboratory data and field observations, however, are needed to test it and various other models for the origin of the Sudbury Igneous Complex and the meaning of the high-precision geochronological results on the various phases of the complex.

CONCLUDING REMARKS

After more than a century of geoscientific investigations, the major geologic features of the Sudbury Structure are well established. Over the last 10–15 yr, since the wide acceptance of its impact origin and since a major academia-industry-government cooperative research program culminating in the publication of the Ontario Geological Survey's volume on the Sudbury Structure (Pye et al., 1984), research on the structure has increased considerably. Many previously controversial field and laboratory observations have been reconciled with an impact origin for the Sudbury Structure. However, despite the impressive amount of research accomplished, new problems and controversies continue to arise. The Sudbury Igneous Complex (as does the Sudbury Structure as a whole) "furnishes a most interesting case of the painfully slow, caterpillar-like, yet logical way in which we grope our way to an understanding of big and intricate geologic bodies" (Collins, 1934, p. 134). The following research papers and A. J. Naldrett's summary included in this Special Paper represent a further step toward a comprehensive understanding of the Sudbury Structure.

ACKNOWLEDGMENTS

This introductory note to the following papers on the Sudbury Structure benefited from discussions with many Sudbury researchers over the last 10–15 yr and from the excellent reviews by B. M. French and G. W. Johns. This is Lunar and Planetary Institute Contribution 935.

REFERENCES CITED

Barlow, A. E., 1904, Report on the origin, geological relations and composition of the nickel and copper deposits in the Sudbury mining district, Ontario, Canada: Geological Survey of Canada Annual Report 873, 236 p.

Barlow, A. E., 1906, On the origin and relations of the nickel and copper deposits of Sudbury, Ontario, Canada: Economic Geology, v. 1, p. 454–466.

Bateman, A. M., 1917, Magmatic ore deposits, Sudbury, Ontario: Economic Geology, v. 12, p. 391–426.

Becker, L., Bada, J. L., Winans, R. E., Hunt, J. E., Bunch, T. E., and French, B. M., 1994, Fullerenes in the 1.85-billion-year-old Sudbury impact structure: Science, v. 265, p. 642–645.

Becker, L., Poreda, R. J., and Bada, J. L., 1996, Extraterrestrial helium trapped in fullerenes in Sudbury impact structure: Science, v. 271, p. 249–252.

Bell, R., 1891a, On the Sudbury mining district: Geological Survey of Canada Annual Report 5, P. 1, p. 5F–95F.

Bell, R., 1891b, The nickel and copper deposits of Sudbury district, Canada: Geological Society of America Bulletin, v. 2, p. 1215–140.

Brockmeyer, P., 1990, Petrographische und geochemische Untersuchungen an polymikten Breccien der Onaping Formation, Sudbury-Distrikt (Ontario, Kanada) [Ph.D. thesis]: Germany, Universität Münster, 228 p.

Chai, G., and Eckstrand, R., 1993, Origin of the Sudbury Igneous Complex, Ontario—differentiate of two separate magmas: Current Research, Pt. E: Geological Survey of Canada Paper 93-1E, p. 219–230.

Chai, G., and Eckstrand, R., 1994, Rare-earth element characteristics and origin of the Sudbury Igneous Complex, Ontario, Canada: Chemical Geology, v. 113, p. 221–244.

Coleman, A. P., 1905, The Sudbury nickel region: Report of the Ontario Bureau of Mines, 1904, v. 14, P. 3, 183 p.

Collins, W. H., 1934, Life history of the Sudbury Nickel Irruptive: Transactions of the Royal Society of Canada, 3rd Ser., v. 28, sec. 4, p. 123–177.

Corfu, F., and Lightfoot, P. C., 1996, U-Pb geochronology of the Sublayer environment, Sudbury Igneous Complex, Ontario: Economic Geology, v. 91, p. 1263–1269.

Cowan, E. J., 1996, Deformation of the eastern Sudbury Basin [Ph.D. thesis]: Toronto, Ontario, Canada, University of Toronto, 341 p.

Dence, M. R., 1972, Meteorite impact craters and the structure of the Sudbury basin, in Guy-Bray, J. V., ed., New developments in Sudbury geology: Geological Association of Canada Special Paper 10, p. 7–18.

Deutsch, A., 1994, Isotope systematics support the impact origin of the Sudbury Structure (Ontario, Canada), in Dressler, B. O., Grieve, R. A. F., and Sharpton, V. L. eds., Large meteorite impacts and planetary evolution: Geological Association of America Special Paper 293, p. 289–302.

Deutsch, A., Grieve, R. A. F., Avermann, M., Bischoff, L., Brockmeyer, P., Buhl, D., Lakomy, R., Müller-Mohr, V., Ostermann, M., and Stöffler, D., 1995, The Sudbury Structure (Ontario, Canada): a tectonically deformed multi-ring impact basin: Geologische Rundschau, v. 84, p. 697–709.

Dickin, A. P., Artan, M. A., and Crocket, J. H., 1996, Isotopic evidence for distinct crustal sources of North and South Ranges ores, Sudbury Igneous Complex: Geochimica et Cosmochimica Acta, v. 60, p. 1605–1613.

Dietz, R. S., 1964, Sudbury Structure as an astrobleme: Journal of Geology, v. 72, p. 412–434.

Dietz, R. S., 1972, Sudbury astrobleme, splash emplaced Sub-Layer and possible cosmogenic ores, in Guy-Bray, J. V., ed., New developments in Sudbury geology: Geological Association of Canada Special Paper 10, p. 29–40.

Dressler, B. O., 1982, Footwall of the Sudbury Igneous Complex, District of Sudbury: Ontario Geological Survey Miscellaneous Paper 106, p. 73–75.

Dressler, B. O., 1984, The effects of the Sudbury event and the intrusion of the Sudbury Igneous Complex on the footwall rocks of the Sudbury Structure, in Pye, E. G., Naldrett, A. J., and Giblin, P. E., eds., The geology and ore deposits of the Sudbury Structure: Ontario Geological Survey Special Volume 1, p. 97–136.

Dressler, B. O., 1987, Precambrian geology of Falconbridge Township: Ontario Geological Survey, Preliminary Map P 3067.

Dressler, B. O., Morrison, G. G., Peredery, W. V., and Rao, B. V., 1987, The Sudbury Structure, Ontario, Canada—a review, in Pohl, J., ed., Research in terrestrial impact structures: Braunschweig/Wiesbaden, Germany, Friedrich Vieweg & Sohn, p. 39–68.

Dressler, B. O., Peredery, W. V., and Muir, T. L., 1992, Geology and mineral deposits of the Sudbury Structure: Ontario Geological Survey Guidebook 8, 33 p.

Dressler, B. O., Weiser, T., and Brockmeyer, P., 1996, Recrystallized impact glasses of the Onaping Formation and the Sudbury Igneous Complex, Sudbury Structure, Ontario, Canada: Geochimica et Cosmochimica Acta, v. 60, p. 2019–2036.

Dressler, V. L., and Sharpton, V. L., 1997, Breccia formation at a complex impact crater: Slate Islands, Lake Superior, Ontario, Canada: Tectonophysics, v. 275, p. 285–311.

Dupuis, L., Whitehead, R. E. S., and Davies, J. F., 1982, Evidence for a genetic link between Sudbury Breccia and fenite breccias: Canadian Journal of Earth Sciences, v. 19, p. 1174–1184.

Faggart, B. E., Basu, A. R., and Tatsumoto, M., 1985, Origin of the Sudbury Structure by meteorite impact: neodymium isotopic evidence: Science, v. 230, p. 436–439.

Fairbairn, H. W., and Robson, G. M., 1942, Breccia at Sudbury, Ontario: Journal of Geology, v. 50, p. 1–33.

French, B. M., 1968, Sudbury Structure, Ontario: some petrographic evidence for an origin by meteorite impact, in French, B. M., and Short, N. M., eds., Shock metamorphism of natural materials: Baltimore, Maryland, Mono Book Corp., p. 383–412.

French, B. M., and Short, N. M., 1968, Shock metamorphism of natural materials: Baltimore, Maryland, Mono Book Corp., 644 p.

Grieve, R. A. F., Stöffler, D., and Deutsch, A., 1991, The Sudbury Structure: Controversial or misunderstood?: Journal Geophysical Research, v. 6, p. 22753–22764.

Howe, E., 1914, Petrographical notes on the Sudbury nickel deposits: Economic Geology, v. 9, p. 505–522.

Johns, G. W., and Dressler, B. O., 1995, The Sudbury Igneous Complex—an impact melt sheet? in Proceedings, 27th, Lunar and Planetary Science Conference, Abstracts: Houston, Texas, Lunar and Planetary Institute, p. 679–680.

Knight, C. W., 1917, Geology of the Sudbury area and description of Sudbury ore bodies: Toronto, Ontario, Canada, Report of the Royal Ontario Nickel Commission, p. 104–211.

Krogh, T. E., Davis, D. W., and Corfu, F., 1984, Precise U-Pb zircon and baddeleyite ages for the Sudbury area, in Pye, E. G., Naldrett, A. J., and Giblin, P. E., eds., The geology and ore deposits of the Sudbury Structure: Ontario Geological Survey Special Volume 1, p. 431–446.

Lakomy, R., 1990, Implications for cratering mechanics from a study of the Footwall Breccia of the Sudbury impact structure, Canada: Meteoritics, v. 25, p. 195–207.

Lightfoot, P. C., Doherty, W., Farell, R. R., Keays, R. R., Moore, M., and Pekeski, D., 1997, Geochemistry of the Main Mass, Sublayer, Offsets, and inclusions from the Sudbury Igneous Complex, Ontario: Ontario Geological Survey Open File Report 5959, 231 p.

Melosh, H. J., 1989, Impact cratering—a geologic process. Oxford, U.K., Oxford University Press, 245 p.

Milkereit, B., Green, A., and 24 others, 1992, Deep geometry of the Sudbury Structure from seismic reflection profiling: Geology, v. 20, p. 807–811.

Muir, T. L., 1984, The Sudbury Structure: considerations and models for an endogenic origin, in Pye, E. G., Naldrett, A. J., and Giblin, P. E., eds., The geology and ore deposits of the Sudbury Structure: Ontario Geological Survey Special Volume 1, p. 449–489.

Müller-Mohr, V., 1992a, Gangbreccien der Sudbury-Struktur; Geologie, Petrographie und Geochemie der Sudbury-Breccie, Ontario, Kanada [Ph.D. thesis]: Münster, Germany, Universität Münster, 139 p.

Müller-Mohr, V., 1992b, Breccias in the basement of a deeply eroded impact structure, Sudbury, Canada: Tectonophysics, v. 216, p. 219–226.

Naldrett, A. J., and Hewins, H. R., 1984, The main mass of the Sudbury Igneous Complex, *in* Pye, E. G., Naldrett, A. J., and Giblin, P. E., eds., The geology and ore deposits of the Sudbury Structure: Ontario Geological Survey Special Volume 1, p. 231–251.

Naldrett, A. J., Hewins, H. R., Dressler, B. O., and Rao, B. V., 1984, The contact Sublayer of the Sudbury Igneous Complex, *in* Pye, E. G., Naldrett, A. J., and Giblin, P. E., eds., The geology and ore deposits of the Sudbury Structure: Ontario Geological Survey Special Volume 1, p. 253–274.

Peredery, W. V., and Morrison, G. G., 1984, Discussion of the origin of the Sudbury Structure, *in* Pye, E. G., Naldrett, A. J., and Giblin, P. E., eds., The geology and ore deposits of the Sudbury Structure: Ontario Geological Survey Special Volume 1, p. 461–511.

Pye, E. G., Naldrett, A. J., and Giblin, P. E., 1984, The geology and ore deposits of the Sudbury Structure: Ontario Geological Survey Special Volume 1, 603 p.

Reimold, W. U., and Colliston, W. P., 1994, Pseudotachylites of the Vredefort Dome and the surrounding Witwatersrand basin, South Africa, *in* Dressler, B. O., Grieve, R. A. F., and Sharpton, V. L., eds., Large meteorite impacts and planetary evolution: Geological Association of America Special Paper 293, p. 177–196.

Riller, U. P., 1996, Tectonometamorphic episodes affecting the southern footwall of the Sudbury Basin and their significance for the origin of the Sudbury Igneous Complex. central Ontario, Canada [Ph.D. thesis]: Toronto, Ontario, Canada, University of Toronto, 135 p.

Schwarz, E. J., and Buchan, K. L., 1982, Uplift deduced from remanent magnetization: Sudbury area since 1250 Ma: earth and Planetary Science Letters, v. 58, p. 65–74.

Shanks, W. S., and Schwerdtner, W. R., 1990, Structural analysis of the central and southwestern Sudbury Structure, Southern Province, Canadian Shield: Canadian Journal of Earth Sciences, v. 28, p. 411–430.

Shanks, W. S., Dressler, B. O., and Schwerdtner, W. R., 1990, New developments in Sudbury geology, *in* International workshop on meteorite impacts on the early Earth, Perth, Australia: Houston, Texas, Lunar and Planetary Institute Contribution 746, p. 46.

Speers, E. C., 1957, The age relationship and origin of common Sudbury Breccia: Journal of Geology, v. 65, p. 497–514.

Spray, J. G., 1997, Superfaults: Geology, v. 25, p. 579–582.

Spray, J. G., and Thompson, L. M., 1995, Friction melt distribution in a multi-ring impact basin: Nature, v. 373, p. 130–132.

Stevenson, J. S., 1972, The Onaping ash-flow sheet, Sudbury, Ontario, *in* Guy-Bray, J. V., ed., New developments in Sudbury geology: Geological Association of Canada Special Paper 10, p. 41–48.

Williams, H., 1957, Glowing avalanche deposits of the Sudbury Basin: Ontario Department of Mines Annual Report for 1956, v. 65, P. 3, p. 57–89.

Wu, J., Milkereit, B., and Boerner, D., 1994, Timing constraints on deformation history of the Sudbury Structure: Canadian Journal of Earth Sciences, v. 31, p. 1654–1660.

Yates, A. B., 1948, Properties of the International Nickel Company of Canada, *in* Structural geology of Canadian ore deposits: Canadian Institute of Mining and Metallurgy, p. 596–617.

Zhou, M., Lightfoot, P. C., Keays, R. R., Moore, M. L., and Morrison, G. G., 1997, Petrogenetic significance of chromium spinels from the Sudbury Igneous Complex, Ontario, Canada: Canadian Journal of Earth Sciences, v. 34, p. 1405–1409.

MANUSCRIPT ACCEPTED BY THE SOCIETY DECEMBER 16, 1998

Geological Society of America
Special Paper 339
1999

Sudbury Breccia distribution and orientation in an embayment environment

John S. Fedorowich
Falconbridge Ltd., P.O. Box 40, Falconbridge, Ontario P0M 1S0, Canada
Don H. Rousell
Department of Earth Sciences, Laurentian University, Sudbury, Ontario P3E 2S4, Canada
Walter V. Peredery
Wallbridge Mining Co. Ltd., 129 Fielding Road, Lively, Ontario P3Y 1L7, Canada

ABSTRACT

The distribution patterns and brecciation intensity of Sudbury Breccia are generally not well constrained within a 5–15-km-wide zone that surrounds the Sudbury Igneous Complex. This zone is punctuated by local areas of abundant Sudbury Breccia. The object of this chapter is to quantify the distribution and orientation within one such area, the Strathcona Embayment.

Estimates of the percentage distribution of Sudbury Breccia are based on detailed field mapping. For example, in a 7-km^2 area, only 4.7% of the total outcrop area (22.8%) consists of Sudbury Breccia. The area was subdivided into subareas varying from 2 to 8% Sudbury Breccia content, with the highest concentration in the immediate vicinity of the Strathcona Embayment. Embayments are interpreted as slump features in the footwall to the Sudbury Igneous Complex and reflect a reworking of ground that has been structurally weakened by meteorite impact brecciation. One of the conclusions from these measurements is that Sudbury Breccia distribution can be used to predict embayment structures, which host nickel–copper–platinum group element (Ni-Cu-PGE) deposits throughout the Sudbury Basin.

A large network of Sudbury Breccia in the Strathcona mine consists of 46% breccia in the center, and only 4–18% in the adjoining hangingwall and footwall. Dike-like bodies and veins of Sudbury Breccia have widely variable orientations that cluster subparallel to the outer margin of the Sudbury Igneous Complex and, like the complex, decrease in dip with depth. Envelopes of Sudbury Breccia are controlled by reverse and normal shears that are interactive with tensional openings in an overall compressional regime with maximum compression approximately orthogonal to the outer contact of the Sudbury Igneous Complex.

INTRODUCTION

Frood breccia, Levack breccia, Sudbury-type breccia, common Sudbury breccia, and pseudotachylyte are names given in the past to a distinctive rock type within the footwall rocks surrounding the Sudbury Igneous Complex (SIC). Although this rock possesses significant textural and distribution variation over a wide area (Fig. 1), the term Sudbury Breccia (SB) has endured. As is typical among breccias, several end-members can be assigned based on textural or compositional variations. Published data on the accurate percentage distribution of SB are lacking. This may be the result of several factors, including apparent irregular distri-

Fedorowich, J. S., Rousell, D. H., and Peredery, W. V., 1999, Sudbury Breccia distribution and orientation in an embayment environment, *in* Dressler, B. O., and Sharpton, V. L., eds., Large Meteorite Impacts and Planetary Evolution II: Boulder, Colorado, Geological Society of America Special Paper 339.

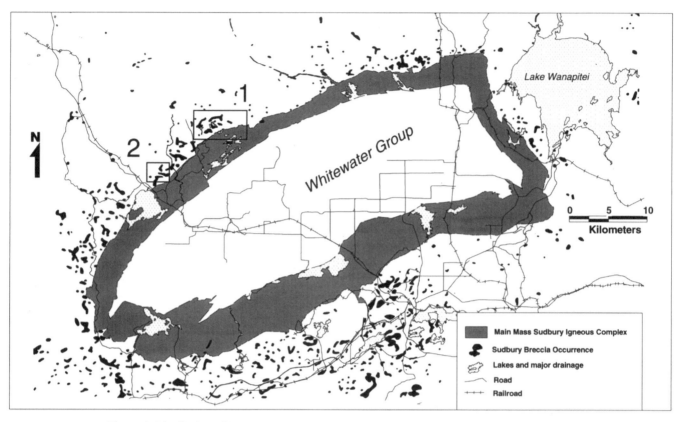

Figure 1. Distribution of Sudbury Breccia occurrences around the Sudbury Basin (modified after Dressler, 1984c). Each of the solid black areas represents an outcrop or series of outcrops in which Sudbury Breccia can be found. These areas are not necessarily entirely composed of Sudbury Breccia.

bution; variation in bedrock exposure; and the erratic nature of the bodies in terms of persistence and connectivity.

This chapter quantifies distribution of Sudbury Breccia on the local scale to determine if significant differences exist, and to relate these differences to local and regional structures. In order to achieve this aim, detailed mapping of the Sudbury Breccia was undertaken underground in the Strathcona mine and on the surface (Fig. 2; see Fig. 1 for location). The Strathcona area in the North Range has a locally high concentration of SB flanked by more sparsely distributed SB, has good exposures on surface, and presents probably the most extensive presently accessible underground exposure of SB. The Strathcona Deep Copper ore zones are largely hosted within SB; therefore, there is a close link between Cu-Ni-PGE mineralization and SB occurrences (Abel, 1981; Coats and Snajdr, 1984; Fedorowich, 1996). Accordingly, a better understanding of the distribution of Sudbury Breccia may prove useful in mineral exploration for this type of ore.

REGIONAL OBSERVATIONS

On the regional scale, Sudbury Breccia occurrences (Fig. 1) are concentrated within a 5-km-wide collar surrounding the outer margin of the Sudbury Igneous Complex, with a scattering of occurrences beyond the collar (Card et al., 1984; Dressler, 1984a,b; Grieve, 1987; Spray, 1997).

SDBX forms irregular-shaped bodies or dikes with either straight, jagged, or wavy contacts. A ramifying network of veins, of varying intensity, extends outward from these bodies, and contacts between the breccia and the host rocks appear sharp at the outcrop scale. Sudbury Breccia bodies tend to follow zones of structural weakness such as lithologic boundaries, faults, joints, and foliations (Fairbairn and Robson, 1942), but they also thoroughly penetrate homogeneous and competent rock. In detail, the breccia may intersect bedding at a large angle, may be associated with small local structures such as folds or faults, or may occur unrelated to any apparent discontinuity (Yates, 1938; Speers, 1957; Dressler, 1984b; Müller-Mohr, 1992). A regional-scale classification of end-member types, observable at the outcrop

Figure 2 (following two pages). A, Strathcona region outcrop distribution map. All footwall rocks to the SIC are shown as unfilled outlines, with Sudbury Breccia highlighted by solid black fill. Lakes and mine shafts (past and present producers) are shown for reference: 1 = Big Levack; 2 = Longvack, 3 = South Longvack; 4 = Coleman; 5 = Strathcona #1; 6 = Strathcona #2. B, Strathcona lithologic contacts interpreted from outcrops, highlighting Sudbury Breccia (SDBX) occurrences. Inset shows an up-plunge view of the Strathcona Embayment in Figure 3D.

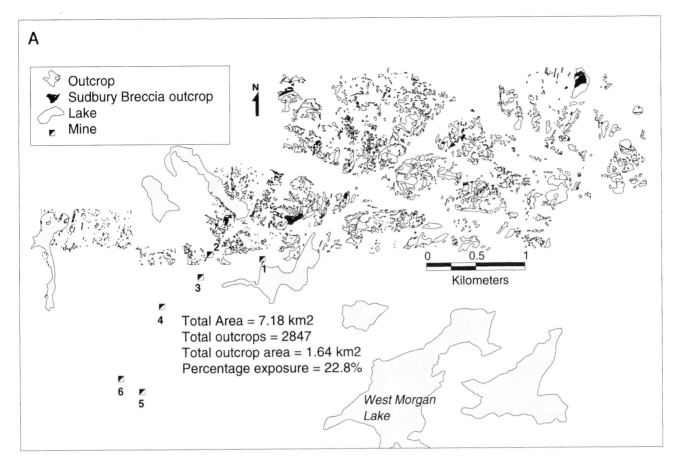

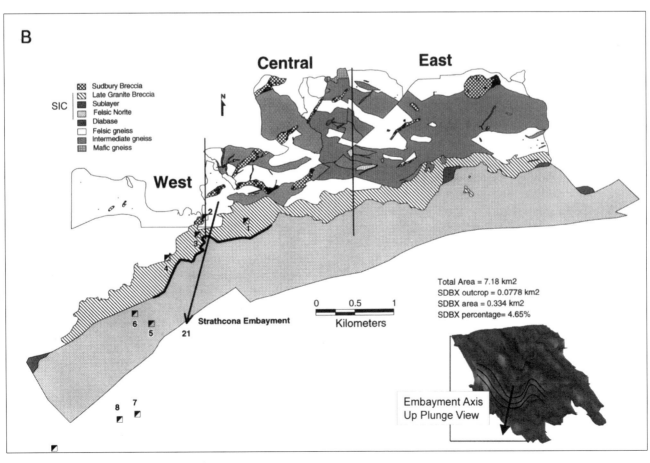

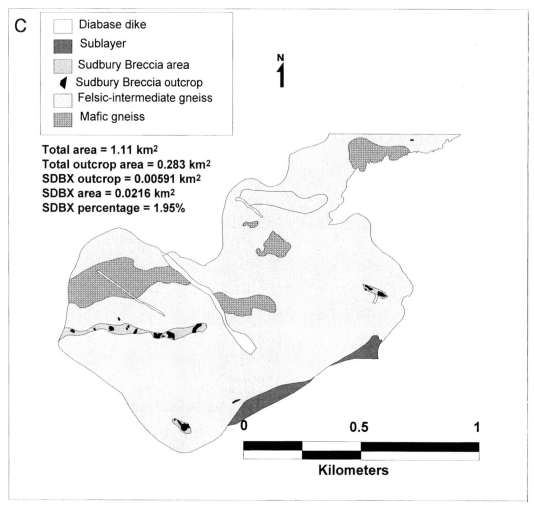

Figure 2 (continued). C, Hardy area map of Sudbury Breccia (SDBX) occurrences.

scale, is given in Table 1, together with their textural characteristics and macroscopic structures.

Sudbury Breccia bodies range in size from millimeter-scale veins to kilometer-scale (Fig. 1), such as the Frood belt in the South Range, which is up to 1.5 km wide and at least 25 km long (Yates, 1938). Yates (1948) later stated that the Frood breccia belt extends an additional 25 km to the southwest, from Copper Cliff to Worthington and Victoria mines. However, this is not universally accepted, and is currently being investigated more fully by Scott and Spray (1996, and this volume). Although good descriptions of local geometry and contact relationships can be found for a few areas (Yates, 1938; Speers, 1957; Card et al., 1971; Dressler, 1984b; Muller-Mohr, 1992; Spray and Thompson, 1995; Thompson, 1996), details about SB distribution are generally qualitative, with no detailed quantitative distribution analysis reported to this date.

ANALYTICAL METHODS

Outcrop mapping was carried out using conventional techniques, and 1:2,500-scale airphotos for control (Tirschmann, 1989; Sweeney and Farrow, 1990; Fedorowich, 1995). All outcrops were entered within the MAPINFO database as closed polygons with attributes, which allowed precise calculation of the outcrop areas and querying of the database. Underground mapping was carried out on cleanly washed walls and ceilings in each of the ramps and drifts. Surveyed outlines of underground openings were used for control, and the position of data points was established by 1-m spaced marks placed along the entire length of the openings from existing survey pin control points. The conventional back projection technique of mapping was performed throughout at a mapping scale of 1:250. Data were then digitized within the Silverscreen-CAMP 3D graphic platform, in which measurements of Sudbury Breccia distribution were obtained.

SURFACE SUDBURY BRECCIA DISTRIBUTION AND ORIENTATION

Sudbury Breccia was mapped in detail (Fig. 2A) in a 6-km-long area of basement exposure in the immediate footwall of the Strathcona Embayment (Tirschmann, 1989; Sweeney and Farrow,

TABLE 1. CHARACTERISTICS OF SUDBURY BRECCIA

Type	Location	Field Relations	Characteristics	Mechanisms
Massive-looking	North and East Range	Sharp contacts, with strain gradients common, ramifying vein networks	Unoriented subrounded polymict fragments, comminuted/milled matrix	Mechanical fragmentation, sliding, rotation, and milling
Flow Texture	South Range Minor in N and E Range	Vein and dike networks, locally overprinted by shear deformation	Subrounded polymict fragments, high aspect ratios, comminuted matrix with flow surfaces	Laminar flow (originally turbulent?); or progressive shear strain with brecciation
Igneous Texture	North Range +	Cryptic texture within massive breccia, faintly distinguishable on outcrop	Similar in appearance to massive SB, but with fine-grained igneous matrix; plagioclase laths, actinolite, and FeTi oxides	Flash melts, slower cooling

TABLE 2. STRATHCONA AND HARDY SURFACE MAPPING SDBX DISTRIBUTION ANALYSIS

Location	Total Area (km^2)	Total Exposure (km^2)	Outcrop (%)	Exposed SDBX (km^2)	Interpreted SDBX Contact Area (km^2)	Mean Outcrop (km^2)	Outcrops	Standard Deviation (km^2)	SDBX (%)	Comments
West	0.68	0.13	19.1	0.0028	0.013	0.0002	20	0.00046	1.9	West of embayment, sparse SDBX
Central	2.56	0.68	26.7	0.0496	0.200	0.0004	118	0.00086	7.8	Central SDBX core; embayment shadow
East	3.94	0.82	20.9	0.0255	0.121	0.0005	49	0.00161	3.0	East of embayment, moderate SDBX
Strathcona total	7.18	1.64	22.8	0.0756	0.334	0.0004	187	0.00109	4.7	Up plunge from Deep Copper
Northwest of Hardy pit	1.11	0.28	25.7	0.0059	0.022	0.0003	18	0.00035	2.0	1 km from SIC contact, sparse SDBX

1990; Fedorowich, 1995). The total area of the map is 7.18 km^2 and the overall outcrop exposure is 22.8% (Table 2).

The Strathcona area has been subdivided, based on the percentage of outcrop that consists of Sudbury Breccia, into three subareas: west (2%), east (2–4%), and central (>4%), as defined by distribution of SB. The central area is immediately adjacent to the Strathcona Embayment, which is a synformal closure of the Sudbury Igneous Complex contact plunging 21° to an azimuth of 195° (Fig. 2B). The embayment has a minimum 3-km plunge length, and has a rake of 37° southwest within the plane of the SIC contact, which dips 32°–35° southeast (Fig. 2B) (Fedorowich, 1997). The central area has an SB distribution of 7.8%. The percentages for the west and east areas, 1.9% and 3.0%, respectively (Table 2), indicate a two- to threefold decrease in percentage distribution of SB away from the embayment.

In order to confirm the Sudbury Breccia "background" percentage within the area immediately adjacent to the SIC, but outside the influence of an embayment, a second area was measured northwest of Hardy pit (Fig. 2C; see Fig. 1 for location). The map area covers 1.11 km^2, and the distribution percentage of SB is 1.95% (Table 2; Fig. 2C). This percentage is similar to the west area in Figure 2B. Comparative analysis thus suggests a general background value of 2–3% SB within the collar proximal to the SIC contact.

Sudbury Breccia bodies in the Strathcona area (Fig. 2B) mainly strike northeast-southwest, with a subsidiary east-west trend. Available mapping and drilling information indicates that contacts dip steeply, and small-scale breccia veins tend to have orientations similar to the larger bodies (Fedorowich, 1995). The steep surface contacts of SB at Strathcona partly confirm the orientation analysis of Dressler (1984b), which suggested a multi-oriented near-vertical orientation distribution.

UNDERGROUND SUDBURY BRECCIA DISTRIBUTION AND ORIENTATION — STRATHCONA MINE

Sudbury Breccia bodies exhibit a wide variation in orientation and width (Fig. 3) throughout the area mapped in Strathcona mine (4.5 km of drifts, ramps, and stopes, as reported in Fedorowich, 1996). Individual dikes are as much as 40 m wide; however, networks form complicated interconnected systems that are much wider (Fig. 4A,B). Shear displacement commonly observed between footwall and hangingwall blocks of the SB (Fig. 5A,B) indicate that development involved displacement and

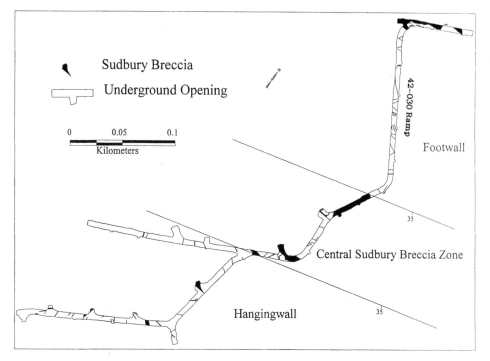

Figure 3. Strathcona underground 4,200- (1,280 m) and 4,400-ft (1,341 m) level ramps illustrating Sudbury Breccia distribution. Sudbury Breccia occurs dominantly within Archean felsic gneisses. Hangingwall and footwall refer to the relative position with respect to the central Sudbury Breccia envelope.

concomitant filling of mechanically fragmented, rolled, and milled rock, which is summarized in Table 1. Microscopic and mesoscopic examination both indicate that cataclastic comminution was widespread and extreme, and where clasts are highly rounded, frictional melting may have taken place (cf. Lin, 1999).

A representative area of ramps connecting the 4,200- (1,280 m) and 4,400-ft (1,341 m) levels is shown in Figure 3. The area is subdivided into a central SB zone, with an overall higher percentage of SB, and footwall (FW) and hangingwall (HW) zones, which have a lower abundance. Most of the Cu-Ni-PGE mineralization in "Deep Copper Zone 38" is within the envelope of the central Sudbury Breccia Zone, but is nonuniformly distributed (Fig. 3). At the 4,200-ft (1,280m) level the zone has an attitude of 068, dipping 35° southeast, a horizontal width of 91 m, and a true width of 53 m. This width varies both up- and down dip. Both the HW and FW to the central zone are locally penetrated by <1-m-wide veins, with several discrete bodies on the order of 2–5 m wide. Detailed measurements of the area enclosed by these veins are shown in Table 3, subdivided according to footwall, central zone, and hangingwall, which have 18.0, 46.4, and 4.1% SB by area, respectively. The entire ramp network has 15.5% SB by area, which is higher than the percentage of area distribution defined from surface mapping. It should be cautioned, however, that the ramp remains relatively close to the central SB zone, whereas the surface map covers a wider area (Fig, 2B).

Details of the SB exposed along a 200-m-long Footwall portion of the ramp are shown in plan and cross-sectional views (Fig. 4A,B). SB bodies that display shear strain gradients along their margins are interpreted as shear veins (Fig. 5A-D), whereas a few other elements interspersed with the shear veins are interpreted as tensional arrays (Fig. 5B–D). Offsets of diabase dikes and gneissic units on the order of 20 cm to 2 m are produced by SB displacements throughout the mapped area (Fig. 4A-B).

A view of the east wall of the ramp (Fig. 4B) illustrates the vertical component of slip for each SB shear vein, where such a component of slip could be established, whereas the horizontal components of slip are displayed in plan view in Figure 4A. In general, displacements tend to be oblique slip, either reverse or normal. All of the significant SB features, and massive Cu-Ni-PGE mineralized veins that occur sparsely in this ramp, are projected a few meters above and below the ramp elevation, and their connectivity has been interpreted accordingly.

SB is formed in both shear and tensional modes of deformation. An example of a SB shear vein in Figure 5A exhibits a well-developed strain gradient and dragfolding of gneissic layering. Both of these indicate reverse-sinistral oblique-slip movement along the shear (Angelier, 1994; Killick, 1992), consistent with the apparent 15 cm offset of the mafic gneiss. At the microscopic scale, the margins of the shear merge almost imperceptibly with the interior, which is composed of comminuted rock particles and fragments up to 2 mm in diameter. The matrix is dominated by very fine grained feldspar, chlorite, and epidote, representing an episode of secondary mineral growth (Fedorowich, 1995, for electron microprobe analyses of matrix material; Dressler, 1984b). Figure 5B shows a 1-cm-wide SB fault; a >2-m normal displacement is deduced from the excellent dragfolding. Minor, ragged extensional veinlets of SB (<1 cm wide) occur within the hangingwall block, and confirm the interactive shear and tensional modes of SB development. (cf. Melosh, 1989; Wood, 1998).

An example of weakly developed flow texture SB is illustrated in Figure 5, C–D. Flow surfaces are defined by the paral-

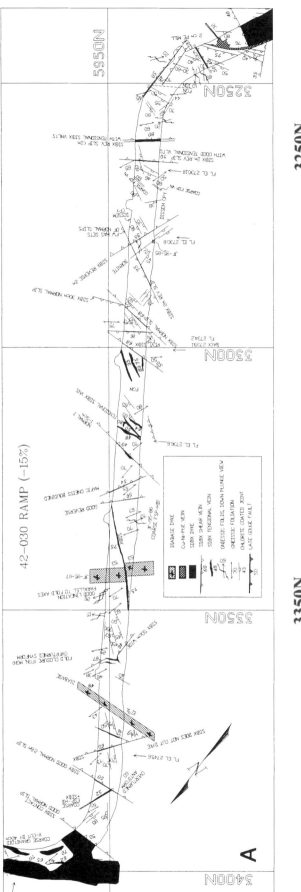
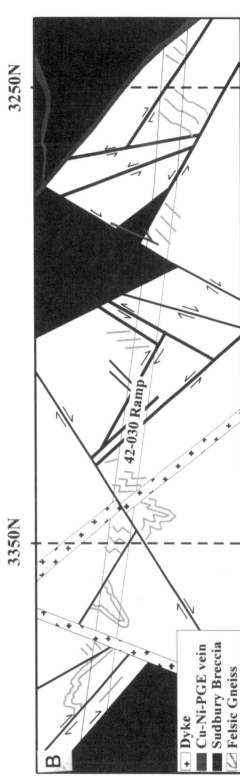

Figure 4. A, Representative portion of 4,200- (1,280 m) and 4,400-ft (1,341 M) level ramps illustrating the detailed mapping of orientation and distribution of Sudbury Breccia within the gneissic host rocks. SB shear veins are shown with barbs pointing in the dip direction; tensional arrays are shown with solid black fill. Several tight fold closures were mapped within the gneisses, and are illustrated in their down-plunge views. B, Cross-section view of the east wall of Figure A, illustrating the distribution and shear sense for Sudbury Breccia features.

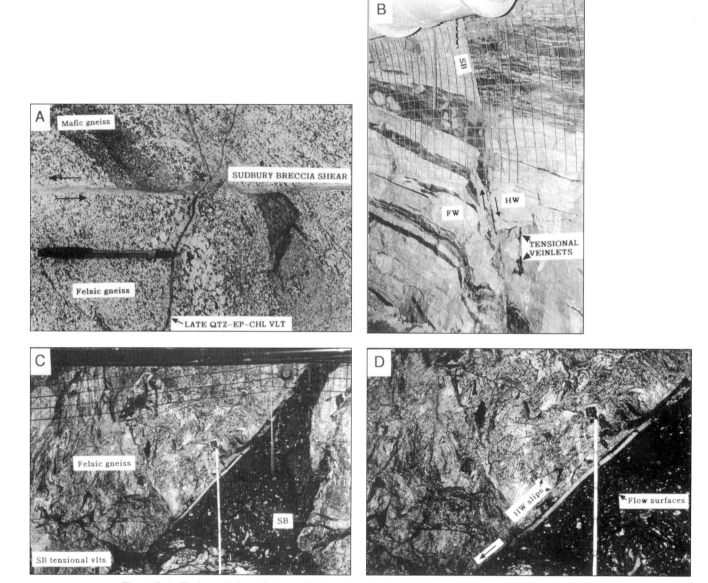

Figure 5. A, Horizontal view of a 1-cm Sudbury Breccia shear from Strathcona area illustrating a 15-cm component of left-lateral horizontal displacement, and later cross-cutting by fibrous epidote-quartz-chlorite veinlets. B, Vertical view of a Sudbury Breccia normal shear with a component of ~2 m of normal slip, and tensional veinlets developed within the hangingwall. C, Wall view of an SB dike with associated tensional veinlets. Hanging-wall felsic gneiss is strongly brecciated. Field of view, 3 × 5 m. D, Close-up view shows series of contact-parallel hanging-wall slips that display normal slip-shear sense. Diffuse flow surfaces and elongated clasts also define a preferred orientation and strain gradient consistent with normal shear.

TABLE 3. STRATHCONA UNDERGROUND 4200–4400 LEVEL RAMPS
SDBX DISTRIBUTION ANALYSIS

	Opening Area	Exposure (%)	SDBX Area (km^2)	SDBX (%)	Comments
Footwall	0.0012	100	0.00022	18.0	0.30 km ramp in FW to main SDBX zone
Central Breccia Zone	0.0008	100	0.00036	46.4	0.12 km ramp through main SDBX zone
Hangingwall	0.0024	100	0.00010	4.1	0.41 km ramp in HW to main SDBX zone
Total Ramp	0.0044	100	0.00068	15.5	Total ramp

lel alignment of fragments, and the planes along which extensional deformation has been accomplished near the hangingwall contact (Fig. 5D). A set of multi-oriented extensional veinlets also penetrates within the hangingwall block. Normal slip sense is found on subsidiary fractures that parallel the HW contact (Fig. 5D), and drag sense for flow surfaces is also consistent with a component of normal shear within and along the margins of this body.

A statistical analysis of the orientations of SB bodies was carried out along the ramps and drifts on the 4,200-ft (1,280 m) and 4,400-ft (1,341 m) levels and combined within an entire set of data for three working levels of the mine (Fig. 4). A contoured stereogram of planes to all tabular SB bodies (n = 430) shows a strong point maximum at >10% (Fig. 6A). The corresponding great circle, 068/35°SE, is subparallel to the SIC contact at this level (053/32°SE). A rose diagram illustrates the strike-distribution of SB along this set of ramps (Fig. 6A). A plot of shears that display good reverse movement (n = 74) yields similar results (Fig. 6B), with an 18% point maximum and a corresponding plane at 074/36°SE. Shears with a normal sense of displacement (n = 67) exhibit a scattered distribution (Fig. 6C), with four possible point maxima. A plot of extension veinlets (n = 193) shows wide variation in orientation with three point maxima <6% (Fig. 6D). Thus, features formed by reverse and normal shear have a stronger preferred orientation, whereas those formed by tensional stresses have more scatter in their orientations. When considered altogether, the orientations overall do not mimic the gneissocity and lithologic trends of the host rocks (Figs. 2B and 4A).

DISCUSSION

Although Sudbury Breccia bodies vary widely in orientation, they do have a preferred orientation. In a plot of all SB bodies and another of reverse shears only, both exhibit a strong preferred orientation subparallel to the SIC contact. We did not expect this result, as the initial impression of SB, both underground and on the surface, is one of chaos. The conclusion from the present study regarding the distribution of SB at Strathcona mine is that 50–200-m wide envelopes of SB are controlled by reverse and normal shears that are interactive with tensional openings in an overall compressional regime. Underground and surface patterns correspond. Previous orientation analysis, carried out on the surface in the Sudbury region, indicate mostly steep dips, no distinct clustering, and an overall concentric distribution extending outward from the SIC contact up to 80 km (Dressler, 1984b; Peredery and Morrison, 1984; Dressler et al., 1987, Thompson and Spray, 1994; Thompson, 1996).

Away from the zones of more abundant SB, the distribution is more diffuse, but the orientations appear to remain sympathetic. The overall shallower orientation of SB bodies found in the present study compared with the steeper orientations documented at surface by Dressler (1984b) indicates a *real* shallowing of the envelopes of SB with depth. This shallowing is thought to reflect the geometry of the impact crater wall that underwent brecciation

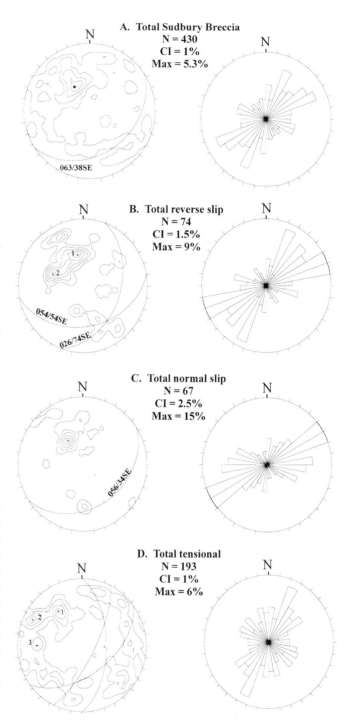

Figure 6. A, Orientation distribution data for all Sudbury Breccia features measured within the 4,200- (1,280 m) and 4,400-ft (1,341 M) level ramps. Data are displayed as a Fisher contoured equal area plot of poles, and a rosette plot, for all SB features. B, Orientation distribution data for all Sudbury Breccia features that display good reverse shear sense. C, Orientation distribution data for all Sudbury Breccia features that display good normal shear sense. D, Equal area plot of poles to tensional veins within the 4200- (1,280 m) and 4,400-ft (1,341 M) level ramps showing considerable scatter in orientation, with several weak point maxima.

and local frictional melting at high strain rate with maximum compression approximately orthogonal to the basin margin.

Thompson and Spray (1994) and Spray and Thompson (1995) studied SB along Highway 144 northeast of the SIC, along an 80-km-long traverse through an area of limited to moderate exposure. They reported shear and tensional modes of emplacement, and suggested that concentric major and minor zones of SB are generated by frictional melting along faults during crater collapse after shock wave dissipation. It has been speculated that the periodicity may perhaps correspond to shock wave periodicity (Grieve, 1994; Golightly, 1994). The present study found that reverse (compressional) and normal (extensional) SB features were interactive, and cannot be separated into episodes. Further, the distribution analysis of the present study indicates a second level of complexity on the regional scale that is orthogonal to the "pseudotachylyte-rich" concentric zones proposed by Thompson and Spray (1994). In other words, the distinct and measurable variations in the intensity of brecciation that are found within the innermost zone of SB really begs the question of how widespread are these variations within the concentric zones that various authors have proposed?

In terms of cross-cutting relationships, SB is overprinted by contact metamorphism, postimpact brecciation, and shear deformation of another type, which is found to be most pronounced in the embayment environment (Fedorowich, 1997). The SB event predates both the emplacement of the Sudbury Igneous Complex, and Cu-Ni-PGE mineralization found within Strathcona Deep Copper deposit (Coates and Snajdr, 1984; Morrison, 1984; Morrison et al., 1994). Deformational structures (penetrative foliation, ductile flow, brecciation, folds, and extensional sulfide vein networks) that postdate SB are found within the SIC contact horizon in the embayment environment (Fedorowich, 1997). Therefore, the higher abundance of SB in the vicinity of an embayment suggests a direct control for its development, but probably as ground preparation and structural weakening, which allowed for progressive post-SIC slumping to be localized.

CONCLUSIONS

Detailed distribution analysis over two areas of the Sudbury Basin indicates that SB near the Sudbury Igneous Complex contact comprises approximately 2% of the exposed bedrock. However, this increases significantly toward the Strathcona Embayment. The embayment has an envelope of SB up to 7.8% of exposed bedrock within a 2.6 km^2 area. The most plausible explanation for this is that the embayment developed where the ground was structurally prepared; an abundance of shattered ground, and (possibly) poorly consolidated breccia facilitated slumping during a "relaxational regime" that followed emplacement of the Sudbury Igneous Complex.

At Strathcona mine, 40–200-m wide envelopes of SB are controlled by reverse and normal shears, which are interactive with tensional openings, in an overall compressional regime. These reflect emplacement under high strain rate and high normal stresses within a shadow zone to a precursor impact crater wall that underwent local frictional melting with maximum compression approximately normal to the crater margin. The brecciation event predates both the Cu-Ni-PGE mineralization and the emplacement of the Main Mass of the SIC, but influenced progressive postemplacement developments.

ACKNOWLEDGMENTS

The mine geology staff at Strathcona, especially P. Johannessen, H. Komarechka, and D. Conroy, are thanked for the insights into Sudbury Breccia relationships within Deep Copper deposit, shared with Fedorowich during the mapping phase of this project. Comments and suggestions by C. Moore, P. Binney, and T. Barnett of Falconbridge were helpful throughout various portions of this study. Ontario Geological Survey regional mapping data were made available by P. Thurston, A. Fyon, and J. Boyd. The patience and coordinating efforts of B. O. Dressler helped achieve submission of this paper to the volume, and thoughtful critiques by J. G. Spray, and V. Muller-Mohr improved its readability.

REFERENCES CITED

Abel, M. K., 1981, The structure of Strathcona mine copper zone: Canadian Institute of Mining and Metallurgy Bulletin, v. 74, p. 89–97.

Angelier, J., 1994, Fault slip analysis and paleostress reconstruction, in Hancock, P. L., ed., Continental deformation: New York, Pergamon Press, p. 53–100.

Card, K. D., Gupta, V. K., McGrath, P. H., and Grant, F. S., 1984, The Sudbury Structure: its regional geological and geophysical setting, in Pye, E. G., Naldrett, A. J., and Giblin, P. E., eds., The geology and ore deposits of the Sudbury Structure, Ontario Geological Survey Special Volume 1, p. 25–43.

Coats, C. J. A., and Snajdr, P., 1984, Ore deposits of the North Range, Onaping-Levack area, Sudbury, in Pye, E. G., Naldrett, A. J., and Giblin, P. E., eds., The geology and ore deposits of the Sudbury Structure: Ontario Geological Survey Special Volume 1, p. 327–346.

Dressler, B. O., 1984a, General geology of the Sudbury area, in Pye, E. G., Naldrett, A. J., and Giblin, P. E., eds., The geology and ore deposits of the Sudbury Structure: Ontario Geological Survey Special Volume 1, p. 57–82.

Dressler, B. O., 1984b, The effects of the Sudbury event and the intrusion of the Sudbury Igneous Complex on the footwall rocks of the Sudbury Structure, in Pye, E. G., Naldrett, A. J., and Giblin, P. E., eds., The geology and ore deposits of the Sudbury Structure: Ontario Geological Survey Special Volume 1, p. 97–136.

Dressler, B. O., 1984c, Sudbury geological compilation; Ontario Geological Survey Map 2491, Precambrian Series, scale 1:50,000.

Dressler, B. O., Morrison, G. G., Peredery, W. V., and Rao, B. V., 1987, The Sudbury Structure, Ontario, Canada — a review, in Pohl, J., ed., Research in terrestrial impact structures: Braunschweig/Wiesbaden, Germany, Friedrich Vieweg & Sohn, p. 39–68.

Fairbairn, H. W., and Robson, G. M., 1942, Breccia at Sudbury: Journal of Geology, v. 50, p. 1–33.

Fedorowich, J. S., 1995, Detailed structural observations at Barnet Property, Strathcona mine area, North Range, Sudbury: Falconbridge Ltd. Exploration Internal Report, Bulletin 477, 23 p.

Fedorowich, J. S., 1996, Structural controls for footwall Cu-Ni-PGE mineralization in the Strathcona Deep Copper deposit: Falconbridge Ltd. Exploration Internal Report, Bulletin 663, 129 p.

Fedorowich, J. S., 1997, Structural controls for footwall Ni-Cu-PGE mineralization within the embayment environment at Fraser Mine, Sudbury: Falconbridge Ltd. Exploration Internal Report, Bulletin 741, 96 p.

Golightly, J. P., 1994, The Sudbury Igneous Complex as an impact melt: evolution and ore genesis, *in* Lightfoot, P. C., and Naldrett, A. J., eds., Proceedings of the Sudbury-Noril'sk Symposium: Ontario Geological Survey Special Volume 5, p. 105–118.

Grieve, R. A. F., 1987, Terrestrial impact structures: Annual Review of Earth and Planetary Science, v. 15, p. 245–270.

Grieve, R. A. F., 1994, An impact model of the Sudbury Structure, *in* Lightfoot, P. C., and Naldrett, A. J., eds., Proceedings of the Sudbury-Noril'sk Symposium: Ontario Geological Survey Special Volume 5, p. 106–119.

Killick, A. M., 1992, Pseudotachylites of the West Rand Goldfield, Witwatersrand Basin, South Africa, [Ph.D. thesis]: Johannesburg, South Africa, Rand Afrikaans University, 273 p.

Lin, A., 1999, Roundness of clasts in pseudotachylytes and cataclastic rocks as an indicator of frictional melting: Journal of Structrual Geology, v. 21, p. 473–478.

Melosh, H. J., 1989, Impact cratering, a geologic process: New York, Oxford University Press, Oxford Monographs on Geology and Geophysics No. 11, 245 p.

Morrison, G. G., 1984, Morphological features of the Sudbury Structure in relation to an impact, *in* Pye, E. G., Naldrett, A. J., and Giblin, P. E., eds., The geology and ore deposits of the Sudbury Structure: Ontario Geological Survey Special Volume 1, p. 513–520.

Morrison, G. G., Jago, B. C., and White, T. L., 1994, Footwall mineralization of the Sudbury Igneous Complex, *in* Lightfoot, P. C., and Naldrett, A. J., eds., Proceedings of the Sudbury-Noril'sk Symposium: Ontario Geological Survey Special Volume 5, p. 119–132.

Müller-Mohr, V., 1992, Breccias in the basement of a deeply eroded impact structure, Sudbury, Canada: Tectonophysics, v. 216, p. 219–226.

Scott, R. G., and Spray, J. G., 1996, The Frood Breccia Belt of the Sudbury Impact Structure: the largest known pseudotachylyte body: Geological Society of America Abstracts with Programs, v. 28–7, p. A–383.

Shand, S. J., 1916, The pseudotachylyte of Parijs (Orange Free State) and its relation to "trapschotten gneiss" and "flinty crush rock": Quarterly Journal of the Geological Society of London, v. 72, p. 198–217.

Speers, E. C., 1957, The age relationship and origin of common Sudbury Breccia: Journal of Geology, v. 65, p. 497–514.

Spray, J. G., and Thompson, L. M., 1995, Friction melt distribution in a multiring impact basin: Nature, v. 373, p. 130–132.

Spray, J. G., 1997, Superfaults: Geology, v. 25, p. 579–582.

Sweeny, M., and Farrow, C., 1990, Geology report, Morgan West Project (6–243), Levack and Morgan Townships; Falconbridge Ltd. Exploration Internal Report, Bulletin, B–270, 35 p.

Thompson, L. M., and Spray, J. G., 1994, Pseudotachylytic rock distribution and genesis within the Sudbury impact structure, *in* Dressler, B. O., Grieve, R. A. F., and Sharpton, V. L., eds., Large meteorite impacts and planetary evolution: Geological Society of America Special Paper 293, p. 275–287.

Tirschmann, P. A., 1989, Barnet geology report: Falconbridge Ltd. Exploration Internal Report, Bulletin 241, 20 p.

Wood, C. R., 1998, Origin and emplacement of the Hess Offset dyke, North Range of the Sudbury Impact Structure, [M.S. thesis]: Fredericton, New Brunswick, Canada, University of New Brunswick, 199 p.

Yates, A. B., 1938, The Sudbury Intrusive: Royal Society of Canada Transactions, Section 4, p. 151–172.

Yates, A. B., 1948, Properties of the International Nickel Company of Canada, *in* Structural geology of Canadian ore deposits: Canadian Institute of Mining and Metallurgy Jubilee Volume, p. 598–617.

MANUSCRIPT ACCEPTED BY THE SOCIETY DECEMBER 16, 1998

Impact diamonds in the Suevitic Breccias of the Black Member of the Onaping Formation, Sudbury Structure, Ontario, Canada

V. L. Masaitis and G. I. Shafranovsky
Karpinsky Geological Institute, St. Petersburg, Russia
R. A. F. Grieve
Geological Survey of Canada, Ottawa, Ontario, Canada
F. Langenhorst
Bayerisches Geoinstitut, University of Bayreuth, Germany
W. V. Peredery
Peredery and Associates, Sudbury, Ontario, Canada
A. M. Therriault
Geological Survey of Canada, Ottawa, Ontario, Canada
E. L. Balmasov and I. G. Fedorova
Karpinsky Geological Institute, St. Petersburg, Russia

ABSTRACT

Eleven samples (mostly of the Onaping Formation) from the 1.85-Ga Sudbury impact structure were examined with respect to their carbon phases. Six impact diamonds, <0.6 mm in diameter, were discovered in two samples of the Black Member of the Onaping Formation. These diamonds occur in a variety of colors, are cubic, and are generally friable due to inclusions and polycrystallinity. In one case, the hexagonal phase lonsdaleite was detected. According to transmission electron microscopy analyses, the mean grain size of individual diamond crystallites is in the range 50–100 nm. The diamonds are pervaded by numerous planar defects parallel to {111} and are organized in 100–200-nm-thick-bands, presumably inherited from precursor graphite. The diamonds were produced by the solid state transformation of graphite. Impact diamonds from the Onaping Formation are similar to other impact diamonds and are yet another expression of the effects of shock metamorphism at the Sudbury impact structure.

INTRODUCTION

Impact diamonds in terrestrial impact lithologies were first recognized at the 100-km diameter Popigai impact structure, Siberia (Masaitis et al., 1972), where they originate from the solid-state transformation of graphite in crystalline target rocks, subjected to shock pressure >35 GPa. At Popigai, impact diamonds most commonly occur together with diaplectic minerals and glasses in allochthonous basement clasts in the impact melt rocks, the so-called tagamites, and suevite breccias filling the crater interior (Masaitis, 1994).

Following their initial discovery at Popigai, impact diamonds were recognized in impactites at other impact structures in Russia (Kara, Puchezh-Katunki), Ukraine (Zapadnaja, Ternovka, Boltysh, Ilintsy), and Germany (Ries) (Rost et al., 1978; Gurov et al., 1985; Yezersky, 1986; Nikolsky, 1991; Valter et al., 1992; Khakhaev et al., 1994). These diamonds are polycrystalline, usually lonsdaleite-bearing, and are regarded as high-pressure carbon polymorphs, which originated from precursor graphite or coal in strongly shocked and melted target rocks. Diamonds have also been found in association with ejecta from Chicxulub, Mexico, in Cretaceous-Tertiary boundary deposits (Carlise and Braman, 1991; Gilmour et al., 1992). Recently, it has been suggested that at least some diamonds and other phases found in impact-related deposits formed from chemical vapor deposition in the ejecta plume (Hough et al., 1995, 1997).

It has recently been argued that impact diamonds should occur at some Canadian impact craters, with the 200–250-km diameter Sudbury impact structure, Ontario, a highly likely candidate (Grieve and Masaitis, 1996). Much has been written about the Sudbury impact structure (e.g., Grieve et al., 1991; Stöffler et al., 1994) since an impact origin was first proposed (Dietz, 1964). Although not all aspects of the impact origin for the structure are without controversy (cf. Grieve et al., 1991; Chai and Eckstrand, 1994), it is generally agreed that it is the tectonized remnant of a very large and ancient impact structure (Boerner and Milkereit, 1999). The most notable rocks of the area are the so-called Sudbury Igneous Complex, which is host to the world-class Ni-Cu-PGE deposits. The Igneous Complex is overlain by the 1.8-km-thick Onaping Formation, which consists of impact melt breccia, suevite, and reworked suevites. The Black Member of the Onaping Formation contains 0.06–>1.4% carbon, a part of which (maximum 0.5%) represents organic carbon (Avermann, 1992). The isotopic composition of organic carbon from the reworked suevites of the Black Member ($\delta^{13}C$ 31.1–31.06‰) indicates that it is similar to carbon of marine carbonate sediments (Avermann, 1992; Bunch et al., this volume; Heymann et al., this volume). The carbon matter is dispersed in the groundmass of the suevites, which contains fragments and particles of shocked Archean and Proterozoic and metasedimentary crystalline rocks and recrystallized impact glasses (e.g., Muir and Peredery, 1984; Dressler et al., 1996). A few hundred parts per million of fullerenes (C_{60} and C_{70}) have been reported within the Onaping Formation (Becker et al., 1994). They form under intense temperatures and pressures, but the source of carbon is unknown, although an origin from the impacting body is speculated (Becker et al., 1994). Recently, identification of fullerenes in the Onaping Formation has been questioned (Heymann et al., 1997).

EXPERIMENTAL TECHNIQUES

Eleven samples from the Sudbury Impact Structure (Table 1) were selected for carbon extraction in the laboratory of the Petrology Division of the Karpinsky Geological Institute, St. Petersburg, Russia. All samples were crushed and subsequently sieved into fractions <2mm. These fractions were washed to remove the <0.07-mm fraction and finally treated by a standard technique for the extraction of carbon phases (Kirikilitsa et al., 1981). The resulting concentrates were studied initially under the stereomicroscope to identify the different mineral phases and, in particular, crystalline carbon minerals. Subsequent characterization included use of scanning and transmisssion electron microscopy (SEM, TEM) and other techniques.

RESULTS

The following types of carbonaceous matter were recovered from the concentrates: (1) Small (~2 mm) tablet-shaped or isometric lumps: consisting of dispersed, fine-grained graphite; aggregates of this type disintegrate very easily and were observed in the early stages of the treatment of the samples. (2) Tiny drops and particles of liquid (yellow to brown) and solid (brown to black) hydrocarbons: observed at the same time as the lumps in (1); these displayed yellow luminescence under ultraviolet light but were not characterized further. (3) More stable aggregates of black carbon matter: similar to coal, grains of this type were less than 0.3 mm in diameter. (4) Monocrystalline graphite grains: plastic, with lustrous surfaces, the largest grains were up to 0.75 mm in diameter. In sample SUD-95-4, a graphite grain with pinacoid faces covered by dense twinning striations was observed. Plasticity and cleavage along (0001) are absent. These characteristics indicate recorded shock pressure up to 35 GPa (Valter et al., 1992). (5) Six small diamonds: extracted from the treatment residues of samples SUD-95-4 and TGA-SUD-6-1995.

Sample SUD-95-4 was divided into three parts; each one was treated separately, and in each one diamond grain was found. Three diamonds of similar dimensions were extracted from the second sample TGA-SUD-6-95. The diamonds are 0.1–0.6 mm

TABLE 1. SAMPLES EXAMINED IN THIS CHAPTER

Sample	Lithology	Weights of Working Fractions (g)
SUD–95–1	Fluidal "glass" from Black Member (Onaping)	1410
SUD–95–2	Aleurolite from the base of Chelmsford	70
SUD–95–3	Suevite from Gray Member (Onaping)	170
SUD–95–4	Suevite from Black Member (Onaping)	120, 310, 545
SUD–95–5	Suevite from Black Member (Onaping)	670
SUD–144–1995, 52847–220	Suevite from Black Member (Onaping)	105
SUD 52848, MS 1061	Suevite from Black Member (Onaping)	270
SUD 141–95, MS 115C	Suevite from Black Member (Onaping)	345
TGA–95, SUD6–1.140–95	Suevite from Black Member (Onaping)	640
S2487–50	Suevite from Black Member (Onaping)	115
TGA–SUD–6–95	Suevite from Black Member (Onaping)	400

across, irregular or flattened in shape, white, dark brown, gray, or black in color, and usually have a blocky appearance (Table 2; Fig.1) similar to the graphite grains recovered. The diamonds are fragile and can be easily disintegrated under weak pressure. For example, white, sugar-like grain (S-1) disintegrated into highly abrasive dust, which striated the carborundum test plate. The gray grain (S-2) decomposed into polyhedral fragments, with a stringy structure. Other gray grains (S-4, S-5) delaminated, mostly along tiny graphite inclusions, producing thin flakes. Dark colored grains (S-3, S-6) were more resistant and disintegrated into polyhedral fragments. All grains and their fragments are harder than the carborundum plate.

As visible by SEM (Fig. 2A), the irregular surfaces of the diamond grains display features of heavy corrosion, which result in a pitted honeycomb-like relief. The corrosion penetrates along fissures, boundaries of blocks, and twinning striations, and is responsible for the fragility of the grains (Fig. 2B). Graphite impurities in some grains (S-3, S-4, S-5, S-6) may be the result of high-temperature formation, as well as thermal corrosion during residence in the suevite breccias.

Under the petrographic microscope, the diamonds are generally not transparent, but some thin fragments exhibit semi-transparency. Thus, a fragment of the gray grain (S-5) showed strong birefringence ($\Delta n = 0.025$) and straight extinction, characteristic of apographitic impact diamonds with ordered crystalline structure (Masaitis et al., 1990). Similar observations were made on thin edges of the dark-brown grain (S-3), which contained very thin graphite flakes. In UV light (l = 300–400 μm), the diamonds exhibited yellow luminescence. Cathodoluminescence was absent.

An x-ray study of the diamond grains, using the diffractometer system RigokuD/max-RC Geigerflex, confirms that they are composed of a cubic phase (Fig. 3). Diffractograms of gray (S-2) and black (S-6) diamonds displayed a broadening of (111) reflections, but this peak itself is very weak. This indicates very low ordering of the crystalline framework in these diamonds. This is in contrast to gray ones (S-4 and S-5), which show an intense (111) peak. The width of this peak confirms the polycrystalline nature of these grains; moreover, the composing crystallites are partly disordered. Lonsdaleite was detected in black diamond (S-6) by the Debye-Scherrer technique. The lonsdaleite is polytype 2H, typical for solid-state transformations (Frondel and Marvin, 1967). In other grains, lonsdaleite and graphite were either below the detection limit (~5%) or these phases were absent.

Transmission electron microscopic analyses also indicated the polycrystallinity of the diamond aggregates, with mean grain sizes

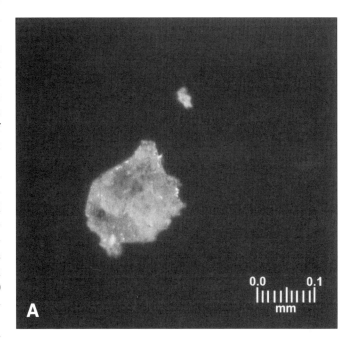

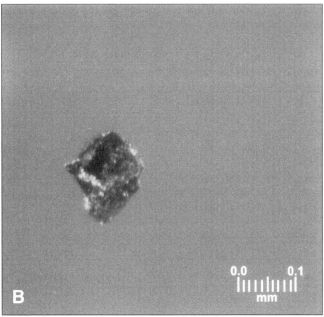

Figure 1. Impact diamonds extracted from the Black Member of the Onaping Formation. A, gray grain with laminated inner structure (S-5); B, irregular black grain (S-6).

TABLE 2. PROPERTIES OF DIAMONDS RECOVERED FROM BLACK ONAPING FORMATION

No.	Size (mm)	Color	Twinning	Graphite	Lonsdaleite
S-1	0.1	White	n.a	n.a.	n.a.
S-2	0.1	Gray	n.a.	Present	n.a.
S-3	0.2 x 0.2	Dark brown	Present	Present	Present
S-4	0.6 x 0.4 x 0.3	Gray (light)	n.a.	Present	n.a.
S-5	0.2 x 0.1	Gray (dark)	Present	Present	n.a.
S-6	0.3 x 0.3	Black	n.a.	Present	n.a.

Note: n.a. = not applicable.

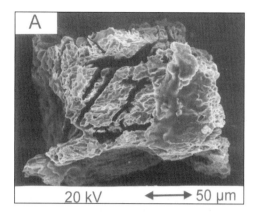

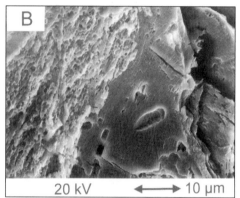

Figure 2. Secondary electron images of impact diamonds from Sudbury. A, Dark brown, lonsdaleite-bearing grain (S-3) with fine-pitted relief. B, Detailed view of the corroded surface of gray grain (S-5).

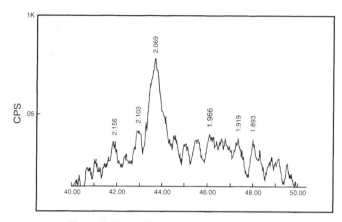

Figure 3. X-ray diagram of gray diamond S-4.

tural characteristics are equivalent to those of so-called diamond paramorphs from other impact craters and indicate a solid-state transformation of preexisting graphite by shock to form diamond (Masaitis et al., 1990). The electron paramagnetic resonance spectra of one diamond grain (S-4) was obtained in Q range at a frequency 37100 MHZ and 20°C temperature. To exclude the influence of paramagnetic centers in atmospheric air, the experiment was carried out in nitrogen. The grain was studied at magnetic field intensities of superhigh frequency. The intensity and form of the spectrum in the field of g-factor (g = 2.0030 ± 0.0003, ΔH = 4.5–4.8 Oe) and the lack of paramagnetic nitrogen were identical to that detected previously in the cubic phase of apographitic impact diamonds (Suchardjevskiy et al., 1992).

of individual crystallites in the 50–100-nm range (Fig. 4). The crystallites are organized in bands 100–200 nm thick, which may be inherited from the precursor graphite. This preferred orientation of crystallites can also be recognized by streakiness in selected area electron diffraction patterns. Individual diamond crystallites are pervaded by numerous planar defects parallel to {111}, which likely represent stacking faults or microtwins. These microstruc-

CONCLUSIONS

1. Six small diamonds were recovered from the reworked suevites of the Black Member of the Onaping Formation of the Sudbury Structure. The morphology, optical and textural properties, polycrystallinity, coexistence with lonsdaleite, absence of paramagnetic nitrogen, and microstructural characteristics are compatible with an impact origin of the diamonds.

Figure 4. A, TEM bright-field image of a fragment of gray diamond aggregate (S-5) mounted on a holey carbon grid. The polycrystalline fragment contains small (50 nm) crystallites, which are arranged in alternating layers. Crystallite in the lower left contains numerous planar defects. B, Selected area electron diffraction pattern of the entire fragment. Lens-shaped 111-reflections indicate the preferred orientation of crystallites. Streaks may be due to the planar defects.

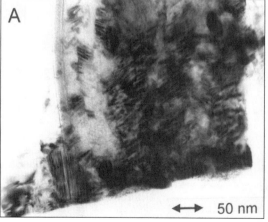
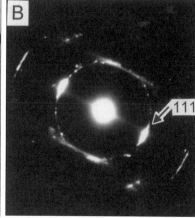

2. The diamonds are accompanied by other forms of carbonaceous matter, including shocked monocrystalline graphite. Similarity in the appearance of the different carbon phases indicates the diamonds formed by the solid-state transformation of polycrystalline and monocrystalline graphite precursors present in the target rocks at the time of impact.

3. The discovery of impact diamonds at the Sudbury Structure confirms, once again, the extraterrestrial origin of this geologic structure. The diamonds from the reworked suevites of the Black Member of the Onaping Formation are the most ancient impact diamonds discovered so far. Their age is coincident with the 1.85-Ga age of the Sudbury impact event (Krogh et al., 1996).

4. Although acknowledging that distant ejecta from the Sudbury impact event may be eroded away or strongly reworked during subsequent geologic events, it is possible that thin layers of this ejecta may be recognized in sequences of Paleoproterozoic rocks by the presence of impact diamonds similar to those found in the reworked suevite breccias of the Black Member of the Onaping Formation.

ACKNOWLEDGMENTS

We thank A. LeCheminant for a review of an early version of this manuscript and I. Gilmour, D. Heymann, and B. Dressler for an informative set of official reviews.

REFERENCES CITED

Avermann, M.-E., 1992, The genesis of the allochthonous polymict breccia of the Onaping Formation, Sudbury Structure, Ontario, Canada [Ph.D. thesis]: University of Munster, Munster, German, 175 p.

Becker, L., Bada, J. L., Winans, R. E., Hunt, J. E., Bunch, T. E., and French, B. M., 1994, Fullerenes at the 1.85 billion-year-old Sudbury impact site: Science, v. 265, p. 642–645.

Boerner, D. E., and Milkereit, B., 1999, Geophysical impact: studies of the Sudbury Structure: Canadian Journal Earth Sciences (in press).

Carlise, D. B., and Braman, D. R., 1991, Nanometre-size diamonds in Cretaceous/Tertiary boundary clays of Alberta: Nature, v. 352, p. 708–709.

Chai, G., and Eckstrand, R., 1994, Rare earth element characteristics and the origin of the Sudbury Igneous Complex: Chemical Geology, v. 113, p. 221–224.

Dietz, R. S., 1964, Sudbury Structure as an astrobleme: Journal of Geology, v. 72, p. 412–434.

Dressler, B. O., Weiser, T., and Brockmeyer, P., 1996, Recrystallized impact glasses of the Onaping Formation and the Sudbury Igneous Complex, Sudbury Structure, Ontario, Canada: Geochimica et Cosmochimica Acta, v. 60, p. 2019–2036.

Frondel, C., and Marvin, U. B., 1967, Lonsdaleite, a hexagonal polymorph of diamond: Nature, v. 214, p. 587–589.

Gilmour, I., Russel, S. S., Arden, J. W., Franchi, I. A., and Pillinger, C. T., 1992, Terrestrial carbon and nitrogen isotopic ratios from Cretaceous-Tertiary boundary nanodiamonds: Science, v. 258, 1624–1626.

Grieve, R. A. F., and Masaitis, V. L., 1996, Impact diamonds, in LeCheminant, A. N., Richardson, D. G., DiLabio, R. N. W., and Richardson, K. A., eds., Searching for diamonds in Canada: Geological Survey of Canada Open-File Report 3228, p. 183–186.

Grieve, R. A. F., Stöffler, D., and Deutsch, A., 1991, The Sudbury structure: controversial or misunderstood? Journal of Geophysical Research, v. 96, p. 22753–22764.

Gurov, E. P., Mielnichuk, E. V., Metalidi, S. V., Rjabenko, V. A., and Gurova, E. P., 1985, The peculiarities of geological structure of the eroded astrobleme in the western part of Ukrainian shield: Doklady AN Ukrainian SSR, Ser. B, no. 1, p. 9–12 (in Russian).

Heymann, D., Buseck, P. R., Knell, J., and Dressler, B. O., 1997, The search for fullerenes in rocks from the White Water Group of the Sudbury Structure; Large meteorite impacts and planetary evolution [Abs.]: Houston, Texas, Lunar and Planetary Institute, Contribution 922, p. 23.

Hough, R. M., Gilmour, I., Pillinger, C. T., Arden, J. W., Gillies, K. W. R., Yuan, J., and Milledge, H. J., 1995, Diamond and silicon carbide in impact melt rock from the Ries impact crater: Nature, v. 378, p. 41–44.

Hough, R. M., Masaitis, V. L., Gilmour, I., and Pillinger, C. T., 1997, Diamond and silicon carbide at the Ries and Popigai impact craters, in Proceedings, 28th, Lunar and Planetary Science Conference, Abstracts: Houston, Texas, Lunar and Planetary Institute p. 605–606.

Khakhaev, B. N., Masaitis, V. L., and Pevsner, L. A., 1994, Deep scientific drilling in the Puchezh-Katunki impact structure, in Proceedings, 7th International Symposium on the Observation of the Continental Crust through Drilling: Santa Fe, New Mexico, April 25-30, 1994, p. 208–210 (in Russian).

Kirikilitsa, S. I., Kashkarov, I. F., and Polkanov, Yu. I., 1981, Systematic recommendations for the recognition and extraction of fine diamonds: Simferopol, Institute of Mineral Resources, 106 p. (in Russian).

Krogh, T. E., Kamo, S. L., and Bohor, B. F., 1996, Shock metamorphosed zircons with correlated U-Pb discordance and melt rocks with concordant protolith ages indicate an impact origin for the Sudbury Structure. Earth processes: Reading the Isotopic Code: Washington, D.C., Geophysical Monograph 95, p. 343–353.

Masaitis, V. L., 1994, Impactites from the Popigai crater, in Dressler, B. O., Grieve, R. A. F., and Sharpton V. L., eds., Large meteorite impacts and planetary evolution: Geological Society of America Special Paper 293, p. 153–162.

Masaitis, V. L., Futergendler, S. I., and Gnevushev, M. A., 1972, Diamonds in impactites of the Popigai meteoritic crater: Zapiski Vsesoyuzngo Mineralogicheskogo Obshchestva, Pt. 101, no. 1, p. 108–112 (in Russian).

Masaitis, V. L., Shafranovsky, G. I., Yezersky, V. N., and Reshetnyak, N. B., 1990, Impact diamonds in ureilites and impactites: Meteoritika, v. 49, p. 180–196 (in Russian).

Muir, T. L., and Peredery W. V., 1984, The Onaping Formation, in Pye, E. G., Naldrett, A. J., and Giblin, P. E., eds., The geology and ore deposits of the Sudbury Structure: Ontario Geological Survey Special Volume 1, p. 139–210.

Nikolsky, A. P., 1991, Geology of the Pervomayskoe iron deposit and transformation of its structure caused by meteoritic impact: Moscow, Nedra Press, 71 p. (in Russian).

Rost, R., Dolgov, Yu. A., and Vishnevsky, S. A., 1978, Gases in inclusions of impact glasses of the Ries crater (Germany) and the discovery of high-pressure polymorphs of carbon: Doklady AN USSR, v. 24, p. 695–698 (in Russian).

Stöffler, D., Deutsch, A., Avermann, M., Bischoff, L., Brockmeyer, P., Buhl, D., Lakomy, R., and Müller-Mohr, V., 1994, The formation of the Sudbury Structure, Canada: towards a unified model, in Dressler, B. O., Grieve, R. A. F., and Sharpton, V. L., eds., Large meteorite impacts and planetary evolution: Geological Society of America Special Paper 293, p. 303–318.

Suchardjevskiy, S. M., Shafranovsky, G. I., and Balmasov, E. L., 1992, The EPR study of impact diamonds from astroblemes, in Proceedings, 23rd, Lunar and Planetary Science Conference, Abstracts: Houston, Texas, Lunar and Planetary Institute, p. 1381–1382.

Valter, A. A., Yeremenko, G. K., Kvasnitsa, V. N., and Polkanov, Yu. A., 1992, Impact metamorphic minerals of carbon: Kiev, Naukova Dumka Press, 171 p. (in Russian).

Yezersky, V. A., 1986, High-pressure polymorphs formed as a result of impact transformation of coal: Zapiski Vsesoyuznog Mineralogicheskogo, pt. 115, p. 26–33 (in Russian).

MANUSCRIPT ACCEPTED BY THE SOCIETY DECEMBER 16, 1998

Printed in U.S.A.

The Green Member of the Onaping Formation, the collapsed fireball layer of the Sudbury impact structure, Ontario, Canada

M. E. Avermann*
Institut für Planetologie, Westfälische Wilhelms–Universität Münster, Wilhelm-Klemm-Str. 10, D-48149 Münster, Germany

ABSTRACT

The Green Member, the chlorite-shard horizon of the older literature of the impact breccias of the Onaping Formation of the Sudbury Structure, is characterized by a unique, cryptocrystalline matrix and small, strongly tempered, and shock metamorphosed mineral and rock fragments. It is distinctly different from the clastic matrix breccias of the Gray Member underlying it and from those of the Black Member overlying it. The member is continuous around the Sudbury Basin, is up to 70 m thick, and is believed to represent the collapsed fireball of the Sudbury impact structure. As such, it originated from condensation of a vapor phase collapsed from high atmospheric regimes onto earlier deposited suevitic breccias of the Gray Member before deposition and redeposition of the Black Member breccias. The mineral and rock fragments of the Green Member represent early excavated fall-back components.

INTRODUCTION

The term Sudbury Structure comprises the central Sudbury Basin, the Sudbury Igneous Complex, as well as brecciated and shocked basement rocks surrounding the Sudbury Igneous Complex. These Archean and Proterozoic rocks of the Superior and Southern Provinces of the Canadian Shield (Card, 1978; Dressler, 1984a) represent the target rocks and occur as fragments in the investigated breccias of the Onaping Formation. The Sudbury Basin is elliptical in shape and contains the Proterozoic rocks of the Whitewater Group. This group consists of, from bottom to top, allogenic breccias of the Onaping Formation, pelagic sediments of the Onwatin Formation, and turbidites of the Chelmsford Formation (Rousell, 1984). The Sudbury event is dated by the 1.85-Ga age of the Sudbury Igneous Complex (Krogh et al., 1984).The South Range of the Sudbury Structure is strongly deformed (Shanks and Schwerdtner, 1991a,b; Milkereit et al., 1992; Cowan and Schwerdtner, 1997; Riller and Schwerdtner, 1997).

Since Dietz's (1964) work, the discussion of the origin of the Sudbury Structure has been controversial in favor of endogenic (Stevenson, 1972; Muir, 1984) or exogenic models (French, 1972; Dence, 1972; Peredery, 1972a; Peredery and Morrison, 1984; Dressler, 1984b; Dressler et al., 1987; Stöffler et al., 1989, 1994; Grieve et al., 1991). The Sudbury Structure is interpreted to represent the deeply eroded remnant of a multi-ring or peak-ring impact basin (Stöffler et al., 1989, 1994). In the same way the origin of the Onaping breccias has been debated for years. Endogenic (Stevenson, 1972; Muir, 1984; Gibbins, 1994) as well as impact-related (French, 1972; Peredery, 1972a,b; Brockmeyer, 1990; Avermann, 1994) models have been presented.

The polymict, allogenic breccias of the Onaping Formation occur in the Sudbury Basin, which is surrounded by the Sudbury Igneous Complex. I have mapped the breccias in detail in the southern East Range (Fig. 1) and in various profiles around the basin. The 1,800-m-thick allogenic breccia sequence of the Onaping Formation consists from bottom to top, the Basal, Gray, Green, and Black Members. Table 1 compares the nomenclatures applied for the Onaping breccias in the last 40 yr. The Green Member, *sensu* Brockmeyer (1990), is considered to be a continous uniform breccia layer on top of the Gray Member. It is the former "chlorite shard horizon" that was regarded by Muir and

---

*Present address: Forsthövel-Rörenstraße 3, 59387 Ascheberg, Germany; marile@gmx.net.

Avermann, M. E., 1999, The Green Member of the Onaping Formation, the collapsed fireball layer of the Sudbury impact structure, Ontario, Canada, *in* Dressler, B. O., and Sharpton, V. L., eds., Large Meteorite Impacts and Planetary Evolution II: Boulder, Colorado, Geological Society of America Special Paper 339.

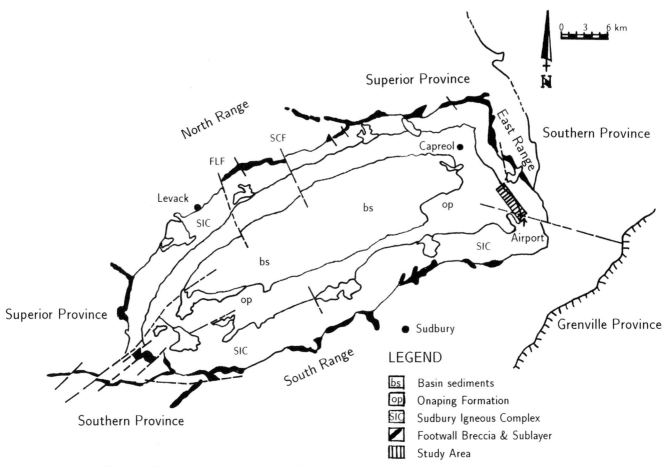

Figure 1. Geologic setting of the Sudbury Structure (i.e., Sudbury Basin, Sudbury Igneous Complex, and brecciated footwall adjacent to the Sudbury Igneous Complex not shown on the map) in the southern Canadian Shield and locality of the study area in the southern East Range of the Sudbury Structure. FLF = Fecunis Lake fault; SCF = Sandcherry Creek fault; numbers 1 through 5 refer to comparative profiles. Modified from Pye et al., 1984.

TABLE 1. VARIOUS STRATIGRAPHY OF ONAPING FORMATION

Present Study	Past Studies					
	Gibbins 1994	Brockmeyer 1990	Muir and Peredery 1984	Muir 1981, 1983	Peredery 1972a, 1972b	Stevenson 1961, 1972
Upper Black Member Lower Black Member	Dowling Member	Black Member	Black Member	Black Member	Black Onaping	Black tuff
Green Member	Contact unit	Green Member	Chlorite shard horizon	(Black Member)	Transition zone	n.d.
Gray Member	Sandcherry Member	Gray Member	Gray Member	Green Member	Gray Onaping	Gray tuff
Basal Member	Basal Intrusion	Basal Member	Melt body	Gray Member	Melt rock Melt body	Pepper-and-salt micropegmatite (chilled)
Basal Member	Basal Member	Basal Member	Basal Member	Basal Member	Basal breccia	Quartzite breccia (tectonic)

Modified from Muir and Peredery, 1984.

Peredery (1984) as the transition zone between the Gray and Black Members. Here the term is not used in the sense of Muir (1981, 1983), who called the Gray Member the "green member." Gibbins (1994) described a "Contact Unit" on the base of the Black Member that is marked by an increase in the proportion of chloritic shards.

The Green Member was mapped in detail in the southern East Range at a scale of 1:15,000. This unit was also investigated along comparative profiles in the northern East Range, as well as in the North and South Ranges. Although this unit is only up to 70 m thick, it is thought to be continuous around the structure (Dressler et al., 1987). It is unique; similar rocks have not been observed in other terrestrial impact structures. Muir and Peredery (1984) first described this 10–70-m-thick unit as a transition zone characterized by a sudden increase in carbon and the occurrence of chloritic shards. In Muir and Peredery (1984), Muir included the "chlorite shard horizon" in the Black Member, and he defined the Black Member as a unit consisting of chloritic shards, minor carbon, and a gray-greenish matrix. In Muir and Peredery (1984), however, Peredery put the transition zone underneath the Black Member. Brockmeyer (1990) mapped the transition zone in detail in the North Range of the Sudbury Structure, named it the "Green Member" because of the green color of its matrix, and reinterpreted the chloritized "shards" as secondary chlorite–filled collapsed cavities. According to Gibbins (1994), stratigraphy of the Onaping Formation is not due to a color change or the presence or absence of carbon, but only to the morphology and size of shards, and the percentage of matrix and lithic fragments.

Peredery (1972a) described a 15–70-m-thick transition zone (the Green Member of this chapter) between Gray and Black Members of the Onaping Formation as characterized by a sudden increase in carbon, and black, chloritized, glassy shards embedded in the matrix. The lower part of the Black Member, overlying the Green Member, contains —as does the Gray Member— many fragments of basement rocks. Upward, the Black Member exhibits a decrease in size and abundance of fragments.

PETROGRAPHY OF THE GREEN MEMBER

In the following description, the macroscopic and microscopic characteristics of the Green Member are described. Table 2 summarizes features for the subdivision of the Onaping Formation into the Gray, Green, and Lower Black Members. In the southern East Range, the attitude of the Green Member (140°, 30–50°SW) is in agreement with seismic reflection data of the granophyre of the Sudbury Igneous Complex and the Onaping Formation as a unit (Milkereit et al.,1992). The contact with the underlying Gray Member is sharp to gradational. Only one sharp contact traceable over 20 m was observed (Fig. 2). At this contact, the fragments in the Green Member are aligned parallel to the contact. Figure 3 shows a typical outcrop of the Green Member. In general, this unit is characterized by a brown weathering surface and (compared to the underlying Gray Member) by a decrease of melt inclusions, a decrease in the size of fragments and by the occurrence of chloritized particles. These chloritized particles were not observed in the Gray Member and make up as much as 27 vol% of the unit. Small (centimeter scale) melt inclusions and fragments of basement rocks are common. Rock fragments consist of metasediments (including dark siltstones), granites, and diabases. Mineral fragments, mainly quartz, and melt inclusions also occur. Fragments commonly are strongly shock metamorphosed and exhibit tempered rims.

The underlying Gray Member is a clastic matrix breccia. Clasts of basement rocks and recrystallized irregularly shaped melt inclusions are common. Fragments of metasediments make up more than 60% of the fragment population and vary in size. The remainder of the fragments are derived from the Archean basement. The contact of the Green Member with the overlying Lower Black Member is, in general, gradational over 2 m. Only one sharp upper contact in a specimen from the North Range was observed, it is shown in Figure 4.

The Lower Black Member above the Green Member is characterized by the onset of the black as well as clastic matrix, an increase of fragment volume and in size of fragments, an increase in melt inclusions, and by the occurrence of chloritized particles. It is 100–200-m thick and exhibits features (larger fragments, higher amount of melt inclusions) comparable to those observed in the Gray Member.

The matrix of the Green Member is gray-green and brownish weathering. Commonly it is schistose. In thin sections (Fig. 5), it is dense and microcrystalline and contains chloritized melt particles up to 0.5 mm in size, as well as chloritized melt inclusions. In images obtained by scanning electron microscopy (SEM) (Fig. 6), the crystalline character of the matrix is obvious. Only rare fragments are visible in the image in Figure 6. Other micrographs of thin sections not shown here exhibit recrystallized melt particles and rock and mineral fragments in the microcrystalline matrix, with streaks of secondary chloritized melt particles.

Avermann (1992) has investigated the various units of the Onaping Formation geochemically and concluded that the chemical composition of the various members of both the North and East Ranges reflect the chemical composition of target rocks. No meteoritic component could be recognized in either the matrix of the impact melt breccia of the Basal Member or the matrix of the Green Member. The average composition of recrystallized impact glasses of the suevitic breccias of the Onaping Formation, however, is similar to the composition of the granophyre of the Sudbury Igneous Complex (Dressler et al., 1996). This is an indication that at least the upper Igneous Complex is an impact melt.

DISCUSSION

Planar deformation features in rock and mineral clasts and recrystallized melt inclusions characterize the Gray Member breccias as suevitic, i.e., formed as the result of an impact of an asteroid or comet. Ignoring its carbon content, the 100–200-m-thick Lower Black Member also shows features of suevitic breccias and is overlain by redeposited Black Member rocks.

TABLE 2. PETROGRAPHIC FEATURES FOR THE SUBDIVISION OF
THE ONAPING FORMATION INTO THE GRAY, GREEN, AND LOWER BLACK MEMBERS

	Gray Member	Green Member	Lower Black Member
Outcrops	100–300 m wide zone, strikes NW–SE	10–25 m wide zone, strikes 140°, dips 30–50° SW	175–200 m wide zone
Upper contact	Sharp	Gradational/sharp	Gradational
Lower contact	Gradational/sharp	Gradational	Gradational
Fragment content	Fine-grained arkoses, coarse-grained metasediments, graywackes, conglomerates, granites, gneisses, volcanics, gabbros, melt inclusions, breccia fragments	Metasediments, granites, diabases, mafic fragments, siltstones, melt inclusions, chloritized particles	Metasediments (arkoses, graywackes, sandstones, cherts), granites, gneisses, migmatites, mafic fragments, breccia fragments, chloritized particles, sulfides, melt inclusions
Fragment volume	As much as 70%	~70%	~53%
Fragment shape	Angular, rounded, with streaks, cut, pointed	Angular, rounded, with streaks, cut, pointed	Angular, rounded
Fragment size	Few mm–2m	Few mm–cm	Few mm–50cm
Matrix	Dark gray, high clast content	Gray, graygreen, dense	Dark, unresolvable
Texture	Clastic	Cryptocrystalline, with streaks	Clastic
Mineralogy	Alkali feldspar-granitic, quartzmonzodioritic	Alkali feldspar-granitic/ granitic-granodioritic	Alkali feldspar-granitic/ tonalitic
Fragments	Recrystallized, tempered, sometimes spherulithic, as well plastically deformed as relict texture	Recrystallized, reaction rims, corrosion rims, common relict texture	Recrystallized, relict texture with decorated, planar elements
Shock stage	Stages 0–IV	Stages I–IV	Stages 0–IV

Figure 2. Sharp contact between the Gray (bottom of photograph) and the Green Members (top of photograph), marked by arrows. Scale bar in centimeters. Location of outcrop is northwest of Sudbury Airport in the southern East Range.

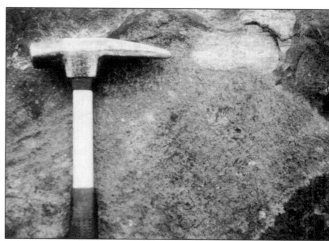

Figure 3. Typical outcrop of the Green Member of the Onaping Formation as it can be seen in the study area. Hammer for scale. Location of outcrop is northwest of Sudbury Airport in the southern East Range.

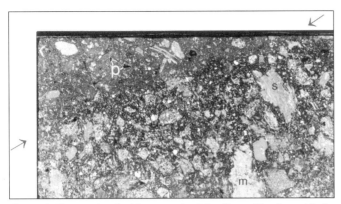

Figure 4. Contact (marked by arrows) of the Green Member with the Black Member. The Green Member (upper left of photograph) is characterized by a dense matrix with small fragments and chloritized particles (p). The matrix of the Black Member (bottom of photograph) contains metasediments (m) and melt inclusions (s). Rock slab = 4 cm long. Locations is North Range, near Fecunis Lake fault.

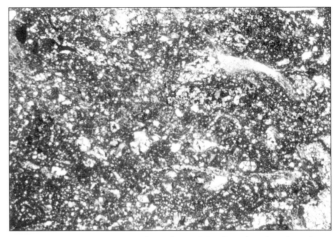

Figure 5. Photomicrograph of the Green Member showing the microcrystalline matrix with chloritized melt particles and melt inclusions. Crossed polarizers; photograph = 5.3 mm wide.

Ames and Gibson (1995) described peperite bodies near the top of the Onaping Formation and concluded that magmatism occurred throughout Onaping time and therefore deposition of the Onaping Formation was not an instantaneous event. After Ames and Gibson (1995, p. 171), "this is evidence against a meteorite fall-back origin for the breccia." This is, however, dubious. It is correct that fall-back breccias are deposited instantaneously, but neither the Black Member nor the whole Onaping Formation is interpreted as a fall-back breccia. Ames and Gibson (1995) argued for a unit near the top of the Onaping Formation; that means 1,000 m above breccias of the Gray Member, which are considered by many authors to be fall-back breccias (French, 1972; Peredery, 1972a,b; Peredery and Morrison, 1984; Dressler et al., 1987; Avermann and Brockmeyer, 1992; Avermann, 1994; Stöffler et al., 1994). The breccias of the Gray Member are equivalent to those of suevitic breccias, but based on its petrographic character, the lower part of the Gray Member is interpreted as a ground-surge deposit that grades into fall-back breccias (Avermann and Brockmeyer, 1992; Avermann, 1994). Most of the excavated material was deposited outside the crater and redeposited to some extent into the central depression of the crater. The suevitic breccias of the Lower Black Member are considered to have been deposited this way. Further on, the rocks of the Upper Black Member are interpreted neither as fall-back breccias nor as having been deposited instantaneously, but instead as reworked impact breccias. Many sedimentary features and many authors (Peredery, 1972a; Beales and Lozej, 1975; Muir, 1981, 1983; Muir and Peredery, 1984; Brockmeyer, 1990) showed that these rocks were deposited under water. This is not in conflict with the impact theory, which tells us that these rocks once were created by a giant impact and then were reworked and preserved in the Sudbury Basin.

The significance of the Green Member between the gray and black suevitic units was not recognized prior to investigations by Brockmeyer (1990) and Avermann (1992). Melosh and Pierazzo

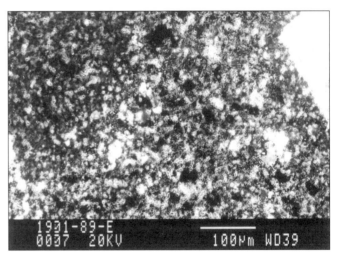

Figure 6. Scanning electron microscope image of the Green Member, which reveals the crystalline character of the matrix. Bright spots indicate quartz; dark spots, sulfides. Location is southern East Range.

(1997) recently argued that plume dynamics may play a significant role in the formation of large impact structures. Specific deposits may result from the collapse of fireball plumes as part of the late-stage impact process. Melosh and Vickery (1991) presented a theory for estimating the sizes of liquid droplets as a function of impactor size and velocity. For a projectile with a 10-km diameter and an impact velocity about 20 km/s these authors estimated the spherule size as $r_\infty = 550$ µm. This is consistent with the presumed diameter of the projectile and with the observed size of melt droplets in the Green Member of the Sudbury Structure. The Green Member rocks are here interpreted to represent the collapsed fireball horizon of the Sudbury Impact Structure. It exhibits a cryptocrystalline matrix suggesting deposition of fine melt particles raining out of the collapsing impact plume. The fragments of the Green Member are strongly shock

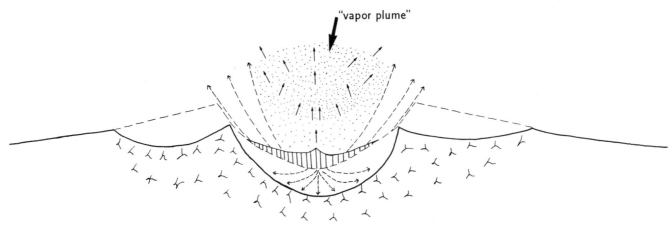

Figure 7. Diagrammatic vapor plume model for the origin of the Green Member from early excavation to high atmospheric regimes. Sketch is not to scale. (Modified from Melosh, 1989).

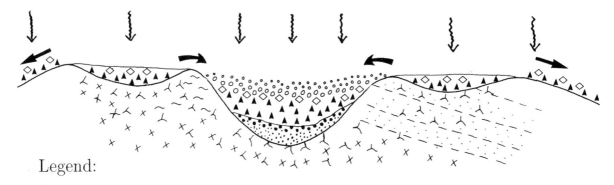

Legend:
- Onwatin slates
- Upper Black Member
- Lower Black Member
- Green Member
- Gray Member
- Basal Member
- Sudbury Igneous Complex
- Sudbury Breccia
- Huronian sediments

Figure 8. Diagrammatic sketch of the deposition of the Black Member units with continuing rain out of melt particles. Sketch is not to scale.

metamorphosed and exhibit evidence that they were subjected to high temperatures. The Green Member originated from early excavation to high atmospheric regimes, condensation of a vapor phase (Fig. 7), and deposition with the final fall-back material. Rainout of melt particles continued into the Lower Black Member (Fig. 8) and is responsible for its gradational contact with the Green Member. Bunch et al. (1997) argued that a conversion of target carbonate to carbon may have happened in the vapor plume. However, carbon is not present in the Green Member rocks but may have resided in the atmosphere for a somewhat longer period of time than it took for melt particles to rain out.

CONCLUDING REMARKS

The Green Member of the Onaping Formation on top of the Gray Member is a melt-rich fall-back layer and is interpreted to represent the termination of the excavation phase of the impact process. Its petrographic characteristics suggest that it was formed by the rainout of melt particles and strongly shocked target rock fragments. This process is in agreement with a vapor-plume model proposed by Melosh (1989). The rainout of melt particles continued into the Lower Black Member, as suggested by the gradational contact between the Green and Black Members. The Gray Member consists of ground-surge deposits and suevitic fall-back material. The rocks of the Lower Black Member have been interpreted as suevitic material redeposited into the central depression of the crater during the modification stage of the impact process. The Upper Black Member represents reworked impact breccias.

Contrary to Gibbins' (1994) interpretation, I consider color changes of breccias and the presence of carbon to be diagnostic and to give important hints to the origin of these breccias. However, a once-ubiquitous distribution of carbon throughout the Onaping Formation and a subsequent oxidation and mobilization upward, as proposed by Ames and Gibson (1995), is not possible for the following reasons:

1. Recent investigations (Avermann, 1994; Heymann et al., this volume) showed that most of the carbon in the Onaping Formation is organic. If the carbon was once distributed ubiquitously, as proposed by Ames and Gibson (1995), it must have originated (besides the interpretation of the Onaping Formation) from melt breccias, tuff breccias, or fall-back breccias.

2. Most of the organic carbon is concentrated in the Black Member of the Onaping Formation, but it is combined only with the matrix of the Black Member and not with fragments or even with Gray Member fragments in the Black Member, which I believe explains the observed absence of carbon in silicified rocks in the Dowling member as described by Ames and Gibson(1995). A hydrothermal mobilization, however, would have touched the whole-rock unit.

3. Carbon is concentrated in the Black Member, but hydrothermal mobilization throughout the sealing cover of the Green Member (Heymann et al., this volume) is not conceivable.

4. A once-ubiquitous distribution of carbon does not explain the origin of the matrix carbon. Heymann and Dressler (1997) showed that there was not enough carbon in the target rocks prior to the Sudbury event.

Therefore, carbonaceous material in the Black Member is interpreted to have formed as a result of biogenetic activity in a restricted, euxinic basin (Avermann, 1994; Heymann et al., this volume).

ACKNOWLEDGMENTS

I am greatly indebted to B. O. Dressler for the introduction to the geology and long-lasting fascination of the Sudbury Structure. Thanks also for the considerable field support by the Ontario Geological Survey. The manuscript was improved by the comments of B. O. Dressler, G. G. Morrison, and an anonymous reviewer.

REFERENCES CITED

Ames, D. E., and Gibson, H. L., 1995, Controls on and geological setting of regional hydrothermal alteration within the Onaping Formation, footwall to the Errington and Vermilion base metal deposits, Sudbury Structure, Ontario, in Current Research 1995-E: Ottawa, Geological Survey of Canada, p. 161–173.

Avermann, M., 1992, Die Genese der allochthonen, polymikten Breccien der Onaping-Formation, Sudbury-Struktur, Ontario, Kanada [Ph.D. thesis]: Münster, Germany, University of Münster, 175 p.

Avermann, M., 1994, Origin of the polymict, allochthonous breccias of the Onaping Formation, Sudbury Structure, Ontario, Canada, in Dressler, B. O., Grieve, R. A. F., and Sharpton, V. L., eds., Large meteorite impacts and planetary evolution: Geological Society of America Special Paper 293, p. 264–274.

Avermann, M., and Brockmeyer, P., 1992, The Onaping Formation of the Sudbury Structure, Canada: an example of allochthonous impact breccias: Tectonophysics, v. 216, p. 227–234.

Beales, F. W., and Lozej, G. P., 1975, Sudbury Basin sediments and the meteoritic impact theory of origin for the Sudbury Structure: Canadian Journal of Earth Sciences, v. 12, p. 629–635.

Brockmeyer, P., 1990, Petrographische und geochemische Untersuchungen an polymikten Breccien der Onaping-Formation, Sudbury-Distrikt (Ontario, Kanada), [Ph.D. thesis]: Münster, Germany, University of Münster, 228 p.

Bunch, T. E., Becker, L., Schultz, P. H., and Wolbach, W. S., 1997, New potential sources for Black Onaping carbon, in Conference on Large Meteorite Impacts and Planetary Evolution (Sudbury 97): Houston, Texas, Lunar and Planetary Institute, Contribution 922, p. 7.

Card, K. D., 1978, Geology of the Sudbury-Manitoulin area, District of Sudbury and Manitoulin: Ontario Geological Survey Report 166, 238 p.

Cowan, E. J., and Schwerdtner, W. M., 1997, Funnel-shaped emplacement geometry of the Sudbury Igneous Complex: a structural perspective, in Conference on Large Meteorite Impacts and Planetary Evolution (Sudbury 1997): Houston, Texas, Lunar and Planetary Institute, Contribution 922, p. 11.

Dence, M. R., 1972, Meteorite impact craters and the structure of the Sudbury Basin, in Guy-Bray, J. V., ed., New developments in Sudbury geology: Geological Association of Canada Special Paper 10, p. 7–18.

Dietz, R. S., 1964, Sudbury Structure as an astrobleme: Journal of Geology, v. 72, p. 412–434.

Dressler, B. O., 1984a, General geology of the Sudbury area, in Pye, E. G., Naldrett, A. J., and Giblin, P. E., eds., The geology and ore deposits of the Sudbury Structure: Toronto, Ministry of Natural Resources, Ontario Geological Survey Special Volume 1, p. 57–82.

Dressler, B. O., 1984b, The effects of the Sudbury event and the intrusion of the Sudbury Igneous Complex on the Footwall rocks of the Sudbury Structure, in Pye, E. G., Naldrett, A. J., and Giblin, P. E., eds., The geology and ore deposits of the Sudbury Structure: Toronto, Ministry of Natural Resources, Ontario Geological Survey Special Volume 1, p. 97–136.

Dressler, B. O., Morrison, G. G., Peredery, W. V., and Rao, B. V.,1987, The Sudbury Structure, Ontario, Canada —A review, in Pohl J., ed., Research in terrestrial impact structures: earth evolution sciences: Braunschweig/Wiesbaden, Germany, Friedrich Vieweg & Sons, p. 39–68.

Dressler, B. O., Weiser, T., and Brockmeyer, P., 1996, Recrystallized impact glasses of the Onaping Formation and the Sudbury Igneous Complex, Sudbury Structure, Ontario, Canada: Geochimica et Cosmochimica acta, vol. 60, p. 2019–2036.

French, B. M., 1972, Shock metamorphic features in the Sudbury Structure, Ontario: a review, in Guy-Bray, J. V., ed., New developments in Sudbury geology: Geological Association of Canada Special Paper 10, p. 19–28.

Gibbins, S. F. M., 1994, Geology, geochemistry, stratigraphy, and mechanism of emplacement of the Onaping Formation, Dowling Area, Sudbury Structure, Ontario, Canada [M.S. thesis]:Sudbury, Ontario, Canada, Laurentian University.

Grieve, R. A. F., Stöffler, D., and Deutsch, A., 1991,The Sudbury Structure: controversial or misunderstood?: Journal of Geophysical Research, v. 96, E5, p. 22,753–22,764.

Heymann, D., and Dressler, B. O., 1997, Origin of carbonaceous matter in rocks from the Whitewater Group of the Sudbury Structure, in Conference on Large Meteorite Impacts and Planetary Evolution (Sudbury 97): Houston, Texas, Lunar and Planetary Institute, Contribution 922, p. 21.

Krogh, T. E., Davis, D. W., and Corfu, F., 1984, Precise U-Pb zircon and baddeleyite ages for the Sudbury area, in Pye, E. G., Naldrett, A. J., and Giblin, P. E., eds., The geology and ore deposits of the Sudbury Structure: Toronto, Ministry of Natural Resources, Ontario Geological Survey Special Volume 1, p. 431–446.

Melosh, H. J., 1989, Impact Cratering—a geologic process: New York, Oxford University Press, 245 p.

Melosh, H. J., and Pierazzo, E., 1997, Impact vapor plume expansion with realistic geometry and equation of state: 28th Lunar and Planetary Science Conference, Houston, p. 935–936.

Melosh, H. J., and Vickery, A. M., 1991, Melt droplet formation in energetic impact events: Nature, v. 350, p. 494–497.

Milkereit, B., and 25 others, 1992, Geometry of the Sudbury Structure from high-resolution seismic reflection profiling: Geology, v. 20, p. 807–811.

Muir, T. L., 1981, Geology of Capreol area, District of Sudbury: Toronto, Ontario Geological Survey Open-File Report 5344, 168 p., with 4 maps, scale 1:15,840.

Muir, T. L., 1983, Geology of Morgan Lake, Nelson Lake area, District of Sudbury: Toronto, Ontario Geological Survey Open-File Report 5426, 203 p., with 3 maps, scale 1:15,840.

Muir, T. L., 1984, The Sudbury Structure: Considerations and models for an endogenic origin, *in* Pye, E. G., Naldrett, A. J., and Giblin, P. E., eds., The geology and ore deposits of the Sudbury Structure: Toronto, Ministry of Natural Resources, Ontario Geological Survey Special Volume 1, p. 450–489.

Muir, T. L., and Peredery, W. V., 1984, The Onaping Formation, *in* Pye, E. G., Naldrett, A. J., and Giblin, P. E., eds., The geology and ore deposits of the Sudbury Structure: Toronto, Ministry of Natural Resources, Ontario Geological Survey, Special Volume 1, p. 139–210.

Peredery, W. V., 1972a, The origin of rocks at the base of the Onaping Formation, Sudbury, Ontario [Ph.D. thesis]:Toronto, Ontario, University of Toronto, 366 p.

Peredery, W. V., 1972b, Chemistry of fluidal glasses and melt bodies in the Onaping Formation, *in* Guy-Bray, J. V., ed., New developments in Sudbury geology: Geological Association of Canada, Special Paper No. 10, p. 49–59.

Peredery, W. V., and Morrison, G. G., 1984, Discussions of the origin of the Sudbury structure, *in* Pye, E. G., Naldrett, A. J., and Giblin, P. E., eds., The geology and ore deposits of the Sudbury Structure: Toronto, Ministry of Natural Resources, Ontario Geological Survey, Special Volume 1, p. 491–511.

Pye, E. G., Naldrett, A. J., and Giblin, P. E., eds., 1984, The geology and ore deposits of the Sudbury Structure: Toronto, Ministry of Natural Resources, Ontario Geological Survey, Special Vol. 1, 603 p.

Riller, U., and Schwerdtner, W. M., 1997, Paleoproterozoic tectonism in the eastern Penokean orogen and its significance for the origin of the Norite of the Sudbury Igneous Complex, *in* Conference on Large Meteorite Impacts and Planetary Evolution (Sudbury 1997): Houston, Texas, Lunar and Planetary Institute, Contribution 922, p. 48.

Rousell, D. H., 1984, Onwatin and Chelmsford Formations, *in* Pye, E. G., Naldrett, A. J., and Giblin, P. E., eds., The geology and ore deposits of the Sudbury Structure: Toronto, Ministry of Natural Resources, Ontario Geological Survey Special Volume 1, p. 211–218.

Shanks, W. S., and Schwerdtner, W. M., 1991a, Structural analysis of the central and southwestern Sudbury Structure, southern province, Canadian Shield: Canadian Journal of Earth Sciences, v. 28, p. 411–430.

Shanks, W. S., and Schwerdtner, W. M., 1991b, Crude quantitative estimates of the original northwest-southeast dimension of the Sudbury Structure, south central Canadian Shield: Canadian Journal of Earth Sciences, v. 28, p. 1677–1686.

Stevenson, J. S., 1961, Recognition of the Quartzite Breccia in the Whitewater Series, Sudbury Basin, Ontario: Transactions of the Royal Society of Canada, v. 55, ser. 3, p. 57–66.

Stevenson, J. S., 1972, The Onaping ash-flow sheet, Sudbury, Ontario, in Guy-Bray, J. V., ed., New developments in Sudbury Geology: Geological Association of Canada Special Paper 10, p. 41–48.

Stöffler, D., Avermann, M., Bischoff, L., Brockmeyer, P., Deutsch, A., Dressler, B. O., Lakomy, R., and Müller-Mohr, V.,1989, Sudbury, Canada: remnant of the only multi-ring(?) impact basin on Earth: Meteoritics, v. 24, p. 328.

Stöffler, D., Deutsch, A., Avermann, M., Bischoff, L., Brockmeyer, P., Buhl, D., Lakomy, R., and Müller-Mohr, V., 1994,The formation of the Sudbury Structure, Canada: towards a unified impact model, *in* Dressler, B. O., Grieve, R. A. F., and Sharpton, V. L., eds., Large Meteorite Impacts and Planetary Evolution: Geological Society of America Special Paper 293, p. 303–318.

Manuscript Accepted by the Society December 16, 1998

Carbonaceous matter in the rocks of the Sudbury Basin, Ontario, Canada

Ted E. Bunch
Space Science Division, Ames Research Center, NASA, Moffett Field, California 94035, USA
Luann Becker
Institute of Geophysics and Planetology, University of Hawaii, Honolulu, Hawaii 96822, USA
David Des Marais, Anne Tharpe
Space Science Division, Ames Research Center, NASA, Moffett Field, California 94035, USA
Peter H. Schultz
Department of Geological Sciences, Brown University, Providence, Rhode Island 02912, USA
Wendy Wolbach
Department of Chemistry, DePaul University, Chicago, Illinois 60614, USA
Daniel P. Glavin, Karen L. Brinton, Jeffrey L. Bada
Scripps Institution of Oceanography, University of California at San Diego, La Jolla, California 92093, USA

ABSTRACT

Petrographic, chemical, and isotopic analyses and field studies have found a probable source for the abundant carbon in the Onaping Black Member. Fragments of a pre-impact, C-rich mudstone, found throughout the breccia deposits, contain an average of ~15 vol% kerogen which has a $\delta^{13}C$ of –30 per mil, comparable to the bulk average of carbonaceous matter in the Black Member and to "poorly graphitized carbon" (PGC) present as fine particulates in the breccias. We suggest that shock processing on impact or in the impact cloud reduced some of the kerogen to PGC, which then fell back into ongoing deposition, particularly in the lower Black Member. PGC decreases in content upward through the deposits and is replaced by unprocessed kerogen as the main contributor to the carbon inventory. Soot of classic shape but smaller size has been identified and quantified in the lower Onwatin. Future investigations may show that the soot extends downward into the upper Black Member. The D/L ratios for detected aminos acids imply that they are both terrestrial in origin and of recent geologic age (<200 k.y.). In addition to the fullerenes reported earlier, other fullerene species of C_{74}, C_{78}, C_{84}, and C_{100} have been identified. Two sources for the fullerenes in the Sudbury impact deposits were that the fullerences synthesized during the impact event and/or in the impact cloud, or that they were present in the bolide and survived impact. Accretionary-mantled clasts (up to 5 cm in size) and accretionary lapilli (≤1 mm) are found mostly in the lower Black Member. The presence of lapilli, in particular, implies that the impact debris cloud was still active over the impact area during crater wall slumping and resurge-driven deposition of the lower Black Member. We attribute most of the carbon inventory in the Black Member to shock processing of preexisting terrestrial carbonaceous matter, although we do not preclude input from the impacting bolide and late depositional stage biotic activity.

Bunch, T. E., Becker, L., Des Marais, D., Tharpe, A., Schultz, P. H., Wolbach, W., Glavin, D. P., Brinton, K. L., and Bada, J. L., 1999, Carbonaceous matter in the rocks of the Sudbury Basin, Ontario, Canada, *in* Dressler, B. O., and Sharpton, V. L., eds., Large Meteorite Impacts and Planetary Evolution II: Boulder, Colorado, Geological Society of America Special Paper 339.

INTRODUCTION

The Sudbury Impact Structure is known for structural and mineralogic impact signatures, rich ore deposits, and an unusual abundance of carbonaceous matter in the Black Member of the Onaping Formation. Although the presence of carbon has been known for some time (Burrows and Rickaby, 1930), few efforts have been made to systematically characterize carbonaceous components. Recently, Heymann and Dressler (1997a) determined that a large portion of the carbonaceous material is highly disordered graphite traditionally referred to as "poorly graphitized carbon" (PGC). Avermann (1994) and Heymann et al. (1997a,b) analyzed Black Member samples for bulk carbon and $\delta^{13}C$ contents. Fullerenes (C_{60} and C_{70}) have been found in the Black Member (Becker et al., 1994), in addition to "caged" He (C_{60}@He) in some fullerenes of possible interstellar origins (Becker et al., 1996). An initial search for the origin of the carbonaceous components of the Black Member was done by Bunch et al. (1997). Whitehead et al., (1990) and Arengi (1977) focused investigations on carbon characteristics in the lower Onwatin and contact unit.

The bulk elemental carbon content in the Sudbury Impact Structure deposits is unique among terrestrial impact craters. Understanding the nature of all carbonaceous components, in addition to knowing the source(s) may give us better insight into (1) survivability/synthesis of exogenous organic compounds on impact; (2) endogenous evolution of impact and postimpact abiotic/prebiotic/carbonaceous chemistry; (3) impact plume chemistry, products of which may occur in fall-back materials; and (4) time constraints on postimpact events, e.g., depositional rates of the Black Member and the lower Onwatin Formation.

A research consortium was formed in 1995 in order to address these and other issues regarding postcratering events. In this chapter we describe preliminary characteristics of several carbonaceous components in the upper and lower Black Member of the Onaping Formation and the lower Onwatin Formation, and a potential source rock for much of the carbon. These results are then used to place constraints on the deposition of the postimpact stratigraphy and the physicochemical pathways taken by carbon during and after impact.

GEOLOGIC SETTING

The Onaping Formation consists of a variety of impact and nonimpact deposits and is subdivided into the Basal, Gray, Green, and Black Members (e.g., Avermann and Brockmeyer, 1992; Stöffler et al., 1994). The Basal Member is presently viewed as a clast-rich impact melt breccia (<300 m thick); it is overlain by the Gray Member (< 600 m), a suevitic fall-back breccia that contains shocked minerals and clasts, in addition to flow-textured, vesiculated melt inclusions. The thin-bedded Green Member (3–<70 m thick) (Brockmeyer, 1990) consists of fine-grained, clastic material embedded in a clastic matrix, both of which are heavily chloritized. Avermann and Brockmeyer (1992) and Avermann (this volume) have concluded that this member represents an agglomeration of melt and clastic debris that was "ballistically" deposited and marks the end of melt particle and rock fall-back within the crater. The Black Member traditionally consists of two units. The lower unit (<200 m thick) is partly suevitic and contains fragments of the underlying members in a fine-grained clastic carbonaceous matrix (organic carbon = 0.2–<2.4 wt%) (Avermann, 1994; Heymann et al., 1997b). The upper unit is roughly gradational over ~600 m with respect to fewer and smaller clasts and increasing matrix, in addition to an increasing ratio of unshocked to shocked clasts; moreover, it displays characteristics consistent with reworking and sedimentation in an aquatic environment (Avermann, 1994). Detailed descriptions of the entire Black Member, as well as other members of the Onaping Formation, are given by Muir and Peredery (1984).

Gibbons et al. (1997) and Ames et al. (1998) suggested a different origin for the Onaping: emplacement by the continuous introduction of a melt of unspecified origin into a subaqueous environment which then produced explosively derived "hyaloclastites." In addition, they subdivided the Onaping on the basis of shard content and morphology into the Sandcherry Member (Basal and Gray Members) and the Dowling Member (Black Member); the contact between the two is a welded unit (Green Member).

The very upper portion (~200 m) of the Black Member is commonly fine grained with particles rarely greater than 2 mm. It signifies the transition from the somewhat chaotic character of the underlying lower and upper units to the very fine grained C-rich argillaceous rocks (up to 5 wt% organic carbon) (Arengi, 1977) of the Onwatin Formation. Whitehead et al. (1990) referred to this portion of the upper Black Member as the transitional unit, a term used in this chapter. In addition, a large section (~200 m) of the rocks between the lower and upper units is sufficiently different in textural characteristics and components to conveniently subdivide the Black Member into lower (~200 m), middle (~200 m), upper (~200 m), and transitional (~200 m) units. For examples of unit rock types, see Figure 1. These provisional units allow us to present data consistent with gradual, but distinct, changes within the vertically extensive Black Member and into the lower Onwatin Formation.

As noted by others (e.g., Fleet et al., 1987), rocks in the South Range and portions of the upper and transitional units in southeast corner of the East Range have been regionally metamorphosed to the amphibolite facies and those in the North Range to the Greenschist Facies.

FIELD SAMPLING AND ANALYTICAL METHODOLGY

Field collections

Samples were taken along the North, East, and South Range exposures to ensure good areal and vertical distributions of components. Collections were made of the lower Black Member at

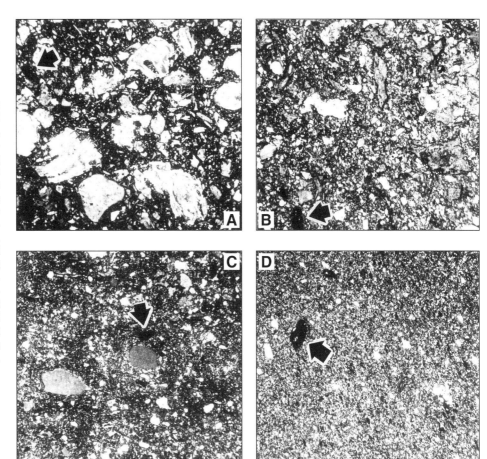

Figure 1. Photomicrographs (plane light) showing textures and grain size typical of each unit of the Black Member. Arrows point to C-rich mudstone fragments. A, Lower unit sample showing light to greenish, altered glass fragments and dark matrix enriched in carbon. Bulk carbon content is 0.83 wt%. B, Sample from the middle unit with slightly smaller grain size and lighter, less carbon-rich matrix (carbon = 0.41 wt%). Glass fragments are less angular and mostly brownish in color. C, Upper unit with fewer glass fragments and numerous metasedimentary fragments. Carbonaceous matter less evenly distributed (carbon = 0.56 wt%). D, Transitional unit showing smaller fragments and reduced size, in addition to much less matrix carbon (carbon = 0.50 wt%). Base width of each 15 mm.

Highfalls (Dowling Township), Nelson and Frenchman Lakes, north and east of Capreol, and along the new Northwest Bypass (Highway 144) south of Chelmsford. These samples were collected with stratigraphic reference to the lower contact where possible. Sampling in the Highfalls area was made at 3–5-m intervals along both sides of the Onaping River, highway, and railroad cuts where continuous exposures afforded good vertical control. Thus the first ~100 m of the lower Black Member were sampled with known stratigraphic position. Other units were collected south and southeast of Capreol and along the Northwest Bypass, but with only estimates of stratigraphic position due to large covered intervals and only inferred contact locations. Because the contact of the Onaping and Onwatin is very poorly exposed, we obtained core drill samples of the gradational contact unit and 10 m on both sides of the contact, courtesy of Inco and Falconbridge mining companies.

Isotopic measurements of carbon

Analyses of samples selected from the entire Black Member with reference to stratigraphic position were carried out using the following procedures for kerogen-bearing rocks. Rock processing for the measurement of carbon abundance and stable isotopic composition is initiated by discarding any weathered surfaces or secondarily emplaced material (e.g., veins), thus creating an aliquot of clean interior rock chips using a chipmunk crusher. To remove any organic contamination, these chips are treated with a 20% HF–10% HCl solution, then rinsed with large amounts of water, and dried in an oven at 75°C. The process is repeated until about 10% dry weight is lost by dissolution. These cleaned rock chips are then pulverized in a ring and puck mill that has been precleaned with water and methanol extraction, followed by grinding of fired Ottawa sand. The pulverized rock is then dissolved in a mixture of concentrated HF and boric acid (Robl and Davis, 1993), which dissolves silicate and carbonate phases without creating insoluble fluoride precipitates. The resulting slurry is centrifuged and decanted, and the insoluble is residue retreated with the HF/boric acid mixture. After recentrifugation and decanting, the insoluble residue is rinsed with distilled water to neutral pH. The residue is then dried at 75°C in an oven. The residue is added to a prefired (at 900°C) 9-mm quartz tube, that is presealed at one end. About 1 g of CuO (previously cleaned of any organic contaminants by heating at 700°C for several hours) and 200 mg of silver foil (precleaned in hydrogen gas at 700°C) are added to the tube which is then evacuated on a vacuum line and sealed using a torch. This tube is then combusted at 850°C for 4 h in a muffle furnace. On cooling, one end of the tube is then attached to a vacuum line, evacuated, and cracked open to retrieve, purify, measure, and package the carbon dioxide for isotopic analysis. Isotopic measurements vs. the PDB standard are performed on a Nuclide 6-60 RMS mass spectrometer. The procedure for isotopic analysis of inorganic carbon in carbonates is achieved by dissolution in concentrated phosphoric acid, as described previously (Strauss et al., 1992).

Analyses of amino acids

Amino acid compositions, as well as their enantiomeric ratios, were investigated in four different Sudbury rock samples using high-performance liquid chromatography (HPLC). Samples were crushed to a powder (9-20-5a, 44 mg; Cap Ong#1, 34 mg; Capreol 67a, 1.2018 g; demineralized Dowling #1, 280 mg; bulk Dowling #1, 2.8125 g). All glassware was cleaned in Chromerge and heated at 500°C for 3 h. A portion of the Dowling sample was first demineralized in HF to dissolve the mineral matrix prior to amino acid extraction. Amino acid extracts were prepared by adding double-distilled water to each sample and heating for 24 h at 100°C in a sealed glass tube. The supernatant was decanted and dried and the residue hydrolyzed in 6N double-distilled HCl for 24 h at 100°C in a sealed glass tube. The hydrolyzed residue was then dried down and passed through a desalting column containing AG 50W-X8 resin (Bio Rad). The amino acid fraction was then eluted with aqueous NH_4OH. The eluate was dried, resuspended in aqueous borate buffer pH 9.4, and dried again to remove ammonia. The amino acid residue was then redissolved in double-distilled water and derivatized by the OPA/NAC (*o*-phthaldialdehyde/*N*-acetyl *L*-cysteine) technique (OPA and NAC both obtained from Fisher). The derivatives were separated by reverse phase HPLC with fluorescence detection and identified by comparison of retention times with standards (Zhao and Bada, 1989, 1995; Brinton and Bada, 1996). A control blank was processed and analyzed in parallel with the samples and used to make blank corrections.

RESULTS

Modal analyses and sample descriptions

Modal analyses. Modal analyses were conducted on six thin sections each from the middle, upper, and transitional units. Measurements were made on sections (25 × 45 mm) where fragments >1 mm were counted at 0.5-mm intervals. Results are summarized in Table 1. Our intent here is to gain insight into the type and characteristics of distal metasedimentary rocks that were washed into the Black Member after impact. The lower Black Member was excluded due to its high abundance (>80%) of altered melt glass and crystalline basement fragments. Observations of 12 thin sections from the lower unit give a rough estimate of the metasedimentary populations in decreasing order of abundance: quartz-rich rocks (quartzites, arkoses, etc.), siltstones, and C-rich mudstone (for abundances and charateristics of glass varieties, see Muir and Peredery, 1984).

TABLE 1. MODAL ANALYSES OF THE BLACK MEMBER ABOVE THE LOWER UNIT

Rock Type	Black Member Unit		
	Middle (wt. %)	Upper (wt. %)	Transitional (wt. %)
Impact produced:			
Melt fragments (clear to brown)	24	16	2
Melt fragments (green)	5	8	2
Mechanical (PDFs, etc.)	1	1	0
Basalts	2	1	0
Granitic rocks	3	1	0
Sedimentary/metasedimentary rocks:			
Quartzite/arenites	19	4	2
Carbonate rock	1	17	54
Claystone	3	3	0
Siltstone	0	2	0
C-mudstone	3	6	8
Microbreccia	1	4	0
Unknown	3	3	0
Chert	10	9	9
Minerals:			
Quartz	20	17	19
Feldspars	2	4	0
Sulfides	4	5	4

From Table 1 it is clear that the middle and upper lithologies have a greater variety of rock types and numerous abundance trends are apparent. The abundance of quartz-rich rocks and melt fragments decrease upward in the Black Member. Carbonate rocks and the C-rich mudstone increase upward and comprise nearly 65 vol% of the components in the transitional unit. In addition to the compositional trends, clast size also decreases upward as noted by Muir and Peredery (1984) and Avermann (1994). Although these trends are gradual over the entire Black Member, there is a general lack of obvious sorting and bedding in the lower, middle, and upper units. Moreover, there are numerous examples of discordant, diamictite-like lenses that range in size from tens of meters areally and meters thick to small lenses a few meters in size. They commonly show fragment concentrations of one rock type, e.g., quartzites and C-rich mudstones, that may be continuous over meters.

Modal analyses have not been made on the recently obtained core samples. Bedding and size sorting are very apparent in samples from 13 m below the transition/contact unit boundary to the topmost sample 2 m above the contact unit/Onwatin boundary (Figs. 2 and 3). Study of the continuous core at the Falconbridge storage site indicates an overall picture of quiescent depositional conditions for these argillaceous sediments, except for periods of energetic disruptions as noted by the presence of autobrecciation and contorted bedding over stratigraphic distances of centimeters (Fig. 3). Distal wash-in fragments are absent; the mudstones (now argillites) are enriched in particulate carbon, sulfides, and authigenic granular and stringer carbonates.

Petrographic and SEM observations. One of our objectives was to locate a previously unidentified source of a C-rich rock,

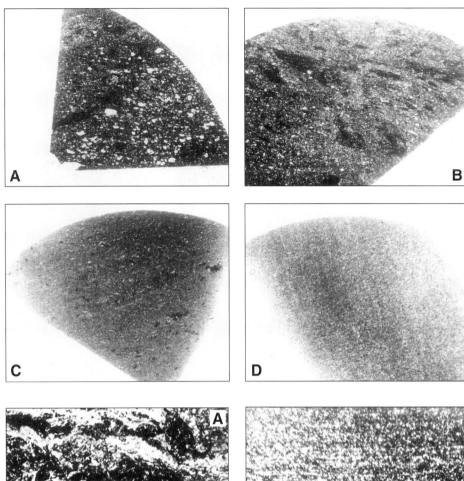

Figure 2. Examples of the transitional, contact, and lower Onwatin argillites (initially mudstones). A, Sample from the transitional unit, 13 m below the contact, showing autobrecciation of mudstones and formation of authigenic carbonates. B, sample taken at the transitional/contact unit boundary; similar to the sample in C, but with less carbonate and fragments and more sulfides. D, Boundary at the contact unit and the Onwatin. Sample devoid of clasts and most carbonate, but is sulfide-rich; this is the start of the well-bedded Onwatin argillite series. All photomicrographs; plane light; scale same, as for Figure 2. All sections parallel to the bedding plane. See Figure 3 for sections normal to the bedding plane.

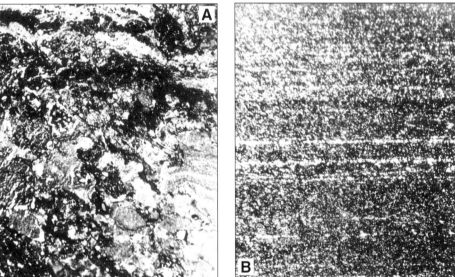

Figure 3. A, Photomicrograph of section cut normal to the bedding plane of a sample 2 m above the transitional/contact unit boundary, indicating small-scale autobrecciation of carbon-, and sulfide-rich mudstones and contemporaneous carbonate deposition. B, Sample from 1 m above the contact unit/Onwatin boundary showing finely spaced quartz layers between thicker layers of carbonaceous matter, sulfides, and quartz.

possibly from the underlying Huronian Supergroup or from unknown preimpact deposits, that could account for some of the unusually high amounts of carbonaceous matter in the Black Member. Regional rocks from the Huronian Supergroup that have been analyzed contain insignificant amounts of intrinsic "organic" carbon, in contrast to carbonate carbon (e.g., Strauss and Moore, 1992). A petrographic search found black, mostly opaque, fragments of a mudstone (now an argillite) in the upper and transitional units (Fig. 4). Subsequent observations indicate that these rock fragments may be pervasive throughout the Black Member (Fig. 1), although they are uncommon in the lower unit. Scanning electron microscope (SEM) and microprobe analyses show that they consist of quartz (41–63 wt%), feldspars and clays (8–30%), and carbonaceous matter (7–24%). They are typically subrounded to blocky in shape in the lower and middle units and rectangular to ribbon-shaped in the upper/transitional units and may show intercalated carbon particle–rich and quartz-rich layers. The abundances of these fragments per unit, together with other rock and mineral fragments, are given in Table 1.

SEM observations indicate that the clastic and compositional

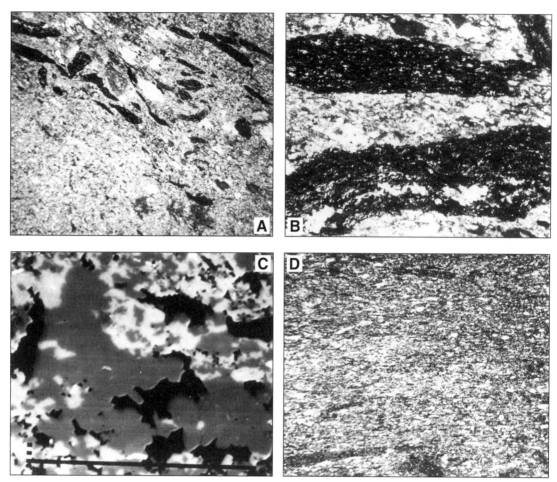

Figure 4. Carbon-rich mudstone. A, Fragments of a C-rich mudstone in an upper unit sample (South Capreol). Photomicrograph, plane light; base, 20 mm. B, Enlargement of fragments showing original layering of carbon and silicate components. Photomicrograph, plane light; base, 3-mm. C, SEM image showing grain relationships in the mudstone: carbonaceous matter (kerogen) is dark, quartz is dark gray, feldspars are light gray, and layer lattice silicates are bright and Fe-rich. D, Sample of the transitional unit (South Range) showing foliation and stretched mudstone fragments from regional metamorphism (amphibolite facies). Photomicrograph, plane light; base, 1 mm.

character of the Black Member is mostly unchanged until well into the upper units. Matrix in the lower and middle units is typically dark from carbon content and is fine-grained with large amounts of alteration products. Matrix in the upper and especially in the transitional units tends to be gray, coarser grained, and dominated by authigenic carbonates and recrystallized quartz. Most of the organic carbon is apparently concentrated in the C-rich mudstone fragments.

Rounded- to ovoid-shaped clasts with thick mantles of dark matrix and angular glass fragments are found throughout the the lower unit (Fig. 5) and are similar to objects described by Peredery (1972) and Muir and Peredery (1984) as "mudballs." They are particularly abundant in discontinuous lenses at ~40 and 70 m above the contact at the Highfalls area. As noted by Peredery (1972), crude alignments of elongated particles form concentric shells. Electron microprobe and SEM analyses show enhancement of carbon in the fine-grained mantle component of up to twice the carbon content in neighboring host rock matrices. These struc-

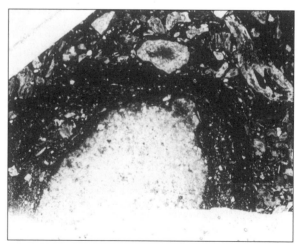

Figure 5. Photomicrograph of accretionary-mantled clast displaying a layered shell (3–5 mm) of coarser material (angular glass shards) closer to the arkosic core (35 × 18 mm) covered by a fine-grained shell of matrix-like material. Plane light; base, 20 mm.

tures, which we refer to as accretionary mantled clasts, are also found in the middle and upper units, although the mantles show evidence of partial erosion, presumably from abrasive action during vigorous reworking of the sediments.

A few round to ovoid accretionary lapilli were also found in the lower and middle units and are similar to those described from the Ries Crater (Graup, 1981; Newsom et al., 1990) with the exception that the average diameter of the Sudbury lapilli is 1.4 mm compared with 4.8 mm in the Ries. The Sudbury lapilli are also enriched in carbon relative to the average matrix of the host rock.

Carbonaceous components, bulk carbon, and carbon isotopic compositions

Fullerenes and soot. Fullerenes, C_{60} at 720 atomic mass units (amu), and C_{70} at 840 amu, initially were found by laser desorption and electron impactor mass spectrometric analyses in samples of the lower Black Member from the Highfalls area (Becker et al., 1994). The C_{60} content is greatly enhanced over C_{70} in these samples, as shown by peak ratios of 10–15:1. We have made a broader search for fullerenes throughout the Black Member by analyzing 26 additional samples with the same techniques of Becker et al. (1994). Of these, fullerenes were found in additional samples from Highfalls where $C_{70} > C_{60}$ with a small mass spectrum peak at 1008 amu, which could signify C_{84}; in the middle unit at Frenchman Lake, where only a peak at 1200 (C_{100}) was found; and in the upper unit, where one sample showed the presence of C_{60} only, and a second sample contained $C_{60} > C_{70}$. In addition, the existence of two other possible fullerenes was noted in one Highfalls sample at 888 amu (C_{74}) and at 936 amu (C_{78}). Analyses of the fine-grained material in the accretionary-mantled clast (shown in Fig. 5) showed the presence of $C_{60} > C_{70}$ fullerenes.

Soot is a form of acid-insoluble carbon that condenses directly from the gas to the solid, usually during combustion of carbonaceous matter, but also by experimental explosive detonation (Kuzentsov et al., 1994) and hypervelocity impact (Bunch et al., 1993), both under nonoxidizing conditions. Soot has been found in association with another impact event in the Cretaceous/Tertiary (K/T) boundary layer clay (Wolbach et al., 1985) as a result of global fires triggered by the Chixculub impact event. In our efforts to isolate and characterize elemental carbon components in the Onaping and Onwatin Formations, samples were treated by the procedures of Wolbach and Anders (1989), which mainly involve acid demineralization and oxidation with acidic dichromate.

A sample of the lower Black Member, known to contain large amounts of PGC (Heymann and Dressler, 1997a), yielded a total elemental carbon concentration of 0.68 ± 0.1 wt%. Of this carbon, none had the classic, aciniform ("clustered like grapes") morphology of soot (Medalia et al., 1983); thus we calculated an upper limit of 0.0029 wt% soot. The lower Onwatin sample was found to contain 2.6 ± 0.3 wt% total elemental carbon. Of this carbon, 1.3 ± 0.2 wt% soot was determined; kerogen and PGC were also identified. The difficulty with soot identification in this sample was the tiny size of what were thought to be soot particles, up to 100 times smaller than those observed at the K/T boundary. The larger of these particles did show classic aciniform morphology. However, morphology of the smaller particles was less clear, due to resolution limits of the SEM. Soot in the lower Onwatin was found to be a factor three to four times greater than soot concentrations observed in seven K/T boundary sites, where concentrations range from 0.0017 to 0.35 wt% (Wolbach et al., 1985).

In order to confirm the identification of soot, we submitted the sample to FTIR (Fourier transform infrared spectroscopy) analysis. Any graphite, soot, or "amorphous" carbon will have surface functional groups (i.e., attachments of molecules like CH_2) when exposed to air and moisture and are unique for any carbon form (O'Reilly and Mosher, 1983; Wolbach and Anders, 1989). The heavy oxidation treatment of the soot after acid demineralization promotes enhancement of certain function groups during oxidation. Figure 6 shows an FTIR spectrum for the Onwatin soot. These frequencies and peak heights after oxidation over 600 h are unique to soot.

Amino acids. The detected amino acids consist of aspartic and glutamic acid, serine, glycine, and alanine (Table 2). A similar distribution of amino acids was observed in all of the Sudbury rock samples, although absolute concentrations range from 0.1 to 8 ppm in the bulk matrix. The demineralized Dowling #1 sample yielded a higher concentration of amino acids (5–23 ppm) than in the bulk sample, which suggests that a large fraction of amino acids is still locked inside the mineral matrix even after the water extraction procedure. A large peak that is not present in the procedural blank is present in both the Capreol 67a and Cap Ong #1 samples. The peak is approximately 10 times more concentrated than any of the identified amino acids and has a retention time similar to methylamine. However, when the samples were spiked with a methylamine standard, separate peaks were observed, which indicates that this peak is not methylamine.

Extraterrestrial amino acids such as α-AIB (α-aminoisobutryric acid) and isovaline were not observed in any of the Sudbury samples above the detection limit (see Table 2). The reaction of these α-dialkyl amino acids with the OPA/NAC requires longer times to reach completion than that of protein amino acids, thus providing us with an additional way to confirm their identification (Zhao and Bada, 1995; Brinton and Bada, 1996). Although a peak with the same retention time as α-AIB was observed in the Capreol 67a sample after the derivatization reaction had proceeded for 1 min, the peak did not grow with longer derivatization (15 min) and was therefore not α-AIB.

Amino acid racemization has been used successfully as a technique for dating geologic samples. The reaction follows reversible first-order kinetics, with half-lives ($D/L = 0.5$) of aspartic acid and alanine on the order of 10^5 yr in most geologic environments. The racemization reaction can be written as:

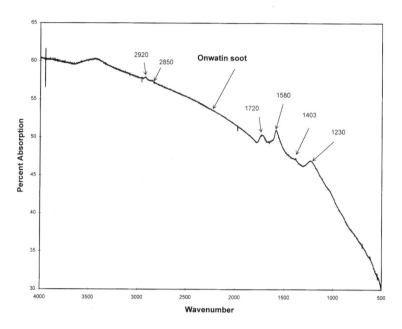

Figure 6. FTIR spectrum of soot in the lower Onwatin Formation. Band assignments from O'Reilly and Mosher (1983) and Wolbach and Anders (1989) are: 2920 and 2850 that represent CH stretching frequencies of methyl groups (CH_2); band 1720 is the C·O stretching frequency of carboxylic acid (COOH); 1580 is the aromatic ring frequency, C·C; 1403 could be due to CH_2, COOH, or CHO; and 1230 is the CH bending frequency.

TABLE 2. SUMMARY OF TOTAL AMINO ACID CONCENTRATIONS

Amino Acid	9-20-5a (ppm)	(D/L)	Cap-Ong #1 (ppm)	(D/L)	Capreol 67a (ppm)	(D/L)	Dowling #1 (Demin.) (ppm)	(D/L)	Dowling #1 (bulk) (ppm)	(DL)
D/L Aspartic Acid	3.7	0.1	3.8	<0.06	0.13	0.3	20.7	0.1	0.18	0.1
D/L Serine	4.0	n.r.	7.9	n.r.	0.10	<0.1	5.4	<0.01	0.22	<0.05
D/L Glutamic Acid	3.5	n.r.	5.1	n.r.	0.21	n..r.	12.3	n.r.	0.46	n.r.
Glycine	4.1	....	7.6	....	0.29	....	23.0	....	0.67	....
D/L Alanine	3.2	0.18	4.3	0.16	0.10	0.2	6.6	0.07	0.25	0.1
AIB	<0.1	....	<0.2	....	<0.4	....	<2	....	<0.02	....
D/L Isovaline	<0.1	n.d.	<0.1	n.d.	<0.01	n.d.	<0.02	n.d.	<0.02	n.d.

Note: Total concentrations represent both free and bound amino acids in the rock matrix and are reported in parts per million (ppm). The uncertainties in the concentrations are roughly ±0.1 ppm for 9-20-5a, Cap Ong #1 and Dowling (Demin.); ±0.02 ppm for Capreol 67a and Dowling. D/L ratios are approximately ±0.1. n.r. = not resolved; n.d. = no data.

$$L\text{-amino acid} \underset{k'_i}{\overset{k_i}{\rightleftarrows}} D\text{-amino acid},$$

where k_i and k'_i are the first-order rate constants for interconversion between the enantiomers. Given the special case of an initial D/L ratio of zero and a single chiral center (as in aspartic acid and alanine), $k_i = k'_i$ and the kinetic equation for this reaction becomes:

$$\ln\left[\frac{1+(D/L)}{1-(D/L)}\right] = k_i t,$$

where D/L is the enantiomeric ratio at time t. For a more complete discussion of amino acid racemization kinetics, see Bada (1985). The D/L ratios for aspartic acid and alanine in the Sudbury samples fall in the range 0.07–0.3. In all the samples, concentrations of D-amino acids were near the detection limit for the technique, and therefore there is greater error associated with these measurements (factor of 2–3), and with the D/L ratios calculated from these values. In contrast to the L amino acids that are associated with terrestrial biota, all known abiotic processes give racemic mixtures of amino acids (D/L = 1). However, the D/L ratios for aspartic acid and alanine are significantly less than 1, which is an indication that these amino acids are both terrestrial in origin and of recent geologic age (<200 ka).

Bulk organic carbon and isotopic compositions. Results of carbon isotopic analyses are given in Table 3 and Figure 7. Bulk organic carbon shows no particular stratigraphic trend. The isotopic data are consistent with the ^{13}C-enriched end of the range of –30 to –40 per mil for Proterozoic kerogens and for carbonaceous matter modified from organic remains (Schidlowski, 1988; Strauss et al., 1992). In addition to our results, other available isotopic data (Heymann et al., 1997a; Avermann, 1994) are also plotted in Figure 7. Both data sets define a trend from more ^{13}C-depleted carbon in the lower unit to heavier carbon in the upper

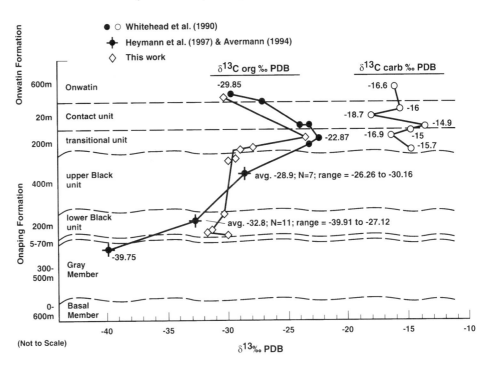

Figure 7. Relationship of bulk δ¹³C compositions with stratigraphic position from the Gray Member through the Black Member and into the Onwatin Formation. Available published isotopic data are given as averages and range where stratigraphic positions are not well known (Heymann et al., 1997a; Avermann, 1994) and approximate positions (Whitehead et al., 1990). The second data set consists of samples of known stratigraphic position (this work). Both show a similar trend: isotopically light carbon at the lowest levels, with increasingly heavier carbon up to the transitional/contact unit boundary and then a reversal to lighter carbon into the Onwatin Formation.

TABLE 3. BULK ORGANIC CARBON AND ISOTOPIC ANALYSES

Sample*	C (wt. %)	δC_{org} (‰) (PDB)	Location	Formation
Onwatin[†]	3.10	−30.44	East Range (core sample)	Onwatin
9-19-3[†]	0.50	−23.82	South Range	Transitional unit
Cap Ong-1[†]	0.56	−28.82	East Range	Upper unit
9-20-1[†]	0.64	−29.24	South Range	Upper unit
9-20-5a[†]	0.85	−30.01	East Range	Upper unit
Cap Ong[†]	n.d.	−30.10	East Range	Middle unit
Dow Onp-1	0.83	−31.93	North Range	Lower unit
9-18-1	0.30	−30.11	North Range	Lower unit
Dow Ong-2	n.d.	−32.54	North Range	Lower unit
9-18-2	0.68	−32.10	North Range	Lower unit

*Listed in descending stratigraphic order.
[†]Kerogen rich.

unit and on to an abrupt change to much heavier carbon as the contact unit is approached. At the transitional/contact unit boundary, another abrupt change to lighter carbon takes place with a return in the lowest Onwatin to isotopic values near the average for the upper Black unit (−30 per mil).

Few Huronian Supergroup rocks contain organic carbon. One of the few rocks that has significant carbon, the Gowganda shale, has an organic content of 4 wt% with a $\delta^{13}C$ of −27.4 per mil (Strauss and Moore, 1992). We have shown above that the Black Member contains an abundance of C-rich mudstone, particularly from the middle unit upward, where the carbonaceous matter is predominately kerogen. Sample 9-20-5a in Table 3 contained hand-picked C-rich mudstone fragments, and its $\delta^{13}C$ value of −30.01 per mil is used as the approximate isotopic composition for the kerogen.

DISCUSSION

Our preliminary data are useful in evaluating the various components in the carbonaceous matter and in interpreting the sedimentary history of the Black Member of the Onaping Formation. Our present understanding of the carbonaceous components, based on field and laboratory data, is presented in Figure 8. We know that much of the carbon in the lower Black Member is present as poorly graphitized carbon (Heymann and Dressler, 1997a). This ultra-fine particulate matter may have been formed in the impact process and was deposited in the breccias of the lower Black Member. Fallout of PGC into the Gray and Green Members may have occurred as well, although most of this carbon was probably oxidized over time in these hot deposits. Bulk carbon content in the Gray Member is <0.1 wt% (Heymann et al., 1997a).

We have also found kerogen-rich mudstone fragments in breccias of the lower Black Member, that increase upward in abundance as PGC decreases in the deposits to a point where, in the upper unit, the kerogen may account for >80% of the carbonaceous material. These fragments undoubtedly washed in from distal sources, together with shocked and unshocked rock fragments and fine-grained detritus. Mudstones (now argillites) on top of the breccias in the upper transitional and contact units were formed from very fine C- and S-bearing sediment that contributed to the general anoxic conditions. In addition, Whitehead et al. (1990) suggested that hydrothermal venting from the hot, underlying Basal and Gray Members may have started in the transitional unit through the contact unit into lower Ontwatin waters and contin-

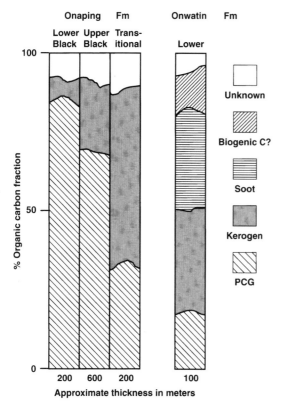

Figure 8. Diagram showing estimated carbon component contents for units of the Black Member and the lower Onwatin.

ued for long periods of time into the Onwatin waters and may have had significant effects on C and S isotopic contents in the contact unit. These mudstones, metamorphosed later to argillites, are different in isotopic composition compared to those that occur in the underlying breccias and to the rocks in the overlying Onwatin Formation (Fig. 7). Thus far soot particles have been found in one sample from the lowest exposed Onwatin; future soot analyses of samples in the Contact unit and into the uppermost portions of the upper Black Member may help to constrain the depositional rate for the Black Member. Impact diamonds and SiC (Masaitis, 1997) also occur in the Black Member.

Avermann (1994) proposed that the carbonaceous matter in the Black Member and the Onwatin Formation was biogenically produced during deposition, based on three lines of evidence: (1) carbon isotopic data from two samples of the Black Member and one from the Onwatin are close to –30 per mil, which is within the isotopic range for biologic remains (Schidlowski, 1988); (2) a few relics of algae were found in both deposits; and (3) contrary to the suggestions of Peredery (1972) and Peredery and Morrison (1984) that the Black Member was deposited in a matter of minutes, Avermann concluded that the rate of sedimentation was slow, possibly over thousands of years. Heymann and Dressler (1997b) concluded that postimpact regional metamorphism in the Greenschist facies "graphitized" the organic matter to PGC. Furthermore, they contended that "super-fast" algal blooms could have kept pace with crater deposition. Thus the carbon is thought to have originated from living organisms during infill of the crater basin that was later processed to particulate graphite (PGC) now present in the Black Member and lower Onwatin deposits.

Algal and bacterial life may have existed in sediments of the lower Onwatin, as shown by Arengi (1977) and Whitehead et al. (1990) and, even in the transitional unit of the Black Member and the contact unit. That algal life accompanied deposition of the Black Member and gave rise to the bulk of the existing carbon in the Black Member can be challenged from several observations. These arguments, together with our conclusions with respect to the origin of Black Member carbon, are discussed below.

1. The conclusion that crater filling occurred over a long period of time is inconsistent with our field and petrographic observations, where little evidence was found for such data as size sorting, clast rounding, and graded bedding that would be expected from long-term sedimentation. Instead, the deposits resemble those of convulsive events, some of which could have immediately followed a giant impact, i.e., giant mass failures from crater wall and central uplift slumping, water-driven avalanches, giant tsunamis, and giant earthquakes that theoretically follow large impacts (e.g., Clifton, 1988). Only in the top 100 m of several core samples from the transitional and contact units that we studied is there evidence for quiescent deposition; even here, there are numerous, small, turbulent interuptions. In essence, the nearly instantaneous and consistent dumping of massive volumes of material overwhelmed the ability of water to establish more orderly sedimentation. Energetic deposition probably continued for most of the Black Member. How long this took is unknown, although it is clear that evidence in support of numerous, long-term ponding episodes for most of the Black Member, necessary for the simultaneous growth of microbial life, has not been found. However, the more quiescent depositional conditions of the transitional and contact units may have been receptive to biologic growth.

2. Avermann (1994) tentatively identified a few algal relics in the Black Member, and used this evidence to support her conclusion that the huge amounts of carbon in the Black Member were the result of algal decomposition. Algal relics should be prolific in view of the very large amounts of carbon. Graphitic carbon (PGC) is evenly distributed in the matrix in each sample examined from the lower unit; it also occurs as mantles around some clasts. Peredery (1972) suggested that these "mudballs" (accretionary-mantled clasts) may have had a sedimentary origin associated with turbulent conditions at the time of deposition. Alternatively, Peredery suggested that these structures could have formed by accretion in the impact dust cloud. Fullerenes found in the fine-grained mantles by us may have accreted to the mantle during either scenario. The PGC in the mantles is indistinct from PGC in the matrix. Clearly, they had the same origin, one that is inconsistent with an algal precursor through later metamorphism.

The abundance of PGC in the matrix commonly decreases vertically in the deposits until it becomes a minor component of the carbon inventory, even though bulk carbon contents remain fairly constant within a range of ~0.30–<1.0 wt% for most of the Black Member lithologies. This may be due to the offsetting, concomitant increase of kerogen-bearing, C-rich mudstone fragments with decreasing PGC. Simple calculations show that for a 6% volume of C-rich mudstones in the upper unit (Table 1), which has an average of 15 wt% carbon with a density of 80% compared to that of the host rock (2.0 vs. 2.5 g/cc), kerogen yields a value of ~0.7 wt% carbon. Compared to the average carbon content (Table 3) for upper unit rocks (0.68 wt%), we can conclude that most of the carbon in the upper unit, in addition to the middle and upper units, derives from kerogens in the C-rich mudstone fragments.

3. It is clear that the isotopic signatures for most of the Black Member are consistent with derivation from biogenic activity. The question is, at what time and where did this activity happen? The presence of carbonaceous matter in the Black Member, with a $\delta^{13}C$ average of –30 per mil and range of –29 to –32.5 per mil (Table 3), compares well (Fig. 7) with the average isotopic compositions of Avermann (1994) and Heymann et al. (1997b). Moreover, we have shown that the –30 per mil value is consistent with the kerogens in the unprocessed C-rich mudstones and with PGC. From this we conclude that kerogens in the mudstone are the precursor for matrix PGC, i.e., the parent C-rich mudstones were present at the time of impact and were partly processed by impact-shock heating into PGC with the unprocessed mudstone fragments ejected outward. Graphitized kerogen products (PGC) fell back into the rapidly growing Black Member deposits and to distal areas outside the immediate crater vicinity. Subsequent water-driven resurgent activity brought much of the ejecta and other materials into the growing Black Member deposits.

The isotopic trend for $\delta^{13}C_{org}$ with stratigraphic position in the Black Member, the contact unit, and the lower Onwatin (Fig. 7) may have resulted from the effects of two contrasting depositional environments: (1) Rapid burial for most of the Black Member deposits, under anoxic conditions, probably preserved the preimpact isotopic compositions of C-rich mudstone kerogens and PGC that resulted from impact processing of these kerogens as discussed above, and (2) The transitional unit shows a return to slower, more ordered sedimentation, concomitant with an increase in carbonate and sulfide contents. The dramatic shift to ^{13}C-enriched organic carbon is consistent with an environment dominated by bacterial populations (e.g., *Chloroflexus* and green bacteria) that synthesized ^{13}C-enriched organic carbon (Des Marais, 1997) and might indicate reduced sulfidic waters. The unusual ^{13}C-depleted carbonates suggest that they may have formed in a basin that sustained extensive oxidation of ^{13}C-depleted organic carbon (Whitehead et al., 1990). The abrupt change back to isotopically lighter organic carbon and large increases in bulk organic carbon (up to 4 wt%) (Arengi, 1977) in the lower Onwatin probably signifies a return to heavy basin infill and burial of organic debris.

Accretionary lapilli probably originated in the impact dust cloud. Newsom et al. (1990) pointed out the similarities of volcanic lapilli that accrete from abundant water and dust in an eruption column to those lapilli found in the Ries Crater impact deposits. Because of the presence of the lapilli, they concluded that the cloud was still present during much of the fall-back suevite deposition. For the same reasons, the Sudbury dust cloud may still have been active over the crater vicinity and neighboring distal regions during initial implacement of the lower Black Member deposits. We cannot rule out that the lapilli were washed-in during deposition, although their delicate structure would have had difficulty surviving in the high-energy wash-in and turbulent depositional conditions.

Implications of soot in the deposits suggest fallout from a global soot-bearing cloud similar to that proposed by Wolbach et al. (1985). Whereas soot found at the K/T boundary arose from the burning of biomass, soot in the Sudbury deposits would not have formed from biomass burning because there was no significant biomass, only simple organisms (e.g., algae and bacteria). However, the aqueous environment of the preexisting organic carbon (C-rich mudstone and any biota present) indicates that very energetic remobilization of this organic carbon would have been required in order to thermally alter it to soot. In the absence of other alternatives, we conclude that the soot formed by evaporation/condensation from the hot impact fireball and/or from a later global cloud. The smaller size of soot particles in the lower Onwatin compared with soot from the K/T is certainly consistent with there being different physical conditions at the time of their formation (e.g., carbon and oxygen abundance; combustion temperature). If the soot is a product of the Sudbury impact, then deposition of the Onaping Formation into the lower Onwatin Formation was extremely rapid.

CONCLUDING REMARKS

The Sudbury Impact Structure has many preserved features essential for determining the effects of impact processing of carbonaceous matter. For example, the Sudbury basin has a complete and exposed stratigraphic record from the melt zone through fall-back and reworked infill deposits into overlying basin filling and shallow marine sediments. The C-rich deposits have not been seriously compromised by metamorphism, although low refractory organic compounds (e.g., amino acids) have not survived with time. Thus the basin has collected shocked matter from impact, fall-back, fallout, and wash-in from distal sources, and recorded the reintroduction of simple life forms. Missing from the record are deposits analogous to the Chicxulub impact ejecta/fallout K/T boundary layer. The two craters are complementary in that what is absent in one is present in the other, which affords the opportunity to study the entire range of impact environments.

In response to our initial quests in this study as outlined in the Introduction, we offer the following summary and conclusions:

1. Survivability/synthesis of exogenous organic compounds on impact. To date, we have focused only on amino acids of extraterrestrial signature and found none. The more refractory organic compounds (e.g., polycyclic aromatic hydrocarbons, or

PAHs) may have in part survived impact and postimpact metamorphism. This issue was not vigorously pursued in this work. However, we have found a few PAHs in several Black Member samples and more in the Onwatin argillites, although we cannot preclude at this time that they are modern contaminants. In view of these results, the detection of exogenous organic compounds has a low probability, although the question of PAH survivability/synthesis remains open. The delivery of fullerenes with an extrasolar He signature is probable. Fullerenes without "caged" He probably formed on impact.

2. Endogenous evolution of impact and postimpact abiotic/prebiotic carbonaceous chemistry. We conclude that the huge PGC component is largely the result of impact processing of preexisting kerogens. Thermal dissociation of biogenic elements (C, H, N, P, and S) from kerogens and other carbonaceous materials may have provided essential nutrients for the apparent fast recharge of life in the basin.

3. The conversion of carbonaceous precursors to PGC and soot mostly occurred in the impact fireball. The cooling impact cloud may have been a site for organic synthesis, given the vast amount of carbon, other biogenic elements, and water vapor that were probably in the cloud. With the possible exception of PAHs, we have not found any evidence of organic compounds.

4. Time constraints on postimpact events. The occurrence of accretionary lapilli in the lower Black Member may indicate that deposition of the Black Member overlapped with late-stage fallout. The presence of soot in the lower Onwatin Formation is enigmatic. It is tempting to speculate that the soot represents fallout over the region from a global impact cloud analogous to that which operated after the Chicxulub event. However, without better knowledge of soot distribution in the stratigraphic record of the basin, soot cannot be used as a time marker.

Note added in proof: Subsequent analyses of core samples from the Onaping transitional unit, through the contact unit, and into the bottom of the Onwatin show various quantities of soot. These soot particles differ from those observed in the upper portions of the Onwatin in being larger and having classic aciniform morphology. Moreover, these new data and information from ongoing analyses of samples from the upper Onaping may indeed suggest that soot can be used as a time marker and further constrain the time period (rapidity) of deposition for the Onaping, transitional, and Onwatin deposits.

ACKNOWLEDGMENTS

We thank B. Dressler, B. French, W. Meyer, T. Muir, G. Morrison, and W. Peredery for fruitful discussions and assistance with the field work. We also thank J. Paque and Max Bernstein for valuable discussions and analytical assistance. Falconbridge, Ltd., and Inco, Ltd., are thanked for their assistance and use of core samples. Comments and suggestions by the reviewers, B. Dressler, R. M. Hough, and J. Kerridge, are much appreciated. Partial support from the National Aeronautics and Space Administration Exobiology Program and the American Chemical Society Petroleum Research Fund are gratefully acknowledged.

REFERENCES CITED

Ames, D. E., Watkinson, D. H., and Parrish, R. R., 1998, Dating of a regional hydrothermal system by the 1850 Ma Sudbury impact event: Geology, v. 26, p. 447–450.

Arengi, J. T., 1977, Evolution of the Sudbury Basin [M. S. thesis]: Toronto, Canada, University of Toronto, 141 p.

Avermann, M., 1994, Origin of the polymict, allochthonous breccias of the Onaping Formation, Sudbury Structure, Ontario, Canada, in Dressler, B. O., Grieve, R. A. F., and Sharpton, V. L., eds., Large meteorite impacts and planetary evolution: Boulder Colorado, Geological Society of America Special Paper 293, p. 265–274.

Avermann, M., and Brockmeyer, P., 1992, The Onaping Formation of the Sudbury Structure, Canada: an example of allochthonous impact breccias: Tectonophysics, v. 216, p. 227–234.

Bada, J. L., 1985, Amino acid racemization dating of fossil bones: Annual Reviews of Earth and Planetary Science, v. 13, p. 241–268.

Becker, L., Bada, J. L., Winans, R. E., Hunt, J. E., Bunch, T. E., and French. B. M., 1994, Fullerenes in the 1.85 billion year old Sudbury Impact Structure: Science, v. 265, p. 642–645.

Becker, L., Poreda, R. J., and Bada, J. L., 1996, Extraterrestrial helium trapped in fullerenes in the Sudbury Impact Structure: Science, v. 272, p. 249–252.

Brinton, K. L. F., and Bada, J. L., 1996, A reexamination of amino acids in lunar soils: Implications for the survival of exogenous organic material during impact delivery: Geochimica et Cosmochimica Acta, v. 60, p. 349–356.

Brockmeyer, P., 1990, Petrographische und geochemische Untersuchungen an polymikten Breccien der Onaping-Formation, Sudbury-Distikt (Ontario, Kanada) [Ph.D. thesis]: Münster, Germany, University of Münster, 228 p.

Bunch, T. E., Becker, L., Bada, J. L., Macklin, J., Radicati, F., Fleming, R. H., and Erlichman, J., 1993, Hypervelocity impact experiments for carbonaceous impactors, in Long-Duration Exposure Facility—2nd Post-Retrieval Symposium: San Diego, California, NASA Conference Publication 3194, p. 453–477.

Bunch, T. E., Becker, L., Schultz, P. H., and Wolbach, W. S., 1997, New potential sources for Black Member Carbon: in Conference on Large Meteorite Impacts and Planetary Evolution: Houston, Texas, Lunar and Planetary Institute, Contribution 922, p. 7.

Burrows, A. G., and Rickaby, H. C., 1930, Sudbury basin area: Ontario Department of Mines Annual Report, v. 38, pt. 3, p. 1–55.

Clifton, H. E., 1988, Sedimentologic relevance of convulsive geologic events: Geological Society of America Special Paper 229, p. 1–5.

Des Marais, D. J., 1997, Isotopic evolution of the biogeochemical carbon cycle during the Proterozoic eon: Organic Geochemistry v. 27, p 185–193.

Fleet, M. E., Barnett, R. L., and Morris, W. A., 1987, Prograde metamorphism of the Sudbury Igneous Complex: Canadian Mineralogist, v. 25, p 499–514.

Gibbins, S. F. M., Gibson, H. L., Ames, D. E., and Jonasson, I. R., 1997, The Onaping Formation: stratigraphy, fragmentation, and mechanisms of emplacement, in Conference on large meteorite impacts and planetary evolution: Houston, Texas, Lunar and Planetary Institute, Contribution 922, p. 16.

Graup, G., 1981, Terrestrial chondrules, glass spherules, and accretionary lapilli from the suevite, Ries Crater, Germany: Earth and Planetary Science Letters, v. 55, p. 407–418.

Heymann, D., and Dressler, B. O., 1997a, Raman study of carbonaceous matter and anthraxolite in rocks from the Sudbury, Ontario, Impact Structure, in Lunar and Planetary Science Conference: Houston, Texas, Lunar and Planetary Institute, Abstracts, p. 563–564.

Heymann, D., and Dressler, B. O., 1997b, Origin of carbonaceous matter in rocks from the Whitewater Group of the Sudbury Structure, in Conference on

Large Meteorite Impacts and Planetary Evolution: Lunar and Planetary Institute, Contribution 922, p. 21.

Heymann, D., Dressler, B. O., Dunbar, R. B., and Mucciarone, D. A., 1997a, Carbon isotopic composition of carbonaceous matter in rocks from the Whitewater Group of the Sudbury Structure, in Conference on Large Meteorite Impacts and Planetary Evolution: Lunar and Planetary Institute, Contribution 922, p. 22.

Heymann, D., Dressler, B. O., and Thiemans, M. H., 1997b, Origin of native sulfur in rocks from the Sudbury Structure, in Conference on Large Meteorite Impacts and Planetary Evolution: Lunar and Planetary Institute, Contribution 922, p. 22–23.

Kuzentsov, V. L., Chuvilin, E. M., Moroz, E. M., Kolomiichuk, V. N., Shaikhutdinov, Sh. K., and Butenko, Yu. V., 1994, Effect of explosion conditions on the structure of detonation soots: ultradisperse diamond and onion carbon: Carbon, v. 32, p. 873–882.

Masaitis, V. L., Shafranovsky, G. L., Grieve, R. A. F., Langenhorst, F., Peredery, W., Balmasov, E. L., Fedorova, I. G., and Therriault, A., 1997, Discovery of diamonds at the Sudbury Structure, in Conference on Large Meteorite Impacts and Planetary Evolution: Lunar and Planetary Institute, Contribution 922, p. 33.

Medalia, A. I., Rivin, D., and Sanders, D. A., 1983, Characteristics of carbonaceous pollutants: Science of the Total Environment, v. 32, p. 1–22.

Muir, T. L., and Peredery, W. V., 1984, The Onaping Formation, in Pye, E.G., Naldrett, A. J., and Giblin, P. E., eds., The geology and ore deposits of the Sudbury Structure: Ontario Geological Survey Special Volume 1, p. 139–210.

Newsom, H. E., Graup, G., Iseri, D. A., Geismann, J. W., and Keil, K., 1990, The formation of the Reis Crater, West Germany: evidence of atmospheric interactions during a larger cratering event: Geological Society of America Special Paper 247, p. 195–206.

O'Reilly, J. M., and Mosher, R. A., 1983, Functional groups in carbon black by FTIR Spectroscopy: Carbon, v. 21, p. 47–51.

Peredery, W., 1972, Chemistry of fluidal glasses and melt bodies in the Onaping Formation, in Guy-Bray, J. V., ed., New developments in Sudbury geology: Geological Association of Canada Special Paper, p. 49–59.

Peredery, W., and Morrison, G. G., 1984, Discussion of the origin of the Sudbury Structure, in Pye, E.G., Naldrett, A. J., and Giblin, P. E., eds., The geology and ore deposits of the Sudbury Structure: Ontario Geological Survey, Special Volume 1, p. 491–511.

Robl, T. L., and Davis, B. H., 1993, Comparison of the HF-HCl and HF-BF$_3$ maceration techniques and the chemistry of resultant organic concentrates: Organic Geochemistry, v. 20, p. 249–255.

Schidlowski, M., 1988, A 3,800-million-year isotopic record of life from carbon in sedimentary rocks: Nature, v. 333, p. 313–318.

Stöffler, D., Deutsch, A., Avermann, M., Bischoff, L., Brockmeyer, P., Buhl, D., Lakomy, R., and Müller-Mohr, V., 1994, The formation of the Sudbury Structure, Canada: Toward a unified model, in Dressler, B. O., Grieve, R. A. F., and Sharpton, V. L., eds., Large meteorite impacts and planetary evolution: Geological Society of America Special Paper 293, p. 303–317.

Strauss, H., and Moore, T. B., 1992, Abundances and isotopic compositions of carbon and sulfur in whole rock and kerogen samples, in Schopf, J. W., and Klein, C., eds, The Proterozoic biosphere, a multidisciplinary study: Cambridge, United Kingdom, Cambridge University Press, p. 711–798.

Strauss, H., Des Marais, D. J., Hayes, J. M., Lambert, I. B., and Summons, R. E., 1992, Procedures of whole rock and kerogen analysis, in Schopf, J. W., and Klein, C., eds, The Proterozoic biosphere, a multidisciplinary study: Cambridge, United Kingdom, Cambridge University Press, p. 699–708.

Whitehead, R. E. S., Davies, J. F., and Goodfellow, W. D., 1990, Isotopic evidence for hydrothermal discharge into anoxic seawater, Sudbury Basin, Ontario, Canada: Chemical Geology (Isotopic Geochemical Section), v. 86, p. 49–63.

Wolbach, W. S., and Anders, E., 1989, Elemental carbon in sediments: determination and isotopic analysis in the presence of kerogen: Geochimica et Cosmochimica Acta, v. 53, p. 1637–1647.

Wolbach, W. S., Lewis, R. S., and Anders., E., 1985, Cretaceous extinctions: evidence for wildfires and search for meteoritic material: Science, v. 230, p. 167–170.

Zhao, M., and Bada, J. L., 1989, Extraterrestrial amino acids in Cretaceous Tertiary boundary sediments at Stevns Klint, Denmark: Nature, v. 339. p. 463–466.

Zhao, M., and Bada, J. L., 1995, Determination of alpha-dialkylamino acids and their enantiomers in geological samples by high performance liquid chromatography after derivatization with a chiral adduct of O-phthaldialdehyde: Journal of Chromatography, v. A690, p. 55–67.

MANUSCRIPT ACCEPTED BY THE SOCIETY DECEMBER 16, 1998

Origin of carbonaceous matter, fullerenes, and elemental sulfur in rocks of the Whitewater Group, Sudbury impact structure, Ontario, Canada

D. Heymann
Department of Geology and Geophysics, Rice University, Houston, Texas 77005, USA

B. O. Dressler
Lunar and Planetary Institute, Houston, Texas 77059, USA

J. Knell
Departments of Geology and Chemistry, Arizona State University, Tempe, Arizona 85287, USA

M. H. Thiemens
Department of Chemistry & Biochemistry, University of California–San Diego, La Jolla, California 92093, USA

P. R. Buseck
Departments of Geology and Chemistry, Arizona State University, Tempe, Arizona 85287, USA

R. B. Dunbar* and D. Mucciarone*
Department of Geology and Geophysics, Rice University, Houston, Texas 77005, USA

ABSTRACT

New carbon determinations of 20 rocks confirm that both the breccias of the Black Member of the Onaping Formation and the mudstones of the Onwatin Formation of the Whitewater Group in the 1.8-Ga Sudbury impact structure contain appreciable amounts of carbonaceous matter. The origin of the carbon has been a long-standing enigma. Carbon-bearing target rocks, either carbonaceous mudstones or carbonate target rocks are not its source, and neither is CO_2 from the atmosphere of the impact area, or carbon from the impacting projectile itself. Fumarolic activity is also ruled out. New isotopic $\delta^{13}C$ results of carbon from 17 Black Member breccias, and 1 each from a Gray Member breccia and an Onwatin Formation range from –35.22 to –26.26‰. Together with Raman spectra of carbon from 11 of these samples, they lead to the conclusion that the Black Member carbon is biogenic. This unit cannot have formed as an "instant clastic sediment," but must have accumulated over several millions of years while prolific prokaryotic activity (archaea and bacteria) occurred. Isotopic $\delta^{13}C$ and Raman results of carbon from an anthraxolite confirm that this carbon is derived from the Onwatin Formation. A search for fullerenes C60 and C70 by high-performance liquid chromatography (HPLC) in 17 rocks from the Black Member of the Onaping Formation, 1 rock from the Gray Member of the Onaping Formation, and by mass spectrometry of 3 breccias from the Black Member failed to find any fullerenes. Elemental sulfur in the breccias from the Gray and Black Members of the Onaping Formation and in one sample of the Onwatin Formation is due to weathering of pyrrhotite. The isotopic composition of the elemental sulfur as well as that of the sulfide-S contains a mass-independent fractionation component whose origin is not understood.

---

*Present addresses: Department of Geology and Environmental Sciences, School of Earth Sciences, Stanford University, Stanford, California 94305.

Heymann, D., Dressler, B. O., Knell, J., Thiemens, M. H., Buseck, P. R., Dunbar, R. B., and Mucciarone, D., 1999, Origin of carbonaceous matter, fullerenes, and elemental sulfur in rocks of the Whitewater Group, Sudbury impact structure, Ontario, Canada, *in* Dressler, B. O., and Sharpton, V. L., eds., Large Meteorite Impacts and Planetary Evolution II: Boulder, Colorado, Geological Society of America Special Paper 339.

INTRODUCTION

The Sudbury multi-ring impact structure in Ontario, Canada, is 1.85 Ga in age (Krogh et al., 1984; Corfu and Lightfoot, 1996). It is located at the boundary of two structural provinces of the Canadian Shield, the Archean Superior Province, and the Proterozoic Southern Province. It lies north of the Grenville Front Boundary Fault (Fig. 1). South of this fault, Grenvillian gneisses and intrusive rocks occur that are characterized by a ~1-Ga metamorphic episode. The Archean footwall rocks in the north are granites, gneisses, and migmatites. South of the structure, the Proterozoic rocks consist of metavolcanics and metasediments of the Huronian Supergroup, intruded by 2.2-Ga Nipissing diabase (Dressler, 1984).

The structure has an estimated diarmeter of ~180–250 km (Peredery and Morrison, 1984; Dressler et al., 1987; Grieve et al., 1991) and is strongly deformed due to thrusting from the southeast (Shanks and Schwerdtner, 1990, Milkereit et al., 1992). Shock-metamorphosed target rocks beneath the Sudbury Igneous Complex (SIC), shatter cones up to ~20 km from the SIC, strong brecciation of target rocks up to 80 km from the SIC, and shock-metamorphosed target rock clasts and recrystallized impact glasses in thick suevitic breccia deposits of the Onaping Formation within the Sudbury Basin provide evidence that the structure was formed as a result of impact. At the base of the SIC, in the so-called Sublayer, in the Footwall Breccias, and in nearby footwall rocks, massive sulfur deposits occur that are rich in nickel, copper, cobalt, platinum group elements, and other

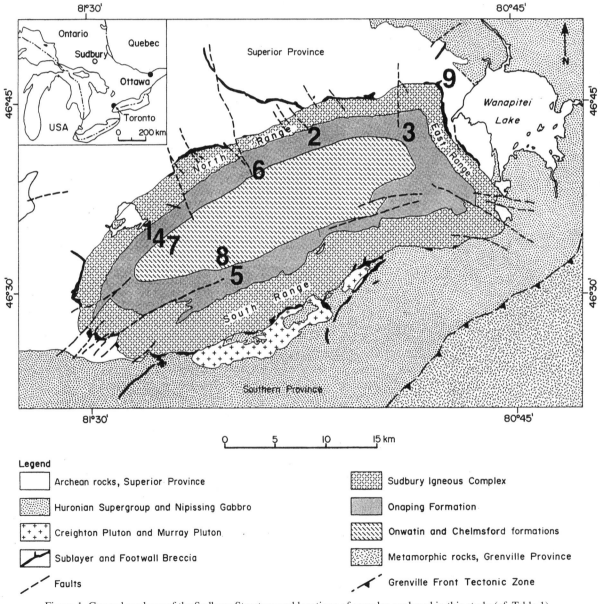

Figure 1. General geology of the Sudbury Structure and locations of samples analyzed in this study (cf. Table 1).

precious metals. They make the Sudbury camp one of the richest mining districts on Earth.

The SIC consists of layers, from bottom to top: the quartz dioritic, inclusion-bearing Sublayer; norite; quartz gabbro; and granophyre. It has been interpreted as a strongly contaminated endogenic, magmatic body (Naldrett and Hewins, 1984), a differentiated impact melt sheet (Faggart et al., 1985; Lakomy, 1989; Brockmeyer, 1990; Grieve et al., 1991; Deutsch et al., 1995), or a combination of a granophyre impact melt and lower SIC units that formed as a result of impact-triggered magmatism, or deep crustal melting (Dence, 1972; Shanks et al., 1990; Chai and Eckstrand, 1993, 1994; Dressler et al., 1966).

Within the Sudbury Basin, enclosed within the canoe-shaped Sudbury Igneous Complex, a sequence ~1,600 m thick of heterolithic breccias of the Onaping Formation (Muir and Peredery, 1984) is overlain by ~600 m of mudstones of the Onwatin Formation (Rousell, 1984a), and ~850 m of wacke-turbidites of the Chelmsford Formation (Rousell, 1984a), all making up the Whitewater Group (Fig. 1). The breccias of the Onaping Formation are traditionally subdivided into four members: the igneous-matrix Basal Member; the Gray Member polymict suevitic breccias; the Green Member; and the Black Member, a redeposited suevitic-breccia unit (Fig. 2). The Gray Member is from 200–700 m thick, and the Black Member is 800–1200 m thick (Muir and Peredery, 1984). The Green Member has been interpreted to represent a fireball collapsed onto the suevitic breccias of the Gray Member before redeposition of the Black Member suevitic breccia occurred (Avermann, 1997). Stöffler et al. (1994) proposed that most of the Black Member is redeposited and reworked clastic breccia material brought into the basin from outside the central crater depression, and that the redeposition occurred under euxinic aquatic conditions. The breccias of all four members contain target rock and mineral fragments of various sizes and shapes (Fig. 2). They commonly exhibit shock metamorphic features. The Gray, Green, and Black Members include recrystallized glass fragments (Muir and Peredery, 1984; Dressler et al., 1996). Masaitis et al. (1997) observed shock-produced diamonds in the breccias of the Black Member. They form polycrystalline aggregates, 0.1–0.6 mm in diameter, of crystallites in the 50–100 nm range. Some aggregates contain up to 20% lonsdaleite.

Two units of the Whitewater Group, namely, the Black Member of the Onaping Formation and the Onwatin Formation, contain appreciable amounts of carbonaceous matter. The origin of the carbon in the Black Member breccias of the Onaping Formation is a long-standing enigma. The regional target rocks contain little carbonaceous matter. Becker et al. (1994a) proposed that the carbon came from the impactor. Bunch et al. (1997) proposed that the carbon formed by the reduction of CO_2 derived from target carbonates. Avermann (1992, 1994) suggested that the carbon is biogenic. To shed more light on this issue, we have determined carbon isotopic compositions and obtained Raman spectra of a large number of carbon samples from the Onaping, and one from the Onwatin Formation. Carbon isotopic composi-

Figure 2. Suevitic breccia of the Black Member of the Onaping Formation. Shock-metamorphosed rock and mineral fragments and recrystallized glass fragments are set in a matrix of rock, mineral, and glass fragments, and fine carbonaceous matter. Plane polarized light. Scale, 0.45 mm.

tions are widely used to distinguish carbon from different sources. Raman spectra are useful because they must be consistent with conditions of greenschist facies metamorphism if the carbon in the Black Member of the Onwatin Formation is of biogenic origin (see Pasteris and Wopenka, 1991).

Becker et al. (1994a, 1996, 1997) reported the occurrence of fullerenes C_{60} and C_{70} in Black Member breccias. Becker et al. (1996, 1997) reported that a small fraction of the fullerenes contained isotopically anomalous helium. The discoveries were made with the analytical technique of laser-desorption mass spectrometry. We felt that a confirmation of these discoveries with HPLC was needed. Also, given the fullerene concentrations reported by Becker et al. (1994a), it seemed possible to purify by high performance liquid chromatography (HPLC) enough C_{60} from these rocks to determine its carbon isotopic composition, which could shed light on its origin.

It was in the course of the search for fullerenes in the Black Member breccias that we discovered the occurrence of elemental sulfur in these rocks. Weathering of pyrrhotite was suspected to be the origin of the sulfur. Tests were carried out to confirm this hypothesis.

CARBONACEOUS MATTER

Samples

Twenty rock samples and one anthraxolite sample were collected mainly from locations in the northern Sudbury Basin. Anthraxolite is a lustrous black, platy material that contains approximately 95% carbon plus quartz and pyrite. It occurs in two large veins in the Onwatin Formation. In the north, the rocks of the Black Member of the Onaping Formation and of the

Onwatin Formation are little deformed and regional greenschist facies overprint is lower than in the south where the deformation is stronger. Sample locations are shown in Figure 1. We determined total elemental carbon content of all but one of the rock specimens, and subjected several samples, including the anthraxolite, to Raman spectrography and carbon isotopic determination.

Determination of total carbon content

Powdered samples of about 0.2 g were analyzed with a standard LECO carbon and sulfur analyzer at the Houston Advanced Research Center (HARC). Carbon was also determined in several samples of 0.5 mg with a Carlo-Erba NA-1500 microanalyzer at the Department of Geology and Geophysics at Rice University. Results are listed in Table 1. The duplicate analyses agree well. The carbon contents of the Black Member of the Onaping Formation have a range of 0.2–2.5 wt% with an average value of 1.0 wt%. These results, as well as the result for the Gray Member of the Onaping Formation, are consistent with earlier reported data on carbon contents by Muir and Peredery (1984) and Avermann (1994). Samples with low carbon contents probably come from the lower portion, and samples with higher carbon contents from the upper portion of the Black Member of the Onaping Formation (see Avermann, 1992, 1994).

Raman spectroscopy

Raman spectra were obtained for carbonaceous matter in nine rocks of the Black Member, in one Onwatin rock, and in the anthraxolite from near the Errington #1 Mine. The rock powders were treated with HCl(concentrated)-HF(49%) to remove most inorganic matter. The anthraxolite was only powdered. The dried residues and other samples were mounted between two circular microscope slide cover glasses and were then fastened in the holder of the Raman spectrometer. Laser radiation at 488 nm was used to obtain the spectra. The scattered radiation was collected at an angle of 45° and analyzed with an SPEX 1403 double monochromator. Representative results are shown in Figure 3. The spectra show peaks at 1354 and 1602 cm^{-1}. These have roughly equal intensity in all of the samples measured, which is consistent with organic matter transformed to poorly organized graphite by greenschist facies metamorphism (Pasteris and Wopenka, 1991).

Carbon isotopic measurements

Powdered rock samples of about 0.2 mg were first treated with toluene to remove soluble organic carbon. The dried samples were then treated with HCl to remove carbonates. Carbon dioxide for the mass spectrometric analysis was generated by O_2 flash-combustion in a Carlo-Erba NA-1500 microanalyzer. Nitrogen oxides were reduced to N_2. Carbon dioxide, N_2, H_2O, and SO_2, which were carried in a stream of He, were separated with a gas-chromatographic column, and the CO_2 was recovered for mass

TABLE 1. TOTAL CARBON AND SULFUR CONTENTS, CARBON ISOTOPIC VALUES, AND ELEMENTAL SULFUR CONTENTS FOR ROCKS FROM THE BLACK AND GRAY MEMBERS OF THE ONAPING FORMATION, ONE ROCK FROM THE ONWATIN FORMATION, AND ONE ANTHRAXOLITE

Sample Catalog Name* Number†	Total Carbon§ (wt. %)	$\delta^{13}C$ Permille**	Total Sulfur§ (wt. %)	Elemental Sulfur§ (ppm)
Black Member Onaping Formation				
High Falls (Location 1, Figure 1)				
SURG1 not known	0.60	-34.52	0.60	97
SUBD4 BO.OF.1	0.31/0.29	-35.22	0.65	238/220
SUBF1 CSF.66.49	0.89	-27.12	1.17	70
SUBF2 CSF.66.44	0.95		0.68	15
SUBF4 CSF.66.43	0.89	-33.08	0.82	12
SUBF7 CSF.66.37	0.73	-29.55	0.69	13
SUBF8 CSF.66.49				32
SUGJ1 BO.OF.3	1.12	-31.60	0.31	33
SUGJ2 BO.OF.2	0.62	-31.89	0.37	28
Dowling Township (Location 4)				
SUBF6 CSF.68.182	0.74		0.41	4.1
Nelson Lake (Location 2)				
SUBD2 BO.NL.2	0.20/0.20		0.17	69
North Capreol (Location 3)				
SUBD5 BO.CA.3	0.78	-31.45	0.10	7.1
South Capreol (Location 3)				
SUBD3 BO.CA.4	1.16/1.13	-30.16	0.46	30
East Capreol (Location 3)				
SUBF3 CSF.94.7A	2.39/2.47	-33.68	0.66/0.67	0.4
SUBF5 CSF.94.5	1.06	-32.77	0.65	3.5
South Chelmsford (Location 5)				
SUGJ3 BO.,CM.1	1.83/1.95	-26.26	0.73/0.88	11
SUGJ5 BO.CM.2	1.14	-29.01	0.27	90
Nickel Offset Road (Location 6)				
SUGJ4 BO.NO.1	1.60/1.67	-30.29	0.56/0.61	
Gray Member Onaping Formation				
High Falls Area (Location 1)				
SUBD1 D95.0104	0.069/0.078	-35.07	0.044/0.045	55
Onwatin Formation				
(Location 7)				
SUGJ6 BO.O1	2.38	-31.31	0.020	28
Anthraxolite				
(Location 8)				
Anthraxolite EM		-31.73		

*SURG = samples donated by R.A.F. Grieve; SUBD = from the collection of B.O. Dressler; SUBF = samples donated by B.M. French; SUGJ = samples donated by G.W. Johns.
†BO samples are from the collection of the Ontario Geological Survey. CSF.xx.yy are samples collected by B.M. French; 19xx denotes year of collection, yy the sample number. EM = Errington Mine #1.
§When two numbers are entered, the first was obtained at Houston Advanced Research Center and the second at Rice University (see text).
**Relative to the international PDB standard.

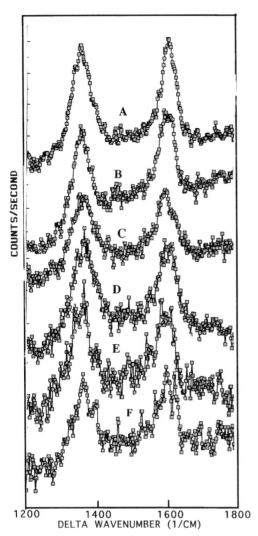

Figure 3. Raman spectra of carbonaceous matter from five Whitewater Group rocks and one anthraxolite from the Sudbury Basin. Primary radiation, 488 nm. The intensity ratios of the peaks at 1,354 and 1,602 cm^{-1} are characteristic of organic matter that has experienced greenschist facies metamorphism (Pasteris and Wopenka, 1991). A = SURG2, High Falls; B = SUBD2, Nelson Lake; C = SUBF3, Capreol; D = SUGJ3, Chelmsford; E = SUGJ6, Onwatin; F = Anthraxolite from Errington #1 Mine.

spectrometry. The mass spectrometer was a semi-automated VG 602E. The performance of the instrument was tested with NBS21 and sulfanilamide (CHNS) standards. The δ^{13}C values listed in Table 1 are reported relative to the international PDB standard.

DISCUSSION

The origin of the carbonaceous matter in the Whitewater Group rocks, especially in the Onaping Formation, has been a long-standing enigma. Muir and Peredery (1984) stated, "If the Black Member is considered to represent a rapidly deposited sequence, then the presence of carbonaceous material throughout the matrix of the member extending laterally over a distance of at least 60 km and vertically about 800 m is most unusual. The suggestion by some investigators (Burrows and Rickaby, 1935) that the carbonaceous matter developed by the reduction of gases during fumarolic activity does not agree with the present distribution pattern nor its abundance in the Black Member. Besides, the numerous feeder structures presumably required for such an exhaustive fumarolic activity are absent or lacking in the Onaping Formation. The origin of the carbonaceous matter is not certain."

Williams (1957) and Thomson (1960) suggested that some or all of the carbonaceous matter of the Whitewater Group might have a biogenic origin. Rousell (1984a) thought that the carbonaceous material of the Onwatin Formation might have been derived from floating algal mats in a deep, restricted basin. A biogenic origin for the carbonaceous matter in the Black Member of the Onwatin Formation was advocated by Beales and Lozej (1975) and Avermann (1992, 1994). Becker et al. (1994a, 1996), whose reported discovery of fullerenes in rocks of the Black Member attracted considerable attention to the problem, suggested that much of the carbon came from the impactor. Bunch et al. (1997) recently suggested that copious amounts of C could be formed by impact-processing of carbonates and also of carbonaceous shales of the Huronian Supergroup.

Let us set the stage by estimating the mass of carbonaceous matter now contained in the Black Onaping Formation. We assumed a circular area of 50-km diameter and ~1-km thickness before deformation: the total volume is roughly 2,000 km^3, somewhat larger than the volume of 1,670 km^3 estimated by Stevenson (1972). With a mean density of 2.73 g/cm^3 and an average C content of 1 wt%, we estimated the mass of Black Member carbon to be on the order of 10^{17} g. These analytical results (Table 1), their interpretations, and other considerations are now applied to select the most probable sources of carbon among five potential sources: target rocks, the impacting projectile, fumarolic activity, the atmosphere above ground zero, and prolific biogenic activity.

Target rocks

One of the remarkable differences between carbon morphologies at the Sudbury and other impact structures is that "shock metamorphosed carbons" such as chaoite and diamond are very rare at Sudbury (Masaitis et al., 1997), but are fairly common at other impact sites (e.g., El Goresy and Donnay, 1968; Masaitis, 1972; Hough et al., 1995; Koeberl et al., 1997; French et al., 1997). This observation already suggests that there was very little carbon of any chemical form present in the target rocks.

Fall-back material in impact craters consists of breccias composed of clasts of variably shock-metamorphosed target rocks and impact glass in a clastic matrix of the same material. Therefore, target rocks are a potential source for the carbon in the Black Member of the Onaping Formation. However, carbonaceous rocks in the basement of the Sudbury Structure are scarce.

We know only of minor and thin carbonaceous mudstone beds, interbedded with metavolcanics of the Huronian Supergroup. They occur south of the SIC, and their volume is much too small to account for the carbon in the rocks of the Whitewater Group. Moreover, only minor carbonaceous target-rock clasts are found in the breccias of the Onaping Formation.

Carbonate rocks of the Espanola Formation of the Huronian Supergroup occur both east and west of the SIC (Card et al., 1977; Dressler, 1982). They constitute another potential source of carbon, but $(CO_3)^{-2}$ must be reduced to neutral carbon. Bunch et al. (1997) have reported that the conversion of carbonate to carbon might have happened in the vapor plume of the Sudbury impact. One problem with this scenario is that most Precambrian carbonates have $\delta^{13}C$ values in the narrow range of about -1 to $+2‰$ relative to PDB standard. Carbon that forms from CO_2 in equilibrium with that gas is not likely to change its $\delta^{13}C$ value by more than about $-10‰$ (Bottinga, 1996). For the nonequilibrium conditions prevailing in the plasma of the plume, it is even less likely that $\delta^{13}C$ would change by that much. Hence it seems unlikely that shock-driven reduction of Espanola carbonate could yield elemental carbon with the $\delta^{13}C$ values reported in Table 1.

Carbonate clasts are extremely scarce in the breccias of the Onaping Formation; hence, carbonate rocks were either absent or relatively scarce themselves in the Sudbury impact area. Moreover, plume processes of terrestrial impacts tend to deposit only small fractions of their products in the crater as was learned, for example, from the global and regional distribution of iridium and spherules from the Chicxulub event (Alvarez et al., 1980; Kring, 1993). Lastly, the very low abundance of carbon in the Gray Member fall-back breccias underlying the Black Member is further evidence that target rocks were almost certainly not the source of the carbon.

Fumarolic activity

One of the first hypotheses for the origin of the carbon is that it formed by the reduction of fumarolic gases in what must have been a megafield of fumaroles (Burrows and Rickaby, 1935). Some objections have already been cited above. Another problem is that fumarolic gases have widely variable compositions depending on the nature of their sources. CO_2, CO, and H_2O are the most common constituents of known fumaroles, with only minor amounts of H_2S, SO_2, H_2, CH_4, HCl, and other compounds present. Such exhalations obviously contain too little reducing matter for the production of much carbon. CH_4-rich fumaroles are known to occur in the vicinity of some oil and gas fields, but no C deposition has been reported at these fumaroles. An additional problem is that the Gray Member must have been in place when the fumarolic activity occurred, yet its breccias have much lower carbon contents than those of the Black Member. Fumarolic activity as the source for the carbonaceous matter in the Whitewater Group rocks can be ruled out.

Projectile

The extraterrestrial, solar system bodies that collide with Earth and are known to contain substantial amounts of carbon, or carbon-bearing compounds, include comets, iron meteorites, and stony meteorites of the carbonaceous chondrite and ureilite classes. Two criteria can be applied here. The first is the required mass and size of the projectile; the second is the isotopic composition of the carbon. Assume that the impactor contained 1% (by mass) of carbonaceous matter, which is a good average for carbonaceous chondrites; Also assume that the density of the impactor was the same as that of the rocks of the Black Member of the Onaping Formation. If all of this carbon became incorporated into the Onaping Formation, then the volume of the impactor must have been at least the same as that of the Onaping Formation, or 2,000 km³, which yields a minimum radius of about 8 km. This is a minimum value because the present volume of the Onaping Formation is smaller than the original one due to erosion, and because the clasts that make up the bulk of the Black Member breccias are derived from well-documented regional formations and not from the projectile. Both factors suggest much larger projectile radii than 8 km.

Whitehead et al. (1990) reported six $\delta^{13}C$ values in the range -29.85 to $-22.8‰$ for the noncarbonate carbon in the upper 5–10 meters of the Black Member, in the overlying 20 m of the so-called "contact unit," defined as the unit in which organic carbon and sulfur contents are much lower than in the uppermost Onaping Formation, and in the lowermost 75 m of the Onwatin Formation. They pointed out that this range overlaps with the range of -0.4–$25.2‰$ for carbon in carbonaceous chondrites (Kerridge, 1985), a conclusion cited in Bunch et al. (1997). However, the results of Whitehead et al. (1990) for the Black Member of the Onaping Formation come from only a single site of that formation, hence cannot be representative for the bulk of the Black Member Breccias. Our $\delta^{13}C$ values are more representative of the Black Member breccias. They fall in the range -35.22 to $-26.26‰$ with an average of $-31.2‰$ in good agreement with the values reported by Avermann (1994) and are outside the range of carbonaceous chondrites. We therefore consider an origin of the carbon from the impactor to be, at least, unproven.

Atmosphere

Another potential source for the carbonaceous matter in the Whitewater Group rocks is a shock-induced reduction affecting the air masses in the target area (Heymann and Dressler, 1997). In this scenario, atmospheric CO_2 dissociates and ionizes in a high-energy plasma above ground-zero. On cooling, oxygen eventually combines with Fe^{2+} in the plume to form Fe^{3+} and with electrically neutral C atoms to form CO. On further cooling, CO disproportionates to CO_2 and C (pure CO at 1 atm disproportionates at 700°C). This scenario could possibly account for the absence of carbon in the Gray Member breccias. The impact would have generated a quasi-vacuum at the target area into which the Gray Member breccias were deposited before the air masses, together with the elemental carbon, returned to ground-zero. The most serious problem is that there simply was not enough CO_2 available in the atmosphere above the area, even 1.85 Ga ago, to account for the mass of carbon in the Whitewater Group rocks. It is also doubtful that this process could produce carbon $\delta^{13}C$ isotopic values more than 10‰ negative. Furthermore, substantial amounts of carbonaceous matter, such as found in the Whitewater Group rocks of the Sudbury basin, have not been observed in other large terrestrial impact structures such as the Chicxulub Structure in Yucatán, where vast masses of platform carbonates occur in the target rocks. We rule out an atmospheric origin for the carbon in the Black Member breccias.

Biogenic activity

Having ruled out or questioned Proterozoic and Archean target rocks, the bolide, fumarolic activity, and the atmosphere as potential sources for carbonaceous matter in the rocks of the Whitewater Group, one important potential source remains, namely, biogenic activity. The theory that carbon in rocks of Precambrian age can be biogenic in origin is not new (see, for example, Hayes et al., 1983). The amounts of biogenic carbon that can form are respectable as some cyanobacteria produce up to 8 g of C/m^2 per day (Schidlowski, 1987). The problem, however, is that the Black Member of the Onaping Formation was considered by some to be an "instant fallback clastic sediment," or an "instant tsunami-transported clastic sediment" (Peredery and Morrison, 1984), whereas the Precambrian rocks in Hayes et al. (1983) are all normal sedimentary or metasedimentary rocks. The environment of an "instant sediment" is not conducive for biogenic activity other than perhaps at its very top when much cooler and quieter conditions returned. For biogenic activity to have permeated the entire Black Member with carbonaceous matter at the level of 1 wt%, the sedimentation rate cannot have overwhelmed the production of organic matter by what we assume were in the main prokaryotes in the water column.

We now present evidence and arguments in support of the conclusion that biogenic activity is the most likely, and perhaps the only, origin of the carbonaceous matter in the Black Member of the Onaping Formation. The isotopic carbon compositions of carbonaceous matter of this work (Table 1) and of Avermann (1994) are consistent with an origin from Precambrian organisms (Fig. 4). Our $\delta^{13}C$ values are in the same range as those of many of the total organic carbon values from Precambrian formations of the same period reported by Hayes et al. (1983). The range of values at the High Falls location, –27.12‰ to –35.22‰ is large. Perhaps this suggests rather chaotic variations of temperature on short distances in the growing Black Member "sediment," or else varying amounts of multiple biogenic components.

For kerogen to become carbonaceous matter, its host rock must experience metamorphism. Pasteris and Wopenka (1991) demonstrated that Raman spectra of the metamorphic graphite can be used as an indicator of the degree of metamorphism, which, in the breccias of the Onaping and Onwatin Formations, was greenschist facies grade. Single-crystal graphite has only one first-order Raman line at 1,575 cm^{-1} (Tuinstra and Koenig, 1970), the so-called "G-line." Many investigators, including Tuinstra and Koenig (1970), reported that the forbidden Raman "D-line" near 1,360 cm^{-1} appears in samples of disordered graphite, and that

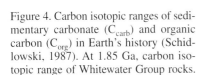

Figure 4. Carbon isotopic ranges of sedimentary carbonate (C_{carb}) and organic carbon (C_{org}) in Earth's history (Schidlowski, 1987). At 1.85 Ga, carbon isotopic range of Whitewater Group rocks.

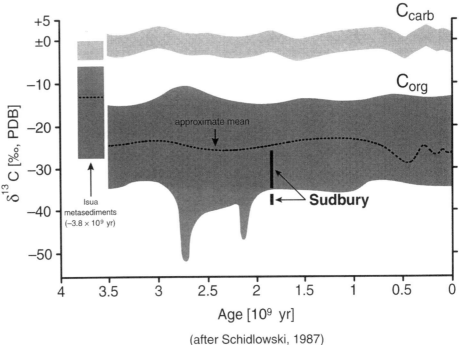

(after Schidlowski, 1987)

the ratio D/G increases as the degree of disorder increases. A simultaneous shift of the G-line toward the blue (1,602 cm^{-1} in Fig. 3) was observed (Nakamizo et al., 1978). Disorder may include increasingly smaller grain size as well as structural defects. The reverse is also the case, namely, that a decrease of disorder, for example, in response to metamorphism of the parent rock, leads to a decrease of the D/G intensity ratio of the graphite in the rock, which is the principle used by Pasteris and Wopenka (1991). Their observed D/G ratios in chlorite rocks of six metamorphic suites vary, but have an average value near 1.0, close to the values seen in Figure 3 for samples of carbonaceous matter from the Onaping Formation. The most salient observation by Pasteris and Wopenka (1991), however, is that graphite from biotite, garnet, staurolite, kyanite, andalusite, and sillimanite zones all have D/G ratios less than 0.5. The Black Member breccia has experienced greenschist facies metamorphism, hence one expects the Raman spectra of carbonaceous matter derived from kerogen in these rocks to be consistent with that degree of metamorphism. Our Raman results taken by themselves do not prove that the origin of the carbon is biogenic, but in context they strongly support that hypothesis.

A biogenic origin was already proposed for carbon in the Onwatin and Chelmsford Formations (Hayes et al., 1983), which are listed in Hayes et al. (1983) as Precambrian biogenic carbons with $\delta^{13}C$ values of –30.7 and –31.6‰, respectively. Avermann (1994) reported $\delta^{13}C$ of –30.24‰ for one Onwatin sample and relics of what she described as algae in another. As Table 1 shows, these and our $\delta^{13}C$ value of the Onwatin sample are in the range of the Onaping samples, and so is the single Onwatin result from Whitehead et al. (1990). Moreover, the Raman spectrum of the Onwatin carbon is very similar to that of the Onaping carbon. The striking similarity of the carbons throughout the entire Whitewater Formation is perhaps the strongest argument in favor of the biogenic origin of the carbon in the Onaping Formation.

How, then, might one depict the filling of the central depression above the Green Member of the Onaping Formation with inorganic material primarily from outside the depression, and the acquisition of organic carbon? Did the infilling matter already contain organic carbon? That is definitely among the possibilities and would shorten the time required to produce the estimated 10^{17} g of carbonaceous matter in the central depression alone. Was the water initially hot and how fast did it cool to more "normal" temperatures? Chances are that the water was initially very hot, conditions that were favorable for the growth of hyperthermophylic archaea. How fast the waters cooled to more normal temperatures is impossible to say, but cooling brought about conditions favorable for other kinds of archaea and bacteria, especially cyanobacteria. The carbonaceous matter now in the Black Member of the Onaping Formation was derived from the inorganic forms of CO_2 and CH_4. We assume that most of it was produced by cyanobacteria.

Biogenic activity continued during the sedimentations of the Onaping, Onwatin, and Chelmsford Formations. If the cyanobacteria produced as much as 8 g of C/m^2 per day (Schidlowski, 1987) over the surface of the undeformed Sudbury basin only, then it would have taken about 20 m.y. to produce the carbon presently in the Whitewater Group rocks. The carbon was probably formed over a considerably larger area around the Sudbury basin and was redeposited in the basin in less than that time. The observation by Avermann (1994) that the C content of the Black Member breccias increases from base to top is consistent with this model because one expects the rate of accumulation of rock powder and clasts in the basin to have decreased in time, thereby presenting the cyanobacteria increasingly greater opportunity to multiply and produce organic carbon before being smothered by the accumulating materials.

Similar conditions favorable for the growth of prokaryotes possibly existed in a shallow-marine impact environment where wave action, turbidity currents and other submarine transport mechanisms were responsible for deposition of Black Member breccia into the Sudbury Crater.

The Gray Member breccias of the Onaping Formation are practically C-free. They were not exposed to organic activity, not even at the top of the formation, where the Gray Member was sealed off by the Green Member, interpreted to represent the collapsed fireball layer (Avermann, 1997).

Anthraxolite

Two large anthraxolite veins are known in the Onwatin Formation. The veins contain about 95% carbon with some quartz and pyrite (Rousell, 1984b). Rousell (1984b) suggested that tectonometamorphism had remobilized, perhaps hydrothermally, and concentrated local carbonaceous material to form the anthraxolite veins. The Raman results (Fig. 3) and carbon isotopic data (Table 1) of this work and of Mancuso et al. (1989) confirm that Onwatin carbonaceous matter could be the parent material of the anthraxolite carbon.

FULLERENES

Introduction

Buseck et al. (1992) reported the discovery of fullerene molecules in shungite from the Kola Peninsula in Russia. Soon thereafter, searches for other natural occurrences of these molecules began in terrestrial samples, in meteorites, and in samples from the Moon. The terrestrial searches were carried out in products from high-energy environments such as fulgurites (Daly et al., 1993), breccias from the Sudbury impact structure (Becker et al., 1994a, 1996, 1997), and soot, presumably from wildfires triggered by the Chicxulub impact (Heymann et al., 1994; Becker et al., 1994a). Fullerenes were discovered in some samples of these materials. Moreover, it was found that C_{60} and C_{70} are the most abundant naturally occurring members of the fullerene family on Earth, presumably because they are thermodynamically the most stable.

Fullerenes are known in the laboratory to form only in the gas phase, either by the condensation of elemental carbon in an inert, hydrogen-poor atmosphere (Kroto et al., 1985; DeVries et al., 1993), by the burning of carbon-bearing matter in oxygen-starved, sooting flames (Howard et al., 1991; Heymann et al., 1994), or by the pyrolysis of naphthalene and possibly other polycyclic aromatic hydrocarbons (PAHs) (Taylor et al., 1993). Saunders et al. (1993) demonstrated that C_{60} is capable of endogenically trapping and then storing He atoms inside its cage when the molecule forms in an He-bearing gas.

Meteorites, especially carbonaceous chondrites, are a potential source of fullerenes on Earth. However, it is unclear how plentiful that source is. Becker et al. (1994b, 1995) reported 100 ppb C_{60} in one sample of the Allende meteorite. Becker and Bunch (1997) and Becker et al (1997) reported contents of ~5 and ~10 ppb in two more Allende samples. However, neither Tingle et al. (1991), Gilmour et al. (1993), DeVries et al. (1993), nor Heymann (1995a,b, 1997) found fullerenes in up to about 300 g of Allende or in other meteorites. Thus, judging from a rough mass balance of all of the Allende samples studied, it appears that C_{60} is very heterogeneously distributed in this meteorite, and its bulk concentration could well be many orders of magnitude smaller than the reported 100 ppb, or even 10 ppb.

Becker et al. (1994a, 1996) reported C_{60} and C_{70} fullerenes in ppm concentrations in rocks from the Black Member of the Onaping Formation. They proposed that these fullerenes are extraterrestrial, perhaps even presolar, on the grounds that a small fraction contains He@C_{60}, with He having an unusual, nonterrestrial isotopic ^{3}He/^{4}He ratio. Their paper prompted the present study, which attempts to verify the findings of Becker et al. (1994a, 1996) and, if feasible, to extract and purify C_{60} and C_{70} in quantities of milligrams for high-precision ^{13}C/^{12}C isotopic measurements. Fullerenes with ^{13}C/^{12}C outside the known range of terrestrial carbon definitely qualify for extraterrestrial provenance.

Experimental procedures and results

The rocks for which carbon and fullerene analyses were done are listed in Table 1. This set included 17 rocks from the Black Member of the Onaping Formation, 1 rock from the Gray Member of the Onaping Formation, and 1 rock from the Onwatin Formation. Figure 1 shows the sample locations. The carbon contents of these rocks, when considered in the context of the work of Avermann (1992, 1994) suggest that the set includes rocks from base, center, and top of the Black Member. All rocks except SUBF8 were studied at Rice University, where the analysis was done by high-performance liquid chromatography. Three rocks (SUGJ1, SUBD3, and SUBF8), were studied at Arizona State University, where the analysis was done by positive-ion electron ionization mass spectrometry, as described in Daly et al. (1993).

C_{60} and C_{70} were extracted from the finely powdered samples by stirring them for 16–25 h in toluene. Separate splits from SURG1, SUBD2, SUBD3, SUBD4, and SUBD5 powders were first demineralized by treatment with HCl-HF and, in some cases, HF–boric acid (Robl and Davis, 1993). Duplicate runs were done with SUBF1 and SUBF2. The initial weights of one SUGJ1 and all SUBD samples treated with toluene were in the range of 20–25 g. When no fullerenes were found in these samples, replicates were run for SUGJ1 and all SUBD samples with initial weights of 100 g. A total of 27 distinct samples (including replicates) were investigated at Rice and 3 samples at Arizona State University.

Following extraction, the contents of the flasks were filtered through Whatman #42 paper to remove the coarsest solid particles. The flask and solid matter on the filter were washed with additional toluene. At this stage, the total volume of the extracts was several hundred milliliters and, in some cases more than 1 L. In order to increase the sensitivity of the analysis, these volumes were reduced to approximately 1 ml with a traditional distillation setup, followed by entrainment of toluene in an Ar stream to less than 0.5 ml. Final solutions were filtered through 0.2 µm PTFE filters prior to HPLC analysis. To test the yield of the extraction and evaporation procedures, 20 g of powdered SURG1 was spiked with 500 ng of C_{60} and was taken through the complete experiment; 461 ng, 92.1% of C_{60}, was recovered.

The HPLC system has been described before (Heymann et al., 1994). However, a Cosmosil column was also used with a mobile phase of 1 ml min^{-1} of toluene. Only 25 ml of extract could be injected into the Nova-Pak C18 column, but up to 500 ml into the Cosmosil column. The identification of C_{60} and C_{70} by HPLC rests on significant absorption signals at the known retention times of these fullerenes in the chromatograms and on their absorption spectra registered by the photo diode array (PDA) at these times. The retention times of the fullerenes as well as the absorption per nanogram of injected fullerene were calibrated by the injection of samples containing known amounts of synthetic C_{60} and C_{70} (see, for example, Heymann et al., 1994). The sensitivity of the method is illustrated in Figure 5, where the signals at the retention time of C_{60} are shown for increasingly smaller amounts of C_{60} injected into the Nova-Pak C18 column. Amounts as small as 0.5 ng injected can be detected.

Both procedural and instrumental blanks were run. Procedural blanks were experiments carried out with a sample of powder from the Black Member that had been washed with toluene. Acceptable procedural blanks yielded less than 1 ng of apparent C_{60}. No extract from a rock was injected unless three successive instrumental blanks showed the absence of any C_{60} memory.

A typical chromatogram obtained from the extraction of a powdered rock is shown in Figure 6. No signals above background are seen at the retention times of C_{60} and C_{70}, hence no detectable fullerenes occur in the solution. In this case, 22% of the extract was used for the analysis. No fullerenes were found in any of the rocks, either at Rice University or at Arizona State University.

Discussion

Fullerenes appear to be heterogeneously distributed in the Black Onaping Formation. Our samples (SUBF1, SUBF2,

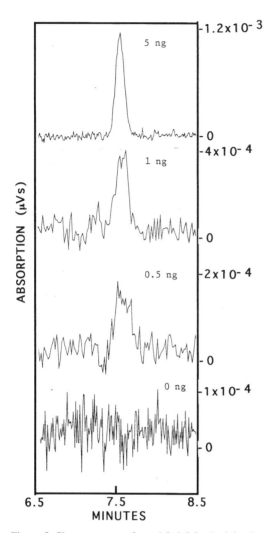

Figure 5. Chromatograms from 6.5–8.5 for the injection of 0.5–5 ng C_{60}. The vertical coordinate is absorption at 330 nm in microvolt-seconds. The response of the PDA is linear up to at least 20,000 ng of C_{60} injected.

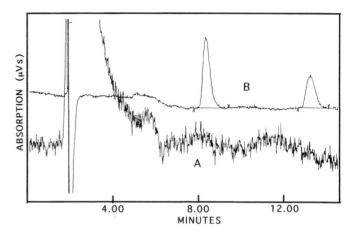

Figure 6. Chromatogram A = 25 µl of SURG1 extract injected into the Nova-Pak C18 column. Mobile phase 0.45 ml/min toluene and 0.55 ml/min methanol. Vertical coordinate is absorption in microvolts. Chromatogram B = injection of 25 ng each of C_{60} and C_{70} into the Nova-Pak C128 column.

SUBF4, SUBF7, and SUBF8), and samples CSF-66-36, and CSF-66-43 of Becker et al. (1994a) all came from the roadcut on the west side of Highway 144; SUBF6 and CSF-68-182 (Becker et al., 1994a) came from an off-road outcrop along a power line about 1 mile west of the road (B. French, personal communication, 1997). Our samples do not contain fullerenes; those of Becker et al. (1994a) do. Given the dimensions of the outcrop, this difference suggests that the scale of fullerene heterogeneity in the Black Member of the Onaping Formation is on the order of tens of meters, or maybe even less. However, in two cases, splits from the same hand specimen were distributed both to our team and to Becker et al. (1994a). SUBF4 is a fragment from CSF-66-43 and SUBF6 is a fragment from CSF-68-182 (B. French, personal communication, 1997). Such proximity suggests that the scale of fullerene heterogeneity may be on the order of centimeters or less.

Two important but unanswered issues are raised by the C_{60} found by Becker et al. (1994a) in the breccias of the Black Member. The first is whether the heterogeneity of fullerenes in the Black Member was established at the time of impact ("originally heterogeneous" or OHET); the second is whether it came about by the transformation of an originally homogeneous distribution to its current distribution, for instance, during the greenschist facies metamorphism of the Black Member ("originally homogeneous" or OHOM).

The OHET hypothesis implies that a small fraction of a fullerene-bearing impactor became scrambled with target rock at depth. An analogous scrambling seems to have happened at the Chicxulub Crater, where carriers of extraterrestrial iridium were found in breccia (Schuraytz et al., 1997). An OHET-type fullerene distribution seems possible, given the notoriously heterogeneous character of the Black Member breccia.

The OHOM hypothesis encounters a quantitative problem. In the earlier section on carbon, the mass of the Black Member was estimated as $\sim 10^{19}$ g. With an average C_{60} content of 5 ppm (bases on data in Becker et al., 1994a), the total mass of C_{60} in the Black Member would be $\sim 10^{14}$ g from an extraterrestrial source (Becker et al., 1996), or locally produced by the impact (Becker et al., 1994a; Bunch et al., 1997), or some of both. The only extraterrestrial matter in which C_{60} was reported is the Allende meteorite, samples of which contained ~ 100 ppb (Becker et al., 1994b) and ~ 5 and ~ 10 ppb (Becker and Bunch, 1997). A 100% extraterrestrial provenance would require an Allende-type impactor with 100 ppb of C_{60} to have had a mass of at least $\sim 10^{21}$ g, hence a volume of about 5×10^4 km^3 (radius, 22.8 km). There are three reasons why this is a minimum value: (1) Heymann (1997) deduced that the bulk C_{60} content of Allende is at least one or two orders of magnitude smaller than 100 ppb, and the recent results of Becker and Bunch (1997) suggest that the bulk C_{60} content may be even less than that. The minimum value of the volume of this impactor would become $>10^6$ km^3 (radius, 62 km). (2) C_{60} and C_{70} are quite volatile (Abrefah et al, 1992;

Matthews et al., 1993); hence, a substantial fraction of both would have been volatilized and become globally distributed, analogous to the global Ir anomaly at the Cretaceous-Tertiary boundary. (3) Some C_{60} was destroyed by ultraviolet (UV) radiation and high temperatures in the transient Sudbury cavity.

Another concern is that the Black Member has experienced greenschist facies metamorphism, probably in the lower part of the range 300°–500°C lasting several 10^5, perhaps 10^6 years (B. French, personal communication, 1997). Several studies on the thermal stability of fullerenes have been published. Vassallo et al. (1991) heated C_{60} stepwise in vacuum to six temperatures in the range of 100°–600°C, holding at each temperature for 60 s. They did not observe any change in the IR spectrum of the fullerene. When a mixture of oxygen and argon (25% v/v O_2) was admitted into the sample chamber, bands of CO_2 and CO began to appear (after 60 s) at 250°C, and by 450°C all evidence of C_{60} was gone. In a separate run, C_{60} was held at a constant 250°C. No fullerene remained after 90 min. Qualitatively, these results are similar to those obtained by Chibante et al. (1993), who found that fullerenes heated in air to 250°C were fully decomposed after 65 h. Heat treatment of fullerenes in an argon atmosphere was carried out at 200°, 400°, 500°, 800°, and 1000°C by Mochida et al. (1995). The yield of C_{60} was 91% for heating during 2 h at 200°C; at 400°C, the yield was 84%. These conditions are perhaps more similar to those in the Black Onaping Formation during metamorphism, and the rates of destruction are smaller than those in the presence of oxygen. Nevertheless, if the thermal decomposition of C_{60} can be treated by first-order kinetics as Chibante et al. (1993) argued, all but one part in 10^6 of the original C_{60} will have been destroyed after heating for only 300 h at 200°C and after about 160 h at 400°C.

The apparent discrepancy between predictions based on these laboratory studies and the observation that fullerenes have survived greenschist facies metamorphism of the Black Member of the Onaping Formation might be explained as follows. The C_{60} molecule, on heating in a chemically inert environment, can either break up into smaller fullerenes, rings, or chains of carbon, or it can coalesce to form larger fullerenes such as C_{118} (Yeretzian et al., 1992). The first reaction path is available both for isolated and "touching" fullerene molecules, but for the second it is always necessary that two or more fullerene molecules are within range of chemical action. Thus, if the fullerenes in the breccias are greatly dispersed, then they can decompose only by breakup. Perhaps nature tells us here that the rate of breakup is much slower than the rate of coalescence, at least in the range of temperatures of greenschist facies metamorphism and that the surviving fullerene molecules were not in contact.

SULFUR

Introduction

The following terms are defined for this section: sulfur, S^0 indicates elemental sulfur; sulfide, S^{2-} indicates the most reduced state; and sulfate, SO_4^{2-}, the oxidized 6+ state. The symbol S represents the sum of all chemical states of the element.

Sulfides are common minerals in the Onaping rocks. They are, in decreasing order of abundance, pyrrhotite, chalcopyrite, pyrite, and sphalerite, with a mean S^{2-} content of 0.53 wt% in the Black Member (Muir and Peredery, 1984). Whitehead et al. (1990) reported an upward trending increase of organic carbon and S contents in the upper 160 m of the Black Member of the Onaping Formation in a drill hole bored approximately in the middle of the South Range. Rousell (1984b) proposed that sulfide mineralization in Onaping rocks involved preexisting sulfides from the country rocks.

Thode et al. (1982) determined isotopic δ^{34}S-values in the range +1.7‰–+9.2‰ in rocks from the Onaping Formation. Whitehead et al. (1990) found δ^{34}S-values up to +7.5‰ in the Onwatin Formation, and Thode et al. (1982) reported six values in the range +15.7‰ to +26‰ in the same formation. Sulfates have not been found in the Onaping Formation, but we report here the occurrence of S^0.

In their report on the discovery of C_{60} and C_{70} fullerenes in breccias from the Black Member of the Onaping Formation, Becker et al. (1994a) noticed peaks at m/z values corresponding to S_2^+ to S_9^+ ions in the laser desorption time-of-flight mass spectrum of matter that had been obtained by toluene extraction of the demineralized High Falls sample CSF-66-43. S^0 dissolves well in toluene at room temperature, but the S^0 observed by Becker et al. (1994a) might also have formed by oxidation of H_2S during the acid demineralization treatment, hence might not have occurred in the rock proper. In order to examine whether S^0 is truly contained in rocks from the Onaping Formation and what its origin might be, a representative suite of rocks (Table 1) was studied. The set includes mostly samples from the Black Member of the Onaping Formation, but also one sample from the Gray Member, and one sample from the Onwatin Formation. Figure 1 shows the locations of the outcrops from which samples were collected.

Experimental procedures and results

Total S was determined with the same instrumentation used for the determination of carbon. The procedures are described in the section on carbon and the results are listed in Table 1. S in the Black Member of the Onaping Formation ranges from 0.10–1.17 wt%, with an average of 0.55 wt%, consistent with data in Muir and Peredery (1984) and Thode et al. (1982). Figure 7 shows that there is no significant correlation between total S and C contents of all samples studied.

All but one of the powdered rock samples for S^0 analysis had masses in the range of 1–2 g; the exception was the sample from the Gray Member of the Onaping Formation (SUBD1) whose mass was 50.0 g. The powders were stirred for 14 h in toluene. Solid and fluid were separated by filtration through Whatman #42 paper, and the fluid volumes were reduced to about 5 ml by vacuum distillation at 60°C. The fluids were transferred to preweighed bottles through 0.2 mm PTFE filters. When needed,

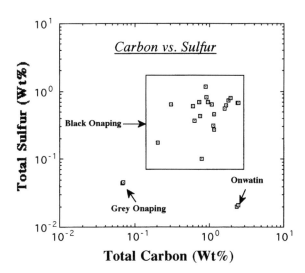

Figure 7. Total carbon vs. total sulfur content of the rocks of Table 1. The samples from the Black Member of the Onaping Formation lie in the box. There is no clear-cut correlation of C and S in these samples.

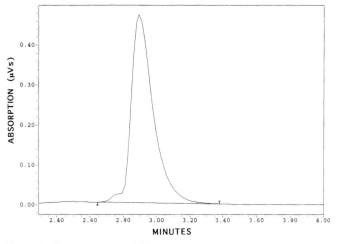

Figure 8. Chromatogram of 25 µl of SUBD4 extract injected into the Cosmosil column. Mobile phase 1 ml/min toluene. The S^0 peak rides on a small background of extracted organic matter.

the volumes could be further reduced by blowing dry argon gas into the bottles.

S^0 was determined in the extracts by HPLC with the system described above in the previous section on fullerenes (the Cosmosil column was used). Figure 8 shows the chromatogram for injection of 25 ml of the extract from SURG1. The peak at 2.88 min is due to S^0 as was demonstrated by the injection of 485 ng of S^0 from a standard solution. Unfortunately, the absorption spectrum of S^0 in the working range of 300–450 nm of the photo diode array is featureless and similar to the absorption spectra of organic substances coextracted from the rocks, hence cannot be used as supporting evidence for the presence of S^0. In order to further test every assignment of the peak at 2.88 min to S^0, a portion of each extract was exposed at room temperature for 14 h to HCl-cleaned copper shavings, which removed S^0 from the solution by forming CuS, and was then reanalyzed. The resulting chromatograms (not shown here) all had only one small and broad peak in the range of 2–4 min due to the coextracted organic matter. These "Cu-cleaned" chromatograms were used to correct the chromatograms of the untreated extracts for absorption by substances other than S^0 at 2.88 min. All results are reported in Table 1. The S^0 contents of the rocks varies greatly, from 4.1–230 ppm. There is no significant correlation between S and S^0 contents.

Since weathering of pyrrhotite was suspected to be the source of the S^0, three samples were taken from the bulk pyrrhotite block ZV15 obtained from the Whistle Mine Sublayer at the northeastern corner of the SIC (Fig. 1). A sample of crust was obtained by gently filing off 4.49 g of crustal matter. By using the conical end of a 1-in. (24mm) diameter drill, 5.53 g were "mined" from below this cleaned surface to a depth of 5 mm. An interior piece at a depth of about 10 cm depth was obtained by sawing. All samples were powdered in an agate mortar. A separate powdered sample of 250 g was exposed to water at 24°C for 10 d, recovered by filtration, washed with acetone, and dried at 40°C. Each of the four samples was refluxed for 16 h with toluene and separated from the extract by filtration through Whatman #42 paper. These extracts were then processed and analyzed like the extracts from rocks. The results are presented in Table 2. By the repeated dry magnetic separation of a 50-g sample of powdered SUBD4, a pyrrhotite separate of 0.371 g was obtained that was analyzed for S^0 in the same manner as all other samples. Its S^0 content is listed in Table 2.

S^0 extracted from SUBD4 and the following samples from pyrrhotite ZV15 were used for isotopic measurements. A pyrrhotite/S^0 sample pair was prepared by refluxing 500 g of powdered pyrrhotite from the Whistle Mine overnight with toluene for the extraction of S^0. Solid and fluid were separated by filtering through Whatman #42 paper. The solid was washed with toluene. About 1 g of the pyrrhotite was further washed with toluene until the fluid was free of S^0. This material was dried and 0.371 g was used for the isotopic determination. The extract was reduced in volume and the S^0 was purified by repeated HPLC separations. The isotopic compositions of S^0 and the S^0 in pyrrhotite were determined by methods described by Gao and Thiemens (1991, 1993a,b). The results of the isotopic measurements are given in Table 2.

Discussion

The total S contents of the rock samples are in the range reported by others (Muir and Peredery, 1984; Thode et al., 1982), but, as demonstrated by the plot in Figure 7, there is no clear-cut increase of S with C among the Onaping rocks of this study. This is different from the tight correlation between the two elements in the data set of from the upper 160 m of the Black Member of Whitehead et al. (1990). The difference means that the prove-

TABLE 2. ELEMENTAL SULFUR CONTENT AND ISOTOPIC COMPOSITION OF SULFUR OF PYRRHOTITE SAMPLES

Sample	S^0
Sulfur Content (Location 9, Figure 1)	
ZV15 Crust	29.4 ppm
ZV15 0-0.15 cm	4.4 ppm
ZV15 interior	None detected
ZV15 "on water"	1.0 %
Pyrrhotite from SUBD4	1.4 %

Sample	$\delta^{33}S$ (‰)	$\delta^{34}S$ (‰)	$\Delta^{33}S$ (‰)	$\delta^{36}S$ (‰)	$\Delta^{36}S$ (‰)
Isotopic Data					
Pyrrhotite	2.62	4.75	0.245	8.6	-0.76
S^0 from Pyrrhotite	2.93	5.19	0.335	9.7	-0.52
S^0 from SUBD4	4.87	9.55	0.095	17.6	-1.21

Note: $\delta^i S = 1000\{([^iS/^{32}S]sample/[^iS/^{32}S]standard)^{-1}\}$. Standard is Canyon Diablo Troilite (CDT) sulfur.
$\Delta^{33}S = \delta^{33}S - 0.5 \times \delta^{34}S$; $\Delta^{36}S = \delta^{36}S - 1.97 \delta^{34}S$.

nances of S and C in the rocks of this work, unlike the samples of Whitehead et al. (1990), were decoupled. Most of the S^{2-}, which represents the bulk of S in our rocks, is from preexisting sulfides and not from bacterial action, and C is from cyanobacteria as explained above.

All rocks contain S^0 in the range of 0.4–238 ppm. The S^0 contents of the Gray Member and Onwatin samples, 55 and 28 ppm respectively, are surprisingly large considering the comparatively low total S contents of these two rocks of 0.045 and 0.020%. There does not seem to be a clear-cut geographic trend among the S^0 contents, but with only a few samples from most locations it is difficult to say whether there is none.

With regard to the origin of the S^0, one may consider CI carbonaceous chondrites, which are known to contain up to a few percent elemental S^0 (Gao and Thiemens, 1993a). Could the impactor have been a CI carbonaceous chondrite? This is unlikely for two reasons. S^0 should not have survived greenschist facies metamorphism in this pyrrhotite-rich environment, furthermore, the isotopic composition of the sulfur extracted from SUBD4, $\delta^{34}S = 9.55$‰, is outside the range of sulfur isotopic compositions in the CI chondrites (Gao and Thiemens, 1993b). A meteoritic origin can be excluded.

Although a small contribution from sulfate-reducing bacteria to S^0 in these rocks cannot be ruled out, the bulk of the S^0 is not likely to come from a bacterial process that normally yields S^0 with negative $\delta^{34}S$ values. A more likely explanation is that the S^0 came from the weathering of a sulfide, most likely pyrrhotite. The S^0 content of pyrrhotite block ZV15 decreases from crust to interior. Pyrrhotite extracted from rock SUBD4 contains 1.4% S^0, even more than the pyrrhotite from ZV15 exposed to water in the laboratory. The isotopic compositions of S^{2-} in ZV15 pyrrhotite and of S^0 extracted from it are very similar. Also, the value of $\delta^{34}S$ of S^0 from SUBD4, although it is at the upper end of the range of Thode et al. (1982), is not inconsistent with an origin by weathering of pyrrhotite in SUBD4. Nearly all pyrrhotite grains in SUBD4 are now covered with an oxide skin, whereas pyrite grains are not. The simplified weathering reaction of pyrrhotite is $2FeS + 3 H_2O \rightarrow Fe_2O_3 + 2S^0 + 6H^+ + 6 e^-$, but the possible participation of H_2S, FeS_2, as well as a variety of iron oxides means that the actual chemistry is likely to be more complex. The absence of a correlation between S^0 and S in our data set probably means that the concentration of S^0 is governed by rock properties and local weathering conditions, but not by total S content.

One remarkable aspect of the isotopic data is that the $\Delta^{33}S$ and $\Delta^{36}S$ values as defined and listed in Table 2, should be zero for perfect mass-dependent fractionations, but they are not zero. This shows that a component of mass-independent fractionation is present, with ^{33}S excess and ^{36}S deficiency relative to the perfect mass-dependent fractionation. Analogous mass-independent fractionations have been reported for the Allende meteorite (Rees and Thode, 1977), in ureilites (Thiemens and Jackson, 1986), in a polymer of carbon disulfide (Colman et al., 1996), and in meteorite sulfonic acids (Cooper et al., 1997). The mass-independent fractionation of the CS_2 polymer is intriguing. The polymer forms when gaseous or liquid CS_2 is irradiated by UV or visible light (Colman and Trogler, 1995; Cataldo, 1995). Spectra of a comet show the presence of CS from which the presence of CS_2 in the comet is implied (Jackson et al., 1986). If the impactor at Sudbury had been a comet, then the volatile CS and CS_2 would have vaporized and possibly ignited. Any polymer from the comet that made it into the breccias would, however, have become decomposed during metamorphism (for decomposition, see Colman et al., 1996). CS, CS_2, and CS_2-polymer are probably too ephemeral to be useful as fingerprints of cometary impacts. The explanation for the mass-independent fractionation that persists is that mineral reactions involving naturally occurring compounds of sulfur result occasionally in this mass-independent fractionation and are perhaps more common than previously thought possible.

SUMMARY

Prior to the crater-forming impact at Sudbury, 1.8 Ga ago, the target rocks apparently contained only very small amounts of the element carbon in any of its naturally occurring chemical forms. Sulfides were quite abundant in the target rocks and the weathering of sulfides would have produced some S^0 in them. Regardless of whether the impactor was a comet, a carbonaceous chondrite, a ureilite, or an iron meteorite, little if any of its carbon appears to have contributed significantly to the bulk of carbonaceous matter in rocks of the Onaping Formation of the Whitewater Group. This conclusion is based in part on the estimated mass of carbonaceous matter now present in the Onaping Formation and in part on the abundant isotopic evidence. Sulfides, S^0, and undoubtedly all other constituents of the impactor were also substantially vaporized and redeposited away from the crater, perhaps even world-wide as is the case of the iridium from the Chicxulub impactor.

As was reported, the impactor contained C_{60} and C_{70} fullerenes, some of which survived the impact in rocks of the Black Member of the Onaping Formation and the later greenschist facies metamorphism of this formation. Exactly how the fullerenes were trapped in the Black Member rocks is unclear. However, given the amounts of fullerenes found in the CV3 chondrite Allende, it is clearly impossible that the incorporation resulted in a homogeneous C_{60} concentration of 1–10 ppm as found in three Black Member samples by Becker et al. (1994a), because that would imply an implausibly large radius for the impactor. The results discussed in this chapter show that it may be difficult to find additional Black Member rocks containing fullerenes.

When preimpact atmospheric conditions returned, there must have been a cavity of >50 km diameter in which the basal, Green, and Gray Members of the Onaping Formation were in place. The bulk of the Black Member breccia clasts were still widely scattered, perhaps on terraces, mountain slopes, and other features around the central depression, but with little, if any carbonaceous matter present. When the masses of rock of the region had cooled sufficiently, a lake began to form in the crater and Black Member breccia clasts and rock flour were transported into it by local streams and sedimented under euxinic aquatic conditions. Almost simultaneously, prokaryotic growth, probably of hyperthermophylic archaea, commenced in still relatively hot waters. Later, the growth of cyanobacteria probably contributed the bulk of the carbonaceous matter now in the Black Member of the Onaping Formation.

Although the lithology of the lake deposits changed from suevitic breccias to mudstones when the sedimentation of the Onwatin Formation began, large concentrations of biogenic organic carbon continued to be generated by algae into the Onwatin Formation.

Similar conditions favorable for the growth of prokaryotes possibly existed in a shallow-marine impact environment where wave action, turbidity currents, and other submarine transport mechanisms were responsible for deposition of the Black Member breccia into the Sudbury Crater.

One, or perhaps several, episodes of metamorphism, deformation, and erosion eventually generated the present-day lithologies of the Whitewater Group of rocks and the structure of the Sudbury basin. The greenschist facies metamorphism recorded in the rocks of the Onaping and Onwatin formations transformed the organic carbon to the carbonaceous matter now present with its characteristic carbon isotopic composition and Raman spectra. Any S^0 present in the rocks prior to metamorphism must have reacted, hence the S^0 now found in the rocks of the Onaping and Onwatin Formations is probably of very recent vintage. The Onaping and Onwatin Formations of the Whitewater Group of the Sudbury Structure are among the world's largest burial grounds of Precambrian biogenic remains.

NOTE ADDED IN PRESS: While our manuscript was in the stages of review and editing, we obtained the following additional carbon contents and carbon isotopic data on carbonaceous matter in rocks from the Onwatin and Chelmsford Formations.

Sample name	Carbon content (wt%)	$\delta^{13}C$ (‰)
ONWATIN FORMATION		
High Falls (Location 1)		
ON1	4.07	−28.21
Vermillion Lake Road		
ON2	3.16	−31.64
ON3	2.07	−29.71
ON4	4.86	−31.26
ON5	4.36	−29.30
CHELMSFORD FORMATION		
West of Chelmsford on highway 144 (Location 6)		
CHEM1	0.52	−31.85
CHEM2	0.61	−31.35
CHEM3	0.44	−31.95

These results strengthen our conclusion that the carbonaceous matter of the Black Member of the Onaping Formation is biogenic.

ACKNOWLEDGMENTS

We thank Bevan French, Richard Grieve, and Glen Johns for donating samples. We also thank Louis Elrod and Ewa Szymczyk of the Houston Advanced Research Center for performing the carbon and sulfur analyses. Appreciation is extended to John Margrave, Robert Hauge, and Judy Chu for assistance with the Raman spectrometry, and to Richard E. Smalley for the use of his HPLC facility. Bevan French has provided salient information and has shared his ideas with us. Last, but not least, we thank M. E. Avermann and I. Gilmour for constructive reviews of this paper. National Aeronautics and Space Administration Grant NAG-5-4308 provided support (to P.R.B.)

REFERENCES CITED

Abrefah, J., Olander, D. R., Balooch, M., and Siekhaus, W. J., 1992, Vapor pressure of buckminsterfullerene: Applied Physics Letters, v. 60, p. 1313–1314.

Alvarez, L. W., Alvarez, W., Asaro, F., and Michel, H. V., 1980, Extraterrestrial cause for the Cretaceous-Tertiary extinction: Science, v. 208, p. 1095–1108.

Avermann, M. E., 1992, Die Genese der allochtonen, polymikten Breccien der Onaping Formation, Sudbury-Struktur, Ontario, Kanada [Ph.D. Thesis]: Münster, Germany, University of Münster 175 p.

Avermann, M. 1994, Origin of the polymict, allochtonous breccias of the Onaping Formation, Sudbury Structure, Ontario, Canada, in Dressler, B. O., Grieve, R. A. F., and Sharpton, V. L., eds., Large Meteorite Impacts and Planetary Evolution: Geological Society of America Special Paper 293, p. 265–274.

Avermann, M. E., 1997, Investigations on the Green Member of the Onaping Formation, Sudbury Structure, Ontario, Canada [abs.]: in conference of Large Meteorite Impacts and Planetary Evolution: Houston, Texas, Lunar and Planetary Institute Contribution No. 922, p. 4–5.

Beales, F. W., and Lozej, G. P., 1975, Sudbury basin sediments and the meteoritic impact theory of origin for the Sudbury Structure: Canadian Journal of Earth Sciences, v. 12, p. 629–635.

Becker, L., and Bunch, T. E., 1997, Fullerenes, fullerranes and polycyclic aromatic hydrocarbons in the Allende meteorite: Meteoritics, v. 32, p. 479–487.

Becker, L., Bada, J. L., Winans, R. E., Bunch, T. E., and French, B. M., 1994a, Fullerenes in the 1.85-billion-year-old Sudbury impact structure: Science, v. 265, p. 642–645.

Becker, L., Bada, J. L., Winans, R. E., and Bunch, T. E., 1994b, Fullerenes in Allende meteorite, Nature, v. 372, p. 507.

Becker, L., Bada, J. L., and Bunch, T. E., 1995, Fullerenes and fulleranes in the Allende meteorite [abs.] *in* Proceedings, 26th Lunar and Planetary Science Conference, Abstracts: Houston, Texas, Lunar and Planetary Institute, p. 87–88.

Becker, L., Poreda, R. J., and Bada, J. L., 1996, Extraterrestrial helium trapped in fullerenes in the Sudbury Impact Structure: Science, v. 272, p. 249–252.

Becker, L., Bada, J. L., Poreda, R. J., and Bunch, T., 1997, Extraterrestrial helium (He@C_{60}) trapped in fullerenes in the Sudbury impact structure [abs.], *in* Conference on Large Meteorite Impacts and Planetary Evolution (Sudbury 97), Houston, Texas, The Lunar and Planetary Institute Contribution 922, p. 5.

Bottinga, Y., 1996, Carbon isotopic fractionation between graphite, diamond, and carbon dioxide: Geochimica et Cosmochimica Acta, v. 33, p. 49–64.

Brockmeyer, P., 1990, Petrographie, Geochemie und Isotopenuntersuchungen an der Onaping Formation im Nordteil der Sudbury-Struktur (Ontario, Kanada) und ein Modell zur Genese der Struktur [Ph.D. Thesis]: Münster, Germany, University of Münster.

Bunch, T. E., Becker, L., Schultz, P. H., and Wolbach, W. S., 1997, New potential sources for Black Onaping carbon [abs.], *in* Conference on Large Meteorite Impacts and Planetary Evolution (Sudbury 97): Houston, Texas, The Lunar and Planetary Institute Contribution 922, p. 7.

Burrows, A. G., and Rickaby, H. C., 1935, Nickel Field restudied: Ontario Department of Mines, Annual Report for 1934, v. 43(2), 49 p.

Buseck, P. R., Tsipurski, S. J., and Hettich, R., 1992, Fullerenes from the geological environment: Science, v. 257, p. 215–217.

Card, K. D., Innes, D. G., and Debicki, R. L., 1977, Stratigraphy, sedimentology, and petrology of the Huronian Supergroup in the Sudbury-Espanola area: Ontario Geological Survey Geoscience Study, v. 16, 99 p.

Cataldo, F., 1995, On the photopolymerization of carbon disulfide: Inorganica Chimica Acta, v. 232, p. 27–33.

Chai, G., and Eckstrand, R. O., 1993, Origin of the Sudbury Igneous Complex, Ontario—differentiate of two different magmas: Current Research, Pt. E; Geological Survey of Canada Paper v. 93-1E, p. 219–230.

Chai, G., and Eckstrand, R. O., 1994, Rareearth element characteristics and origin of the Sudbury Igneous Complex, Ontario, Canada: Chemical Geology, v.113, p. 221–244.

Chibante, L. P. F., Pan, C., Pierson, M. L., Haufler, R. E., and Heymann, D., 1993, Rate of decomposition of C_{60} and C_{70} heated in air and the attempted characterization of the products: Carbon, v. 31, p. 185–193.

Colman, J. J., and Trogler, W. C., 1995, Photopolymerization of carbon disulfide yields the high-pressure phase $(CS_2)_x$: Journal of the American Chemical Society, v.117, p. 11270–11277.

Colman, J. J., Xu, X., Thiemens, M. H., and Trogler, W. C., 1996, Photopolymerization and mass-independent sulfur isotopic fractionations in carbon disulfide: Science, v. 273, p. 774–776.

Cooper, G. W., Thiemens, M. H., Jackson, T. L., and Chang, S., 1997, Sulfur and hydrogen isotope anomalies in meteorite sulfonic acids: Science, v. 277, p. 1072–1074.

Corfu, F., and Lightfoot, P. C., 1996, U-Pb geochronology of the sublayer environment, Sudbury Igneous Complex, Ontario: Economic Geology, v. 91, p. 1263–1269.

Daly, T. K., Buseck, P. R., Williams, P., and Lewis, C. F., 1993, Fullerenes from a fulgurite: Science, v. 259, p. 1599–1601.

Dence, M. R., 1972, Meteorite impact craters and the structure of the Sudbury Basin, *in* Guy-Bray, J. V., ed., New developments in Sudbury geology: Geologic Association of Canada Special Paper 120, p. 7–18.

Deutsch, A., Grieve, R. A. F., Avermann, M., Bischoff, L., Brockmeyer, P., Buhl, D., Lakomy, R., Müller-Mohr, V., Ostermann, M., and Stöffler, D., 1995, The Sudbury Structure (Ontario, Canada): a tectonically deformed multiring impact basin: Geologische Rundschau, v. 84, p. 697–709.

DeVries, M. S., Reihs, K., Wendt, H. R., Golden, W. G., Hunziker, H. E., Fleming, R., Peterson, E., and Chang, S., 1993, A search for C_{60} in carbonaceous chondrites: Geochimica et Cosmochimica Acta, v. 57, p. 933–935.

Dressler, B. O., 1982, Geology of the Wanapitei Lake area, District of Sudbury: Ontario Geological Survey, Report 213, 131 p.

Dressler, B. O., 1984, General geology of the Sudbury area, *in* Pye, E. G., Naldrett, A. J., and Giblin, P. E., eds., The geology and ore deposits of the Sudbury Structure: Ontario Geologic Survey Special Volume 1, p. 57–82.

Dressler, B. O., Weiser, T., and Brockmeyer, P., 1966, Recrystallized impact glasses of the Onaping Formation and the Sudbury Igneous Complex, Sudbury Structure, Ontario, Canada: Geochimica et Cosmochimica Acta, v. 60, p. 2019–2036.

Dressler, B. O., Morrison, G. G., Peredery, W. V., and Rao, B. V., 1987, The Sudbury Structure, Ontario, Canada—A review, *in* Pohl, J., ed., Research in terrestrial impact structures: Braunschweig/Wiesbaden, Germany, Friedrich. Viehweg & Sohn, p. 39–68.

El Goresy, A. and Donnay, G., 1968, A new allotropic form of carbon from the Ries crater: Science, v. 161, p. 363–364.

Faggart, B. E., Basu, A. R., and Tatsumoto, M., 1985, Origin of the Sudbury Structure by meteorite impact: neodymium isotopic evidence: Science, v. 230, p. 436–439.

French, B. M., Koeberl, C., Gilmour, I., Shirey, S. B., Dons, J. A., and Naterstad, J., 1997, The Gardnos impact structure, Norway: petrology and geochemistry of target rocks and impactites: Geochimica et Cosmochimica Acta, v. 61, p. 873–904.

Gao, X., and Thiemens, M. H., 1991, Systematic study of sulfur isotopic composition in iron meteorites and the occurrence of excess ^{33}S and ^{36}S: Geochimica et Cosmochimica Acta, v. 55, p. 2671–2679.

Gao, X., and Thiemens, M. H., 1993a, Isotopic composition and concentration of sulfur in carbonaceous chondrites: Geochimica et Cosmochimica Acta, v. 57, p. 3159–3169.

Gao, X., and Thiemens, M. H., 1993b, Variations of the isotopic composition of sulfur in enstatite and ordinary chondrites: Geochimica et Cosmochimica Acta, v. 57, p. 3171–3176.

Gilmour, I., Russell, S. S., Newton, J., Pillinger, C. T., Arden, J. W., Dennis, T. J., Hare, J. P.,

Kroto, H. W., Taylor, R., and Walton, D. R. M., 1993, A search for the presence of C_{60} as an interstellar grain in meteorites, *in* Proceedings, 22nd, Lunar and Planetary Science Conference, Abstracts: Houston, Texas, The Lunar and Planetary Institute, p. 445–446.

Grieve, R. A. F., Stöffler, D., and Deutsch, A., 1991, The Sudbury Structure: controversial or misunderstood?, Journal of Geophysical Research, v. 96, p. 22753–22764.

Hayes, J. M., Kaplan, I. R., and Wedeking, K. W., 1983, Precambrian organic geochemistry, preservation of the record, *in* Schopf, J. W., ed., Earth's earliest biosphere, its origin and evolution, Chap. 5, Princeton, New Jersey, Princeton University Press, p. 93–134.

Heymann, D., 1995a, Search for extractable fullerenes in the Allende meteorite [abs.]. *in* Proceedings, 25th Lunar and Planetary Science Conference, Abstracts: Houston, Texas, The Lunar and Planetary Institute, p.595–596.

Heymann, D., 1995b, Search for extractable fullerenes in the Allende meteorite: Meteoritics, v. 30, p. 436–438.

Heymann, D., 1997, Fullerenes and fulleranes in meteorites revisited: Astrophysical Journal, v. 489, p. L111–L114.

Heymann, D. and Dressler, B. O., 1997, Raman study of carbonaceous matter and anthraxolite in rocks from the Sudbury, Ontario, impact structure, *in* Proceedings, 28th Lunar and Planetary Science Conference, Abstracts: Houston, Texas, The Lunar and Planetary Institute, p. 1265–1266.

Heymann, D., Chibante, L. P. F., Brooks, R. R., Wolbach, W. S., and Smalley, R. E., 1994, Fullerenes in the K/T boundary layer: Science, v. 256, p. 645–647.

Hough, R. M., Gilmour, I., Pillinger, C. T., Arden, J. W., Gilkes, K. W. R., Yuan, J., and Milledge, H. J., 1995, Diamond and silicon-carbide in impact melt rock from the Ries impact crater: Nature, v. 378, p. 41–44.

Howard, J. B., McKinnon, J. T., Makarovsky, Y., LaFleur, A. L., and Johnson, M. E., 1991, Fullerenes C_{60} and C_{70} in flames: Nature, v. 352, p. 139–141.

Jackson, W. M., Butterworth, P. S., and Ballard, D., 1986, The origin of CS in comet IRAS-ARAKI-ALCOCK 1983d: Astrophysical Journal, v. 304, p. 515–518.

Kerridge, J. K., 1985, Carbon, hydrogen, and nitrogen in carbonaceous chondrites: isotopic abundances and isotopic compositions in bulk samples: Geochimica et Cosmochimica Acta, v. 49, p. 1701–1714.

Koeberl, C., Masaitis, V. L., Shafvranovsky, G. I., Gilmour, I., Langenhorst, F., and Schrauder, M., 1997, Diamonds from the Popigai impact structure, Russia: Geology, v. 25, p. 967–970.

Kring, D. A., 1993, The Chicxulub impact event and possible causes of K/T boundary extinctions, *in* Boaz, D., and Dornan, M., eds., Proceedings 1st Annual Symposium of Fossils of Arizona: Mesa, Arizona, Mesa Southwest Museum and Southwest Paleontological Society, p. 63–79.

Krogh, T. E., Davis, D. W., and Corfu, F., 1984, Precision U-Pb zircon and baddeleyite ages for the Sudbury area, *in* Pye, E. G., Naldrett, A. J., and Giblin, P. E., eds., The geology and ore deposits of the Sudbury Structure, Ontario Geologic Survey Special Volume 1, p. 431–446.

Kroto, H. W., Heath, J. R., O'Brien, S. C., Curl, R. F., and Smalley, R. E., 1985, Buckminsterfullerene: Nature, v. 318, p. 162–163.

Lakomy, R., 1989, Petrographie, Geochemie und Sr-Nd-Untersuchungen an der Footwall-Breccie im Nordteil der Sudbury-Struktur: [Ph.D. Thesis]: Münster, Germany, University of Münster, 165 p.

Mancuso, J. J., Kneller, W. A., and Quick, J. C., 1989, Precambrian vein pyrobitumen: evidence for petroleum generation and migration 2 Ga ago: Precambrian Research, v. 44, p. 137–146.

Masaitis, V. L., Furtergondler, S. I., and Gnevushev, M. A., 1972, Diamonds in impactites of the Popigai meteor crater: Proceedings of the All-Union Mineralogical Society, v. 1, p. 108–112. [in Russian]

Masaitis, V. L., Shafranovsky, G. I., Grieve, R. A. F., Langenhorst, F., Peredery, W. V., Balmasov, E. L., Federova, I. G., and Therriault, A., 1997, Discovery of impact diamonds at the Sudbury Structure [abs.]; *in* Conference on Large Meteorite Impacts and Planetary Evolution: Houston, Texas, The Lunar and Planetary Institute Contribution 922, p. 33.

Matthews, C. K., Sai Baba, M., Lakhsmi P., Narasimhan, T. S., Balasubramanian, R., Siavaraman, N., Srinivasan, T. G., and Vasudeva Rao, P. R., 1993, Vapor pressure and enthalpy of sublimation of C_{70}: Journal of Fullerene Science and Technology, v.1, p. 101.

Milkereit, B., and 43 others, 1992, Deep geometry of the Sudbury structure from seismic reflection profiling: Geology, v. 20, p. 807–811.

Mochida, I., Egashira, M., Koura, H., Dakeshita, K., Yoon, S.-H., and Korai, Y., 1995, Carbonization of C_{60} and C_{70} fullerenes to fullerene soot: Carbon, v. 33, p. 1186–1188.

Muir, T. L., and Peredery, W. V., 1984, The Onaping Formation: *in* Pye, E. G., Naldrett, A. J., and Giblin, P. E., eds., The geology and ore deposits of the Sudbury Structure: Ontario Geologic Survey Special Volume 1, p. 130–210.

Nakamizo, M., Honda, H., and Inagaki, M., 1978, Raman spectra of ground natural graphite: Carbon, v. 16, p. 281–283.

Naldrett, A. J., and Hewins, R. H., 1984, The main mass of the Sudbury Igneous Complex, *in* Pye, E. G., Naldrett, A. J., and Giblin, P. E., eds., The geology and ore deposits of the Sudbury Structure: Ontario Geologic Survey Special Volume 1, p. 235–251.

Pasteris, J. D., and Wopenka, B., 1991, Raman spectra of graphite as indicators of degree of metamorphism: Canadian Mineralogist, v. 29, p. 1–9.

Peredery, W. V., and Morrison, G. G., 1984, Discussion of the origin of the Sudbury Structure, *in* Pye, E. G., Naldrett, A. J., and Giblin, P. E., eds., The geology and ore deposits of the Sudbury Structure: Ontario Geologic Survey Special Volume 1, p. 491–511.

Rees, C. E., and Thode, H. G., 1977, A ^{33}S anomaly in the Allende meteorite: Geochimica et Cosmochimica Acta, v. 41, p. 1679–1682.

Robl, T. A., and Davis, B. H., 1993, Comparison of the HF-HCl and HF-BF_3 maceration techniques and the chemistry of the resultant organic concentrates: Organic Geochemistry, v. 20, p. 249–255.

Rousell, D. H., 1984a, Onwatin and Chelmsford Formations, *in* Pye, E. G., Naldrett, A. J., and Giblin, P. E., eds., The geology and ore deposits of the Sudbury Structure: Ontario Geologic Survey Special Volume 1, p. 211–217.

Rousell, D. H., 1984b, Mineralization in the Whitewater Group, *in* Pye, E. G., Naldrett, A. J., and Giblin, P. E., eds., The geology and ore deposits of the Sudbury Structure: Ontario Geologic Survey Special Volume 1, p. 219–232.

Saunders, M., Jiménez-Vaquez, R., Cross, J., and Poreda, R. J., 1993, Stable compounds of helium and neon He@C_{60} and Ne@C_{60}: Science, v. 259, p. 1428–1430.

Schidlowski, M., 1987, Application of stable carbon isotopes to early biochemical evolution on earth: Annual Review of Earth and Planetary Science, v. 15, p. 47–72.

Schuraytz, B. C., Lindstrom, D. J., and Sharpton, V. J., 1997, Constraints on the nature and distribution of iridium host phases at the Cretaceous-Tertiary boundary: implications for projectile identity and dispersal on impact [abs.], *in* Conference on Large Meteorite Impacts and Planetary Evolution (Sudbury 97): Houston, Texas, Lunar and Planetary Institute Contribution 992, p. 50.

Shanks, W. S., and Schwerdtner, W. R., 1990, Structural analysis of the central and southwestern Sudbury Structure, Southern Province, Canadian Shield: Canadian Journal of Earth Science, v. 28, p. 411–430.

Shanks, W. S., Dressler, B. O., and Schwerdtner, W. R., 1990, New developments in Sudbury geology [abs.]: *in* International Workshop on Meteorite Impacts on the Early Earth, Perth, Australia: Houston, Texas, The Lunar and Planetary Institute Contribution 746, p. 46.

Stevenson, J. S., 1972, The Onaping ash-flow sheet, Sudbury, Ontario, *in* Guy-Bray, ed., New developments of Sudbury geology, Geological Association of Canada Special Paper 10, p. 41–48.

Stöffler, D., Deutsch, A., Avermann, M., Bischoff, L., Brockmeyer, P., Buhl, D., Lakomy, R., and Müller-Mohr, V., 1994, The formation of the Sudbury Structure, Canada: toward a unified impact model: *in* Dressler, B. O., Grieve, R. A. F., and Sharpton, V. L., eds., Large meteorite impacts and planetary evolution: Geologic Society of America Special Paper 293, p. 303–318.

Taylor, R., Langley, G. J., Kroto, H. W., and Walton, D. R. M., 1993, Formation of C_{60} by pyrolysis of naphthalene: Nature, v. 366, p. 728–731.

Thiemens, M. H., and Jackson, T., 1986, Further measurements of sulfur isotopes in ureilites: evidence for a ^{33}S isotopic anomaly [abs.]: *in* Lunar and Planetary Science XXVI, The Planetary Institute, Houston, Texas, USA, p. 1405–1406.

Thode, H. G., Dunford, H. B., and Shima, M., 1982, Sulfur isotopic abundances in rocks of the Sudbury District and their geologic significance: Economic Geology, v. 57, p. 565–578.

Thomson, J. E., 1960, On the origin of algal-like forms and carbon in the Sudbury Basin, Ontario: Transactions of the Royal Society of Canada, v. 54, p. 65–75.

Tingle, T. N., Becker, C. H., and Malhotra, R., 1991, Organic compounds in the Murchison and Allende carbonaceous chondrites studied by photoionization mass spectrometry: Meteoritics, v. 26, p. 117–127.

Tuinstra, F., and Koenig, J. L., 1970, Raman spectrum of graphite: Journal of Chemical Physics, v. 53, p. 1126–1130.

Vasallo, A. M., Pang, L. S. K., Cole-Clarke, P. A., and Wilson, M. A., 1991, Emission FTIR study of C_{60} thermal stability and oxidation: Journal of the American Chemical Society, v. 113, p. 7820–7821.

Whitehead, R. E. S., Davies, J. F., and Goodfellow, W. D., 1990, Isotopic evidence for hydrothermal discharge into anoxic seawater, Sudbury Basin, Ontario, Canada: Chemical Geology, v. 86, p. 49–63.

Williams, H., 1957, Glowing avalanche deposits of the Sudbury Basin: Ontario Department of Mines Annual Report for 1956, v. 65 (3), p. 57–89.

Yeretzian, C., Hansen, K., Diederich, F., and Whetten, R. L., 1992, Coalescence reactions of fullerenes: Nature, v. 359, p. 44–47.

MANUSCRIPT ACCEPTED BY THE SOCIETY DECEMBER 16, 1998

Isotopic evidence for a single impact melting origin of the Sudbury Igneous Complex

A. P. Dickin, T. Nguyen, and J. H. Crocket
School of Geography and Geology, McMaster University, Hamilton, Ontario L8S 4M1, Canada

ABSTRACT

An impact-related origin for the Sudbury Igneous Complex is now generally accepted, but the sources of its component parts and the relationships between them remain in doubt. In this chapter we use multiple isotopic techniques to help identify the source rocks of the respective units and to better understand relationships between them.

Initial osmium isotope compositions for sulfide ores from the Sublayer of the complex generally fall within the same range as do published data for local crustal rocks. Most significantly, disseminated ores from two mafic inclusions from South Range mines have initial osmium isotope ratios as radiogenic as any other samples, and are strongly suggestive of a totally crustal origin. Apparent low initial osmium ratios seen in some ores, particularly from the North Range, may be due to later disturbance of the rhenium-osmium (Re-Os) system.

Initial Pb isotope compositions for the Main Mass of the complex fall between the extremes of isotopic composition previously identified in the Sublayer and are attributed to shock melting of distinct crustal rocks on the two sides of the complex. South Range norites preserve the distinct isotopic signature of the local Proterozoic country rocks, whereas norites from the composite southwest–north–east limb of the complex appear to result from mixing between melts of Archean and Proterozoic crust. Granophyres from the upper parts of the complex are intermediate between all of these signatures. Initial Sr isotope ratios are indicative of the same process, with quite variable signatures in the Sublayer, due to incomplete mixing between different crustal lithologies, but very uniform signatures in the norite, gabbro, and granophyre units of the Main Mass.

Given the isotopic and trace element evidence for a cogenetic origin for the norite and granophyre, mineral composition data are used to refine a crystal fractionation model for the complex. This model provides a good fit to the observed petrology of the complex, supporting its origin as a single impact melt body.

INTRODUCTION

An impact-related origin for the Sudbury Igneous Complex (SIC) is now generally accepted, but the sources of its component parts and the relationships between them remain in doubt. Isotopic evidence provides powerful constraints on magma genesis, and isotopic investigations have therefore provided the focus of several studies on the origin of the complex. Sr isotope systematics were studied by Gibbins and McNutt (1975), Rao et al. (1984), and Deutsch (1994); oxygen isotope systematics were studied by Ding and Schwarcz (1984); Nd isotope systematics were studied by Faggart et al. (1985) and Deutsch (1994); and Os isotope systematics were studied by Walker et al. (1991) and Dickin et al. (1992). The initial isotopic compositions of these elements were

Dickin, A. P., Nguyen, T., and Crocket, J. H., 1999, Isotopic evidence for a single impact melting origin of the Sudbury Igneous Complex, *in* Dressler, B. O., and Sharpton, V. L., eds., Large Meteorite Impacts and Planetary Evolution II: Boulder, Colorado, Geological Society of America Special Paper 339.

all indicative of an essentially crustal origin for the complex. Hence, Faggart et al. (1985), Dickin et al. (1992), and Deutsch (1994) attributed the complex entirely to impact-induced crustal melting, whereas Rao et al. (1984) and Walker et al. (1991) argued for a limited mantle-derived contribution.

Isotopic investigations have also revealed systematic compositional differences between the two sides of the Sudbury Igneous Complex, termed the North and South Ranges. Thus, Ding and Schwarcz (1984) observed higher $\delta^{18}O$ signatures in norites from the Main Mass of the South Range, while Rao et al. (1984) and Dickin et al. (1996) observed more radiogenic initial Sr and Pb isotope ratios, respectively, for samples of the Contact Sublayer from the South Range. In the first two studies, these differences were attributed to local contamination effects by the wallrocks of the complex. However, in the latter study, the distinct Pb isotope ratios of the North and South Range Sublayer were attributed to wholesale melting of distinct crustal source rocks on the two sides of the complex. These source rocks are represented by the 2.7-Ga Levack gneisses of the Superior Province on the North Range, and the 2.3–2.4-Ga metasediments and metavolcanics of the Huronian Supergroup (Southern Province) on the South Range. We also showed (Dickin et al., 1996; Dickin and Crocket, 1998) that initial Pb and Nd isotope ratios for Sublayer magmas and inclusions from the North and South Ranges overlapped completely with the compositions of nearby Sudbury Breccias, formed by shock melting of local crustal units. From these observations and inferences, we supported a model of impact-induced shock melting for the origin of the complex as a whole.

An origin for SIC as a single impact-derived melt sheet was advocated by Grieve et al. (1991) on the basis of comparisons between the SIC and other impact sites. However, in other recent studies, distinct origins for the upper granophyric unit and lower noritic unit of the complex have been suggested. For example, Chai and Eckstrand (1994) argued that the upper part of the complex was produced by impact melting, whereas the lower part was mantle-derived. Cohen et al. (1997) supported this model to the extent that they argued for a significant mantle-derived component in the Sublayer of the complex, particularly its mafic and ultramafic inclusions. On the other hand, Dressler et al. (1996) and Dressler and Sharpton (1998) developed a somewhat different model; in it, the granophyre was attributed to a true impact melt sheet, whereas the lower part of the complex was attributed to a distinct (but still impact-triggered) deep crustal melt.

In this study, our objective was to use isotopic evidence to assess the origin of the magmas that crystallized to form the Sublayer and Main Mass of the SIC. We used osmium isotope data for Sublayer ores from several mines to further examine the question of whether an entirely crustal origin for the complex is reasonable. We then used Pb and Sr isotope data, coupled with petrologic evidence, to examine relationships between the upper and lower parts of the complex.

SAMPLE SELECTION AND EXPERIMENTAL METHODS

Ore samples studied in this chapter are aliquots of the samples previously analyzed for Pb isotope composition by Dickin et al. (1996), and come from three mines in the North

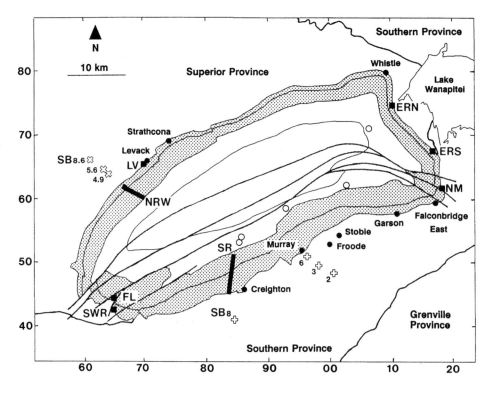

Figure 1. Geological setting of the Sudbury Igneous Complex relative to the three nearby geologic provinces. The Main Mass of the complex (shaded) is divided into a lower gabbro-norite unit and an upper granophyre unit. Mines (named) are located near the footwall of the complex. Heavy bars indicate traverses through the Main mass; black squares indicate basal norite samples; open circles indicate Onaping Formation; open crosses indicate Sudbury breccia (SB). FL = Fairbank Lake; NM = Norduna Mine; ERN = East Range (North); ERS = East Range (South); NRW = North Range (West); LV = Levack; SWR = Southwest Range; SR = South Range. Numbers on figure are Sudbury breccia sample numbers.

TABLE 1. OSMIUM ISOTOPE DATA FOR SULFIDE ORES

Sample Number	Mine	Sample Size (g)	Re (ppb)	Os (ppb)	$\frac{^{187}Re}{^{188}Os}$	$\frac{^{187}Os}{^{188}Os}$	$\frac{^{187}Os}{^{188}Os_I}$
South Range							
C1	Creighton	~0.5	336	156	11.24	0.863	0.52
C2	Creighton	0.65	265	156	8.86	0.805	0.54
C5	Creighton	0.85	342	145	12.34	0.849	0.47
C13	Creighton	0.51	324	259	6.46	0.707	0.51
		0.51	303	234	6.68	0.710	0.51
		0.18	799	593	6.97	0.730	0.52
1650–41	Murray	0.49	216	60	19.52	1.187	0.60
		0.43	233	67	19.02	1.174	0.60
		0.69	163	47	19.09	1.197	0.62
FD2	Froode	0.48	356	77	25.29	1.183	0.42
FD7	Froode	0.43	352	84	22.53	1.086	0.40
STB6	Stobie	0.62	246	70	18.61	0.956	0.39
		0.79	173	49	19.05	0.996	0.42
G6	Garson	0.55	230	75	16.41	1.031	0.53
		0.63	222	69	17.28	1.064	0.54
F2	Falconbridge E	~1.0	236	68	18.31	1.014	0.46
		~1.0	246	68	19.43	1.031	0.44
F3	Falconbridge E	~0.5	748	172	23.49	1.160	0.45
F5	Falconbridge E	~1.0	480	94	28.37	1.395	0.53
North Range							
S20–22	Strathcona	0.60	301	35	50.65	1.845	0.31
		0.69	278	32	51.27	1.806	0.25
S25–13	Strathcona	0.56	207	32	37.15	1.608	0.48
		0.65	176	26	38.64	1.620	0.45
L13	Levack	0.61	635	201	16.56	0.832	0.33
		0.40	828	271	16.00	0.829	0.34
W4	Whistle	0.25	319	33	54.80	1.549	-0.11
		0.23	344	35	56.03	1.581	-0.12
		0.16	519	55	53.45	1.596	-0.03
W5	Whistle	0.36	244	39	35.77	1.568	0.48
W9	Whistle	0.45	940	221	23.01	1.133	0.44
		0.20	2378	537	23.97	1.123	0.40
W18	Whistle	0.60	204	37	30.54	1.419	0.49
		0.66	201	37	30.47	1.409	0.48

Range and six mines in the South Range (Table 1). They consist of pyrrhotite that was separated by hand magnet from a variety of disseminated and massive Sublayer ores. Details of locality and sample type were given by Dickin et al. (1996).

Silicate rocks from the Main Mass of the complex were collected along Highway 144 in a traverse across both North and South Ranges of the complex. These transects were made in order to examine isotopic variation as a function of structural height above the footwall. The transects are indicated by the heavy bars in Figure 1, and sample localities are given as UTM grid references (to the nearest 100 m) in Table 2. In addition, samples of norite from the Sublayer and lowest parts of the Main Mass were collected from the southwest, northern, and eastern parts of the complex to examine lateral isotopic variations around the rim of the complex (Fig. 1).

Samples of the Onaping Formation were collected from a variety of localities, but those on the North Range were found to have very radiogenic Pb and Sr isotope ratios. The very large age corrections on these samples prohibited the calculation of reliable initial ratios. Hence only samples from the South and East Ranges are included (Fig. 1).

For each Re-Os analysis, approximately 1 g of ore was mixed with 5 mg of a mixed Re-Os spike in solid solution in a nickel sulfide matrix (Richardson et al., 1989). This mixture was dissolved in warm 5M nitric acid in a covered beaker, and after the reaction was complete (about 3 days later), the undissolved silicate residue was filtered off and weighed. The solution was split into two aliquots for Re and Os analysis. After addition of 10% periodic acid (HIO_3), osmium was distilled from the first aliquot at ca. 95°C and trapped in 9M HBr. The HBr solution was refluxed in a Teflon beaker at 100°C, then dried down. The sample was then purified by dissolution in water and adsorption onto one 20-mesh anion exchange bead in water. After rinsing with 10 µl of water, the sample was stripped off the bead with

TABLE 2. Pb ISOTOPE DATA FOR SUDBURY SILICATE ROCKS

Sample Number	Rock Type	UTM Grid Reference	$\frac{^{206}Pb}{^{204}Pb}$	$\frac{^{207}Pb}{^{204}Pb}$	$\frac{^{208}Pb}{^{204}Pb}$	$\frac{^{207}Pb}{^{204}Pb_i}$
South Range transect						
SR0.8	Norite	838 455	16.737	15.534	36.451	15.46
SR2.0	Norite	840 467	17.707	15.644	38.410	15.45
SR2.5	Gabbro	841 472	18.129	15.630	37.545	15.36
SR2.5F*			17.786	15.618	37.713	15.40
SR2.9	Gabbro	842 476	20.599	15.847	40.930	15.39
SR2.9F*			19.122	15.755	38.203	15.38
SR3.4	Mixed rock	842 482	22.498	16.085	41.319	15.30
SR3.9	Granophyre	843 487	22.315	16.039	38.397	15.26
SR4.9	Granophyre	844 497	25.488	16.450	40.522	15.34
SR5.3	Granophyre	845 501	26.284	16.510	41.120	15.29
SR5.7	Granophyre	847 505	23.516	16.177	39.134	15.27
SR5.7F*			21.591	15.974	38.990	15.29
SR6.0	Granophyre	847 508	20.588	15.867	37.811	15.30
North Range transect						
NLV4	Norite	700 646	16.538	15.418	36.174	15.31
NRW1	Norite	672 623	17.518	15.501	36.435	15.27
NRW1F*			16.674	15.398	36.024	15.27
NRW2	Gabbro	682 626	17.283	15.472	36.541	15.27
NRW2F*			17.912	15.557	36.681	15.28
NRWC	Gabbro	685 624	19.942	15.806	41.123	15.32
NRWCF*			16.280	15.406	36.376	15.34
NRWB	Mixed rock	685 623	23.019	16.131	45.014	15.28
NRWBF*			18.052	15.596	37.992	15.32
NRWAM*	Granophyre	685 622	26.096	16.456	43.221	15.25
NRWAF*			27.227	16.599	45.595	15.27
NRW5M*	Granophyre	707 612	28.354	16.709	40.800	15.25
NRW5F*			24.916	16.361	41.394	15.30
Basal norite, Southwest-North-East Range						
SWR0	Norite	644 427	18.727	15.653	37.692	15.29
SWR4	Norite	649 445	18.691	15.602	37.991	15.22
NLV1	Norite	702 651	16.879	15.446	36.327	15.29
ERN1	Norite	099 740	18.184	15.571	37.034	15.26
ERS1	Norite	167 661	16.508	15.429	36.039	15.34
NM1	Norite	181 607	18.793	15.665	37.132	15.30
Onaping Formation						
OT2	Matrix	027 618	23.735	16.129	40.023	15.16
OT3	Matrix	061 708	21.701	15.954	37.954	15.24
OT9	Matrix	853 522	22.636	16.101	38.451	15.30
OT10	Matrix	858 533	22.486	15.992	39.223	15.17
OT11	Matrix	929 579	23.350	16.132	38.839	15.23

*Separated mineral phases: F = feldspar; M = magnetite.

9M HBr. Rhenium was separated from the second aliquot on a 1-ml anion exchange column in 1M HCl, then purified by adsorption onto a few 20-mesh anion beads in 1M HCl, before stripping with 2M HNO_3.

Re and Os were loaded on zone-refined Pt filaments, which were pre-leached in nitric acid to lower the Re blank (Walczyk et al., 1991). Both Re and Os were analyzed as negative ions on a VG Isolab 54 instrument at McMaster University using a Gallileo channeltron connected via capacitance isolation to an Ortec pulse counting system. For Os analysis, oxygen was bled into the source, and both isotope dilution and isotope ratio analysis were performed in a single run with off line iterative deconvolution of fractionation. Typical within run precision was 0.1% (1 σ).

Blanks measured during the course of the work were 30 pg for osmium and 700 pg for rhenium. This is a high Re blank (attributed to Re contamination in the mass spectrometer). However, since the minimum size of Re sample was about 100 ng (Table 1), the blank is expected to contribute less than 1% to the analysis and, therefore, no correction was made. The Os blank is negligible, relative to the sample size.

Most samples were analyzed in duplicate dissolutions to test for the reproducibility of initial ratios (Table 1). This procedure evaluates a combination of analytical reproducibility and geologic closed-system reproducibility. It does not actually quantify the approach to closed-system conditions, but may be somewhat indicative. For all samples except the two with highest Re/Os ratios (from Strathcona and Whistle), the scatter of initial ratio duplicates averaged 4%, much less than the spread between different samples. Below we argue that the two samples displaying marked scatter were probably geologically disturbed.

Three samples from Creighton (C1, 2, 13) represent aliquots of material previously analyzed for Re-Os by ICP-MS (Dickin et al., 1992). There are systematic offsets between the old and new data of 4–7% for isotope ratios and 8–18% for elemental ratios and initial ratios. This is not particularly surprising, in view of the relatively large errors associated with ICP analysis at low count rates. Nevertheless, the new initial ratios data for Creighton are within the total range of initial ratios previously measured at Creighton. On the other hand, the large variations in absolute Re and Os concentrations between duplicate dissolutions, both within the new study and between the two studies, are attributed to the "nugget" effect commonly observed in PGE analysis.

Pb isotope ratios were measured principally on whole-rock samples of the Main Mass. For the North Range transect and selected samples from the South Range, feldspar separates were also analyzed (nonmagnetic fraction at 2 A on a Frantz separator). Whole-rock granophyres from the North Range gave Pb isotope ratios that were too radiogenic to yield accurate initial ratios, so the magnetic fraction (hand magnet) was analyzed, with reasonable success. All samples for Pb analysis were powdered, then leached with warm 6M HCl. Analysis was performed in the same way as in our previous work on Sudbury (Dickin et al., 1996). All data are fractionation corrected (0.08–0.1% per atomic mass unit), based on frequent analysis of the NBS 981 standard. Model initial $^{207}Pb/^{204}Pb$ ratios (Table 2) were calculated by the same method as Dickin et al. (1996). Reproducibility of initial ratios is estimated as about 0.15% (2 σ), based on agreement between whole-rock and mineral fractions.

Sr isotope analysis was performed on whole-rock samples only, since Rb-Sr mineral systems at Sudbury are disturbed (Gibbins and McNutt, 1975). Sr isotope ratios were analyzed by routine methods, fractionation corrected to $^{86}Sr/^{88}Sr = 0.1194$, with average within-run precision of 0.003% (2 σ). Rb/Sr elemental concentrations and ratios were measured by x-ray fluorescence using a calibration line of international standards and in-house standards previously analyzed using isotope dilution by the first author. Based on replicate analysis of samples and standards, precision of Rb/Sr ratios is estimated as about 1% (2 σ), and based on the miss-fit of standards to the calibration line, the estimated accuracy of Rb/Sr ratios is 2% (2 σ). Since the age correction is the prime source of uncertainty in determination of the initial ratio, the average ^{87}Rb/^{86}Sr ratio of 0.5 yields an uncertainty of 0.0003 (2 σ) on initial ratios. Samples with ^{87}Rb/^{86}Sr over 1 were excluded, as these give less accurate initial ratios due to the effects of geologic disturbance of the Rb-Sr system (Gibbins and McNutt, 1975). Such exclusion was not possible for the Sudbury Breccia, since most samples are radiogenic. However, these data were not used for detailed comparisons.

Microprobe analysis of plagioclase from sample SR2.0 was performed at the University of Western Ontario with a 20-kV accelerating voltage and 20-nA beam current, using count times of 20 s on peaks and backgrounds.

RESULTS

Rhenium-osmium data

Rhenium-osmium data are plotted on an isochron diagram in Figure 2, along with a 1.85-Ga reference line (Krogh et al., 1984). Samples with ^{187}Re/^{188}Os below 40 appear to define several short arrays with slope ages of ca. 1.85 Ga, but with different intercepts. The best explanation of this pattern is that ores were formed at the same time as the rest of the complex, but with variation in initial Os isotope ratio. However, there is no evidence for consistently different initial ratios on the two sides of the complex, as seen in Pb isotope data (Dickin et al., 1996). One sample from Strathcona mine lies below the main array, and an inclusion sample from the Whistle pit yields data that are even further removed from the other samples, with an apparent negative initial ratio (Table 1). We attribute both of these results to geologic disturbance since 1.85 Ga. Two-point regression lines through the Strathcona samples (S20-22 and S25-13) and through two Whistle samples (W4 and W9) yield approximate 1 b.y. slopes (not shown), which might imply an effect by the Grenville orogeny. However, other Whistle samples do not fit such an array, so it should not be given too much significance.

In order to compare initial Os isotope ratios with possible source rocks at 1.85 Ga, initial ratios are plotted as histograms for the North and South Ranges in Figure 3. Based on the reproducibility of duplicate dissolutions, the uncertainties on initial ratios are significantly less than the x axis intervals. Stars show the composition of the mantle and local crustal units at 1.85 Ga (Walker et al., 1991; Dickin et al., 1992). Disseminated ores from mafic inclusions are marked by letters, signifying the mine from which they were sampled. The range of initial ratios in the ores overlaps strongly with the estimated composition of local crustal units at 1.85 Ga, supporting the argument of Dickin et al. (1992) that the ores are 100% crustally derived. Most notably, inclusion samples from the Creighton and Murray mines have initial ratios near the upper limit of the range. Since these are from mafic rocks, they would be the most likely to reveal a mantle-derived

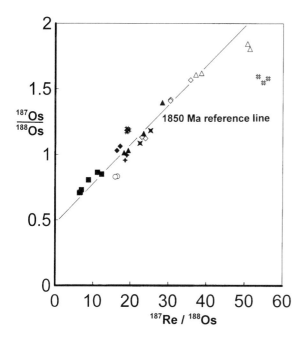

Figure 2. Re-Os isochron diagram for Sublayer ore samples from nine Sudbury mines. Symbols distinguish mines; in South Range: black squares indicate Creighton; tilted black crosses indicate Froode; vertical black crosses indicate Stobie; black cross indicates Murray; black diamonds indicate Garson; black triangles indicate Falconbridge East; in North Range: open circles indicate Levack; open triangles indicate Strathcona; open diamonds, open crosses indicate Whistle.

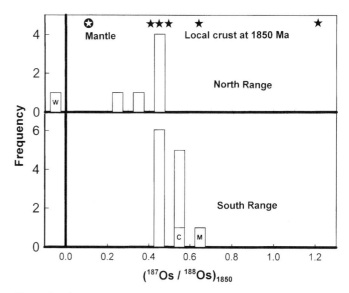

Figure 3. Histograms compare initial Os isotope ratios for North Range and South Range ores with published country rock data. Disseminated ores from mafic-ultramafic inclusions are marked with letters: M = Murray mine; C = Creighton; W = Whistle.

component, if such a component were present in the complex. Other authors have reported low initial Os ratios in inclusions from the Whistle pit (Cohen et al., 1997), but these may represent Re-poor mafic crustal rocks.

The initial ratio data in Figure 3 can be further evaluated by comparison with the Re-Os data of Walker et al. (1991). From that study a range of initial $^{187}Os/^{188}Os$ ratios from 0.40 to 0.58 can be calculated for six samples from Levack West, with an average of 0.51 ± 0.04. This suggests that our low initial ratio for Levack may result from geologic disturbance. On the South Range, five samples analyzed by Walker et al. from the main Falconbridge mine yield a range of initial ratios of 0.58–0.62, with an average of 0.60 ± 0.01. Strathcona data analyzed by Walker et al. had scattered initial ratios of 0.09–1.02, again probably due to disturbance, but the median initial ratio of 0.55 is close to the North Range peak in Figure 3.

The data of Walker et al. (1991) for Levack and Falconbridge, on the North and South Ranges, respectively, yield distinct initial ratios outside of error. However, after removal of samples we consider to be disturbed (S20-22, L13, W4), our grand means of initial ratios for the North and South ranges are 0.47 ± 0.03 and 0.50 ± 0.02 (2 SDM), respectively, showing that our initial ratios for the two sides of the complex are not distinguishable. Since our data represent a wider sampling from several mines, it seems that the two ranges have essential the same initial osmium signature, within the uncertainties allowed by closed system behavior of the Re-Os system.

In conclusion, we believe that osmium isotope data continue to support a 100% crustal origin for the sulfide ores. In some other proposed impact sites, evidence has been found of unradiogenic osmium from the impactor body (e.g., Koeberl and Shirey, 1997); however, we see no convincing evidence for a significant component of this material at Sudbury. Alternatively, it has been suggested (Zhou et al., 1997) that the Sudbury Structure resulted from the impact of a chondritic body that contributed material to some of the mafic inclusions within the Sublayer. However, all mafic inclusions analyzed thus far have ϵ_{Nd} values of about –5 to –10, far away from the chondritic point, but in close agreement with Sudbury Breccia matrices (Dickin and Crocket, 1998), which were clearly produced by shock melting of local crustal rocks. Thus, it appears that the contribution of the impactor body to the Sublayer rocks was overwhelmed by the immense amount of crustal melting that resulted from the impact.

Pb isotope data

Model initial Pb isotope ratios (Table 2) are plotted in Figure 4 as a composite traverse across the whole complex. This is composed of the transects across the North and South Ranges (Fig. 1), along with new data for the Onaping Formation and published data for the Sublayer from mines (Dickin et al., 1996).

The Sublayer data for the North and South Ranges define the limit of isotopic variation within the complex, while the Onaping samples fall in the gap between the ranges of Pb isotope ratio seen in the Sublayer. This is consistent with an origin for the Onaping by variable mixing of two distinct sources on either side of the complex during the impact event. From the limited data available, there seems to be no geographic correlation with isotope ratio within the Onaping Formation.

The granophyre also has an intermediate Pb isotope ratio, but with a narrower range than the Onaping Formation. Dickin et al. (1996) suggested that such an intermediate composition would be expected in the granophyre, due to homogenization of distinct North and South Range melts within a long-lived magma chamber.

The norites from the South Range transect show no tendency for isotopic homogenization with less radiogenic melts from the North Range, but instead fall within the limits of Sublayer compositions for the South Range. This shows that the Sublayer Pb signature cannot be simply a result of local crustal assimilation, and supports the impact melting model of Dickin et al. (1996), which called for North Range and South Range melts that reflect the distinct Pb isotope signatures of the local target rocks. The more complex Pb isotope signature of the North Range norite is discussed below.

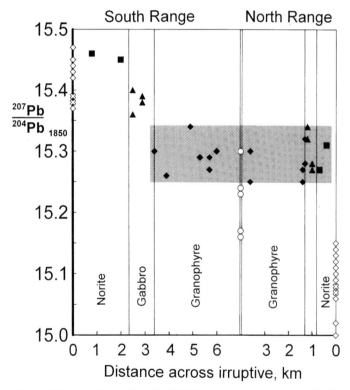

Figure 4. Composite profile of initial Pb isotope ratios across the North and South Ranges of the SIC. Black squares indicate Main Mass norite; black triangles indicate oxide-rich gabbro; black diamonds indicate granophyre; open circles indicate Onaping tuff; open diamonds, open crosses indicate Sublayer norite and inclusions from Dickin et al. (1996). Shaded band represents range of initial ratios in the granophyre.

Sr isotope data

Initial Sr isotope data (Table 3; Fig. 5) show that the Sublayer has a much wider range of initial ratios than the Main Mass, as reported in previous studies (e.g., Rao et al., 1984; Deutsch, 1994). The Sublayer on the South Range has a more radiogenic initial ratio, and also shows much more isotopic variation than on the North Range, consistent with the origin of the former by melting of a mixture of radiogenic Huronian sediments and unradiogenic mafic rocks. Average initial ratios for the South and North Range in the present study (0.7076 and 0.7043, respectively) compare closely with averages from Rao et al. (0.7074 and 0.7045, respectively). Compared with these values, the Main Mass has an intermediate and homogeneous initial isotope ratio, from which the departure of sample SR 3.4 can be attributed to metamorphic disturbance. Excluding this sample, the averages

TABLE 3. Sr ISOTOPE DATA FOR ROCKS OF THE SIC

Sample Number	Mine/Rock Type	Rb (ppm)	Sr (ppm)	$^{87}Rb/^{86}Sr$	$^{87}Sr/^{86}Sr$	$^{87}Sr/^{86}Sr_I$
South Range, Sublayer						
1650.03	Murray	10.7	90.6	0.342	0.71516	0.7061
1650.14	Murray	13.0	47.6	0.792	0.72762	0.7066
1650.16	Murray	13.7	72.1	0.551	0.72148	0.7068
1650.41	Murray	36.0	448	0.233	0.71266	0.7065
FD2	Froode	46.1	160.9	0.831	0.73269	0.7106
FD7	Froode	46.5	184.8	0.730	0.73088	0.7115
G6	Garson	70.2	327	0.623	0.72235	0.7058
South Range transect, Main Mass						
SR0.8	Norite	57.8	429	0.391	0.71695	0.7065
SR2.0	Norite	29.9	391	0.222	0.71267	0.7068
SR2.5	Gabbro	39.8	458	0.252	0.71329	0.7066
SR2.9	Gabbro	34.0	502	0.196	0.71214	0.7069
SR3.4	Mixed rock	74.6	285	0.760	0.72812	0.7079
SR5.3	Granophyre	57.6	196.1	0.852	0.72938	0.7067
North Range, Sublayer						
L13	Levack	24.7	347	0.206	0.70927	0.7038
S14.7	Strathcona	25.4	381	0.193	0.70915	0.7040
S20.22	Strathcona	15.0	438	0.099	0.70684	0.7042
S25.13	Strathcona	15.6	198.1	0.228	0.71092	0.7049
W4	Whistle	31.5	289	0.316	0.71271	0.7043
W14	Whistle	51.0	229	0.646	0.72155	0.7044
North Range, Main Mass						
NLV4	Norite	47.8	482	0.288	0.71465	0.7070
NRW1	Norite	41.7	487	0.248	0.71351	0.7069
NRW2	Gabbro	37.0	492	0.218	0.71244	0.7066
NRWC	Gabbro	51.4	416	0.358	0.71635	0.7068
NRWB	Mixed rock	52.5	315	0.483	0.71953	0.7067
Basal norite, Southwest-North-East Range						
SWR0	Norite	21.0	417	0.146	0.71097	0.7071
SWR4	Norite	62.1	414	0.435	0.71714	0.7055
NLV1	Norite	50.6	470	0.312	0.71599	0.7077
ERN1	Norite	23.4	515	0.132	0.71026	0.7067
ERS1	Norite	93.6	422	0.643	0.72539	0.7083
NM1	Norite	44.1	400	0.320	0.71544	0.7069
Sudbury Breccia						
North Range						
SB4.9	Matrix	81.3	788	0.299	0.71212	0.7042
SB5.6	Matrix	41.1	404	0.295	0.71428	0.7064
SB8.6	Matrix	170.8	186.6	2.654	0.77620	0.7056
South Range						
SB2	Matrix	97.8	215.5	1.316	0.75337	0.7184
SB3	Matrix	210	294	2.071	0.76930	0.7142
SB6	Matrix	75.6	142.8	1.535	0.74895	0.7081
SB8	Matrix	83.7	206	1.178	0.74086	0.7095

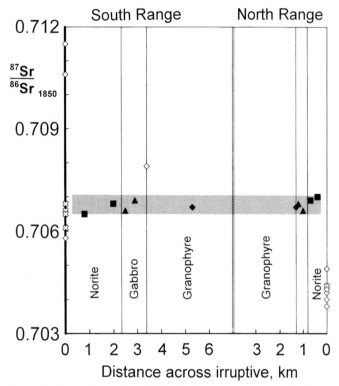

Figure 5. Composite profile of initial Sr isotope ratios across the North and South Ranges of the SIC. Symbols as in Figure 4. Shaded band represents range of initial ratios in the Main Mass.

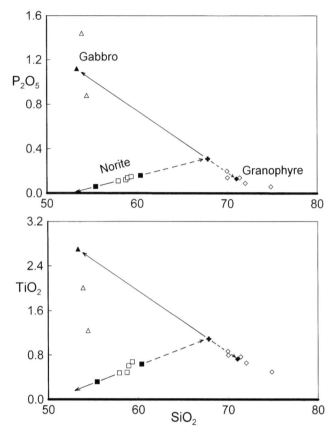

Figure 6. Variation diagrams for phosphorus and titanium against silica, in weight-percent oxides. Solid symbols are samples used for modeling (Table 4), from the South Range transect. Solid lines are control lines for cumulate phases of the norite and gabbro. Dashed lines are proposed liquid evolution lines. Open symbols represent other samples from the Main Mass used in this study.

for the South and North Range transects in the present study (0.7067, 0.7068) are almost identical to the South and North Range averages of Gibbins and McNutt (1975) for norite, gabbro, and one granophyre sample with $^{87}Rb/^{86}Sr <1$ (0.7069, 0.7068). Given that the results of the present Sr isotope study essentially duplicate those of previous studies, the principal reason for analyzing Sr isotope ratios herein was to allow direct comparison between initial Pb and Sr isotope compositions measured on the same samples. These are discussed below.

MAJOR ELEMENT CHEMISTRY

Data presented by Lightfoot et al. (1997) show that the norite and granophyre of the Main Mass have identical trace element abundance patterns (spidergrams). This is striking, in view of the wide range of elemental ratios in the crustal source rocks of the SIC (e.g., Ce/Yb ratios) (Dickin and Crocket, 1998). When combined with the isotopic homogeneity of the complex discussed above, this strongly suggests a cogenetic relationship between the upper and lower parts of the complex.

Dickin and Crocket (1998) used major element data to test a least squares mixing model in which the gabbro-granophyre system was a residue of crystal fractionation of the norite magma. It was notable that noritic rocks with a variety of modal plagioclase/pyroxene ratios lay on a single trend on variation diagrams of phosphorus and titanium against silica. This suggested that the principal cumulus phases of the norite (plagioclase, clinopyroxene, orthopyroxene) had similar silica contents around 52.5 wt% (Fig. 6). Since full mineral analyses were not available for the norite, minerals with the appropriate composition were used for the modeling from the Marginal Border Group of the Skaergaard intrusion (Hoover, 1989).

The least-squares mixing calculations have now been repeated (Table 4) using average pyroxene microprobe analyses from the South Range calculated from the data of Naldrett et al. (1970). Two mixing calculations were done, the first using new microprobe data from the cores of plagioclase grains from sample SR2.0. The second calculation used the Skaergaard plagioclase used in Dickin and Crocket (1998). Plagioclase, orthopyroxene, clinopyroxene, and whole-rock gabbro (Table 4) were removed as phases from a starting composition that is the most silica-rich South Range norite (SR 0.8). The latter is regarded as a proxy for the "original liquid" of the complex. Both models reproduce the observed composition of the South Range granophyre with low (or very low) squared residuals (1.00 and 0.15, respectively) (Table 4). Most of the misfit using the first model (analyzed Sudbury plagioclase) is due to sodium, and can

TABLE 4. MAJOR ELEMENT ANALYSES USED FOR LEAST-SQUARES MODELING

	W.R. SR0.8	Plagioclase SR2.0	Plagioclase Rim	KT26	Opx 2WO	Cpx 2WO	W.R. SR2.9	W.R. SR5.3	Model 1 (SR2) Result	Model 1 (SR2) Misfit	Model 2 (KT26) Result	Model 2 (KT26) Misfit
SiO_2	60.36	52.72	56.03	52.59	53.28	52.33	53.32	71.01	71.01	+0.00	71.03	−0.02
TiO_2	0.64	0.00	0.00	0.00	0.41	0.70	2.70	0.73	0.69	+0.04	0.68	+0.05
Al_2O_3	15.85	29.70	27.60	29.82	1.34	1.48	14.23	12.06	12.21	−0.15	12.08	−0.02
FeO*	6.83	0.59	0.31	0.32	20.08	10.12	12.02	5.31	5.38	−0.07	5.34	−0.03
MnO	0.11	0.00	0.00	0.00	0.00	0.30	0.16	0.07	0.20	−0.13	0.21	−0.14
MgO	4.47	0.00	0.00	0.07	23.21	15.26	3.95	0.25	0.18	+0.07	0.18	+0.07
CaO	6.74	11.96	9.36	13.04	1.68	19.81	8.31	2.37	2.37	+0.00	2.36	+0.01
Na_2O	3.02	4.78	6.31	3.90	0.00	0.00	3.00	4.04	3.08	+0.96	3.70	+0.34
K_2O	1.81	0.06	0.03	0.25	0.00	0.00	1.20	4.02	4.16	−0.14	3.95	+0.07
P_2O_5	0.16	0.00	0.00	0.00	0.00	0.00	1.12	0.13	0.13	+0.00	0.12	+0.01
Total[†]	100	99.8	99.6	100	100	100	100	100	99.4		99.7	
Mode[1]		31.1			13.9	4.9	9.8		40.3	$r^2 = 1.00$		
Mode[2]				30.7	15.0	3.3	10.1				40.9	$r^2 = 0.15$

W.R. = Whole-rock analyses from Dickin and Crocket, 1998.
Plagioclase SR2.0 = average of four cores (new data), and one rim.
Plagioclase KT26 is from Hoover, 1989.
Pyroxenes 2WO opx and cpx are averages from Naldrett et al., 1970.
*Total iron as FeO, due to the reduced nature of the complex.
[†]Totals normalized to 100% anhydrous.
Mineral modes are the percent abundances that yield the best fit to models 1 and 2.

---

probably be explained by metamorphic redistribution of sodium from the rims of plagioclase grains into the cores, since the rock has suffered greenschist-facies metamorphism. The modeled proportions of norite cumulus phases (Table 4) are in good accord with the modal mineral observations of Naldrett et al. (1970). In addition, the modeled proportions of norite, gabbro, and granophyre are in quite good agreement with their estimated thicknesses in the complex.

It is therefore concluded that crystal fractionation represents an excellent model to explain the relationships between the major element compositions of the different units of the complex. Isotopic mixing relationships between these units will now be discussed.

DISCUSSION

The SIC is believed to have been originally circular or approximately circular in shape, and to have reached its present form in response to tectonic compression, principally by brittle faulting in an east-west zone across the complex (Shanks and Schwerdtner, 1991). The principal zones of faulting are believed to run from near Fairbank Lake in the west to the Norduna mine in the east (Fig. 1), separating the South Range, as a single fault-bounded block, from the remainder of the complex.

The structural break in the complex may have developed very early in its history, possibly when the complex was still semi-liquid, since it was emplaced during Penokean deformation. The location of this structural break, relative to the distribution of the crustal source rocks of the complex, may have controlled the process of isotopic homogenization that we believe occurred during the cooling of the complex.

We can attempt to understand this process by comparing initial Pb and Sr isotope ratios from geographically different parts of the complex (Fig. 7). Sublayer rocks from the North and South Ranges define different-shaped distributions on this diagram, with much more Sr isotope variation on the South Range, but less Pb isotope variation. This reflects the fact that the South Range Sublayer is in contact only with Proterozoic rocks that had a very wide range of $^{87}Sr/^{86}Sr$ ratios at 1.9 Ga, but a relatively smaller range of $^{207}Pb/^{204}Pb$ ratios, with a radiogenic signature. The composition of the basal Main Mass norite on the South Range is consistent with homogenization of the observed ranges of Sublayer $^{87}Sr/^{86}Sr$ composition (0.7058–0.7115) and $^{207}Pb/^{204}Pb$ composition (15.37–15.47). However, this homogenization must have been somewhat more weighted toward unradiogenic Sr than the mean of our sample suite.

The scenario for the North Range is more complex, since basal norites collected from widely separated localities (Fig. 1) form an array trending between the Sublayer compositions of the North and South Ranges. However, the North Range is structurally continuous with the East Range and a segment near Fairbank Lake that we term the "Southwest Range." These latter two segments of the complex are in close proximity to the Huronian rocks of the Southern Province at the present erosion level (Fig. 1), suggesting that these crustal units may have contributed locally to the basal norite melt. Hence, the Main Mass norite on the North Range probably results from homogenization of impact melts of Archean and Proterozoic crust. The apparent

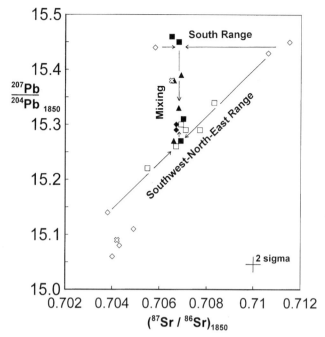

Figure 7. Plot of initial Pb against initial Sr isotope ratio for samples from the Sublayer and Main Mass of the SIC. Symbols as in Figure 4; in addition, open squares indicate basal norite samples from the southwest–north–east limb of the complex. Arrows show proposed mixing lines (see text). Estimated average 2 σ error bars are shown.

straightness of the diagonal mixing line in Figure 7 implies similar Pb/Sr ratios in the two end-members, which is reasonable in the circumstances.

The partially homogenized Pb isotope signatures of the South Range gabbro suggest that, after a period of isolated evolution, the two limbs of the complex regained communication part way through their crystallization. This could have been caused by collapse of the central part of the complex, caused by hydrothermal cooling in this vicinity, resulting in the forcible reinjection of magmas from the center of the complex toward its outer limbs (which are the parts of the complex we now see). Such a disturbance may actually have initiated the crystallization of the oxide-rich gabbro on the South Range. The process of isotopic homogenization then continued as the system evolved further, finally producing the residual granophyre with a Pb isotope signature that is homogeneous between the two sides of the complex. However, the average initial $^{207}Pb/^{204}Pb$ ratio of the granophyre is much closer to the North Range norite, which is consistent with the fact that the southwest–north–east limb of the complex probably made up 70% of its perimeter.

CONCLUSIONS

Based on the available isotopic and geochemical data, it is suggested that the SIC resulted from a single impact melting event that gave rise to the whole complex entirely from crustal sources. However, the distinct isotopic signatures of the North and South Range Sublayer indicate the involvement of different crustal target rocks on the two sides of the complex. It also appears that the complex was separated into two limbs early in its history, allowing separate magmatic evolution on the South Range and the Southwest–North–East Range, before further movements reestablished connection and allowed additional magma mixing. This reinjection model is consistent with the proposal of Peredery and Naldrett (1975) that the Main Mass granophyre was injected laterally into a part of the complex already roofed by plagioclase-rich granophyre. Such a model, while complex in detail, is typical of the kinds of magma interactions that occur during magma evolution and caldera collapse of modern volcanic systems (e.g., Condomines et al., 1982).

ACKNOWLEDGMENTS

We are most grateful to the Inco and Falconbridge companies for access to operating mines in order to collect ore samples from Sudbury. We thank Steve Beneteau, Colin Coates, Bob McNutt, Walter Peredery, and Jean Richardson for generously supplying samples and assisting with fieldwork. Catharina Jager is thanked for expert assistance in the clean chemistry lab. We also thank C. Koeberl and M. Ostermann for detailed reviews that helped to significantly improve the manuscript, and C. Hemmond for earlier helpful comments on the Sublayer Pb data. Finally, we acknowledge financial support from Natural Sciences and Engineering Research Council of Canada Research Grants (to A.P.D. and J.H.C.).

REFERENCES CITED

Chai, G., and Eckstrand, R., 1994, Rare-earth element characteristics and origin of the Sudbury Igneous Complex, Ontario, Canada: Chemical Geology, v. 113, p. 221–244.

Cohen, A. S., Burnham, O. M., Hawkesworth, C. J., Lightfoot, P. C., and Cooper, M., 1997, Os isotope study of ultramafic inclusions and separated sulphides in the Sublayer, Sudbury Igneous Complex, Ontario, in The origin and fractionation of highly siderophile elements in the Earth's mantle: European Association of Geochemistry Meeting, Mainz, Germany, Abstracts, p. 29–30.

Condomines, M., Tanguy, J. C., Kieffer, G., and Allegre, C. J., 1982, Magmatic evolution of a volcano studied by ^{230}Th–^{238}U disequilibrium and trace element systematics: the Etna case: Geochimica et Cosmochimica Acta, v. 46, p. 1397–1416.

Deutsch, A., 1994, Isotope systematics support the impact origin of the Sudbury Structure (Ontario, Canada): in Dressler, B. O., Grieve, R. A. F., and Sharpton, V. L., eds., Large meteorite impacts and planetary evolution: Geological Society of America Special Paper 293, p. 289–302.

Dickin, A. P., and Crocket, J. H., 1998, Reply to Dressler, B. O., and Sharpton, V. L., Isotopic evidence for distinct crustal sources of North and South Range ores, Sudbury Igneous Complex: a discussion: Geochimica et Cosmochimica Acta, v. 62, p. 319–322.

Dickin, A. P., Richardson, J. M., Crocket, J. H., McNutt, R. H., and Peredery, W. V., 1992, Osmium isotope evidence for a crustal origin of platinum group elements in the Sudbury nickel ore: Geochimica et Cosmochimica Acta, v. 56, p. 3531–3537.

Dickin, A. P., Artan, M. A., and Crocket, J. H., 1996, Isotopic evidence for distinct crustal sources of North and South Range ores, Sudbury Igneous Complex: Geochimica et Cosmochimica Acta, v. 60, p. 1605–1613.

Ding, T. P., and Schwarcz, H. P., 1984, Oxygen isotopic and chemical compositions of rocks of the Sudbury Basin, Ontario: Canadian Journal of Earth Science, v. 21, p. 305–318.

Dressler, B. O., and Sharpton, V. L., 1998, Isotopic evidence for distinct crustal sources of North and South Range ores, Sudbury Igneous Complex: a discussion: Geochimica et Cosmochimica Acta, v. 62, p. 315–317.

Dressler, B. O., Weiser, T., and Brockmeyer, P., 1996, Recrystallised impact glasses of the Onaping Formation and the Sudbury Igneous Complex, Ontario, Canada: Geochimica et Cosmochimica Acta, v. 60, p. 2019–2036.

Faggart, B. E., Basu, A. R., and Tatsumoto, M., 1985, Origin of the Sudbury Complex by meteorite impact: Science, v. 230, p. 436–439.

Gibbins W. A., and McNutt, R. H., 1975, Rubidium-strontium mineral ages and polymetamorphism at Sudbury, Ontario: Canadian Journal of Earth Sciences v. 12, p. 1990–2003.

Hoover, J. D., 1989, Petrology of the Marginal Border Series of the Skaergaard intrusion: Journal of Petrology, v. 30, p. 399–439.

Koeberl, C., and Shirey, S. B., 1997, Re-Os isotope systematics as a diagnostic tool for the study of impact craters and distal ejecta: Palaeogeography, Palaeoclimatology, Palaeoecology, v. 132, p. 25–46.

Krogh, T. E., Davis, D. W., and Corfu, F., 1984, Precise U-Pb zircon and baddeleyite ages for the Sudbury area, in Pye, E. G., Naldrett, A. J., and Giblin, P. E., eds., The geology and ores deposit of the Sudbury Structure: Ontario Geological Survey Special Volume 1, p. 431–446.

Lightfoot P. C., Keays, R. R., Morrison, G. G., Bite, A., and Farrell, K. P., 1997, Geochemical relationships in the Sudbury Igneous Complex: origin of the Main Mass and offset dykes: Economic Geology, v. 92, p. 289–307.

Naldrett, A. J., Bray, J. G., Gasparrini, E. L., Podolsky, T., and Rucklidge, J. C., 1970, Cryptic variation and the petrology of the Sudbury Nickel Irruptive: Economic Geology, v. 65, p. 122–155.

Peredery, W. V., and Naldrett, A. J., 1975, Petrology of the upper Irruptive rocks, Sudbury, Ontario: Economic Geology, v. 70, p. 164–175.

Rao, B. V., Naldrett, A. J., and Evensen, N. M., 1984, Crustal contamination in the Sublayer, Sudbury igneous Complex: a combined trace element and strontium isotope study: Ontario Geological Survey Miscellaneous Paper 121, p. 128–146.

Richardson, J. M., Dickin, A. P., McNutt, R. H., McAndrew, J. I., and Beneteau, S. B., 1989, Analysis of a rhenium-osmium solid-solution by inductively coupled plasma mass spectrometry: Journal of Analytical Atomic Spectrometry, v. 4, p. 465–471.

Shanks, W. S., and Schwerdtner, W. M., 1991, Crude quantitative estimates of the original northwest-southeast dimension of the Sudbury Structure, south central Canadian Shield: Canadian Journal of Earth Sciences, v. 28, p. 1677–1686.

Walczyk, T., Hebeda, E. H., and Heumann, K. G., 1991, Osmium isotope ratio measurements by negative thermal ionization mass spectrometry (NTI-MS): Fresenius Journal of Analytical Chemistry, v. 341, p. 537–41.

Walker, R. J., Morgan, J. W., Naldrett, A. J., Li, C., and Fassett, J. D., 1991, Re-Os isotope systematics of Ni-Cu sulfide ores, Sudbury Igneous Complex, Ontario: Earth and Planetary Science Letters, v. 105, p. 416–429.

Zhou, M.-F., Sun, M., and Malpas, J., 1997, Origin of the Sudbury Igneous Complex (northern Ontario, Canada): a new interpretation: Proceedings, 30th International Geological Congress, Beijing, v. 9, p. 429–440.

MANUSCRIPT ACCEPTED BY THE SOCIETY DECEMBER 16, 1998

Sudbury Igneous Complex: Simulating phase equilibria and in situ differentiation for two proposed parental magmas

Alexey A. Ariskin
Vernadsky Institute, Kosygin St. 19, Moscow 117975, Russia; ariskin@geokhi.ru
Alexander Deutsch
Institut für Planetologie, Universität Münster, D-48149 Münster, Germany; deutsca@uni-muenster.de
Markus Ostermann[*]
Institut für Planetologie, Universität Münster, D-48149 Münster, Germany; markus.ostermann@irmm.jrc.be

ABSTRACT

One of the major problems in the geology of the 1.85-Ga Sudbury impact structure, Ontario, Canada, is the origin and the evolution of the Sudbury "Igneous" Complex. The 2.5-km-thick Main Mass of this complex consists of from bottom to top, mafic to felsic norites, a layer of oxide- and apatite-rich quartz-gabbro, and granophyres. Genetically related to these lithologies are fragment-laden breccias on top (Basal Member of the Onaping Formation), and below the complex (Sublayer). The Sudbury Complex is interpreted to represent either a differentiated impact melt sheet, a hybrid endogenic magma that intruded in several pulses, or an endogenic magma topped by impact-generated granophyres.

Using the COMAGMAT–3.5 phase equilibria model, equilibrium and fractional crystallization were computed at 1 atm with the two initial liquids SIC (granophyres and norites in the ratio 2:1) and QRN (quartz-rich norite from the South Range). The calculated liquidus temperatures are 1,120°C for the SIC, and 1,180°C for the QRN parental melts. The modeled crystallization sequence for the SIC parent is independent of f_{O2}: Opx → Opx + Pl → Opx + Pl + Aug → Opx + Pl + Aug + Mt. For QRN, we observed Ol → Ol + Pl → Opx + Pl → Opx + Pl + Aug → Opx + Pl + Aug + Ilm (followed by Mt at the QFM buffer), and Opx + Pl + Aug + Mt (at the NNO buffer). The calculated liquid lines of descent show that the SIC composition potentially can produce about 70% of a residual melt containing more than 67.5 wt% SiO_2. In contrast, the quartz-rich norite magma can generate only 30% of such a "granophyric" liquid. These results would imply that fractionation of either the SIC or the QRN parent can produce the huge observed mass of granophyres, currently constituting up to two-thirds of the Main Mass.

Dynamic calculations using the INTRUSION subroutine of COMAGMAT, however, yield a more realistic and different picture. We numerically simulated convective-cumulative in situ magma differentiation with regard to a silica-rich melt trapped in natural rocks as intercumulus material. Results of these computations indicate that, even in the case of the silica-enriched SIC starting composition, the total amount of "granophyres" does not exceed 12 vol% of the modeled Main Mass. In the case of the

---

[*]Present address: EC Joint Research Centre-IRMM, B-2440 Geel, Belgium.

Ariskin, A. A., Deutsch, A., and Ostermann, M., 1999, Sudbury Igneous Complex: Simulating phase equilibra and in situ differentiation for two proposed parental magmas, *in* Dressler, B. O., and Sharpton, V. L., eds., Large Meteorite Impacts and Planetary Evolution II: Boulder, Colorado, Geological Society of America Special Paper 339.

QRN liquid, the granophyres amount to only 1–3 vol%. These results indicate that, under the assumption of a closed melt system evolution of the Main Mass, *granophyres in their present volume did not originate by differentiation of the norites in their presently known volume.* The lower "mafic" part of the Main Mass, however, is modeled perfectly with the dynamic COMAGMAT calculations, using a ferroandesitic melt compositionally similar to QRN, as starting composition. The modeled series of cumulates exactly duplicates the natural sequence from mafic and quartz-rich norites to the quartz-gabbro as observed in drill core 70011 of the North Range. These findings substantiate the interpretation that the norites, the quartz-gabbro, and a small cap of granophyres are genetically linked by differentiation. On the other hand, the results contradict any interpretation of the quartz-gabbro as the product of mixing between a "granophyric" and a "noritic" magma.

INTRODUCTION

The 1.850-g.y.-old (Krogh et al., 1984) Sudbury Structure, Ontario, lies within a Proterozoic supracrustal sequence (Huronian Supergroup) in the South, and Archean basement rocks (Abitibi Subprovince) of the Superior Province in the North (Fig. 1) (Dressler, 1984). The structure covers an area of roughly 15,000 km² and includes brecciated country rocks, the Sudbury "Igneous" Complex, and the Sudbury Basin. The Sudbury Structure is now widely acknowledged as tectonized erosional remnant of a more than 200-km sized multi-ring impact basin, yet the origin and evolution of the >2.5-km-thick Igneous Complex in the central depression of the structure remains controversial (see, Deutsch et al., 1995, for review). Models interpreting the Igneous Complex as coherent impact melt sheet (e.g., Grieve et al., 1991; Masaitis, 1993; Deutsch et al., 1995) compete with mixed endogenic-exogenic views ("two-magma model") (e.g., Chai and Eckstrand, 1993, 1994; Dressler et al., 1996), and, at the other extreme, a purely endogenic magmatic origin of the Igneous Complex has also been proposed (e.g., Naldrett 1984; Dressler et al., 1987). The impact melt nature of the Sudbury Igneous Complex is supported by evidence including isotope systematics (e.g., Ding and Schwarcz, 1984; Faggart et al., 1985; Deutsch, 1994; Ostermann, 1996; Ostermann and Deutsch, 1998), constraints from impact mechanics (e.g., Lakomy, 1990; Grieve et al., 1991; Deutsch and Grieve, 1994), and geophysical data (e.g., Milkereit et al., 1992).

As shown in Figure 2, the Main Mass of the Igneous Complex includes three major units: the Lower Zone of mafic to felsic norites; the Middle, or Transition, Zone, consisting of an oxide-rich gabbro with high modal amounts of Ti-Fe oxides (up to 10%) and apatite (the so-called quartz-gabbro); and the grano-

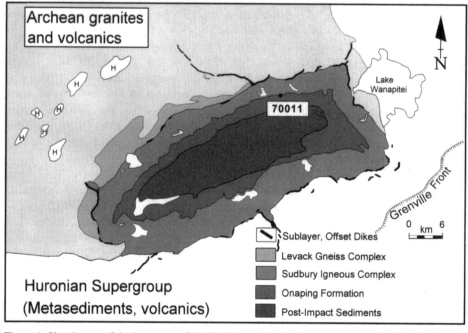

Figure 1. Sketch map of the inner part of the Sudbury multi-ring impact structure showing location of drill core 70011.

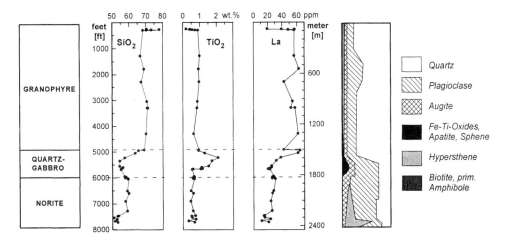

Figure 2. Variation of selected major elements and La over drill core 70011, North Range. To compare geochemical and mineralogic compositions, the simplified mineralogic composition of the Main Mass of the Sudbury melt system in the North Range is shown (after Naldrett and Hewins, 1984). Geochemical and mineralogic features are distinctly related between, for example, the noritic and quartz-gabbroic parts of the schematic stratigraphic column. Modified after Ostermann (1996).

phyric Upper Zone. Based on field evidence, Collins (1934) estimated a 1:2 ratio between norite and granophyre. Due to this odd volume ratio, the genetic relation between granophyres, the quartz-gabbro, and the norites still represents an unsolved problem, which is independent of the ongoing discussion of exogenic vs. endogenic origin.

A detailed geochemical study of the continuously sampled INCO core 70011 drilled in the North Range (Fig. 1) revealed for the first time rather smooth changes of elemental concentrations in the petrographically defined transitions from one to the other lithology of the Main Mass. Ostermann (1996), and Ostermann and Deutsch (1998) concluded that their geochemical data indicate a link of the three major units by differentiation. This interpretation prompted the present study to test whether petrologic modeling supports the proposed differentiation mechanism. Progress in theoretical igneous petrology over the past decades allows petrologists to study most of these effects numerically by means of numerical phase equilibria and dynamic models designed to simulate magma differentiation processes over a wide range of compositions and conditions.

Several different programs exist, which are based on data from melting experiments with igneous rocks. This empirical approach extends the available experimental data base to poorly studied natural systems and/or to a wider range of P-T-f_{O_2} (oxygen fugacity) conditions (e.g., Nielsen and Dungan, 1983; Frenkel and Ariskin, 1984; Ghiorso, 1985; Ariskin et al., 1987, 1993; Weaver and Langmuir, 1990; Ghiorso and Sack, 1995). These models allow calculation of equilibrium and/or fractional crystallization in terrestrial basaltic to andesitic magmas, and they have been successfully applied to genetic interpretation of tholeiitic and calc-alkaline volcanic suites (Ghiorso and Carmichael, 1985; Nielsen, 1990; Ariskin et al., 1988, 1990, 1995). All these models except COMAGMAT (Ariskin et al., 1993) fail in the solution of thermal and dynamic problems, e.g., the movement of solidification fronts in magma chambers, effects of crystal settling on volumetric relations, as well as mineralogy and chemistry of rocks formed during differentiation (Frenkel et al., 1989).

In this chapter, we present the results of combined phase equilibria and dynamic calculations carried out with COMAGMAT software in order to simulate the proposed in situ differentiation of the impact melt layer (the so-called Sudbury Igneous Complex) in the central part of the Sudbury multi-ring impact structure (e.g., Grieve et al., 1991; Deutsch et al., 1995; Ostermann, 1996). The results of the simulation, which have been carried out for two different "initial" compositions of the Main Mass of the Sudbury Igneous Complex (Collins, 1934; Naldrett, 1984) set new constraints on the evolution of this controversely discussed silicate melt with an estimated total volume of about 12,500–25,000 km³ (Deutsch and Grieve, 1994).

See Table 1 for a list of abbreviations used in this chapter.

COMAGMAT PROGRAM

General description

The COMAGMAT software consists of a series of linked programs developed to simulate a variety of igneous processes including both simple crystallization of volcanic suites and in situ differentiation of tabular intrusions (Ariskin et al., 1993). COMAGMAT is designed for melts, ranging from basaltic to dacitic compositions. The modeled elements include Si, Ti, Al,

TABLE 1. LIST OF ABBREVIATIONS

Ab	albite
An	anorthite
Aug	cpx with an augitic composition
Cpx	clinopyroxene
En	enstatite
F_{crit}	maximal possible fraction of solid phases in a magma, which can accumulate to form a primary cumulate
F_{crys}	fraction of solid phases suspended in a magma
F_{int}	fraction of solid phases suspended in a magma directly after emplacement in the chamber
f_{O2}	oxygen fugacity
Fa	fayalite
Fo	forsterite
Fs	ferrosilite
Ilm	ilmenite
MGN	magnesium number (100xMg/(Mg+Fe))
Mt	magnetite
NNO	nickel-bunsenite oxygen buffer
Ol	olivine
Opx	orthopyroxene
P	pressure
Pl	plagioclase
QFM	quartz-fayalite-magnetite oxygen buffer
QRN	average chemical composition of the fine-grained Quartz-Rich Norites of the South Range according to Naldrett (1984)
SIC	average chemical composition of the Main Mass of the Sudbury "Igneous" Complex according to Collins (1934)
t	time
t*	chilling duration
T	temperature
Ulv	ulvöspinel
Wo	wollastonite

Fe_{tot} (divided into Fe^{3+} and Fe^{2+}) ±Mn, Mg, Ca, Na, K, P, as well as 20 trace elements (see below). In the last version of COMAGMAT–3.5 used for this work (Ariskin, 1997[1]), the modeled minerals include olivine (Fo - Fa solution), plagioclase (An - Ab solution plus K), three pyroxenes (augite, pigeonite, and orthopyroxene: En - Fs - Wo solutions plus Al and Ti), ilmenite and magnetite (Fe^{2+} - Fe^{3+} - Ti - Al - Mg solutions). The program can be used at low to moderate pressures, approximately up to 12 kb, and allows calculations of open (12 oxygen buffers) and closed system differentiation with respect to oxygen (Ariskin, 1987, 1993).

Thermodynamic background

Mineral-melt geothermometers. The basic components of the COMAGMAT model are a system of empirically calibrated equations that describe mineral-melt equilibria as function of temperature and melt composition. These geothermometers have been calibrated using a data base, which consists of published results of melting experiments at 1 atm (Ariskin et al., 1987, 1996). For this calibration, the activities of the mineral components were assumed to equal the mole fractions of the cations in a single site (ideal solution). This model implies that the effects of nonideality on mineral-melt equilibria are attributed primarily to the liquid phase. Applying the calibrated geothermometers to liquid compositions of the initial data base, one can invert the calculations to estimate mineral-melt equilibria temperatures. Calculated and experimentally observed temperatures coincide with an accuracy of ±10°–15°C (Ariskin et al., 1987, 1993). A similar comparison of calculated and experimentally produced mineral compositions shows that Fo, An, En, and Wo contents can be predicted within 1–3 mol%. It is important to note that the best fit of calculated mineral-melt equilibria with experimental data is obtained for tholeiitic and transitional systems ranging from magnesian-basalts to dacites (45–60 wt% SiO_2, $Na_2O + K_2O < 5$ wt%).

Trace element partitioning. One of the major advantages of COMAGMAT is the linkage between major and trace element systematics. Computation of major element mineral-melt equilibria controlled parameters, such as temperature and fractionating mineral proportions, set constraints for the behavior of trace elements, based on empirically determined distribution coefficients. COMAGMAT calculates the distribution of Mn, Ni, Co, Cr, V, Sc, Sr, Ba, Rb, Cu, and 10 rare earth elements (REE) in an evolving melt system (Barmina et al., 1989b, 1992).

Algorithm to solve the equilibrium problem. To simulate the crystallization of multiple saturated magmas, an algorithm must be developed for the calculations of mineral-melt equilibrium at a given set of independent parameters of state, such as temperature and pressure. To address the problem, COMAGMAT takes advantage of numerical solutions of nonlinear empirical equations that describe mineral-melt equilibria and mass action law, coupled with the mass balance constraints for the whole system composition (Frenkel and Ariskin, 1984). This approach combines basic empirical and thermodynamic considerations into a hybrid algorithm. The simulation of the differentiation of natural magmas is thus performed step by step with increasing crystallization (Ariskin et al., 1993).

Dynamic background

To model dynamics of in situ differentiation of magmas, we have developed the INTRUSION subroutine. This part of the COMAGMAT program is placed in the crystallization algorithm so that after determination of the phase equilibria for a given crystallinity, the modeled magma composition can be modified by subtraction of the calculated equilibrium (cumulative) minerals and the melt, trapped in the cumulates (Frenkel et al., 1989). This enables the user to simulate a variety of igneous differentiation processes, accounting for changes in phase proportions and compositions during the system's evolution.

The INTRUSION subroutine simulates a number of physical processes taking place simultaneously during in situ magma differentiation. These processes include (1) loss of the heat of crys-

---

[1] The COMAGMAT program and its description are available on request from A. A. Ariskin.

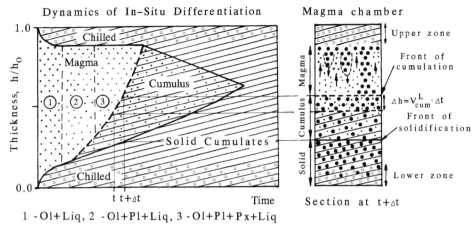

Figure 3. Dynamics of in situ magma differentiation based on the convective-cumulative model of Frenkel et al. (1989). High efficiency of vigorous convection is assumed to maintain uniform temperature and composition in the magma body up to the point where the magma chamber is filled with cumulates and the chemical differentiation is completed.

tallization through the upper and lower margins of a tabular magma chamber; (2) movement of solidification fronts into the magma body; (3) crystal settling, together with the formation of crystal-bonded (cumulus) and nonbonded (porphyric) crystal aggregates; and (4) the evolution of the primary mineral and the trapped liquid compositions. The basic framework for this model and its full mathematical description are given in Frenkel et al. (1989) and summarized in Ariskin et al.(1993).

Convective-cumulative model. The main problem in simulating mineral-melt equilibria combined with heat-mass transfer in a magma chamber is to define the relationship between the increment of crystallization and the time needed to cool and crystallize the melt to a specific crystallinity. The development of an algorithm that will accurately model this link depends on the constraints chosen for the responsible physical mechanisms. The convective-cumulative model used in COMAGMAT was originally designed for simulating the in situ differentiation of Siberian sills (Frenkel et al., 1989). This model is constrained by more than 20 yr of field work, petrographic and geochemical analysis, as well computer simulations.

Perhaps the most important result from this multi-disciplinary work is the obviously high efficiency of convection in 10^2–10^3-m-thick magma bodies. The convection yields a nearly uniform temperature distribution and quite homogeneous melt composition up to the point where the magma chamber becomes filled with cumulates. This will result in fractionation of the magma, but not perfect fractional crystallization due to effects caused by suspended crystals. These solid phases are assumed to form near the upper boundary of chamber, then descend through the convecting magma into cumulates at the base of the magma body. In this model, convection does not prevent settling of the suspended crystals. Figure 3 schematically illustrates the convective-cumulative model of magma evolution.

Frenkel et al. (1989) proposed a number of semi-empirical constraints that approximate the convective-cumulative process in order to link the dynamic equations with the basalt crystallization algorithm. Calculation of the equilibrium crystallization for a given initial magma composition always starts under the assumption of an entirely liquid system ($F_{crys} = 0$). This does not imply that the magma at the beginning of its evolution in the magma chamber indeed is free of crystals but simply that these calculations proceed incrementally up to the value $F_{crys} = F_{int}$, which is the assumed fraction of solid phases suspended in the magma at the time of emplacement. Beginning from F_{int}, the INTRUSION subroutine of COMAGMAT is used iteratively, after calculations of phase equilibria at each degree of crystallization.

Given the bulk composition and thickness H of a sheet-like magma body, the INTRUSION subroutine starts to compute the differentiation process at the initial time $t = 0$, with an assumed chilling duration t^*. This t^* defines the thickness of the upper and lower chilled zones (Frenkel et al., 1989). The formation of the chilled zones along the margins for the period $0 < t < t^*$, as well the formation of cumulates in the lower zone at $t > t^*$, results in a reduced volume of the model magma system for both the liquid (trapped melt) and the equilibrium solid (cumulative) phases. After correcting for this volume loss, the remaining melt is incrementally crystallized further in terms of ΔF. The phase equilibria at the new and higher crystallinity is then calculated. Each forthcoming iteration through the INTRUSION subroutine begins with the determination of the heat and crystal mass flux and the time interval corresponding to the given ΔF_{crys}.

In summary, the COMAGMAT algorithm simulates at the same time crystallization and the thermal history of the magma body, including the dynamics of crystal settling and the movement of the crystallization or accumulation fronts within the magma chamber. The simulation ends when the calculated fraction of solids suspended in the magma reaches the value of an assumed critical crystallinity F_{crit}. This F_{crit} is defined as the maximal possible fraction of minerals, which can be accumulated in the magma to form a primary cumulate. If the value of F_{crit} is

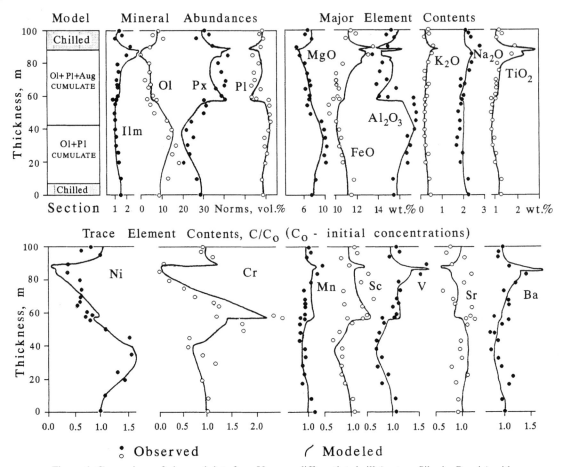

Figure 4. Comparison of observed data from Vavucan differentiated sill (eastern Siberia, Russia) with dynamic simulations using the COMAGMAT model. The phase boundary, which is separating Ol + Pl and Ol + Pl + Aug "cumulates" (porphyric rocks), predicted by the model, corresponds to boundary between poikilophitic and taxiophitic dolerites observed in nature (Frenkel et al., 1989; Ariskin et al., 1993).

not reached during these calculations, this modeling will proceed up to the point (height) where the upper crystallization front encounters the front of crystal accumulation.

Application of the convective-cumulative model to the natural case. The test of the proposed in situ differentiation mechanism is whether observed major and trace element distributions in tabular intrusions are quantitatively reproduced by the proposed model (Frenkel et al., 1989; Ariskin et al., 1993; Chalokwu et al., 1996). Figure 4 compares measured geochemical characteristics with results of the model calculations for one Siberian differentiated sill, the 100-m-thick Vavukan intrusion (Frenkel et al., 1989). Given the large number of data, particularly major and trace element analyses, and the relatively small number of variable parameters, the solution explaining all observational data is by no means overdetermined.

The successful application of the model to this sill indicates that the convective-cumulative mechanism represents a plausible physical explanation for the differentiation of sheet-like magma bodies, 10^2–10^3-m-thick. With this background, we applied the algorithm to the ~2,500-m-thick Main Mass of the Sudbury Igneous Complex in order to evaluate the evolution of this impact melt sheet. In principle, this silicate melt body cooled slowly enough (i.e., on the order of 0.5 Ma) to differentiate (Ivanov et al., 1997; Ivanov and Deutsch, this volume).

MODELING PHASE EQUILIBRIA FOR THE SUDBURY IGNEOUS COMPLEX

Defining parental melts

The principal problem in modeling a complex silicate melt system such as the Sudbury Igneous Complex lies in the definition of the parental melt (or starting) composition. There exist several approaches for this question. The classic petrologic-geologic approach uses bulk chemical analyses of fine-grained lithologies from the margins of layered intrusions, which are assumed to represent chilled initial liquids (Wager and Brown, 1967). This technique works only if thermal or chemical processes at the margin did not alter the composition of such rocks. The calculation of weighted average compositions from

TABLE 2. AVERAGE COMPOSITION OF THE MAIN ROCK TYPES OF THE SUDBURY "IGNEOUS" COMPLEX IN DRILL 70011 OF THE NORTH RANGE*

Rock Type	Granophyre	Quartz-gabbro	Norite	Sublayer	Calculated Average for the Main Mass	Analytical Error
Drillcore	70011	70011	70011	70011	70011	
% of drillcore	59.6	13.7	20.2	6.5	100	
(wt. %)						
SiO_2	69.5	55.2	57.3	51.7	63.9	±0.54
TiO_2	0.80	1.44	0.59	0.53	0.83	±0.01
Al_2O_3	12.6	14.7	15.4	9.00	13.22	±0.18
$Fe_2O_{3\,(tot)}$[†]	5.49	10.4	8.42	12.15	7.19	±0.21
MnO	0.08	0.14	0.13	0.21	0.11	±0.01
MgO	1.01	3.66	5.70	12.27	3.05	±0.15
CaO	2.24	7.25	6.39	10.17	4.28	±0.10
Na_2O	4.15	3.86	3.19	1.69	3.76	±0.14
K_2O	2.76	1.45	1.41	0.55	2.16	±0.05
P_2O_5	0.17	0.46	0.15	0.12	0.20	±0.02
Total	99.2	99.1	99.6	98.4	99.86	
LOI[§]	1.04	1.08	1.31	1.62	1.14	

*Ostermann, 1996.
[†]$Fe_2O_{3\,(tot)}$ = total iron.
[§]LOI = Loss on ignition.

well-documented vertical sections of tabular intrusion bodies, e.g., in differentiated traps, yields more accurate estimates of parental magma composition (Frenkel et al., 1989). This approach is less sensitive to processes at the contact. Such calculations, however, may result in an initial melt composition enriched in certain components that are suspended or accumulated in the magma as crystal phases.

A constructive way to escape both these problems is to combine calculated weighted average compositions with experimental studies or phase equilibria simulations. In this concept, known as geochemical thermometry (Frenkel et al., 1988), the equilibrium crystallization is calculated for an assumed parent melt up to the temperature corresponding to the estimated percentage of primary crystallization or crystal accumulation. This approach has been successfully applied to, for example, to the genetic interpretation of Siberian differentiated traps, hypabyssal bodies from eastern Kamchatka, and some mafic layered intrusions (Barmina et al., 1989a, b; Chalokwu et al., 1993, 1996).

Estimates of parental melt compositions for the Sudbury Igneous Complex

Application of the geochemical thermometry technique to the Sudbury Igneous Complex presents serious difficulties. The main problem rests in the fact that a successful approach requires strongly defined contacts to the country rocks, and correct estimates of major rock types in terms of volumetric proportions (thickness), geochemical composition, and density. In the case of the Sudbury Igneous Complex, two facts hinder a sound assessment of the parental melt by the geochemical thermometry technique. First the Sublayer rocks and melt breccias of the Basal Member (Onaping Formation) reliably belong to the impact melt body (Brockmeyer, 1990; Grieve et al., 1991; Deutsch and Grieve, 1994; Deutsch et al., 1995), and, hence, should be combined with the Main Mass for calculation of an weighted average composition of the whole Sudbury Igneous Complex. Volumetric proportions and geochemical composition of both formations, however, are not well constrained. Second, principal differences occur in the stratigraphic columns of the North and South Ranges of the Sudbury Igneous Complex, such as the diversity of norites, and different volumetric proportions between the three major lithologies, i.e., norites, quartz-gabbro, and granophyres (Naldrett and Hewins, 1984). Due to these difficulties, any estimated "bulk composition of the initial Sudbury melt" could represent only a first approximation of the true initial melt. Results of computer-based interpretations of the further evolution of such a melt, therefore, provide simply generalized conclusions on the possible evolution of the Sudbury Igneous Complex in nature.

In this study, two melt compositions have been used for phase equilibria calculations; they are assumed to be parental for the Main Mass of the Sudbury Igneous Complex. As proposed by Collins (1934) on the basis of field studies, the first average composition corresponds to a mix of granophyres and norites in the canonical ratio 2:1; for this initial melt, we use the abbreviation SIC. This average composition is close to the average composition of the Main Mass, calculated recently by Ostermann and Deutsch (1998) by weighting average compositions of the major lithologies in core 70011 (Table 2) drilled in the North Range

TABLE 3. PROPOSED PARENTAL SUDBURY MAGMAS
USED IN COMAGMAT CALCULATIONS

Oxide	Average SIC (Collins, 1934)	Quartz-rich Norite QRN (Naldrett, 1984)
SiO_2	64.19	57.00
TiO_2	0.75	1.34
Al_2O_3	14.89	16.40
FeO_{tot}	5.57	7.33
MnO	0.08	0.13
MgO	2.90	6.40
CaO	4.10	7.28
Na_2O	3.35	2.41
K_2O	2.96	1.55
P_2O_5	0.23	0.16

Note: Normalized to water-free basis (100 wt. %).

(Fig. 1). The second parent used in our computations corresponds to the fine-grained Quartz-Rich Norites (QRN) of South Range; this lithology was proposed by Naldrett (1984) and Naldrett et al. (1986) as the initial, mantle-derived and contaminated magma for the Sudbury Igneous Complex. Table 3 lists these two assumed "parental" compositions.

Compared to the QRN composition, the SIC is high in SiO_2 (64 vs. 57 wt%) and alkalies yet depleted in the "mafic" and "plagioclase" components MgO (2.9 vs. 6.4 wt% for QRN), FeO, CaO (4.1 vs. 7.3 wt%), and Al_2O_3 (see Table 3). This significant compositional difference reflects the large proportion of exposed granophyre (Collins, 1934). It is noteworthy that drill core 70011 displays relative to the estimate of Collins (1934) a similar volumetric ratio between norite and granophyre (Ostermann, 1996; A. Therriault, personal communication). The SIC parental melt is "ferrodacitic" and the QRN is "ferroandesitic" composition. Using both given "initial" compositions, we applied COMAGMAT to study whether the three major lithologies of the Main Mass, i.e., norite, quartz-gabbro, and granophyre, are related by differentiation (e.g., Ostermann, 1996) or belong to separate silicate melt systems (e.g., Chai and Eckstrand, 1993, 1994). In this context, and as already pointed out by researchers including Kuo and Crocket (1979) and Naldrett (1984), we should emphasize that the odd ratio between silica-rich and mafic members of the Mass Mass represents a fundamental problem in all previously published models for the origin and evolution of the Sudbury Igneous Complex. This ratio contrasts with the observations on all true (i.e., endogenic) layered magmatic bodies.

Phase equilibria calculation for Sudbury parental melts

The COMAGMAT–3.5 program (Ariskin, 1997) was used for the modeling of equilibrium crystallization and fractional crystallization of the SIC and QRN parental magmas. Advantages of the new COMAGMAT version include equations describing mineral-melt equilibria for Ti-magnetite and ilmenite, which are significant components in the quartz-gabbro (Naldrett and Hewins, 1984) (see Fig. 2). In addition, a new orthopyroxene-melt model computes more accurately the orthopyroxene crystallization (Bolikhovskaya et al., 1996). Moreover, using the INFOREX experimental database (Ariskin et al., 1996), more realistic molar mineral-melt distribution coefficients for TiO_2 in clino- and orthopyroxenes, as well for K_2O in plagioclase, are calculated. These coefficients are also integrated in the COMAGMAT–3.5 software.

Crystallization sequence. The calculations of equilibrium and fractional crystallization were carried out at 1 atm total pressure and with a 1 mol% increment of solids from one step to the next. In addition, two different oxygen buffers were used, QFM (quartz-fayalite-magnetite) and NNO (nickel-bunsenite), in order to better constrain the f_{O2} conditions at which magnetite starts to crystallize on the calculated liquidus simultaneously with high-Ca clinopyroxene. This assemblage occurs in norites and the quartz-gabbro (Naldrett and Hewins, 1984). The maximum compositional range encountered is about 70 mol%, which corresponds to the final calculated temperatures of 1,025°C for SIC, and 1,050°C for QRN compositions, respectively.

Figure 5 illustrates the results for the simulated equilibrium crystallization of the two proposed parental magmas. As shown in these plots, the calculated liquidus temperatures are 1,120°C for the SIC and 1,180°C for the QRN composition, respectively. For the SIC parent, the modeled crystallization sequence is Opx (orthopyroxene) → Opx + Pl (plagioclase) → Opx + Pl + Cpx [Aug] (augitic clinopyroxene) → Opx + Pl + Aug + Mt (magnetite). This sequence is independent of f_{O2}. The main differences between calculations at QFM, and NNO buffers lie in the higher crystallization temperature for *Mt* at more oxidizing conditions.

For the QRN composition of the initial melt, calculated phase relations turned out to be more complex. The modeled crystallization sequence for silicate minerals is also independent of f_{O2}, whereas the appearance of iron-titanium oxides at the liquidus displays small, yet important, differences: Ol (olivine) → Ol + Pl → Opx + Pl → Opx + Pl + Aug → Opx + Pl + Aug + Ilm (ilmenite) → Opx + Pl + Aug + Ilm + Mt (at QFM); and Ol → Ol + Pl → Opx + Pl → Opx + Pl + Aug + Mt (at NNO). This variance in the onset crystallization temperatures for ilmenite and Ti-magnetite at QFM and NNO is in good agreement with available experimental data (Toplis and Carroll, 1995). In addition, we note that during progressive crystallization of QRN, olivine is completely dissolved after the first appearance of orthopyroxene. This orthopyroxene is partly dissolved with the onset of augite crystallization, implying peritectic (reaction) relations of these phases at the modeled range of temperatures and compositions.

The disappearance of *Ol* with commencement of Opx crystallization in the QRN system deserves a more detailed discussion. According to petrologic observations, olivine is absent in the norites of the Main Mass, even in the most magnesian norites of the North Range (Naldrett and Hewins, 1984). This observation might be considered as argument against the QRN composition representing the real parental melt. Yet, the phase equilibria calculations presented in Figure 5 demonstrate that such a con-

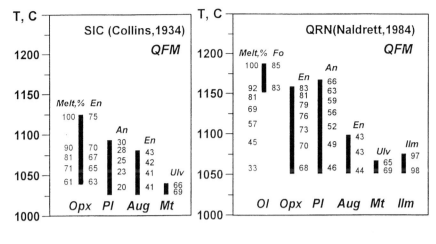

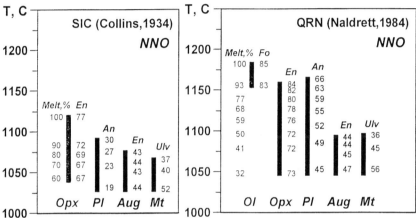

Figure 5. Mineral crystallization sequences calculated with the COMAGMAT model for the SIC and QRN parents at 1 atm and QFM to NNO oxygen buffer conditions. Values to the right of the magnetite and ilmenite lines represent the percent of ulvöspinel and ilmenite component in solid solutions, calculated using the equations of Stormer (1983).

clusion would be erroneous. After magma fractionation and complete accumulation of primary minerals, the first Ol crystals at the liquidus do not survive solidification of the crystal-trapped melt system, due to a highly efficient reaction of primary solids with the silica-enriched trapped liquid.

In general, both crystallization sequences, i.e., for the SIC or QRN initial melt, calculated at NNO conditions correspond well to observations in the Main Mass of the Sudbury Igneous Complex. Calculated mineral compositions provide additional genetic information.

Opx compositions. Naldrett and Hewins (1984) present data on the change in hypersthene and augite mineral chemistry in the norites of the South, and North Ranges. The most magnesian Opx crystals with the magnesium number $MGN = 100 \times Mg/(Mg + Fe)$ of 74–76 occur in mafic norites of the North Range; whereas, most of the felsic and quartz-rich norites contain Opx with MGN of 60–65. Comparing these natural Opx compositions with the calculated ones, shown in Figure 5, we conclude that the SIC composition crystallizes liquidus Opx with MGN = 75–77. This initial composition, hence, may accurately duplicate the parental melt. We want to emphasize, however, that the seemingly good accordance in Opx mineral chemistry could also be due to re-equilibration of primary more magnesian Opx crystals with a trapped liquid having a low MGN. Such postcrystallization processes occur generally both in differentiated sills (Barmina et al., 1989a), and mafic layered intrusions (Chalokwu et al., 1993).

Pl compositions. According to Naldrett and Hewins (1984), the anorthite content decreases upward from An_{61} at the base of the South Range norite to An_{50} halfway to the quartz-gabbro. Plagioclase in the quartz-gabbro and granophyres are altered to a high degree, so that only the bulk normative An content may be used to estimate the evolution of plagioclase compositions. The observed normative An content of Pl decreases abruptly from An_{50} in the quartz-gabbro to An_{15-20} in the granophyres (Naldrett and Hewins, 1984). In this respect, the QRN composition matches observation better than the SIC starting liquid (Fig. 5).

Liquid lines of descent. Figure 6 illustrates the evolution of SiO_2 contents in the residual melts of the SIC and QRN starting compositions, calculated at NNO conditions for equilibrium, and fractional crystallization. Figure 6a was constructed in terms of degree of crystallization, Figure 6b gives silica contents as function of the modeled temperature. The liquid lines of descent for both parent melts exhibit only minor variations related to the type of crystallization process (equilibrium vs. fractional). The initial composition, SIC and QRN, however, causes significant differences in the respective trajectories.

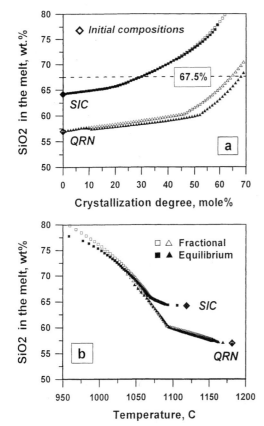

Figure 6. Liquid lines of descent during equilibrium and fractional crystallization of the SIC and QRN parents calculated at 1 atm and NNO oxygen buffer conditions. The SiO_2 contents in the melts are displayed as a function of the crystallization degree and the equilibrium temperature.

Figure 6a shows to what extent the SIC and QRN parents should be crystallized to produce differentiates with a silica content similar to that of the granophyre, i.e., >67.5 wt% (Collins, 1934). In the simulation, the SIC composition produces about 70 mol% of a residual melt with SiO_2 in excess of 67.5 wt%. In contrast, the QRN melt generates only 30 mol% of such a granophyric liquid. Figure 6b gives evidence that for both initial compositions, all calculated T-SiO_2 trajectories converge in the temperature range of 1,070°–1,025°C.

The results of our modeling indicate that, in principle, fractionation of the SIC parent could explain formation of enough granophyre to constitute two-thirds of the Main Mass in sections in the North Range. In contrast, differentiation of the QRN magma could result in the less thick granophyre horizon observed in the South Range (Naldrett and Hewins, 1984). Such a straightforward mass balance, however, seems to be of limited value for the natural differentiation process of the Sudbury melt body. In fact, due to the porosity of natural cumulates, a part of the silica-enriched liquid will be trapped as intercumulus material.

Although the norites of the Main Mass are not typical cumulates (Warren et al., 1996), they may contain a certain amount of trapped melt, which in turn results in a significant decrease of the total volume of granophyric melt that can be generated from the parents. In conclusion, both parental melts can yield, in principle, "granophyric" residuals. This fact strongly supports the concept that the Sudbury Igneous Complex represents a differentiated single melt body.

Constraints set by our phase equilibria calculations, however, do not define conditions of melt evolution in detail, and questions remain. For example, the structure of the lower zones of the Main Mass constitutes a serious problem. Any parental melt proposed for the Sudbury Igneous Complex must also generate differentiates compositionally similar to the quartz-rich and mafic norites observed in the lower zones of the Main Mass. The following gives a more quantitative treatment of this problem.

SIMULATING IN SITU MAGMA DIFFERENTIATION OF THE MAIN MASS

Using the INTRUSION subroutine of the COMAGMAT-3.5 software, we performed a set of calculations for the two proposed parental melts, SIC and QRN, simulating the convective-cumulative style of in situ differentiation as described above. The simulation technique for the Sudbury problem, which has been successfully applied to Siberian sills (Fig. 4), is a typical example of a forward model. Beginning with computer simulations to calculate the response of a modeled system to a set of assumed thermodynamic and dynamic parameters (initial and boundary conditions), the calculated results are then compared with observations of the natural case. The model may be used iteratively by modifying the parameters until a satisfactory fit is obtained. After defining a set of optimal parameters, this information is used to gain further insight into how crystallization will change other system parameters. They include the temperature path, changes in chemistry of liquidus phases, and changes in phase chemistry of the model cumulates and coexisting magma at every level in the body.

The detailed geochemical study of drill core 70011 (Ostermann, 1996) in the North Range (Fig. 1) served as natural data set to control and optimize parameters. The weighted average compositions for the three major lithologies of the Main Mass in the core samples are listed in Table 2.[2]

Modeling the whole Main Mass section

Initial and boundary conditions. Based on drill core 70011 (Fig. 2) (A. Therriault, personal communication), the thickness H of the magma body was set to 2,224 m. Initial crystallinity of the parental magmas was varied in the range $0 < F_{int} < 10\%$, with the chilling duration t^* being 720 hr. Thermophysical parameters for country rocks, such as thermal conductivity, density, and heat capacity, were used from the Siberian sills simulations

---

[2] Detailed data are available by request from the authors, and as EXCEL file under <http://saturn.uni-muenster.de/>.

(Table 4), the initial temperature gradient at the upper crystallization front was assumed to be $\Delta T = 1,000°C$. The main variables include the critical crystallinity of modeled cumulates, which is $40 < F_{crit} < 80$ vol% (i.e., the portion of trapped melt F_m is 100% - F_{crit}). The *efficient* crystal settling velocities for the minerals Ol, Opx, Pl, Aug, and Mt varied in the range from 0.1 to 50 m a^{-1}. About 50 such simulations were performed. Figure 7 shows two typical results.

Main results. Figure 7 illustrates the distribution of SiO_2 and TiO_2 in the modeled sequence, which results from dynamic calculations of convective-cumulative differentiation of the Main Mass at $F_{crit} = 40\%$, and 60%; other parameters are listed in Table 4. The results for the SIC, and the QRN parent are displayed in Figure 7. Each of the two plots shown includes two curves. The solid one displays the evolution for $F_{crit} = 40\%$, the dotted one stands for $F_{crit} = 60\%$. We note that even in the case of the silica-enriched Collins' (1934) SIC composition, the total amount of "granophyres" generated does not exceed 4–12% of the total thickness. For the QRN starting composition, this amount is decreased to less than 3%. This is no surprise, as a part of the silica-enriched liquid is assumed to be trapped as intercumulus material.

These results indicate the following conclusion, which is independent on the initial melt composition, and boundary conditions. The convective-cumulative magma fractionation processes was neither able to produce the 60–70% of granophyre observed in the North Range nor the about 40% of granophyre occurring in the South Range. We assume this statement to be valid also for other magma fractionation mechanisms, such as directional crystallization from the bottom. In addition, neither

TABLE 4. PARAMETERS OF DYNAMIC CALCULATIONS SIMULATING THE FORMATION OF THE SUDBURY MAIN MASS BY THE CONVECTIVE-CUMULATIVE MODEL†

Initial conditions	
Parental magma compositions	SIC and QRN, see Table 2
Thickness of the Main Mass magma body, H	2224 m
Initial crystallinity of the parental magmas, F_{int}	5 vol. %
The chilling duration, t*	720 hours
Boundary conditions	
Temperature gradient at the upper crystallization front, ΔT	1000°C
Density of country rocks	2.700 g/cm^3
Heat conduction of country rocks	0.006 cal/cm*sec*grad
Heat capacity of country rocks	0.250 cal/g*grad
Assumed dynamic parameters	
Critical primary crystallinity of modeled cumulates, F_{crit}	40 and 60 vol. %
Crystal settling velocities, m/year	
Olivine	20.0
Orthopyroxene	0.5
Plagioclase	0.5
Augite	6.0
Magnetite	1.0
Calculated	
Time needed to fill the magma chamber with cumulates, years	
$F_{crit} = 40\%$ (SIC-QRN)	(2.22–2.90) x 10^3
$F_{crit} = 60\%$ (SIC-QRN)	(3.68–5.72) x 10^3

†For the results shown in Figures 7, 8, and 9. For more details concerning the convective-cumulative model parameters, see Frenkel et al., 1989, and Ariskin et al., 1993.

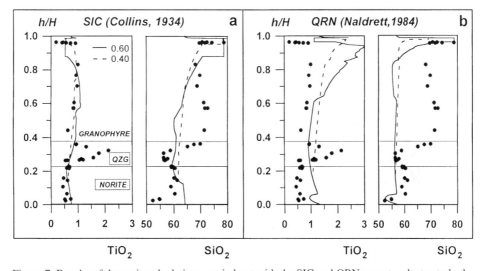

Figure 7. Results of dynamic calculations carried out with the SIC and QRN parent melts to study the effect of the convective-cumulative style of in situ magma differentiation on the composition of rocks, which could be formed in the Sudbury Main Mass as a result of crystal settling. The solid lines represent SiO_2, TiO_2, and other major element contents modeled at 1 atm and NNO oxygen buffer conditions using the INTRUSION subroutine of COMAGMAT (Frenkel et al., 1989; Ariskin et al., 1993). The assumed critical crystallinities are 40 and 60 vol%, respectively; for other main thermal and dynamic parameters, see Table 3. Natural data are from Ostermann and Deutsch (1998).

modeled parental melt can generate the sharp increase in TiO_2 occurring in the quartz-gabbro of the lower part of the Main Mass (Fig. 7; see also Table 2 and Fig. 2). This basic conclusion holds only in the case of closed system evolution, and under the assumption that the Main Mass indeed represents a complete section of the differentiated Sudbury melt body.

Formation of norites and the quartz-gabbro

The modeled results plotted in Figure 7b for TiO_2 and SiO_2 reproduce the titanium and silica distributions in the norites and quartz-gabbro of drill core 70011 in most details. The elemental distributions modeled for the lower zone appear to be "stretched" in the Main Mass along the vertical axis. The differentiation of a melt similar to the QRN parent could, therefore, be responsible for the formation of the lower part of the Main Mass including both norites and the quartz-gabbro. To test this hypothesis, we normalized the major element distributions, calculated for the whole Main Mass section at $F_{crit} = 60\%$, to the total relative thickness of both these "mafic" lithologies $H = 823$ m. The newly obtained modeled distributions are again compared with geochemical data for norites and gabbro of drill core 70011 in Figure 8.

Evolution of major element distributions. In the normalized height sections of Figure 8, the calculated concentrations of major element oxides match those in the natural rocks at the same height levels, except TiO_2, and Na_2O. The somewhat strange behavior of sodium could be related to the significant postimpact alteration, which is documented for the Main Mass by reopening of isotope systems (e.g., Deutsch, 1994), for example. But far more important are the calculated major element evolution trends in Figure 8, which are similar to those observed in the drill core. MgO reaches its maximum close to the contact with the Sublayer, whereas the highest FeO_{tot} and TiO_2 concentrations occur in the middle part of the quartz-gabbro. The decrease in CaO and Al_2O_3 is coupled with the increase in SiO_2 and K_2O toward the boundary with granophyres. Given the large number of geochemical analyses, and the relatively small number of variable parameters, this calculation reproduced the natural observations to a sufficient degree; this result is not a casual artifact of the model.

Calculated cumulative assemblages. An additional output of the COMAGMAT program allows the quantitative interpretation of the modeled major element distributions in terms of modal equilibrium phase occurrences. Figure 9 summarizes data for the calculated evolution of trapped melt temperature and volume, porosity F_m of the modeled cumulates, and the amount of "primarily accumulated" (Chalokwu et al., 1993) phases as result of efficient crystal settling.

Interpreting these data, the maximum MgO concentrations in the lower part of the modeled sequence are caused by accumulation of olivine crystals initially suspended in the parental QRN magma (Fig. 5; Table 4). Enrichment of FeO_{tot} and TiO_2 in the quartz-gabbro is correlated with the maximum of cumulative

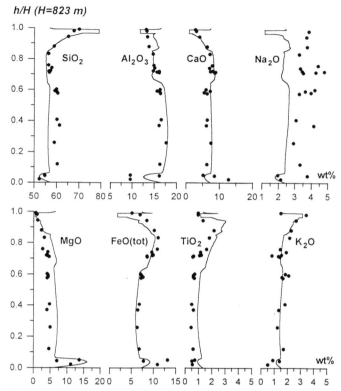

Figure 8. Diagrams show results for the QRN parent as given in Figure 7, but normalized to the total thickness of "mafic" rocks observed in the North Range (Ostermann and Deutsch, 1998). These plots demonstrate that the QRN parental magma is a plausible source for the observed upward sequence of rocks, from the mafic and quartz-rich norites to oxide gabbro.

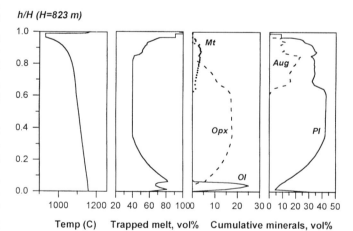

Figure 9. Height normalized distribution of the temperature of the trapped melt and "primarily accumulated" mineral phases (cumulates) in the rocks as result of crystal settling. Normalization for true thickness is for the North Range, drill core 70011, see text for further explanations. These data, coupled with the calculated liquid lines of descent, define all chemical features of modeled rocks shown in Figure 8 (Frenkel et al., 1989; Ariskin et al., 1993).

magnetite, a feature already described in detail (e.g., Naldrett and Hewins, 1984). The increase in CaO within the lowermost quartz-gabbro is due to the onset of high-Ca clinopyroxene crystallization. Cumulative plagioclase controls the behavior of Al_2O_3, whereas the increase in K_2O toward the boundary with the granophyre reflects the amount and composition of the trapped melt (Fig. 9). The behavior of SiO_2 is mainly controlled by primary phase proportions in the Opx + Pl and Aug + Mt + Pl cumulates. Enrichment in silica within the uppermost quartz-gabbro is caused by the abrupt decrease in the total amount of cumulative phases toward the boundary with the overlying granophyre (Fig. 9). The calculated compositions of the extreme differentiates duplicate the granophyre, which rest on the more "mafic" part of the Main Mass. Thus, the results shown in Figures 8 and 9 suggest the following:

1. A ferroandesitic melt compositionally similar to the QRN of Naldrett (1984) might be considered as the parent melt for the lower part of the Main Mass of the Sudbury melt system.

2. The convective-cumulative process (Frenkel et al., 1989) is a plausible physical mechanism to explain differentiation of this melt into the naturally observed sequence from mafic and quartz-rich norites to the quartz-gabbro with a small cap of silica-rich granophyres.

3. This lithologic sequence corresponds to the series of rocks consisting of accumulated crystals and trapped melt.

Undoubtedly, the quartz-gabbro belongs to the rock series, which differentiated from one single parent melt. The quartz-gabbro lithology, which certainly does not represent an "intermediate" magma, originated from mixing between impact-derived granophyres and endogenic norites (Chai and Eckstrand, 1993, 1994). The most important conclusion from points 1 to 3, however, is that differentiation of the ferroandesitic/QRN melt body cannot explain the huge mass of granophyres overlying the quartz-gabbro at present.

CONCLUSIONS

The results of modeling the Sudbury melt system with the COMAGMAT–3.5 program yield one important conclusion: If the Main Mass in its currently known (or estimated) volume evolved as a closed magmatic system, the observed volume ratio of 2:1 for granophyres to norites cannot be obtained by differentiation. This conclusion already has been drawn, based on petrologic-geologic (Masaitis, 1993; Warren et al., 1996) and geochemical data (Chai and Eckstrand, 1993, 1994). The latter authors, although based on a strongly biased data set, have argued for the existence of a batch of an "impact-triggered" endogenic magma overlain by a silica-rich impact melt, which finally resulted in formation of the Sudbury Main Mass. New detailed geochemical data for drill core 70011 (Ostermann, 1996), on the other hand, provide evidence for a continous evolution of trace element concentrations over the Main Mass (Fig. 10). The observed distribution patterns corresponds to those known for fractionally crystallized single parent melts

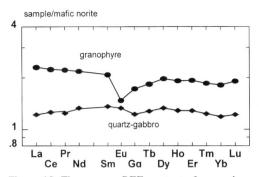

Figure 10. The average REE content of granophyre (black circles) and quartz-gabbro (black diamonds), normalized to the average content of mafic norite (data after Ostermann, 1996), indicate a close relation of the three major lithologies of the Main Mass.

(Ostermann and Deutsch, 1998). We purpose two different scenarios to solve this dilemma:

1. The ferroandesite QRN magma indeed represents the parent melt for the Main Mass, but the initial volume of the impact melt was much larger than the presently estimated volume of the Sudbury Igneous Complex. This postulate implies the existence of cumulative rocks enriched in mafic phases, such as olivine and orthopyroxene (Naldrett and Hewins, 1984), which are complementary to the observed volume of granophyres. Geophysical data do not totally exclude the presence of such cumulate layers at the bottom of the Sudbury melt system (B. Milkereit, personal communication).

2. The second, more speculative hypothesis focuses on the high initial temperature for the impact melt on the order of 2,000°C, and the high initial formation temperature of the overlying suevite layer, estimated at about 550°C (Ivanov et al., 1997; Ivanov and Deutsch, this volume). Due to these extreme conditions, melting of the already impact-melt rich breccias may have occurred. Such molten material could have been assimilated at a large scale by the underlying impact melt layer to form part of the granophyres. If such a process indeed took place, the smooth trends for all elements in the continuously analyzed drill core 70011 indicate a high efficiency in mixing between silica-rich differentiates of the impact melt body and the silica-rich melts, generated from the suevites during slow cooling of the underlying impact melt layer. In addition, we generally would not expect to see significant differences in trace element contents, as both the impact melt body and the breccias have, to a large degree, the same precursor lithologies.

Both scenarios deserve a careful study, and both may have contributed to this unique feature, which is known as the Sudbury melt system. Finally we observe that current understanding of the evolution of large-scale impact melts is still in its infancy. Moreover, kinetics and dynamics of assimilation and mixing at such high temperatures have not been studied in detail. We hope that thermophysical modeling (Ivanov et al., 1997; Ivanov and Deutsch, 1998) will provide more accurate constraints to understand the evolution of the Sudbury impact melt in greater detail.

ACKNOWLEDGMENTS

We appreciate stimulating discussions with R. A. F. Grieve (Geological Survey of Canada, Ottawa), B. A. Ivanov (Russian Academy of Science, Moscow), E. V. Koptev-Dvornikov (Moscow State University, Moscow), V. L. Masaitis (Karpinsky Institute, St. Petersburg), St. Tait (Institut de Physique du Globe, Paris), and O. I. Yakovlev (Vernadsky Institute, Moscow), as well as the careful reviews by B. Marsh and B. O. Dressler. This study was supported by DFG Grants 436 RUS 17/56/97 and DE 401/9-1 (to A.D.), and by the Russian Foundation of Basic Research Grant N96-05-64231 (to A.A.A.). A.A.A acknowledges travel support from the Organizing Committee of the Sudbury '97 meeting.

REFERENCES CITED

Ariskin, A. A., 1997, COMAGMAT–3.5: a renewed program for the calculations of magma differentiation processes: Moscow, Russia, Vernadsky Institute.

Ariskin, A. A., Barmina G. S., and Frenkel, M. Ya., 1987, Computer simulation of basalt magma crystallization at a fixed oxygen fugacity: Geochemistry International, v. 24, p. 92–100.

Ariskin, A. A., Barmina G. S., Frenkel, M. Ya., and Yaroshevsky, A. A., 1988, Simulating low-pressure tholeiite-magma fractional crystallization: Geochemistry International, v. 25, p. 21–37.

Ariskin, A. A., Frenkel, M. Ya., and Tsekhonya T. I., 1990, High-pressure fractional crystallization of tholeiitic magmas: Geochemistry International, v. 27, p. 10–20.

Ariskin, A. A., Frenkel, M. Ya., Barmina, G. S., and Nielsen, R. L., 1993, COMAGMAT: a FORTRAN program to model magma differentiation processes: Computers and Geosciences, v. 19, p. 1155–1170.

Ariskin A. A., Barmina, G. S., Ozerov, A. Yu., and Nielsen, R. L., 1995, Genesis of high-alumina basalts from Klyuchevskoi Volcano: Petrology, v. 3, p. 449–472.

Ariskin, A. A., Barmina, G. S., Meshalkin, S. S., Nikolaev, G. S., and Almeev, R. R., 1996, INFOREX–3.0: a database on experimental phase equilibria in igneous rocks and synthetic systems. II. Data description and petrological applications: Computers and Geosciences, v. 22, p. 1073–1082.

Barmina, G.S., Ariskin, A.A., Koptev-Dvornikov, E.V., and Frenkel, M.Ya., 1989a, Estimates of the primary compositions of cumulate-minerals in differentiated traps: Geochemistry International, v. 26, p. 32–42.

Barmina, G. S., Ariskin, A. A., and Frenkel, M. Ya., 1989b, Petrochemical types and crystallization conditions of the Kronotsky Peninsula plagiodolerite (eastern Kamchatka): Geochemistry International, v. 26, p. 24–37.

Barmina, G. S., Ariskin, A. A., and Kolesiov, G. M., 1992, Simulating the REE spectra of hypabyssal rocks in the Kronotsky series, eastern Kamchatka: Geochemistry International, v. 29, p. 45–54.

Bolikhovskaya, S. V., Vasil'yeva, M. O., and Koptev-Dvornikov, E. V., 1996, Simulating low-Ca pyroxene crystallization in basite systems: new geothermometer versions: Geochemistry International, v. 33, p. 1–19.

Brockmeyer P., 1990, Petrographische und geochemische Untersuchungen an polymikten Breccien der Onaping-Formation, Sudbury-Distrikt (Ontario, Kanada) [Ph.D. thesis]: Münster, Germany, Universität Münster, 228 p.

Chai, G., and Eckstrand, O. R., 1993, Origin of the Sudbury Igneous Complex, Ontario—differentiate of two separate magmas: Geological Survey Canadian Papers, v. 93–1E, p. 219–230.

Chai, G., and Eckstrand, O. R., 1994, Rare-earth element characteristics and origin of the Sudbury Igneous Complex, Ontario, Canada: Chemical Geology, v. 113, p. 221–244.

Chalokwu, C. I., Grant N. K., Ariskin, A. A., and Barmina G. S., 1993, Simulation of primary phase relations and mineral compositions in the Partridge River intrusion, Duluth Complex, Minnesota: implications for the parent magma composition: Contributions to Mineralogy and Petrology, v. 114, p. 539–549.

Chalokwu, C. I., Ariskin, A. A., and Koptev-Dvornikov, E. V., 1996, Magma dynamics at the base of an evolving mafic magma chamber: incompatible element evidence from the Partridge River intrusion, Duluth Complex, Minesota, USA: Geochimica et Cosmochimica Acta, v. 60, p. 4997–5011.

Collins, W. H., 1934, The life-history of the Sudbury Nickel Irruptive. 1, Petrogenesis: Transactions of the Royal Society of Canada, v. 28, p. 123–177.

Deutsch, A., 1994, Isotope systematics support the impact origin of the Sudbury Structure (Ontario, Canada), in Dressler, B. O., Grieve, R. A. F., and Sharpton, V. L., eds., Large meteorite impacts and planetary evolution: Geological Society of America Special Paper, v. 293, p 289–302.

Deutsch, A., and Grieve, R. A. F., 1994, The Sudbury Structure: constraints on its genesis from Lithoprobe results: Geophysical Research Letters, v. 21, p. 963–966.

Deutsch, A., Grieve, R. A. F., Avermann, M., Bischoff, L., Brockmeyer, P., Buhl, D., Lakomy, R., Müller-Mohr, V., Ostermann, M., Stöffler, D., 1995, The Sudbury Structure (Ontario, Canada): a tectonically deformed multi-ring basin: Geologische Rundschau, v. 84, p. 697–709.

Ding, T. P., and Schwarcz, H. P., 1984, Oxygen isotopic and chemical compositions of rocks of the Sudbury Basin, Ontario: Canadian Journal of Earth Science, v. 21, p. 305–318.

Dressler, B. O., 1984, General geology of the Sudbury area, in Pye, E. G., Naldrett, A. J., and Giblin, P. E., eds., The geology and ore deposits of the Sudbury Structure: Ontario Geological Survey Special Volume 1, p. 57–82.

Dressler B. O., Morrison G. G., Peredery W. V., and Rao, B. V., 1987, The Sudbury Structure Ontario, Canada—A review, in Pohl, J., ed., Research in terrestrial impact structures: Braunscheig/Wiesbaden, Germany, Friedrich Vieweg & Sons, p. 39–68.

Dressler, B. O., Weiser, T., and Brockmeyer, P., 1996, Recrystallized impact glass of the Onaping Formation and the Sudbury Igneous Complex, Sudbury Structure, Ontario, Canada: Geochimica et Cosmochimica Acta, v. 60, p. 2019–2036.

Faggart, B. E., Basu, A. R., and Tatsumoto, M., 1985, Origin of the Sudbury Complex by meteoritic impact neodymium isotopic evidence: Science, v. 230, p. 436–439.

Frenkel, M. Ya., and Ariskin, A. A., 1984, A computer algorithm for equilibration in a crystallizing basalt magma: Geochemistry International, v. 21, p. 63–73.

Frenkel, M. Ya., Ariskin, A. A., Barmina, G. S., Korina, M. I., and Koptev-Dvornikov, E. V., 1988, Geochemical thermometry of magmatic rocks—principles and example: Geochemistry International, v. 25, p. 35–50.

Frenkel, M. Ya., Yaroshevsky, A. A., Ariskin, A. A., Barmina, G. S., Koptev-Dvornikov, E. V., and Kireev, B. S., 1989, Convective-cumulative model simulating the formation process of stratified intrusions, in Bonin, B., ed., Magma-crust interactions and evolution: Athens, Greece, Theophrastus Publications, S.A., p. 3–88.

Ghiorso, M. S., 1985, Chemical mass transfer in magmatic processes. I, Thermodynamic relations and numeric algorithms: Contributions to Mineralogy and Petrology, v. 90, p. 107–120.

Ghiorso, M. S., and Carmichael, I. S. E, 1985, Chemical mass transfer in magmatic processes. II, Applications in equilibrium crystallization, fractionation and assimilation: Contributions to Mineralogy and Petrology, v. 90, p. 121–141.

Ghiorso, M. S., and Sack, R. O., 1995, Chemical mass transfer in magmatic processes. IV, A revised and internally consistent thermodynamic model for the interpolation and extrapolation of liquid-solid equilibria in magmatic systems at elevated temperatures and pressures: Contributions to Mineralogy and Petrology, v. 119, p. 197–212.

Grieve, R. A. F., Stöffler, D., and Deutsch, A., 1991, The Sudbury Structure—controversial or misunderstood: Journal of Geophysical Research, v. 96, p. 22753–22764.

Ivanov, B. A., Deutsch, A., and Ostermann, M., 1997, Solidification of the Sudbury impact melt body and nature of offset dikes, *in* Proceedings, 28th, Lunar Planetary Science Conference, Abstracts: Houston, Texas, Lunar and Planetary Institute, p. 633–634

Krogh, T. E., Davis, D. W., and Corfu, F., 1984, Precise U-Pb zircon and baddeleyite ages for the Sudbury area, *in* Pye, E. G., Naldrett, A. J., and Giblin, P. E., eds., The geology and ore deposits of the Sudbury structure: Ontario Geological Survey Special Volume 1.

Kuo, H. Y., and Crocket, J. H., 1979, Rare earth elements in the Sudbury Nickel Irruptive —comparison with layered gabbros and implications for Nickel Irruptive petrogenesis: Economic Geology, v. 79, p. 590–605.

Lakomy, R., 1990, Implications for cratering mechanics from a study of the Footwall Breccia of the Sudbury impact structure, Canada: Meteoritics, v. 25, p. 195–207.

Masaitis, V. L., 1993, Origin of the Sudbury Structure from the points of new petrographic, mineralogical and geochemical data: Transactions of All-Russian Mineralogical Society, v. 122, p. 1–17 (in Russian).

Milkereit, B., Green, A., Berrer, E., Boerner, D., Broome, J., Cosec, M., Cowan, J., Davidson, A., Dressler, B., Fueten, F., Grieve, R. A. F., James, R., Kraus, B., McGrath, P., Meyer, W., Moon, W., Morris, W., Morrison, G., Naldrett, A. J., Peredery, W., Rousell, D., Salisbury, M., Schwerdtner, W., Snajdr, P., Thomas, M., and Watts, A., 1992, Geometry of the Sudbury Structure from high-resolution seismic reflection profiling: Geology, v. 20, p. 807–811.

Naldrett, A. J., 1984, Summary, discussion and synthesis, *in* Pye, E. G., Naldrett, A. J., and Giblin, P. E., eds., The geology and ore deposits of the Sudbury structure: Ontario Geological Survey Special Volume 1, p. 533–569.

Naldrett, A. J., and Hewins, R. H., 1984, The main mass of the Sudbury igneous complex, *in* Pye, E. G., Naldrett, A. J., and Giblin, P. E., eds., The geology and ore deposits of the Sudbury structure: Ontario Geological Survey Special Volume 1, p. 235–251.

Naldrett, A. J., Rao, B. V., and Evensen, N. M., 1986, Contamination at Sudbury and its role in ore formation, *in* Gallagher, M. J., Ixer, R. A., Neary, C. R., and Pritchard, H. M., eds., Metallogeny of basic and ultrabasic rocks: London, Institute of Mining and Metallurgy Special Publication, p. 75–91.

Nielsen, R. L., 1990, Simulation of igneous differentiation processes, *in* Nicholls, J., and Russell, J. K., eds., Modern methods of igneous petrology: understanding magmatic processes: Reviews in Mineralogy, v. 24, p. 63–105.

Nielsen, R. L., and Dungan, M. A., 1983, Low-pressure mineral-melt equilibria in natural anhydrous mafic systems: Contributions to Mineralogy and Petrology, v. 84, p. 310–326.

Ostermann, M., 1996, Die Geochemie der Impaktschmelzdecke (Sudbury Igneous Complex) im Multiring-Becken Sudbury [Ph.D. thesis]: Münster, Germany, Universität Münster, 168 p.

Stormer, J. C., Jr., 1983, The effects of recalculation on estimates of temperature and oxygen fugacity from analyses of multicomponent iron-titanium oxides: American Mineralogist, v. 68, p. 586–594.

Toplis, M. J., and Carroll, M. R., 1995, An experimental study of the influence of oxygen fugacity on Fe-Ti oxide stability, phase relations, and mineral-melt equilibria in ferro-basaltic systems: Journal of Petrology, v. 36, p. 1137–1170.

Wager, L. R., and Brown, G. M., 1967, Layered igneous rocks: San Francisco, W. H. Freeman, 588 p.

Warren, P. H., Claeys, P., and Cedillo-Pardo, E., 1996, Mega-impact melt petrology (Chicxulub, Sudbury, and the Moon): effects of scale and other factors on potential for fractional crystallization and development of cumulates, *in* Ryder, G., Fastovsky, D., and Gartner, S., eds., The Cretaceous-Tertiary event and other catastrophes in Earth history: Geological Society of America Special Paper 307, p. 105–124.

Weaver, J. S., and Langmuir, C. H., 1990, Calculation of phase equilibrium in mineral-melt systems: Computers and Geosciences, v. 16, p. 1–19.

MANUSCRIPT ACCEPTED BY THE SOCIETY DECEMBER 16, 1999

Sudbury impact event: Cratering mechanics and thermal history

B. A. Ivanov
Institute for Dynamics of Geospheres, Russian Academy of Sciences, Moscow 117939, Russia;
e-mail: baivanov@glasnet.ru

A. Deutsch
Institut für Planetologie, Universität Münster, Wilhelm-Klemm-Str. 10, D-48149 Münster, Germany;
e-mail: deutsca@uni-muenster.de

ABSTRACT

The Sudbury Igneous Complex (SIC) is interpreted as the solidified impact melt body of the 1.850-g.y.-old Sudbury impact structure. First results of cratering and thermal modeling for this ~250-km sized multi-ring structure are presented. The numerical calculations were done for the vertical impact of a stony (granite) body (cylindrical projectile, 12.5 km in diameter and height) impacting at a granite target with a velocity of 20 km s^{-1}. These simulations yield estimates of the transient cavity dimensions and the temperature field below the impact structure just after the modification stage. One-dimensional heat transfer modeling sets constraints for the thermal history of the impact melt. Cooling of the melt sheet, the present SIC, from the initial temperature of 2,000°K to the liquidus at 1,450°K lasted several 100 k.y., and below the solidus at 1,270°K, about 300 k.y. to 2 m.y., depending on the initial melt thickness H(SIC). The cooling sequence was modeled for H(SIC) of 2.5, 4, and 6 km. Given this long duration of cooling, postimpact tectonic processes during the Penokean orogeny may well have deformed the melt sheet prior to its final solidification. Prolonged cooling as well as this large-scale deformation may explain the present structural position and the composition of the Offset Dikes, consisting of differentiated impact melt.

INTRODUCTION

It is now largely acknowledged that the 1.850-g.y.-old (Krogh et al., 1984) Sudbury Structure, in Ontario, represents the tectonized erosional remnant of a multi-ring impact basin about 250 km in diameter. For an introduction to the geologic setting, see Dressler (1984), and papers in Pye et al. (1984); for the impact mechanical interpretation of the lithologic units of the Sudbury Structure, we refer to Deutsch et al. (1995). Within the frame of this impact model, the Sudbury Igneous Complex (SIC), together with the clast-rich sequences on top (Basal Member of the Onaping Formation) and bottom (Sublayer) are interpreted as solidified impact melt body (Brockmeyer, 1990; Grieve et al., 1991; Deutsch et al., 1995). Postimpact tectonism resulted in deformation of the impact structure, and hence this melt body, causing overthrusting in the South Range (e.g., Shanks and Schwerdtner, 1991; Milkereit et al., 1994) and finally yielding an elliptically shaped bowl, the SIC. According to Lithoprobe investigations, the maximum depth of this bowl is 6 km below the present surface (e.g., Milkereit et al., 1994; Deutsch and Grieve, 1994).

Reconstruction of the postimpact deformation of the Sudbury Structure (Roest and Pilkington, 1994) allows restoration of the initial geometry of the SIC as a melt sheet with a thickness of at least 2.5 km, with a minimum diameter of 60 km, covering the inner depression of the Sudbury Crater. About 3

Ivanov, B. A., and Deutsch, A., 1999, Sudbury impact event: Cratering mechanics and thermal history, *in* Dressler, B.O., and Sharpton, V. L., eds., Large Meteorite Impacts and Planetary Evolution II: Boulder, Colorado, Geological Society of America Special Paper 339.

km of impact, breccias and postcrater deposits form the overburden on this melt sheet (Grieve et al., 1991; Deutsch et al., 1995). With an estimated volume of $1-2.5 \times 10^4$ km^3, the SIC is by far the largest known terrestrial impact melt sheet. The essential difference of the melt pool at Sudbury compared to impact melt layers of smaller craters is that, due to its large size, solidification took a much longer time. This time was sufficient to allow chemical differentiation of the initially rather homogeneous melt (Ariskin et al., this volume; Ostermann, 1996) into the three main lithologies of the SIC: a thick upper layer of granophyres, underlain by quartz-gabbro and quartz-rich norites (Naldrett and Hewins, 1984).

In this context, we emphasize that geochemical and isotopic characteristics of the material solidified in the Offset Dikes (Grant and Bite, 1984) around the main SIC body match that of the norite, i.e., it is a differentiated impact melt (e.g., Ostermann, 1996; Wood and Spray, 1998). This observation implies that the Offset Dikes did not originate simultaneously with the cratering event, but only after the onset of differentiation during late-stage adjustments of the crater basement. We can, however, imagine an alternate origin of the Offset Dikes. They may represent fractures in the crater floor, filled from above with impact melt. This possibility would imply a melt pool initially much larger than the SIC at its present erosional level. In favor of the latter hypothesis is the presence of concentric Offset Dikes, which strike parallel to the outer margin of the SIC, and which lack at present physical connection to the Main Mass of the SIC. To choose between the two alternatives requires additional geochemical work on samples from Offset Dikes in combination with proper modeling of the original crater morphology and the cooling history of the melt pool (see discussion by Ostermann, 1996). Both goals are difficult to achieve.

The intent of this chapter is to estimate the cooling history of the SIC based on numerical simulation of the Sudbury cratering event. The mechanics of large-scale impacts include several topics that are still under investigation. The style of the transient cavity collapse defines displacement of the crustal material at the target site, which in turn confines the temperature field below the crater directly after modification of the transient cavity. A realistic modeling of the cooling history should take into account the effects of local convection and/or differentiation of the impact melt. Before we can address these topics, however, we need to construct a preliminary time sequence of events. For that, some approximations are used to evaluate the following effects: (1) shock wave propagation and evolution of the transient cavity, (2) the temperature field below the structure due to the combined action of shock heating and material displacement, and (3) the rate of cooling in the central part of the structure.

CRATERING MECHANICS

The numerical simulation of the cratering event was conducted to assess the temperature field at the Sudbury impact site just after modification of the transient crater. We used the SALE 2D hydrocode (Amsden et al., 1980) with some modifications described by Ivanov (1994) and Ivanov et al. (1996a, 1997). The computational grid consists of 120×120 cells in vertical and horizontal directions. The space step is 1.25 km for the first 80×100 cell region around the point of impact, and increases progressively with a factor of 1.15 for the next row or column. The grid boundaries are approximately 300 km away from the point of impact. The Eulerian mode of the SALE code with the stress remapping was used to present the target deformation from the very beginning of the process to the end of a transient crater collapse. From 5,000 to 10,000 marked Lagrangian particles were placed in the computational area to trace displacement of the material. A tabulate equation of state (EOS) for a granite containing 5 wt% of water (Zamyshliaev and Evterev, 1990) was used to describe the thermodynamics of both the target and projectile.

Numerical modeling was done for a stony (granitic) body impacting the granite target vertically at a velocity of 20 kms^{-1}. After several code runs, a cylindrical projectile, 12.5 km in diameter and in height (or an equivalent spherical projectile, 14 km in diameter), turned out to fit the estimated volume of impact melt of 8,000 km^3 at the Sudbury Structure (Grieve and Cintala, 1992). Recent numerical calculations by Pierazzo et al. (1997) show that estimates of the melt production may depend on the resolution of the modeling. The resolution used in this study gives a reasonable compromise of the ability to calculate both the early stage of cratering, i.e., the shock compression, and the late stage of the transient crater collapse.

The description of the mechanical behavior of the target rock material during large-scale cratering events needs a particular discussion. Previous investigations of impact mechanics indicate that the late stage of crater formation—the collapse of the transient cavity in a gravity field—may be modeled with traditional rock mechanics if one ascribes very specific mechanical properties to the rocks in the vicinity of the transient crater. The effective strength of the rocks lies around 30 bar (Melosh, 1977); the effective angle of internal friction is less than 5° (McKinnon, 1978). Rock media with such properties are termed as "temporarily fluidized." The nature of this "fluidization," however, is still poorly understood (for a review, see Melosh, 1989). Melosh (1979, 1989) suggested an acoustic (vibration) nature of this fluidization. This model obviously represents the best approach to the problem. An open question, however, is how to implement Melosh's model (or other possible models) in the hydrocode for numerical simulation of the dynamic collapse of the transient crater.

The peculiar feature of rock deformation during this process is that the rock medium deforms not like a plastic metal-like continuum, but as system of discrete rock blocks. Deep drilling of impact craters has revealed that the subcrater basement and the central mount consist of a system of rock blocks ranging in size from 50 to 200 m (Ivanov et al., 1996b). The model describing oscillations of these blocks allows the formulation of the appropriate rheologic law for the subcrater flow during the modification stage. This approach yields a realistic model of the mechanical behavior of rocks during cratering (Ivanov and Kostuchenko, 1997). In large-scale cratering, however, block vibrations should be suppressed at a depth of several tens of kilometers due to the high lithostatic pressure.

To estimate material displacement during cratering, two simple mechanical models were applied: Model 1, the standard rock description (internal friction, Hugoniot Elastic Limit, and thermal softening close to the melting point); and the modified test Model 2 with artificially decreased internal friction to simulate the possible acoustic fluidization.

Figure 1 illustrates the general succession of events after the impact, i.e., transient cavity growth and collapse. One can see uplifting and overturning of isotherms due to both structural uplift and additional shock heating. The residual mantle uplift turned out as the parameter most sensitive to the model used. In general, however, models with all assumed friction coefficients result in an essentially similar modification of the geothermal field.

Figure 2 shows the shock pressure isobars, plotted at the initial projectile/target configuration. The EOS used here does not explicitly describe melting; for a recent extensive discussion of impact melting, we refer to Pierazzo et al. (1997). However, the 50-GPa isobar in Figure 2 gives a realistic approximation of the initial distribution of shock melted material.

To simulate the possible acoustic fluidization in the target below the shock melted material, several calculations were performed in which the friction coefficient varied from "normal" values of 1 to 0.5 to 0 (a completely fluid target). Figure 3 compares three variants with various assumed friction coefficients, ranging from 0.5 to 0.0625. In all cases, the zone finally heated is relatively small compared to the final diameter of the structure, estimated to be on the order of 200–300 km. Moreover, thermal softening (the plastic limit decrease with temperature to zero at the melting point) makes the target in all cases weak enough to produce a fluid-like splash of the central mount before the final collapse (cf. Fig. 1, time frames T = 200; T = 250). The collapse of the transient central uplift produces the folding of isotherms at the central part of the structure (Fig. 3, right-hand side). The low spatial resolution of the computer model (cell dimension, 1.25 km) prevents any conclusions about the exact morphology of the final crater. Nevertheless, the well-resolved transient crater diameter is in the range of 90–100 km, in keeping with estimates by Grieve et al. (1991). A residual mantle uplift of the order of 10 km is observed only for smallest friction coefficients used (Fig. 3).

Despite the fact that applied mechanical models are still immature, our numerical simulations provide a qualitative estimate of the geothermal field under the "just formed" modified Sudbury structure. This situation is illustrated in Figure 4 (left diagram) showing the distribution of the geotherms (temperature vs. depth) at T = 400 s, and radial distances R from the point of impact, with R being 5, 10, 20, and 30 km. At R of 30 km, the temperature profile matches the initial one, except for the uppermost 10 km, due to a relatively thin layer of splashed out impact melt. At an R of 5 km, however, we see a thermal disturbance down to about 50 km below the pre-impact surface. However, the resolution of the numerical simulation presented here was not fine enough to reproduce the geometry of the impact melt sheet. Such task is a challenge for future modeling. Moreover, due to the water content of 5 wt% in the model granite, not a simple granite melt is produced in the cratering event but instead a mixture of shock-melted granite and water vapor. In reality, such a mixture would be separated, a process hard to address in simple numerical calculations. Nevertheless, the main goal of the numerical modeling was to estimate the residual heating of the target, providing the input parameters for the thermal modeling. We estimate the distribution of temperature in the final crater at T = 400 s with a resolution of approximately four cells, or 5 km. This number rests on the size of computational cells, which is 1.25 km in our modeling of the cratering event.

THERMAL MODELING

Simple estimates (1D [one dimensional] implicit numerical code, see Appendix 1) were done to evaluate the cooling history of the SIC body. The geometric constraints of the model are three flat layers: (1) overburden material with a thickness of 2.5 km, resting on (2) a 2.5-km-thick melted layer, which in turn is underlain by (3) rocks of the crust, uplifted due to the transient cavity collapse (Fig. 4, right diagram). The surface boundary conditions of layer (1) are held constant at a temperature of 300°K (see comments in Appendix 1); temperature within layer (1) ranges from 300°K ("cold breccia") to 850°K ("hot suevite"). In nature, this layer is represented by the Onaping breccia Formation. Melt layer (2), in geologic terms, the Sudbury "Igneous" Complex, has a conservatively estimated initial temperature of 1,800°–2,000°K. For layer (3), which corresponds to lithologic units such as the Footwall Breccia and the megabreccias below, we used the temperature profile derived from numerical simulations presented above (Fig. 4). Thermal constants used in our calculations were those that have been applied by Onorato et al. (1978) in the thermal modeling of the Manicouagan Crater.

Figures 5 and 6 illustrate the time-related decay of the maximum temperature in the melt layer (lines labeled 1), the evolution of the temperature at the lower boundary of the melt (lines labeled 2), and at 1.2 km below the bottom of the melt pool (lines labeled 3). The standard case of the impact melt pool (SIC) thickness H = 2.5 km is compared in Figure 5 with H = 4 km, and in Figure 6, with H = 6 km. An original thickness of 6 km for the melt pool is discussed by Ariskin et al. (this volume) to explain the differention of the SIC. The illustrated comparisons allow to estimate variations in the thermal history if the SIC pool was initially deeper than it appears to be at present, after extensive tectonic deformation and erosion.

The calculations yielded the following results for the impact melt layer of the Sudbury Structure: the initial temperature decreases below the liquidus point (assumed at 1,450°K; see Ariskin et al., this volume) in about 250 k.y., and below the solidus point (assumed at 1,270°K), in about 500 k.y. In contrast, this time span is only 1 k.y. for the 200-m-thick melt sheet of the Manicouagan structure (Onorato et al., 1978).

This result simply reflects the [length]^2/[time] scaling (e.g., Zel'dovich and Raizer, 1967). A factor of 2 is assigned as a minimum uncertainty to the solidification time of the SIC due to uncertainties in thermal properties and boundary conditions. Two- or three- dimensional (2D, 3D) thermal modeling and convective heat transfer inside the melted body also can modify the numbers; however, the order of magnitude from our simple estimate should remain unchanged. For the most extreme case, in which the initial SIC thickness is estimated to be 6 km (Fig. 6), the time to solidify the melt increases three times, amounting between 1 and 2 Ma. The enormous thickness of the melt layer (SIC), compared to other terrestrial impact structures, and its prolonged cooling, explain why the initially homogeneous impact melt underwent differentiation at Sudbury (Ariskin et al., this volume). In addition, the calculated cooling sequence set constraints for the formation of the Offset Dikes. Their emplacement should have taken place no later than 0.25–0.5 m.y. after the impact event; these numbers rise for an increase in the initial thickness of the SIC. We emphasize that these estimates are much tighter than recent high-precision dating results. U-Pb crystallization ages of zircon and baddeleyite, separated from quartz-dioritic lithologies of the Foy Offset Dike (Ostermann et al., 1996), and in the main body of the SIC (Krogh et al., 1984), are identical within the given 2σ error limits of +4/–3 Ma. Presently available geochronologic methods are insufficient for a better resolution of the succession of events, whereas thermal modeling yields a more detailed time frame.

Other useful outcomes of the modeling are temperature estimates at the contact of the melt pool with crater floor lithologies, and for rocks beneath this contact. Mineralogic estimates, based on geothermometry, indicate a maximum temperature of as much as 1300°K near the contact (Lakomy, 1990). Thermal modeling (Figs. 5 and 6) gives a temperature of about 1200°K for the lower SIC/footwall contact. The temperature 1.2 km below this contact gradually grows and reaches a maximum of 950°K only 400 k.y. after the impact. This temperature compares well with geothermometric measurements, yielding approximately 850°K in the footwall rocks 1.2 km below the contact to melt lithologies (Lakomy, 1990). Future comparison of the model and its fitting to observational data will allow us to set strict constraints for the entire scenario of the formation of the Sudbury Crater. In this context, evaluating the maximum temperatures in country rocks from the Offsets represent an important future goal of petrologic investigations. Such temperatures will help to confirm or reject the aforementioned possibility of the formation of the Offset Dikes at the bottom of a much larger impact melt pool.

CONCLUSIONS AND OUTLOOK

Numerical modeling of the Sudbury impact event demonstrates that, for an assumed impact melt volume of 8,000 km^3, the melt zone does not reach to the lower crust (below approximately 30 km) and the upper mantle. This result is in accordance with geochemical and isotopic data for the SIC, indicating that material from the 1.85 g.y. subcontinental mantle did not contribute to the impact melt (e.g., Faggart et al., 1985; Deutsch, 1994; Walker et al., 1991; Dickin et al., 1992, 1996; Ostermann, 1996). The maximum depth of the transient cavity for a cratering event such as Sudbury amounts to about 40–50 km (see, Lakomy, 1990; Deutsch and Grieve, 1994). This implies that the impact melt zone will be highly distorted during transient cavity growth. The possible difference in motion along the transient cavity walls enhances mixing of the melt with highly shocked but not melted material. Despite this huge maximum depth of the transient cavity, the mantle (assumed to be below 45 km) is not highly deformed. The residual mantle uplift depends strongly on the assumed rheology of the target rocks. The mantle uplift may be only of the order of several kilometers if the acoustic fluidization is suppressed by lithostatic pressure close to the mantle depth.

According to our calculations, the region of enhanced temperatures under the structure is concentrated within a restricted zone near the center of impact. Shock heating and structural uplift give a "neck" of hot rocks with a radius of 50 km. Beyond this radial distance, hot ejected material lies on relatively cold target rocks, yet the pre-impact temperature field of the target lithologies remains nearly unchanged.

The observationally derived model of the vertical cross

Figure 1. General sequence of impact crater formation, simulated with the modified SALE hydrocode for the Sudbury event. The cylindrical projectile with height and diameter both equal to 12.5 km impacts the granite surface at the velocity 20 km s^{-1}. Selected time frames correspond to 0, 20, 60, 100, 140, 200, 300, 350, and 400 s after impact (left to right, top to bottom). Vertical (depth) and horizontal (radial distance) scales are in kilometers. The axis of symmetry (zero radial distance) divides each frame in two halves. The left side presents distortion and displacement of originally horizontal marked layers in the target. Thick lines correspond to layers below 45 km, an estimated Moho depth. In the right half, displacement and distortion of isotherms are shown. The variant used to construct this picture was computed for artificially decreased internal friction (Model 2, see text for further explanation). For the given variant, isotherms from 300°K to 1,200°K are shown. To outline the boundary of a condensed matter, the light gray tone corresponds to a density above 2,000 kg m^{-3}; the dark gray tone marks cells with a density less than 2,000 kg m^{-3} but above 200 kg m^{-3}. Cells where the density is less than 200 kg m^{-3} are white (blank). Only the central part of the computational zone is shown. The computational boundaries are approximately at 300 km from the point of impact. The assumed friction coefficient is equal to 0.125.

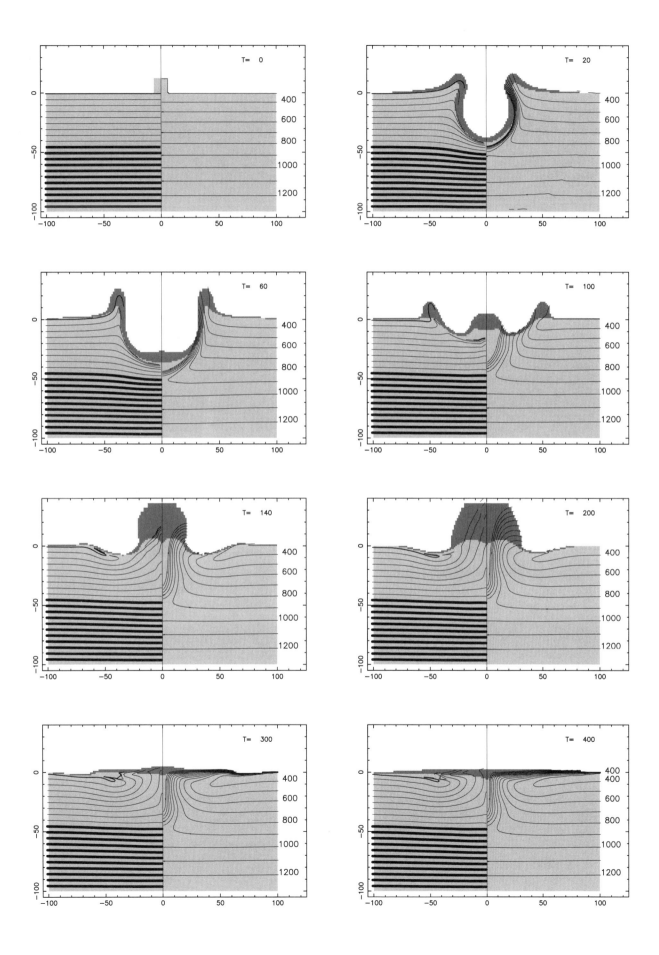

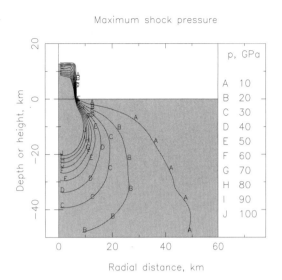

Figure 2. Maximum shock pressure isobars overlapping the initial geometry of the target and the projectile. Isobars from 10 to 100 GPa are marked with letters A–J (see the legend to the right). The 50-GPa isobar (E) is close to the threshold for impact melting of the granitic target. The maximal melt depth is close to 30 km, which is less than the mantle/crust boundary estimated at 45 km (see text). Zero radial distance is the axis of symmetry.

section of suevite, melt, and basement rocks (Brockmeyer, 1990; Grieve et al., 1991; Deutsch et al., 1995) was used for the thermal modeling. The cooling of this typical vertical section, modeled in the one-dimensional approximation, gives a solidification time for the SIC on the order of several hundred kiloyears which depends mostly on the initial thickness of the melt layer. Due to this prolonged cooling, the observed tectonic deformation of the impact structure (e.g., Shanks and Schwerdtner, 1991) can occur well before the final solidification of the impact melt. Such a deformation, which acted during the cooling, could explain several well-known features of the SIC. Among those are the so-called intrusional contacts, as well as textural and magnetic features, which show significant variations not only in different lithologies but also between North and South Range (e.g., Morris, 1984; Naldrett and Hewins, 1984). We are convinced that improvements of the thermal model, for example using 3D (three-dimensional) geometry, and accounting for convection (e.g., Warren et al., 1996), would not shift the estimated cooling times outside the range defined by the values given for an initial melt sheet thickness of 2.5–6 km.

Heat is removed from the melt mostly through the free surface. The temperature of rocks at depths of a few kilometers below the melt may rise for the first several hundred kiloyears but never reached temperatures above 1,100°K–1,200°K. Several questions arise from this outlined thermal scenario:

1. How is the overburden material, i.e., the Onaping Formation, stabilized on top of the >2.5-km liquid impact melt layer, with a diameter in excess of 60 km?

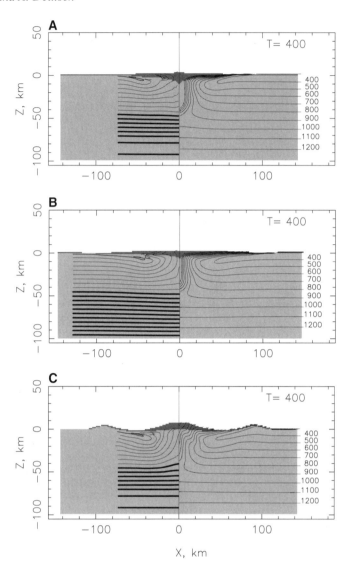

Figure 3. The position of the marked, initially horizontal layers (left side) and isotherms (right side) at T = 400 s after the impact for model friction coefficients k_p, equal to (a) 0.5, (b) 0.125, and (c) 0.0625. Color code (dark/light gray) and line thickness correspond to those of Figure 1. Isotherms are labeled in degrees Kelvin.

2. If the Offset Dikes were indeed formed only during SIC differentiation (i.e., delayed by several thousands of kiloyears), what was the mechanical reason for the opening of existing breccia-filled fractures and the injection of the dike-forming melt?

3. Are there field indications and/or petrologic data to prove that Offset Dikes are really dikes, not melt-filled fractures, in the basement of a larger, now eroded part of the melt sheet?

Some of these topics are discussed by Ostermann (1996); others should be addressed in future investigations.

ACKNOWLEDGMENTS

Part of this work was accomplished during a scientific visit of Ivanov to Münster. We thank R. A. F. Grieve, M. Pilk-

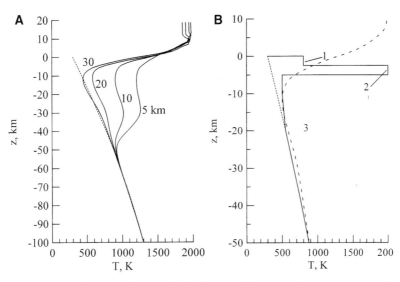

Figure 4. a, Thermal profiles along vertical columns at distances 5, 10, 20, and 30 km from the impact center, 400 s after the impact (end of the numerical simulation of the cratering event) for the variant shown in Figure 3B. The short dashed line corresponds to the initial thermal gradient. The high-temperature mixture of melted granite and water vapor extends above the surface. The deposition of the granite melt droplets will later create the melt pool, overburden with suevites (not simulated in this chapter). b, The initial geotherm (short dashed line), the computed geotherm at the distance of 25 km from the center (long dashed line), and the geotherm assumed for the 1D (one dimensional) thermal modeling. According to geologic observations, the model geotherm included layers of suevite (1) and impact melt (2) at the top of the granite basement (3).

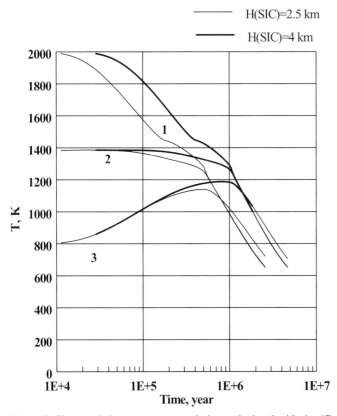

Figure 5. Characteristic temperature evolution, calculated with the 1D (one dimensional) heat conductivity model for the initial thermal profile shown in Figure 4B; assumed initial melt pool thickness H(SIC) is 2.5, and 4 km. 1 - maximum temperature inside the melt layer (2 in Fig. 4B); 2 - temperature at the contact of the melt pool and the basement (2 and 3 in Fig. 4B); 3 - temperature in the basement (3 in Fig. 4B) at the depth of 1.2 km below the melt/basement contact.

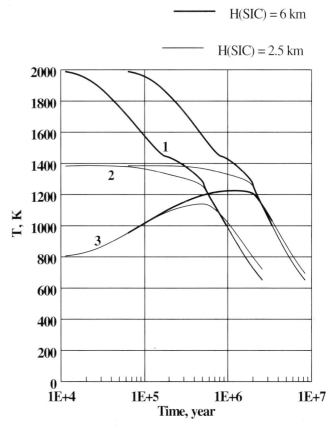

Figure 6. The same as in Figure 5 for initial melt pool thickness H(SIC) of 2.5 and 6 km.

ington, and V. Masaitis for constructive discussions during the joint field trip to the Popigay impact crater in Siberia during July and August 1997. Additional discussions with A. A. Ariskin and M. Ostermann, as well as constructive reviews by E. Pierazzo and an anonymous colleague, are highly appreciated. Much of the of numerical modeling was done with the help of staff of the Institute for Planetary Exploration, DLR, Berlin. This work was supported by grants from the German Science Foundation (Grants De 401/9 and 436, both to A.D.). Additional support from the Russian Foundation for Basic Research, Grant 96-05-64167, was also provided (to B.A.I.).

APPENDIX 1

The one-dimensional heat flow equation:

$$\frac{\partial T}{\partial_t} = K(T)\frac{\partial^2 T}{\partial z^2},$$

where T is the temperature, t is the time, $K(T)$ is the thermal diffusivity as a function of temperature, and z is the depth, has been solved numerically. The computational mesh presented the vertical column of the target material as a set of cells with the step $\Delta z = 100$ m. The discrete analogue of the heat transfer equation allows the calculation of the temperature at the next time step $n + 1$ from the temperature at the time step n in the implicit form:

$$\frac{T_i^{n+1} - T_i^n}{\Delta t} = \frac{K_{i+1/2}\left(T_{i+1}^{n+1} - T_i^{n+1}\right) - K_{i-1/2}\left(T_i^{n+1} - T_{i-1}^{n+1}\right)}{\left(\Delta z\right)^2},$$

where

$$K_{i+1/2} = \left(K\left(T_{i+1}^n\right) + K\left(T_i^n\right)\right)/2$$

$$K_{i-1/2} = \left(K\left(T_i^n\right) + K\left(T_{i-1}^n\right)\right)/2.$$

This well-known absolutely stable algorithm has the first order of approximation in time and the second order of approximation in space. The standard subroutine "tridiag.for" from Numerical Recipies by Press et al. (1992) was used to solve this equation at each time step. The melting/solidification process introduces the latent heat into the equation. However, the approximation by Jaeger (1968) (see also Onorato et al., 1978) allows inclusion of the latent heat production/consumption between liquidus and solidus into the function $K(T)$.

The suevite layer of the model was assumed to have a temperature of 800°K. To simplify calculations we assumed the standard 300°K temperature. This gives an artificially large heat flux (500°K at 100 m). A better boundary condition should constrain not temperature at the boundary point, but the heat flux. In the case of a planet without an atmosphere, it may be done by keeping the boundary flux equal to the radiation of the black body ($\sim \sigma T^4$, where σ is a Stephan-Boltzman constant). At terrestrial conditions, however, one should take into account such factors as water vaporization, cold water delivery, and atmospheric circulation. In the simplified model presented here, we assume the heat flux above the surface, which is enough large to keep the boundary temperature at the constant level of 300°K.

REFERENCES CITED

Amsden, A. A., Ruppel, H. M., and Hirt, C. W., 1980, SALE: a simplified ALE computer program for fluid flow at all speeds: Los Alamos, New Mexico, Los Alamos National Laboratory Report LA-8095, 101 p.

Brockmeyer, P., 1990, Petrographische und geochemische Untersuchungen an polymikten Breccien der Onaping-Formation, Sudbury-Distrikt (Ontario, Kanada) [Ph.D. thesis]: Münster, Germany, University of Münster, 228 p.

Collins, W. H., 1934, The life–history of the nickel irruptive (1) petrogenesis: Transactions of the Royal Society of Canada, v. 28, sec. 4., p. 123–177.

Deutsch, A., 1994, Isotope systematics support the impact origin of the Sudbury Structure (Ontario, Canada): in Dressler, B. O., Grieve, R. A. F., and Sharpton, V. L., eds., Large meteorite impacts and planetary evolution: Geological Society of America Special Paper 293, p. 289–302.

Deutsch, A., and Grieve, R. A. F., 1994, The Sudbury Structure: constraints on its genesis from Lithoprobe results: Geophysical Research Letters, v. 21, p. 963–966.

Deutsch, A., Grieve, R. A. F., Avermann, M., Bischoff, L., Brockmeyer, P., Buhl, D., Lakomy, R., Müller-Mohr, V., Ostermann, M., Stöffler, D., 1995, The Sudbury Structure (Ontario, Canada): a tectonically deformed multi-ring basin: Geologische Rundschau, v. 84, p. 697–709.

Dickin, A. P., Richardson, J. M., Crocket, J. H., McNutt, R. H., and Peredery, W. V., 1992, Osmium isotope evidence for a crustal origin of platinum group elements in the Sudbury nickel ore, Ontario, Canada: Geochimica et Cosmochimica Acta, v. 56, p. 3531–3537.

Dickin, A. P., Artan, M. A., and Crocket, J. H., 1996, Isotopic evidence for distinct crustal sources of North and South Range ores, Sudbury Igneous Complex: Geochimica et Cosmochimica Acta, v. 60, p. 1605–1613.

Dressler, B. O., 1984, General geology of the Sudbury area, in Pye, E. G., Naldrett, A. J., and Giblin, P. E., eds., The geology and ore deposits of the Sudbury Structure: Ontario Geological Survey Special Volume 1, p. 57–82.

Faggart, B. E., Basu, A. R., and Tatsumoto, M., 1985, Origin of the Sudbury Complex by meteoritic impact neodymium isotopic evidence: Science, v. 230, p. 436–439.

Grant, R. W., and Bite, A., 1984, Sudbury quartz diorite offset dikes, in Pye, E. G., Naldrett, A. J., and Giblin, P. E., eds., The geology and ore deposits of the Sudbury Structure: Ontario Geological Survey Special Volume 1, p. 275–301.

Grieve, R. A. F., and Cintala, M. J., 1992, An analysis of differential of melt-crater scaling and implications for the terrestrial impact record: Meteoritics, v. 27, p. 526–538.

Grieve, R. A. F., Stöffler, D., and Deutsch, A., 1991, The Sudbury Structure: controversial or misunderstood? Journal of Geophysical Research, v. 96, p. 22753–22764.

Jaeger, J. C., 1968, Cooling and solidification of igneous rocks in basalts, in Hess, H. H., and Poldervaart, A., eds., The Poldervaart treatise on rocks of basaltic composition: New York, John Wiley, p. 503–533.

Ivanov, B. A., 1994, Geomechanical models of impact cratering: Puchezh-Katunki structure, in Dressler, B. O., Grieve, R. A. F., and Sharpton, V. L., eds., Large meteorite impacts and planetary evolution: Geological Society of America Special Paper 293, p. 81–91.

Ivanov, B. A., Badukov, D. D., Yakovlev, O. I., Gerasimov, M. V., Dikov, Yu. P., Pope, K. O., and Ocampo, A. C., 1996a, Degassing of sedimentary rocks due to Chicxulub impact: hydrocode and physical simulations: in Ryder, G., Fastovsky, D., and Gartner, S., eds., The Cretaceous-Tertiary event and other catastrophes: Geological Society of America Special Paper 307, p. 125–139.

Ivanov, B. A., Kocharyan, G. G., Kostuchenko, V. N., Kirjakov, A. F., and Pevzner, L. A., 1996b, Puchezh-Katunki impact crater: preliminary data on recovered core block structure, in 27th Lunar and Planetary Science Conference, Abstracts: Houston, Texas, Lunar and Planetary Science Institute, p. 589–590.

Ivanov, B. A., and Kostuchenko, V. N., 1997, Block oscillation model for impact crater collapse, *in* 28th Lunar and Planetary Science Conference, Abstracts: Houston, Texas, Lunar and Planetary Science Institute, p. 631–632.

Ivanov, B. A., DeNiem, D., and Neukum, G., 1997, Implementation of dynamic strength models into 2D hydrocodes: applications for atmospheric breakup and impact cratering: International Journal of Impact Engineering, v. 20, p. 411–430.

Krogh, T. E., Davis D. W., and Corfu, F., 1984, Precise U-Pb zircon and baddeleyite ages for the Sudbury area, *in* Pye, E. G., Naldrett, A. J., and Giblin, P. E., eds., The geology and ore deposits of the Sudbury Structure: Ontario Geological Survey Special Volume 1, p. 431–446.

Lakomy, R., 1990, Implications for cratering mechanics from a study of the Footwall Breccia of the Sudbury impact structure, Canada: Meteoritics, v. 25, p. 195–207.

McKinnon, W. B., 1978, An investigation into the role of plastic failure in crater modification, *in* Proceedings, 9th Lunar and Planetary Science Conference: Houston, Texas, Lunar and Planetary Institute, p. 3965–3973.

Melosh, H. J., 1977, Crater modification by gravity: a mechanical analysis of slumping, *in* Roddy, J., Pepin, R. O., and Merill, R. B., eds., Impact and explosion cratering: New York, Pergamon Press, p. 1245–1260.

Melosh, H. J., 1979, Acoustic fluidization: A new geologic process? Journal of Geophysical Research, v. 84, p. 7513–7520.

Melosh, H. J., 1989, Impact cratering: a geologic process: New York, Oxford University Press, and Oxford, United Kingdom, Clarendon Press, 245 p.

Milkereit, B., Green, A., Wu, J., White, D., and Adam, E., 1994, Integrated seismic and borehole geophysical study of the Sudbury Igneous Complex: Geophysical Research Letters, v. 21, p. 931–934.

Morris, W. A., 1984, Paleomagnetic constraints on the magmatic, tectonic, and metamorphic evolution of the Sudbury basin region, *in* Pye, E. G., Naldrett, A. J., and Giblin, P. E., eds., The geology and ore deposits of the Sudbury Structure: Ontario Geological Survey Special Volume 1, p. 411–427.

Naldrett, A. J., and Hewins, R. H., 1984, The main mass of the Sudbury igneous complex, *in* Pye, E. G., Naldrett, A. J., and Giblin, P. E., eds., The geology and ore deposits of the Sudbury Structure: Ontario Geological Survey Special Volume 1, p. 235–251.

Onorato, P. I. K., Uhlmann, D. R., and Simonds, C. H., 1978, The thermal history of the Manicouagan impact melt sheet, Quebec: Journal of Geophysical Research, v. 83, p. 2789–2798.

Ostermann, M., 1996, Die Geochemie der Impaktschmelzdecke (Sudbury Igneous Complex) im Multiring-Becken Sudbury [Ph.D. thesis]: Münster, Germany, Univiersity of Münster, 168 p.

Ostermann, M., Schärer, U., and Deutsch, A., 1996, Impact melt dikes in the Sudbury multi-ring basin (Canada): implications from uranium-lead geochronology on the Foy Offset Dike: Meteoritics and Planetary Science, v. 31, p. 494–501.

Pierazzo, E., Vickery, A. M., and Melosh, H. J., 1997, A reevaluation of impact melt production: Icarus, v. 127, p. 408–423.

Press, W. H., Teukolsky, S. A., Vettering, W. T., and Flannery, B. P., 1992, Numerical recipes in fortran: the art of scientific computing, 2nd ed. New York, Cambridge University Press, 963 p.

Pye, E. G., Naldrett, A. J., and Giblin, P. E., eds., 1984, The geology and ore deposits of the Sudbury Structure: Ontario Geological Survey Special Volume 1, p. 381–410.

Roest, W. R., and Pilkington, M., 1994, Restoring post-impact deformation at Sudbury: a circular argument: Geophysical Research Letters, v. 21, p. 959–962.

Shanks, W. S., and Schwerdtner, W. M., 1991, Crude quantitative estimates of the original northwest-southwest dimension of the Sudbury Structure, south-central Canadian Shield: Canadian Journal of Earth Science, v. 28, p. 1677–1686.

Walker, R. J., Morgan, J. W., Naldrett, A. J., Li, C., and Fassett, J. D., 1991, Re-Os isotope systematics of Ni-Cu sulfide ores, Sudbury Igneous Complex, Ontario evidence for a major crustal component: Earth and Planetary Science Letters, v. 105, p. 416–429.

Warren, P. H., Claeys P., and Cedillo-Pardo, E., 1996, Mega-impact melt petrology (Chicxulub, Sudbury, and the moon): effects of scale and other factors on potential for fractional crystallization and development of cumulates: Geological Society of America Special Paper 307, p. 105–124.

Wood, C. R., and Spray, J. G., 1998, Origin and emplacement of Offset Dykes in the Sudbury impact structure: constraints from Hess: Meteoritics and Planetary Science, v. 33, p. 337–347.

Zamyshliaev, B. V., and Evterev, L. S., 1990, Models of dynamic deforming and failure for ground media: Moscow, Nauka Press, 215 p., [*in Russian*].

Zel'dovich, Ya. B., and Raizer, Yu. P., 1967, The physics of shock waves and high-temperature hydrodinamic phenomena. New York, Academic Press.

MANUSCRIPT ACCEPTED BY THE SOCIETY DECEMBER 16, 1998

Emplacement geometry of the Sudbury Igneous Complex: Structural examination of a proposed impact melt-sheet

E. J. Cowan, U. Riller, W. M. Schwerdtner*
Department of Geology, University of Toronto, Toronto, Ontario M5S 3B1, Canada

ABSTRACT

The main mass of the 1.85-Ga Sudbury Igneous Complex (SIC) has been recently interpreted as a 2.5-km-thick impact melt-sheet that differentiated into norite, gabbro, and granophyre layers. This interpretation requires the SIC to have been emplaced as a horizontal sheet ponded in a complex impact crater whereby orogenic folding is regarded as the cause of its synformal geometry. However, three independent lines of structural evidence from the SIC and its Huronian host rocks indicate that the SIC was not a horizontal sheet at the time of its consolidation.

1. Planar mineral fabrics of the unstrained norite and gabbro layers are subparallel to the synformal base of the SIC and mineral lineation plunges toward the center of the SIC. This radial lineation pattern is inconsistent with an initial horizontal sheet geometry of the SIC, but is consistent with preconsolidation strain caused by gravitational reorientation of crystals on inclined magma chamber walls.

2. Anisotropy of magnetic susceptibility (AMS) of the granophyre reveals a magnetic lineation that is orthogonal to the basal contact of the SIC. This fabric is correlated to acicular plagioclase crystals and is thus interpreted as a wall-orthogonal crystallization texture. Fold-induced strain is expected to overprint such textures most severely where the curvature of the SIC is greatest, e.g., in the North Lobe. However, the angular departure of this lineation from its initial contact-orthogonal orientation is minimal in this area. Shortening strains estimated from AMS numerical modeling and microstructural analysis are significantly lower than strains expected from orthogonal flexural folding, <15 vs. 50%, respectively. The observed low strain levels are in agreement with a primary parabolic geometry of the SIC, but are inconsistent with folding of a consolidated horizontal melt-sheet.

3. Structural analysis of Huronian host rocks shows that deformation of these rocks can be explained without invoking rotation of Huronian strata as a consequence of impact cratering. Moreover, the absence of pervasive, post-SIC ductile strain in Huronian rocks and the adjacent norite, the uniform northwest-southeast–directed, late-orogenic compression, and the regional tectonometamorphic correlation suggest that shortening of the SIC was not accomplished by noncylindrical buckle folding but rather by imbrication of the southern SIC on thrust surfaces of the South Range Shear Zone.

---

*Present address: Cowan, SRK Consulting, 25 Richardson Street, West Perth, WA 6005, Australia; e-mail: juncowan@yahoo.com; Riller, Institut für Geowissenschaften und Lithosphärenforschung, Universität Giessen, Senckenbergstraße 3, 35390 Giessen, Germany.

Cowan, E. J., Riller, U., and Schwerdtner, W. M., 1999, Emplacement geometry of the Sudbury Igneous Complex: Structural examination of a proposed impact melt-sheet, *in* Dressler, B. O., and Sharpton, V. L., eds., Large Meteorite Impacts and Planetary Evolution II: Boulder, Colorado, Geological Society of America Special Paper 339.

If the SIC is accepted as an impact melt body, a nonhorizontal initial configuration of the SIC has far-reaching implications for the emplacement mechanism of impact melts in large craters. Published theoretical models of crater formation invariably predict a horizontal emplacement geometry for the impact melt-sheet. In the light of our structural results, such models and the fractionation mechanism of the SIC require revision.

INTRODUCTION

The 1.85-Ga Sudbury Igneous Complex (SIC), in central Ontario, Canada is a 2.5-km-thick layered igneous body whose main mass consists of a lower norite layer overlain by gabbro and granophyre sheets, respectively (Figs. 1 and 2). The main mass is underlain by a discontinuous sulfide-rich noritic unit called the Sublayer, and radial and concentrically oriented quartz dioritic Offset Dikes extend out from the Sublayer, transecting the basement lithologies. Although studied for more than a century, the mechanism of SIC emplacement has been debated since the first quarter of this century when Coleman (1905, 1907) interpreted the SIC as an intrusion that differentiated in situ. Coleman's interpretation was followed by a hypothesis in which the norite and the granophyre were regarded as separate intrusions (Phemister, 1925). Subsequent studies have supported either, or a combination, of these two views (Naldrett et al., 1970; Peredery and Naldrett, 1975; Naldrett and Hewins, 1984).

The interpretation that Sudbury was a site of extraterrestrial impact was first suggested by Dietz (1964), but he regarded the SIC as a mantle-derived igneous body triggered by the impact. Recently, however, a radically new interpretation, i.e., an impact melt origin, was introduced for the entire SIC (Faggart et al., 1985; Grieve et al., 1991). Prior to these publications, the SIC

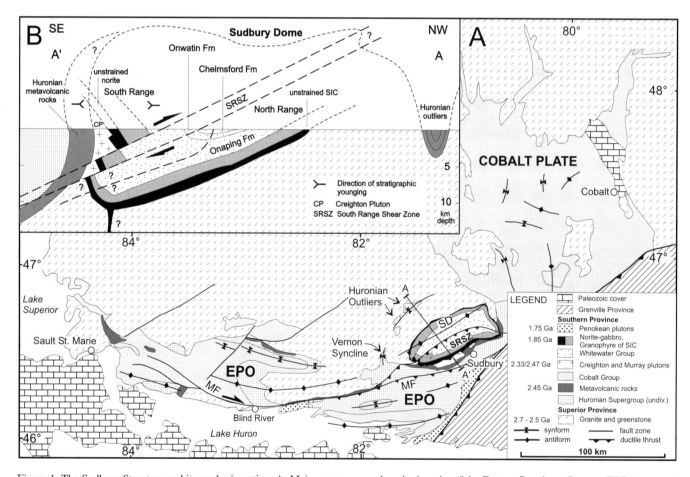

Figure 1. The Sudbury Structure and its geologic setting. A, Major structures and geologic units of the Eastern Penokean Orogen (EPO) in central Ontario. SD = Sudbury Dome (Riller and Schwerdtner, 1997); MF = Murray Fault; SRSZ = South Range Shear Zone (Shanks and Schwerdtner, 1991a). B, Northwest-southeast profile of the Sudbury Basin and the Sudbury Dome based on the Lithoprobe seismic profile (Milkereit et al., 1992). Y-shaped arrows indicate the reversal in stratigraphic younging between the Whitewater Group and the Huronian strata.

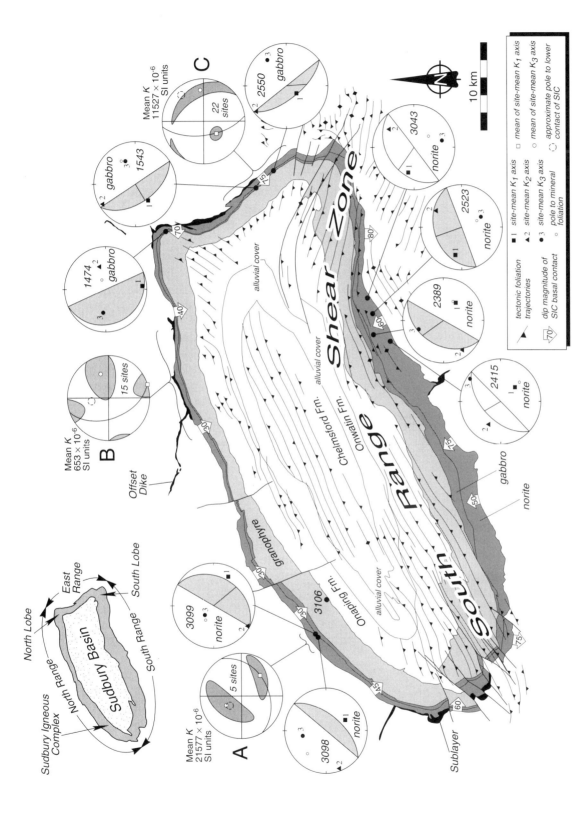

Figure 2. Sites and mean fabric data from image analysis plotted on lower hemisphere equal-area stereonets (shaded great circle indicates mean mineral lineation) of gabbro and norite samples. Mean principal axes of the AMS ellipsoid K_1, K_2 and K_3 are labeled 1, 2 and 3 on the stereoplots. Small open circles indicate poles to the mean mineral foliation determined from image analysis as described in text. Mineral L-S fabric determined from sites 2415 and 3043 were visually estimated from hand samples without the aid of AMS (see text; unshaded great circle shows mean mineral foliation; line shows mean mineral lineation). Lower hemisphere equal-area stereoplots (A), (B) and (C) are magnetic fabrics obtained from hand samples at several stations (numbers of sites indicated), with each sample site consisting of two to eight core samples: (A) is from sample domain A, (B) from domains B and C, and (C) from domains H and I of Figure 5. The mean of the site–mean magnetic foliation is shown as a great circle, as well as its pole as an open circle. The mean of the site–mean magnetic lineation is indicated by an open square. Both are shown with shaded 95% confidence cones, and their mean site–mean susceptibilities (K) are indicated. The magnetic foliation poles in (A) and (C) are subparallel to the inferred basal contact of the SIC (large dashed open circles are approximate poles). The departure of magnetic foliation pole from the SIC contact, and the large 95% confidence cones of (B) is due to the very low susceptibility of the samples. The mean site–mean magnetic lineation nevertheless point in a radial orientation with respect to the center of the Sudbury Basin. Dips of the outer contact of the SIC are from Rousell (1984, Fig. 5.1). The foliation trajectories of strained lithologies were constructed from more than 2,500 measurements of individual foliation data, and objectively interpolated by a spatial averaging method (Cowan, 1996). The location of site 3106 (Fig. 3B) is shown in the North Range.

was interpreted as a magmatic intrusion, regardless of whether it was emplaced as a consequence of endogenetic volcanism or impact-induced magmatic activity. More recent geochemical studies (Chai and Eckstrand, 1993a,b; Norman, 1994), in situ magmatic fractionation modeling of the SIC (Ariskin, 1997), field observations (Dressler et al., 1996), and radiometric data from the Sublayer (Corfu and Lightfoot, 1996) cast doubt on the validity of the impact melt model for all or part of the SIC. However, geochemical and impact theoretical arguments in support for the entire SIC as an impact melt are nevertheless favored by many workers (e.g., Grieve et al., 1993; Deutsch and Grieve, 1994; Grieve, 1994; Deutsch et al., 1995; Ostermann and Deutsch, 1997).

Prior to the impact melt hypothesis, the discussion centered frequently on evidence for the predeformational geometry of the SIC. Proponents of the differentiation model assumed the SIC as a horizontal sheet, which was modified by folding during the Penokean Orogeny (e.g., Collins and Kindle, 1935). By contrast, workers favoring the multiintrusion model viewed locally steep contacts of the SIC to be original, and thus supportive evidence for their hypothesis (Thomson and Williams, 1959). The impact melt model of Grieve et. al. (1991) requires that the SIC formed in a horizontal melt-lake, a scenario that is geometrically equivalent to those in which the SIC intruded as a horizontal sheet. Horizontal geometry is required because the tripartite compositional variation of the SIC is interpreted by Grieve et al. (1991) to be the result of magmatic differentiation. Large, complex impact craters are also known to possess a subhorizontal crater floor in which the impact-melt body ponds as a flat sheet (Grieve, 1975; Floran et al., 1978); in addition, numerical models of large impact craters predict a horizontal sheet geometry for impact melts (O'Keefe and Ahrens, 1994; Ivanov and Deutsch, 1997). Only noncylindrical folding can produce the present geometry of the SIC if the initial horizontal configuration is accepted, regardless of its origin (Cowan and Schwerdtner, 1994).

The rivalry between the various emplacement hypotheses of SIC continues today. Still, it may be useful if the emplacement of the SIC can be examined independent of geochemical data and arguments based on impact theory. Knowledge of the predeformational geometry of the SIC may provide independent tests of proposed emplacement mechanisms, and consequently, of the origin of the SIC. It is therefore our intention to summarize in this chapter recently obtained structural information on the primary geometry of the SIC, without expounding on the technical details of the structural analysis that are presented elsewhere (Cowan, 1996, 1999; Riller, 1996; Riller et al., 1996, 1998; Riller and Schwerdtner, 1997). We present three sets of structural evidence data: (1) on the structural petrology of the norite and the gabbro (Cowan, 1996), (2) on the strain levels in the granophyre (Cowan, 1996, 1999), and (3) on the deformation of Huronian host rocks (Riller, 1996). All the data sets, together with published field observations made by previous workers, point to a parabolic, or dish-shaped, primary geometry for the SIC.

GEOLOGIC OVERVIEW

The synformal SIC is part of the Paleoproterozoic Eastern Penokean Orogen, which lies at the southern margin of the Archean Superior Province (Fig. 1). The orogen formed by folding and thrusting of Archean granitoid rocks and volcano-sedimentary sequences of the Huronian Supergroup mainly during the Blezardian (ca. 2.47–2.2 Ga) and Penokean (ca. 1.9–1.8 Ga) tectonic pulses (Card et al., 1972; Zolnai et al., 1984; Riller and Schwerdtner, 1997). Strained Archean greenstone, granitoids, and 2.7-Ga-old high-grade gneisses of the Levack Gneiss Complex (Krogh et al., 1984) underlie the area north of the SIC (Fig. 1). By contrast, southward overturned Huronian metavolcanic strata underlie the South Range and form the cover to Archean granitoid basement rocks west of the SIC (Card, 1965; Card and Palonen, 1976). Synformal keels of Huronian rocks within Archean basement rocks, known as Huronian outliers (Pye et al., 1984), occur at a distance of approximately 20 km to the west and north of the SIC (Fig. 1) (Dressler, 1984). East of the SIC, folded Huronian strata of the Cobalt plate exhibit an east-dipping fold enveloping surface (Fig. 1). These structural relationships suggest that the synformal SIC is superimposed on a crustal dome structure, herein called the Sudbury Dome (Fig. 1), which is cored by high-grade metamorphic Archean basement rocks and which is larger than the SIC at surface.

The Sudbury Basin consists of the synformal SIC that encloses, in the map pattern, folded sediments of the Whitewater Group (Fig. 2) (Clendenen et al., 1988; Hirt et al., 1993). Shock-metamorphic structures, such as shatter cones, pseudotachylytes, and planar deformation features in quartz, feldspar, and zircon, are documented from the Archean Superior and Proterozoic Southern Province lithologies that envelope the 60- × 27-km elliptical outline of the Sudbury Basin at the surface (Dietz and Butler, 1964; French, 1967; Dressler et al., 1991; Lakomy, 1990; Krogh et al., 1984, 1996; Spray and Thompson, 1995). The presence of these structures in the host rocks of the SIC suggests shock-induced deformation associated with the hypervelocity impact of an extraterrestrial mass.

In addition to the shock metamorphic features, it has been known for some time that devitrified glass of impact origin is preserved in the Onaping Formation of the Whitewater Group (Peredery, 1972a; Muir and Peredery, 1984; Dressler et al., 1996). This unit consists of heterolithic breccia fragments derived from nearby Archean and Proterozoic rocks and is interpreted as an impact-generated suevite deposit (Peredery, 1972a,b; Avermann, 1994), or an impact-generated pseudovolcanic sequence of pyroclastic and hydroclastic deposits (Ames and Gibson, 1995; Ames et al., 1998). In agreement with impact melting, recent studies of the SIC indicate high levels of crustal contamination of its igneous rocks (Faggart et al., 1985; Grieve et al., 1991; Grieve, 1994). Accordingly, Grieve et al. (1991) have argued that the SIC differentiated into its tripartite composition by in situ differentiation of an impact-generated melt-lake. The impact melt model for the SIC consequently constrains the initial geometry of the SIC to a horizontal sheet (Cowan and Schwerdtner, 1994). This is because an impact melt-sheet with a thickness of several kilometers such as the SIC can be generated only by a large impact that results in a multi-ring complex crater with a horizontal floor (Cintala and Grieve, 1994). Thus, the impact-melt hypothesis as presented by Grieve et al. (1991) excludes magma chamber geometries and emplacement mechanisms such as lopolithic intrusion of the SIC along steep walls (cf. Hamilton, 1960; Peredery and Morrison, 1984).

STRUCTURAL PETROLOGY OF THE GABBRO AND NORITE

Fabric studies of granitic plutons have shown that the last increments of magma strain are recorded by the fabric of magmatic minerals (Paterson et al., 1989; Cruden and Aaro, 1992; Bouchez et al., 1992; Nicolas, 1992). Thus, the orientation, shape, and symmetry of these mineral fabrics can yield information on the flow of a magma just prior to its consolidation in a pluton, which, in turn, may provide information on the geometry of the pluton (Cruden and Launeau, 1994). Similarly, as the mineral fabric data obtained from sheet-like igneous bodies may constrain its geometry, attitude, and magmatic flow characteristics (Cruden, 1998). Mineral fabrics in portions of the SIC, which were not affected by solid-state deformation, can be used to unravel its original geometry and aid in discriminating among rival emplacement models of the SIC. Mineral fabrics of the gabbro and norite units were obtained using the latest methods available in digital image analysis. The results are discussed in terms of fabric patterns expected in a melt-lake, and in an igneous body that intruded along parabolic contacts with the host rock.

Analytical methods

Nine oriented block samples of gabbro and norite devoid of solid-state tectonic deformation features were collected in the field (sample sites 3098, 3099, 1474, 1543, 2550, 3043, 2523, 2389, and 2415 in Fig. 2). Two samples were taken from both the North and East Ranges, four from the South Range, and one from the North Lobe of the SIC (Fig. 2). Six to 10 cores (2.2 cm long, with a 2.54-cm diameter) were drilled from each sample. The anisotropy of magnetic susceptibility (AMS) was determined for each core at low magnetic field strength using a Sapphire SI-2 induction coil instrument. Statistical site averaging of the magnetic fabric was done using the matrix averaging routines of Hext (1963) and Jelinek (1978).

Determination of AMS is an established physical method used in the field of petrofabric studies of igneous rocks (Guillet et al., 1983; Cruden and Launeau, 1994; Archanjo et al., 1995). Mineral fabrics are rarely measurable with conventional tools such as a compass but can be rapidly estimated by measuring the AMS. AMS is a symmetric second-rank tensor that relates the intensity of the applied magnetic field (H) to the acquired magnetization (M) of a material:

$$M_i = K_{ij} H_j,$$

The dimensionless susceptibility tensor K_{ij} has the principal components $K_1 \geq K_2 \geq K_3$, which correspond to the principal radii of a triaxial magnetic fabric ellipsoid (Hrouda, 1982). The K_1 axis and the K_3 axis are commonly found to parallel the mineral lineation axis and the pole to mineral foliation plane, respectively. Accordingly, the rock samples were cut parallel to the principal planes of the magnetic fabric resulting in three orthogonal sections for each sample. If the magnetic fabric was subconcordant to the mineral fabric, then polished, large thin-sections were made parallel to the principal planes for further petrofabric analysis. The thin-sections were labeled according to the principal axes of the magnetic fabric (Fig. 2) (K_1 = magnetic lineation axis; K_2 = intermediate axis; or K_3 = magnetic foliation–normal axis).

The determination of the mineral foliation (S) and lineation (L) of the gabbro and norite samples was done by digital image analyses conducted on multiple electron microprobe x-ray maps of the thin-sections (Cowan, 1996). Multi-channel image analysis technique utilized in this study is based on multi-spectral classification using principal component analysis, extensively used in remote sensing (Drury, 1993), but can be used on all types of multi-channel digital images (Launeau et al., 1994). This technique allowed the identification of all mineral phases within the scanned area, and mineral alignment data for each orthogonal face was quantitatively resolved using the intercept method of Launeau and Robin (1996). Figure 3A shows such a well-defined L-S fabric in a norite sample of the North Range, characterized by the alignment of tabular plagioclase crystals (in white) on three orthogonal sections that were assembled to a block diagram (other identified mineral phases were omitted for clarity). The reader is referred to Cowan (1996) for details of electron microprobe data acquisition and digital image processing.

The magnetic fabric can be substantially discordant to the principal petrofabric planes (sites 1474, 3098, 2389, 2415, and 3043 in Fig. 2), but in some of these cases the mineral L-S fabric can be visually estimated in the rock specimen from preferred orientation of plagioclase laths on orthogonal sections (sites 2389, 2415, and 3043). The orientation of the mineral foliation and the lineation were determined with these methods at each site, with the exception of sites 1474 and 3098, where mineral lineation could not be resolved with confidence.

Results

The mineral foliation is effectively parallel to the basal contact of SIC at most localities (Fig. 2) and is consistent with all of the five reliable samples analyzed with the image analysis technique (samples 3099, 1543, 2550, 2523, and 2389). Similar fabric orientations are also obtained by measuring AMS in other samples from the East and North Ranges, as well as visually determined L-S fabrics in samples 2415 and 3043 from the South Range (Fig. 2). Although the magnetic fabrics are not concordant to the mineral fabric at every site, visual examination and image analysis of samples obtained from apparently unstrained gabbro and norite in the East and North Ranges confirms that AMS yields reliable results. Both magnetic and mineral foliations of the norite and gabbro dip toward the center of the Sudbury Basin (Fig. 2). These foliations, however, appear to be shallower than the basal contact of the SIC at most localities, as also noted from the South Range norite by Naldrett and Hewins (1984). Both magnetic and mineral lineations plunge consistently toward the center of the Sudbury Basin (Fig. 2) and so, form a radial lineation pattern. It is important to recognize that the center of symmetry of this fabric pattern coincides with the center of the elliptical Sudbury Basin.

Interpretation

The mineral foliation is dipping shallower than the basal contact of the SIC at most localities (Fig. 2). Thus the formation of the L-S fabric of the gabbro and norite as a result of noncylindrical folding of a semi-consolidated sheet can be ruled out as consistently steep S fabrics are not observed (Fig. 4A). Emplace-

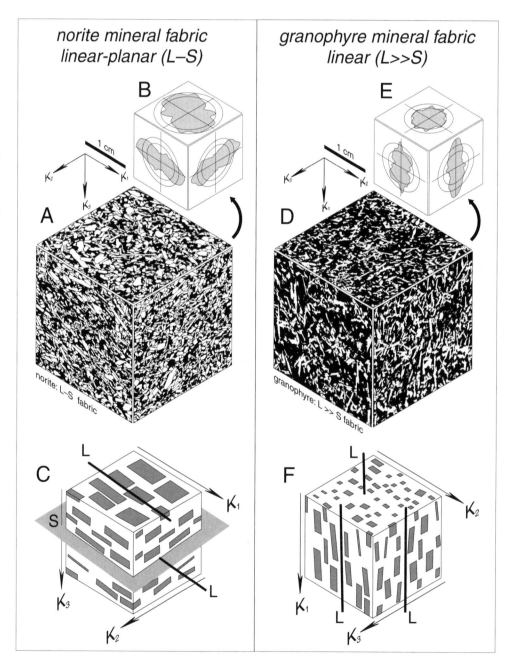

Figure 3. A, Plagioclase fabric of norite from site 3099 in the North Range (Fig. 2, UTM 467240E, 5162260N). As shown in Figure 2, the mineral foliation is subparallel to the K_1–K_2 plane, which in turn is subparallel to the contact of the SIC. D, The plagioclase fabric of granophyre from site 3106 from the North Range near Onaping Falls (Fig. 2; UTM 470700E, 5161060N). The mineral lineation is defined by alignment of acicular plagioclase crystals that parallel the K_1 direction, which in turn is orthogonal to the contact of the SIC. Smoothed histograms of sectional plagioclase long-directions in both (B) and (E) are shown with the inner circle in the histogram plots representing uniform distribution, the outer circle is plus one standard deviation from uniform distribution. Block diagrams in (C) and (F) schematically show the mineral fabric represented in the L-S fabric of the norite (A) and L>>S fabric of the granophyre (D).

ment models for the SIC that may be distinguished by the mineral L-S fabric pattern in the gabbro and the norite include: (1) emplacement as an impact-induced melt-lake (Fig. 4B), (2) viscous drag of magma on intrusive contacts (Fig. 4C), (3) expulsion of magma as a consequence of roof subsidence (Fig. 4D), and (4) postemplacement adjustment of magma due to inclined magma chamber walls (Fig. 4E).

The radial lineation pattern in the gabbro and the norite is difficult to reconcile with a ponded impact melt-sheet. Such a melt-sheet would most likely undergo gravitational settling and compaction of early crystallized mineral phases. This would result in a dominantly foliated mineral fabric devoid of mineral lineation (S>>L fabric) (Fig. 4B), rather than in the observed L-S fabric. Viscous drag along the lower and upper contacts during magma intrusion provides a possible explanation for the obliquity between the contact and the mineral foliation and the radial mineral lineation (Fig. 4C). However, this early formed fabric will be short-lived, and will be destroyed by subsequent magma strain. The presence of identical down-plunging L fabrics in mineral phases of both the gabbro and the norite indicates that fabric formation persisted during the process of igneous fractionation and consolidation. Similar down-plunge L-S fabric geometry is known from large mafic intrusions with inclined basal contacts (e.g., Nicolas, 1992; Quadling and Cawthorn, 1994). In such intrusions, the lineation exhibits a radial symmetry with respect to the intrusive body in map view, comparable to the mineral fabric pattern seen at Sudbury. Gravitationally induced flow and drag within the magma chamber appears to be the most plausible explanation for the L-S mineral fabrics of the gabbro and norite (Fig. 4E). Such a process would occur throughout the fractionation process, thus explaining the presence of identical L-S fabrics in the gabbro and norite (Fig. 2).

In summary, the L-S mineral fabrics of the gabbro and norite are inconsistent with an initial horizontal geometry of these units. The contact-parallel mineral foliation and down-plunging mineral lineation of the norite and gabbro are best explained by igneous strain resulting from gravitational settling of magmatic mineral phases on inclined magma chamber walls. This process occurred over a protracted period of time, commensurate to that of the fractionation process.

MAGNITUDE OF SOLID-STATE STRAIN IN THE GRANOPHYRE

Although the South Range of the SIC is transected by south-dipping thrust surfaces of the South Range Shear Zone (Shanks and Schwerdtner, 1991b), the SIC is reported to be unaffected by solid-state strain elsewhere (Rousell, 1984; Muir, 1984). This is incompatible with the standard impact melt model for the SIC, as solid-body rotation during folding of an initially horizontal, and solidified impact melt-sheet, is required to produce dips of 40°–70° in the North and East Ranges, respectively (Fig. 2). Such rotations must invariably impart solid-state strain to the hinge zones of folds, for example, in the North and South Lobes

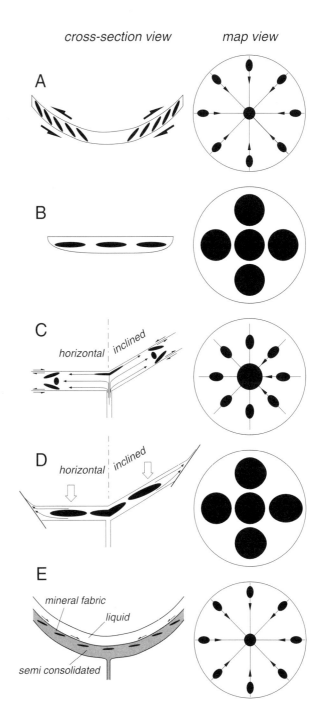

Figure 4. Kinematic and emplacement models for the gabbro and norite bodies and expected fabric patterns. Expected strain fabric resulting from (A) noncylindrical folding; (B) ponded impact melt, which is expected to produce a planar mineral fabric due to gravitational settling of crystallizing species; (C) frictional drag of magma on its wall during intrusion; D largely planar fabrics produced as a result of roof subsidence; and E, layer-parallel L-S fabrics resulting from gravitational strain effects and sedimentation within an inclined magma chamber. Mineral lineation (lines in map view, and arrow indicating plunge direction) is radial and down-plunging with respect to the parabolic shaped igneous magma chamber in the last scenario.

(Fig. 2). The strain magnitudes in rocks of these areas are very low (e.g., Dressler, 1987) but have never been estimated quantitatively. The aim of the following study is to quantify strain intensity of the granophyre in the North Lobe, and to compare these estimates with shortening strains expected to result from folding of a consolidated igneous sheet with the curvature equivalent to that of the North Lobe.

Analytical methods

The level of tectonic strain is difficult to estimate in igneous rocks that have been affected by a minor solid-state overprint. Magnetic fabric can, however, be used to quantify the tectonic strain. The magnetic fabric in weakly strained igneous rocks arises from the combination of strain acquired in the igneous state before consolidation and superposed solid-state strain. The igneous component of the AMS fabric is destroyed only if the imposed solid-state strain is large enough; otherwise, the igneous fabric will be detectable by AMS. The amount of shortening strain required to modify, and eventually obliterate, an AMS fabric due to igneous strain can be modeled numerically, assuming magnetic grain reorientation due to distortion of the host rock mass (Richter, 1992; Benn, 1994). Such a numerical method was used to estimate the amount of shortening strain resulting in the measured magnetic fabric anisotropy of the granophyre (Cowan, 1996, 1999). In order to compare these strain estimates with strains expected to result from folding of a consolidated igneous sheet, the interlimb angle of the SIC in the North Lobe was calculated by down-plunge projection. As in the fabric study of the gabbro and norite samples, the mineral fabrics of the granophyre were first estimated with AMS and supplemented by digital image analysis.

Results

Combined magnetic fabric and image analyses indicate that the microstructure of the granophyre is characterized by *comb-layering* or *crescumulate* crystalline texture (Moore and Lockwood, 1973; McBirney and Noyes, 1979). Crescumulate texture forms by wall-orthogonal crystal growth due to a temperature gradient and preferential crystal nucleation in the magma chamber (Lofgren and Donaldson, 1975; Lofgren, 1983). In the SIC, this fabric is distinguished by a mineral lineation oriented orthogonal to its basal contact and is best seen by eye in the plagioclase-rich granophyre (Peredery, 1972b) close to the upper contact with the Onaping Formation (Fig. 3B). The fabric is also readily identified by the contact-orthogonal magnetic lineation determined from nine sampling domains of the North and East Ranges (Fig. 5).

Interpretation

The crescumulate texture in the granophyre serves as an ideal natural strain gauge, as the initial orthogonality between its linear fabric element and the SIC contact will be destroyed if subjected to even minor tectonic shortening (Fig. 6). This is particularly relevant to the North Lobe if its curvature formed by flexural folding. This fold mechanism seems unlikely because the mean magnetic lineation is perpendicular to the basal contact of the SIC in all domains (Fig. 5). The orthogonality may, however, have been preserved during folding if strain axes were coaxial to the petrofabric, i.e., parallel to the axis of mineral alignment and the contact plane of the SIC such as by orthogonal flexure (Twiss and Moores, 1992, p. 240–241). In such a scenario, large shortening strains in the order of 30% in the outer arc and 50% in the inner arc of the fold are expected to be present in the host rock of the granophyre (Fig. 7).

Microstructural examination of host rocks revealed only minor distortion of quartz by dislocation creep (Cowan, 1999, Fig. 13). This indicates a low level of solid-state overprint and is therefore inconsistent with high shortening strains expected from folding by orthogonal flexure in the North Lobe (Fig. 7B). These observations are consistent with a maximum of 15% shortening in the granophyre, but are most likely much less based on numerical modeling of the magnetic fabric (Cowan, 1996, 1999). Further, the absence of layer-parallel stretching in the outer arc negates the orthogonal flexure mechanism in the North Lobe (Fig. 7).

In summary, the preservation of contact-orthogonal mineral and magnetic fabrics in the granophyre and overall low strain magnitudes in the North Lobe are inconsistent with a fold origin of the SIC in this area.

PALEO-PROTEROZOIC TECTONISM IN THE SUDBURY AREA

Three features have been attributed to either the formation of a multi-ring basin by meteorite impact (Grieve et al., 1991; Deutsch et al., 1995; Spray and Thompson, 1995) or crustal doming due to orogenic deformation (Cooke, 1948; Card et al., 1984; Riller, 1996; Riller and Schwerdtner, 1997) (Fig. 1): (1) circumferencial distribution of Huronian synforms around the SIC; (2) southward overturned Huronian strata exposed south of the South Range norite and (3) the presence of high-grade metamorphic rocks below, and north of, the SIC in the North Range. Dietz (1964) attributed the reversal of stratigraphic younging between the Whitewater Group and Huronian strata immediately south of the South Range SIC (Fig. 1B) to recumbent folding, analogous to overturning of strata at the collars of impact and explosion craters (Shoemaker, 1960; Jones, 1977; Roddy, 1977; Shoemaker and Kieffer, 1978). Consequently, the age and mechanism of tilting of Huronian strata are critical for deciphering impact-induced from orogenic deformation and thus for constraining the primary geometry of the SIC. In the impact model, the SIC was emplaced as a subhorizontal melt-sheet that cooled at the surface and acquired its synformal geometry by noncylindrical buckle folding prior to northwestward thrusting of the South Range (Grieve et al., 1991; Deutsch et al., 1995). Synformal buckling must have

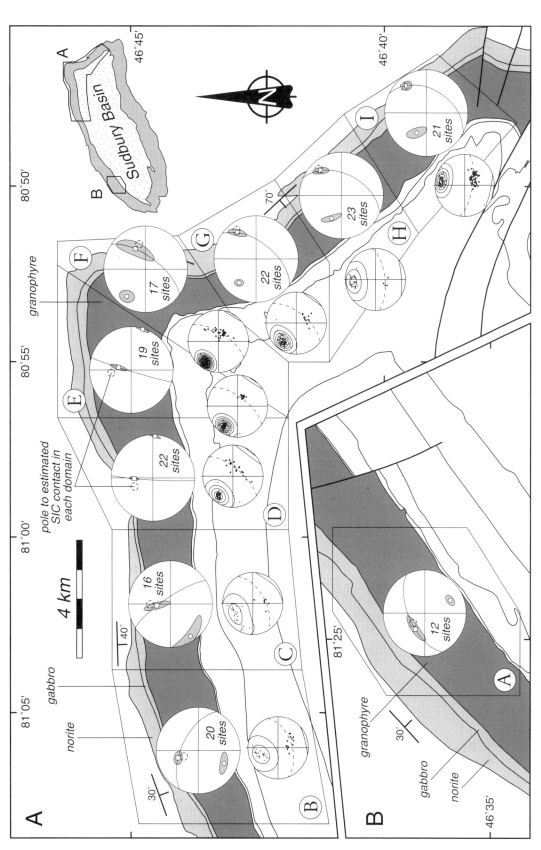

Figure 5. Mean magnetic foliation and lineation from the granophyre (dark shading; light shading indicates gabbro/norite). Each domain consists of as many as 23 sample sites, and each site has 3–12 core samples in which AMS fabrics were determined. Domain means of site-means were calculated according to statistical site averaging method of Jelinek (1978). Mean magnetic lineation (K_1) is shown in open square, and mean foliation pole (K_3) as open circle on the lower hemisphere stereonets with shaded 95% confidence cones about the mean values. Mean magnetic foliation also shown as solid great circle. Stereoplots are divided into Domains A through I, and the number of station means used to calculate the domain mean is shown in the stereonets. The dips of the North and East Ranges are also shown in map view and their estimated pole orientations are plotted as large dashed open circles for each domain on the stereonets. Note that the pole to the basal contact of the SIC in each domain is subparallel to the direction of the magnetic lineation in the same domain (see text). Smaller stereoplots show the tectonic foliation in the Onaping Formation (contoured) and lineation (not contoured). Mean tectonic foliation is drawn as a dashed great circle.

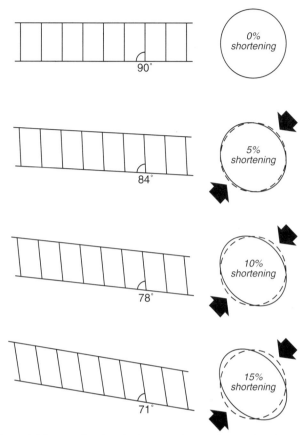

Figure 6. Diagram showing the departure from orthogonality of line elements for different amounts of coaxial homogeneous strain. At 0% total shortening, the vertical lines are analogous to the crescumulate fabric of the granophyre, whereas the horizontal lines represent the contact planes of the SIC. Note that orthogonality between line elements is lost after small amounts of shortening strain (see text for explanation).

imparted ductile strain to the SIC and its host rock unless rotation of these rocks was accomplished bodily. Nevertheless, the attitude of the Huronian strata south of the SIC should differ from that exposed along strike to the southwest of the SIC. Alternatively, orogenic crustal doming occurred prior to emplacement of the SIC. In this scenario, ductile strain in Huronian rocks should be larger and older than those in the SIC. Moreover, northwestward translation of the South Range would be the dominant shortening mechanism of the SIC and consequently ductile strain fabrics in the SIC should be associated spatially with the South Range Shear Zone.

In our previous studies (Riller, 1996; Riller et al., 1996, 1998; Riller and Schwerdtner, 1997), we presented structural evidence strongly supporting the formation of a Sudbury Dome during Blezardian tectonism and a primary synformal geometry of the SIC, which was modified by Penokean thrusting along the South Range Shear Zone. Based on these results, an outline of the effects of individual orogenic pulses on Huronian rocks and the SIC are presented below and complemented with results of a shape fabric and a paleostress analysis.

Blezardian tectonism (2.47–2.2 Ga)

Huronian rocks of the Elliot Lake Group acquired their pervasive tectonic foliation and lineation (Fig. 8B) during intrusion of the 2.33-Ga granitoid Creighton Pluton (Frarey et al., 1982) and the 2.47-Ga Murray Pluton (Krogh et al., 1996). Syntectonic intrusive relationships indicate that the pervasive strain fabrics in Huronian rocks formed under lower amphibolite facies conditions in the Blezardian tectonic pulse. The elliptical outline of the plutons at surface and low shape fabric intensities in granitoid rocks are best explained by emplacement of granitoid magma into a dilation zone between Archean basement and Huronian cover rocks during formation of the Sudbury Dome (Riller and Schwerdtner, 1997). Dilation at the basement-cover interface was most likely induced by fold detachment in its hinge zone during regional horizontal contraction. Since 2.2-Ga Nipissing gabbro bodies truncate the Vernon syncline, which is a section of the rim synform to the Sudbury Dome (Fig. 1), crustal doming occurred during Blezardian tectonism (Roscoe and Card, 1992).

Pervasive shape fabrics in ductilely deformed Huronian metavolcanic and metasedimentary rocks are defined by elongate primary markers such as vesicles, lapilli, pebbles, and preferentially oriented metamorphic mineral aggregates (Fig. 8B). Assuming initial sphericity, these markers can be used for estimating total ductile strain imparted to Huronian rocks during orogenic deformation. Aspect ratios of 20–30 markers per station were measured in oriented rock samples cut parallel to the A–B and B–C principal planes of the shape fabric ellipsoid (A ≥ B ≥ C). Since the mineral lineation is subvertical and parallel to A, mean sectional shape fabric ellipse diameter ratios (R) on A–C planes are generally larger than those on the horizontal B–C planes that are shown in Figure 8B (for original data, see Riller and Schwerdtner, 1997). Apart from high diameter ratios (R ≈ 5) and variable orientation of ellipse long axes near the margin of the Creighton Pluton due to contact strain, ellipses are characterized by moderate diameter ratios (2 < R < 4). The preferred orientation of ellipse long axes ($\phi \approx 60°$) parallel to the strike of the pervasive planar strain fabrics (Fig. 8B,C) suggests that most of the ductile strain seen in Huronian rocks accumulated during Blezardian tectonism and is therefore related to crustal doming.

Penokean tectonism (ca. 1.9–1.8 Ga)

Penokean ductile deformation occurred under middle greenschist facies metamorphic conditions but is not pervasive in the South Range of the SIC and its Huronian host rocks. A concentration of ductile strain due to competency contrast should be expected at the interface between Huronian supracrustal rocks, Proterozoic granite plutons and the norite if the SIC and its enveloping rocks were subjected to Penokean buckle folding. However, this is not apparent in the field. Kilometer-scale folds in Huronian strata, identified as Blezardian due to their axial-planar foliation formed under amphibolite-facies conditions, are truncated by effectively unstrained South Range norite (e.g., at station 3

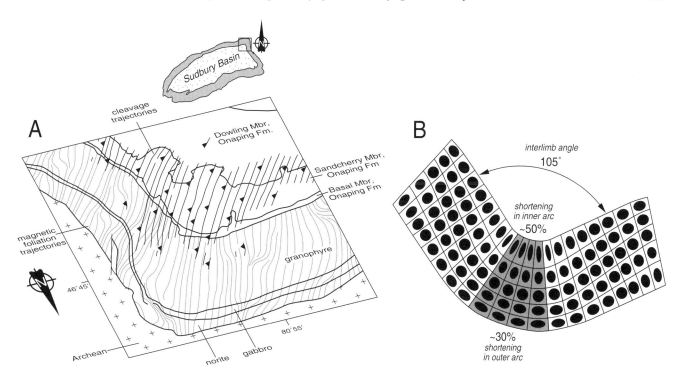

Figure 7. A, Down-plunge projection view of the North Lobe. The boundary between Sandcherry and Dowling Members (formerly Grey and Black Members) are marked by the appearance of shard layers, and not the absence or presence of carbon (see Ames and Gibson, 1995). Solid lines with dip-barbs are cleavage trajectories of the Onaping Formation and parts of the strained granophyre. Most of the SIC is unstrained, however, and the magnetic foliation is the only evidence for deformation (shown as the dashed trajectories). Note that these two sets of trajectories are subparallel to each other (both computer interpolated), and are axial planar to the North Lobe. B, Plane strain buckle fold showing orthogonal flexure in the hinge region (shaded) with the same interlimb angle as the North Lobe (105°). The outer arc must undergo layer-parallel stretch in order to preserve the orthogonality of lines, and conversely, axial planar shortening in the inner arc. This layer-parallel stretch is absent from the North Lobe shown in (A). Strain levels are also incompatible between the North Lobe (A) and the fold model (B), since shortening of ~50 and ~30% is theoretically required in the inner and outer arcs, respectively. Evidence for large degree of solid-state strain, however, is clearly lacking in the North Lobe.

in Fig. 8B). Similarly, hinge zones of mesoscopic Blezardian folds were destroyed by largely unstrained Sudbury Breccia bodies (see below). Finally, undeformed 1.85-Ga quartz-diorite Offset Dikes (Krogh et al., 1984) transect Blezardian shape fabrics in Huronian rocks (Fig. 8B).

Since the formation of pseudotachylyte (Sudbury Breccia) is regarded as penecontemporaneous with emplacement of the SIC (e.g., Spray and Thompson, 1995), shape fabrics of breccia fragments should record post-SIC Penokean strains. As vertical outcrop surfaces are rarely exposed, the mean diameter ratio (R) of these strains in horizontal section was estimated by applying the inertia tensor method (P.-Y. F. Robin, unpublished data) to the orientation and aspect ratio of 30–70 fragments per station (Fig. 8). Fabric ellipse diameter ratios of breccia fragments are considerably lower ($1 \leq R \leq 1.5$) than those of primary markers ($1.4 \leq R \leq 5.3$) in the same area (Fig. 8B,C). This confirms that most of the bulk ductile strain was imparted on Huronian rocks prior to emplacement of the SIC and is in agreement with static growth of greenschist-metamorphic mineral assemblages in Huronian metavolcanic rocks during Penokean tectonism (Riller and Schwerdtner, 1997). Consequently, the South Range and its Huronian host rocks were not affected by large amounts of shear-induced rotation due to buckle folding, or shearing on listric faults, during the later stages of Penokean ductile deformation.

By contrast, higher values of fabric ellipse diameter ratios are found on or close to the South Range Shear Zone (Fig. 8A). However, there is no preferred alignment of breccia fragments in Archean rocks north and west of the SIC (e.g., station 2047 in Fig. 8A). Thus, the South Range Shear Zone, and the spatially related ductile strain recorded in the Whitewater Group within the Sudbury Basin, appears to be the northernmost locus of Penokean ductile deformation (Clendenen et al., 1988; Shanks and Schwerdtner, 1991a,b; Hirt et al., 1993). Ductile strain fabrics of the South Range Shear Zone transect the southwestern SIC where they are concordant to the interface of Archean basement with Huronian cover rocks for about 15 km before turning southwest, at location 2310, to join with the Murray fault (Fig. 8A). This indicates that the shear zone is part of a regional thrust system, and is unrelated to local deformation associated with rotation of the North and South Ranges, as invoked by Cowan and Schwerdtner (1994). The sum of the above structural relationships suggest that Penokean shortening of the SIC and the Sudbury Dome was

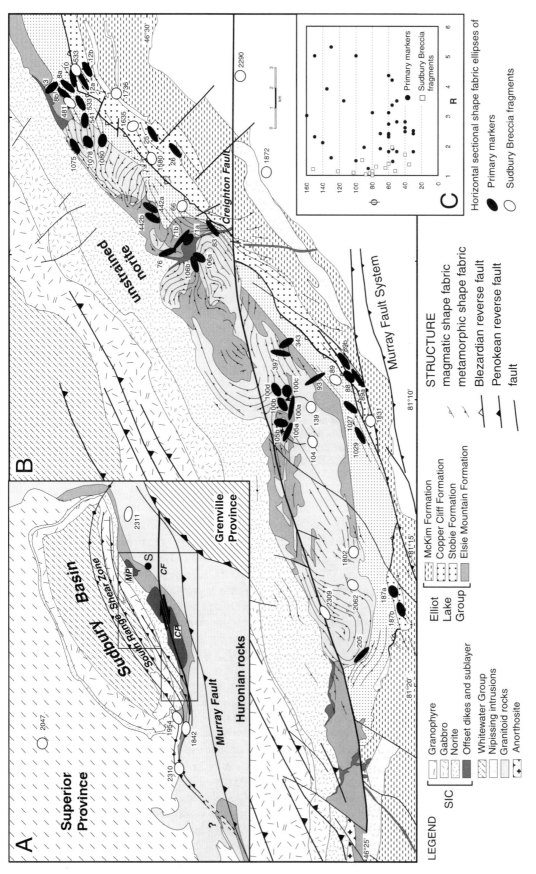

Figure 8. A and B, Ductile strain and shape fabrics in the South Range of the SIC and its host rocks (S = town of Sudbury; CF = Creighton fault; MP = Murray Pluton). Foliation trajectories of the South Range Shear Zone are from Shanks and Schwerdtner (1991a); those in Huronian rocks are from Riller et al. (1996) and Riller and Schwerdtner (1997). Ellipses represent the arithmetic mean of the normalized horizontal sectional aspect ratio of measured primary markers and Sudbury Breccia fragments per station. Unless indicated by arrows, the center of each ellipse is located at the station of data acquisition. C, Plot showing the diameter ratio (R) versus the azimuth (φ) of horizontal sectional shape fabric ellipses. Note that primary markers reflect the total strain, whereas shape fabrics of Sudbury Breccia fragments record Penokean strain accumulated after 1.85 Ga.

accomplished chiefly by imbrication of the South Range and its Huronian host rocks. Ductile distortion and rotation of these rocks due to large-scale Penokean buckle folding can be ruled out as an operative mechanism.

Late-Penokean contraction

Folding of the SIC and its host rocks may have been accomplished by rigid-body rotation under brittle conditions, i.e., at a late stage of Penokean deformation. This requires the presence of mainly layer-parallel discontinuities with substantial offsets (Cowan and Schwerdtner, 1994), which should be particularly apparent where the curvature of the SIC is greatest (i.e., the North and South Lobes and the southwestern part of SIC). Faults transecting the SIC in the North Lobe occur only in its eastern limb (Fig. 2), and strain patterns associated with the faults in the South Lobe are consistent with dip-slip displacements (Cowan, 1996). These strain patterns, cannot explain the sharp curvature of the SIC at these localities. Similarly, large strike separations on faults transecting the southwestern SIC are attributed to northwestward translation of the South Range on thrust surfaces of the South Range Shear Zone (Shanks and Schwerdtner, 1991a).

To further explore the possibility of late-orogenic, noncylindrical buckling of the SIC, we analyzed the principal paleostress directions using small-scale shear fractures with known slip sense (slickensides). The orientation of 1,100 fault surfaces and their respective slip sense was recorded at 55 stations in the Sudbury area (Fig. 9). The orientation of compressive principal paleostresses ($\sigma_1 \geq \sigma_2 \geq \sigma_3$) was calculated using the algorithm by Sperner and Ratschbacher (1994) based on the numerical dynamic analysis of calcite twin lamellae by Spang (1972). Assuming that deformation was coaxial and occurred under a homogeneous stress field and that the striation on a fault surface is parallel to the direction of the maximum resolved shear stress, the direction of the calculated maximum compressive stress (σ_1) will be close to the incremental shortening direction (Bott, 1959; Pollard et al., 1993). Taking into account field observations on mineralization and cross-cutting relationships, successive brittle deformation increments can then be distinguished by separating a given fault population into homogeneous subsets. Noncylindrical buckling of the SIC requires the presence of at least two such subsets, which are expected to differ substantially in their respective σ_1 directions.

Calculated σ_1 directions in Huronian rocks and the SIC are subhorizontal and show a strong preferred orientation in northwest-southeast direction, whereas σ_2 and σ_3 directions delineate a great circle perpendicular to the mean σ_1 direction (Fig. 9C). By contrast, σ_1 in the Grenville Province indicates east-west compression (Fig. 9A). The paucity of such σ_1 directions north of the Grenville Front and the fact that mineralized fault surfaces have not been observed in 1.23-Ga Sudbury dikes (Krogh et al., 1987) suggest that brittle deformation in Huronian rocks and the SIC is unrelated to Grenvillian tectonism and likely to be late Penokean in age. Moreover, fault surfaces in the Sudbury area are generally decorated by single sets of chlorite fibers that are parallel to occasional calcite or quartz fibers on the same surfaces. This indicates a constant direction of shear stress during fault reactivation and is in agreement with uniform northwest-southeast orientation of σ_1. The calculated stress states are correlate well with late Penokean reverse shear on the South Range Shear Zone and dextral strike-shear on the Murray and Creighton faults. However, uniform northwest-southeast–directed compression is incompatible with noncylindrical folding of the SIC.

Grenvillian tectonism (ca. 1.2–1.1 Ga)

Owing to the parallelism of structures in Huronian rocks with those in the adjacent Grenville Front in the Sudbury area, shortening of the Sudbury Structure has also been attributed to Grenvillian contraction (e.g., Wynne-Edwards, 1972). However, the structural and thermal effects of the Grenvillian orogeny on Huronian rocks were insignificant on a regional scale. The strike of first-order structures in the eastern Penokean Orogen changes progressively from east-southeast between Sault St. Marie and Blind River to north-northeast near Sudbury and to dominantly north in the Cobalt Plate (Fig. 1A). This structural grain is truncated by, but not deflected near, the northeast-striking Grenville Front. Furthermore, dikes of the 1.23-Ga Sudbury Swarm are effectively undeformed and unmetamorphosed and transect Penokean ductile strain fabrics such as the South Range Shear Zone. Metamorphic temperatures due to Grenvillian overprint of Huronian rocks, exposed about 60 km northeast of Sudbury, decrease to less than 400°C beyond a distance of 2 km from the Grenville Front (Hyodo et al., 1986).

In summary, the structure and distribution of Huronian rocks as seen at the current level of erosion in the Sudbury area can be explained without invoking rotation of strata as a consequence of impact-related deformation. The lack of structural evidence for ductile and brittle folding strain in the SIC and its host rocks after 1.85 Ga during the Penokean and Grenvillian tectonism suggests that the parabolic geometry of the SIC is mostly primary.

SYNTHESIS

Historical interpretations

Several lines of independent structural evidence presented herein lead us to regard the shape of the SIC as parabolic or funnel-shaped at the time of its consolidation (Fig. 10): (1) the presence of radial igneous fabrics of the gabbro and norite dipping toward the center of the synformal SIC, (2) the preservation of crescumulate petrofabrics of the granophyre in the North Lobe, (3) steep attitudes of Huronian strata prior to the emplacement of the SIC, and (4) the absence of structures indicating strain imparted to the SIC and its Huronian host rocks by buckle folding. Such interpretation of structural field relationships with regard to the primary shape of the SIC is not new. About 40 yr ago, in situ differentiation was commonly accepted as the origin

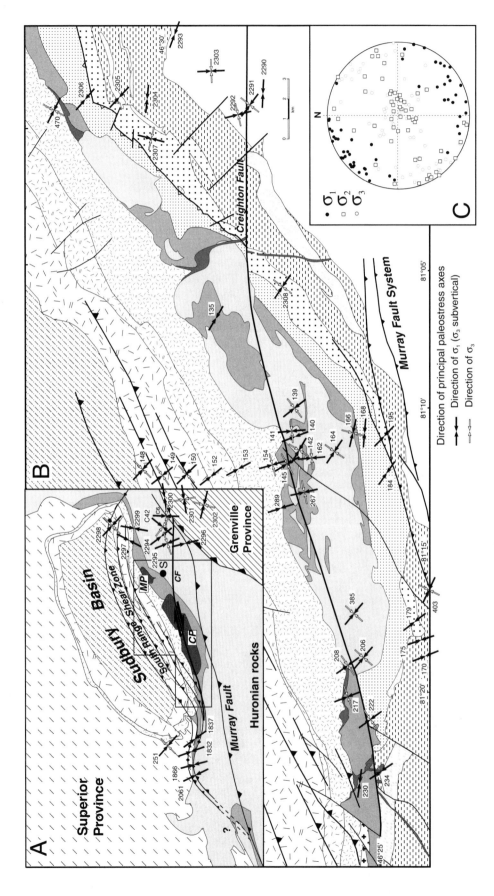

Figure 9. A and B, Map showing the direction of principal paleostress axes ($\sigma_1 \geq \sigma_2 \geq \sigma_3$) derived from analysis of late orogenic, brittle shear fractures in rocks from the Sudbury area and the Grenville Front. Legend explained in Figure 8. B, Lower hemisphere equal-area projection showing orientation of principal paleostress axes in the SIC and Huronian rocks.

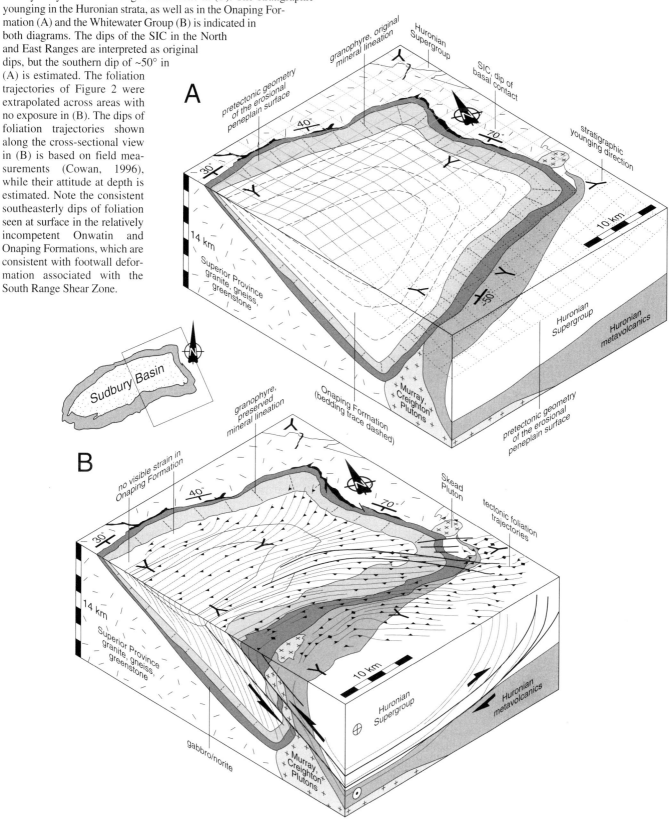

Figure 10. A, Scaled isometric block diagram of the eastern Sudbury Structure at the time of consolidation of SIC, after the deposition of the Onwatin Formation (bedding trace in dashed lines); and B, diagram of the present structural configuration of the Sudbury Structure. The upper paleohorizontal plane of the block diagram in (A) is tangential to the present erosional peneplain surface (mesh surface) in the north, but does not represent the upper depositional surface of the Onaping Formation. The funnel-shaped SIC necessarily implies that the Onaping Formation filled a parabolic depression on impact. The lineated and contact-orthogonal mineral fabric of granophyre is schematically shown in both diagrams, but its delicate fabric is destroyed by the South Range Shear Zone in (B). The stratigraphic younging in the Huronian strata, as well as in the Onaping Formation (A) and the Whitewater Group (B) is indicated in both diagrams. The dips of the SIC in the North and East Ranges are interpreted as original dips, but the southern dip of ~50° in (A) is estimated. The foliation trajectories of Figure 2 were extrapolated across areas with no exposure in (B). The dips of foliation trajectories shown along the cross-sectional view in (B) is based on field measurements (Cowan, 1996), while their attitude at depth is estimated. Note the consistent southeasterly dips of foliation seen at surface in the relatively incompetent Onwatin and Onaping Formations, which are consistent with footwall deformation associated with the South Range Shear Zone.

of the tripartite main mass of the SIC. Thomson and Williams (1959, p. 61) stated that "Those who persist in believing that the eruptive is a folded sill must explain why much of its south side is overturned toward the north whereas the beds beyond are 'right side up' and facing south at high angles. They must also explain why most of the east side of the eruptive is very steep…," and, "In fact, gravitational magmatic segregation of the type generally visualized at Sudbury is *physically impossible* [their emphasis] in those places where the eruptive dips vertically or outward, when it is known that the eruptive was never extensively folded after its emplacement." While reverse shearing of the southern SIC may well have affected its attitude (Shanks and Schwerdtner, 1991a), the comments of Thomson and Williams (1959) still apply to the eastern SIC, which is characterized by dips of 70° (Fig. 2), in some places reported to be vertical (Dressler, 1982, p. 94). Similarly, Cooke (1948) explained the subvertical attitude of south-facing Huronian strata south of the SIC by folding prior to emplacement of the SIC, which is also corroborated by our structural results (for details, see Riller and Schwerdtner, 1997; Riller et al., 1998). Thus, the above observations, which have been part of the geologic literature for a long time, must be incorporated into models of the emplacement and primary geometry of the SIC.

Contrasting SIC petrofabrics

A common characteristic of terrestrial impact melts is that they cool as a single unit (Grieve, 1975; Floran et al., 1978). The pattern of mineral fabric in the norite and the gabbro differs from that of the granophyre and indicates that these units could not have shared the same crystallization history. Unlike the granophyre, which does not record a petrofabric modified by gravitationally induced flow, the gabbro and norite exhibit well-developed contact-parallel L-S mineral fabrics. This may indicate that the crystallization period of these phases relative to the granophyre was longer. The contrasting petrofabrics of the granophyre and the gabbro/norite may indicate, but do not necessarily prove, different origins for both units and appear to conflict with the hypothesis that the entire SIC was created by in situ differentiation of an impact melt. Chai and Eckstrand (1993a,b) suggested that two chemically distinct magmas contributed to the formation of the SIC. Their interpretation is corroborated by the phase equilibria calculations of Ariskin (1997), who could not account for the high volumetric proportion of granophyre by in situ differentiation of a single magmatic body. Unusually high volumetric proportion of the granophyre in relation to the mafic phase of the SIC suggests either, the presence of a large hidden parental source, regardless of whether the SIC is interpreted as an impact melt, mantle-derived body, or a mixture of the two; or it suggests that the granophyre phase represents an impact melt, whereas the gabbro, norite, Sublayer, and Offset Dikes were derived mainly from an impact-induced magmatic source (Dence, 1972; Shanks et al., 1990; Dressler et al., 1996). We cannot rule out either model on the basis of our structural data. However, both scenarios are consistent with the observed chemical discontinuity between the granophyre and norite/gabbro layers (Chai and Eckstrand, 1993a,b) and the contrasting petrofabric character of the granophyre and the norite/gabbro phases (Fig. 3).

Initial geometry of the SIC

Seismic profiles show that the SIC is bound by two parallel contact surfaces (Milkereit et al., 1992). This excludes emplacement models in which the primary geometry of the SIC is characterized by a concave base and a horizontal top, i.e., a ponded impact sheet (e.g., Grieve et al., 1991, Fig. 8; Wu et al., 1994, Fig. 4A; Thompson and Spray, 1994, Fig. 8; Wood and Spray, 1998, Fig. 7). Such an initial configuration of nonparallel surfaces requires substantial tectonic strain to generate the observed parallelism of the base and top of the SIC. However, there is no field evidence for large strains in the SIC of the East and North Ranges (Rousell, 1984; Muir, 1984). Consequently, theoretical models that invoke large degrees of tectonic strains in the entire the SIC are implausible (e.g., Morrison, 1984, Fig. 23.3; Golightly, 1994, Fig. 10.3; Roest and Pikington, 1994, Fig. 3). We argue that the upper surface of the SIC was initially nonhorizontal and that the SIC consolidated as a parabolic body. Strain data suggests that the SIC was never circular along a horizontal plane (i.e., not the present erosion plane, since this plane was not originally horizontal: see Fig. 10A). Thus, the North Lobe and the South Lobe are most likely primary features (Fig. 10A).

Knowledge on the original dips of the gabbro/norite and granophyre sheets is most critical for discriminating between emplacement models. In situ differentiation of the SIC, characterized by primary contacts dipping between 30° and 90°, would likely lead to subhorizontal phase boundaries between the norite, gabbro, and the granophyre. Based on the seismic image, however, the phases boundaries are subparallel to the inclined base of the SIC (Milkereit et al., 1992). Phase boundaries that are parallel to steeply inclined contacts are known from large ultramafic intrusions (e.g., references in Cawthorn, 1996). While it is possible that the relatively thin norite and gabbro sheets may have formed by in situ differentiation along steep contacts, such origin appears to be implausible for the granophyre, given its large volume in relation to the total volume of the SIC (Ariskin, 1997).

Conflicts with theoretical impact models

Our structural results of the SIC and its host rocks are best explained in terms of a primary parabolic geometry of the SIC, but this conflicts with popular impact cratering models. Finite-element models of large impact structures predict a horizontal impact melt-sheet ponded in a complex crater (e.g., references in Melosh, 1989; Ivanov and Deutsch, 1997) and are consistent with the geometry of impact-melt rocks associated with some terrestrial impacts (Grieve, 1975; Floran et al., 1978). Assuming that our interpretation of structural relationships is correct and that geochemical data are consistent with at least the granophyre representing impact melt (Chai and Eckstrand, 1993b), numerical models predicting impact cratering dynamics, melt generation, and emplacement may not readily apply to the SIC. In particular, such models do not account for the inferred emplacement of the granophyre sheet in a parabolic geometry after deposition of the Onaping Formation. They are compatible with an origin of the granophyre only as a melt sheet ponded in the impact crater before emplacement of the suevite (assuming that the Onaping Formation represents a suevite [Ames and Gibson, 1995]). If the granophyre is regarded as an impact melt, a hitherto unknown physical process must have injected the granophyre magma after emplacement of the Onaping Formation. Any model explaining the origin of the SIC must account for its synformal geometry during consolidation.

Numerical models of impact cratering predict large differential displacements on faults at ultra-high strain rates and a pervasive disaggregation of target rocks by large amounts of rigid-body rotation and discontinuous deformation (e.g., O'Keefe and Ahrens, 1994; Ivanov and Deutsch, 1997). Such deformation has been explained by Melosh (1979) in terms of acoustic fluidization. It is inferred that the process of acoustic fluidization can be accomplished only by distributed fracturing of the crust, yet there is little field evidence for the presence of an extensive network of brittle faults separating rotated blocks in Archean or Huronian target rocks. Pseudotachylyte dikes in the Sudbury area have been recently interpreted as structures marking fault surfaces that formed during the latest stage of crater modification (Thompson and Spray, 1994; Spray and Thompson, 1995; Spray, 1997). The lateral continuity of these features, however, remains to be substantiated with detailed fieldwork away from road exposures (Spray and Thompson, 1995, Fig. 1A). Moreover, the continuity of pre-SIC structures in Huronian host rocks bordering on the southern SIC suggests little differential rotation within these rocks following the impact (Cowan and Schwerdtner, 1994; Riller, 1996). We contend that much of the geologic history of the Sudbury area can be explained without invoking impact-related deformation. The conspicuous lack of extensive translation and rotation of target rocks upon impact (cf. Melosh, 1979) must be included in impact cratering models.

CONCLUSIONS

Results of our structural analysis of the Sudbury Igneous Complex and its host rocks suggest that the complex had a parabolic geometry and was noncircular in plan view at the time of its consolidation (Fig. 10). Structural evidence for these conclusions include the following points:

1. The presence of radial igneous fabrics of the gabbro and norite dipping toward the center of the synformal SIC, interpreted to be a product of inwardly dipping initial SIC contacts.

2. The preservation of contact-orthogonal petrofabrics of the granophyre in the North Lobe, implying that folding was not responsible for the tight curvature after consolidation of the granophyre.

3. Steep attitudes of Huronian strata prior to the emplacement of the SIC, consistent with a formation of a structural dome prior to the Sudbury impact event, but inconsistent with impact-related overturning of strata.

4. The lack of evidence for large strains imparted to the SIC and its Huronian host rocks by noncylindrical buckle folding, which is required if the SIC was a horizontal sheet prior to deformation.

We concur with the original interpretation by Dietz (1964), in which the Sudbury region was the site of a hypervelocity impact. Our structural evidence appears, at least indirectly, to partly affirm Dietz's interpretation that the SIC is an impact-induced magmatic body (see also French, 1970), or more precisely, a mixture of impact melt and magmatic component generated by impact-induced decompression melting (cf. Dence, 1972; Dressler et al., 1996; Rousell et al., 1997). While recent geochemical evidence for the contribution of impact-induced crustal melt appears persuasive to some (e.g., Lightfoot et al., 1997), the geometric state of the SIC is inconsistent with presently accepted models of impact cratering. The chemical signature of crustal contamination of the SIC will remain a contentious issue until our structural data are incorporated into a unifying geologic model explaining the compositional and geometric peculiarities of the SIC.

There are many geologic features of the Sudbury Structure that cannot be explained with current models of impact cratering: (1) the primary basinal geometry of the SIC, (2) the inferred nontabular initial geometry of the Onaping Formation (Fig. 10A), (3) the paucity of large rotated blocks bound by pseudotachylyte dikes, and (4) the inferred late intrusion of the gabbro/norite (Dressler, 1987). In the light of our structural results, modes of impact melt formation and theoretical impact cratering models may have to be reassessed. Geologic features predicted from these models have led to a strongly biased interpretation of geologic relationships in the Sudbury area (e.g., Huronian outliers representing down-faulted blocks due to crater wall collapse) and to models that purport the existence of features that are not substantiated (e.g., interpretation of the SIC as a ponded melt, interpretation of pseudotachylytes delineating normal faults). Notwithstanding the fact that we may not immediately have satisfactory models to explain our observations, we view that careful description of field data is essential to interpreting and modeling observed features of the Sudbury Structure.

ACKNOWLEDGMENTS

We thank INCO, Ltd., and Falconbridge, Ltd., for permission to access their properties. Magnetic fabric statistics were computed using software AMS-PLOT written by John Dehls. Doreen Ames and Christina Wood are thanked for providing copies of their publications in press. The manuscript greatly benefited from comments by Stuart Brown, and from helpful reviews by Burkhard Dressler, Roy Kligfield, and Peter Schultz. This research was funded by an Ontario Geological Survey Special

Grant and Natural Enviroment Research Council operating grants (to W.M.S); by support from the Canadian Commonwealth Scholarship and Fellowship Plan (to E.J.C.); by Grant HSPII-AUFE from the German Academic Exchange Service and Grants Ri 916/1-1 and Ri 916/2-1 from the German Science Foundation (to U.R.). Computer support and funding during the preparation of the manuscript was provided by the Tectonics Special Research Centre, University of Western Australia. This is Lithoprobe publication 938.

REFERENCES CITED

Ames, D. E., and Gibson, H. L., 1995, Controls on and geological setting of regional hydrothermal alteration within the Onaping Formation, footwall to the Errington and Vermilion base metal deposits, Sudbury Structure, Ontario: Geological Survey of Canada Current Research, v. 1995-E, p. 161–173.

Ames, D. E., Watkinson, D. H., and Parrish, R. R., 1998, Dating of regional hydrothermal system induced by the 1850 Ma Sudbury impact event: Geology, v. 26, p. 447–450.

Archanjo, C. J., Launeau, P., and Bouchez, J. L., 1995, Magnetic fabric vs. magnetite and biotite shape fabrics of the magnetite-bearing granite pluton of Gameleiras (northeast Brazil): Physics of the Earth and Planetary Interiors, v. 89, p. 63–75.

Ariskin, A. A., 1997, Simulating phase equilibria and in situ differentiation for the proposed parental Sudbury magmas; *in* Conference on Large Meteorite Impacts and Planetary Evolution: Houston, Texas, Lunar and Planetary Institute, Contribution 922, p. 3–4.

Avermann, M., 1994, Origin of the polymict, allochthonous breccias of the Onaping Formation, Sudbury Structure, Ontario, Canada, *in* Dressler, B. O., Grieve, R. A. F., and Sharpton, V. L., eds., Large meteorite impacts and planetary evolution: Geological Society of America Special Paper 293, p. 265–274.

Benn, K., 1994, Overprinting of magnetic fabrics in granites by small strains: numerical modeling: Tectonophysics, v. 233, p. 153–162.

Bott, M. P. H., 1959, The mechanisms of oblique slip faulting: Geological Magazine, v. 96, p. 109–117.

Bouchez, J. L., Delas, C., Gleizes, G., and Nédélec, A., 1992, Submagmatic microfractures in granites: Geology, v. 20, p. 35–38.

Card, K. D., 1965, Geology of Hyman and Drury Townships: Ontario Department of Mines Geological Report, v. 34, p. 38.

Card, K. D., Church, W. R., Franklin, J. M., Frarey, M. J., Robertson, J. A., West, G. F., and Young, G. M., 1972, The Southern Province, *in* Price, R. A., and Douglas, R. J. W., eds., Variations in tectonic styles in Canada: Geological Association of Canada Special Paper 11, p. 335–380.

Card, K. D., Gupta, V. K., McGrath, P. H., and Grant, F. S., 1984, The Sudbury Structure: its regional geological and geophysical setting, *in* Pye, E. G., Naldrett, A. J., and Giblin, P. E., eds., The geology and ore deposits of the Sudbury Structure: Ontario Geological Survey Special Volume 1, p. 25–43.

Card, K. D., and Palonen, P. A., 1976, Geology of the Dunlop-Shakespeare area, district of Sudbury: Ontario Division of Mines Geoscience Report, v. 139, p. 1–52.

Cawthorn, R. G., 1996, Layered intrusions: New York, Elsevier Scientific, Developments in Petrology, p. 531.

Chai, G., and Eckstrand, O. R., 1993a, Origin of the Sudbury Igneous Complex—differentiate of two separate magmas: Geological Survey of Canada Paper 93-1E, p. 219–230.

Chai, G., and Eckstrand, O. R., 1993b, Rare-earth element characteristics and origin of the Sudbury Igneous Complex, Ontario, Canada: Chemical Geology, v. 112, p. 221–244.

Cintala, M. J., and Grieve, R. A. F., 1994, The effects of differential scaling of impact melt and crater dimensions on lunar and terrestrial craters: some brief examples, *in* Dressler, B. O., Grieve, R. A. F., and Sharpton, V. L., eds., Large meteorite impacts and planetary evolution: Geological Society of America Special Paper 293, p. 51–59.

Clendenen, W. S., Kligfield, R., Hirt, A. M., and Lowrie, W., 1988, Strain studies of cleavage development in the Chelmsford Formation, Sudbury Basin, Ontario: Tectonophysics, v. 145, p. 191–211.

Coleman, A. P., 1905, The Sudbury Nickel region: Ontario Bureau of Mines Annual Report, v. 14(3), p. 1–188.

Coleman, A. P., 1907, The Sudbury laccolithic sheet: Journal of Geology, v. 15, p. 759–782.

Collins, W. H., and Kindle, E. D., 1935, The life history of the Sudbury nickel irruptive, pt. 2: Intrusion and deformation: Transactions of the Royal Society of Canada 3rd ser., v. 29, p. 27–47.

Cooke, H. C., 1948, Regional structure of the Lake Huron–Sudbury area, structural geology of Canadian ore deposits: Canadian Institute of Mining and Metallurgy Publication, Jubilee Volume, p. 580–589.

Corfu, F., and Lightfoot, P. C., 1996, U-Pb geochronology of the Sublayer environment, Sudbury Igneous Complex: Economic Geology, v. 91, p. 1263–1269.

Cowan, E. J., 1996, Deformation of the eastern Sudbury Basin [Ph.D. thesis]: Toronto, Ontario, University of Toronto, 332 p.

Cowan, E. J., 1999, Magnetic fabric constraints on the initial geometry of the Sudbury Igneous Complex: A folded sheet or a basin-shaped igneous body? Tectonophysics, v. 307, P. 135–162.

Cowan, E. J., and Schwerdtner, W. M., 1994, Fold origin of the Sudbury Basin, *in* Lightfoot, P. C., and Naldrett, A. J., eds., Proceedings of the Sudbury-Noril'sk symposium: Ontario Geological Survey Special Volume 5, p. 45–55.

Cruden, A. R., 1998, On the emplacement of tabular granitoids: Journal of the Geological Society of London, v. 155, P. 853–862.

Cruden, A. R., and Aaro, S., 1992, The Ljugaren granite massif, Dalarna, central Sweden: Geologiska Föreningens i Stockholm Förhandligar, v. 114, p. 209–225.

Cruden, A. R., and Launeau, P., 1994, Structure, magnetic fabric and emplacement of the Archean Lebel Stock, SW Abitibi Greenstone Belt: Journal of Structural Geology, v. 16, p. 677–691.

Dence, M. R., 1972, Meteorite impact craters and the structure of the Sudbury Basin, *in* Guy-Bray, J. V., ed., New developments in Sudbury geology: Geological Association of Canada Special Paper 10, p. 7–18.

Deutsch, A., and Grieve, R. A. F., 1994, The Sudbury Structure: constraints on its genesis from Lithoprobe results: Geophysical Research Letters, v. 21, p. 963–966.

Deutsch, A., Grieve, R. A. F., Avermann, M., Bischoff, L., Brockmeyer, P., Buhl, D., Lakomy, R., Müller-Mohr, V., Ostermann, M., and Stöffler, D., 1995, The Sudbury Structure (Ontario, Canada): a tectonically deformed multiring impact basin: Geologische Rundschau, v. 84, p. 697–709.

Dietz, R. S., 1964, Sudbury Structure as an astrobleme: Journal of Geology, v. 72, p. 412–434.

Dietz, R. S., and Butler, L. W., 1964, Shatter-cone orientation at Sudbury, Canada: Nature, v. 204, p. 280–281.

Dressler, B. O., 1982, Geology of the Wanapitei Lake area, district of Sudbury: Ontario Geological Survey Report, v. 213, 131 p.

Dressler, B. O., 1984, Sudbury geological compilation: Ontario Geological Survey, Geological Series, Map 2491, scale 1:50,000.

Dressler, B. O., 1987, Precambrian geology of the Falconbridge Township, District of Sudbury: Ontario Geological Survey, Geological Series, Preliminary Map P. 3067, scale 1:15,840.

Dressler, B. O., Gupta, V. K., and Muir, T. L., 1991, The Sudbury Structure, *in* Thurston, P. C., Williams, H. R., Sutcliffe, R. H., and Stott, G. M., eds., Geology of Ontario: Ontario Geological Survey Special Volume 4, p. 593–625.

Dressler, B. O., Weiser, T., and Brockmeyer, P., 1996, Recrystallized impact glasses of the Onaping Formation and the Sudbury Igneous Complex, Sudbury Structure, Ontario, Canada: Geochemica et Cosmochimica, v. 60, p. 2019–2036.

Drury, S. A., 1993, Image interpretation in geology: London, Chapman & Hall, 283 p.

Faggart, B. E., Basu, A. R., and Tatsumoto, M., 1985, Origin of the Sudbury Complex by meteoritic impact: neodymium isotropic evidence: Nature, v. 230, p. 436–439.

Floran, R. J., Grieve, R. A. F., Phinney, J. L., Warner, J. L., Simonds, C. H., Blanchard, D. P., and Dence, M. R., 1978, Manicouagan impact melt, Quebec. 1, Stratigraphy, petrology, chemistry: Journal of Geophysical Research, v. 83, no. B6, p. 2737–2759.

Frarey, M. J., Loveridge, W. D., and Sullivan, R. W., 1982, A U-Pb zircon age for the Creighton granite, Ontario: Geological Survey of Canada Paper, 81-1C, p. 129–132.

French, B. M., 1967, Sudbury structure, Ontario: some petrographic evidence for origin by meteorite impact: Science, v. 156, p. 1094–1098.

French, B. M., 1970, Possible relations between meteorite impact and igneous petrogenesis, as indicated by the Sudbury Structure, Ontario, Canada: Bulletin Volcanologique, v. 34, p. 466–517.

Golightly, J. P., 1994, The Sudbury Igneous Complex as an impact melt: evolution and ore genesis, in Lightfoot, P. C., and Naldrett, A. J., eds., Proceedings of the Sudbury-Norl'sk Symposium: Ontario Geological Survey Special Volume 5, p. 105–117.

Grieve, R. A. F., 1975, Petrology and chemistry of the impact melt at Mistastin Lake crater, Labrador: Geological Society of America Bulletin, v. 86, p. 1617–1629.

Grieve, R. A. F., 1991, Terrestrial impact: the record in the rocks: Meteoritics and Planetary Science, v. 26, p. 175–194.

Grieve, R. A. F., 1994, An impact model of the Sudbury Structure, in Lightfoot, P. C., and Naldrett, A. J., eds., Proceedings of the Sudbury-Noril'sk Symposium: Ontario Geological Survey Special Volume 5, p. 119–132.

Grieve, R. A. F., Stöffler, D., and Deutsch, A., 1991, The Sudbury Structure: controversial or misunderstood?: Journal of Geophysical Research, v. 96, no. E5, p. 22753–22764.

Grieve, R. A. F., Stöffler, D., and Deutsch, A., 1993, Clarification to "The Sudbury Structure: controversial or misunderstood?": Journal of Geophysical Research, v. 98, no. E11, p. 20903–20904.

Guillet, P., Bouchez, J.-L., and Wagner, J.-J., 1983, Anisotropy of magnetic susceptibility and magmatic structures in the Guerande Granite Massif (France): Tectonics, v. 2, p. 419–429.

Hamilton, W., 1960, Form of the Sudbury lopolith: Canadian Mineralogist, v. 6, p. 437–447.

Hext, G. R., 1963, The estimation of second-order tensors, with related tests and design: Biometrika, v. 50, p. 353–373.

Hirt, A. M., Lowric, W., Clendenen, W. S., and Kligfield, R., 1993, Correlation of strain and the anisotropy of magnetic susceptibility in the Onaping Formation: evidence for a near-circular origin of the Sudbury Basin: Tectonophysics, v. 225, p. 231–254.

Hrouda, F., 1982, Magnetic anisotropy of rocks and its application in geology and geophysics: Geophysical Surveys, v. 5, p. 37–82.

Hyodo, H., Dunlop, D. J., and McWilliams, M. O., 1986, Timing and extent of Grenvillian magnetic overprint near Temagami, Ontario, in Moore, J. M., Davidson, A., and Baer, A. J., eds., The Grenville Province: Geological Association of Canada Special Publication 31, p. 119–126.

Ivanov, B. A., and Deutsch, A., 1997, Sudbury impact event: cratering mechanics and thermal history; in Conference on Large Meteorite Impacts and Planetary Evolution: Houston, Texas, Lunar and Planetary Institute, Contribution 922, p. 26–27.

Jelinek, V., 1978, Statistical processing of anisotropy of magnetic susceptibility measured on groups of specimens: Studia Geophysica et Geodaetia, v. 22, p. 50–62.

Jones, G. H. S., 1977, Complex craters in alluvium, in Roddy, D. J., Pepin, R. O., and Merrill, R. B., eds., Impact and explosion cratering: New York, Pergamon Press, p. 163–183.

Krogh, T. E., Davis, D. W., and Corfu, F., 1984, Precise U-Pb zircon and baddeleyite ages for the Sudbury area, in Pye, E. G., Naldrett, A. J., and Giblin, P. E., eds., The geology and ore deposits of the Sudbury Structure: Ontario, Canada, Ontario Geological Survey Special Volume 1, p. 431–446.

Krogh, T. E., Corfu, F., Davis, D. W., Dunning, G. R., Heaman, L. M., Kamo, S. L., Machado, N., Greenough, J. D., and Nakamura, E., 1987, Precise U-Pb isotopic ages of diabase dykes and mafic to ultramafic rocks using trace amounts of baddeleyite and zircon, in Halls, H. C., and Fahrig, W. F., eds., Mafic dyke swarms: Geological Association of Canada Special Paper 34, p. 147–152.

Krogh, T. E., Kamo, S. E., and Bohor, B. F., 1996, Shock metamorphosed zircons with correlated U-Pb discordance and melt rocks with concordant protolith ages indicate an impact origin for the Sudbury Structure, earth processes: reading the isotopic code: Washington, D.C., Americal Geophysical Union, p. 343–353.

Lakomy, R., 1990, Implications for cratering mechanics from a study of the Footwall Breccia of the Sudbury impact structure, Canada: Meteoritics and Planetary Science, v. 25, p. 195–207.

Launeau, P., and Robin, P.-Y. F., 1996, Fabric analysis using the intercept method: Tectonophysics, v. 267, p. 91–119.

Launeau, P., Cruden, A. R., and Bouchez, J.-L., 1994, Mineral recognition in digital images of rocks: a new approach using multichannel classification: Canadian Mineralogist, v. 32, p. 919–933.

Lightfoot, P. C., Keays, R. R., and Doherty, W., 1997, Can impact-generated melts have mantle contributions? Geochemical evidence from the Sudbury Igneous Complex; in Conference on Large Meteorite Impacts and Planetary Evolution: Houston, Texas, Lunar and Planetary Institute, Contribution 922, p. 30.

Lofgren, G. E., 1983, Effect of heterogeneous nucleation on basaltic textures: a dynamic crystallization study: Journal of Petrology, v. 24, p. 229–255.

Lofgren, G. E., and Donaldson, C. H., 1975, Curved branching crystals and differentiation in comb-layered rocks: Contributions to Mineralogy and Petrology, v. 49, p. 309–319.

McBirney, A. R., and Noyes, R. M., 1979, Crystallization and layering of the Skaergaard intrusion: Journal of Petrology, v. 20, p. 487–554.

Melosh, H. J., 1979, Acoustic fluidization: a new geologic process?: Journal of Geophysical Research, v. 84, p. 7513–7520.

Melosh, H. J., 1989, Impact cratering: a geological process, Oxford, United Kingdom, Oxford University Press, Oxford Monographs on Geology and Geophysics, v. 11, 245 p.

Milkereit, B., Green, A., and the Sudbury Working Group, 1992, Geometry of the Sudbury Structure from high-resolution seismic reflection profiling: Geology, v. 20, p. 807–811.

Moore, J. G., and Lockwood, J. P., 1973, Origin of comb layering and orbucular structure, Sierra Nevada batholith, California: Geological Society of America Bulletin, v. 84, p. 1–20.

Morrison, G. G., 1984, Morphological features of the Sudbury Structure in relation to an impact origin, in Pye, E. G., Naldrett, A. J., and Giblin, P. E., eds., The geology and ore deposits of the Sudbury Structure: Ontario Geological Survey Special Volume 1, p. 513–521.

Muir, T. L., 1984, The Sudbury Structure; considerations and models for an endogenic origin, in Pye, E. G., Naldrett, A. J., and Giblin, P. E., eds., The geology and ore deposits of the Sudbury Structure: Ontario Geological Survey Special Volume 1, p. 449–490.

Muir, T. L., and Peredery, W. V., 1984, The Onaping Formation, in Pye, E. G., Naldrett, A. J., and Giblin, P. E., eds., The geology and ore deposits of the Sudbury Structure: Ontario Geological Survey Special Volume 1, p. 139–210.

Naldrett, A. J., and Hewins, R. H., 1984, The main mass of the Sudbury Igneous Complex, in Pye, E. G., Naldrett, A. J., and Giblin, P. E., eds., The geology and ore deposits of the Sudbury Structure: Ontario Geological Survey Special Volume 1, p. 235–251.

Naldrett, A. J., Bray, J. G., Gasparini, E. L., Podolski, T., and Rucklidge, J. C., 1970, Cryptic variation and the petrology of the Sudbury Nickel Irruptive:

Economic Geology, v. 65, p. 122–155.

Nicolas, A., 1992, Kinematics in magmatic rocks with special reference to gabbros: Journal of Petrology, v. 33, p. 891–915.

Norman, M. D., 1994, Sudbury Igneous Complex: impact melt or endogenous magma? Implications for lunar crustal evolution, *in* Dressler, B. O., Grieve, R. A. F., and Sharpton, V. L., eds., Large meteorite impacts and planetary evolution: Geological Society of America Special Paper 293, p. 331–341.

O'Keefe, J. D., and Ahrens, T. J., 1994, Impact-induced melting of planetary surfaces, *in* Dressler, B. O., Grieve, R. A. F., and Sharpton, V. L., eds., Large meteorite impacts and planetary evolution: Geological Society of America Special Paper 293, p. 103–109.

Ostermann, M., and Deutsch, A., 1997, The Sudbury Igneous Complex (SIC) as impact melt layer: geochemical evidence for in situ differentiation: *in* Conference on Large Meteorite Impacts and Planetary Evolution: Houston, Texas, Lunar and Planetary Institute, Contribution 922, p. 38–39.

Paterson, S. R., Vernon, R. H., and Tobisch, O. T., 1989, A review of criteria for the identification of magmatic and tectonic foliations in granitoids: Journal of Structural Geology, v. 11, p. 349–363.

Peredery, W. V., 1972a, Chemistry of fluidal glasses and melt bodies in the Onaping Formation, *in* Guy-Bray, J. V., ed., New developments in Sudbury geology: Geological Association of Canada Special Paper 10, p. 49–59.

Peredery, W. V., 1972b, The origin of rocks at the base of the Onaping Formation, Sudbury, Ontario [Ph.D. thesis]: Toronto, Ontario, University of Toronto, 366 p.

Peredery, W. V., and Morrison, G. G., 1984, Discussion of the origin of the Sudbury Structure, *in* Pye, E. G., Naldrett, A. J., and Giblin, P. E., eds., The geology and ore deposits of the Sudbury Structure: Ontario Geological Survey Special Volume 1, p. 491–512.

Peredery, W. V., and Naldrett, A. J., 1975, Petrology of the upper irruptive rocks, Sudbury, Ontario, Canada: Economic Geology, v. 70, p. 164–175.

Phemister, T. C., 1925, Igneous rocks of Sudbury and their relation to the ore deposits: Report of the Ontario Department of Mines, v. 34, p. 1–61.

Pollard, D. D., Saltzer, S. D., and Rubin, A. M., 1993, Stress inversion methods: are they based on faulty assumptions?: Journal of Structural Geology, v. 15, p. 1045–1054.

Pye, E. G., Naldrett, A. J., and Giblin, P. E., 1984, The geology and ore deposits of the Sudbury Structure: Ontario Geological Survey Special Volume 1, 603 p.

Quadling, K., and Cawthorn, R. G., 1994, The layered gabbronorite sequence, Main Zone, eastern Bushveld Complex: South African Journal of Geology, v. 97, p. 442–454.

Richter, C., 1992, Particle motion and the modeling of strain response in magnetic fabrics: Geophysical Journal International, v. 110, p. 451–464.

Riller, U., 1996, Tectonometamorphic episodes affecting the southern footwall of the Sudbury Basin and their significance for the origin of the Sudbury Igneous Complex, central Ontario, Canada [Ph.D. thesis]: Toronto, Ontario, University of Toronto.

Riller, U., and Schwerdtner, W. M., 1997, Midcrustal deformation at the southern flank of the Sudbury Basin, central Ontario: Geological Society of America Bulletin, v. 109, p. 841–854.

Riller, U., Cruden, A. R., and Schwerdtner, W. M., 1996, Magnetic fabric, microstructure and high-temperature metamorphic overprint of early Murray granite pluton, central Ontario: Journal of Structural Geology, v. 18, p. 1005–1016.

Riller, U., Schwerdtner, W. M., and Robin, P.-Y. F., 1998, Low-temperature deformation mechanisms at a lithotectonic interface near the Sudbury Basin, Eastern Penokean Orogen, Canada: Tectonophysics, v. 287, p. 59–75.

Roddy, D. J., 1977, Large-scale impact and explosion craters: comparisons of morphological and structural analogs, *in* Roddy, D. J., Pepin, R. O., and Merrill, R. B., eds., Impact and explosion cratering: New York, Pergamon Press, p. 185–246.

Roest, W. R., and Pikington, M., 1994, Restoring post-impact deformation at Sudbury: a circular argument: Geophysical Research Letters, v. 21, p. 959–962.

Roscoe, S. M., and Card, K. D., 1992, Early Proterozoic tectonics and metallogeny of the Lake Huron region of the Canadian Shield: Precambrian Research, v. 58, p. 99–119.

Rousell, D. H., 1984, Structural geology of the Sudbury Basin, *in* Pye, E. G., Naldrett, A. J., and Giblin, P. E., eds., The geology and ore deposits of the Sudbury Structure: Ontario Geological Survey Special Volume 1, p. 83–95.

Rousell, D. H., Gibson, H. L., and Jonasson, I. R., 1997, The tectonic, magmatic and mineralization of the Sudbury Structure: Exploration Mining Geology, v. 6, p. 1–22.

Shanks, W. S., and Schwerdtner, W. M., 1991a, Structural analysis of the central and southwestern Sudbury Structure, Southern Province, Canadian Shield: Canadian Journal of Earth Sciences, v. 28, p. 411–430.

Shanks, W. S., and Schwerdtner, W. M., 1991b, Crude quantitative estimates of the original northwest-southeast dimension of the Sudbury Structure, south-central Canadian Shield: Canadian Journal of Earth Sciences, v. 28, p. 1677–1686.

Shanks, W. S., Dressler, B., and Schwerdtner, W. M., 1990, New developments in Sudbury geology; *in* International Workshop on Meteorite Impacts on the Early Earth: Perth, Australia, Lunar and Planetary Institute, Contribution 746, p. 46.

Shoemaker, E. M., 1960, Penetration mechanics of high velocity meteorites, illustrated by Meteor Crater, Arizona: Report, 21st, International Geological Congress, v. pt. 7, p. 418–434.

Shoemaker, E. M., and Kieffer, S. W., 1978, Guide to the astronaut trail at Meteor Crater, *in* Shoemaker, E. M., and Kieffer, S. W., eds., Guidebook to the geology of Meteor Crater, Arizona: Tempe, Arizona, Center for Meteorite Studies, p. 34–62.

Spang, J. H., 1972, Numerical method for dynamic analysis of calcite twin lamellae: Geological Society of America Bulletin, v. 83, p. 467–472.

Sperner, B., and Ratschbacher, L., 1994, A Turbo Pascal program package for graphical presentation and stress analysis of calcite deformation: Zeitung der deutschen geologischen Gesellschaft, v. 145, p. 414–423.

Spray, J. G., 1997, Superfaults: Geology, v. 25, p. 579–582.

Spray, J. G., and Thompson, L. M., 1995, Friction melt distribution in a multi-ring impact basin: Nature, v. 373, p. 130–132.

Thompson, L. M., and Spray, J. G., 1994, Pseudotachylytic rock distribution and genesis within the Sudbury impact structure, *in* Dressler, B. O., Grieve, R. A. F., and Sharpton, V. L., eds., Large meteorite impacts and planetary evolution: Geological Society of America Special Paper 293, p. 275–287.

Thomson, J. E., and Williams, H., 1959, The myth of the Sudbury Lopolith: Canadian Mining Journal, v. 80, p. 57–62.

Twiss, R. J., and Moores, E. M., 1992, Structural geology: New York, W. H. Freeman, 532 p.

Wood, C. R., and Spray, J. G., 1998, Origin and emplacement of Offset Dykes in the Sudbury impact structure: constraints from Hess: Meteoritics and Planetary Science, v. 33, p. 337–347.

Wu, J., Milkereit, B., and Boerner, D., 1994, Timing constraints on deformation history of the Sudbury Structure: Canadian Journal of Earth Sciences, v. 31, p. 1654–1660.

Wynne-Edwards, H. R., 1972, The Grenville Province, *in* Price, R. A., and Douglas, R. J. W., eds., Variations in tectonic style in Canada: Geological Association of Canada Special Paper 11, p. 263–334.

Zolnai, A. I., Price, R. A., and Helmstaedt, H., 1984, Regional cross section of the Southern Province adjacent to Lake Huron, Ontario: implications for the tectonic significance of the Murray Fault Zone: Canadian Journal of Earth Sciences, v. 21, p. 447–456.

MANUSCRIPT ACCEPTED BY THE SOCIETY DECEMBER 16, 1998

Structural evolution of the Sudbury impact structure in the light of seismic reflection data

D. E. Boerner
Geological Survey of Canada, 601 Booth Street, Ottawa, Ontari K1A 0E9, Canada
B. Milkereit
Institute of Geophysics, Christian Albrechts University, Olshausenstrasse 40, 24098 Kiel, Germany

ABSTRACT

Rock property data, in situ logging results, and surface-based seismic reflection data all show that strong reflection coefficients characterize the contact between the Sudbury Igneous Complex and its footwall rocks. This ability of seismic methods to map the crater floor, coupled with the pristine state of the Sudbury Igneous Complex immediately following impact, provides essential geometric data for obtaining a more complete understanding of the Sudbury Structure deformation. In particular, we seek to classify seismic reflection events as arising from impact processes, contacts between different rock units, or pre- or postimpact tectonic deformation. Seismic reflection events observed in the relatively undeformed Archean foreland north of the Sudbury Structure are difficult to attribute to impact processes, or to postimpact tectonic events, and are thus best interpreted as representing the effects of pre-impact orogenesis. Impact-induced deformation is notably absent from the seismic data, yet geophysical interpretations from outside the Sudbury Igneous Complex lack sufficient constraining geologic information and hence are necessarily speculative, demanding further work to achieve resolution. The one prominent exception is an apparent buckling of the North Range Igneous Complex that may represent a response to early folding of the entire structure. Within the South Range of the Sudbury Structure, the seismic data are dominated by structures related to postimpact ductile/brittle failure. Projecting surface observations of the South Range deformation to depth using the seismic reflection data suggests that faults nucleated beneath the folds, likely from within the buried Sudbury Igneous Complex. This interpretation suggests an episodic deformation history, beginning with large-scale, noncylindrical folding, and followed by ductile/brittle failure.

INTRODUCTION

The genesis of the 1,850-Ma (Krogh et al., 1984) Sudbury Structure has been the subject of extensive geologic, geochemical, and geophysical studies for nearly a century (Giblin, 1984). A first-order observation is the widespread and diverse evidence of shock deformation features that document the occurrence of a high-energy meteorite impact (Dietz, 1964; French, 1969; Grieve et al., 1991; Deutsch and Grieve, 1994; Joreau et al., 1996). While it is clear that the structure has undergone tectonic shortening (Milkereit et al., 1992), conflicting genetic implications of detailed structural studies argue that the database of observational evidence regarding Sudbury deformation is, as yet, incomplete. Perhaps the principal reason

---

Boerner, D. E., and Milkereit, B., 1999, Structural evolution of the Sudbury impact structure in the light of seismic reflection data, *in* Dressler, B. O., and Sharpton, V. L., eds., Large Meteorite Impacts and Planetary Evolution II: Boulder, Colorado, Geological Society of America Special Paper 339.

the Sudbury Structure has remained so inscrutable derives from its location astride the southern margin of the Superior Province, a region with a long history of multiple orogenesis. In this chapter, we summarize the available seismic reflection data from Sudbury and its environs to discuss and speculate about the causative geologic deformation mechanisms. Drill results and geologic ground verification strongly confine the interpretation of the North Range footwall contact in the seismic reflection data. However, the interpretation of other seismic reflection events is less certain, and may be equivocal. Thus, we seek to devise a consistent framework to discuss the implications of these geophysical data so that they can be properly tested with future work.

The Sudbury Structure is located near the contact between Paleoproterozoic Huronian supracrustal rocks of the Southern Province and Archean basement rocks of the Superior Province (Fig. 1) and comprises the entire region of crust affected by the impact. The Structure includes the basinal Whitewater Group sediments cradled by the Sudbury Igneous Complex (SIC) and the encircling shock- and impact-induced metamorphic features (Dietz, 1964; French, 1969; Dressler, 1984a; Grieve et al., 1991; Spray and Thompson, 1995) seen throughout the country rock. The SIC, consisting of the unusually siliceous Granophyre, Quartz Gabbro, and Norite, is a primary feature related to the impact event (Fig. 1) (Faggart et al., 1985; Grieve et al., 1991; Walker et al., 1991). Although the genetic details of the SIC tripartite stratification are still debated (e.g., Naldrett, 1984; Chai and Eckstrand, 1993; Riller and Schwerdtner, 1997), it is clear that the molten SIC cooled soon after the impact. Thus, the SIC was a pristine geologic unit at approximately 1,850 Ma, which can serve to help distinguish between pre- and postimpact deformation. Excellent descriptions of the detailed geologic setting appear in Dressler (1984a,b) and the collection of papers edited by Pye (1984) and Naldrett (1994). Background information on the geophysical view of Sudbury is provided in the series of papers introduced by Boerner et al. (1994b).

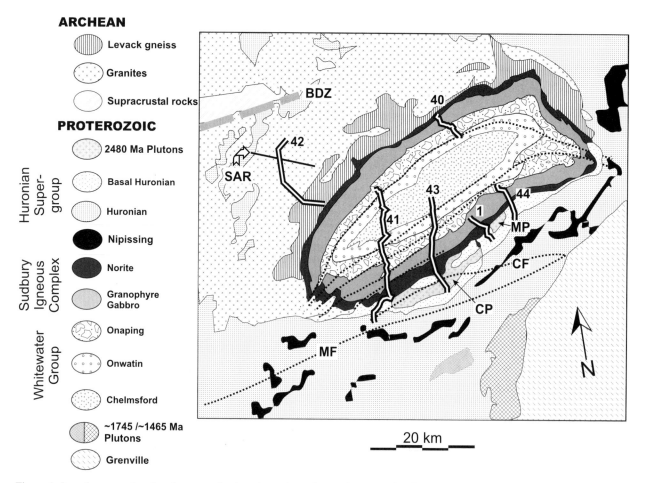

Figure 1. Location map showing the generalized geology along the southern margin of the Superior Province. The Sudbury Structure includes the Sudbury Igneous Complex, the Whitewater Group basinal sediments, and the environs subjected to shock by the meteorite impact at 1,850 Ma. SAR indicates the location of a major remote sensing lineament. BDZ = Benny deformation zone; MF = Murray fault; CF = Creighton fault; MP = Murray pluton; CP = Creighton pluton (modified after Dressler, 1984b).

MULTIPLE OROGENESIS OF THE SOUTHERN PROVINCE

The exposed southern margin of the Superior Province was affected by at least four distinct orogenic events. Uplift of the Levack Gneiss Complex (Fig. 1) occurred at approximately 2,711 Ma (James et al., 1992) and was followed by emplacement of the Cartier Granite around 2,642 Ma (Meldrum et al., 1997). Initial deformation of the Huronian Supergroup marginal sedimentary prism (2,500–2,220 Ma) (Bennett et al., 1994) occurred before intrusion of the Nipissing dikes at 2,220 Ma (the Blezardian orogeny) (Card, 1978; Zolnai et al., 1984; Riller and Schwerdtner, 1997). The margin was likely compressed again during by the 1,890–1,830-Ma Penokean orogeny. The actual eastward extent of the Penokean orogeny into Ontario is equivocal (Davidson et al., 1992; Card, 1992), yet there is isotopic evidence of a transition between Archean to Proterozoic crustal ages buried beneath the Grenville Province in Ontario (Dickin and McNutt, 1989; Dickin et al., 1990). It was during the waning stages of the Penokean orogeny that the meteorite impact occurred. The SIC is thought to have developed from an impact-generated crustal melt sheet (Grieve et al., 1991), possibly with later intrusive contributions (Chai and Eckstrand, 1993; Riller and Schwerdtner, 1997). At some point following solidification, the SIC was deformed.

Minor plutonism around the Grenville Front occurred in discrete intervals from 1,700 to 1,450 Ma (Fig. 1) (Davidson et al., 1992), perhaps representing the far-field effects of orogeny (see also Fueten and Redmond, 1997). The most recent major tectonic event near Sudbury was the 1,300–1000-Ma Grenvillian orogeny characterized by polyphase deformation, intense metamorphism, and igneous activity. Despite its vigor and the proximity to Sudbury, there is little evidence that Grenville orogeny affected the SIC (Brocoum and Dalziel, 1974; Davidson, 1992). In particular, unstrained Sudbury diabase dikes (1,235 Ma) (Dudás et al., 1994) transect the Sudbury Structure, but are highly disturbed where continuations can be traced into the Grenville Province (Bethune, 1997). Today, the Grenville Front (Fig. 1) is a steeply south-dipping deformed hanging wall above a crustal scale ramp (Green et al., 1988; Rivers et al., 1993) that defines the northernmost limit of the orogenic metamorphic effects.

Despite a history of repeated margin formation and destruction, there is surprisingly little evidence preserved in the sedimentary record of foreland basins related to plate flexure and tectonic shortening. Postimpact sedimentary rocks within the Sudbury area include the Whitewater Group, made up of impact breccia of the Onaping Formation; pelagic metasedimentary rocks of the Onwatin Formation (Muir and Peredery, 1984; Deutsch and Grieve, 1994; Gibbins et al., 1997); and metawackes of the Chelmsford Formation. The Onwatin and Chelmsford Formations reflect quiescent tectonic conditions (isolated basins with restricted sediment input), even while being deposited amid the orogenic paroxysm that consolidated most of Laurentia between 2,000 and 1,750 Ma (Hoffman, 1989). These formations have no known preserved equivalents in Ontario.

Given the extent of deformation experienced by the southern margin of the Superior Province, it is important to note that geologic, geochemical, and geophysical data suggest the North Range of the Sudbury Structure is relatively undeformed. Deformation is concentrated in the southern half of the Sudbury Structure, implying that the North Range was somehow mechanically decoupled from collision and subsequent relaxation. These observations, combined with the past century of geologic mapping, provide unprecedented control on identifying exposed structures and determining lateral movement sense. The seismic data are similarly clear in defining the character of some buried structures, particularly the SIC/footwall contact. Geologic interpretation of the seismic images, however, requires distinguishing the relative timing of the composite tectonic events that are represented in the contemporary seismic images.

SEISMIC IMAGING AND GEOLOGIC VERIFICATION

Six profiles of seismic reflection data were acquired across the Sudbury Structure (Fig. 1). Line 42 crossed the Cartier Granite and Levack Gneiss to the edge of the North Range, whereas Line 40 (Milkereit et al., 1992, 1994b) is the only seismic profile to cross the North Range. Lines 1, 41, 43, and 44 were acquired across the deformed South Range (Boerner et al., 1998).

In situ logging and borehole seismic experiments have been conducted at multiple locations within the SIC using full waveform sonic and density logging tools (Fig. 2). Narrow diameter drill holes were surveyed to depths of about 2,000 m, providing superb control on the origin of reflections and the basic velocity structure of the SIC, particularly at the SIC/footwall contact (Fig. 3). Figure 4 shows density and velocity logs, aligned along the SIC/footwall contact. A narrow range of velocities character-

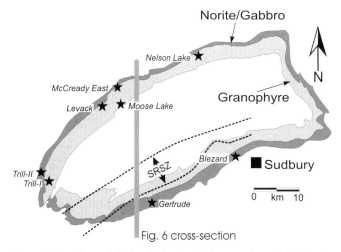

Figure 2. Locations of the boreholes (stars) that were logged for density and seismic velocity (results shown in Fig. 4) relative to the Sudbury Igneous Complex (compare with Fig. 1). The dashed lines indicate the approximate location of the South Range Shear Zone (SRSZ) (Shanks and Schwerdtner, 1991a). Also shown is the location of the cross section shown in Figure 6.

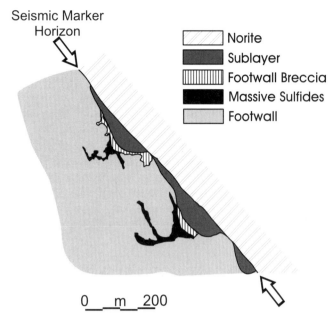

Figure 3. Detail of Norite footwall showing the geologic structure of the crater floor derived from drilling and mining operations (simplified after Morrison, 1984).

extend the logging results to show that the Onaping/Granophyre contact in the SIC is likely transparent to seismic reflection surveys. Prominent reflections occur from within the Whitewater Group (at the Chelmsford-Onwatin and Onwatin-Onaping contacts), the Granophyre-Norite contact of the SIC, and at the contact with the footwall complex (Figs. 3,4). No along-strike variation of the physical properties is observed within the SIC. Similarly, there is no noticeable difference between the velocities or densities of the North Range and South Range rocks (Salisbury et al., 1994).

The Sudbury region is characterized by a wealth of geologic map information, geophysical logs of existing deep drill holes, and physical rock property determinations from core samples. Consequently, interpretation of the surface seismic data from the Sudbury Structure is well supported by basic velocity/density information and is closely coupled to geologic observations. In terms of seismic exploration, the contact between the generally "transparent" SIC Norite and the "reflective" footwall complex define a clear regional marker horizon. Seismic methods can map the bottom of the SIC and topography at the crater floor (see Fig. 3, the principal location for the rich Sudbury Ni-Cu deposits) (Morrison, 1984).

ize the SIC Norite, varying from 6,200 to 6,400 m/s, whereas the Sublayer and footwall complex display more extreme velocity variations (6,000–6,700 m/s). Density logs follow the same trend as the P-wave velocities: relatively uniform densities between 2.75 and 2.8 g/cm^3 are associated with the Norite; Sublayer and footwall complex densities scatter between 2.75 and 3.0 g/cm^3.

Laboratory studies of drill-core sample density and compressional wave velocity (Salisbury et al., 1994) corroborate and

The ability to correlate seismic reflection events with contacts between different rock units in the North Range provides a dramatic contrast with the South Range (Lines 1, 41, 43, and 44) (Milkereit et al., 1992, unpublished data) where reflections can arise only from tectonically imposed structures. Deformation-induced features, such as shear zones and faults, are thought to exhibit strong seismic reflectivity (e.g., Jones and Nur, 1984; Law and Snyder, 1997), a result corroborated by extensive observational evidence in the South Range of the Sudbury Structure (Milkereit et al., 1992). The transition from pristine North Range

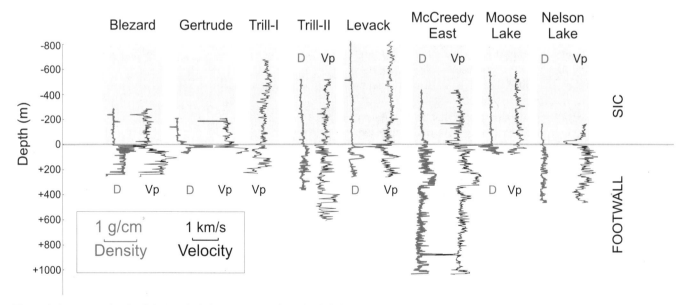

Figure 4. Summary sketch of the borehole log data that relates the SIC footwall contact to a strong seismic discontinuity. Drill-hole locations are shown in Figure 2.

to deformed South Range SIC occurs over a distance of less than 20 km and is compatible with the ground observations of metamorphic grade and deformation magnitude (Cowan and Schwerdtner, 1994). The geomorphologic feature corresponding to the boundary between extensive deformation and intact foreland (buttress) is the ductile-brittle southeast- over northwest-shortening features characteristic of the South Range Shear Zone (SRSZ in Fig. 2) (Shanks and Schwerdtner., 1991a). Consequently, any seismic reflection response occurring structurally below the SRSZ is unlikely to represent brittle deformation structures, unless they predate the impact event.

PRE-IMPACT DEFORMATION: NORTH RANGE FOOTWALL

Pre-impact deformation should be most apparent in the seismic data from Line 42 (Miao et al., 1993; W. Moon, personal communication) (Fig. 5) across the Cartier Granite and Levack Gneiss Complex exposed north of the SIC (Fig. 1). Interpretation of these preliminary images is complicated, but can be aided by extrapolating the seismic data from Lines 40 and 41 westward along the strike of the SIC to construct a combined profile across the North Range and into the Whitewater Group. Limited North Range deformation and the lack of apparent spatial variation in SIC velocities and densities (Salisbury et al., 1994) help to justify this extrapolation. Reflections from the Levack Gneiss on Line 42 are characteristically subhorizontal but are sufficiently chaotic to make migration difficult (W. Moon personal communication), perhaps reflecting the discontinuous nature of the geology (Fueten et al., 1992; Card, 1994). In contrast, Line 40 shows reflections from within the Levack Gneiss that are generally subparallel to the SIC. Several prominent south-dipping reflection events can be identified in the unmigrated data (R1–R3 in Fig. 5).

Following migration, these linear features will be displaced laterally and have a steeper dip, but should project to approximately the same surface location. Although somewhat incoherent, feature R1 apparently projects into ductile shear zone mapped at the contact with the Cartier Granite (Fueten et al., 1992). Farther to the north, two subparallel features (R2 and R3 in Fig 5) within the Cartier Granite display more continuous seismic reflection responses, suggesting they are later, perhaps tectonically induced, features. Further detailed processing followed by depth migration is required to fully elucidate the relationship of these structures to the smaller, more discrete reflection signature in their immediate vicinity.

The surface projection of R2 has been associated (W. Moon, personal communication) with a northwest-southeast lineament observed in synthetic aperture radar (SAR) maps (location indicated in Fig. 1). Although magnetic and electromagnetic data show the lineament is not a dike, its geologic significance is otherwise unknown. The general northwest-southeast strike of the SAR lineament is consistent with the orientation of the brittle structures in the region, but is oblique to the east-west gneissic foliation and major ductile deformation features (Fueten et al., 1992). This lineament strikes obliquely to both the axis of the Sudbury Structure and to the postimpact, minor (>1 km) sinistral movement on the north-trending Onaping fault system (Buchan and Ernst, 1994; Card, 1994). Consequently, it is difficult to associate R2 with impact processes, or with postimpact deformation.

Although not constrained in the near-surface, R3 projects approximately into the location of the Benny Greenstone Belt deformation zone (BDZ) (Card, 1994) located north of the seismic profile (Fig. 1). Shallow, south-dipping foliation and stretching lineations observed in the BDZ reminded Card (1994) of the style and orientations of the structures of the SRSZ, suggesting a Proterozoic age. Kellett and Rivard (1994) have placed bounds on the age of the BDZ by noting that the Huronian sedimentary outliers bordering the Benny Greenstone Belt are deformed while cross-cutting 1,235-Ma Sudbury dikes are undisturbed. Sedimentologic observations by Rousell and Long (1998) further indicate the Huronian outliers were deposited (before 2,200 Ma) in topographic depressions, supporting the contention that the Benny deformation zone, and hence reflection event R3, predate the impact event.

The limited geologic evidence of postimpact North Range footwall deformation supports the contention that R2 and R3 are pre-impact structures. The Cartier Granite is mostly undeformed, containing only minor shear zones and alteration along late faults (see also Buchan and Ernst, 1994). Minor greenschist alteration, predominantly along Proterozoic faults and shear zones, characterizes a late stage in the polymetamorphic history of the Levack Gneiss (James et al., 1992). Lacking evidence to the contrary, we interpret the reflection responses observed in the SIC footwall to be Archean- or Blezardian-age shear or fault zones. Although these seismic reflection events possess apparent dips similar to the SIC footwall contact, their true dips are steeper, and our ten-

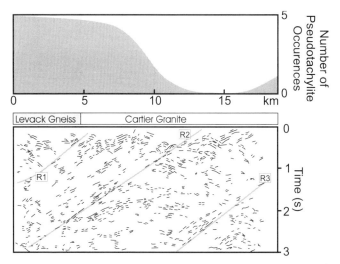

Figure 5. Unmigrated line drawing of the Line 42 seismic data (after Miao et al., 1993) with a schematic representation of the pseudotachylite distribution histogram mapped by Spray and Thompson (1995) to show the lack of correlation between reflection events and pseudotachylite mapping.

tative correlation with surface geology suggests these features have distinct strike directions, and hence different true dips. The apparent correspondence of the seismic reflection event dips may signify only a tectonic vergence toward the foreland characteristic of the different orogenies affecting the southern margin of the Superior Province.

IMPACT-INDUCED DEFORMATION

If evidence of impact-induced deformation is present in the seismic data, it should be most clearly seen in the footwall of the North Range SIC where there has been minimal postemplacement disturbance. The primary geologic evidence of impact-induced deformation in this region consists of localized pockets of Sudbury breccia found north of the SIC (Dressler, 1984a,b; Card, 1994). Mapping by Spray and Thompson (1995) suggests the pseudotachylite occurrences north of the SIC define zones of high-velocity slip generated at impact. Footwall breccia may substantially contribute to the observed reflection response defining the contact between the SIC and the Levack Gneiss complex. However, at greater distances from the SIC, the slip zones postulated by Spray and Thompson (1995) do not correlate with the surface projections of the seismic reflectors within the Cartier Granite (Fig. 5). As the pseudotachylite bodies are small and are distributed as pervasive anastomosing networks (Spray and Thompson 1995), they may be undetectable using surface-based seismic reflection methods. Thus, it seems that seismic data from within the Cartier Granite does not provide any indication of structures created by the impact event. Remote sensing methods studied by Kellett and Rivard (1995) were also unable to resolve the presence of friction melt rings, suggesting that any large-scale morphologic signature attributable to impact-induced fracturing is absent or extremely subtle.

POSTIMPACT DEFORMATION

Sudbury Igneous Complex Norite

Seismic reflection data from the relatively undeformed North Range (Line 40) (Milkereit et al., 1994a; White et al., 1994) show that the Norite and underlying highly reflective footwall can be traced southward from outcrop, to a depth of at least 5 km beneath the Sudbury Basin (Fig. 6). While the Granophyre-Norite interface is also reflective and can be traced to depth, scattered energy seen on Line 41 beneath the Sudbury Basin signifies a dramatic change in the dip of this contact (e.g., Milkereit et al.,

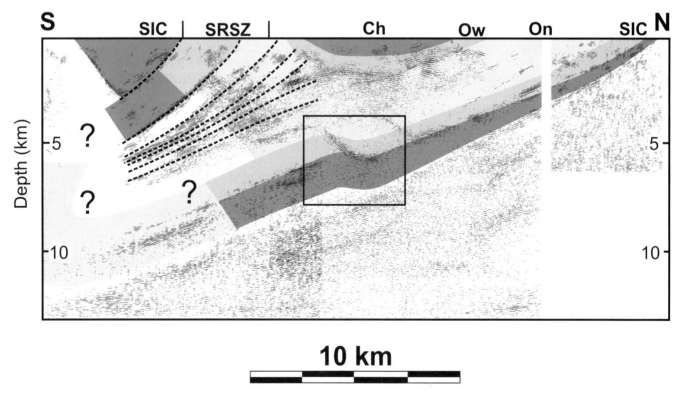

Figure 6. Cross section of the Sudbury Structure based on migrated seismic data (modified from Milkereit et al., 1992). The rectangle outlines the region where a dramatic change in the dip of the Norite layer occurs. The heavy dashed lines are meant to indicate schematically the locations of brittle subsurface fault zones. The shading illustrates the interpreted geometry of the SIC at depth. SRSZ = South Range Shear Zone; Ch = Chelmsford Formation; Ow = Onwatin Formation; On = Onaping Formation; SIC = Sudbury Igneous Complex.

1992). Hints of similar reflection patterns are seen on Lines 1, 43, and 44, although the individual profile lengths are too short (relative to the depth) to allow unambiguous definition. However, data from these profiles indicate the causative structure apparently extends laterally across most of the SIC. Down-dip geologic projections using the Line 41 seismic data (with auxiliary control from drilling) indicate that the disturbed Granophyre-Norite contact is located structurally below the most intense deformation associated with the SRSZ.

Interpreting the seismic data at depth requires some understanding of the initial SIC geometry. Global comparative studies of terrestrial impact sites make it clear that at least some portion, if not all, of the SIC represents an impact melt sheet (e.g., Grieve et al., 1991). A melt pond implies gravitational leveling, and is consistent with the observation that the composition (Naldrett, 1984) and isotopic signature of the SIC requires only contributions from crustal rocks (Grieve et al., 1991). Under this assumption, large-scale, non cylindrical folding may be the only means of obtaining a doubly plunging syncline exposing oblique sections through the thickness of the SIC (e.g., Cowan and Schwerdtner, 1994; Wu et al., 1994). This requirement for folding is not obviated by the structural mapping data that suggest some portion of the SIC might be intrusive (Cowan and Schwerdtner, 1997; Riller and Schwerdtner, 1997). If only one component of the SIC were intrusive, the remaining portion must have originated as a horizontal melt sheet and must have been folded. A wholly intrusive origin for the SIC components does not place any restrictions on the emplacement geometry, but would require multiple injection pulses of crustal melts into a funnel-like geometry to account for the different concentric rock units. Intrusion-only models also need to explain the absence of an impact-induced melt sheet. Accordingly, our preferred interpretation is that at least part of the SIC is an impact melt sheet, suggesting that folding is required to create the primary funnel-shaped geometry of the SIC.

The change in dip of the Norite-Granophyre interface seen in Figure 6 could be interpreted to mean that the SIC acted as a competent layer during initial folding. Compression would have occurred against a foreland buttress (Levack Gneiss), inducing layer parallel shortening and buckle folding. This interpretation is consistent with the observation that the deformed contact is indicated in all the seismic profiles that traverse the axis of the SIC. Note that the chaotic reflection pattern thought to identify the footwall Levack Gneiss complex continues to somewhat greater depths, and might be similarly folded, although it is somewhat indistinct in this region. Interpreting the geometry of the SIC as resulting from overturned folding is also consistent with the notion that brittle faults may have nucleated in the competent layer (the SIC) near the fold axis (Fig. 6; see also Liu and Dixon, 1991).

Sediments of the Whitewater Group

Concretions in the Chelmsford are deformed and show 30% shortening (Brocoum and Dalziel, 1974), in accordance with Shanks and Schwerdtner (1991b) and shortening estimates for the Onaping Formation (38–67%) by Clendenen et al. (1988) and Hirt et al. (1993). Perhaps 40% of the Onaping compression occurred as layer parallel shortening prior to buckling and the formation of the SRSZ (Hirt et al., 1993). While the layer parallel shortening is not visible in the seismic data, Wu et al. (1994) showed strong evidence that brittle thrust faulting offsets the contact between the Onwatin and Onaping Formations (Fig. 6). In general, however, the brittle deformation is blind, ending within the incompetent Onwatin slates. These seismic observations do provide a relative timing constraint by suggesting SIC deformation postdates the lithification of the Whitewater Group, compatible with the development of the SRSZ (e.g., Cowan and Schwerdtner, 1994). Postdepositional deformation is also suggested by the foreland basin–type setting required for sedimentation of the Onwatin and Chelmsford Formations, which precludes the possibility of a high-energy environment typical of syndepositional deformation. These results suggest that the Whitewater Group was involved in the deformation, but that the details of overthrust shear seen in the SRSZ are not well preserved in the incompetent Onwatin and Chelmsford Formations.

South Range Shear Zone

Extensive ductile shearing and brittle faulting associated with the SRSZ characterize the South Range seismic data and likely delineate the most northerly extent of substantive postimpact deformation (Boerner, unpublished data). Based on field and drill-hole observations, the primary deformation associated with the SRSZ is localized within the Onaping and SIC Granophyre, but it does cross-cut lithology in places. The consequence of late tectonic features imposed on the South Range is that contacts between adjacent SIC units are difficult to identify from the seismic data, and the interpretations must be predominantly structural, and not based on lithology. In some rare instances, however, it is possible to identify the contact between the Norite and South Range footwall rocks (Fig. 7) (Milkereit et al., 1996). The Whitewater Group and the South Range exhibit most of the shortening, implying a rather complete mechanical decoupling from the Archean foreland during deformation. Based on electromagnetic, geologic and rock property data, Boerner et al. (1994a) proposed that graphite within the Onaping provided the lubrication on the primary glide plane that accommodated some of the postfailure shortening.

The seismically imaged brittle faults down-dip are listric and merge at depth just south of the exposed SIC, near the fold hinge suggested by the seismic data (Fig. 6). A simple interpretation of these seismic data is that the faults nucleated beneath the existing fold (Liu and Dixon, 1991). As the development of the SRSZ follows lithification of the Whitewater Group (Cowan and Schwerdtner, 1994), a period of tectonic quiescence is required between the initial folding and ductile/brittle-thrusting event to allow the deposition of the Whitewater Group. This episodic

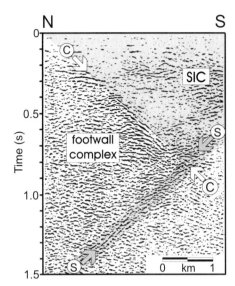

Figure 7. South Range footwall contact displayed in a migrated seismic section showing the Norite-Footwall contact (C) and the superimposed shear zone (S) that apparently truncates the SIC.

pause in tectonic development need not have lasted long. Cowan and Schwerdtner (1994) suggest that Whitewater Group deposition might have taken place within 6 m.y.

Axial deformation geometry

The seismic data portray a complex image of a northwest vergent shear zone that traverses virtually the entire South Range of the SIC (Boerner, unpublished data). Structural mapping evidence (Shanks and Schwerdtner, 1991a) document the surface expression of this deformation and demonstrates that some South Range rocks were tectonically transported to the north. The deformation zones and fault systems that accommodated the compression are convex in the direction of vergence (antiaxial) and are also concordant with metamorphic trends observed within the Whitewater metasediments and SIC (Fleet et al., 1987; Cowan and Schwerdtner, 1994). Together, these observations suggest the overall elliptical geometry of the SIC resulted from northwest directed shortening (e.g., Shanks and Schwerdtner, 1991a; Cowan and Schwerdtner, 1994), although the amount of shortening is not uniform across the SIC (Boerner et al., 1998). The seismically determined true dips of the South Range Shear Zone, ranging between 28° (east) and 35° (west), are relatively shallow and consistent with the range of dips observed for the frontal thrusts in sand models (Huiqi et al., 1992). Such shallow dips on the SRSZ indicate it may be near the deformation front (Marshak et al., 1992), an interpretation supported by the eccentricity of the faults and mineral schistosity within the SIC (Cowan and Schwerdtner, 1994). Yet the SRSZ extends beyond the boundaries of the SIC and the dips become more steep on the westernmost seismic profiles. The dip variation of the deformation zones and the observed offsets in lithologic contacts in the western part of the SIC (Fig. 1) are suggestive of greater tectonic thickening to the southwest.

Near their lateral terminus, the net displacements accommodated by thrust faults must be translated into other forms of shortening (Fig. 8) (Liu and Dixon, 1991; Cowan and Schwerdtner, 1994). In the Sudbury Structure, this progression is seen in the transition from southeast over northwest thrusting in the center of the basin, to steep strike-slip at the southeast corner of the SIC (Cowan and Schwerdtner, 1994; Ramsay and Huber, 1987). That there is little lateral movement apparent in the lithologic offsets at the southeast corner of the SIC is interpreted as indicating layer parallel shortening at the lateral terminus of fold-induced thrusts (Cowan and Schwerdtner, 1994), consistent with the patterns observed in centrifuge models (Liu and Dixon, 1991).

IMPLICATIONS

Successful future exploration of the Sudbury Structure relies on proper appraisal and interpretation of the existing data. To this end, it is quite clear that the Norite/footwall contact can be mapped with confidence using seismic methods, providing a number of possibilities for mapping the deformation of the Sudbury Structure. In particular, the surface-mapped geology can be firmly extrapolated to depth along a flat plane dipping at approximately 25° south starting from the North Range. This one concrete observation has profound implications since the postulated origin (at least partially) of the SIC as an impact induced crustal melt sheet (Grieve et al., 1991; Lightfoot et al., 1997) requires a subhorizontal attitude at formation. Adopting the assumption of initial large-scale, noncylindrical folding (Cowan and Schwerdtner, 1994) offers several ramifications when integrated into the interpretation of the seismic data.

1. Solid body rotation likely contributed to the current 25° south dip of the North Range SIC. An immediate consequence of this rotation is that the current erosional surface exposes an oblique view of the impact crater and its surroundings. This tilting should be considered in comparative studies with other terrestrial and extraterrestrial impact sites.

2. The curious reflection pattern indicating a change in dip of the Granophyre/Norite contact seen at depth beneath the center of the Whitewater Group appears to be a buckling that occurred along the entire length of the SIC. Since it lies structurally below the SRSZ and parallels the axis of the SIC, the buckling is most easily attributed to an initial folding event, followed by later brittle failure. The notion of folding, followed by brittle thrusting, is consistent with the basin-filling low-energy sediments of the Onwatin and Chelmsford Formations. These formations were clearly deformed, but only after lithification. While the exact shortening mechanism can still be debated, it seems clear that the SIC acted as a relatively competent layer within the deformation field.

3. The seismic evidence documents extensive deformation of the competent South Range SIC and the relatively incompe-

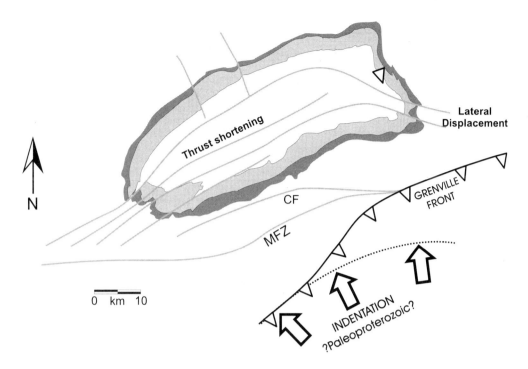

Figure 8. Cartoon interpretation of the major deformational features within the Sudbury Structure showing the transition from thrusting to steep strike-slip (geology and fault locations based on Dressler, 1984b). MFZ = Murray Fault Zone; CF = Creighton Fault

tent Whitewater Group, suggestive of rotation, elongation, and axial alignment during northest-southeast compression. Shortening may thus bias paleocurrent indicators (Brocoum and Dalziel, 1974; Rousell, 1972).

4. Surface faults and the subsurface seismic events are convex in the direction of vergence, suggesting lateral displacement on the eastern flank of the Sudbury Structure and implying indentation of the colliding terrain during the late stages of the Penokean orogeny. However, the exact relationship of this deformation to the earlier folding remains equivocal.

5. The true dip of the latest brittle faults increases to the west, implying greater tectonic thickening in this direction.

6. The paucity of Norite deformation in the North Range requires an almost complete decoupling of the Archean foreland from the advancing orogenic belt. This geologic inference is confirmed by the transition from the tectonically disturbed seismic images of the South Range to the lithologically defined images of the North Range. This decoupling may have been accomplished in zones and could be related to the high-metamorphic grade contrasts across the Murray Fault Zone (Fig. 1) (Zolnai et al., 1984; Riller and Schwerdtner, 1997).

The seismic images obtained from the Sudbury Structure offer an intriguing glimpse into the internal structure of an ancient orogenic belt. However, interpretation of the deep data require initial assumptions about the formation of the SIC. While we favor a genetic interpretation that includes an impact melt sheet, other possibilities remain open. In this regard, more work is required to understand and limit the range of plausible models for SIC genesis.

ACKNOWLEDGMENTS

F. Schwerdtner, G. Stott, and B. Dressler provided careful and thought-provoking reviews of an earlier version of this manuscript. Funding for the Sudbury seismic studies came from INCO Exploration and Technical Services, Inc., Falconbridge, Ltd., the Ontario Geological Survey, the Geological Survey of Canada, and the Natural Sciences and Engineering Research Council of Canada through a grant to Lithoprobe. This is Lithoprobe Contribution 1041, and Geological Survey of Canada Contribution 1998045.

REFERENCES CITED

Bennett, G., Dressler, B. O., and Robertson, J. A., 1994, The Huronian supergroup and associated intrusive rocks, in Geology of Ontario: Ontario Geological Survey Special Volume 4, p. 549–592.

Bethune, K. M., 1997, The Sudbury dyke swarm and its bearing on the tectonic development of the Grenville Front, Ontario, Canada: Precambrian Research, v. 85, p. 117–146.

Boerner, D. E., Kellett, R., and Mareschal, M., 1994a, Inductive source EM sounding of the Sudbury Structure: Geophysical Research Letters, v. 21, p. 943–946.

Boerner, D. E., Milkereit, B., and Naldrett, A. J., 1994b, Introduction to the Special Section of the Lithoprobe Sudbury Project: Geophysical Research Letters, v. 21, p. 919–922.

Brocoum, S. J., and Dalziel, I. W. D., 1974, The Sudbury Basin, the Southern Province, the Grenville Front, and the Penokean Orogeny: Geological Society of America Bulletin, v. 85, p. 1571–1580.

Buchan, K. L., and Ernst, R. E., 1994, Onaping fault system; age constraints on deformation of the Kapuskasing structural zone and units underlying the Sudbury Structure: Canadian Journal of Earth Sciences, v. 31, p. 1197–1205.

Card, K. D., 1978, Metamorphism of the Middle Precambrian (Aphebian) rocks of the eastern Southern Province, *in* Fraser, J. A., and Heywood, W. W., eds., Metamorphism in the Canadian Shield: Geological Survey of Canada Paper 78-10, p. 269–282.

Card, K. D., 1992, Circa 1.75 Ga ages for plutonic rocks from the Southern Province and adjacent Grenville Province: what is the expression of the Penokean orogeny? Discussion, *in* Radiogenic age and isotopic studies, Report 6: Geological Survey of Canada Paper 92-2, p. 227–228.

Card, K. D., 1994, Geology of the Levack Gneiss complex, the Northern footwall of the Sudbury Structure, Ontario, *in* Current Research, pt. C: Geological Survey of Canada, p. 269–278.

Chai, G., and Eckstrand, O. R., 1993, Origin of the Sudbury Igneous Complex, Ontario—differentiate of two separate magmas, *in* Current research, pt. E: Geological Survey of Canada Paper 93-1E, p. 219–230.

Clendenen, W. S., Kligfield, R., Hirt, R. M., and Lowrie, W., 1988, Strain studies of cleavage development in the Chelmsford Formation, Sudbury Basin, Ontario: Tectonophysics, v. 145, p. 191–211.

Cowan, E. J., and Schwerdtner, W. M., 1994, Fold origin of the Sudbury basin, *in* Lightfoot, P. C., and Naldrett, A. J., eds., Proceedings of the Sudbury–Noril'sk symposium: Ontario Geological Survey Special Volume 5, p. 45–55.

Davidson, A., 1992, Relationship between faults in the Southern Province and the Grenville front south-east of Sudbury, Ontario, *in* Current research, pt. C.: Geological Survey of Canada Paper, 92-1C, pp. 121–127.

Davidson, A., van Breemen, O., and Sullivan, R. W., 1992, Circa 1.75 Ga ages for plutonic rocks from the Southern Province and adjacent Grenville Province: what is the expression of the Penokean orogeny? *in* Radiogenic age and isotopic studies, Report 6: Geological Survey of Canada Paper 92-2, p. 107–118.

Deutsch, A., and Grieve, R. A. F., 1994, The Sudbury Structure; constraints on its genesis from Lithoprobe results: Geophysical Research Letters, v. 21, p. 963–966.

Dickin, A. P., and McNutt, R. H., 1989, Nd model age mapping of the southeast margin of the Archean foreland in the Grenville province of Ontario: Geology, v. 17, p. 299–302.

Dickin, A. P., McNutt R. H., and Gifford, P. M., 1990, A neodymium isotope study of plutons near the Grenville Front in Ontario, Canada: Chemical Geology, v. 83, p. 315–324.

Dietz, R. S., 1964, Sudbury Structure as an astrobleme: Journal of Geology, v. 72, p. 412–434.

Dressler, B. O., 1984a, The effects of the Sudbury event and the intrusion of the Sudbury Igneous Complex on the footwall rocks of the Sudbury structure, *in* Pye, E. G., Naldrett, A. J., and Giblin, P. E., eds., The geology and ore deposits of the Sudbury Structure: Ontario Geological Survey Special Volume 1, p. 97–136.

Dressler, B. O., 1984b, Sudbury geological compilation, Ontario Geological Survey Map 2491, Scale 1:50,000.

Dudás, F. Ö., Davidson, A., and Bethune, K. M., 1994, Age of the Sudbury diabase dikes and their metamorphism in the Grenville Province, Ontario, *in* Radiogenic age and isotopic studies, Report 8: Geological Survey of Canada, p. 97–106.

Faggart, B. E., Jr., Basu, A. R., and Tatsumoto, M., 1985, Origin of the Sudbury Complex by meteorite impact: neodymium isotopic evidence: Science, v. 230, p. 426–429.

Fleet, M. E., Barnett, R. L., and Morris, W. A., 1987, Prograde metamorphism of the Sudbury Igneous Complex: Canadian Mineralogist, v. 25, p. 499–514.

French, B. M., 1969, Distribution of shock-metamorphic features in the Sudbury Basin, Ontario, Canada: Meteoritics, v. 4, p. 173–174.

Fueten, F., and Redmond, D. J., 1997, Structural history between the Sudbury Structure and the Grenville Front, Ontario, Canada: Geological Society of America Bulletin, v. 109, p. 268–279.

Fueten, F., Seabright, R., and Morris, W., 1992, A structural transect across the Levack Gneiss–Cartier Batholith Complex, northwest of the Sudbury Structure, *in* Lithoprobe Report; Abitibi-Grenville Project; Abitibi-Grenville Transect: Lithoprobe Report 33, p. 11–15.

Gibbins, S. F. M., Gibson, H. L., Ames, D. E., and Jonasson, I. R., 1997, The Onaping Formation: Stratigraphy, fragmentation, and mechanisms of emplacement, *in* Conference on Large Meteorite Impacts and Planetary Evolution: Houston, Texas, Lunar and Planetary Institute, Contribution 922, p. 16.

Giblin, P. E., 1984, History of exploration and development, of geological studies and development of geological concepts, *in* Pye, E. G., Naldrett, A. J., and Giblin, P. E., eds., The geology and ore deposits of the Sudbury Structure: Ontario Geological Survey Special Volume 1, p. 3–23.

Green, A. G., Milkereit, B., Davidson, A., Spencer, C., Hutchinson, D. R., Cannon, W. F., Lee, M. W., Agena, W. F., Behrendt, F. C., and Hinze, W. J., 1988, Crustal structure of the Grenville Front and adjacent terranes: Geology, v. 16, p. 788–792.

Grieve, R. A. F., Stöffler, D., and Deutsch, A., 1991, The Sudbury Structure: controversial or misunderstood?: Journal of Geophysical Research, v. 96, p. 22753–22764.

Hirt, A. M., Lowrie, W., Clendenen, W. S., and Kligfield, R., 1993, Correlation of strain and anisotropy of magnetic susceptibility in the Onaping Formation: evidence for a near-circular origin of the Sudbury Basin: Tectonophysics, v. 225, p. 231–254.

Hoffman, P., 1989, Precambrian geology and tectonic history of North America, *in* Bally, A. W., and Palmer, A. R., eds., Geology of North America—An Overview: Boulder, Colorado, Geological Society of America, Geology of North America, v. A, p. 447–512.

Huiqi, L., McClay, K. R., and Powell, D., 1992, Physical models of thrust wedges, *in* McClay, K. R., ed., Thrust Tectonics: London, Chapman & Hall, p. 71–82.

James, R. S., Sweeney, J. M., and Peredery, W., 1992, Thermobarometry of the Levack gneisses–footwall rocks to the Sudbury Igneous Complex: Lithoprobe Report; Abitibi–Grenville Transect, v. 25, p. 179–182.

Jones, T. D., and Nur, A., 1984, The nature of seismic reflections from deep crustal fault zones: Journal of Geophysical Research, v. 89, p. 3153–3171.

Joreau, P., French, B. M., and Doukhan, J. C., 1996, A TEM investigation of shock metamorphism in quartz from the Sudbury impact structure (Canada): Earth and Planetary Science Letters, v. 138, p. 137–143.

Kellett, R. L., and Rivard, B., 1994, Characterization of the Benny deformation zone, Sudbury, Ontario: Canadian Journal of Earth Sciences, v. 33, p. 1256–1267.

Krogh, T. E., Davis D. W., and Corfu, F., 1984, Precise U-Pb zircon and baddeleyite ages for the Sudbury area, *in* Pye, E. G., Naldrett, A. J., and Giblin, P. E., eds., The geology and ore deposits of the Sudbury Structure: Ontario Geological Survey Special Volume 1, p. 431–446.

Law, A., and Snyder, D. B., 1997, Reflections from a mylonitized zone in central Sweden: Journal of Geophysical Research, v. 102, p. 8411–8425.

Lightfoot, P. C., Keays, R. R., and Doherty, W., 1997, Can impact-generated melts have mantle contributions? Geochemical evidence from the Sudbury Igneous complex: Houston, Texas, Lunar and Planetary Institute, Contribution 922, p. 30.

Liu, S., and Dixon, J. M., 1991, Centrifuge modelling of thrust faulting: structural variation along strike in fold-thrust belts: Tectonophysics, v. 188, p. 39–62.

Marshak, S., Wilkerson, M. S., and Hsui, A. T., 1992, Generation of curved fold-thrust belts: Insights from simple physical and analytical models, *in* McClay, K. R., ed., Thrust Tectonics: London, Chapman and Hall, p. 83–92.

Meldrum, A., Abel-Rahman, A. F. M., and Wodicka, N., 1997, The nature, age and petrogenesis of the Cartier Batholith, northern flank of the Sudbury Structure, Ontario, Canada: Precambrian Research, v. 82, p. 265–285.

Miao, X., Moon, W. M., and Milkereit, B., 1993, Crustal structures in the northwest of the Sudbury basin from seismic reflection line 42, Lithoprobe Seismic Processing Facility Newsletter, v. 6, p. 25–28.

Milkereit, B., and the Sudbury Working Group, 1992, Deep Geometry of the Sudbury Structure from seismic reflection profiling: Geology, v. 20, p. 807–811.

Milkereit, B., Green, A., Wu, J., White, D., and Adam, E., 1994a, Integrated seismic and borehole geophysical study of the Sudbury Igneous Complex:

Geophysical Research Letters, v. 21, p. 931–934.

Milkereit, B., White, D., Adam, E., Boerner, D. E., and Salisbury, M., 1994b, Implications of the Lithoprobe seismic reflection transect for Sudbury geology, *in* Lightfoot, P. C., and Naldrett, A. J., eds., Proceedings of the Sudbury-Noril'sk symposium, Ontario Geological Survey Special Volume, v. 5, p. 11–20.

Milkereit, B., Eaton, D., Wu, J., Perron, G., Salisbury, M., Berrer, E. K., and Morrison, G., 1996, Seismic imaging of massive sulfide deposits: Pt. II., Reflection seismic profiling: Economic Geology, v. 91, p. 829–834.

Morrison, G. G., 1984, Morphological features of the Sudbury Structure in relation to an impact origin, *in* Pye, E. G., Naldrett, A. J., and Giblin, P. E., eds., The geology and ore deposits of the Sudbury Structure: Ontario Geological Survey Special Volume 1, p. 513–520.

Muir, T. L., and Peredery, W. V., 1984, The Onaping Formation, *in* Pye, E. G., Naldrett, A. J., and Giblin, P. E., eds., The geology and ore deposits of the Sudbury Structure: Ontario Geological Survey Special Volume 1, p. 142–197.

Naldrett, A. J., 1984, Summary, discussion and synthesis, *in* Pye, E. G., Naldrett, A. J., and Giblin, P. E., eds., The geology and ore deposits of the Sudbury Structure: Ontario Geological Survey Special Volume 1, p. 533–570.

Naldrett, A. J., 1994, The Sudbury-Noril'sk Symposium, an overview, *in* Lightfoot, P. C., and Naldrett, A. J., eds., Proceedings of the Sudbury-Noril'sk symposium: Ontario Geological Survey Special Volume, v. 5, p. 3–8.

Pye, E. G., 1984, The origin of the Sudbury Structure; preface, *in* Pye, E. G., Naldrett, A. J., and Giblin, P. E., eds., The geology and ore deposits of the Sudbury Structure: Ontario Geological Survey Special Volume 1, p. v.

Ramsay, J. G., and Huber, M. I., 1987, The techniques of modern structural geology; Volume 2; Folds and fractures: London, Academic Press, p. 462.

Riller, U., and Schwerdtner, W. M., 1997, Mid-crustal deformation at the southern flank of the Sudbury Basin, central Ontario, Canada: Geological Society of America Bulletin, v. 109, p. 841–854.

Rivers, T., van Gool, J. A. M., and Connelly, J. N., 1993, Contrasting tectonic styles in the northern Grenville Orogen: implications for the dynamics of orogenic fronts: Geology, v. 21, p. 1127–1130.

Rousell, D. H., 1972, The Chelmsford Formation of the Sudbury Basin—a Precambrian turbidite: Geological Association of Canada Special Paper 10, p. 79–91.

Rousell, D. H., and Long, D. G. F., 1997, Are outliners of the Huronian Supergroup preserved in structures associated with the collapse of the Sudbury Impact Crater?: Journal of Geology, v. 106, p. 407–419.

Salisbury, M. H., Iuliucci, R., and Long, C., 1994, Velocity and reflection structure of the Sudbury Structure from laboratory measurements: Geophysical Research Letters, v. 21, p. 923–926.

Shanks, W. S., and Schwerdtner, W. M., 1991a, Structural analysis of the central and southwestern Sudbury Structure, Southern Province, Canadian Shield: Canadian Journal of Earth Sciences, v. 28, p. 411–430.

Shanks, W. S., and Schwerdtner, W. M., 1991b, Crude quantitative estimates of the original northwest-southeast dimension of the Sudbury Structure, south-central Canadian Shield: Canadian Journal of Earth Sciences, v. 28, p. 1677–1686.

Spray, J. G., and Thompson, L. M., 1995, Friction melt distribution in a multiring impact basin: Nature, v. 373, p. 130–132.

Walker, R. J., Morgan, J. W., Naldrett, A. J., Li, C., and Fassett, J. D., 1991, Re-Os isotope systematics of Ni-Cu sulfide ores, Sudbury Igneous Complex, Ontario: evidence for a major crustal component: Earth and Planetary Science Letters, v. 105, p. 416–429.

White, D. J., Milkereit, B., Wu, J., Salisbury, M. H., Mwenifumbo, J., Berrer, E. K., Moon, W., and Lodha, G., 1994, Seismic reflectivity of the Sudbury Structure, North Range, from borehole logs: Geophysical Research Letters, v. 21, p. 935–938.

Wu, J., Milkereit, B., and Boerner, D. E., 1994, Timing constraints on deformation history of the Sudbury Impact Structure: Canadian Journal of Earth Sciences, v. 31, p. 1654–1660.

Zolnai, A. I., Price, R. A., and Helmstaedt, H., 1984, Regional cross section of the Southern Province adjacent to Lake Huron, Ontario, Implications for the tectonic significance of the Murray Fault Zone: Canadian Journal of Earth Sciences, v. 21, p. 447–456.

MANUSCRIPT ACCEPTED BY THE SOCIETY DECEMBER 16, 1998

… Geological Society of America
Special Paper 339
1999

Summary: Development of ideas on Sudbury geology, 1992–1998

A. J. Naldrett
Department of Geology, University of Toronto, Toronto, Ontario M5S 3B1, Canada

ABSTRACT

Evidence favoring an impact origin for the Sudbury Structure continues to accumulate, particularly the recent discovery in the Black Member of the Onaping Formation of shocked microdiamonds resembling those found at other impact structures. The 1990 Lithoprobe Vibroseis survey of the Sudbury basin and its subsequent interpretation indicated that the original diameter of the Sudbury Structure was of the order of 200 km, as opposed to previous estimates of 60–100 km. This new size estimate made it possible to interpret the Sudbury Igneous Complex (SIC) as an impact melt with no requirement for a mantle-derived contribution.

The very felsic compositions of all of the exposed components of the SIC make it difficult to account for the origin of the dunite, harzburgite, wehrlite, websterite, and melanorite inclusions in the Sublayer at Sudbury, without calling upon their derivation from a preexisting layered intrusion in the target area. Recent U-Pb dating of baddeleyite and zircon from melanorite inclusions, giving a 1.85-Ga age, indicated that these are related to the development of the SIC, and are not from an older intrusion. However, new Re-Os isotopic data on variably mineralized melanorite inclusions indicate that the ore component of these inclusions has interacted with a dominantly crustal source with an average age of about 2.6 Ga, while the silicate component is younger. The solution to this enigma is probably that the mafic/ultramafic inclusion suite is composite, the more olivine-rich inclusions being from a preexisting source and the melanorites being of Sudbury age, with a major component of the mineralization also predating the SIC.

Modeling the fractional crystallization of hypothetical SIC magmas indicates that the rock types, and the sequence of cumulus phases, is consistent with a fractionation origin, but that the relative proportions of the rock types is not; there is a great excess of granophyre relative to norite exposed at surface. The application of computer models designed to simulate the impact of large bolides to the Sudbury Structure leads to the conclusion that the resulting impact melt would have been superheated for the first 100,000–200,000 yr of its existence, as it cooled from about 1,700°C to its liquidus temperature (about 1,200°C). During this time a massive amount of the overlying basal breccia of the Onaping Formation would have been melted to accumulate at the top of the underlying impact melt. Sulfur solubility in silicates melts is strongly temperature-dependent, so that cooling through this 500°C temperature interval of superheat would also have favored liquation of a sulfide liquid if the initial impact melt had been anywhere close to sulfide saturation, and the very complete settling of this liquid to the base of the complex.

It is proposed that the bolide impacted an area of the crust underlain in part by a preexisting mafic intrusion, and that the initial melt was more mafic than average

Naldrett, A. J., 1999, Summary: Development of ideas on Sudbury geology, 1992–1998, *in* Dressler, B. O., and Sharpton, V. L., eds., Large Meteorite Impacts and Planetary Evolution II: Boulder, Colorado, Geological Society of America Special Paper 339.

crust. As increasing amounts of felsic melt derived from the overlying fall-back breccia (Basal Member of the Onaping Formation) accumulated and became partly incorporated at the top of the impact melt, a two-layer igneous system developed, with convection occurring simultaneously and separately in an upper felsic and lower, more mafic layer. Limited diffusion of alkalies occurred downward from the upper layer, across the boundary layer separating it from the lower layer, increasing the felsic component of the latter as time progressed. Convection currents within the lower layer streamed down the cooler footwall rocks, giving rise to the observed igneous foliation parallel with the contact, and the igneous lineation oriented toward the center of the Sudbury Structure. The upper granophyric layer cooled, and crystallized primarily from the top down, so that the igneous lineation in the portions of this that are preserved is orthogonal to the roof. Thus, observations as to the contrasting igneous fabrics of granophyre and norite are explicable in terms of the cooling of a single, two-layered body of magma.

It is suggested that the Sublayer consists primarily of initial melt, enriched by sulfides and inclusions that settled out of the overlying magma. The magma did not become involved in the convective mixing affecting the Main Mass norite, and thus preserved a more "local" geochemical signature.

INTRODUCTION

The most recent comprehensive volumes concerned with the origin of the Sudbury Structure are a book, *The Geology and Ore Deposits of the Sudbury Structure* (Pye et al., 1984), which was a landmark summary of Sudbury geology as it was understood at the time, and two 1992 symposia: the proceedings of the symposium, The Sudbury-Noril'sk symposium (Lightfoot and Naldrett, 1994), and the Large Meteorite Impacts and Planetary Evolution (Dressler et al., 1994). This chapter summarizes many of the key questions (and ideas about their answers) in the minds of Sudbury geologists at the time of the 1992 symposia, traces the evolution of ideas from this stage to the 1997 symposium (the subject of this volume), and then summarizes the main points made in the latter symposium. It concludes with a statement of my own views as to how much of the present information and ideas can be synthesized into a comprehensive model, albeit one that still has many shortcomings.

NEW CONCEPTS ABOUT SUDBURY GEOLOGY, 1992–1997

The two 1992 Sudbury symposia

In the Proceedings of the Sudbury-Noril'sk Symposium and in other papers published at about the same time, it was pointed out that the results of the initial Sudbury Lithoprobe transects (Milkereit et al., 1992) showed that a reflector corresponding to the interpreted extent of the norites and gabbros of the North Range extended south, with very little change in the 20°–30° southerly dip to about the southern limit of the Sudbury Igneous Complex (SIC). This implies that the original diameter of the crater was of the order of 150–200 km. New gravity, density, and seismic measurements allowed McGrath and Broome (1994) to develop a gravity model that accounted for the excess of mass present at Sudbury, without calling on a second layered intrusion hidden at depth, as had been required by previous models (Gupta et al., 1984). Magnetic susceptibility measurements on the same borehole material, together with the seismic constraints, enabled Hearst et al. (1994) to propose that the broad regional magnetic high that lies within and immediately northwest of the Sudbury Structure is produced by a strongly magnetic variant of the Levack Gneiss Complex, possibly the result of contact metamorphism against the SIC.

The knowledge that the original, final (i.e., not transient) diameter of the Sudbury Structure was of the order of 200 km, much larger than previous estimates of 60–80 km, has raised a critical point with respect to the origin of the SIC within it. Grieve et al. (1991), Grieve (1994), and Golightly (1994), using extrapolations of the relationship between crater diameter and volume of impact melt produced, argued that the Sudbury Impact Structure was more than large enough to account for all of the SIC as an impact melt, thus providing support for the earlier conclusion reached by Faggart et al. (1985) on the basis of Nd and Sr isotopic data. Cowan and Schwerdtner (1994) argued that the basinal shape of the Sudbury Structure was not the consequence of impact alone, but was due in part to synclinal folding. Morrison et al. (1994) and Jago et al. (1994) pointed to the close relationship between zones of Sudbury Breccia (pseudotachylite) and Cu-rich ore deposits 500–1,000 m within the footwall of the North and East Ranges at Sudbury. Naldrett et al. (1994) interpreted these ores to be the final residual liquid from the fractional crystallization of sulfide liquid that had given rise to deposits along the contact. This suggests that surface tension effects between this liquid and silicates had caused the liquid to migrate out from the monosulfide solid solution (mss) cumulates constituting the contact ore bodies into the country rocks.

In the symposium on Large Metorite Impacts and Planetary

Evolution, Thompson and Spray (1994) identified four types of pseudotachylite (Sudbury Breccia): one type (ductile-margined) they suggested was related to passage of the shock wave; the other three types (sharp-margined, aphanitic to crystalline matrixed [the most common] injection and true cataclastic [relatively rare]) they concluded postdated the shock wave and were related to recovery. Thompson and Spray found that the distribution of the pseudotchylite zones defined three zones of major development separated by zones with minor development, as had been described originally by Muir and Peredery (1984) and Dressler (1984); this point was used to argue that the original diameter was at least 200 km, in agreement with the estimates from other lines of evidence (see above). A similar conclusion as to the size of the original structure was reached by Butler (1994) on the basis of the interpretation of Landsat imagery.

Avermann (1994) reviewed earlier information and ideas about the Onaping Formation and subdivided it into the Basal Member (0–300 m thick), Grey Member (300–500 m thick), Green Member (5–70 m thick), Lower Black Member (100–200 m thick), and Upper Black Member (800–900 m thick). She interpreted the Basal Member as melt rock forming the upper part of the impact melt and containing inclusions of melt breccia and suevitic breccia, the Grey Member as a suevitic breccia resulting from ground-surge following impact, the Green Member as the final fall-back into the crater, the Lower Black Member as suevitic breccia redeposited during adjustments within the crater, and the Upper Black Member as reworked breccia material that forms a transition into the overlying slates of the Onwatin Formation.

Deutsch (1994) interpreted isotopic ratio data for the SIC, suevitic breccias, and country rocks as indicating that the SIC was entirely an impact melt sheet, although he acknowledged that his principal argument that *no* mantle-derived melt was involved relied on the difficulty that such a melt would have in assimilating the required 50%+ of crustal rocks required to account for the isotopic data. Norman (1994) noted that the composition of the SIC is very similar to that of lunar samples, which previously had been interpreted as primary, not impact, melts, and which had, therefore been used to give insight into the thermal history of the Moon. Norman pointed out that if the SIC is an impact melt, the lunar samples could also be impact in origin, with no relevance to the Moon's thermal evolution. He cautioned that an entirely impact, as opposed to endogenous, origin for Sudbury should be treated with caution.

Development of ideas following the 1992 symposia

Interpretation of the SIC as being entirely due to the impact melt raised some new problems. The bulk composition of the SIC is very siliceous, and it was not clear how the harzburgites, wehrlites, websterites, and melanorites that occur as inclusions in the ore-bearing Sublayer (Naldrett, 1994; Scribbins et al., 1984) could have crystallized from this magma. The suggestion was raised (D. Peck, personal communication to A. J. N., cited in Naldrett, 1994) that they might have been derived from a preexisting layered intrusion in the target area of the impactor, which was one of a series that included the East Bull Lake body 100 km west of Sudbury. This latter body is part of the 2.4-Ga mafic magmatism accompanying the rifting that gave rise to the Huronian ocean that developed along the southern margin of the Superior continent at this time. Subsequently, Keays et al. (1995) suggested that the Sudbury Ni-Cu sulfides were also derived from a preexisting layered intrusion that was disrupted by the impact.

Data apparently in conflict with this interpretation were presented by Corfu and Lightfoot (1996), who had dated zircons and baddeleyite from inclusions of melanorite, pyroxenite, and diabase from the Whistle mine at the northeastern corner of the SIC. All dates correspond to the Sudbury age of 1.85 Ga, which links the origin of these rocks very closely with that of the SIC. In view of the importance of their findings, the nature of their samples requires comment. The diabase cuts across the contact between Sublayer and country rock at the northern end of the Whistle pit; in their article, Corfu and Lightfoot suggested it could be a fragment of disrupted chilled margin. However, many of the large bodies of diabase at the footwall of the SIC are disrupted and recrystallized, and occur in both igneous textured and metamorphic textured breccias. An alternate, and more likely interpretation (P.C. Lightfoot, personal communication, June 1998), is that they are are samples of older diabase for which the zircon age has been re-set by the impact.

The melanorites are described as pods within a Sublayer matrix, without sharp margins and in one case gradational over 5 m into leuconorite. While this material is not typical of the inclusion suite described by Scribbins et al. (1984), Lightfoot has pointed out to the author (P. C. Lightfoot, personal communication, to A. J. N., 1998) that inclusions of the type documented by Scribbins et al. are rare. Furthermore, melanorites of the kind dated in the Corfu and Lightfoot study, which appear to belong to a spectrum of geochemically similar rocks, are the most common melanocratic inclusion type. In discussing their data, Corfu and Lightfoot (1997) concluded that the melanorite samples appear to have crystallized from magma in which the major elements indicated a much more magnesian bulk composition than any magma responsible for the Main Mass of the SIC, but in which the trace elements ratios were enriched in LREE (light rare earth element) and other incompatible elements to an extent inconsistent with their crystallization from primitive mantle melts.

Chai and Eckstrand (1994) documented a marked compositional break between the quartz gabbro and the overlying granophyre, and argued that this implied derivation from two different magmas originating from different sources. They postulated that the norite and quartz gabbro were the product of a primary mantle melt that had become contaminated by Archean granulites of the lower crust, whereas the granophyre was an upper crustal impact melt. This view was challenged by Lightfoot et al. (1997a), who pointed out that, with the exception of Sr, P, Eu, and Ti (which are very dependent on addition or removal of plagioclase, apatite, or Fe-Ti oxides), the principal components of the Main Mass of the SIC, namely the felsic norite, quartz gabbro, and granophyre, have extremely similar trace element patterns. In particular, Lightfoot et al. (1997a) drew attention to the similarity in Th/Zr ratios (0.04–0.05) between the SIC and granophyre which, they argued, would be an extraordinary coincidence if any of these units had been contaminated by, or derived from, different crustal reservoirs. They concluded that the Main Mass is the consequence of a pulse of mantle-derived picritic magma rising into and mixing with impact melt in a proportion of 20 picrite to 80 impact melt. Using mixing calculations involving a primitive picrite and average crust, Lightfoot et al. (1997a) showed that the composition of the felsic norite is consistent with this explanation. They interpreted the granophyre as the product of fractionation of the Main Mass magma.

Lightfoot et al. (1997b) commented that the relative proportions of trace elements in the Sublayer is very different to that in Felsic Norite, in contrast to the broad similarities that they observed between the Felsic Norite and other components of the Main Mass (Quartz-rich Gabbro, Granophyre). The sublayer rocks are poorer in LREE (light rare earth elements) and LILE (large ion lithophile elements) but have HREE (heavy rare earth elements) and HFSE (high field strength elements) similar to the felsic norite. They argued that these differences cannot be explained by closed system fractional crystallization or partial melting. They found that the Sublayer from different embayments, specifically the Levack-McCreedy West, Fraser mine, Whistle, Little Stobie, and Crean Hill embayments had similar compositions within themselves, but somewhat different compositions from embayment to embayment. They modeled the Whistle sublayer as a mix of mafic norite, diabase, and footwall granite and speculated that the differences in sublayer composition between different locations and with the Main Mass of the SIC reflect different inclusion and/or country rock compositions.

The Quartz Diorite occupies dike-like structures that project away from the SIC, in places extending several tens of kilometers from the contact, and has traditionally been included with the Sublayer along the basal contact (Souch, et al., 1969; Naldrett et al., 1984). However, the data of Lightfoot et al. (1997a) indicated that the trace element concentrations are much closer to those of the felsic norite than the sublayer. These authors argued that this implies a close genetic relationship, closer than is the case for the Sublayer. However, they noted that the Sr, Eu, P and Ti negative anomalies that characterize the Main Mass are either not present or are less pronounced in the offset Quartz Diorite, which implies that there was less fractionation of plagioclase, apatite, and Fe-Ti oxides from the quartz diorite magma. They suggested that the offsets may have been formed from the original, unfractionated Main Mass magma that penetrated fractures and, through reaction with country rocks, had its composition slightly modified on a local scale.

Following the report of Buseck et al. (1992) on fullerenes (structural variants of carbon in which the carbon atoms occur at the nodes of structures resembling geodesic domes) in Shungite from the Kola Peninsula of Russia, Becker et al. (1994) reported the discovery of concentrations of fullerenes (in parts per million) in samples from the Onaping Formation. These authors (Becker et al., 1996) also reported the presence of helium with a distinctive "extraterrestrial" $^3He/^4He$ isotopic composition, which led them to suggest that the fullerenes were extraterrestrial in origin, and were derived from the impactor.

SUMMARY OF SUDBURY PAPERS IN THIS VOLUME

Structure of the Sudbury Basin

Elsewhere in this volume, Boerner and Milkereit (chapter 32) summarize the seismic reflection data presently available for the Sudbury Structure, and confirm the earlier conclusion (see above) that the base of the SIC is well defined to a depth of 10–11 km, at which depth it is directly below the southern margin of the South Range. As before, the data support the conclusion that the South Range has been transported tectonically northward. They point out that two orogenies have affected the area. The first, the Penokean, is transected by the impact structure and thus occurred largely prior to the impact, whereas the second, the Grenvillian, has not greatly affected rocks of the Sudbury Structure, although Boerner and Milkereit suggest that loading of the southern margin of the Superior craton by overthrusting from the southeast during both late Penokean and Grenvillian times may have caused a broad regional south-southeasterly rotation, which accounts for some of the southerly dip shown by the North Range of the SIC.

Cowan et al. (Chapter 31) reverse some of the conclusions of Cowan and Schwerdtner (1994), particularly that calling for the SIC to have been folded subsequent to emplacement. Study of the folding and deformation of the Huronian shows that this was accomplished prior to emplacement of the SIC, and that all of the deformation experienced by the latter is attributable to deformation along the South Range Shear Zone. These authors note that measurements of remanant magnetic susceptibility show that the granophyre is characterized by a mineral lineation (which they interpret as magmatic) that is orthogonal to the base of the SIC. Mineral fabrics in the norite and quartz gabbro are parallel to the synformal base of the SIC, with a magnetic lineation plunging toward the center of the structure. Cowan et al. suggest that this is consistent with the gravitational strain experienced by crystals settling on inclined walls. They make the point that the fabrics that they have observed are not strained where the curvature of the body is greatest, indicating that folding is not responsible for the curvature. Their conclusions are that an early Blezardian age (2.4 Ga) dome was truncated by Nipissing magmatism (2.2 Ga), and that northwest compression accompanying the Penokean (1.9–1.8 Ga) orogeny accounts for subsequent folding. The SIC was emplaced as two events, the granophyre and the norite-gabbro, along a prefolded structure which is essentially the present site of the SIC. Subsequently, further compression, possibly during the waning, post-SIC phases of the Penokean orogeny, gave rise to the South Range Shear Zone. Finally, they note that the steep dips along the East Range cannot have resulted from folding, and they therefore preclude in situ differentiation of a melt sheet.

Fedorowich et al. (Chapter 23) present for the first time a statistically valid examination of the development of Sudbury Breccia. Using the footwall in the vicinity of the Strathcona embayment (a bulge of the SIC into the footwall), they show that the Sudbury Breccia comprises about 8% of the country rocks close to the embayment, as opposed to about 2% away from it. Although the breccia zones are very irregular, dips of the zones are generally parallel to the contact. Envelopes of breccia appear to be controlled by reverse and normal shears that were interactive with tensional openings in an overall compressive regime. They attribute the relationship between the Sudbury Breccia and the embayment to the breccia as an aspect of the ground preparation, which allowed the country rocks to slump into the developing impact crater, and thus give rise to the embayment.

Origin of the Onaping Formation

Interpretation of the Sudbury Structure as an astrobleme continues to gain strength; most would argue that the evidence is almost incontrovertible. In their study of samples from the Onaping Formation, Masaitis et al (Chapter 24) found six diamonds, 50–100 nm in diameter, similar to those found at other impact sites; this further reinforces the conclusion that major shock accompanied the formation of the Sudbury Structure. Avermann (Chapter 25), following on her recognition of the Green Member of the Onaping Formation as the final fall-back breccia, interprets it to be the "collapsed fire ball horizon" of the Sudbury Structure. The source of the high noncarbonate carbon content of the Onaping (0.3–0.85 wt%) and Onwatin (~3 wt%) formations is discussed in two chapters in this volume (Heyman et al., Chapter 27; Bunch et al., Chapter 26).

Both papers report that the $\delta^{13}C$ content is of the order of −30‰, which is consistent with the carbon being derived from plant or animal remains. Heyman et al. looked for, but could not find, fullerenes, even in samples collected proximal to those used in the studies of Becker et al. (1994, 1996). The failure to detect fullerenes indicates the extremely inhomogeneous distribution of this material; and these authors state that the results of their work show that it may be difficult to find additional Black Member [Onaping] rocks containing fullerenes. Heyman et al. argue that the concentration of fullerenes in the principal extraterrestrial body in which their abundance has been documented, the Allende meteorite, is so low that to account for the concentrations reported by Becker et al. in the Onaping would require an impossibly large chondritic impactor. They also conclude that other extraterrestrial bodies, such as comets, could not have provided the carbon in the Onaping, and that this must be terrestrial. They consider a number of different sources for the carbon and conclude that the most probable is that it formed as very rapid algal blooms developing as the Black Member of the Onaping was being deposited. Bunch et al. report the tentative identification of other fullerene species, C_{74}, C_{78}, C_{84} and C_{100}, in addition to the C_{60}, and C_{70} reported by Becker et al. (1994, 1996). Bunch et al. also disagree with the (algal bloom) explanation of Heyman et al. for the carbon throughout the Black Member of the Onaping Formation on the grounds that these rocks were formed too quickly for organic matter to grow during deposition: if the carbon had resulted in this way, algal remains would surely have been preserved—which appears not to be the case. They point to the high proportion of kerogen-bearing mudstones in the upper part of the Black Onaping, and identify C-rich mudstones in the target area as responsible for the carbon throughout the Black Member.

Origin of the SIC

Ariskin et al. (Chapter 29) take the approach of modeling the crystallization of the SIC, using the computer program COMAGMAT–3.5 that had been developed with respect to other differentiated mafic bodies. Of course, their conclusions are only as good as the programs are applicable, and it is difficult to evaluate this from the present paper, but, assuming its applicability, they show conclusively that using either Collins' (1934) weighted average for the bulk composition of the SIC, or Naldrett and Hewins' (1984) Quartz-rich Norite as the initial composition of the SIC magma cannot account for the high proportion of granophyre to norite exposed at surface. On the other hand, a Quartz-rich Norite starting composition results in a model that gives a very close approximation to both the chemical variations and appearance and disappearance of phase in the lower, mafic portion of the SIC.

Ariskin et al. conclude that the SIC, including the granophyre, appears to be the result of differentiation of a single magma. If this were so, the cumulates that should have resulted from this differentiation do not correspond to the norite, quartz-rich gabbro, and granophyre in the proportions in which they are exposed; there must be a larger body of mafic cumulates hidden somewhere in the structure. Ariskin et al. point out that the superheat generated by the impact may have resulted in the incorporation of a large proportion of the overlying fall-back breccia into the upper part of the differentiating melt sheet. They suggest that, since the initial impact melt and the breccias forming the roof of the SIC have been derived from the same target rocks, the continuity of all elements from norite through quartz gabbro into the granophyre that they document in their study is not incompatible with this late stage of assimilation of roof rocks.

Ivanov and Deutsch (Chapter 30) have used a modified version of the SALE computer program of Amsden et al. (1980) for fluid flow to accomplish the following: (1) construct a time-dependent model for the formation of the transient crater, and thus show the progression in the distortion of layers in the target area and position of isotherms close to and beneath the crater following impact; (2) model the maximum shock pressures experienced in the target area; and (3) model the temperature evolution within and beneath the melt pool from 10^4 to 10^7 yr after impact for an impactor of 14-km diameter traveling at 20 km/s. They examine two models, one in which standard rock mechanics are applied to the rocks of the target area, and a second in which acoustic fluidization (Melosh, 1989) decreases internal friction of the target rocks.[1]

Given the applicability of computer modeling, a number of points appear from this work that have not found their way fully into much of the geologic literature about Sudbury. The first is that the transient crater reached a depth of 40–50 km, but that the zone of impact melting did not extend this deep and would not have affected the subcontinental mantle. Nevertheless, Ivanov and Deutsch's diagrams show that a substantial amount of lower crustal material could have been brought to surface (as the result of fluid flow and not merely as inclusions or impact melt). (My personal observation is that I suspect that the high-grade gneisses of the Levack complex, which so faithfully rim the outer margin of the North Range of the SIC, may have been uplifted in this way.) Ivanov and Deutsch state that the amount to which the mantle would have been uplifted beneath the structure depends on which of their models is correct—20 km for the acoustically fluidized model, about 2 km for no contribution from acoustic fluidization, and an intermediate amount if acoustic fluidization occurred but was damped out by lithostatic load at depths approaching those of the mantle. Impact melting, according to Ivanov and Deutsch, should be restricted to target rocks within a radius of about 15 km of the center of impact (i.e., beneath only the very central part of the final, 75–100-km-radius crater), but extending to a depth of 20–30 km beneath the surface. One of the least appreciated possibilities (at least by me) that this study raises is the degree of superheat generated within the impact melt, and the length of time over which this would have persisted. According to Ivanov and Deutsch's estimates, the initial temperature of the impact melt would have been about 1,727°C, and superliquidus temperatures (>1,177°C) would have persisted for 100,000–250,000 yr after impact. Approximately 500,000 yr would have been required for the whole of the impact sheet to cool to the solidus (997°C).

WHERE DO WE STAND NOW, AND WHAT FOR THE FUTURE?

The foregoing summary indicates that some problems with respect to the origin of the Sudbury Structure are close to being solved and others are far from it. It is clear that the impact origin of the structure has been established beyond almost all reasonable doubt. Some questions remain concerning the origin of the carbon in the Black Member of the Onaping Formation, the presence or absence of fullerenes, and their implications with respect to the nature of the impactor, but the discovery of impact diamonds places yet another nail in the coffin of impact disbelievers!

The original shape of the structure and the extent to which it has been folded are more open to question. While the structural studies have shown that through-going strain related to folding does not appear to have developed in the rocks of the SIC, it is possible that some components of the present basinal structure owe their origin to flexural-slip folding. For example, the steeply dipping East Range could possibly have been given its present orientation as a result of the Wanapitei impact, just to the east, with uplift occurring along a series of co-planar reverse faults (e.g., the Waddell Lake fault system) (J. Fedorowich, personal communication, 1998). Alternatively, the dips here may originally have been steep, with the layering developing at a high angle. Certain aspects of the Sudbury Structure that have previously been attributable to folding have now been established as original intrusive structures, e.g., the sharp corner at the northeastern extremity of the SIC. The original attitude of the South Range is more difficult to establish, and rotation may have been imposed on it during the period of deformation that gave rise to the South Range Shear Zone. Cowan et al. (Chapter 31) attribute significance to the difference in igneous fabrics between the granophyre and norite; they suggest that this difference implies different ages of emplacement. I, however, propose a scenario (below) that I believe could explain this without calling on two ages of emplacement.

Some of the greatest uncertainties remain in fitting the different phases of the SIC into a unified model. Particularly difficult are these: Concerning the origin of the mafic and ultramafic inclusions in the Sublayer—are the inclusions derived in part from a preexisting layered intrusion or did they form during cooling after impact? If the former is true, why do they have zircon and baddeleyite ages of 1.85 Ga? If the latter is the case, how could dunites and pyroxenites have crystallized from a magma siliceous enough to give rise to the SIC? Another difficulty is the question of the granophyre —Ariskin et al. (Chapter 29) point out that this rock type is too abundant to be explained easily by fractionation of any reasonable estimate (they use a 2:1 mixture of granophyre and norite and also the Quartz-rich Norite composition of Naldrett and Hewin [1984]) for the original magma composition. Chai and Eckstrand (1994) have identified the granophyre as representing direct impact melt and the norite as a mixture of primary basalt with this, but Lightfoot et al. (1997a) have said that the continuity in ratios of incompatible elements between norite and granophyre implies a single, not a multiple, source of magma.

The application of rhenium-osmium (Re-Os) isotope systematics to Sudbury has given a new insight into the origin of the Ni-Cu ores. Walker et al. (1991) found that the high initial $^{187}Os/^{186}Os$ ratios in many of the ore deposits required a minimum of 50% and possibly all of the Os to have been derived from a crustal source. Dickin et al. (1996) interpreted their Os isotope data to indicate a totally crustal origin. The recent work on sulfide-bearing melanorite inclusions at Sudbury (Hawkesworth et al., 1997), which has been reported since the 1997 symposium, indicates that these lie on a mixing line between a younger, low Re/Os end-member (the silicates of the inclusions) and an older (2.6 Ga), high Re/Os end-member (the sulfides). The latter com-

---

[1]On the basis of drilling, Ivanov and Deutsch propose that blocks of rocks 50–200 m in size comprise the units which were fluidized. They point out that lithostatic pressure should dampen out this kind of fluidiziation at depths greater than several tens of kilometers.

ponent (which could itself be a well-homogenized mixture of two components) is undoubtedly crustal. Thus, it seems that the origin of the osmium (and by implication, the remainder of the PGE [platinum group elements], since the proportion of Os to the other PGE is typical for all deposits of mafic association, with no evidence that the Sudbury ores are enriched in Os), and thus the ores—which one would have thought to have best preserved evidence of a primary mantle contribution—is crustal. Much of the available isotopic evidence is beginning to point to an exclusively crustal origin as the most reasonable explanation for the SIC.

For the moment, let us examine the Sudbury Structure from an exclusively crustal viewpoint. If this is correct, how does one explain the 1.85-Ga ultramafic inclusions? While Corfu and Lightfoot (1997) could obtain only baddeleyite and zircon from diabase (which is not part of the suite of ultramafic inclusions), melanorite pods, and one completely altered pyroxenite, the geochemical data of Lightfoot et al. (1997b) on an average of three pyroxenite inclusions have many similarities with Sublayer norite; this suggests that some of the more melanocratic inclusions have Sublayer affinities. Scribbins et al. (1984) described many of their pyroxenite and peridotite inclusions in the Sublayer as having a distinctly shattered appearance. It is possible that the mafic/ultramafic inclusions, which thus far have been grouped together (I exclude from this group the diabase inclusions and the inclusions of Nipissing-like metagabbro from the South Range, which have always been distinguished separately), come from more than one provenance. For example, the melanorite pods and some pyroxenites (new work has shown that the melanorites are the predominant inclusion type of as-yet unknown provenance at Sudbury) appear on the basis of Corfu and Lightfoot's (1997) dating to have formed synchronously with the SIC. Some dunites, olivine pyroxenites, and pyroxenites, particularly those described from Strathcona mine by Scribbins et al., 1984), may be much older, possibly derived

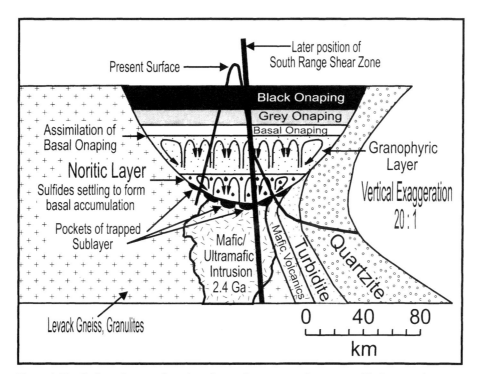

Figure 1. The Sudbury Igneous Complex after readjustments to the crater, and before complete crystallization of norite and granophyre. The granophyre and norite have separated into two independently convecting layers. Assimilation of the overlying Onaping Formation is occurring at the upper contact of the "granophyric Layer," greatly assisted by the superheat within the impact melt. Sulfide immiscibility develops in the "noritic Layer," as this cools, but before it reaches its liquidus, so that the sulfides are unimpeded by silicates as they settle through the hot, low viscosity magma to collect near the lower contact. Once the noritic Layer reaches its liquidus, crystals accumulate on the walls from cool currents of magma sweeping down them to produce a lineated fabric down-dip and parallel to the walls. Magma resulting from initial melting collects in pockets along the base of the SIC, and sulfides and inclusions settle into these to produce the inclusion and sulfide-bearing Sublayer. As the granophyre reaches its liquidus, crystallization occurs with an igneous fabric developing orthogonal to isotherms with the magma, that, near the center of the structure, are parallel with the roof. Subsequent tilting of the whole region to the south, coupled with deformation along the South Range Shear Zone, gives rise to the present form of the Sudbury Structure and the erosion level as shown by the line marked "Present Surface."

from a preexisting 2.45-Ga intrusion, similar to the East Bull Lake body west of Sudbury.[2]

The deep level to which impact melting extended, as suggested by Ivanov et al. (Chapter 30), means that the paroxysm of the impact may not have resulted in perfect homogenization of all of the impact melt; in particular, localized zones of melting may have existed within the wall rocks and have retained a distinctly "local" flavor, perhaps only partially influenced by the Main Mass magma. Where the wallrocks were mafic/ultramafic rocks, the melt, while still retaining a distinctly primitive composition (and thus being capable of giving rise to the melanorites that are preserved as "inclusions") and containing shattered inclusions of the local mafic/ultramafic country rocks, could also have acquired a distinctive trace element profile as a result of this limited mixing with the Main Mass.

Given the excess heat in the area, zones rich in this material could have persisted at depth for relatively long periods of time (tens to hundreds of thousands of years), crystallizing to give rise to additional ultramafic cumulates, and then have worked their way up along the floor of the crater as postimpact readjustments continued, exploiting particularly the embayments marked by, and resulting from, zones in the footwall with a very high proportion of Sudbury Breccia (see the discussion of Fedorowich et al., above). Further incomplete mixing with Main Mass magma during this late-stage emplacement would account for the heterogeneous nature of the Sublayer, especially the presence of melanorite pods as noted by Lightfoot et al. (1997b). The Sublayer geochemistry would then tend to have a distinctly different, local character, consistent with the observations of Lightfoot et al. (1997b), but nevertheless an unusually high concentration of incompatible elements for mantle-derived magma. A portion of any sulfides that had previously settled out of the main body to concentrate along the crater floor would have been picked up by and have become intermixed with the Sublayer magmas as they moved into position along the floor.

This model, in which the sulfides settled out of the main body of the SIC and were then picked up by the Sublayer, has some distinct advantages over one requiring them to have been developed within and introduced along with the Sublayer, since sulfide ores are not exclusively associated with the Sublayer. Mafic norite, which is included with the Main Mass of the SIC, contains appreciable concentrations of sulfide (cf. Naldrett and Kullerud, 1967), as do the offset dikes, which Lightfoot et al. (1997a) have shown to be much closer on geochemical grounds to the Main Mass than to the Sublayer. Furthermore, the distinctive character of the Sublayer has been documented principally on the North Range, where it appears in sharp contact with the Felsic norite (for example, on the 2,500-m level of the Strathcona mine), which lacks both the sulfides and inclusions (Naldrett et al., 1984) of the immediately adjacent Sublayer. This distinction between Sublayer and Main Mass norites is much less apparent along much of the South Range, where the South Range and Quartz-rich Norites appear to be transitional with the Sublayer in many localities.

It is very rare for a terrestrial magma to intrude with sufficient superheat that 250 Ka will pass before the solidus is reached. Given the demonstration of Ivanov et al. (Chapter 30) that this was very possibly the case at Sudbury, a much greater degree of country-rock assimilation, particularly melting of the roof rocks, is to be expected here than is the case with other intrusions. I suggest that a magma comprising cumulates represented by the South Range Norite + Felsic Norite + Quartz Gabbro[3] represents the initial composition of the impact melt. Given the depth to which Ivanov et al. have indicated impact melting might have extended, the source for this melt would have included the relatively felsic Huronian strata, mafic early Huronian volcanic rocks, and perhaps a layered intrusion(s) of the same age, and Archean tonalites, and possibly mafic to komatiitic Archean greenstones. If the ejecta from the crater contained a higher proportion of felsic Huronian strata and a lower proportion of the proposed Huronian mafic intrusion than the impact melt that remained in the crater, the roof rocks to the impact melt (which comprised the blanket formed by the fall-back of the ejecta) would have had a more felsic composition than the impact melt itself. As Ariskin et al. (Chapter 29) propose, the hot impact breccias forming the roof would have been subjected to very considerable melting as a result of the superheat of the impact melt.

I suggest that the melted suevite was incorporated into the upper part of the impact melt, giving it a lower density than the lower part. Very quickly, this separated as a density-stabilized, low-density layer above the norite, separated by a boundary layer (Fig. 1). Both layers would have been convecting, but convecting separately (Campbell and Turner, 1986). Heat from the cooling norite would have been convected up to the boundary layer, conducted across, it and then convected up again through the granophyre layer to augment the melting of the roof rocks. Convection currents would have swept down the relatively cool margins of the norite, and once the solidus was reached and silicates began to crystallize, would have caused the downward and inward lineation that Cowan et al. (Chapter 31) have documented in this unit. The same would likely have occurred in the granophyric layer, but portions of the SIC in which the granophyre was in contact with the walls are no longer preserved, the only border with which the granophyre can be observed to be in con-

---

[2]It is relevent to note here that sheet-like bodies of pyroxenite have been intersected in drill holes penetrating the footwall in the Strathcona-Fecunis Lake area of the North Range (geological staff, personal communication, Falconbridge Ltd.).

[3]Note that both Chai and Eckstrand (1996) and Lightfoot et al. (1997a), when they computed the average composition of the Main Mass of the SIC or the composition of the Norite Layer omitted the South Range Norite. This has distinctly less siliceous composition than the Felsic Norite on which they based their calculations, and yet it is by far more abundant than the Felsic Norite on the South Range. It is difficult to estimate which is the more abundant throughout the original, undeformed SIC, but, if one includes the South Range Norite in equal proportion to the Felsic Norite in computing the average composition of the norite layer, it will be distinctly less siliceous and more dense than the computations that Lightfoot et al. and Chai and Eckstrand indicated.

tact is the roof. Once the granophyric layer reached its liquidus temperature, it is likely that the igneous fabric developed orthogonal to the isotherms prevailing within it, orthogonal to the roof, and thus accounting for the observation of Cowan et al.

At first sight, this model—which has much in common with that of Chai and Eckstrand (1994)—would seem to be in conflict with the trace element data of Lightfoot et al. (1997a). However, the initial homogenization of the impact melt into the layer of "norite" magma would have involved a large proportion of the felsic components in the target area, which would have been the principal source of the incompatible elements in the "norite" magma. These constituents would have also been the principal source of these elements in the suevite breccia, so it is to be expected that the ratios of incompatible elements in the two layers, "norite" and granophyre (melted basal Onaping Formation), are very similar—Lightfoot et al. have admitted this possibility.

The boundary layer between the two principal layers of magma proposed here (i.e., norite and granophyre layers) would not only have been a zone of heat interchange; diffusion of chemical constituents would also have been occurring across this layer. Watson and Jurewicz (1984) have shown that alkalis diffuse along chemical gradients much more quickly than Al_2O_3 or SiO_2, so that, given a higher alkali content in the granophyre than the norite layer, the upper part of the norite layer would have been subjected to a constant influx of alkalis; this would have been picked up by the convecting norite magma and rapidly become homogenized with that part that remained unsolidified.

Calling on the "norite" layer in Figure 1 to be a well-homogenized mixture of felsic and mafic rocks in the target area also provides a clue, as Golightly (1994) suggested, to the formation of the high proportion of rich accumulations of sulfide ore at Sudbury. If the mafic rocks of the target area had been close to saturation with sulfide, melting them and immediately mixing the melt with melted felsic rocks would produce the right conditions for sulfide immiscibility (Li and Naldrett, 1993). The immiscibility might not have occurred immediately, when the melt was several hundred degrees above its liquidus, but on cooling, sulfides could well have become immiscible at a temperature that was still well above the liquidus of the hybrid magma. If this did occur, the time that the magma remained above its liquidus temperature would have allowed very complete settling of the sulfides to the floor of the chamber, unobstructed by the development of silicates, and thus have contributed to the accumulation of a layer of sulfide along the crater floor. These sulfides would have been accessible to incorporation in Sublayer magmas that were exploiting this contact, as discussed above. Since the sulfides had settled out of what was a melt of crustal rocks ranging from Archean to Proterozoic, they would be characterized by very radiogenic Os.

Lightfoot et al. (1997a) pointed out that problems exist with respect to the Ni, Cu, and S contents of the SIC if this is the result of melting of "average crust." This point is illustrated in Table 1. Average felsic norite contains 126 ppm Ni and 155 ppm Cu, not counting the Ni and Cu in the ores that should be back-calculated into any source magma. This is very much higher than the 25 ppm Ni and 30 ppm Cu, which are characteristic of average crust. The lowering of Ni and Cu, which is accomplished by including the granophyre in calculating an average for the initial magma of the SIC, is offset to some extent by inclusion of the metal and sulfur in the ores. Clearly, one has to account for the excess of both metals.

The scenario I developed above involving a lower layer of norite magma and upper zone of granophyre magma requires the noritic layer to contain a much greater proportion of melted mafic/ultramafic rock, and thus a higher Ni and Cu content, than the granophyric layer (whose composition I suggest is greatly influenced by melted fall-back breccia). As emphasized above, this requires the impact melt that remained in the crater to have had a different, more mafic composition to that of the ejecta thrown up out of the crater. If this requirement is shown to be impossible or unlikely in terms of impact dynamics, then a mantle contribution, such as that called for by Lightfoot et al. (1997a) and Naldrett et al. (1986), becomes very necessary.

The preceding scenario can account for most of the new observations on Sudbury geology that have been made over the last 6 yr, particularly in regard to the origin of the SIC. A key question is, does one have to devise a scenario along these lines? Can any mantle contribution coincident with the Sudbury event be positively ruled out? The answer is no. Models calling on an endogenous component to the SIC, particularly that argued so well by Lightfoot et al. (1997a), which calls for up to 20% picrite magma to fountain into the chamber of the SIC, remain valid at the present time.[4]

In summary, the scenario that I propose for Sudbury contains the following points:

1. A preexisting terrain, consisting of Huronian sediments and underlying volcanic rocks, rests unconformably on Archean gneisses. The gneises had been intruded by a mafic intrusion, which is a feeder to the Huronian volcanics and is part of the

TABLE 1. Ni AND Cu CONTENTS OF VARIOUS ROCK TYPES

	Felsic Norite (1)*	Granophyre (1)*	Gray Onaping (2)†	Average Crust (1)*
(ppm)				
Ni	126	31	51	25
Cu	155	5	35	30

*Lightfoot et al., 1997a.
†Muir and Peredery, 1984.

---

[4]If the question is asked, What do I believe to be more likely? the answer is the scenario involving just impact melting. This is a reversal of my previous opinion, made with some reluctance, and which is irrelevant to this summary because, at present, both interpretations remain valid. [Note added during revision: Peter Lightfoot, one of the referees of this paper, asked to have his name mentioned as endorsing this viewpoint very strongly.]

2.45-Ga mafic magmatism that is known to have accompanied the opening of the Huronian ocean.

2. Impact of a meteorite is followed by excavation of a transient crater 40 km deep, and 60 km in diameter. Formation of some of the Sudbury Breccia, ejection of country rocks in the target area, melting of target rocks, including a high proportion of the mafic/ultramafic intrusion to form the "norite" layer, leaves pockets of molten rock within the footwall target rocks.

3. Rapid rebound occurs at the center of crater, and lower crustal high-grade gneisses of the Levack Complex are brought to surface beneath the melt rock filling the crater.

4. A part of the ejecta falls back into the crater to form a 2 km blanket over the melt rock, with the Green Member of the Onaping Formation as the uppermost unit.

5. The crater starts slumping around the margins and the initial crater becomes enlarged to a diameter of about 200 km. Other types of Sudbury Breccia form during this readjustment. Zones around the margin of the crater where Sudbury Breccia is abundant remain as zones of weakness. The Onaping Formation is reworked during this phase of tectonic adjustment with the formation of the Lower and Upper Black Members, which have incorporated clasts of a black mudstone of unknown provenance.

6. Extensive remelting of the Basal Member of the Onaping, as a result of superheat within the melt sheet within the crater, results in a strong chemical gradient developing within the impact melt from a more felsic top to more mafic base. Rapidly this compositional heterogeneity becomes stabilized through the development of two layers, an upper "granophyric" layer and a lower "noritic" layer, which convect independently. Once this occurs, continued melting of the breccias of the overlying Onaping Formation increases the felsic components of the Granophyric layer (Fig. 1).

7. As the impact sheet cools, sulfide immiscibility develops in the lower "noritic" layer, and the resulting sulfides settle rapidly in the low viscosity, superheated magma, unimpeded by silicates, to collect in pools at the base of the sheet, along with mafic and ultramafic inclusions (Fig. 1). The superheated environment existing in the SIC during this stage in its development provided the perfect conditions for sulfide settling. During this stage, magma from the noritic layer was forced out laterally and downward into fractures in the country rocks, particularly exploiting zones of Sudbury Breccia, to give rise to the "Offsets." This magma picked up sulfide and inclusions that were transported as "pods" within it, giving rise to the pod-like ore zones within the offsets.

8. With further cooling, the "noritic" layer reached its liquidus temperature, and pyroxene, followed very quickly by plagioclase, started to crystallize. Since the walls were cooler than the central part of the layer, currents of magma swept down them (Fig. 1), giving rise to the lineated fabric documented by Cowan et al. (Chapter 31). The upper "granophyric" layer reached its liquidus at a later stage. The high viscosity of this very siliceous zone may have damped out convection after the temperature fell to a certain level, and the fabric of the crystallizing minerals developed at right angles to the isotherms within the melt sheet (i.e., orthogonal to the roof).

9. The most difficult component, and one of the most important, to fit into this scenario is the Sublayer. At present, I interpret it as an igneous breccia composed in part of remnants of a very mafic component of the impact melting, a component that failed to become involved in the convection of the lower mafic sheet of magma, and which thus retained more of a "local," less felsic signature. It is possible that the sulfides that had settled out of the overlying sheet, together with inclusions that had settled there, so increased the overall density of the Sublayer that it remained at the contact and did not convect and become mixed with the Main Mass norite. This appears to have occurred particularly in embayments in the lower contact of the SIC. Possibly the physical environment of an embayment is able to shelter material lying within it from being incorporated in the SIC-wide convection affecting the overlying layer. Some later mobility of the Sublayer would seem to be called for in view of conflicting age relationships that have been reported with respect to the Main Mass norites (Naldrett et al., 1984).

The above scenario has resulted from the request that I summarize and synthesize information available when this was written (1998). In no sense is it my work; it is instead an attempt to bring the work of others together into a possible model, and thus it combines aspects of the ideas developed by many scientists recently concerned with the Sudbury Structure.

I believe that key questions requiring answers in the future are:

1. What really occurs in the unusual physical and chemical environment of impact? Do lessons learned from the condensation of matter from a solar gas have applications at Sudbury? Can zircons and baddeleyite really be reset during impact, and if not, how does one explain the 1.85-Ga age of the apparently older diabase?

2. Is the recent geochemical data of Lightfoot et al. (1997a,b) on which the above scenario is so heavily based, of general application to the Sudbury basin as a whole, particularly on the South and East Ranges? So often in the history of the earth sciences, errors have arisen because a scientist has attempted to extrapolate ideas developed in a limited geologic area into a general theory (e.g., The proponents of the flat earth theory; Werner and the Neptunists; Buckland and the Diluvialists).

3. Despite what is suggested here—the origin of the Sublayer, the relationship between it and the mineralization, and the structures controlling its location have no conclusive explanation as yet.

4. The inclusions in the Sublayer are telling us something either about the early development of the SIC or about the nature of the target area of the impactor. The amount of modern geochemical information about them is relatively small, and further samples from all around the Sudbury Structure require serious study.

5. What is the cause of the steep dips along the East Range. Partly due to the paucity of active mines in this area over the past 20 yr, up-to-date information is lacking about this area. If, as Cowan et al. (Chapter 31) claim, the dips are original and not due

to folding, they must represent the steep walls of the original magma chamber and the nature of the rocks adhering to them, and the fabric of these rocks may be expected to be different to that elsewhere. If the dips have resulted from the effect of the Wanapitei impact, the structures along which the deformation have occurred should be apparent. As operations begin at INCO's Victor and Falconbridge's Nickel Rim mines, we may learn more about this environment.

I cannot conclude without commenting that Sudbury differs in so many respects from other world-class accumulations of magmatic sulfides. Sudbury was used for so long as the archetype for these deposits, yet it has turned out to be a very poor model. The superheat, which appears to have existed for so long, and the formation of magma as the result of a process originating from above rather than below, produced an unusual environment in which sulfides formed and accumulated in a manner quite different to the way in which deposits have developed elsewhere. It remains to be seen just how "unique" this unusual environment is. It is possible that it required not only the impact of a very large (10–15-km diameter) impactor, but also a special arrangement of rocks in the target area, comprising felsic rocks close to the surface and mafic/ultramafic rocks at depth.

ACKNOWLEDGMENTS

While this summary is concerned primarily with data and ideas originating during the 1990s, I have also drawn on my Sudbury-related experience, which began at the Falconbridge East Mine in 1957. I am grateful to all of those with whom I have interacted on the geology of Sudbury: colleagues from the early field-trip days; colleagues at the Engineers' Club in Falconbridge; J. E. Hawley of Queen's University, Toronto; Gunnar Kullerud of the Geophysical Laboratory, Washington D.C.; fellow faculty from more than 31 years at the University of Toronto; geologists at INCO and Falconbridge; and above all, of my students at the University of Toronto. Finally, I thank Peter Lightfoot, INCO, Ltd., and John Fedorowich, Falconbridge, Ltd., for their helpful reviews and suggestions concerning this manuscript.

REFERENCES CITED

Amsden, A. A., Ruppel, H. M., and Hirt, C. W., 1980, SALE: a simplified ALE computer program for fluid flow at all speeds: Los Alamos, New Mexico, Los Alamos National Laboratory Report LA-8095, 101 p.

Avermann, M., 1994, Origin of the polymict, allochthonous breccias of the Onaping Formation, Sudbury Structure, Ontario, Canada, in Large meteorite structures and planetary evolution: Dressler, B. O., Grieve, R. A. F., and Sharpton, V. L, eds, Geological Society of America Special Paper 293, v. 319–329.

Becker, L., Bada, J. L., Winans, R. E., Hunt, J. E., Bunch, T. E., and French, B. M., 1994, Fullerenes in the 1.85 billion year old Sudbury impact structure: Science, v. 265, p. 642–645.

Becker, L., Poreda, R. J., and Bada, J. L., 1996, Extraterrestrial helium trapped in fullerenes in the Sudbury impact structure. Science, v. 272, p. 249–252.

Buseck, P. R., Tsipurski, S. J., and Hettich, R., 1992, Fullerenes from the geological environment: Science, v. 257, p. 215–217.

Butler, H. R., 1994, Lineament analysis of the Sudbury multiring impact structure, in Dressler, B. O., Grieve, R. A. F., and Sharpton, V. L., eds., Large meteorite structures and planetary evolution: Geological Society of America Special Paper 293, p. 319–329.

Campbell, I. H., and Turner, J. S., 1986. a laboratory investigation of assimilation at the top of a basaltic magma chamber: Journal of Geology, v. 95, p. 155–172.

Chai, G., and Eckstrand, O. R., 1994, Rare earth element characteristics of the Sudbury igneous complex and its country rocks: new constraints on genesis: Chemical Geology, v. 113, p. 221–244.

Collins, W. H., 1934, Life history of the Sudbury nickel irruptive (i): Transactions of the Royal Society of Canada, sect. IV, v. 28, p. 123–177.

Corfu, F., and Lightfoot, P. C., 1996, U-Pb geochronology of the Sublayer environment, Sudbury Igneous Complex, Ontario: Economic Geology, v. 91, p. 1263–1269.

Cowan, E. J., and Schwerdtner, W. M., 1994, Fold origin of the Sudbury Basin, in Lightfoot, P. C., and Naldrett, A. J., eds., Proceedings of the Sudbury Noril'sk Symposium: Ontario Geological Survey Special Publication 5, p. 45–56.

Deutsch, A., 1994, Isotope systematics support the impact origin of the Sudbury Structure (Ontario, Canada), in Dressler, B. O., Grieve, R. A. F., and Sharpton, V. L., eds., Large meteorite structures and planetary evolution: Geological Society of America Special Paper 293, p. 289–329.

Dickin, A. P., Artan, M. A., and Crocket, J. H., 1996, Isotopic evidence for distinct crustal sources of North and South Range ores, Sudbury Igneous Complex: Geochimica et Cosmochimica Acta, v. 60, p. 1605–1613.

Dressler, B. O., 1984. The effects of the Sudbury Event and the intrusion of the Sudbury Igneous Complex on the footwall of the Sudbury Structure, in Pye, E. G., Naldrett, A. J. and Giblin, P. E., eds., The geology and ore deposits of the Sudbury Structure: Ontario Geological Survey Special Volume 1, p. 97–138.

Dressler, B. O., Grieve, R. A. F., and Sharpton, V. L., eds.,1994, Large meteorite impacts and planetary evolution: Geological Society of America Special Paper 293, 348 p.

Faggart, B. E., Basu, A. B., and Tatsumoto, M., 1985, Origin of the Sudbury Complex by meteorite impact: neodymium isotopic evidence: Science, v. 230, p. 436–439.

Golightly, J. P., 1994, The Sudbury Igneous Complex as an impact melt: evolution and ore genesis, in Lightfoot, P. C., and Naldrett, A. J., eds., Proceedings of the Noril'sk Symposium: Sudbury Ontario Geological Survey Special Publication 5, p. 105–118.

Grieve, R. A. F., 1994, An impact model of the Sudbury Structure, in Lightfoot, P. C., and Naldrett, A. J., eds., Proceedings of the Sudbury-Noril'sk Symposium: Ontario Geological Survey Special Volume 5, p. 119–132.

Grieve, R. A. F., Stöffler, D., and Deutsch, A., 1991, The Sudbury Structure: controversial or misunderstood?; Journal of Geophysical Research, v. 96, p. 22753–22764.

Gupta, V. K., Grant, F. S., and Card, K. D., 1986. Gravity and magnetic characteristics of the Sudbury Structure, in Pye, E. G., Naldrett, A. J. and Giblin, P. E., eds., The geology and ore deposits of the Sudbury Structure: Ontario Geological Survey Special Volume 1, p. 381–410.

Jago, B. C., Morrison, G. G., and Little, T. L., 1994, Metal zonation patterns and micro-textural and micro-mineralogical evidence for alkali and halogen-rich fluids in the genesis of the Victor Deep and McCreedy East Footwall copper ore bodies of the Sudbury Igneous Complex, in Lightfoot, P. C., and Naldrett, A. J., eds., Proceedings of the Sudbury-Noril'sk symposium. Ontario Geological Survey Special Publication 5, p. 65–76.

Hawkesworth, C. J., Cohen, A. S., Cooper, M., Lightfoot, P. C., and Burnham, O. M., 1997, Significant differences in the ore forming processes at Sudbury and Noril'sk: Constraints from Os isotopes: Fall Meeting, American Geophysical Union, San Francisco, 1997, Volume of Abstracts, V41F–1.

Hearst, R. B., Morris, W. A., and Thomas, M. D., 1994, Magnetic interpretation along the Sudbury Structure Lithoprobe transect, in Lightfoot, P. C., and Naldrett, A. J., eds., Proceedings of the Sudbury Noril'sk Symposium: Ontario Geological Survey Special Publication 5, p. 33–44.

Keays, R. R., Vogel, D. C., James, R. S., Peck, D. C., Lightfoot, P. C., and Prevec, S. A., 1995, Metallogenic potential of the Huronian-Nipissing magmatic province, *in* Programme and Abstracts, Symposium on the Northern Margin of the Southern Province: Canadian Mineralogist, v. 33, p. 932–933.

Li, C., and Naldrett, A. J., 1993, Sulfide capacity of magma: a quantitative model and its application to the formation of sulfide ores at Sudbury: Economic Geology, v. 88, p. 1253–1260.

Lightfoot, P. C., and Naldrett, A. J., eds., 1994, Proceedings of the Sudbury-Noril'sk Symposium, Ontario Geological Survey Special Volume 5, 423 p.

Lightfoot, P. C., Keays, R. R., Morrison, G. G., Bite, A., and Farrell, K. P., 1997a, Geochemical relationship in the Sudbury Igneous Complex: origin of the Main Mass and Offset Dikes: Economic Geology, v. 92, p. 289–307.

Lightfoot, P. C., Keays, R. R., Morrison, G. G., Bite, A., and Farrell, K. P., 1997b, Geologic and geochemical relationships between the Contact Sublayer, inclusions and the Main Mass of the Sudbury Igneous Complex: a case study of the Whistle mine embayment. Economic Geology, v. 92, p. 647–673.

McGrath, P. H., and Broome, H. J., 1994, A gravity model for the Sudbury Structure, *in* Lightfoot, P. C., and Naldrett, A. J., eds., Proceedings of the Sudbury Noril'sk Symposium: Ontario Geological Survey Special Publication 5, p. 21–33.

Melosh, H. J., 1989, Impact cratering: A geologic process: New York, Oxford University Press, and Oxford, United Kingdom, Clarendon Press, 245 p.

Milkereit, B., Green, A., Berrer, E., Boerner, D., Broome, J., Cosec, M., Cowan, J., Davidson, A., Dressler, B., Fueten, F., Grieve, R., James, R., Kraus, B., McGrath, P., Meyer, W., Moon, W., Morris, W., Morrison, G., Naldrett, A., Peredery, W., Rousell, D., Salisbury, M., Schwerdtner, W., Snajdr, P., Thomas, M., and Watts, A., 1992, Geometry of the Sudbury Structure from high resolution seismic reflection profiling. Geology, v. 20, p. 807–811.

Milkereit, B., White, D., Adam, E., Boerner, D. and Salisbury, M., 1994, Implications of the Lithoprobe seismic reflection transect for Sudbury geology, *in*, Pye, E. G., Naldrett, A. J., and Giblin, P. E., eds, The geology and ore deposits of the Sudbury Structure: Ontario Geological Survey Special Volume 1, p. 11–20.

Morrison, G. G., Jago, B. C., and White, T. L., 1994, Footwall mineralisation of the Sudbury Igneous Complex, *in* Lightfoot, P. C., and Naldrett, A. J., eds., Proceedings of the Sudbury Noril'sk Symposium, Ontario Geological Survey Special Publication 5, p. 57–64.

Muir, T. L., and Peredery, W. V., 1984, The Onaping Formation, *in* Pye, E. G., Naldrett, A. J., and Giblin, P. E., eds., The geology and ore deposits of the Sudbury Structure: Ontario Geological Survey Special Volume 1, p. 139–210.

Naldrett, A. J., 1994, The Sudbury Noril'sk Symposium, An Overview, *in* Lightfoot, P. C., and Naldrett, A. J., eds., Proceedings of the Sudbury Noril'sk Symposium: Ontario Geological Survey Special Publication 5, p. 3-8.

Naldrett, A. J., and Hewins, R. H., 1984. The Main Mass of the Sudbury Igneous Complex, *in* Pye, E. G., Naldrett, A. J., and Giblin, P. E., eds., The geology and mineral deposits of the Sudbury Structure: Ontario Geological Survey Special Publication 1, p. 235–251.

Naldrett, A. J., and Kullerud, G., 1967, A study of the Strathcona mine and its bearing on the origin of the nickel-copper ores of the Sudbury district, Ontario: Journal of Petrology, v. 8, p. 453–531.

Naldrett, A. J., Hewins, R. H., Dressler, B. O., and Rao, B. V., 1984, The Contact Sublayer of the Sudbury Igneous Complex, *in* Pye, E. G., Naldrett, A. J., and Giblin, P. E., eds., The geology and mineral deposits of the Sudbury Structure: Ontario Geological Survey Special Publication 1, p. 253–274.

Naldrett, A. J., Rao, B. V., and Evensen, N. M., 1986, Contamination at Sudbury and its role in ore formation, *in* Gallagher, M. J., Ixer, R. A., Neary, C. R., and Pritchard, H. M., eds., Metallogeny of Basic and Ultrabasic Rocks: Special Publication of the Institute of Mining and Metallurgy, London, p. 75–92.

Naldrett, A. J., Pessaran, A., Asif, M., and Li, C., 1994, Compositional variation in the Sudbury ores and prediction of the proximity of footwall copper-PGE ore bodies, *in* Lightfoot, P. C., and Naldrett, A. J., eds., Proceedings of the Sudbury Noril'sk Symposium, Ontario Geological Survey Special Publication 5, p. 133–146.

Norman, M. D., 1994, Sudbury Igneous Complex: Impact melt or endogenous magma? Implication for lunar crustal evolution, *in* Dressler, B. O., Grieve, R. A. F., and Sharpton, V. L., eds., Large meteorite structures and planetary evolution: Geological Society of America Special Paper 293, p. 331–341.

Peredery, W. V., and Naldrett, A. J., 1975, Petrology of the Upper Irruptive rocks, Sudbury, Ontario, Canada: Economic Geology, v. 70, p. 164–175.

Pye, E., Naldrett, A. J., and Giblin, P. E., eds., The geology and ore deposits of the Sudbury Structure: Ontario Geological Survey Special Volume 1, 603 p.

Scribbins, B., Rae, D. R., and Naldrett, A. J., 1984, Mafic and ultramafic inclusions in the sublayer of the Sudbury Igneous Complex: Canada. Mineralogist, v. 22, p. 67–75.

Souch, B. E., Podolsky, T., and the geological Staff of the International Nickel Co. of Canada, Ltd., 1969, The sulfide ores of Sudbury: their particular relation to a distinctive inclusion-bearing facies of the Nickel Irruptive. Econonmic Geology, Monograph 4, p. 252–261.

Stöffler, D., Deutsch, A., Avermann, M., Bischoff, L., Brockmeyer, P., Buhl, D., Lakomy, R., and Müller-Mohr, V., 1994, The formation of the Sudbury Structure, Canada: towards a unifed model, *in* Dressler, B. O., Grieve, R. A. F., and Sharpton, V. L., eds., Large meteorite structures and planetary evolution: Geological Society of America Special Paper 293, p. 303–318.

Thompson, L. M., and Spray, J. G., 1994, Pseudotachylitic rock distribution and genesis within the Sudbury impact structure, *in* Dressler, B. O., Grieve, R. A. F., and Sharpton, V. L., eds., Large meteorite structures and planetary evolution: and Geological Society of America Special Paper 293, p. 275–285.

Walker, R. J., Morgan, J. W., Naldrett, A. J., Li, C., and Fassett, J. D., 1991, Re-Os isotopic systematics of Ni-Cu sulfide ores, Sudbury Igneous Complex, Ontario: Evidence for a major crustal component. Earth and Planetary Science Letters, v. 105, p. 416–429.

Watson, E. B., and Jurewicz, S. R., 1984, Behavior of alkalies during diffusive interaction of granitic xenoliths with basaltic magma: Journal of Geology, v. 92, p. 121–131.

MANUSCRIPT ACCEPTED BY THE SOCIETY DECEMBER 16, 1998

Index

[Italic page numbers indicate major references]

A

accessory minerals
 Morokweng impact structure, 71–72
 See also specific minerals
accretionary-mantled clasts
 Black Member, Onaping Formation, Sudbury Basin, 336–337, 340
 enriched in carbon, 337
 fullerenes in, 340
 Lower Black Member, Onaping Formation, Sudbury Basin, 331, 341–342
acid rain, 243
acoustic data, Lake Karikkoselkä impact structure, 136, 144
acoustic fluidization
 Sudbury Structure, 436
 of target rocks, 277–278, 390–391, 415
Acraman impact structure, South Australia, 258
AEM (airborne electromagnetic data). See electromagnetic data
aerial photograph
 BP impact structure, 179
 Lake Karikkoselkä impact structure, 133
 Oasis impact structure, 179
aeromagnetic data, 63
 Lake Karikkoselkä impact structure, 139–142
 Lycksele structure, 126–127
 Mjölnir impact structure, 195
 Morokweng impact structure, 61–62, 64, 85, 87, 89, 92–93
age of pseudotachylite, 121
age of structure
 Berriasian, early, 62
 Chesapeake Bay impact crater, 149–150
 Jurassic-Cretaceous boundary, 61–62
 Kara impact structure, Russia, 206
 Lake Karikkoselkä impact structure, 131, 139, 145–146
 Lycksele structure, 125–126, 130
 Lycksele structure, dated dikes (post-Jotnian, ca. 1.26 Ga), 126
 Mjölnir impact structure, 193, 195, 202
 Morokweng impact structure, 61–62, 72, 83–85, 91–92
 Morokweng impact structure, model ages (Sm-Nd), 82
 pitfalls in determination of, 121
 Popigai impact structure, 1, 20, 47–48
 Slate Islands impact structure, 109, 118, 120–122
 Steinheim impact structure, Germany, 206
 Sudbury (Swarm) dikes, 411, 423
 Sudbury Structure, 299–300, 302, 317, 345–346, 389, 419
 Teague Ring impact structure, 168
 Toms Canyon impact crater, 150
 Vredefort structure, South Africa, 62
age of Sudbury Igneous Complex, 323, 400, 431, 433
age of target or related rocks, Lake Karikkoselkä impact structure, 133, 138
airborne electromagnetic anomalies (AEM). See electromagnetic data
aleurolite, from Chelmsford Formation, Whitewater Group, Sudbury Basin, 318

algae
 Black Member, Onaping Formation, Sudbury Basin, 340, 435
 Onwatin Formation, Sudbury Basin, 340, 352
algorithm to solve equilibrium problem, COMAGMAT-3.5 use of, 376
alkalies, limited diffusion of, Sudbury Igneous Complex, 432
Allende meteorite
 fullerenes in, 353–354, 358, 435
 mass-independent fractionation of sulfur isotopes in, 357
alteration of melt rock, Morokweng impact structure, 66, 72, 84, 88
alteration of minerals
 Morokweng impact structure, 72, 97
 Popigai impact structure, 39, 41
 Slate Islands impact structure, 122
amino acids
 in Black Member, Onaping Formation, Sudbury Basin, 331, 337
 concentrations in Sudbury Basin rocks, 338
 extraterrestrial, not observed in Sudbury Basin rocks, 337, 341
AMS. See anisotropy of magnetic susceptibility
Anabar Group, Popigai impact structure, Russia, chemical composition, 42
analytical methods, 41–42, 99
 acid-resistant residue, 216
 age of impact structure, 121
 amino acids, 334
 anisotropy of magnetic susceptibility (AMS), 403, 406
 BP impact structure, 181
 carbon isotopes, 333–334
 fullerenes in Sudbury Structure rocks, 353
 impact diamonds, 216
 Morokweng impact structure, 68
 Oasis impact structure, 181
 pitfalls in determination of, 121
 X-ray powder diffraction measurements, 206–208
anatectic glass
 chemical composition of, 45
 Popigai impact structure, 35, 45, 55
anatectic rocks of the earliest Earth, 53
angular dependency on the diffraction angle, X-ray powder diffraction measurements, 207
Animikie Group (1.8–1.9 Ga), Slate Islands impact structure, Canada, 112
anisotropy of magnetic susceptibility (AMS)
 gabbro, Sudbury Igneous Complex, 403, 434
 granophyre, Sudbury Igneous Complex, 399, 407, 434
 Levack Gneiss Complex, and Sudbury Structure, 432
 mathematical relationships, 403
 norite, Sudbury Igneous Complex, 403, 434
Anna's Rust Sheet (Vredefort impact structure), similarities to, Morokweng impact structure, 75
annular trough
 Chesapeake Bay impact crater, 149, 151–152, 154–155, 157
 Popigai impact structure, 7–13
 Slate Islands impact structure, 109, 114, 122

Antarctica, Permian-Triassic boundary, shocked quartz at, 244
anthraxolite, 345, 347–349, 352
apatite
 chemical composition, 99
 Morokweng impact structure, 75–76, 78–79, 97
apparent polar wander paths (APWPs) of Baltica
 Burakovskaja gabbro, 140
 Kolvitsa porphyry, 140
 Lake Karikkoselkä impact structure, 140
 Russian Karelia, 140
 Suoperä, Finland, 140
apparent resistivity, Lake Karikkoselkä impact structure, 143
APWP. See apparent polar wander path
Araguainha impact structure, Brazil, 245
archaea (hyperthermophylic) as Sudbury carbon source, 345, 352, 358
Archean (late), spherules, 249
Archean target rocks, Popigai impact structure, Russia, chemical composition, 43–44
archetype of world-class accumulations of magmatic sulfides (Sudbury), 441
Arctic Siberia, Russia, Popigai impact structure, 1, 19–20
Arrecife Alacrañ, 296
Arroyo El Mimbral, Mexico
 carbonaceous components at, 217
 Cretaceous-Tertiary boundary, 216
 diamonds at, 217–219
 impact diamonds, 216
asteroid impact, Slate Islands impact structure, 109
asteroids, size of, 246
Australasian strewn field, 190
Australia, Permian-Triassic boundary, shocked quartz at, 244
autoclastic breccia
 seismic vibration attending crater-formation processes, 116
 Slate Islands impact structure, 117, 122
Avak impact structure, United States, 245
Azuara impact structure, Mexico, gravity anomaly vs. erosional level, 237

B

B.P. Exploration Company, 178
bacteria
 and Black Member, Onaping Formation, Sudbury Basin, 340–341
 and Onwatin Formation, Sudbury Basin, 340–341
 as Sudbury carbon source, 345, 352, 358
baddeleyite, 216
 diabase (Sudbury), 437
 Libyan Desert Glass (LDG), 179
 melanorite inclusions in Sublayer, Sudbury Igneous Complex, 431, 433, 436–437
 Morokweng impact structure, 72, 97
 pyroxenite (Sudbury), 437
 reset by Sudbury impact?, 440
 Sudbury Igneous Complex, 302
 Sudbury Structure, 392
 Whistle Mine, Sublayer mafic inclusions, Sudbury Igneous Complex, 433

443

Baden-Württemberg, Steinheim impact structure, Germany, 206
Balagan River basin, Popigai impact structure, 7, 9–11, 15–16
ballen structure
    Morokweng impact structure, 74, 77
    Popigai impact structure, 32, 34–35
Baltimore Canyon trough, Chesapeake Bay impact crater, 151, 154
Bangemall Group, 167
Barents Sea, 173, 244
barometry. *See* pressure of impact
Barremian, Mjølnir impact structure, 195
Barringer, Colonel Paul, dedication to, iii
Barringer Crater, Arizona, United States, 270
    See also Meteor Crater
Basal Breccia, Onaping Formation, Sudbury Basin, 324
Basal Member, Onaping Formation, Sudbury Basin, 323–324, 332, 347, 373, 389, 433
    cross section, after readjustments to crater and before complete solidification, 437
    melting of by superheated impact melt, 432
    and parent-magma calculation, 379
    remelting of, by superheated impact melt, 440
basement surface, Chesapeake Bay impact crater, 151
basins, Late Cretaceous, Morokweng impact structure, 92
bathymetry
    Deep Bay impact structure, Canada, 134
    Lake Karikkoselkä impact structure, 135, 146
Bee Gorge Member, Western Australia, 251
Benny deformation zone, Sudbury Structure, 423
    map (geologic), 420
Berriasian (lower Berriasian–middle Berriasian) boundary, 244
Berwind Canyon, Colorado
    carbonaceous components at, 217
    Cretaceous-Tertiary boundary, 215–216
    diamonds at, 217, 221
    impact diamonds, 215
Bigach impact structure, Kazakhstan, 245
Big Levack mine, Sudbury Structure, 306–307
biogenic origin of carbon in impact-related rocks. *See* carbonaceous matter in Sudbury Basin rocks
biota, postimpact resurgence of, 342
BIRPS. *See* British Institutions Reflection Profiling Syndicate
Bjarmeland Platform, Barents Sea
    Mjølnir impact structure, 193
    target rocks, 198
Black Member, Onaping Formation, Sudbury Basin, 323–329, 331–342, 433
    algal blooms, 435
    amino acids in, 331, 339–341
    biogenic origin of carbon in, 345, 351–352, 435–436
    carbon isotopes, 339–341, 345, 348–356, 358
    carbon-rich, 336
    cross section, after readjustments to the crater and before complete solidification, 437
    fullerenes in, 332
    fullerenes not found in, 345, 352–355
    impact diamonds, 317–321, 340, 431, 435

kerogen in, 341, 435
modal analyses, 334
petrography, 334–336
poorly graphitized carbon (PGC), 341
silicon carbide in, 340
soot in, 331, 337, 340
sulfur in, 348, 355
sulfur isotopes, 355
textures, devitrification, 333
See also carbonaceous matter in Sudbury Basin rocks
Black Onaping Formation, Sudbury Basin, 324. *See* Black Member, Onaping Formation, Sudbury Basin
Black Reef Quartzite, Morokweng impact structure, 63, 85, 87
Blezard borehole, and Sudbury Igneous Complex, 421–422
Blezardian tectonic pulse, and Sudbury Structure, 402, 408–410, 421, 423, 434
Blind River, Ontario, and eastern Penokean orogen, 411
    map (geologic), 400
block faulting
    Chesapeake Bay impact crater, 151
    Popigai impact structure, 51
bolide. *See* impactor
Boltysh impact structure, Ukraine, 245
    gravity anomaly vs. erosional level, 237
    impact diamonds, 317
Boreal paleogeographic realm, and Jurassic-Cretaceous boundary, 244
Boreal province, extinctions in, Morokweng impact structure, 88
boreholes
    Chesapeake Bay impact crater, 151
    Morokweng impact structure, 65–66, 89, 93, 95
    Sudbury Igneous Complex, 421
bornite, Morokweng impact structure, 91, 97
Bosumtwi impact structure, Ghana, 245
    See also Lake Bosumtwi, Ghana
Bothitong-Reivilo fault, Morokweng impact structure, 92–93
Bottom Breccia Formation, Popigai impact structure, 26–27, 29, 32, 37–39, 48–49, 55
BP impact structure, Libya, 177–192
    chemical composition, 190
    description of samples and locations, 181
    isotopic compositions, 190
    Libyan Desert Glass (LDG), 179
    location of structure, 178
    size of structure, 190
    summary description, 190
    trace elements, 190
breccia
    BP impact structure, 181, 183
    Chesapeake Bay impact crater, 149
    Chicxulub Crater, 221, 266, 273–274, 277
    extraterrestrial craters, 162
    Lake Karikkoselkä impact structure, 131, 136, 144–146
    Mjølnir impact structure, 195
    Morokweng impact structure, 61, 64, 66–68, 77, 80, 82, 85, 87–88, 95, 104–105
    Oasis impact structure, 181–183
    Popigai impact structure, 1–3, 5–6, 8–10, 12, 14–16, 21–22, 24–26, 28–32, 48–50
    Slate Islands impact structure, 109–110, 116–118, 122–123

Steinheim impact structure, Germany, 206
See also Black Member, Gray Member, *and* Green Member (Onaping Formation, Sudbury Basin); Onaping Formation, Whitewater Group, Sudbury Basin; Sudbury breccia
breccia-hosted zinc deposit, Pering Mine, South Africa, 254
brecciation stage of impact process, 114
Brent impact structure, impact melt rocks, 105
Chicxulub Crater, 263
British Institutions Reflection Profiling Syndicate (BIRPS), 292
    Chicxulub Crater, 264, 270, 282
Brockman Iron Formation, Western Australia, 251, 258
Brownie Butte, Montana, United States
    carbonaceous components at, 217–219
    Cretaceous-Tertiary boundary, 216
    diamonds at, 217, 221
    impact diamonds, 215
Buckland and the Diluvialists, 440
bunsenite (NiO), Morokweng impact structure, 72, 91, 99, 106
bunte breccia, as at Ries Crater, Germany
    Chicxulub Crater, 291
    Slate Islands impact structure, 109–110, 116, 122–123
Bunte Breccia, Ries Crater, Germany, 48
Burgstall, near Steinheim impact structure, Germany, 207
burial by crater-fill deposits, structural floor of crater, 163
Burträsk shear zone, Lycksele structure, 126, 128, 130

C

calcite
    crystal structure of, 213
    Hugoniot data on, 205
    shock experiments on, 205
calcite unit-cell dimensions, Ries Crater, Germany, 205
Callovian, Mjølnir impact structure, 195
camera, high-speed frame, experimental fracturing model, 224
Campbellrand Supergroup, South Africa, 250–251
Campeche Bank, Gulf of Mexico, seismic velocity data, 270
Canadian Arctic, affected by Mjølnir impact, 202
Cape Charles, Virginia, Chesapeake Bay impact crater, 149–150, 160
Cape Charles harbor, Chesapeake Bay impact crater, 157, 159
Capreol sampling site, Black Member, Onaping Formation, Sudbury Basin, 333, 336, 348
Carawine Dolomite, Western Australia, 255, 258
carbon
    conversion to poorly graphitized carbon (PGC), 342
    conversion to soot, 342
    total elemental, in Sudbury Basin rocks, 348, 358
carbon (noncarbonate) source, Sudbury Basin rocks, 435
carbonaceous chondrites
    diamonds in, 215
    as Sudbury carbon source, 350
    as Sudbury sulfur source, 357

carbonaceous component, Cretaceous-Tertiary boundary samples, 221
carbonaceous matter in Sudbury Basin rocks, 331–342, 345–358
  analytical methods, 318, 348, 353
  atmosphere as source, 301–302, 345, 349–351
  of biogenic origin, 301, 329, 331, 340, 342, 345, 347, 349, 351–352, 358, 435–436
  carbon-bearing sediment, 339, 345, 349–350
  carbon matter similar to coal, 318
  carbon minerals, 318
  components of, 332
  due to fumarolic or hydrothermal activity, 301
  fumarolic activity as source, 345, 349–350
  graphite, 318, 332
  hydrocarbons, 318
  impact diamonds in Black Member, Onaping Formation, Sudbury Structure, 317
  impact diamonds in Black Member, Onaping Formation, Sudbury Structure, 318–321
  kerogen in, 331, 336–337, 339
  Onaping Formation, Sudbury Basin, 347–358
  Onwatin Formation, Sudbury Basin, 331–332, 345–353, 355–358
  projectile as source, 301–302, 345, 347, 350, 353–354, 357–358
  target rocks as source, 301–302, 331, 345, 347, 349–350
  Whitewater Group, Sudbury Basin, 299–302, 345–347, 349–351, 357–358
carbonate, sedimentary, ranges of carbon isotopes in, 351
carbon availability, and postimpact resurgence of biota, 342
carbon dioxide
  in impact diamonds, 220
  released by carbonate-bearing target rocks, 205
carbon disulfide, inferred in comets, 357
carbon disulfide polymer, mass-independent fractionation of sulfur isotopes in, 357
carbon for Onaping Formation, Sudbury Basin, Gowganda Shale, 339
carbon in impact vapor plume, 327
carbon isotopes
  analytical methods, 333–334, 348
  in Basal Member, Onaping Formation, Sudbury Basin, 339
  in Black Member, Onaping Formation, Sudbury Basin, 332, 338–341, 345, 348–356, 358
  in Chelmsford Formation, Sudbury Basin, 358
  Chicxulub Crater, 215–222
  Cretaceous-Tertiary boundary diamonds, 219
  in diamonds, Cretaceous-Tertiary boundary, 220
  in fireball layer, Cretaceous-Tertiary boundary, 220
  in Gray Member, Onaping Formation, Sudbury Basin, 339
  and hydrothermal venting, Sudbury Basin, 342
  impact diamonds, 215–222
  in kerogen, in Onaping Formation, Sudbury Basin, 331
  in Onwatin Formation, Sudbury Basin, 339
  ranges in organic carbon, 351
  ranges in sedimentary carbonate, 351
  in Sudbury Basin rocks, 435
carbon matter similar to coal, in Sudbury Structure, 318
carbon sources
  carbon-rich mudstone and carbon of Sudbury Structure, 331, 440
  graphite, 221
  impact diamonds, 318
  target rocks, Cretaceous-Tertiary boundary, 221
Carnian-Norian boundary and Manicouagan impact structure, Canada, 243–244
Carswell impact structure, Canada, 245
  gravity anomaly vs. erosional level, 237
Carswell structure, Canada, 174
Cartier Granite and Sudbury Structure, 421, 423–424
cataclasite, Morokweng impact structure, 64
cenotes (sinkholes), Chicxulub Crater, 263, 266, 270, 282, 296
central European strewn field, 190
central peak
  BP impact structure, 179–180
  Chesapeake Bay impact crater, 149–151, 153, 158–159, 162
  extraterrestrial craters, 162
  Mars, craters on, 163
  Mercury, craters on, 162
  Moon, craters on the, 162
  Venus, craters on, 163
central uplift
  decompression, 116
  stages of impact process, 113, 115
central uplift collapse and modification, stages of impact process, 113
chalcedony, Popigai impact structure, 39
chalcocite, Morokweng impact structure, 72, 91, 97, 99
chalcopyrite, Morokweng impact structure, 80, 91, 97, 99
chaoite
  Popigai impact structure, 34
  Sudbury Structure, 349
Charlevoix impact structure, Quebec, similarities to
  age of impact structure, 120
  Slate Islands impact structure, 120
charnockite, Morokweng impact structure, 68, 95, 102
charnoenderbite, Morokweng impact structure, 95, 99, 102
checkerboard texture, Morokweng impact structure, 73
Chelmsford Formation, Whitewater Group, Sudbury Basin, 323
  aleurolite from, 318
  biogenic origin for carbon in, 346
  carbon isotopes, 358
  description of, 346, 421
  foreland-basin depositional setting, 425–426
  map (geologic), 346, 420
  map (structure), 401
  seismic reflection profile, 424
  shortening estimates, 425
  timing of deformation, 426
  total carbon in, 358
chemical composition
  anatectic glass in tagamite, Popigai impact structure, 45
  Archean target rocks, Popigai impact structure, 23–24, 43–44
  Bon Accord nickel deposit, PGEs in, 105
  BP impact structure, 184–185, 187
  East site, relative to Libyan impact craters, 186–187
  Fe^{2+}/Fe^{3+} ratio, 3, 52
  glass, Popigai impact structure, 43–44
  glass in tagamite, Popigai impact structure, 45
  greenstone PGEs, 105
  impactite, Popigai impact structure, Russia, 46
  impact melt rocks, 119
  impact melt rocks, Popigai impact structure, Russia, 41
  komatiite PGEs, 105
  Libyan Desert Glass (LDG), 177, 187
  Morokweng impact structure, 71–72, 82, 91, 96–100, 103
  Oasis impact structure, 186–187
  PGEs in Bon Accord nickel deposit, 105
  PGEs in greenstones and komatiites, 105
  PGEs in nickel-rich veins, Morokweng impact structure, 105
  Popigai impact structure, 23, 37–39, 41, 52–53
  proposed parental Sudbury magmas, 380
  quartz norite, 91
  residual glass in tagamite, Popigai impact structure, 45
  Slate Islands impact structure, 109, 118–119
  Sudbury Igneous Complex units, 379
  suevite, Popigai impact structure, 41–42, 46
  tagamite, Popigai impact structure, 41, 43, 46
  target rocks, 104, 119
  whole-rock analyses, 99
chemical remanent magnetization (CRM), Lake Karikkoselkä impact structure, 139
chemical vapor deposition (CVD), diamond formation, 215, 220–221, 318
Chesapeake Bay impact crater, United States, 150, 243–245
  age of structure, 160
  central peak, 160–161
  comparison to Ries Crater, 161
  cross section, 161
  inner basin, 160
  map (structure), 161–162
  morphology of, 163
  outer rim, 161
  peak ring, 160–161
  seismic expression, 149–164
  seismic reflection method, 195
  size of structure, 160
Chicxulub Crater, Yucatán Peninsula, Mexico, 54, 89, 243–245, 258
  age of structure, 282
  basement uplift?, 297–298
  environments of deposition, 296
  extraterrestrial iridium, 354
  hydrocarbon exploration, 62
  impact diamonds, 318
  impact ejecta, 55, 288, 350
  impact melt rocks, 105
  lack of carbonaceous matter in, 351

Lycksele structure, similarities to, 128–129
morphology of, 281–290
multi-ring basin, 106, 281–289, 291
ocean-bottom seismograph data, 291–298
outermost (third) ring, 263, 266
peak-ring crater, 281–289, 291
potential field data, 281
reefal buildup, 296–297
seismic data, 270–277, 281–289
seismic expression, near-surface, 269–279
seismic expression, Tertiary section, 287, 289, 294–297
seismic reflection profiles (deep), 263–268
size of structure, 158, 282
and Slate Islands impact structure, 112
soot due to impact at, 352
target rocks, 88
upper-crustal structure, 291–298
Chicxulub structure. See Chicxulub Crater
chilled texture of rocks, Morokweng impact structure, 91, 94, 97, 99
chlorite-shard horizon. See Green Member, Onaping Formation, Sudbury Basin
Chloroflexus
Black Member, Onaping Formation, Sudbury Basin, 341
Onwatin Formation, Sudbury Basin, 341
chondritic impactor
Morokweng impact structure, 105
Popigai impact structure, 54
chondritic platinum-group-element (PGE) pattern
Morokweng impact structure, 82, 102
quartz norite, 102
chondritic values, Morokweng impact structure, 91
chromium, Morokweng impact structure, 82, 91, 100, 102, 105
chromium contents, Morokweng impact structure, 97
circular structure of
Lake Bosumtwi, Ghana, 135
Lake Karikkoselkä impact structure, 135, 142, 146
classification of impact rocks. See nomenclature of impact rocks
clay minerals
Morokweng impact structure, 72
Popigai impact structure, 52
claystones, Cretaceous-Tertiary boundary, diamonds in, 221
Clear Creek North, Colorado
carbonaceous components at, 217–219
Cretaceous-Tertiary boundary, 216
diamonds at, 217
ilmenite, 216
Clearwater East impact structure, Canada
gravity anomaly vs. erosional level, 237
Clearwater West impact structure, Canada, 53, 173
density variations, 232
gravity anomaly vs. erosional level, 237
impact melt rocks, 103, 105
clinopyroxene, Sudbury Igneous Complex, major element analyses, 369
coal, lower-temperature sulfide deposits, 106
coastal erosion associated with impact, Mjølnir impact structure, 202
cobalt
Libyan Desert Glass (LDG), 189
Morokweng impact structure, 82, 100
Cobalt Group and Sudbury Structure, 400

Cobalt plate and Sudbury Structure, 400, 402, 411
Coconino Sandstone, open fractures observed in, 188
coesite, 221
Popigai impact structure, 37
Coleman mine, Sudbury Structure, 306–307
COMAGMAT-3.5 simulation program, 373
advantages of current version, 380
description of, 375–378
initial and boundary conditions, Sudbury Igneous Complex, 382–383
INTRUSION subroutine, 376–377, 382–383
Main Mass, Sudbury Igneous Complex, 378–385, 435
results compared to Sudbury Igneous Complex rocks, 382–385
Vavucan differentiated sill, eastern Siberia, Russia, 378
comb layering, granophyre, Sudbury Igneous Complex, 406
cometary orbit, perturbations of, 242
comet impact, Slate Islands impact structure, 109
comets
not source of Sudbury carbon, 357, 435
from Oort cloud, 242, 245, 247
from Oort cloud, and periodicity of extinctions, 241
peak flux, 242
pulses of flux, 242
Shoemaker-Levy, Jupiter, viii
spectra of, 357
compression phase, stages of impact process, 114
condensation of a vapor phase, Green Member, Onaping Formation, Sudbury Basin, 323
condensation of matter from a solar gas, 440
contact and compression, stages of impact process, 110, 113
contact metamorphism in target rocks, Morokweng impact structure, 79
continental shelf, Chesapeake Bay impact crater, 149–150
convective-cumulative model
application to natural case, 378
COMAGMAT-3.5 simulation program, 377
results compared to Sudbury Igneous Complex rocks, 382–385
cooling history, Sudbury Igneous Complex, 378, 389–390, 431
Copper Cliff Formation, Elliot Lake Group, Sudbury Structure, map, structure, 410
Copper Cliff mine, Sudbury Structure, 308
copper contents
average crust, 439
granophyre, Sudbury Igneous Complex, 439
Gray Member, Onaping Formation, Sudbury Basin, 439
norite, Sudbury Igneous Complex, 439
copper sulfides
Morokweng impact structure, 91
See also sulfide ore deposits
coptoclastite, Popigai impact structure, 2–3, 5, 8–11, 16
cordierite, Popigai impact structure, 37, 40, 52
cordierite-olivine aggregates, Popigai impact structure, 35–36
coronas, Slate Islands impact structure, 118

coronas on mineral grains, Morokweng impact structure, 75
correlations with impacts and extinction peaks, 241, 244
crater density, Africa, 62
Crater Galle, Mars, 275
cratering processes
fracturing, 223–226
Mjølnir impact structure, 198, 201
crater modification, Slate Islands impact structure, 116
crater morphology, Popigai impact structure, 22
Crean Hill embayment, Sublayer, Sudbury Igneous Complex, 434
Creighton fault, Sudbury Structure, 410–411
map (geologic), 420
map (structure), 412, 427
Creighton Mine, Sudbury, 362
osmium isotopes, 364
osmium isotopes in sulfide ores, 363, 365
Creighton pluton and Sudbury Igneous Complex, 346, 400, 408
block diagram, 413
map (geologic), 420
crescumulate crystalline texture
granophyre, Sudbury Igneous Complex, 406, 408
model of deformation of, 408
Cretaceous (Late) mass extinction, 241–242, 245
Cretaceous-Tertiary boundary, 263
Chicxulub Crater, 263, 269, 272, 281, 291
impact diamonds, 318
impact doublet, 215
and mass extinctions, 242
Morokweng impact structure, 86, 88
Cretaceous-Tertiary boundary, impact site of. See Chicxulub Crater
Cretaceous-Tertiary boundary diamonds, carbon isotopes, 219
Cretaceous-Tertiary boundary ejecta layer, diamonds in, 217
Cretaceous-Tertiary extinction event, 244
crises due to impact, 241–248
cristobalite-quartz aggregates. See quartz-cristobalite aggregates
CRM. See chemical remanent magnetization
cross section
Chesapeake Bay impact crater, 151, 154
Mjølnir impact structure, 200
Mjølnir impact structure, and geophysical data, 196
Morokweng impact structure, 85, 87
Popigai impact structure, 6, 9, 14–16, 25
Strathcona mine, Sudbury Structure, 311
Sudbury Basin, 400
Sudbury Dome, 400
Sudbury Igneous Complex, after readjustments to the crater and before complete solidification, 437
Teague Ring impact structure, 172
crustal doming at Sudbury Structure
vs. multi-ring basin structures, 406
See also Sudbury Dome
cryptovolcanic objects, Popigai impact structure, 27
crystal accumulation, lack of evidence for in impact melt rocks, 105
crystal fractionation. See fractional crystallization
crystallization sequence, modeled for Sudbury Igneous Complex, 380–381

CVD. *See* chemical vapor deposition
cyanobacteria, as Sudbury carbon source, 352, 357–358

D

Daldyn Formation, Popigai impact structure
    chemical composition, 42
    suevite, 22, 26, 28, 30–33, 36–39, 50, 55
Daldyn River, Siberia, 28, 30
Dales Gorge Member, Brockman Iron Formation–S4, Western Australia, 258
dark matter component of galactic disk, 247
    consisting of very cold molecular clouds, 242
    and periodicity of extinctions, 241
dating. *See* age of structure
decarbonization of carbonates, lack of, 213
decompression, central uplift, 116
Deep Bay impact structure, Canada, deep bathymetry, 134
definitions of impact rocks
    impactite, 49
    Manson impact structure, 49
    Popigai impact structure, 26
    suevite, 49, 114
    suevitic breccia, 114
deformation, postimpact
    Mjölnir impact structure, 193
    vs. original orientation, Sudbury Structure, 399–415
    Sudbury Igneous Complex, 437
    Sudbury Structure, 346, 369–370, 389
    timing, Sudbury Structure, 369
Delaute Island, Canada, Slate Islands impact structure, 110, 114, 116
Dellen impact structure, Sweden, 245
    erosion, effect of on magnetic signature, 238
Delmarva Peninsula, Chesapeake Bay impact crater, 149–151, 154
density, rock
    Lake Karikkoselkä impact structure, 140
    North Range, Sudbury Igneous Complex, 422
    South Range, Sudbury Igneous Complex, 422
    Sudbury Igneous Complex, 421–422, 432
density contrast. *See* gravity model
density data, Ries Crater, Germany, 173
density-stabilized impact melt, Sudbury Igneous Complex, 438–440
depth of melting, modeled, Sudbury Structure, 392
development of ideas on Sudbury Structure, 431–441
diamictite, Permian?, Morokweng impact structure, 85, 87, 92
diamonds
    in carbonaceous chondrites, 215
    grain-size grading of, Cretaceous-Tertiary boundary sites, 221
    paramorphs, 320
    and silicon carbide intergrowths, Ries Crater, 215
    See also impact diamonds
diaplectic feldspar glass, Popigai impact structure, 34–37, 40
diaplectic pyroxene, Popigai impact structure, 39
diaplectic quartz glass
    Morokweng impact structure, 70–71, 74, 77, 79, 82, 84, 103
    Popigai impact structure, 32, 34–35, 37, 40

diatremes, impact, Popigai impact structure, 24–25, 27
Dietz, R. S., dedication to, iii
differentiation of impact melt rocks, Morokweng impact structure, 91–92
Diluvialists and Buckland, 440
dimensions of structure. *See* size of structure
disequilibrium textures, Slate Islands impact structure, 118
Ditshipeng, road to, Morokweng impact structure, 64
Dogger target rocks, Steinheim impact structure, Germany, 206
dolomite, shock experiments on, 205
dolomixtite, 258–259
Dowling Member, Onaping Formation, Sudbury Basin, 324, 329, 332
    See also Black Member, Onaping Formation, Sudbury Basin
Dowling Township sampling site, Black Member, Onaping Formation, Sudbury Basin, 348
drainage-pattern analysis, Lycksele structure, 126
drill core
    INCO drill core 70011, Sudbury Igneous Complex, 374–375, 379–380, 382, 384–385
    Morokweng impact structure, 66–67, 70–71
drill hole(s) in vicinity of
    Chesapeake Bay impact crater, 150–151, 159
    Chicxulub Crater, 263, 274, 291, 296
    Mjölnir impact structure, 194
drilling opportunity
    Chicxulub Crater, 289
    Mjölnir impact structure, 202
dunelike bedforms in spherulite layer, 254
Dupuis Island, Canada, Slate Islands impact structure, 110, 114, 122
Dwyka Group, Morokweng impact structure, 61, 64, 85, 87, 89
dynamic barrier in expanding explosion cloud, Popigai impact structure, 19, 50–51, 55

E

Earaheedy Group (Early Proterozoic), Western Australia, 165, 167–169, 171–172, 174
Earth's geologic history, comparison with other planets, 163
Earth-crossing objects, viii, 243, 246
    flux of, 241
earthquake associated with impact, Mjölnir impact structure, 202
East Bull Lake layered intrusion and Sudbury Igneous Complex, 433, 438
East Range (North), Sudbury Igneous Complex, 362
East Range (South), Sudbury Igneous Complex, 362
East Range, Sudbury Igneous Complex, 414
    application of geochemical data to, 440
    map (structure), 401
    mineral fabrics, 404
    steep dips not from folding, 435
    steep dips possibly from Wanapitei impact structure, 436, 440–441
    Sudbury Breccia (pseudotachylite) and copper-rich ore deposits, 432

East site, relative to Libyan impact craters petrography, 183
    See also BP impact structure; Oasis impact structure
Ed'en-Yurege impact diatreme, Popigai impact structure, 24
Edmonds Island, Canada, Slate Islands impact structure, 110
Egypt, Libyan Desert Glass (LDG), 178
ejecta
    carbonaceous component, Cretaceous-Tertiary boundary layer, 221
    claystone, Cretaceous-Tertiary boundary, 215
    Cretaceous-Tertiary boundary layer, 215
    dense clouds of fine, 243
    diamonds in Cretaceous-Tertiary boundary layer, 219
ejecta, distal
    Chicxulub Crater, 55, 215–222
    Cretaceous-Tertiary boundary sites, 215–222
    lack of, Morokweng impact structure, 86
    Mexico, northeastern, 215–222
    Mjölnir impact structure, 202
    Popigai impact structure, 25
    Siberia, 86
    Sudbury impact event, 320
    Western Interior, United States, and Chicxulub impact, 215–222
ejecta, flight time, Popigai impact structure, 50
ejecta, lack of preserved, Slate Islands impact structure, 114
ejecta, proximal
    Mjölnir impact structure, 201–202
    Mjölnir impact structure, Barents Sea, 193
    Popigai impact structure, 25, 55
ejecta, velocities of, Popigai impact structure, 49–50, 55
El'gygytgyn impact structure, Russia, 37, 245
electromagnetic anomalies, airborne, Lake Karikkoselkä impact structure, 131
electromagnetic data, Lake Karikkoselkä impact structure, 140–141, 146
electromagnetic model, Lake Karikkoselkä impact structure, 145
electromagnetic profiles, Lake Karikkoselkä impact structure, 143
electromagnetic survey, Lake Karikkoselkä impact structure, 134, 137, 142
elemental mobilization, Popigai impact structure, 55
Elliot Lake Group, Huronian Supergroup, 408
Elsie Mountain Formation, Elliot Lake Group, Sudbury Structure, map (structure), 410
"Eltanin asteroid," 249
emplacement geometry of Sudbury Igneous Complex, 399–415
energy estimations
    Mjölnir impact magnitude, 193, 201–202
    Popigai impact structure, 49
    See also autoclastic breccia; seismic vibration attending crater-formation processes
Englehardt and Bertsch's universal stage method and planar deformation features (PDFs), 114
environmental effects of impacts, 241
late Eocene extinction event, 244
late Eocene impacts, 242
EOS. *See* equation of state

epithermal veins, lower-temperature sulfide deposits, 106
equation of state (EOS) for target and projectile, 390–391
equilibrium crystallization, 373, 375, 379–381
erosion, amount of, 125, 194
    BP impact structure, 178, 188
    Lake Karikkoselkä impact structure, 146
    Lycksele structure, 128, 130
    Morokweng impact structure, 106
    Oasis impact structure, 178, 188
    Popigai impact structure, 14
    Slate Islands impact structure, 116, 118
    Sudbury Structure, 300
    Sudbury Structure, cross section, after readjustments to the crater and before complete solidification, 437
    Teague Ring impact structure, 173
erosional levels, variations in, Lycksele structure, 128
erosion, effect of on gravity and magnetic signatures, modeled results, 235–236
Errington #1 Mine, Sudbury Structure, 348
Espanola Formation, Huronian Supergroup, Sudbury Structure, 350
excavation and central uplift, stages of impact process, 114
excavation process, model of, Popigai impact structure, 51
excavation process, stages of impact process, 114
Exmore breccia, Chesapeake Bay impact crater, 149–152, 154–157, 159, 162
experimental fracturing model
    ejecta-plume formation, 225
    fracturing, 223–226
    material, 224
    results compared to meteorite craters, 223, 225–226
    spall phenomena, 225–226
    velocity of crack propagation, 225
explosion cloud, Popigai impact structure, 50, 52
explosion cloud formations, Popigai impact structure, 25, 27, 30, 37, 48–50, 55
    See also fireball layer, Cretaceous-Tertiary boundary; fireball layer, Sudbury impact
extinction peaks and correlations with impacts, 241, 244
extinction rates, 243
extraterrestrial causes of mass extinctions, 241
extraterrestrial craters, impossibility of seeing structural floor of, 162
extraterrestrial helium, in fullerenes in Sudbury Structure, 301–302
extraterrestrial material, Mjölnir impact structure, 195

F

Fairbank Lake, Sudbury Igneous Complex, 362, 369
Falconbridge East Mine, Sudbury, 362
    osmium isotopes in sulfide ores, 363, 365–366
Falconbridge Nickel Rim Mine, 441
fall-back breccia incorporated into Sudbury melt sheet, 435
fall-back from high atmosphere, Green Member, Onaping Formation, Sudbury Basin, 323

faulting
    Chesapeake Bay impact crater, 151, 154, 156–157
    Popigai impact structure, 51
    Slate Islands impact structure, 122
    Sudbury Structure, 305–306, 310–313, 369
faulting, circular, Lycksele structure, 126
faults. See structure
fault surfaces, Sudbury Structure, 411
Fecunis Lake area, Sudbury Structure, pyroxenite sheets in drill holes, 438
Fecunis Lake fault, Sudbury Structure, 324, 327
feldspar inclusions in diaplectic quartz, Popigai impact structure, 34
feldspar ternary diagram, Morokweng impact structure, 72
Felsic Norite, Sudbury Igneous Complex. See norite, Sudbury Igneous Complex
Fennoscandia. See specific countries
    geophysical model, 229
Fennoscandia, northern, affected by Mjölnir impact, 202
ferrihypersthene (non-Ca), Popigai impact structure, 35
ferrohypersthene, Popigai impact structure, 35
field observations, Kara (and Ust-Kara) impact structure, Russia (reference for), 206
Finland, central, Lake Karikkoselkä impact structure, 131
Finland, recognized impact structures, 132
    Iso-Naakkima impact structure, 132
    Lappajärvi impact structure, 132
    Lumparn, 132
    Passelkä, 132
    Sääksjärvi, 132
    Saarijärvi, 132
    Söderfjärden, 132
    Suvasvesi N, 132
fireball layer. See also explosion cloud formations, Popigai impact structure
fireball layer, Cretaceous-Tertiary boundary, 215
    carbonaceous component, 221
    diamonds in, 219
fireball layer, Sudbury impact
    Green Member, Onaping Formation, 323, 327–328, 347
    thickness of, 323
Fladen-type glassy masses, 49
flat earth theory, 440
flattened subsurface fracture zone, 223
fluid inclusions
    Morokweng impact structure, 74, 77
    Popigai impact structure, 52
fluidization of target rocks, 277–278, 390–391, 415, 436
foreland-basin depositional setting
    Chelmsford Formation, Sudbury Basin, 425
    Onwatin Formation, Sudbury Basin, 425
formation of impact structure
    Slate Islands impact structure, 113, 122
    See also stages of impact process
Fortescue Group, Western Australia, 251
Fort Monroe well, Chesapeake Bay impact crater, 155
Fourier analysis of extinctions, 245–246
Foy Offset Dike, Sudbury Igneous Complex, 302, 392
fractional crystallization
    lack of evidence for crystal accumulation in impact rocks, 105
    modeling of, 375

    simulation of Sudbury Igneous Complex crystallization, 373, 380–381, 431, 435
    Sudbury Igneous Complex, 369
fracture front, 225
fracture mechanism, target rocks and, 226
fracturing, Meteor Crater, Arizona, 223
fracturing, and gravity effects on, 226
Franz Josef Land, and Mjölnir impact structure, 194
Fraser Mine, Sublayer, Sudbury Igneous Complex, 434
Frenchman Lake sampling site, Black Member, Onaping Formation, Sudbury Basin, 333, 337
Frere Formation, 166–169, 174
Frere Range, 166, 174
frictional melting, Morokweng impact structure, 61
friction melt rock. See pseudotachylite; Sudbury Breccia (pseudotachylite)
Frood breccia, Sudbury Structure, 308
    Sudbury Breccia (pseudotachylite), 305
Frood Mine, Sudbury, 362
    osmium isotopes in sulfide ores, 363, 365
    strontium isotopes, 367
Frood-Stobie zone, Sudbury Structure, 301
fulgurites, search for fullerenes in, 352
fullerenes, He inside, 353
fullerenes, in shungite, Kola Peninsula, Russia, 352, 434
fullerenes in Sudbury Structure rocks, 299–302
    in accretionary-mantled clasts, 340
    analytical methods, 353
    in Black Member, Onaping Formation, 347, 349
    bulk concentration, 353
    extraterrestrial helium in, 301–302, 342, 347, 434
    in Lower Black Member, Onaping Formation, 332, 337
    not found in Black Member, Onaping Formation, 345, 352–355, 435
    in Onaping Formation, Sudbury Basin, 331–332, 434
    original distribution of, 354
    presence or absence, 436
    questioned, 318

G

gabbro, Sudbury Igneous Complex, 361, 367, 390, 399–400
    compositional break with overlying granophyre, 433
    emplaced separately from granophyre, 434
    initial geometry of, 414
    map (geologic), 407, 420
    map (structure), 401, 410, 413
    mineral fabrics, 401, 403–405, 411, 414–415
    and norite, trace elements compared, 434
    oxide-rich (quartz gabbro), South Range, 373–375, 379–385
    oxide-rich, South Range, 366, 368, 370
    strain measurements, 402
    structural petrology of, 403–405
    See also anisotropy of magnetic susceptibility (AMS)
galactic comet shower models, 241
galactic dynamics, 241–248

galactic-plane crossing hypothesis for mass
    extinctions, 245–246
  evaluation of, 247
  and mean crossing period, 241
  and periodic extinctions, 241
gamma-radiation data, Lake Karikkoselkä
    impact structure, 136, 141–142
Gamohaan Formation, South Africa, 251, 258
Ganyesa dome, South Africa, 62, 250, 257
  map (geologic), 65
  Morokweng impact structure, 63–64,
    84–85, 87, 92–93
garben texture, Morokweng impact structure,
    69–70, 73, 82
Garson Mine, Sudbury, 362
  osmium isotopes in sulfide ores, 363, 365
  strontium isotopes, 367
geochemical data, Kara (and Ust-Kara) impact
    structure, Russia (reference for), 206
geochemical thermometry, 379
geochronological methods, comparison with
    Sudbury Igneous Complex
    emplacement modeling, 392
geochronology. *See* age of structure
geologic heritage site, Popigai impact structure,
    20
geophysical data
  airborne surveys, Lake Karikkoselkä
    impact structure, 137
  Kara (and Ust-Kara) impact structure,
    Russia (reference for), 206
  Mjölnir impact structure, 195, 202
geophysical modeling
  applications, 229–239
  density variations, 231–232
  effect of deformation, reference to, 230
  geometry of the model, 231
  gravity and magnetic signatures and
    erosion, 229–239
  magnetic variations, 232
  methods, 232–234
  physical properties, modeled, 231
  results, 234
  three-dimensional, 230
geothermometers and COMAGMAT-3.5
    simulation program, 376
Gertrude borehole, and Sudbury Igneous
    Complex, 421–422
Ghaap Group, South Africa, 251–252, 257
ghost clasts, Morokweng impact structure, 75
glass
  chemical composition, Popigai impact
    structure, 43–45
  geochemistry and mineralogy of, Popigai
    impact structure, 41–46
  inclusions, Morokweng impact structure,
    98, 102
  petrography, Popigai impact structure,
    32–41
  Popigai impact structure, 19, 52–53, 55
  shard-like fragments of, Slate Islands
    impact structure, 114
  Slate Islands impact structure, 109, 118,
    122–123
GLIMPCE (Great Lakes International
    Multidisciplinary Program on Crustal
    Evolution), Slate Islands impact
    structure, 109–112, 116
gold, Vredefort structure, South Africa, 62
Gosses Bluff impact structure, Australia, 245
  gravity anomaly vs. erosional level, 237
  Jurassic-Cretaceous boundary, 244

Gowganda Shale
  carbon for Onaping Formation, Sudbury
    Basin, 339
  Huronian Supergroup, Sudbury Structure,
    339
granophyre, Sudbury Igneous Complex, 390,
    399–400
  anisotropy of magnetic susceptibility
    (AMS), 407
  block diagram, 413
  comb layering, 406
  compositional break with underlying
    gabbro, 433
  crescumulate crystalline texture, 406, 411
  cross section, after readjustments to the
    crater and before complete
    solidification, 437
  density-stabilized impact melt, 438
  emplaced separately from norite and
    gabbro, 434
  formation as a residual melt, model for, 373
  formed as upper-crustal impact melt, 433
  formed by fractionation of Main Mass
    magma, 434
  formed with norite from single magma, 436
  impact melt, direct, 436
  initial geometry of, 414
  initial magma composition proposed,
    438–439
  injected after Onaping Formation
    deposition, 415
  isotopic signatures of, 361
  lead isotopes, 366
  map (geologic), 420
  map (structure), 401, 407, 410
  mineral fabrics, 405–406, 413–415, 432,
    434, 436, 440
  and norite, trace elements compared, 434
  not a residual melt from mafic Main Mass
    magma, 374
  and numerical simulation of magmatic
    differentiation, 373–375, 379–380,
    382–383, 385
  and Onaping Formation, Sudbury Basin,
    325
  roofed by preexisting plagioclase-rich
    granophyre, 370
  and South Range Shear Zone, 425
  strain measurements, 402
  strontium isotopes, 367–368
  structural petrology of, 404
  trace elements, 368
  See also anisotropy of magnetic
    susceptibility (AMS)
graphite
  description of, from Sudbury Structure, 318
  as fault lubrication, Onaping Formation,
    Sudbury Basin, 425
  transformed to diamond, 317, 319–320
graphite-bearing quartz, Popigai impact
    structure, 34
gravimetric survey, Lake Karikkoselkä impact
    structure, 134, 137
gravitational collapse, 193
gravity anomaly vs. erosional level
  Azuara impact structure, Mexico, 237
  Boltysh impact structure, Ukraine, 237
  Carswell impact structure, Canada, 237
  Clearwater East impact structure, Canada,
    237
  Clearwater West impact structure, Canada,
    237

Gosses Bluff impact structure, Australia,
    237
  Haughton impact structure, Canada, 237
  Lappajärvi impact structure, Finland, 237
  Mistastin impact structure, Canada, 237
  Ries Crater, Germany, 237
  Rouchouart impact structure, France, 237
  Saint Martin impact structure, Canada,
    237
  Steen River impact structure, Canada, 237
gravity data
  Mjölnir impact structure, 194–195
  Sudbury Structure, 432
gravity effects on fracturing, 226
gravity modeling
  Lake Karikkoselkä impact structure, 136,
    139, 145
  no second layered intrusion required at
    Sudbury, 432
  Teague Ring impact structure, 171–173
gravity signature
  causes, 238
  Chesapeake Bay impact crater, 149, 154,
    157, 160
  Chicxulub Crater, 263, 269–270, 275, 277,
    282, 288–289, 291, 295–297
  and erosion, 229–239
  Lake Karikkoselkä impact structure, 131,
    144
  Lycksele structure, 125–130
  Mjölnir impact structure, 196, 199
  modeled results, 235
  Morokweng impact structure, 62, 85, 92
  peak ring, Chicxulub Crater, 278
  Popigai impact structure, 3–4, 24
  Teague Ring impact structure, 165,
    168–173
Gray Member, Onaping Formation, Sudbury
    Basin, 323–329, 332, 347, 355, 358,
    433
  carbon in, 348, 350, 352
  carbon isotopes, 345, 348
  ground-surge deposit, 327–328
  planar deformation features (PDFs), 325
  suevitic breccias of, 327–328
  sulfur (elemental) in, 348, 355–357
Great Lakes International Multidisciplinary
    Program on Crustal Evolution
    (GLIPMCE), Slate Islands impact,
    109–112, 116
Great Sand Sea, Egypt, Libyan Desert Glass
    (LDG), 178
greenhouse gases, 243
  calcite, stability of, 213
  and carbonate-bearing target rocks, 205
Greenland, northeast, affected by Mjölnir
    impact, 202
Green Member, Onaping Formation, Sudbury
    Basin, 323–329, 332, 347, 352, 358
  end of excavation phase of impact process,
    328
  fireball layer, 323, 327–328, 433, 435, 440
  petrography of, 325
  seismic reflection data, 325
  thickness of, 323, 325
Grenville Front tectonic zone, 374, 411
  map (structure), 412, 427
  and plutonism, 421
  and Sudbury Structure, 346
Grenville orogeny
  and Sudbury Igneous Complex, 365
  and Sudbury Structure, 421, 434

Gries limestone breccia
    Kara impact structure, Russia, 206
    Steinheim impact structure, Germany, 207
gries texture, Popigai impact structure, 24, 28, 30
Grieve and Robertson's scaling relationship, Chesapeake Bay impact crater, 157
Griqualand West basin, South Africa, 249–252, 256, 258
Griqualand West Sequence, Morokweng impact structure, 63–64, 85–86
Griqualand West Supergroup (Proterozoic), Morokweng impact structure, 92
ground zero, Chicxulub Crater, 264–265
Gweni Fada impact structure, Chad, 178

H

Hamersley basin, Western Australia, 249–253, 257–259
Hardy area, Sudbury Structure
    distribution of Sudbury Breccia (pseudotachylite), 309
    map (geologic), 308
Haughton impact structure, Canada, 245
    gravity anomaly vs. erosional level, 237
Heidenheim, near Steinheim impact structure, Germany, 206–207
Hekpoort Andesite (Proterozoic), Morokweng impact structure, 65
helium, extraterrestrial, in fullerenes in Sudbury Structure, 301–302, 332, 342, 434
hematite
    chemical composition, 99
    Morokweng impact structure, 76
hemispherical subsurface fracture zone, 223
hercynite, Popigai impact structure, 35, 37, 44
Herodotus (historical reference), 178
heterogeneous original distribution of fullerenes in Sudbury Structure rocks, 354
Heuningvlei, village of, Morokweng impact structure, 64, 85, 87
Highbury structure, size of structure, 62
high-explosive modeling, 223–226
High Falls (Dowling Township) sampling site, Black Member, Onaping Formation, Sudbury Basin, 333, 336–337, 348, 351, 355
High Falls (Dowling Township) sampling site, Onwatin Formation, Sudbury Basin, 358
high-pressure polymorphs of calcite, 212
high-pressure polymorphs of quartz, lack of
    BP impact structure, 188
    Oasis impact structure, 188
high-pressure stability, calcite, crystal structure of, 213
high-speed frame camera, experimental fracturing model, 224
high-temperature origin, Morokweng impact structure, 72, 84
high-velocity strata, Chicxulub Crater, 266
Highway 144 road-cut sample site, Sudbury Structure, 354
homogeneous original distribution of fullerenes in Sudbury Structure rocks, 354
Horace Cove, Canada, Slate Islands impact structure, 110
Hugoniot elastic limit, 391
Huronian ocean's opening south of Superior province, 433, 440
Huronian Supergroup, Sudbury Structure, 362, 367, 374, 399, 409
    block diagram, 413
    carbon for Onaping Formation, Sudbury Basin, 335, 339
    deformation of, 402, 408, 411, 421, 434
    Elliot Lake Group, 408
    Gowganda Shale, 339
    map (geologic), 420
    map (structure), 412
    not related to Grenville deformation, 411
    outliers of, 400, 402, 406, 415, 423
    and Penokean deformation, 411
    preimpact folding, 414–415
    source of melt for Sudbury basal norite, 369
    stratigraphic youngening direction, 400, 406, 413
    Sudbury target rocks, 346, 349–350, 439
"hyaloclastites," source of Onaping Formation, Sudbury Basin, 332
hydrocarbon exploration, Chicxulub Crater, 62
hydrocarbons, in Sudbury Structure, 318
hydrodynamic model of Melosh, 163
hydrogen availability and postimpact resurgence of biota, 342
hydrothermal (lower-temperature) sulfide deposits, 106
hydrothermal cooling, Sudbury Igneous Complex, and reinjection of magmas, 370
hydrothermal mobilization of carbon, Onaping Formation, Sudbury Basin, 329
hydrothermal venting, Sudbury Basin, 339
hygrogenic hypothesis, Popigai impact structure, 52
hypabyssal bodies, eastern Kamchatka, and geochemical thermometry, 379
"hypercane," impact-induced, 257
hyperthermophylic archaea as Sudbury carbon source, 345, 352, 358
hypothetical impact melt, Slate Islands impact structure, 118, 121

I

Iceland spar, X-ray powder diffraction measurements, 206, 208–212
identification strategy, impact structures, 125
igneous foliation, Sudbury Igneous Complex, 432
igneous lineation, Sudbury Igneous Complex, 432
Ilintsy impact structure, Ukraine, impact diamonds, 317
ilmenite
    Clear Creek North, Colorado, 216
    Morokweng impact structure, 71, 75–76, 78–81
    Popigai impact structure, 35, 37, 39–40, 44
ilmenite, nickel-rich, Morokweng impact structure, 72, 99, 106
ilmenite-magnetite, Morokweng impact structure, 97
immiscibility of silicate melts, Popigai impact structure, 52–53
immiscibility of sulfides, Sudbury Igneous Complex
    cross section, after readjustments to crater and before complete solidification, 437
    sulfide ore deposits, 431
    and superheated silicate magma, 439–440
impact, oblique. See oblique impact
impact anatexis
    glass, Popigai impact structure, 38–39
    Popigai impact structure, 20, 52–53
impact breccia, extraterrestrial craters, 162
impact catastrophe characteristics, 243
impact cloud formations. See explosion cloud formations; fireball layer, Cretaceous-Tertiary boundary; fireball layer, Sudbury impact
impact clusters and global correlation difficulties, 244
impact diamonds
    Berwind Canyon, Colorado, 215
    Boltysh impact structure, Ukraine, 317
    Brownie Butte, Montana, 215
    carbon isotopes, 215–222
    carbon sources, 215, 318
    chemical vapor deposition (CVD), origin by, 215, 220–221, 318
    Chicxulub Crater, 215–222, 318
    Cretaceous-Tertiary boundary deposits, 215, 318
    description of, 317–321
    diamond paramorphs, 320
    Ilintsy impact structure, Ukraine, 317
    Kara (and Ust-Kara) impact structure, Russia, 317
    meteoritic component in, 215
    nitrogen, paramagnetic, absence of, 320
    oldest so far recovered, 320
    Popigai impact structure, 20, 25–26, 34, 215, 317
    Puchezh-Katunki impact structure, Russia, 317
    Ries Crater, Germany, 215, 317
    shock origin of, 215–216
    Sudbury Structure, 317–321, 347, 349, 435–436
    Ternovka impact structure, Ukraine, 317
    Zapadnaja impact structure, Ukraine, 317
    See also lonsdaleite
impact doublet, Cretaceous-Tertiary boundary, 215
impact ejecta and extinctions, 243
impact glasses
    isotopic compositions of, 189
    Slate Islands impact structure, 109
    Sudbury Structure, 346
impact-induced decompression melting and Sudbury Igneous Complex, 415
impactites
    chemical composition, Popigai impact structure, 23, 46
    definitions of impact rocks, 49
    glass-bearing, Popigai impact structure, 19–59
    Popigai impact structure, 2–3, 6, 8–10, 50
    projectile contamination of, 54
    See also impact melt rocks; impact melt sheet; tagamite
impact melt rocks
    age, 83–84
    breccia, 32
    Chicxulub, compared to Sudbury Igneous Complex, 297
    Chicxulub Crater, 277, 291–292, 294, 297
    at Chicxulub Crater, seismic reflections due to, 266
    as flows, Popigai impact structure, 10
    Morokweng impact structure, 61, 73–74, 78, 82–84, 87–88, 92–93, 106

Popigai impact structure, 19, 29, 49, 317
  seismic reflections due to, 263
  similarities among, Morokweng impact structure, 71
  Slate Islands impact structure, 113, 117–118, 120–121, 123
  See also impactites; impact melt sheet; Sudbury Igneous Complex; tagamite
impact melt rocks, lack of
  BP impact structure, 188
  Mjølnir impact structure, 195
  Oasis impact structure, 188
  Slate Islands impact structure, 109
impact melt sheet
  emplacement geometry of, 282
  extraterrestrial craters, 162
  Morokweng impact structure, 91, 103
  Popigai impact structure, 7, 22, 25
  Slate Islands impact structure, 122
impact melt sheet, Sudbury Igneous Complex
  conflicting formation theories of, 414–415
  emplacement geometry of, 399–415
  modeled melt production, 390
  nonhorizontal initial configuration of, 400
  not homogeneous because of depth of melting, 438
impactoclastic air-fall layer, Popigai impact structure, 20, 48, 55
impactoclastic fill, Popigai impact structure, 19
impactoclastic mass, Popigai impact structure, 50
impactor
  carbon in, 350
  Mjølnir impact structure, 201
  Popigai impact structure, 54
  size of, 327, 354
  Sudbury Structure, 366
impactor, contamination by
  Morokweng impact structure, 105–106
  Popigai impact structure, 20, 46, 53–55
  See also under meteoritic component
impactor, size of
  Slate Islands impact structure, 114
  Sudbury Structure, 389–390, 435, 441
impactors, secondary, Chicxulub Crater, 266
impact structures
  and extinctions, 243
  identification strategy, 125
  See also specific impact structures
impact–wave height relationship, Mjølnir impact structure, 202
impossibility of seeing structural floor of extraterrestrial craters, 162
incidence angle of impactor, 163
inclusions, lithic, Morokweng impact structure, 94
inclusion trails, Morokweng impact structure, 75, 77
INCO drill core 70011, Sudbury Igneous Complex, 374–375, 379–380, 382, 384–385
INCO Victor Mine, 441
infilling of crater, 193
INFOREX experimental mineral-melt database, 380
initial and boundary simulation conditions, Sudbury Igneous Complex, 382–383
initial geometry, Sudbury Igneous Complex, 437
initial magma composition proposed, Sudbury Igneous Complex, 438

inner basin, Chesapeake Bay impact crater, 155
in situ differentiation
  dynamics, model of, 377
  of melt sheet precluded by Sudbury East Range steep dips, 435
  Morokweng impact structure, 92
  Sudbury Igneous Complex, 375, 400, 402, 412, 414
INTRUSION subroutine, COMAGMAT-3.5 simulation program, 373, 376–377, 382–383
iridium, 221, 250, 253
  BP impact structure, 184–185
  Chicxulub Crater, 354, 358
  Cretaceous-Tertiary boundary samples, 221, 350
  East site, relative to Libyan impact craters, 186
  Libyan Desert Glass (LDG), 189
  Mjølnir impact structure, 193, 195, 202
  Morokweng impact structure, 82, 91–92, 100–101, 104–106
  Oasis impact structure, 186
  Popigai impact structure, 46, 48, 54–55, 106
iridium contamination, Popigai impact structure, 20
iridium spikes, 244
  and extinctions, 243
iron. *See* native iron
Isabella multi-ring basin, Venus, and peak rings, 282
Iso-Naakkima impact structure, Finland, 125
  density variations, 232
  similarities to Lake Karikkoselkä impact structure, 144
isotherms
  and granophyre, Sudbury Igneous Complex, 437, 439–440
  modeled, Sudbury Igneous Complex, 392–394, 435
isotopically light components
  Berwind Canyon, Colorado, 220
  carbon isotopes, 219
  Clear Creek North, Colorado, 220
isotopic compositions
  BP impact structure, 188
  Oasis impact structure, 188
  Sudbury Igneous Complex, 300
  tektites, 190
  See also specific isotopes
isotopic disequilibrium, Morokweng impact structure, 61, 84
Ivory Coast strewn field, 190
Ivory Coast tektites, 54

J

Jacobsville Formation (ca. 800 Ma), Slate Islands impact structure, Canada, 111–112, 120
James River, Chesapeake Bay impact crater, 149–150, 155
James River borehole, Chesapeake Bay impact crater, 151
Jebel Dalma structure. *See* BP impact structure, Libya
Jeerinah Formation and impact layer, Western Australia, 251, 258
Jurassic-Cretaceous boundary
  and Mjølnir impact structure, 244

Morokweng impact structure, 83, 85–86, 88
Siberia, 86
Jurassic-Cretaceous extinction event, 244

K

Kalahari basin, Morokweng impact structure, 93
Kalahari Group (Late Cretaceous–Cenozoic), Morokweng impact structure, 61, 63–64, 66, 72, 85–86, 89, 92
Kamensk impact structure, Russia, 245
  compared with BP and Oasis impact structures, 189
Kara (and Ust-Kara) impact structure, Russia, 245
  impact diamonds, 317
  suevite, 207
Kara impact structure, Russia, 205–214
  suevite, 206
  target rocks, calcite-rich, 205–206
  Ust-Kara structure, part of, 206
Kara River, Kara (and Ust-Kara) impact structure, Russia, 207
Kara Sea
  and the Mjølnir impact, 202
  and Mjølnir impact structure, 194
Kärdla, Estonia, 125
  density variations, 232
Karla impact structure, Russia, 245
Karoo Supergroup (Permian-Jurassic), Morokweng impact structure, 64–65, 75, 84, 87
Kathu drill core, Griqualand West basin, South Africa, 253, 256–259
kerogen, 351–352
  in Black Member, Onaping Formation, Sudbury Basin, 340–341, 435
  in Lower Black Member, Onaping Formation, Sudbury Basin, 339
  in Onaping Formation, Sudbury Basin, 331, 336
  in Onwatin Formation, Sudbury Basin, 337, 340
  and postimpact resurgence of biota, 342
Keweenawan Supergroup (1.1–1.2 Ga), Slate Islands impact structure, 112, 115, 118–121
Khapchan Series (Group), Popigai impact structure, 21, 23–24
  chemical composition, 42
Khara-Khaia Mountain, 24
Khonde-Yaga River, Kara (and Ust-Kara) impact structure, Russia, 207
"kill curve" relationships, 241, 243
Klenova multi-ring basin, Venus, and peak rings, 282
Klippen Breccia Formation, Popigai impact structure, 22, 26–29, 55
Knudsen's Farm, Alberta, Canada
  Cretaceous-Tertiary boundary, 215
  impact diamonds, 215
Kola Peninsula, Russia, fullerenes in shungite in, 352, 434
Kraaipan graben, Morokweng impact structure, 63
Kraaipan Group (Archean), Morokweng impact structure, 64, 75, 92–93
Kufra Oasis, Libya, 178, 181
Kuiper belt comets, 242
Kuruman Iron Formation, South Africa, 250–251

L

Lake Bosumtwi, Ghana
  Bosumtwi impact structure and Lake Karikkoselkä impact structure, 141
  deep bathymetry, 134
Lake Karikkoselkä impact structure, 131–147
  deep bathymetry, 134
  location of structure, 131
Lake Nabberu, 166–167
Lake Superior, Canada, Slate Islands impact structure, 109
Lake Suvasvesi, Finland, deep bathymetry, 134
Lake Teague, 166–167
Landsat imagery
  Lake Karikkoselkä impact structure, 132
  Morokweng impact structure, 92
  and size of Sudbury, 433
lanthanum, Sudbury Igneous Complex, 375
Lappajärvi impact structure, Finland, 125, 245
  density variations, 232
  erosion, effect of on gravity signature, 238
  gravity anomaly vs. erosional level, 237
  and Lake Karikkoselkä impact structure, 144
  and modeled results, 234
Lapten Sea, and Mjölnir impact structure, 194
latitude dependency of magnetic signatures, geophysical model, 229, 232, 237
LDG. See Libyan Desert Glass
lead isotopes
  analytical methods, 364
  Onaping Formation, Sudbury Basin, 366
  Sudbury Igneous Complex, 361–362, 365–366, 369
  Sudbury silicate rocks, 364
Leadman Islands, Canada, Slate Islands impact structure, 110, 116
lechatelierite
  Libyan Desert Glass (LDG), 179
  Morokweng impact structure, 98
  Popigai impact structure, 32, 34, 37–40, 53
lechatelierite, lack of
  BP impact structure, 189
  Oasis impact structure, 189
Levack borehole and Sudbury Igneous Complex, 421–422
Levack Gneiss Complex and Sudbury Structure, 362, 402, 421
  cross section, after readjustments to the crater and before complete solidification, 437
  as a foreland buttress during compression, 423, 425
  interpreted as lower-crustal material brought to surface by fluid flow, 436, 440
  magnetic susceptibility measurements, 432
  map (geologic), 374, 420
  preimpact deformation, 423
  Sudbury Breccia (pseudotachylite), 305
Levack-McCreedy West embayment, Sublayer, Sudbury Igneous Complex, 434
Levack Mine, Sudbury, 362
  osmium isotopes in sulfide ores, 363, 365
  strontium isotopes, 367
Levack West Mine, Sudbury, osmium isotopes in sulfide ores, 366
Levänsaaret, Lake Karikkoselkä impact structure, 132–133
Levänsaari, Lake Karikkoselkä impact structure, 139

Libyan Desert Glass (LDG), 177–178
  age, 178
  BP impact structure, 189
  chemical composition, 190
  cometary collision hypothesis, 178
  compared with BP and Oasis impact structures, Libya, 189
  formation in situ, 179
  Great Sand Sea, Egypt, 178
  impact origin for, 178
  isotopic compositions, 190
  lunar glassy spherules as source of, 178
  mass, 178
  meteoritic component in, 189
  Oasis impact structure, 189
  organic material asserted to be in, 178
  reference to chemical analyses of, 189
  reference to detailed description of, 189
  sedimentary origin, 178
  sol-gel formation hypothesis, 178
  summary description, 190
  trace elements, 190
liebenbergite (Ni-rich olivine), Morokweng impact structure, 72, 91, 99, 106
light components. See isotopically light components
limestone, shock experiments on, 205
Linopen (Loopeng), road to, Morokweng impact structure, 64
liquidization of target rock, Chicxulub Crater, 277
liquid lines of descent, Sudbury Igneous Complex, 381–382
lithic contents, impact melt rocks, 105
Lithoprobe Vibroseis investigations, Sudbury Structure, 389, 400
  and size of crater, 431–432
Little Stobie embayment, Sublayer, Sudbury Igneous Complex, 434
location of structure
  BP impact structure, 178
  Chesapeake Bay impact crater, 149–150
  Chicxulub Crater, 264, 270, 282
  Kara impact structure, Russia, 206
  Lake Karikkoselkä impact structure, 132
  Lycksele structure, 125
  Mjölnir impact structure, 194
  Morokweng impact structure, 62, 92
  Oasis impact structure, 178
  Popigai impact structure, 20
  Slate Islands impact structure, 109–110
  Steinheim impact structure, Germany, 206
  Teague Ring impact structure, 165–166
  Toms Canyon impact crater, 150
Logoi Crater, 53
Logoisk impact structure, Belarus, 245
Lokammona Formation, South Africa, 251
Longvack Mine, Sudbury Structure, 306–307
lonsdaleite, 215, 217, 319–320
  Black Member, Onaping Formation, Sudbury Basin, 317, 347
  Popigai impact structure, 34
lonsdaleite, lack of, in Cretaceous-Tertiary boundary diamonds, 220
Loopeng (Linopen), road to, Morokweng impact structure, 64
lopolithic intrusion of Sudbury Igneous Complex, 403
Lower Black Member, Onaping Formation, Sudbury Basin, 324–328, 433
  kerogen in, 339
  reworking of impact breccias, 440
  stages of impact process, 328

lower-crustal material brought to surface by fluid flow, Sudbury impact event, 394
  interpreted as Levack Gneiss Complex, 436
lunar samples, and Sudbury Igneous Complex, 433

M

Machavie dike, Morokweng impact structure, 92–93
Maddina Basalt, Western Australia, 251
magnetic signature
  causes, 238
  Chicxulub Crater, 263, 291, 297
  and erosion, 229–239
  Lycksele structure, 125, 127, 130
  Mjölnir impact structure, 194–196, 198–199
  modeled results, 235
  Morokweng impact structure, 61–62, 92
  Popigai impact structure, 24
  Sudbury Structure, 423
  target rocks, 198
magnetic susceptibility, Lake Karikkoselkä impact structure, 140
magnetic susceptibility measurements. See anisotropy of magnetic susceptibility (AMS)
magnetite, Morokweng impact structure, 71, 75–79, 81, 97
magnetite, nickel-rich, Morokweng impact structure, 97
magnitude of impact
  Mjölnir impact, 193, 201–202
  See also autoclastic breccia; energy estimations; seismic vibration attending crater-formation processes
Main Mass, Sudbury Igneous Complex, 361
  COMAGMAT-3.5 simulation program applied to, 379–385
  comparison with melanorite inclusions in Sublayer, 433
  crystallization modeling, 374
  description of, 373
  distinguished from Sublayer, 438
  emplacement geometry of, 399–415
  formed by mantle-derived picritic magma mixed with impact melt, 434, 439
  fractionation to produce granophyre, 434
  lack of olivine in, 380–381
  lead isotopes, 364, 366, 369–370
  mineral contents, 375
  and parent-magma calculation, 379
  sampling localities, 363
  source of melt for, 369
  strontium isotopes, 367–370
  Sudbury melt not completely homogenized with, 438
  trace elements, 368, 434
major element analyses
  clinopyroxene, Sudbury Igneous Complex, 369
  mixing calculations, 369
  orthopyroxene, Sudbury Igneous Complex, 369
  plagioclase, Sudbury Igneous Complex, 369
  Sudbury Igneous Complex, 375
  whole rocks, Sudbury Igneous Complex, 369
major element distribution
  Sudbury Igneous Complex, 375, 383–384

Vavucan differentiated sill, eastern Siberia, Russia, 378
Malmian age, Steinheim impact structure, Germany, 207
Malmian limestone (unshocked), X-ray powder diffraction measurements, 206
Manicouagan impact structure, Canada, 103, 243–245
 impact melt rocks, 103, 105
 and Lycksele structure, 128
 and Lycksele structure, gravity signature of, 129
 and Slate Islands impact structure, 118
 thermal modeling, 391–392
Manson impact structure, Iowa, United States, 173, 245
 definitions of impact rocks, 49
 seismic reflection method, 195
mantle
 contribution to Sudbury Igneous Complex, 439
 not highly deformed during Sudbury impact, 392
 not melted by Sudbury impact, 394, 436
map, geologic
 Anabar Group, Popigai impact structure, 42
 Balagan River basin, Popigai impact structure, 10
 Benny deformation zone, Sudbury Structure, 420
 Chelmsford Formation, Sudbury Basin, 420
 Creighton fault, Sudbury Structure, 420
 Creighton pluton and Sudbury Igneous Complex, 420
 gabbro, Sudbury Igneous Complex, 407, 420
 Ganyesa dome, 65
 granophyre, Sudbury Igneous Complex, 420
 Hardy area, Sudbury Structure, 308
 Huronian Supergroup, Sudbury Structure, 420
 Lake Karikkoselkä impact structure, 134
 Levack Gneiss Complex, Sudbury Structure, 420
 Lycksele structure, 126–127
 Mayachika Upland, Popigai impact structure, 5
 Morokweng impact structure, 63, 87, 93
 Murray fault, Sudbury Structure, 420
 Nipissing diabase, 420
 norite, Sudbury Igneous Complex, 407, 420
 Onaping Formation, Sudbury Basin, 420
 Onwatin Formation, Sudbury Basin, 420
 Popigai impact structure, 2, 5, 10, 16, 20, 22, 25
 Slate Islands impact structure, 110
 Strathcona Embayment, Sudbury Structure, 306–308
 Sublayer, Sudbury Igneous Complex, 307–308
 Sudbury Basin, 410
 Sudbury Igneous Complex, 306, 362
 Sudbury Structure, 324, 346, 362, 374, 400–401, 420
 Teague Ring impact structure, 167
 Whitewater Group, Sudbury Basin, 306, 420
map, isopach
 Balagan River basin, Popigai impact structure, 11
 Mayachika Upland, Popigai impact structure, 8
 Morokweng impact structure, 86
 Popigai impact structure, 8, 11
map, location
 Morokweng impact structure, 62
 paleogeography of Morokweng impact, 88
 Slate Islands impact structure, 110
 South Range Shear Zone, Sudbury Igneous Complex, 421
 Teague Ring impact structure, 166
map, relief
 Balagan River basin, Popigai impact structure, 12
 Popigai impact structure, 4, 7, 12
map, structure
 Balagan River basin, Popigai impact structure, 13
 Chesapeake Bay impact crater, 153
 Copper Cliff Formation, Elliot Lake Group, Sudbury Structure, 410
 Creighton fault, Sudbury Structure, 412, 427
 Creighton pluton, Sudbury Structure, block diagram of, 413
 ductile strain, 410
 Elsie Mountain Formation, Elliot Lake Group, Sudbury Structure, 410
 gabbro, Sudbury Igneous Complex, 410, 413
 granophyre, Sudbury Igneous Complex, 407, 410
 granophyre, Sudbury Igneous Complex, block diagram of, 413
 Grenville Front tectonic zone, 412, 427
 Huronian Supergroup, block diagram of, Sudbury Structure, 413
 Huronian Supergroup, Sudbury Structure, 412
 McKim Formation, Elliot Lake Group, Sudbury Structure, 410
 Morokweng impact structure, 63
 Murray fault, Sudbury Structure, 427
 Murray pluton, Sudbury Structure, block diagram of, 413
 norite, Sudbury Igneous Complex, 410, 413
 Offset Dikes, Sudbury Igneous Complex, 410
 Onaping Formation, Sudbury Basin, block diagram of, 413
 paleostress axes, Sudbury area, 412
 shape fabrics, Sudbury Igneous Complex, 410
 Skead pluton, Sudbury Structure, 413
 South Range Shear Zone, Sudbury Igneous Complex, 410, 412
 South Range Shear Zone, Sudbury Structure, block diagram of, 413
 Stobie Formation, Elliot Lake Group, Sudbury Structure, 410
 Sublayer, Sudbury Igneous Complex, 410
 Sudbury Basin, 407, 412
 Sudbury Igneous Complex, 407
 Sudbury Igneous Complex, block diagram of, 413
 Sudbury Structure, 410, 427
 Sudbury Structure, block diagram of, 413
map, underground, Strathcona Mine, Sudbury Structure, 310–311
mapping methods, underground, 308
marble, shock experiments on, 205
Marginal Border Group, Skaergaard intrusion, trace elements, 368
marine impact structures. See submarine impact structures
marine sections
 diamonds often lacking, 221
 diamonds present at Arroyo El Mimbral, Mexico, 221
Marquez impact structure, United States, 245
Marra Mamba Iron Formation, Western Australia, 251
martensitic-type transformation, 220
maskelynite, Slate Islands impact structure, 113, 122
mass extinction, 241–248
 Morokweng impact, 86, 89
 See also times of individual extinctions
mass extinction effects, lack of
 calcite stability and greenhouse gases, 213
 greenhouse gases and calcite stability, 213
 Mjölnir impact structure, 193, 202–203
mass-independent fractionation of sulfur isotopes in Allende meteorite, 357
mass of impactor, Mjölnir impact structure, 201
material-deposition modes, Popigai impact structure, 49
material-transport modes, Popigai impact structure, 49
mathematical relationships
 angular dependency on the X-ray diffraction angle, 207
 anisotropy of magnetic susceptibility (AMS), 403
 central uplift's height and diameter, 230
 gravity anomaly vs. diameter, 237
 gravity anomaly vs. erosional level, 237
 Grieve and Robertson's scaling relationship, 157
 morphometric determination of structural uplift, 173–174
 one-dimensional heat-flow equation, 396
 transient crater's diameter, 230
 velocity of crack propagation, 225
Matsap Quartzite, Morokweng impact structure, 86
Mayachika Upland, Popigai impact crater, 12, 16
McCready East borehole, and Sudbury Igneous Complex, 421–422
McKim Formation, Elliot Lake Group, Sudbury Structure, map, structure, 410
Mead multi-ring basin, Venus, 282
Megabreccia Formation, Popigai impact structure, 22, 25–29, 49, 55
Meitner multi-ring basin, Venus, 282
melanorite inclusions in Sublayer, Sudbury Igneous Complex
 baddeleyite, 431, 433
 and inhomogeneity of melt, 438
 and late-magmatic effects of superheating, 438
 osmium isotopes, 436
 zircon, 431, 433
Melosh, H. J., hydrodynamic model of, 163
melting after shock deformation, Morokweng impact structure, 71
melting depth, modeled, Sudbury Structure, 392
melting evidence, Morokweng impact structure, 82, 84
melt-matrix breccia, Chicxulub Crater, 291, 297
melt phase, Morokweng impact structure, 78–80, 88
melt rocks
 Chicxulub Crater, 221, 274, 291
 Morokweng impact structure, 66–69

Melt Sheet Formation, Popigai impact structure, 22, 24–30, 32, 37, 48, 50–51, 55
melt volume
  Popigai impact structure, 49, 54
  Sudbury Igneous Complex, 390
Meteor Crater, Arizona, United States
  fracture study of, 223
  impact melt rocks, 106
  See also Barringer Crater, Arizona, United States
meteoritic component
  and carbon isotopes, 221
  lack of in BP impact structure, 188
  lack of in Oasis impact structure, 188
  lack of in Onaping Formation, Sudbury Basin, 325
  in Libyan Desert Glass (LDG), 179, 189
  in melt rock, Morokweng impact structure, 62, 82, 88
  minor, in Cretaceous-Tertiary boundary diamonds and other samples, 221
  See also impactor, contamination by
meteoritic sulfonic acid, mass-independent fractionation of sulfur isotopes in, 357
microkrystites, definition of, 252
microkrystites at Cretaceous-Tertiary boundary, 249, 251
micropegmatite, Morokweng impact structure, 69
microspherules. *See* spherules
microtektites. *See* tektites
microtektites, definition of, 252
Midcontinent rift system (ca. 1.1 Ga), 112
Middle Neck Peninsula, Chesapeake Bay impact crater, 154
Mien impact structure, Sweden, 245
millerite (NiS), Morokweng impact structure, 72, 91, 97, 99, 106
Mimbral. *See* Arroyo El Mimbral, Mexico
mineral compositions, accessory. *See* individual minerals
mineral compositions, rock-forming. *See* chemical composition
mineral contents
  Main Mass, Sudbury Igneous Complex, 375
  modeled results, Sudbury Igneous Complex, 384–385
  Vavucan differentiated sill, eastern Siberia, Russia, 378
mineral deposits. *See* sulfide ore deposits
mineral fabrics
  and emplacement of Sudbury Igneous Complex, 399, 401, 403
  plagioclase fabric in granophyre, Sudbury Igneous Complex, 404
mineral lineations, and emplacement of Sudbury Igneous Complex, 399
mineral-melt database (INFOREX), 380
mineral-melt geothermometers and COMAGMAT-3.5 simulation program, 376
Miocene (middle) mass extinction, 242
Mistastin impact structure, Canada, 245
  gravity anomaly vs. erosional level, 237
  impact melt rocks, 103, 105
mixing calculations, Sudbury Igneous Complex, 368, 370
Mjölnir impact structure, Barents Sea, 173, 193–204, 245, 249
  age of structure, 158
  drill hole in vicinity of, 194
  ejecta of, 88
  Jurassic-Cretaceous boundary, 244
  size of structure, 158
model ages (Sm-Nd), Morokweng impact structure, 82
modeling
  cratering mechanics, Sudbury Structure, 389–396
  fracture zone formation, 223
  impact effects, 243
  impact effects, Popigai impact structure, 49
  physical properties for, Lake Karikkoselkä impact structure, 135
  thermal history, Sudbury Structure, 389–396, 435–436
models
  cross section, Popigai impact structure, 51
  excavation process, Popigai impact structure, 51
  impact process, Slate Islands impact structure, 114
  terrestrial impact cratering, 1
  three-dimensional, Teague Ring impact structure, 171
  two-dimensional, Lake Karikkoselkä impact structure, 131, 144
  See also map, relief
modification of structure, postimpact, 193
molecular clouds as dark matter component of galactic disk, 242
Molopo River, Morokweng impact structure, 85, 87
monazite, Morokweng impact structure, 72, 79, 97
Montagnais Crater, Scotian Shelf, Canada, 245, 249
  age of structure, 158
  compared with Mjölnir impact structure, 201–202
  seismic reflection method, 195
  size of structure, 158
Monteville farm location, Griqualand West basin, South Africa, 253–254, 257–258
Monteville Formation, South Africa, spherule layer in, 249–252, 259
  age, 258
  comparison with Hamersley basin, Western Australia, 258–259
  description of, 253–257
  interpretation of, 257–258
Moon, and Sudbury Igneous Complex, 433
Moose Lake borehole, and Sudbury Igneous Complex, 421–422
Morokweng impact structure, South Africa, 61–89, 91–108, 178, 243–245, 250
  age, 61
  basins, Late Cretaceous, 92
  Jurassic-Cretaceous boundary, 244
  map (location), 62
  Morokweng fault, 92–93
  Morokweng Granophyre, 68
  Morokweng impact melt complex, 94
  multi-ring basin, 106
  quartz norite, 66, 68, 91, 93–95, 97, 99–106
  Slate Islands impact structure, 112
morphologic floor of
  Chesapeake Bay impact crater, 151
  extraterrestrial craters, 162
morphology
  Chesapeake Bay impact crater, 149, 151
  Chicxulub Crater, 270

morphometric determination of structural uplift, Teague Ring impact structure, 173–174
Mortimer Island, Canada, Slate Islands impact structure, 109–110, 116, 122
Moshaweng streambed, Morokweng impact structure, South Africa, 64
Motiton Member, Monteville Formation, South Africa, 256, 258
Motley Rock, 24
  Popigai impact structure, 28, 30, 50
Mount Bruce Supergroup, Western Australia, 250
Mount McRae Shale, Western Australia, 251
Mount Sylvia Formation, Western Australia, 251
multi-ring basins
  Chicxulub Crater, 158–159, 263, 266, 268, 288–289
  vs. crustal doming at Sudbury Structure, 406
  Lycksele structure, 130
  Morokweng impact structure, 85
  Sudbury Structure, 301, 374–375, 389, 403
  on Venus, 282
Muong Nong–type tektites, 37
Murray fault, Sudbury Structure, 400, 409–412
  map (geologic), 420
  map (structure), 427
Murray Mine, Sudbury, 362
  osmium isotopes in sulfide ores, 363, 365
  strontium isotopes, 367
Murray pluton and Sudbury Igneous Complex, 346, 400, 408, 410
  block diagram, 413
mushroom-like microstructures, Popigai impact structure, 38–39

N

Nabberu basin, 167, 169, 172, 174
NASA Space Shuttle image, Oasis impact structure, 179
NASA Space Shuttle radar studies, Oasis impact structure, 180
National Oceanic and Atmospheric Administration, and Morokweng impact structure, 92
native iron, Popigai impact structure, 39, 44, 53
native nickel, Popigai impact structure, 53
native platinum, Morokweng impact structure, 72, 91, 99, 106
Nelson Lake borehole, and Sudbury Igneous Complex, 421–422
Nelson Lake sampling site, Black Member, Onaping Formation, Sudbury Basin, 333, 348
neodymium isotopes, Sudbury Igneous Complex, 361–362, 432
Neptunists and Werner, 440
Newark basin, 243
nickel. *See* native nickel
  Libyan Desert Glass (LDG), 189
  Morokweng impact structure, 82, 91–92, 102, 105
nickel contents
  average crust, 439
  granophyre, Sudbury Igneous Complex, 439
  Gray Member, Onaping Formation, Sudbury Basin, 439
  Morokweng impact structure, 71, 97
  norite, Sudbury Igneous Complex, 439

Nickel Offset Road sampling site, Black Member, Onaping Formation, Sudbury Basin, 348
nickel-rich ilmenite, Morokweng impact structure, 91, 99
nickel-rich oxides, Morokweng impact structure, 91
nickel-rich silicates, Morokweng impact structure, 91
nickel-rich spinels, 244
nickel-rich veins, impact melt rocks, 105
nickel-rich veins and nodules, Morokweng impact structure, 94
Nickel Rim Mine, Falconbridge, 441
nickel sulfides
 Morokweng impact structure, 91
 See also sulfide ore deposits
Nipissing diabase, 346, 408, 421, 434
 inclusions, South Range, Sudbury Igneous Complex, 437
 map (geologic), 420
nitric oxides, 243
nitrogen, paramagnetic, absence of, 320
nitrogen availability, and postimpact resurgence of biota, 342
nitrogen isotopes, Cretaceous-Tertiary boundary, 215
nomenclature of impact rocks, 32, 49, 114
 Popigai impact structure, 2, 26
Nördlingen, Germany, Ries Crater, 159
Norduna Mine, Sudbury Igneous Complex, 362, 369
Norfolk, Virginia, Chesapeake Bay impact crater, 150
Norian-Rhaetian boundary, and Manicouagan impact structure, Canada, 244
norite, Sudbury Igneous Complex, 373–375, 379–385, 390, 399–400
 cross section, after readjustments to the crater and before complete solidification, 437
 cross section of footwall, 422
 density-stabilized impact melt, 438
 emplaced separately from granophyre, 434
 formed with granophyre from single magma, 436
 and gabbro and granophyre, trace elements compared, 434
 initial geometry of, 414
 initial magma composition proposed, 438–439
 interpreted as mixture of granophyre and primary basalt, 436
 map (geologic), 407, 420
 map (structure), 401, 410, 413
 mineral fabrics, 401, 403–405, 411, 414–415, 432, 434, 436, 440
 seismic reflection profiles, 424
 strain measurements, 402
 structural petrology of, 403–405
 and Sublayer, trace elements compared, 434
 sulfide contents, high, 438
 unstrained, 408
 See also anisotropy of magnetic susceptibility (AMS)
normative mineral composition, quartz norite, 102
North American strewn field, 190
North Atlantic rift basins, affected by Mjölnir impact, 202

North Lobe, Sudbury Igneous Complex, 401, 409, 411, 414
 mineral fabrics, 405–406
North Range, Sudbury Igneous Complex, 306, 361–362, 364–370, 375, 379, 409, 414
 buckling of, 419, 426
 dip of, 426
 footwall contact, 420
 impact-induced deformation, 424
 Levack Gneiss Complex as lower-crustal material, 436
 mineral fabrics, 404–405
 preimpact deformation, 423
 relatively undeformed, compared to South Range, 421, 424, 427
 seismic data, 420–421
 and Sublayer characterization, 438
 Sudbury Breccia (pseudotachylite) and copper-rich ore deposits, 432
North Range (West), Sudbury Igneous Complex, 362
Northwest Bypass (Highway 144) sampling site, Black Member, Onaping Formation, Sudbury Basin, 333
Northwest Province, South Africa, Morokweng impact structure, 61–90
Novaya Zemlya
 affected by Mjölnir impact, 202
 and Mjölnir impact structure, 194
Nubia Group (Lower Cretaceous), 179–181, 188, 190
 BP impact structure, 177–178
 deformation of, 179
 metamorphism of, 179
 Oasis impact structure, 177–178
nuclear tests
 Prairie Flat, 174
 Snowball, 174
numerical simulation of magmatic differentiation
 COMAGMAT-3.5 simulation program, 373–378
 and fractional crystallization, 377
 Siberian sills, 377
 Sudbury Igneous Complex, 375

O

Oaktree Formation, South Africa, 258
Oasis impact structure, Libya, 177–192
 chemical composition, 190
 description of samples and locations, 181
 isotopic compositions, 190
 Libyan Desert Glass (LDG), 179
 size of structure, 190
 summary description, 190
 trace elements, 190
Oasis Oil Company, Libya, 178
oblique impact
 Popigai impact structure, 16
 Slate Islands impact structure, 114
OBS. See ocean-bottom seismograph data
ocean-bottom seismograph data
 Chicxulub Crater, 264, 291–298
 methods, 293
ocean impacts, 249
"odd" ratio of silica-rich to mafic-rich rocks, Sudbury Igneous Complex, 375, 380, 385, 414, 431, 435–436
Offset Dikes, Sudbury Igneous Complex, 389–390, 400, 414
 emplacement timing, 394, 440

 geochemically closer to Main Mass than to Sublayer, 438
 map (structure), 410
 and possibly much larger impact melt sheet, 394
 sulfide contents, high, 438
 thermal modeling and emplacement timing, 392
 trace elements, 434
Olifantshoek Sequence (Supergroup) (Proterozoic), Morokweng impact structure, 63–64, 85, 92
olivine-cordierite aggregate, Popigai impact structure, 36
olivine-cordierite aggregates, Popigai impact structure, 35
Onaping Falls, North Range, Sudbury Basin, plagioclase fabric in granophyre, 404
Onaping fault system, Sudbury Basin, 423
Onaping Formation, Whitewater Group, Sudbury Basin, 299, 374
 Basal Member, 323–325, 433
 biogenic origin of carbon in, 329
 Black Member, 323–329, 331–342, 433, 435
 block diagram, 413
 carbon (noncarbonate) source, 435
 carbonaceous matter in Sudbury Basin rocks, 347–358
 carbon from Huronian Supergroup, 335
 carbon in Black Member, 318
 chemical composition, 325
 contact with granophyre, Sudbury Igneous Complex, 406
 contact with Onwatin Formation, 333, 335
 cross section, after readjustments to the crater and before complete solidification, 437
 description of, 421
 devitrified glass in, 402
 distribution of carbon in, 328–329
 Dowling Member (Black Member), 324, 329, 332, 409
 faulting along contact with Onwatin Formation, 425
 fullerenes in, 318, 331
 fullerenes in, questioned, 318
 fullerenes in Black Member, 301
 and granophyre of Sudbury Igneous Complex, 325
 Gray Member, 318, 323–329, 433
 Green Member, 323–329, 433
 "hyaloclastites" as source of, 332
 hydrothermal mobilization of carbon, 329
 impact diamonds in Black Member, 317–321, 435
 as an impact-generated pseudovolcanic sequence, 402
 initial geometry of, 413, 415
 lead isotopes, 363, 366
 Lower Black Member, 433
 map (geologic), 420
 map (structure), 401
 melting of by superheated impact melt, 431, 438–439
 meteoritic component, lack of, 325
 and modeled cratering mechanics, Sudbury Structure, 391
 organic carbon in Black Member, 318, 329
 peperite bodies, 327
 petrography of, 326
 poorly graphitized carbon (PGC) in, 331

reworking of impact breccias, 440
Sandcherry Member (Gray Member), 324, 332, 409
seismic reflection profile, 424
shortening estimates, 425
soot in, 331
and South Range Shear Zone, Sudbury Igneous Complex, 425
stability above liquid layer, 394
stratigraphic nomenclature, 324–325, 433
stratigraphic younging direction, 413
strontium isotopes, 363
and Sudbury Igneous Complex, 362
as a suevite, 415
sulfur isotopes, 355
tectonic foliation and lineation, 407
transitional unit to Onwatin Formation, 333, 335
Upper Black Member, 433
See also Basal Member, Black Member, Gray Member, Green Member, Lower Black Member, and Upper Black Member (Onaping Formation, Sudbury Basin)
Ongeluk Lava (Proterozoic), Morokweng impact structure, 65, 75, 87
Onwatin Formation, Whitewater Group, Sudbury Basin, 301, 323
carbon (noncarbonate) source, 435
carbon, total, in, 348, 356
carbon isotopes in, 339–341
contact with Onaping Formation, 333, 335
description of, 421
faulting along contact with Onaping Formation, 425
foreland-basin depositional setting, 425–426
kerogen, 340
map (geologic), 420
map (structure), 401
seismic reflection profile, 424
soot in, 331
sulfur, total, in, 356
sulfur isotopes, 355
timing of deformation, 426
and Upper Black Member, Onaping Formation, 433
Oort cloud comets, 242, 245, 247
and periodicity of extinctions, 241
opaque minerals, Popigai impact structure, 35, 37, 39–40
open fractures observed
at BP impact structure, 188–189
in Coconino Sandstone, 188
at Meteor Crater, Arizona, 188
at Oasis impact structure, 188–189
Ordovician (Late) severe mass extinctions in, 245
organic carbon
in Black Member, Onaping Formation, Sudbury Basin, 318, 329, 338–342
in Onwatin Formation, Sudbury Basin, 339
ranges of carbon isotopes in, 351
origin, high-temperature, Morokweng impact structure, 72
origin of impact melt, 1
origin of impact structures
Chicxulub Crater, 291
Lake Karikkoselkä impact structure, 138, 145
Lycksele structure, 125–130
Morokweng impact structure, 84, 106

Slate Islands impact structure, 109
Sudbury Structure, 299–302, 318, 320, 323, 421
Teague Ring impact structure, 168
origin of Sudbury Igneous Complex, 299, 361–370, 373, 400, 402, 421
conflicts with theoretical models of impact melt sheets, 414–415
cross section, after readjustments to crater and before complete solidification, 437
fundamental problem caused by rock volumes, 380
theories on, 362, 425–426
origin of Sudbury Structure, 431
orthopyroxene compositions, Sudbury Igneous Complex, 369, 381
osmium isotopes
analytical methods, 363–364
melanorite inclusions in Sublayer, Sudbury Igneous Complex, 431
Sublayer, Sudbury Igneous Complex, 361–362, 431
in sulfide ores, 363, 365–366, 431, 436, 439
osmium not enriched at Sudbury, 437
outer rim, Chesapeake Bay impact crater, 149–151, 153, 155, 158, 162
oxides, Morokweng impact structure, 79
oxides, nickel-bearing, chemical composition, 99
oxygen fugacity, Sudbury Igneous Complex, crystallization model, 373, 380, 382
oxygen isotopes, Sudbury Igneous Complex, 361–362

P

PAHs. See polycyclic aromatic hydrocarbons
Pai-Khoi Range, Russia, Kara impact structure, Russia, 206
paleogeographic map of Morokweng impact, 88
paleomagnetic studies
dating, Lake Karikkoselkä impact structure, 139
Lake Karikkoselkä impact structure, 131–132, 136–138, 146
poles, Finland, 140
Sudbury Structure, 300
paleostress axes, Sudbury area, map, structure, 412
Paraburdoo Member, Western Australia, 251
Parchanai Formation, Popigai impact structure, 22, 26, 30, 32, 50
Parchanai River, Siberia, 30
parental magmas of Sudbury Igneous Complex, 378–381, 435
parental rocks. See target rocks
Pastakh impact diatreme, Popigai impact structure, 24–25
Patterson Island, Canada, Slate Islands impact structure, 110, 113–118, 120–122
peak flux of comets, 242
peak formation, hydrodynamic model of Melosh, 163
peak rings
arguments for and against formation on Earth, 281–282
Chesapeake Bay impact crater, 149–154, 156, 158–159, 162
Chicxulub Crater, 159, 263, 266, 269, 273–277, 288, 291

extraterrestrial craters, 162, 274
formation of, 269, 276–278
Jupiter's moons, craters on, 281, 289
Mars, craters on, 163
Mercury, craters on, 162
Mjölnir impact structure, 159
Moon, craters on the, 162, 277, 281, 289
Popigai impact structure, 159, 277
Ries Crater, Germany, 159, 277
Venus, craters on, 163, 289
Pechora Basin
affected by Mjölnir impact, 202
and Mjölnir impact structure, 194
PEMEX drill holes, Chicxulub Crater, 291
PEMEX reflection profiles of Chicxulub Crater, 270, 272, 274–275, 281–289
Penokean orogeny, 408–410
and Sudbury Structure, 369, 389, 400, 402, 421, 427
tilt of Sudbury Structure, 434
pentlandite, Popigai impact structure, 39, 44, 53
perigalactic revolution of Solar System, 245
Pering Mine (Peringmyn), South Africa, 250, 253–257
periodicity of extinctions, 244–245
correlations with impacts, 241
and dark matter component of galactic disk, 241
Oort cloud comets, 241
Permian (Late) severe mass extinctions in, 245
Permian-Triassic boundary, shocked quartz at, 244
Petriccio, Italy, diamonds lacking at Cretaceous-Tertiary boundary, 221
petrography
BP impact structure, 182–183
Green Member, Onaping Formation, Sudbury Basin, 325
Morokweng impact structure, 68–71, 73–74, 77–79
Oasis impact structure, 182–183
Slate Islands impact structure, 118
See also minerals (specific); textures; textures (specific)
petrologic data, Kara (and Ust-Kara) impact structure, Russia (reference for), 206
petrophysics. See physical properties
Petäjälahti road, Lake Karikkoselkä impact structure, 139
Petäjävesi watershed area, Lake Karikkoselkä impact structure, 131–132, 134
PGC. See poorly graphitized carbon
PGEs. See platinum-group elements (PGEs)
phase-equilibria calculations, Sudbury Igneous Complex, 380–382, 414
phosphorus availability and postimpact resurgence of biota, 342
phosphorus-bearing phases, Morokweng impact structure, 79
physical properties
determination of, Lake Karikkoselkä impact structure, 135, 139
for modeling, Lake Karikkoselkä impact structure, 137
plagioclase
compositions in Sudbury Igneous Complex, 381
laths, 252
major element analyses, Sudbury Igneous Complex, 365, 369
zoned, in impact melt rocks, 96, 105

planar deformation features (PDFs), 256
  BP impact structure, 177, 189
  Gray Member, Onaping Formation, Sudbury Basin, 325
  lack of at BP impact structure, 183, 188
  lack of at Oasis impact structure, 183, 188
  Lake Karikkoselkä impact structure, 131, 138, 146
  Mjølnir impact structure, 195
  Morokweng impact structure, 64, 66, 70–72, 74–75, 77, 82, 84–85, 88, 92
  Oasis impact structure, 177, 189
  Popigai impact structure, 31–32, 34–37, 40
  and pressure of formation, 212–213
  Slate Islands impact structure, 109, 113, 116, 118, 122–123
  Steinheim impact structure, Germany, 206
  Sudbury Structure, 402
  Teague Ring impact structure, 168
planar fractures (PFs)
  BP impact structure, 181–183, 188–189
  Mjølnir impact structure, 195
  Oasis impact structure, 181–183, 188–189
  Popigai impact structure, 32, 35–37, 40
  Slate Islands impact structure, 113, 122
plasmas
  diamonds produced in, 220
  high-energy, of atmosphere, 350
platinum. See native platinum
platinum-group elements (PGEs), 250, 346
  deposits, Sudbury structure, 62
  enrichment, Morokweng impact structure, 91–108
  Morokweng impact structure, 82, 86, 101–102, 105–106
  nugget effect on analysis of, 364
  osmium not enriched at Sudbury, 437
  See also sulfide ore deposits
Pliocene (late) ejecta and "Eltanin asteroid" identification, 249
Pliocene extinction event, 242, 244
polycyclic aromatic hydrocarbons (PAHs), 341, 353
  in Black Member, Onaping Formation, Sudbury Basin, 342
  in Onwatin Formation, Sudbury Basin, 342
polymorphs of quartz, lack of
  BP impact structure, 188
  Oasis impact structure, 188
poorly graphitized carbon (PGC), 332
  in Black Member, Onaping Formation, Sudbury Basin, 337, 340–341
  conversion of carbon to, 342
  in Gray Member, Onaping Formation, Sudbury Basin, 339
  in Green Member, Onaping Formation, Sudbury Basin, 339
  in Lower Black Member, Onaping Formation, Sudbury Basin, 339
  in Onaping Formation, Sudbury Basin, 331
  in Onwatin Formation, Sudbury Basin, 337, 340
  and postimpact resurgence of biota, 342
Popigai impact structure, Russia, 1–17, 19–59, 243–245
  age of structure, 1, 19, 47–48, 159–160
  breccia, 7, 48
  central peak, 160–162
  chemical composition, 52
  and Chesapeake Bay impact crater, 149
  comparison with Chesapeake Bay impact crater, 160–161
  description of suevite, 3
  impact diamonds, 317
  impact melt rocks, 103, 105
  map (structure), 161
  multi-ring basin, 106
  outer rim, 161
  peak ring, 160–161
  silicon carbide and diamond in, 220
  size of structure, 159–160
  and Slate Islands impact structure, 114
  suevite, 1–16, 19–22, 26, 28–32, 37–39, 48–50, 52, 54–55
  tagamite, 48
  volatile contents, 52
  water contents of impact-melt rocks, 44
porosity of natural cumulates and Sudbury Igneous Complex, 382
postcrystallization processes, Sudbury Igneous Complex, 381
postimpact processes, Sudbury Igneous Complex, 389, 437
postimpact resurgence of biota, 342
potential field data, Chicxulub Crater, 281, 291
Precambrian correlations across continents, 249–250, 259
preimpact stratigraphy, Chicxulub Crater, 287
pre-Kalahari Group, Morokweng impact structure, South Africa, 87
pressure of impact, 391, 394
  and carbonate stability, 205
  and graphite characteristics, 318
  graphite transformed to diamond, 317
  Morokweng impact structure, 103
  Popigai impact structure, 5, 7, 10, 27, 34, 36, 49–50
  and shatter cones, 212
  Slate Islands impact structure, 109, 114–115
  Sudbury Structure, 318
Prieska facies, Monteville Formation, South Africa, 252, 256
Princess Ranges Quartzite, 167–168, 170
profile. See cross section; seismic reflection profiles
progressive compositional changes with depth, lack of evidence for, impact melt rocks, 105
protobasins
  on Mars, 163
  on Mars, Mercury, and the Moon, similarities to Chesapeake Bay impact crater, 149
pseudotachylite
  as a descriptive term, 113
  distribution in Sudbury Structure, 423
  Morokweng impact structure, 68, 85, 88, 95
  Slate Islands impact structure, 109–110, 113, 116, 121–123
  and stages of impact process, 114
  Sudbury Structure, 402, 424
  See also Sudbury Breccia (pseudotachylite)
pseudotachylite dikes
  Morokweng impact structure, 61, 64
  Popigai impact structure, 27
  Sudbury Structure, 415
pseudotachylite veins
  Morokweng impact structure, 80, 94
  Slate Islands impact structure, 117–118
  Teague Ring impact structure, 168
Puchezh-Katunki impact structure, Russia, 243–245
  impact diamonds at, 317
pulses of comet flux, 242
pyrite, Morokweng impact structure, 82
pyroxene compositions, Morokweng impact structure, 72
pyrrhotite, weathering of, elemental sulfur source, Sudbury Basin, 345, 356–357

Q

quartz, shocked, Popigai impact structure, Russia, 36
quartz-cristobalite aggregates, Popigai impact structure, Russia, 32, 35, 37
Quartz Diorite and Sudbury Igneous Complex. See Offset Dikes, Sudbury Igneous Complex
quartz norite
  chemical composition, 91
  Morokweng impact structure, 66, 68, 91, 93–95, 97, 99–106
quasi-vacuum in target area, 350

R

Ragozinka impact structure, Russia, 245
random orientations of shatter cones, Lake Karikkoselkä impact structure, 137
Rappahannock River, Chesapeake Bay impact crater, 149–151
rare earth elements (REEs)
  BP impact structure, 177, 188–189
  and COMAGMAT-3.5 simulation program, 376
  East site, relative to Libyan impact craters, 188–189
  Libyan Desert Glass (LDG), 177, 188
  Morokweng impact structure, 91, 103–104
  Oasis impact structure, 177, 188–189
  Popigai impact structure, 46
  Sudbury Igneous Complex, 434
Rassokhe River valley, Popigai impact structure, Russia, 24
Raton Pass, New Mexico, United States
  carbonaceous components at, 217–219
  Cretaceous-Tertiary boundary, 216
  diamonds at, 217
reaction rims
  Morokweng impact structure, 75
  Slate Islands impact structure, 118
recommendations for future study, Mjølnir impact structure, 202
REEs. See rare earth elements (REEs)
reinjection of magma, Sudbury Igneous Complex, 370
Reivilo Formation, South Africa, 251, 258
relevance to tectonic stress change, Morokweng impact structure, 88
remote-sensing analysis, Lycksele structure, 126, 130
remote-sensing image
  Lake Karikkoselkä impact structure, 132
  Teague Ring impact structure, 166
Re-Os isotopes. See osmium isotopes
residual glass in tagamite, Popigai impact structure, Russia, chemical composition, 45
resistivity, Lake Karikkoselkä impact structure, 143–144
Rhaetian-Hettangian boundary, 243
Ries Crater, Germany, 21, 24, 28, 55, 125, 163, 245, 270
  accretionary-mantled clasts, 337

age of structure, 159
Bunte Breccia, 29, 48
calcite unit-cell dimensions, 205
and Chesapeake Bay impact crater, 149, 160–161
cross section, 161
diamond and silicon carbide, 215, 220–221
glasses, 53
gravity anomaly vs. erosional level, 237
impact diamonds, 317
impact melt bombs, 103
inner basin, 160
and Morokweng impact structure, 92
silicon carbide and diamond, 215, 220–221
size of structure, 116, 159
and Slate Islands impact structure, 114, 116
suevite, 49, 53
velocity and density data, 173
X-ray powder diffraction measurements, 208
Ries-Type Suevite Formation, Popigai impact structure, 30, 32–33, 37, 39, 50
rim delineation, gamma-radiation data, 136, 141
ring formation, hydrodynamic model of Melosh, 163
ring-tectonic model, 282
Ripon Hills, Western Australia, 259
Rochechouart impact structure, France, 245
rock fracturing, Lake Karikkoselkä impact structure, 137–140, 146
rock types, Sudbury Igneous Complex, 300
rosin, experimental fracturing model, 224
Rouchouart impact structure, France, gravity anomaly vs. erosional level, 237
rutile, 216
  inclusions in diaplectic quartz, Popigai impact structure, 34
  masking diamond's presence, 219
  Morokweng impact structure, 71–72, 75–76, 79, 97
  Popigai impact structure, 39, 44

S

Sääksjärvi, Finland, similarities to, Lake Karikkoselkä impact structure, 144
Saa-Yaga River, Kara (and Ust-Kara) impact structure, Russia, 207
Saint Martin impact structure, Canada, gravity anomaly vs. erosional level, 237
SALE 2D hydrocode and impact modeling, 390, 392–393, 435
SAMPO EM profiles, Lake Karikkoselkä impact structure, 142–144
Sandcherry Creek fault, Sudbury Structure, 324
sanidine, Popigai impact structure, 35
Sault Ste. Marie, Ontario, and eastern Penokean orogen, 411
  map (geologic), 400
Schmidtsdrif Supergroup, South Africa, 251
scientific drilling opportunity
  Chicxulub Crater, 289
  Mjölnir impact structure, 202
SDBX. See Sudbury Breccia (pseudotachylite)
secondary minerals, Morokweng impact structure, 66
seismic character and/or expression
  Chelmsford-Onwatin contact, Sudbury Basin, 422
  Chesapeake Bay impact crater, 149–164
  Chicxulub Crater, deep structure, 263–268
  Chicxulub Crater, morphology, 281–290
  Chicxulub Crater, near-surface, 269–279
  Chicxulub Crater, upper-crustal structure, 291–298
  contact between the Sudbury Igneous Complex and the Levack Gneiss Complex, 424
  granophyre-norite contact, Sudbury Igneous Complex, 422, 424–426
  Mjölnir impact structure, 194–199
  norite-footwall contact, Sudbury Igneous Complex, 424–426
  Onaping-granophyre contact, Sudbury Basin, 422
  Onwatin-Onaping contact, Sudbury Basin, 422
  shear zones and faults, Sudbury Igneous Complex, 422
seismic data
  Chesapeake Bay impact crater, 149–150
  Chesapeake Bay impact crater, collection of, 149, 151
  Chicxulub Crater, 264–265, 267, 282
  Green Member, Onaping Formation, Sudbury Basin, 325
  Lake Karikkoselkä impact structure, 136
  Mjölnir impact structure, 194–195
  preimpact deformation, Sudbury Structure, 423
  Slate Islands impact structure, 109–112
  Sudbury Igneous Complex, 432
  Sudbury Structure, 300, 419–427, 434
seismic reflection method
  Chesapeake Bay impact crater, 195
  Manson impact structure, 195
  Montagnais Crater, Scotian Shelf, Canada, 195
seismic reflection profiles
  availability of, 268
  Chelmsford Formation, Sudbury Basin, 424
  Chesapeake Bay impact crater, 152, 155–159
  Chicxulub Crater, 263–268, 283–287
  granophyre, Sudbury Igneous Complex, 424
  Mjölnir impact structure, 196–199
  norite, Sudbury Igneous Complex, 424
  North Range footwall, Sudbury Structure, 423
  Onaping Formation, Sudbury Basin, 424
  Onwatin Formation, Sudbury Basin, 424
  peak ring, Chicxulub Crater, 277
  Slate Islands impact structure, 111
  South Range Shear Zone, Sudbury Igneous Complex, 424, 426
  Sudbury Igneous Complex, 414
  Sudbury Structure, 419–427, 434
seismic velocity analyses, Chicxulub Crater, 269–270
seismic velocity contrasts, not observed, Chicxulub Crater, 277
seismic velocity data
  Campeche Bank, Gulf of Mexico, 270
  Chicxulub Crater, 265, 270–276, 278, 292–297
  Mjölnir impact structure, 195
  North Range, Sudbury Igneous Complex, 422
  Ries Crater, Germany, 173
  South Range, Sudbury Igneous Complex, 422
  Sudbury Igneous Complex, 421–422
  target rocks, 198
seismic vibration attending crater-formation processes
  Slate Islands impact structure, 123
  See also autoclastic breccia; energy estimations
selenides, copper-bearing, Morokweng impact structure, 91, 97, 106
shape fabrics, Sudbury Igneous Complex, map, structure, 410
shatter cones
  BP impact structure, lack of at, 188
  Kara (and Ust-Kara) impact structure, Russia, 207
  Kara impact structure, Russia, 206
  Lake Karikkoselkä impact structure, 131, 134, 136–137, 139–140, 145–146
  largest known on Earth, 114
  Oasis impact structure, lack of at, 188
  Popigai impact structure, 28, 30
  pressure of formation, 212
  and pressure of formation, 213
  random orientations, 137
  Slate Islands impact structure, 109, 113–116, 122–123
  Steinheim impact structure, Germany, 206
  Sudbury Structure, 300, 346, 402
  Teague Ring impact structure, 168, 174
Shiva hypothesis, 242
shock-attenuation plan, Slate Islands impact structure, 109, 114–115, 123
shock-attenuation studies, Sudbury Structure, 300
shock experiments, application of results, Slate Islands impact structure, 109, 113–114, 123
shock experiments, diamonds produced in, 220
shock melting of target rocks
  Sudbury Breccia (pseudotachylite), 366
  Sudbury Igneous Complex, 361
shock metamorphism, 244, 321
  BP impact structure, 188
  BP impact structure, lack of at, 189
  Cretaceous-Tertiary boundary samples, 221
  Green Member, Onaping Formation, Sudbury Basin, 323, 325, 327–328
  Lake Karikkoselkä impact structure, 131, 137–138
  Mjölnir impact structure, 193, 195, 202
  Morokweng impact structure, 61–62, 66, 70–72, 75, 82, 84, 88, 91, 102–103
  Oasis impact structure, 188
  Oasis impact structure, lack of at, 189
  Popigai impact structure, 27, 29–30, 32, 35–36, 53
  Slate Islands impact structure, 113–114, 116, 122–123
  Sudbury Structure, 346–347, 402, 420
  target rocks, calcite-rich, 205–214
  timing of effects, 113
  X-ray powder diffraction measurements of effects, 205–214
  of zircon, 216
shock pressure
  and coesite, 189
  and lechatelierite, 189
  and mosaicism, 189
  and planar microstructures, 189
  and quartz melting, 189
  and stishovite, 189
  See also pressure of impact
shock remanent magnetization (SRM), 232
  Lake Karikkoselkä impact structure, 139

shock transformation
    graphite to diamond, 221
    quartz to coesite, 221
    quartz to stishovite, 221
    and silicon carbide, 221
shock-wave effects
    at Meteor Crater, Arizona, 188
    Slate Islands impact structure, 109, 113, 123
    stages of impact process, 112
    time of action, Popigai impact structure, 49
shock-wave propagation, Sudbury Structure, 390
shock waves, Slate Islands impact structure, 114
Shoemaker, Eugene
    dedication to, iii
    and Teague Ring impact structure, 165, 168–169
Shoemaker, Carolyn, and Teague Ring impact structure, 165
Shoemaker impact structure. *See* Teague Ring impact structure, Western Australia
Shoemaker-Levy comet fragments, Jupiter, viii
shot gathers, Chicxulub Crater, 293
    inside and outside crater, 272
    peak ring, 276
Siberian differentiated traps, and geochemical thermometry, 379
Siberian Platform, iridium and Jurassic-Cretaceous boundary, 244
Siberian sills, simulation of differentiation of, 377
siderophile element enrichment
    Libyan Desert Glass (LDG), 189
    Morokweng impact structure, 62, 82, 88, 91, 100, 105
    quartz norite, 102
siderophile enrichment, 253
side-scan sonar profiles, Mjölnir impact structure, 195
silicon carbide
    in Black Member, Onaping Formation, Sudbury Basin, 340
    in claystones, Cretaceous-Tertiary boundary, 221
    in Cretaceous-Tertiary boundary samples, 221
    and diamond intergrowths, Ries Crater, 215
Siljan impact structure, Sweden, 125, 173
    density variations, 232
    size of structure, 125
Sinamwenda impact structure, Zimbabwe, 178
single-impact-melting origin of Sudbury Igneous Complex, 361, 370
size of impactor, 163
    Mjölnir impact structure, 201–202
    Slate Islands impact structure, 114
size of structure, 163, 173
    and amount of extinctions, 243
    Chesapeake Bay impact crater, 149, 154–155
    Chicxulub Crater, 158, 263, 266, 291
    geophysical model, 229
    Highbury structure, 62
    impact diamonds, 215
    Kara impact structure, Russia, 206
    Lake Karikkoselkä impact structure, 131, 146
    Lycksele structure, 125–126
    Mjölnir impact structure, 193, 200, 202
    Morokweng impact structure, 61–62, 84–85, 89, 91–92, 106

Oasis impact structure, 180
Popigai impact structure, 1, 3, 19–20, 49, 215
Ries Crater, Germany, 116, 215
Slate Islands impact structure, 109, 114
Steinheim impact structure, Germany, 206
Sudbury Structure, 89, 300, 431, 433
Teague Ring impact structure, 169, 173
Skaergaard intrusion, Marginal Border Group, trace elements, 368
skarn, lower-temperature sulfide deposits, 106
Skead pluton, Sudbury Structure, map, structure, 413
Skellefte ore district, Lycksele structure, 130
Slate Islands Crater. *See* Slate Islands impact structure, Canada
Slate Islands impact structure, Canada
    erosion, effect of on magnetic signature, 238
    Lake Superior, 109–124
    suevite, 109–110, 116, 122–123
smoke clouds, 243
Sogdoky Upland, Popigai impact structure, 2, 14–16
solar gas, condensation of, 440
solidification time, modeled, Sudbury Igneous Complex, 394
Solnhofen limestone (unshocked), X-ray powder diffraction measurements, 206–212
soot
    in Black Member, Onaping Formation, Sudbury Basin, 331, 337, 340
    conversion of carbon to, 342
    in Cretaceous-Tertiary boundary samples, 221, 337, 341, 352
    in Onaping Formation, Sudbury Basin, 341
    in Onwatin Formation, Sudbury Basin, 331, 337–338, 340–341
    source of, 337
    as a time marker, 342
South Africa, spherules, 249
South Chelmsford sampling site, Black Member, Onaping Formation, Sudbury Basin, 348
South Lobe, Sudbury Igneous Complex, 401, 411, 414
    mineral fabrics, 405
South Longvack mine, Sudbury Structure, 306–307
South Range, Sudbury Igneous Complex, 308, 361–362, 364–369, 379
    application of geochemical data to, 440
    base of Sudbury Igneous Complex directly below, 424, 434
    faulting and folding of, 419
    lead isotopes, 370
    mineral fabrics, 404–405
    norites transitional with Sublayer, 438
    northward transport of, 434
    original attitude of, 436
    seismic data, 420–421
    strontium isotopes, 370
South Range Shear Zone, Sudbury Igneous Complex, 399–400, 405, 408–411, 436
    block diagram, 413
    cross section, after readjustments to crater and before complete solidification, 437
    as deformation boundary, 423
    and late Penokean orogeny, 434
    map (location), 421
    map (structure), 401, 410, 412

most intense deformation in, 425, 434
    seismic reflection profile, 424, 426
Southwest Range, Sudbury Igneous Complex, 362, 369
spherule marker bed, 258
spherules, 244, 251
    Archean (late), 249–250
    Australasian strewn field, 190
    clinopyroxene, late Eocene, 251
    Cretaceous-Tertiary boundary, 250, 350
    Ivory Coast strewn field, 190
    Ivory Coast tektites, 54
    layer of, 249–261
    North American strewn field, 190
    South Africa, 249
    Western Australia, 249
    Yukatite strewn field, Popigai impact structure, 26
spinel, chromium-rich, Morokweng impact structure, 71, 76
spinel, Morokweng impact structure, 97
spinel, nickel-rich, Morokweng impact structure, 71
SRM. *See* shock remanent magnetization
stability at high pressure, calcite, crystal structure of, 213
stacking velocities. *See* seismic velocity data
stages of impact process
    and accretionary-mantled clasts, 342
    central uplift, 109, 114, 116, 440
    and chemical composition of melt sheet, 439
    contact and compression, 112
    crater modification, 109, 112, 328, 390–391, 435, 440
    end of excavation phase of impact process, 328
    excavation, 116
    excavation and central uplift, 112
    excavation and ejection, 114
    impact debris cloud, 331, 440
    postimpact processes, 340
    reworking of impact breccias, 328, 440
    Slate Islands impact structure, 112–113, 123
    slumping, 331
    time for excavation process, 114, 122
    transient crater, 114, 440
    and types of Sudbury Breccia (pseudotachylite), 433
Steen River impact structure, Canada, 245
    gravity anomaly vs. erosional level, 237
Steinheim impact structure, Germany, 205–214
    Ries Crater, same age as, 206
    target rocks, calcite-rich, 205–206
stepped combustion
    analytical sensitivity, 220
    determination of diamond, 217–219
    diamonds, Cretaceous-Tertiary boundary, 220
stishovite, 221, 244
    Popigai impact structure, 28
Stobie Formation, Elliot Lake Group, Sudbury area, map (structure), 410
Stobie Mine, Sudbury, 362
    osmium isotopes in sulfide ores, 363, 365
Strathcona Deep Copper ore zones, 306, 310, 314
Strathcona Embayment, Sudbury Structure
    deformational structures, 314
    distribution of Sudbury Breccia (pseudotachylite), 309–314, 435

ground preparation to form, 435
map (geologic), 306–308
pyroxenite sheets in drill holes, 438
structure of, 309
Sudbury Breccia (pseudotachylite), 305–313
Strathcona Mine, Sudbury, 306–307, 362
map of, 310
osmium isotopes, 364, 366
osmium isotopes in sulfide ores, 363, 365
strontium isotopes, 367
Sublayer, Sudbury Igneous Complex, 438
ultramafic components, 437
stratigraphy, in drill cores, Morokweng impact structure, 67
stratigraphy, seismic, Mjölnir impact structure, 195
strewn field. See spherules; tektites
strontium isotopes
analytical methods, 364
Sudbury Igneous Complex, 361–362, 367–369, 432
structural evolution, Sudbury Structure, 419–427
structural features
Chicxulub Crater, 266
Mjölnir impact structure, 195–202
in submarine impact structures, 199–200
Sudbury Breccia (pseudotachylite), 311
structural floor of crater
buried by crater-fill deposits, 163
Chesapeake Bay impact crater, 151–152
impossibility of seeing in extraterrestrial craters, 162
Sudbury Igneous Complex, 419, 422
structural petrology
of gabbro, Sudbury Igneous Complex, 403–405
of norite, Sudbury Igneous Complex, 403–405
structural relationships, Sudbury Igneous Complex, 300
structure
Balagan River basin, Popigai impact structure, 13
of basement surface, Chesapeake Bay impact crater, 151
Chesapeake Bay impact crater, 149, 151, 153–154
Lycksele structure, 125
Morokweng impact structure, 86, 92–93, 106
Popigai impact structure, 3, 5–7, 12–13, 15–17, 22, 24
Slate Islands impact structure, 113
See also cross section; map, structure; seismic reflection profiles
Sublayer, Sudbury Igneous Complex, 299–302, 373, 389, 400, 414
controversy regarding, 302
cross section, 422
cross section, after readjustments to crater and before complete solidification, 437
heterogeneous nature of, 438
and incorporation of sulfides, 439–440
lead isotopes, 361–362, 366, 369–370
map (geologic), 307–308, 346, 374
map (structure), 401, 410
melanocratic inclusions, geochemical similarities, 437
melanorite inclusions in, 438
no conclusive explanation yet found, 440

and norite, trace elements compared, 434
not convectively mixed with Main Mass norite, 432, 440
osmium isotopes, 362, 365–366
and parent-magma calculation, 379
and preexisting layered intrusion, 431, 433, 438
sampling localities, 363
and sheltering of embayments, 440
strontium isotopes, 361, 367–370
ultramafic components, 431, 433, 436
submarine impact structures
Chesapeake Bay impact crater, 158, 160
Chicxulub Crater, 158, 160, 263–268
compared to subaerial impact structure, 193
Mjölnir impact structure, 158, 160, 193
Montagnais Crater, Scotian Shelf, Canada, 158, 160
percentage of all impact structures, 193
special aspects of, and Mjölnir impact structure, 199–200, 202
Toms Canyon Crater, New Jersey Shelf, 158, 160
Sudbury, archetype of world-class accumulations of magmatic sulfides, 441
Sudbury, summary of development of ideas on, 1992–1998, 431–442
Sudbury (Swarm) dikes, age and lack of deformation of, 411, 423
Sudbury Basin, 323, 374
cross section, 400
definition of, 402
map (geologic), 306, 410
map (structure), 401, 407, 412
Sudbury Breccia (pseudotachylite), 305–314, 373, 402, 409, 415
background abundance percentage, 309
classification of end-member types, 309
and copper-rich ore deposits, 432
distribution of, 299–301, 305, 309, 313–314, 435
flow texture, 310, 312
formation in shadow zone, 314
formation of different types of, 433, 440
formed in both shear and tension, 310–311, 314
Frood breccia, 305, 308
and late-magmatic effects of superheating, 438
lead isotopes, 362
Levack breccia, 305
modeled temperature, 391
neodymium isotopes, 362, 366
nickel–copper–platinum-group-element deposits, 305, 314
north of Sudbury Igneous Complex, 424
orientation change with depth, 313
orientations of, relative to host rocks and Sudbury Igneous Complex, 313
origin of, 299–301
overprinting of, 314
photographs of, 313
predated Main Mass of Sudbury Igneous Complex, 314
predated sulfide ore deposition, 314
shape fabric of, 410
and shock-wave periodicity, 314
similarity to Vredefort impact structure, 301
Strathcona Embayment, 305
Strathcona mine, 305–306, 309, 312–314

Strathcona mine, map of, 310–311
strontium isotopes, 367
structural control of, 305–306
and Sudbury Igneous Complex, 362
time of emplacement, 314
Sudbury dome, 400, 408–409
definition of, 402
Sudbury Igneous Complex
age of, 323, 373, 400
age of mafic inclusions in Sublayer, 433
chemical compositions of various units, 379
Chicxulub, compared to, 297
cross section, after readjustments to crater and before complete solidification, 437
density-stabilized impact melt, 438–440
depth of melting, modeled, 438
description of, 347
diabase and Sublayer, Whistle Mine, 433
emplaced as two magmas, 434
emplacement geometry of, 399–415
fall-back breccia incorporated into melt sheet, 435
folding no longer interpreted subsequent to emplacement, 434
formed by primary mantle melt and Archean granulites, 433
formed from single parent magma, 435
geochronological methods, comparison with emplacement modeling, 392
and hidden mafic cumulates, 435
high felsic proportion, 380, 385, 414, 431, 435–436
Huronian Supergroup folded prior to impact, 434
hydrothermal cooling causing magma reinjection, 370
impact-induced decompression melting, 415
impact melt rocks, 105, 297
initial geometry of, 414–415, 425–426, 436–437
initial magma composition proposed, 438–439
isotopes and question of mantle involvement, 433
isotopes indicate exclusively crustal origin, 437
lower-crustal material brought to surface by fluid flow, 394, 436
Main Mass, 362
mantle, not melted by Sudbury impact, 394, 436, 438
map (geologic), 306, 346, 374
map (structure), 407, 413
melt system, 385
mixing calculations, 368, 370
modeled crystallization sequence, 380–381, 435
modeled temperature, 391–392
modeled with two proposed parental magmas, 373–386, 435
and Morokweng impact structure, 92
parental magmas of, 378–381, 435
phase equilibria, 373–386
postcrystallization processes, 381
and preexisting layered intrusion (product of 2.4 Ga mafic magmatism), 433, 438–440
pseudotachylite. See Sudbury Breccia
rock types, 300
and similarity to lunar samples, 433

in situ differentiation to form, 92, 373–386, 390
size implies impact melt sheet, 403, 431
size of, 389–390
solidification time, modeled, 394
superheating of impact melt, 385, 431, 435–436
theories on formation of, 347, 361, 373–374, 415, 435
totally due to impact melting, 432
volume of, 390
See also granophyre; impact melt sheet; Main Mass; norite; Onaping Formation; origin of Sudbury Igneous Complex; *under* seismic; shock metamorphism; Strathcona Embayment; *under* structural; Sublayer; Sudbury impact event; Sudbury Structure; sulfur (elemental); *under* thermal

Sudbury impact event
amount of mantle uplift, 436
cratering mechanics, 389–391, 435–436
ejecta, distal, 320
lower-crustal material brought to surface by fluid flow, 394, 436
thermal history, 389
thermal modeling, 391–392, 435–436

Sudbury Structure, 323
acoustic fluidization of, 436
age of structure, 282, 299, 302, 374, 389
basinal shape of, 432
block diagram of, 413
breccia. *See* Onaping Formation; Sudbury Breccia
cross section, after readjustments to crater and before complete solidification, 437
definition of, 299, 323
deformation, postimpact, 282
depth of melting, modeled, 392
development of ideas on, 431–441
Fecunis Lake fault, 324, 327
listric faults, 425
map (geologic), 307–308, 324, 374, 400, 420
map (structure), 401, 410
map (underground), 310–311
maximum compression direction, 314
multi-ring basin, 106, 374, 389, 403
nickel–copper–platinum-group-element deposits, 305–306, 310–311
PGE (platinum-group element) deposits, 62
questions and disagreements on, 299–302
regional metamorphism, 332
seismic data, 300
shape of, 299
shatter cones, 299
size of, 89, 125, 282, 299–300, 431
and Slate Islands impact structure, 112
stratigraphy, 341
synthesis of present knowledge and future directions, 436–441
See also amino acids; carbonaceous matter; fullerenes; impact diamonds; kerogen; Onaping Formation; Onwatin Formation; organic carbon; *under* origin; poorly graphitized carbon (PGC); *under* seismic; shock metamorphism; soot; Strathcona Embayment; *under* structural; Sudbury Igneous Complex; sulfide ore deposits; sulfur (elemental); Whitewater Group

suevite
carbon in, Onaping Formation, Sudbury Basin, 318
chemical composition, Popigai impact structure, 41–42, 46
definitions of impact rocks, 49, 114
description, 3
diamond and silicon carbide in, Ries Crater, Germany, 220
Kara (and Ust-Kara) impact structure, Russia, 207
Kara impact structure, Russia, 206
Onaping Formation, Sudbury Basin, as a, 415, 433
Popigai impact structure, 1–16, 19–22, 26, 28–32, 37–39, 44, 48–50, 52, 54–55, 317
silicon carbide and diamond in, Ries Crater, Germany, 220
Slate Islands impact structure, 109–110, 116, 122–123
water contents, Popigai impact structure, 44
Suevite Megabreccia Formation, Popigai impact structure, 30–31, 51
Suevite Sand Formation, Popigai impact structure, 30–31, 50
suevitic breccia
Chicxulub Crater, 291, 296–297
definitions of impact rocks, 114
Popigai impact structure, 317
Slate Islands impact structure, 114
suevitic breccias, Onaping Formation, Sudbury Basin, 346
sulfide ore deposits
cross section, Sudbury Structure, 422
cross section, Sudbury, after readjustments to crater and before complete solidification, 437
crustal derivation, based on osmium isotope data, 437
crustal origin for, 366
Deep Copper deposit, Sudbury Structure, 306, 310, 314
derived from pre-Sudbury layered intrusion, 433
formation of, Sudbury Igneous Complex, 432
high-temperature, at Sudbury Structure, 106
Lycksele structure, 130
osmium isotopes, 363, 365–366, 431, 436, 439
and Sudbury Breccia (pseudotachylite), 432
Sudbury Igneous Complex, 318, 361, 400, 437–439
Sudbury not good archetype for magmatic sulfides, 441
Sudbury Structure, 299, 301, 305–306, 310–311, 346
sulfur immiscibility, Sudbury Igneous Complex, 431, 437, 439–440
sulfides
chemical composition, 99
Morokweng impact structure, 80
sulfur (elemental) in Sudbury Basin rocks, 339, 341, 348, 356, 358
analytical methods, 355–356
mass-independent fractionation of, 345, 357
and weathering of pyrrhotite, 347, 356–357
sulfur availability and postimpact resurgence of biota, 342

sulfur isotopes, Sudbury Basin rocks, 355–357
summary of development of ideas on Sudbury, 1992–1998, 431–442
summary of present knowledge and future directions, Sudbury Structure, 436–441
Sunday Harbour, Canada, Slate Islands impact structure, 110, 114–116
Suon-Tumul Upland and Popigai impact structure, 15
superheating of impact melt
allowing complex differentiation relationships, 438
Sudbury Igneous Complex, 385, 431, 435–438
Sudbury not good archetype for magmatic sulfides, 441
Superior province, multiple orogenies, 420–421, 424
Suvasvesi N impact structure, Finland, similarities to, Lake Karikkoselkä impact structure, 146
Svalbard, and Mjölnir impact structure, 194
Sverdrup Basin
affected by Mjölnir impact, 202
and Mjölnir impact structure, 194
Sweden, northern, Lycksele structure, 125
Sweden, Siljan impact structure, 125, 173
synthesis of present knowledge and future directions, Sudbury Structure, 436–441
synthetic aperture radar lineament, and Sudbury Structure, 420, 423

T

tagamite
description, 3
Popigai impact structure, 1–3, 5, 8–16, 19, 22, 26, 28–40, 48–55, 317
tagamite, chemical composition, Popigai impact structure, 23
tagamite, Popigai impact structure, Russia
chemical composition, 41–43, 46
water contents, 44
Tagamy hills, Siberia, 32
Talasniemi, Lake Karikkoselkä impact structure, 139
target rocks, 163
acoustic fluidization of, 277–278, 390–391, 415, 436
BP impact structure, 177, 179, 183, 189
calcite-rich, characteristics of, 207
calcite-rich, shock metamorphism of, 205–214
carbon in, 329
carbon-rich mudstone and Sudbury Structure carbon, 331
chemical composition, Popigai impact structure, 23
Chesapeake Bay impact crater, 154
Chicxulub Crater, 269
disaggregation and displacement of, 415
fracture mechanism, 226
and fracture mechanism, 226
geophysical model, 229
Green Member, Onaping Formation, Sudbury Basin, 327
hydrocarbon-generating, 202
and impact diamonds, 318
isotopic compositions of, 189
Lake Karikkoselkä impact structure, 131, 133, 146

and lead isotope signature of Sudbury Igneous Complex, 366
Libyan Desert Glass (LDG), source of, 179
Libyan impact structures, 177
Lycksele structure, 125–126, 128, 130
and Main Mass, Sudbury Igneous Complex, 361
mechanical behavior of, 390
melting of in Sudbury area, 440
Mjölnir impact structure, 195
modeled temperature, 391–392
Morokweng impact structure, 64–65, 75, 78–82, 84–85, 88, 102
Oasis impact structure, 177, 179, 183, 189
and Onaping Formation, Sudbury Basin, 325
Popigai impact structure, 21–24, 30, 32, 35, 50–52, 54–55
preexisting layered intrusion, Sudbury area, 439
sequence boundaries, 194
Slate Islands impact structure, 109, 112–113, 116, 118–119, 121–123
stages of impact process, 114
Sublayer, Sudbury Igneous Complex, 370
Sudbury Igneous Complex, 301, 438
Sudbury impact event, 323
Sudbury Structure, 302, 346, 402, 431, 441
Teague Ring impact structure, 167–168
Teague Ring impact structure, Western Australia, 165–175
geologic setting, 167
origin, 168
structure, 167–168
tektites, 54, 244, 249
Australasian strewn field, 190
and BP impact structure, 190
central European strewn field, 190
isotopic compositions of, 189–190
Ivory Coast strewn field, 190
Ivory Coast tektites, 54
and Libyan Desert Glass (LDG), 190
Muong Nong–type, 37
North American strewn field, 190
and Oasis impact structure, 190
Popigai impact structure, 48
Yukatite strewn field, Popigai impact structure, 26
temperature
of melt, 216
Morokweng impact structure, 84, 103
Popigai impact structure, 49, 51–52, 55
Sudbury Structure, modeled, 391, 395
Vredefort structure, South Africa, 103
ternary feldspar diagram, Morokweng impact structure, 72
Ternovka impact structure, Ukraine, impact diamonds, 317
Terrace Bay, Canada, Slate Islands impact structure, 109–110
Tethyan paleogeographic realm, and Jurassic-Cretaceous boundary, 244
Tethyan province, extinctions in, Morokweng impact structure, 88
texture, checkerboard, Morokweng impact structure, 73
texture, chilled
Morokweng impact structure, 91, 94, 97, 99
texture, devitrification
Black Member, Onaping Formation, Sudbury Basin, 333

Cretaceous-Tertiary boundary, 251
Moon, 251
texture, garben
Morokweng impact structure, 69–70, 73, 82
texture, symplectitic
Morokweng impact structure, 103
Sudbury Igneous Complex, 103
Vredefort structure, South Africa, 103
textures
BP impact structure, 182
in impact rocks, BP impact structure, 181
in impact rocks, Oasis impact structure, 181
Morokweng impact structure, 74–75, 88, 94
Oasis impact structure, 182
petrographic, Morokweng impact structure, 69, 71
theories on origin of impact structures. *See* origin of impact structures
thermal alteration. *See* thermal overprinting
thermal history, Sudbury impact event, 389
thermal metamorphism, Morokweng impact structure, 91
thermal modeling
Sudbury Igneous Complex, 435
Sudbury impact event, 391–392, 394–395
thermal overprinting
Morokweng impact structure, 68, 72, 75, 77, 79, 84–85
Sudbury Igneous Complex, 302
thermodynamic calibration of COMAGMAT-3.5 simulation program, 376
thermodynamics of modeled target and projectile, crater modification, 390
thermophysical modeling, 385
parameters used for Sudbury Igneous Complex, 383
thermoremanent magnetization (TRM)
Lappajärvi, Finland, 232
Manicouagan impact structure, Canada, 232
Thunder Bay, Canada, Slate Islands impact structure, 109, 121
Th/Zr ratios, Sudbury Igneous Complex, 434
time for contact and compression, Slate Islands impact structure, 122
time for crater-modification process, Slate Islands impact structure, 116, 122
time for excavation and central uplift, Slate Islands impact structure, 114, 122
time for long-term readjustment, 122
time of emplacement, units of Sudbury Igneous Complex, 302
timing of shock-metamorphic effects, 113
titanomagnetite, Morokweng impact structure, 71
Tithonian-Berriasian boundary, 244
Morokweng impact structure, 86
tomographic inversion, Chicxulub Crater, 292–294
Toms Canyon Crater, New Jersey Shelf
age of structure, 158
size of structure, 158
Tookoonooka impact structure, Australia, 245
topographic data, Kara (and Ust-Kara) impact structure, Russia (reference for), 206
trace elements
BP impact structure, 184–185, 187, 189
and COMAGMAT-3.5 simulation program, 376
comparison of Main Mass units, Sudbury Igneous Complex, 433–434

East site, relative to Libyan impact craters, 186–187, 189
Libyan Desert Glass (LDG), 187, 189
Main Mass, Sudbury Igneous Complex, 368
Main Mass units and Sublayer compared, Sudbury Igneous Complex, 434
melanorite inclusions in Sublayer, Sudbury Igneous Complex, 433
Morokweng impact structure, 79, 91, 100–101, 103–104
norite vs. Sublayer, Sudbury Igneous Complex, 434
Oasis impact structure, 186–187, 189
Popigai impact structure, 21, 23, 46, 53–55
quartz norite, 102
Skaergaard intrusion, Marginal Border Group, 368
Slate Islands impact structure, 119
South Range, Sudbury Igneous Complex, 368
Sublayer vs. norite, Sudbury Igneous Complex, 434
Sudbury Igneous Complex, 375, 385, 439
See also impactor, contamination by
trace elements, target rocks, Popigai impact structure, 24
transient crater
Chesapeake Bay impact crater, 155
Chicxulub Crater, 269, 275–276, 288, 291–292, 297–298
Chicxulub Crater, diameter of, 263, 266, 270
Mjölnir impact structure, size of, 195
and peak ring, 277
and peak-ring formation, 278
Slate Islands impact structure, 116
stages of impact process, 112
Sudbury Structure, 301, 389–392, 435–436, 440
Transvaal basin, South Africa, 258
Transvaal Supergroup (Proterozoic), Morokweng impact structure, 63, 65, 75, 84–85
Transvaal Supergroup, South Africa, 250, 259
trevorite, Morokweng impact structure, 72, 76–77, 91, 97, 99, 106
Triassic (Late) extinction event, 244
Triassic-Jurassic boundary, 243
trichites, Popigai impact structure, 35
tridymite, Popigai impact structure, 35
Trill-I borehole, and Sudbury Igneous Complex, 421–422
Trill-II borehole, and Sudbury Igneous Complex, 421–422
troilite-pyrrhotite, Popigai impact structure, 40, 44
troilite-pyrrhotite globules, Popigai impact structure, 37, 39, 53–54
tsunami model of peak-ring formation, 277
tsunamis, 243, 257
Mjölnir impact structure, 202
turbidite deposition associated with impact, Mjölnir impact structure, 202
two-layer igneous system, Sudbury Igneous Complex, 432
two-magma model, Sudbury Igneous Complex, 374
type D craters, Venus, 163
type D craters on Venus, similarities to, Chesapeake Bay impact crater, 149

U

unified theory of impacts, mass extinctions, and galactic dynamics, 241–248
  Earth-crossing objects, 241
  Milky Way Galaxy, 241, 247
unit-cell dimensions, calcite
  Kara (and Ust-Kara) impact structure, Russia, 211–212
  Steinheim impact structure, Germany, 207, 211–212
unit-cell dimensions, calcite, Kara (and Ust-Kara) impact structure, Russia, 207
universal stage method of Englehardt and Bertsch, planar deformation features (PDFs), 114
Upheaval Dome, Utah, United States, 174
Upper Black Member, Onaping Formation, Sudbury Basin, 324–325, 327–328, 433, 435
  reworking of impact breccias, 440
upper-crustal structure, Chicxulub Crater, 291–298
uranium, Vredefort structure, South Africa, 62
ureilite
  mass-independent fractionation of sulfur isotopes in, 357
  as Sudbury carbon source, 350
Ust-Kara structure. *See* Kara impact structure, Russia

V

Vanha-Karikko, Lake Karikkoselkä impact structure, 139
vapor plume dynamics, 327–328
Vavucan differentiated sill, eastern Siberia, Russia, observed and modeled parameters, 378
velocity of asteroid, Slate Islands impact structure, 114
velocity of ejecta, Popigai impact structure, 49, 54–55
velocity of impact, 163, 327
  stages of impact process, 112
  Sudbury Structure, 389–390, 435
Ventersdorp Supergroup (Archean), Morokweng impact structure, 65, 75
Verkhne-Anabar Series, Popigai impact structure, 21, 23–24
Vermillion Lake Road sampling site, Onwatin Formation, Sudbury Basin, 358
Vernon syncline, and Sudbury Structure, 400, 408
vertical jet vapor model, Popigai impact structure, 54–55
Victoria mine, Sudbury Structure, 308
Victor Mine, INCO, 441
volatile contents
  Morokweng impact structure, 98
  Popigai impact structure, 52
volatiles
  Morokweng impact structure, 72, 79–80
  Popigai impact structure, 53
Volgian-Berriasian age (141–149 Ma), Mjölnir impact structure, 193, 195, 201–202
Volgian-Ryazanian boundary, Morokweng impact structure, 88
Volop Group (Proterozoic), Morokweng impact structure, 92
volume of ejecta, Popigai impact structure, 54–55
volume of impact-deformed strata, Mjölnir impact structure, 202
volume of melt
  Popigai impact structure, 49, 54
  Sudbury Igneous Complex, 390
volume work on rocks during impact, 189
Vorstershoop, road to, Morokweng impact structure, 63
Vredefort Granophyre, similarities to, Morokweng impact structure, 68, 71–72, 82, 84, 86, 88
Vredefort impact structure, South Africa
  age, 62
  age of structure, 282
  erosion, amount of, 282
  erosion, effect of on gravity signature, 238
  gold, 62
  impact melt rocks, 103
  and Lycksele structure, 128–129
  modeled results, 234
  and Morokweng impact structure, 77, 92
  multi-ring basin, 106
  size of structure, 89, 125, 178, 282
  and Slate Islands impact structure, 112
  uranium, 62
  Witwatersrand basin, 62
Vredefort impact structure, and Sudbury Breccia (pseudotachylite), 301
Vryburg, South Africa, Morokweng impact structure, 92–93
Vryburg Formation, South Africa, 258

W

Wabar Crater, Saudi Arabia, impact melt rocks, 106
Wabar impact structure, compared with BP and Oasis impact structures, 189
Waddell Lake fault system, and Sudbury Structure, 436
Wanapitei impact glasses, 55
  Sudbury Structure, 53
Wanapitei impact structure, Canada, 245
  and Sudbury Structure, 436
Wandel Sea Basin, and Mjölnir impact structure, 194
Wandiwarra Formation, 167–168, 170
water, role in crystallization rate, Popigai impact structure, 52
water, role in melt generation, Popigai impact structure, 19
water-bearing minerals, Popigai impact structure, 52
water-column disturbance, Mjölnir impact structure, 202
water contents
  impact melt rocks, Popigai impact structure, 44
  suevite, Popigai impact structure, 44
  tagamite, Popigai impact structure, 44
Werner and the Neptunists, 440
West Clearwater Lake Crater, Canada. *See* Clearwater West impact structure
Western Australia, spherules, 249
West Siberian Basin
  and the Mjölnir impact, 202
  and Mjölnir impact structure, 194
Whistle Mine, Sublayer, Sudbury Igneous Complex, 356, 358, 362, 434
  baddeleyite from mafic inclusions, 433
  osmium isotopes, 364
  osmium isotopes in sulfide ores, 363, 365–366
  strontium isotopes, 367
  zircon from mafic inclusions, 433
Whitewater Group, Sudbury Basin
  carbonaceous matter, 299–301
  description of, 421
  devitrified glass in, 402
  ductile strain, 409
  folded, 402
  folding of, 427
  formations of, 323
  map (geologic), 306, 420
  Onaping Formation, as an impact-generated pseudovolcanic sequence, 402
  stratigraphic younging direction, 400, 406, 413
  timing of deformation, 425–426
whole-rock analyses
  Morokweng impact structure, 99–101
  Sudbury Igneous Complex, major elements of, 369
  target rocks, 104
willemseite (nickel-rich talc mineral), Morokweng impact structure, 72, 91, 99, 106
Windidda Formation, 167
Wittenoom Formation, Western Australia, 251, 255, 258
Witwatersrand basin, Vredefort structure, South Africa, 62
Witwatersrand Supergroup, Morokweng impact structure, 84
Wongawol Formation, 167–168, 173
Worthington Mine, Sudbury Structure, 308

X

X-ray powder diffraction measurements
  analytical methods, 207–208
  angular dependency on the diffraction angle, 207
  discussion, 207–213
  Kara (and Ust-Kara) impact structure, Russia, 208–212
  results for carbonate samples, 209–212
  Ries Crater, Germany, 208
  Steinheim impact structure, Germany, 208–212

Y

Yelma Formation, 167–168
York-James Peninsula, Chesapeake Bay impact crater, 154
York River, Chesapeake Bay impact crater, 149–150
Yucatán. *See* Chicxulub Crater, Yucatán Peninsula, Mexico
Yukatite strewn field, Popigai impact structure, 26

Z

Zapadnaja impact structure, Ukraine, impact diamonds, 317
Zhamanshin Crater, Kazakhstan, 37, 53, 245
zircon, 258
  chemical composition, 99
  diabase (Sudbury), 437
  masking diamond's presence, 219

melanorite inclusions in Sublayer, Sudbury
    Igneous Complex, 431, 433, 436–437
Morokweng impact structure, 72, 75,
    78–79, 82–83, 88, 97
planar deformation features (PDFs) in, 402
pyroxenite (Sudbury), 437
reset by Sudbury impact, 433, 440
shock metamorphism of, 216
Sudbury Igneous Complex, 302
Sudbury Structure, 392
Whistle Mine, Sublayer mafic inclusions,
    Sudbury Igneous Complex, 433